Student Solutions Manual

Doreen Kelly

Elementary and Intermediate Algebra

for College Students

Second Edition

Allen R. ANGEL

PEARSON
Prentice Hall

Upper Saddle River, NJ 07458

Senior Acquisitions Editor: Paul Murphy
Supplement Editor: Elizabeth Covello
Editorial Assistant: Kerri-Ann O'Donnell
Assistant Managing Editor: John Matthews
Production Editor: Jeffrey Rydell
Supplement Cover Manager: Paul Gourhan
Supplement Cover Designer: Joanne Alexandris
Manufacturing Buyer: Ilene Kahn

Pearson Prentice Hall
Pearson Education, Inc.
Upper Saddle River, NJ 07458

Printed in the United States of America

10 9 8 7 6 5 4 3

ISBN 0-13-141123-3

Pearson Education Ltd., *London*
Pearson Education Australia Pty. Ltd., *Sydney*
Pearson Education Singapore, Pte. Ltd.
Pearson Education North Asia Ltd., *Hong Kong*
Pearson Education Canada, Inc., *Toronto*
Pearson Educación de Mexico, S.A. de C.V.
Pearson Education—Japan, *Tokyo*
Pearson Education Malaysia, Pte. Ltd.
Pearson Education, *Upper Saddle River, New Jersey*

Table of Contents

Chapter 1

Exercise Set 1.1

11. To prepare properly for this class, you need to do all the homework carefully and preview the new material that is to be covered in class.

13. At least 2 hours of study and homework time for each hour of class time is generally recommended.

15. a. You need to do the homework in order to practice what was presented in class.

b. When you miss class, you miss important information, Therefore it is important that you attend class regularly.

17. Answers will vary.

Exercise Set 1.2

1. Understand, translate, calculate, check, state answer

3. Substitute smaller or larger numbers so that the method becomes clear.

5. Rank the data. The median is the value in the middle.

7. The mean is greater since it takes the value of 30 into account.

9. A mean average of 80 corresponds to a total of 800 points for the 10 quizzes. Pat's mean average of 79 corresponds to a total of 790 points for the 10 quizzes. Thus, he actually missed a B by 10 points.

11. a. $\frac{78+97+59+74+74}{5} = \frac{382}{5} = 76.4$

The mean grade is 76.4.

b. 59, 74, 74, 78, 97
The middle value is 74.
The median grade is 74.

13. a. $\frac{204.83+153.85+210.03+119.76+128.38}{5} = \frac{816.85}{5} = 163.37$

The mean bill is \$163.37.

b. \$119.76, \$128.38, \$153.85, \$204.83, \$210.03
The middle value is \$153.85.
The median bill is \$153.85.

15. a. $\frac{10.63+10.67+10.68+10.83+11.6+11.76+11.87+12.18+12.8+12.91}{10}$

$= \frac{115.93}{10} \approx 11.593$

The mean inches for rainfall for the 10 years is 11.593.

b. 10.63, 10.67, 10.68, 10.83, 11.6, 11.76, 11.87, 12.18, 12.8, 12.91
The middle values are 11.6 and 11.76.

$\frac{11.6+11.76}{2} = \frac{23.36}{2} = 11.68$

The median inches for rainfall for the 10 years is 11.68.

17. Barbara's earnings = 5% of sales
Barbara's earnings = 0.05(9400)
= 470
Her week's earnings were \$470.

19. a. sales tax = 7% of price
sales tax = 0.07(16,700)
= 1169
The sales tax was \$1,169.

b. Total cost = price + tax
Total cost = 16,700 + 1,169
= 17,869
The total cost was \$17,869.

21. a. total cost with payments = down payment + (number of months)(monthly payment)
total cost with payments = 200 + 24(33)
= 200 + 792
= 992
Making monthly payments, it costs \$992.

b. savings = total cost with payments – total cost at purchase
savings = 992 - 950
= 42
He saves \$42 by paying the total at the time of purchase.

23. Women enlisted in the army = 45% of total enlisted women = .45(91,600) = 41,220

Women enlisted in the navy = 24% of total enlisted women = .24(91,600) = 21984

Army Women – Navy Women

= 41,220 – 21,984 = 19,236

There are approximately 19,236 more women enlisted in the army than the navy.

25. $$\text{miles per gallon} = \frac{\text{number of miles}}{\text{number of gallons}} = \frac{16,935.4 - 16,741.3}{10.5} = \frac{194.1}{10.5} \approx 18.49$$
His car gets about 18.49 miles per gallon.

27. savings = local cost – mail order cost
$$\text{local cost} = \$425 + (0.08)(\$425) = \$425 + \$34 = \$459$$
$$\text{mail cost} = 4(\$62.30 + \$6.20 + \$8) = 4(\$76.50) = \$306$$
$$\text{savings} = \$459 - \$306 = \$153$$
Eric saved \$153.

29. a. taxes = 15% of income
taxes = 0.15(\$34,612)
= \$5191.80
Their taxes were \$5191.80

b. taxes
$$= \$6780 + 27.5\% \text{ of (income } - \$45,200) \cdot \text{ taxes}$$
$$= \$6780 + 0.275(\$53,710 - \$45,200)$$
$$= \$6780 + 0.275(\$8510)$$
$$= \$6780 + \$2340.25$$
$$= \$9120.25$$

Their taxes were \$9120.25.

31. a. gallons per year = 365(gallons per day)
gallons per year = 365(11.25 gallons)
= 4106.25 gallons

additional money spent
= (cost)(gallons wasted)
additional money spent
$$= \frac{\$5.20}{1000 \text{ gallons}} \cdot 4106.25 \text{ gallons} \approx \$21.35$$

33. Cost = Flat Fee + .30(each quarter mile traveled)
+ .20(each 30 seconds stopped in traffic)
= 2.00 + .30(12) = .20(3)
= 6.20
His ride cost \$6.20.

35. a. difference in cost = cost in Santa Monica – cost in Austin
= 480(number of weeks) + 109(number of weeks)
= 480(20) – 109(20)
= 9600 – 2180
=7420
It costs \$7420 more for daycare in Santa Monica.

b. $\text{hour of evening babysitting} = \dfrac{\text{total amount one can spend}}{\text{cost per hour of babysitter}}$

$= \dfrac{132}{6}$

$= 22$

You can obtain 22 hours of babysitting.

37. A single green block should be placed on the 3 on the right.

39. $\text{mean} = \dfrac{\text{total cost}}{\text{number of nights spent in a hotel}}$

$\text{mean} = \dfrac{1470.72}{8}$

$= \$183.84$ per night

41. a. $\text{mean} = \dfrac{\text{sum of grades}}{\text{number of exams}}$

$60 = \dfrac{50+59+67+80+56+\text{ last}}{6}$

$360 = 312 + \text{ last}$
$\text{last} = 360 - 312$
$= 48$
Lamond needs at least a 48 on the last exam.

b. $70 = \dfrac{312 + \text{ last}}{6}$

$420 = 312 + \text{ last}$
$\text{last} = 420 - 312$
$= 108$
Lamond would need 108 points on the last exam, so he cannot get a C.

43. Answers will vary.

Exercise Set 1.3

1. a. Variables are letters that represent numbers.

b. Letters often used to represent variables are x, y, and z.

3. $5(x)$, $(5)x$, $(5)(x)$, $5x$, $5 \cdot x$

5. Divide out factors that are common to both the numerator and the denominator.

7. a. The least common denominator is the smallest number divisible by he two denominators.
b. Answers will vary.

9. Part b) shows simplifying a fraction. In part a) common factors are divided out of two fractions.

11. Part a) is incorrect because you cannot divide out common factors when adding.

13. c); $\frac{4}{5} \cdot \frac{1}{4} = \frac{1}{5} \cdot \frac{1}{1} = \frac{1}{5}$. Divide out the common factor, 4. This process can be used only when multiplying fractions and so cannot be used for a) or b). Part d) becomes $\frac{4}{5} \cdot \frac{4}{1}$ so no common factor can be divided out.

15. Multiply numerators and multiply denominators.

17. Write fractions with a common denominator, add or subtract numerators, keep the common denominator.

19. Yes, it is simplified because the greatest common divisor of the numerator and denominator is 1.

21. The greatest common factor of 3 and 12 is 3.
$\frac{3}{12}=\frac{3\div 3}{12\div 3}=\frac{1}{4}$

23. The greatest common factor of 10 and 15 is 5.
$\frac{10}{15}=\frac{10\div 5}{15\div 5}=\frac{2}{3}$

25. The greatest common factor of 17 and 17 is 17.
$\frac{17}{17}=\frac{17\div 17}{17\div 17}=\frac{1}{1}=1$

27. The greatest common factor of 36 and 76 is 4.
$\frac{36}{76}=\frac{36\div 4}{76\div 4}=\frac{9}{19}$

29. The greatest common factor of 40 and 264 is 8.
$\frac{40}{264}=\frac{40\div 8}{264\div 8}=\frac{5}{33}$

31. 12 and 25 have no common factors other than 1. Therefore, the fraction is already simplified.

33. $2\frac{3}{5}=\frac{10+3}{5}=\frac{13}{5}$

35. $2\frac{13}{15}=\frac{30+13}{15}=\frac{43}{15}$

37. $4\frac{3}{4}=\frac{16+3}{4}=\frac{19}{4}$

39. $4\frac{13}{19}=\frac{76+13}{19}=\frac{89}{19}$

41. $\frac{7}{4}=1\frac{3}{4}$ because $7\div 4 = 1 \text{ R } 3$

43. $\frac{15}{4}=3\frac{3}{4}$ because $15\div 4=3 \text{ R } 3$

45. $\frac{110}{20}=5\frac{10}{20}=5\frac{1}{2}$ because $110\div 20=5 \text{ R } 10$

47. $\frac{32}{7}=4\frac{4}{7}$ because $32\div 7=4 \text{ R } 4$

49. $\frac{2}{3}\cdot\frac{4}{5}=\frac{2\cdot 4}{3\cdot 5}=\frac{8}{15}$

51. $\frac{5}{12}\cdot\frac{4}{15}=\frac{1}{12}\cdot\frac{4}{3}=\frac{1}{3}\cdot\frac{1}{3}=\frac{1\cdot 1}{3\cdot 3}=\frac{1}{9}$

53. $\frac{3}{4}\div\frac{1}{2}=\frac{3}{4}\cdot\frac{2}{1}=\frac{3}{2}\cdot\frac{1}{1}=\frac{3}{2}$ or $1\frac{1}{2}$

55. $\frac{3}{8}\div\frac{3}{4}=\frac{3}{8}\cdot\frac{4}{3}=\frac{1}{2}\cdot\frac{1}{1}=\frac{1\cdot 1}{2\cdot 1}=\frac{1}{2}$

57. $\frac{5}{12}\div\frac{4}{3}=\frac{5}{12}\cdot\frac{3}{4}=\frac{5}{4}\cdot\frac{1}{4}=\frac{5\cdot 1}{4\cdot 4}=\frac{5}{16}$

59. $\frac{10}{3}\div\frac{5}{9}=\frac{10}{3}\cdot\frac{9}{5}=\frac{2}{3}\cdot\frac{9}{1}=\frac{2}{1}\cdot\frac{3}{1}=\frac{2\cdot 3}{1\cdot 1}=\frac{6}{1}=6$

61. $\left(2\frac{1}{5}\right)\frac{7}{8}$
$2\frac{1}{5}=\frac{10+1}{5}=\frac{11}{5}$
$\left(2\frac{1}{5}\right)\frac{7}{8}=\left(\frac{11}{5}\right)\frac{7}{8}=\frac{11\cdot 7}{5\cdot 8}=\frac{77}{40}$ or $1\frac{37}{40}$

63. $5\frac{3}{8}\div 1\frac{1}{4}$
$5\frac{3}{8}=\frac{40+3}{8}=\frac{43}{8}$
$1\frac{1}{4}=\frac{4+1}{4}=\frac{5}{4}$
$$\begin{aligned}5\frac{3}{8}\div 1\frac{1}{4}&=\frac{43}{8}\div\frac{5}{4}\\&=\frac{43}{8}\cdot\frac{4}{5}\\&=\frac{43}{2}\cdot\frac{1}{5}\\&=\frac{43\cdot 1}{2\cdot 5}\\&=\frac{43}{10}\text{ or }4\frac{3}{10}\end{aligned}$$

65. $\frac{1}{4}+\frac{3}{4}=\frac{1+3}{4}=\frac{4}{4}=1$

67. $\frac{5}{12}-\frac{1}{12}=\frac{5-1}{12}=\frac{4}{12}=\frac{1}{3}$

69. $\frac{8}{17}+\frac{2}{34}$
$\frac{8}{17}=\frac{8}{17}\cdot\frac{2}{2}=\frac{16}{34}$
$\frac{8}{17}+\frac{2}{34}=\frac{16}{34}+\frac{2}{34}=\frac{16+2}{34}=\frac{18}{34}=\frac{9}{17}$

71. $\frac{4}{5}+\frac{6}{15}$

$\frac{4}{5}=\frac{4}{5}\cdot\frac{3}{3}=\frac{12}{15}$

$\frac{4}{5}+\frac{6}{15}=\frac{12}{15}+\frac{6}{15}=\frac{12+6}{15}=\frac{18}{15}=\frac{6}{5}$ or $1\frac{1}{5}$

73. $\frac{1}{6}-\frac{1}{18}$

$\frac{1}{6}=\frac{1}{6}\cdot\frac{3}{3}=\frac{3}{18}$

$\frac{1}{6}-\frac{1}{18}=\frac{3}{18}-\frac{1}{18}=\frac{3-1}{18}=\frac{2}{18}=\frac{1}{9}$

75. $\frac{5}{12}-\frac{1}{8}$

$\frac{5}{12}=\frac{5}{12}\cdot\frac{2}{2}=\frac{10}{24}$

$\frac{1}{8}=\frac{1}{8}\cdot\frac{3}{3}=\frac{3}{24}$

$\frac{5}{12}-\frac{1}{8}=\frac{10}{24}-\frac{3}{24}=\frac{10-3}{24}=\frac{7}{24}$

77. $\frac{7}{12}-\frac{2}{9}$

$\frac{7}{12}=\frac{7}{12}\cdot\frac{3}{3}=\frac{21}{36}$

$\frac{2}{9}=\frac{2}{9}\cdot\frac{4}{4}=\frac{8}{36}$

$\frac{7}{12}-\frac{2}{9}=\frac{21}{36}-\frac{8}{36}=\frac{21-8}{36}=\frac{13}{36}$

79. $\frac{5}{9}-\frac{4}{15}$

$\frac{5}{9}=\frac{5}{9}\cdot\frac{5}{5}=\frac{25}{45}\frac{5}{9}=\frac{5}{9}\cdot\frac{5}{5}=\frac{25}{45}$

$\frac{4}{15}=\frac{4}{15}\cdot\frac{3}{3}=\frac{12}{45}$

$\frac{5}{9}-\frac{4}{15}=\frac{25}{45}-\frac{12}{45}=\frac{25-12}{45}=\frac{13}{45}$

81. $6\frac{1}{3}-3\frac{1}{5}$

$6\frac{1}{3}=\frac{18+1}{3}=\frac{19}{3}=\frac{19}{3}\cdot\frac{5}{5}=\frac{95}{15}$

$3\frac{1}{5}=\frac{15+1}{5}=\frac{16}{5}\cdot\frac{3}{3}=\frac{48}{15}$

$6\frac{1}{3}-3\frac{1}{5}=\frac{95}{15}-\frac{48}{15}=\frac{95-48}{15}=\frac{47}{15}$ or $3\frac{2}{15}$

83. $5\frac{3}{4}-\frac{1}{3}$

$5\frac{3}{4}=\frac{20+3}{4}=\frac{23}{4}=\frac{23}{4}\cdot\frac{3}{3}=\frac{69}{12}$

$\frac{1}{3}=\frac{1}{3}\cdot\frac{4}{4}=\frac{4}{12}$

$5\frac{3}{4}-\frac{1}{3}=\frac{69}{12}-\frac{4}{12}=\frac{65}{12}$ or $5\frac{5}{12}$

85. $55\frac{3}{16}-46\frac{1}{4}$

$55\frac{3}{16}=\frac{880+3}{16}=\frac{883}{16}$

$46\frac{1}{4}=\frac{184+1}{4}=\frac{185}{4}=\frac{185}{4}\cdot\frac{4}{4}=\frac{740}{16}$

$53\frac{3}{16}-46\frac{1}{4}=\frac{883}{16}-\frac{740}{16}=\frac{143}{16}=8\frac{5}{16}$

Kim has grown $8\frac{5}{16}$ inches. Wrong Solution

87. $1-\frac{25}{36}=\frac{36}{36}-\frac{25}{36}=\frac{36-25}{36}=\frac{11}{36}$

About $\mathbf{\frac{11}{36}}$ of all U.S. employees were not online.

89. $1-\frac{39}{50}=\frac{50}{50}-\frac{39}{50}=\frac{50-39}{50}=\frac{11}{50}$

About $\frac{11}{50}$ of sales were for imported vehicles in 2001.

91. $4\frac{1}{2}=\frac{8+1}{2}=\frac{9}{2}=\frac{9}{2}\cdot\frac{6}{6}=\frac{54}{12}$

$1\frac{1}{6}=\frac{6+1}{6}=\frac{7}{6}=\frac{7}{6}\cdot\frac{2}{2}=\frac{14}{12}$

$1\frac{3}{4}=\frac{4+3}{4}=\frac{7}{4}=\frac{7}{4}\cdot\frac{3}{3}=\frac{21}{12}$

$4\frac{1}{2}+1\frac{1}{6}+1\frac{3}{4}=\frac{54}{12}+\frac{14}{12}+\frac{21}{12}=\frac{89}{12}=7\frac{5}{12}$

The total weight is $7\frac{5}{12}$ tons.

93. 15 feet $2\frac{1}{2}$ in. - 3 feet $3\frac{1}{4}$ in.

$= 14$ feet $14\frac{2}{4}$ in. - 3 feet $3\frac{1}{4}$ in.

$= 11$ feet $11\frac{1}{4}$ in. or

$11(12) + 11\frac{1}{4} = 132 + 11\frac{1}{4} = 143\frac{1}{4}$ in. or

$143\frac{1}{4} \div 12 = \frac{572+1}{4}\cdot\frac{1}{12} = \frac{573}{48} \approx 11.9$ ft.

95. $3\frac{1}{8} = \frac{24+1}{8} = \frac{25}{8}$

$3\frac{1}{8} \div 2 = 3\frac{1}{8} \div \frac{2}{1} = \frac{25}{8}\cdot\frac{1}{2} = \frac{25}{16}$ or $1\frac{9}{16}$

Each piece is $\frac{25}{16}$ or $1\frac{9}{16}$ inches long.

97. $5\frac{1}{2} = \frac{10+1}{2} = \frac{11}{2}$

$5\frac{1}{2}\cdot\frac{1}{4} = \frac{11}{2}\cdot\frac{1}{4} = \frac{11\cdot 1}{2\cdot 4} = \frac{11}{8}$ or $1\frac{3}{8}$

$1\frac{3}{8}$ cups of chopped onions are needed.

99. $15 \div \frac{3}{8} = \frac{15}{1}\cdot\frac{8}{3} = \frac{5}{1}\cdot\frac{8}{1} = \frac{5\cdot 8}{1\cdot 1} = \frac{40}{1} = 40$

Tierra can wash her hair 40 times.

101. $\frac{1}{4}+\frac{1}{4}+1 = \frac{1}{4}+\frac{1}{4}+\frac{4}{4} = \frac{6}{4} = \frac{3}{2}$ or $1\frac{1}{2}$

The total thickness is $1\frac{1}{2}$ inches.

103. $4\frac{2}{3} = \frac{12+2}{3} = \frac{14}{3}$

$28 \div \frac{14}{3} = \frac{28}{1}\cdot\frac{3}{14} = \frac{2}{1}\cdot\frac{3}{2} = \frac{6}{1} = 6$

There will be 6 whole strips of wood.

105. a. Total height of computer + monitor

$= 7\frac{1}{2}$ in. $+ 14\frac{3}{8}$ in.

$7\frac{1}{2} = \frac{15}{2} = \frac{15}{2}\cdot\frac{4}{4} = \frac{60}{8}$

$14\frac{3}{8} = \frac{112+3}{8} = \frac{115}{8}$

$7\frac{1}{2}+14\frac{3}{8} = \frac{60}{8}+\frac{115}{8} = \frac{175}{8}$ or $21\frac{7}{8}$

Total height of computer and monitor is $\frac{175}{8}$ or $21\frac{7}{8}$ inches, so there is sufficient room.

b. $22\frac{1}{2} = \frac{44+1}{2} = \frac{45}{2} = \frac{45}{2}\cdot\frac{4}{4} = \frac{180}{8}$

$22\frac{1}{2} - 21\frac{7}{8} = \frac{180}{8} - \frac{175}{8} = \frac{5}{8}$

There will be $\frac{5}{8}$ inch of extra height.

c. $22\frac{1}{2} = \frac{44+1}{2} = \frac{45}{2} = \frac{45}{2}\cdot\frac{2}{2} = \frac{90}{4}$

$26\frac{1}{2} = \frac{52+1}{2} = \frac{53}{2} = \frac{53}{2}\cdot\frac{2}{2} = \frac{106}{4}$

$2\frac{1}{2} = \frac{4+1}{2} = \frac{5}{2} = \frac{5}{2}\cdot\frac{2}{2} = \frac{10}{4}$

$1\frac{1}{4} = \frac{4+1}{4} = \frac{5}{4}$

$22\frac{1}{2}+26\frac{1}{2}+2\frac{1}{2}+1\frac{1}{4}$

$= \frac{90}{4}+\frac{106}{4}+\frac{10}{4}+\frac{5}{4}$

$= \frac{211}{4}$ or $52\frac{3}{4}$

The height of the desk is $52\frac{3}{4}$ in.

107. a. $\frac{*}{a}+\frac{?}{a} = \frac{*+?}{a}$

b. $\frac{\odot}{?}-\frac{\square}{?} = \frac{\odot-\square}{?}$

c. $\frac{\Delta}{\square}+\frac{4}{\square} = \frac{\Delta+4}{\square}$

d. $\frac{x}{3}-\frac{2}{3} = \frac{x-2}{3}$

e. $\frac{12}{x}-\frac{4}{x} = \frac{12-4}{x} = \frac{8}{x}$

109. number of pills

$$= \frac{(\text{mg per day})(\text{days per month})(\text{number of months})}{\text{mg per pill}}$$

$$\text{number of pills} = \frac{(450)(30)(6)}{300} = 270$$

Dr. Highland should prescribe 270 pills.

111. Answers will vary.

112. $\frac{9+8+15+32+16}{5}=\frac{80}{5}=16$

The mean is 16.

113. In order, the values are: 8, 9, 15, 16, 32. The median is 15.

114. Variables are letters used to represent numbers.

Exercise Set 1.4

1. A set is collection of elements.

3. Answers will vary. One possible answer is the set of all natural numbers less than 0.

5. The set of whole numbers contains the natural numbers and zero which is not a natural number.

7. a. A rational number is any number that can be expressed as a quotient of two integers, denominator not 0.

b. Any integer can be expressed as a quotient of two integers by writing it with a denominator of 1. Therefore every integer is a rational number.

9. a. yes

b. no

b. no

c. yes

11. The integers are {..., –3, –2, –1, 0, 1, 2, 3, ...}.

13. The whole numbers are {0, 1, 2, ...}.

15. The negative integers are {..., –3, –2, –1}.

17. True; the whole numbers are {0, 1, 2, ...}.

19. True; any number that can be represented on a real number line is a real number.

21. False; the integers are {..., –2, –1, 0, 1, 2, ...}.

23. False; $\sqrt{2}$ cannot be expressed as the quotient of two integers.

25. True; $-\frac{1}{5}$ is a quotient of two integers, $\frac{-1}{5}$.

27. True; 0 can be expressed as a quotient of two integers, $\frac{0}{1}$.

29. False; $4\frac{5}{8}$ is rational since it can be expressed as a quotient of two integers.

31. False; $-\frac{5}{3}$ is rational since it is a quotient of two integers.

33. True, either $\varnothing$ or { } is used.

35. False; irrational numbers are real but not rational.

37. True; any rational number can be represented on a real number line and is therefore real.

39. True; irrational numbers are real numbers which are not rational.

41. True; the counting numbers are {1, 2, 3, ...}, the whole numbers are {0, 1, 2, ...}.

43. True; the symbol R represents the set of real numbers.

45. False; every number greater than zero is positive but not necessarily an integer.

47. True; the integers are

$$\left\{\underbrace{\ldots,-2,-1}_{\text{negative integers}},\ \underbrace{0}_{\text{zero}},\ \underbrace{1,2,\ldots}_{\text{positive integers}}\right\}.$$

49. a. 0 is an integer.

b. 0 and $2\frac{1}{2}$ are rational numbers.

c. 0 and $2\frac{1}{2}$ are real numbers.

51. a. 3 and 77 are positive integers.

b. 0, 3, and 77 are whole numbers

c. 0, –2, 3, and 77 are integers.

d. $-\frac{5}{7}$, 0, – 2, 3, $6\frac{1}{4}$, 1.63, and 77 are rational numbers.

e. $\sqrt{7}$ and $-\sqrt{3}$ are irrational numbers.

f. $-\frac{5}{7}$, 0, – 2, 3, $6\frac{1}{4}$, $\sqrt{8}$, $-\sqrt{7}$, 1.63, and 77 are real numbers.

For Exercises 53–63, answers will vary. One possible answer is given.

53. 0, 3, 1

55. $\sqrt{3}, \sqrt{7}, \pi$

57. $\frac{1}{2}$, 6.4, –2.6

59. –1, –7, –24

61. 0, $\sqrt{2}$, –5

63. –7, 1, 5

65. {8, 9, 10, 11, ..., 94}
94 – 8 + 1 = 86 + 1 = 87
The set has 87 elements.

67. a. $A = \{1, 3, 4, 5, 8\}$

b. $B = \{2, 5, 6, 7, 8\}$

c. A and $B = \{5, 8\}$

d. A or $B = \{1, 2, 3, 4, 5, 6, 7, 8\}$

69. a. Set B continues beyond 4.

b. Set A has 4 elements.

c. Set B has an infinite number of elements.

d. Set B is an infinite set.

71. a. There are an infinite number of fractions between any 2 numbers.

b. There are an infinite number of fractions between any 2 numbers.

73. $5\frac{2}{5} = \frac{5 \cdot 5 + 2}{5} = \frac{25 + 2}{5} = \frac{27}{5}$

74. $\frac{16}{3} = 5\frac{1}{3}$ because $16 \div 3 = 5$ R 1

75. $\frac{3}{5} + \frac{5}{8}$
$\frac{3}{5} = \frac{3}{5} \cdot \frac{8}{8} = \frac{24}{40}$
$\frac{5}{8} \cdot \frac{5}{5} = \frac{25}{40}$
$\frac{3}{5} + \frac{5}{8} = \frac{24}{40} + \frac{25}{40} = \frac{49}{40}$ or $1\frac{9}{40}$

76. $\left(\frac{5}{9}\right)\left(4\frac{2}{3}\right)$
$4\frac{2}{3} = \frac{12 + 2}{3} = \frac{14}{3}$
$\left(\frac{5}{9}\right)\left(4\frac{2}{3}\right) = \frac{5}{9} \cdot \frac{14}{3} = \frac{70}{27}$ or $2\frac{16}{27}$

Exercise Set 1.5

1. a. 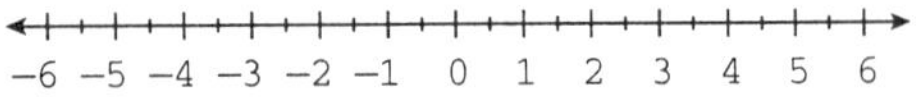

b. 

c. –2 is greater than –4 because it is farther to the right on the number line.

d. $-4 < -2$

e. $-2 > -4$

3. a. 4 is 4 units from 0 on a number line.

b. –4 is 4 units from 0 on a number line.

c. 0 is 0 units from 0 on a number line.

5. Yes; for example, $5 > 3$ and $3 < 5$. Also, $-2 > -5$ and $-5 < -2$.

7. No, $-3 > -4$ but $|-3| < |-4|$.

9. No, $|-4| > |-3|$ but $-4 < -3$.

11. $|7| = 7$

13. $|-15| = 15$

15. $|0| = 0$

17. $-|-5| = -(5) = -5$

19. $-|21| = -(21) = -21$

21. $5 > 2$; 5 is to the right of 2 on a number line.

23. $-6 < 0$; –6 is to the left of 0 on a number line.

25. $\frac{1}{2} > -\frac{2}{3}$; $\frac{1}{2}$ is to the right of $-\frac{2}{3}$ on a number line.

27. $0.7 < 0.8$; 0.7 is to the left of 0.8 on a number line.

29. $-\frac{1}{2} > -1$; $-\frac{1}{2}$ is to the right of –1 on a number line.

31. $3 > -3$; 3 is to the right of –3 on a number line.

33. $-2.1 < -2$; -2.1 is to the left of -2 on a number line.

35. $\frac{4}{5} > -\frac{4}{5}$; $\frac{4}{5}$ is to the right of $-\frac{4}{5}$ on a number line.

37. $-\frac{3}{8} < \frac{3}{8}$; $-\frac{3}{8}$ is to the left of $\frac{3}{8}$ on a number line.

39. $0.49 > 0.43$; 0.49 is to the right of 0.43 on a number line.

41. $5 > -7$; 5 is to the right of -7 on a number line.

43. $-0.006 > -0.007$; -0.006 is to the right of -0.007 on a number line.

45. $\frac{5}{8} > 0.6$ because $\frac{5}{8} = 0.625$ and 0.625 is to the right of 0.6 on a number line.

47. $-\frac{2}{3} < -\frac{1}{3}$; $-\frac{2}{3}$ is to the left of $-\frac{1}{3}$ on a number line.

49. $-\frac{1}{2} > -\frac{3}{2}$; $-\frac{1}{2}$ is to the right of $-\frac{3}{2}$ on a number line.

51. $0.3 < \frac{1}{3}$; 0.3 is to the left of .333... on a number line.

53. $\frac{13}{15} < \frac{8}{9}$; $\frac{39}{45}$ is to the left of $\frac{40}{45}$ on a number line.

55. $-(-6) > -(-5)$; 6 is to the right of 5 on a number line.

57. $3 > |-2|$ since $|-2| = 2$

59. $|-4| > \frac{2}{3}$ since $|-4| = 4$

61. $|0| < |-4|$ since $|0| = 0$ and $|-4| = 4$

63. $4 < \left|-\frac{9}{2}\right|$ since $\left|-\frac{9}{2}\right| = \frac{9}{2}$ or $4\frac{1}{2}$

65. $\left|-\frac{4}{5}\right| < \left|-\frac{5}{4}\right|$ since $\left|-\frac{4}{5}\right| = \frac{4}{5} = \frac{16}{20}$ and $\left|-\frac{5}{4}\right| = \frac{5}{4} = \frac{25}{20}$

67. $|-4.6| = \left|-\frac{23}{5}\right|$ since $|-4.6| = 4.6$ and $\left|-\frac{23}{5}\right| = \frac{23}{5} = 4.6$

69. $\frac{2}{3} + \frac{2}{3} + \frac{2}{3} + \frac{2}{3} = 4 \cdot \frac{2}{3}$ since $\frac{2}{3} + \frac{2}{3} + \frac{2}{3} + \frac{2}{3} = \frac{2+2+2+2}{3} = \frac{8}{3}$ and $4 \cdot \frac{2}{3} = \frac{4}{1} \cdot \frac{2}{3} = \frac{8}{3}$

71. $\frac{1}{2} \cdot \frac{1}{2} < \frac{1}{2} \div \frac{1}{2}$ since $\frac{1}{2} \cdot \frac{1}{2} = \frac{1 \cdot 1}{2 \cdot 2} = \frac{1}{4}$ and $\frac{1}{2} \div \frac{1}{2} = \frac{1}{2} \cdot \frac{2}{1} = \frac{1}{1} \cdot \frac{1}{1} = 1$

73. $\frac{5}{8} - \frac{1}{2} < \frac{5}{8} \div \frac{1}{2}$ since $\frac{5}{8} - \frac{1}{2} = \frac{5}{8} - \frac{4}{8} = \frac{1}{8}$ and $\frac{5}{8} \div \frac{1}{2} = \frac{5}{8} \cdot \frac{2}{1} = \frac{10}{8}$

75. $-|-8|$, .38, $\frac{4}{9}$, $\frac{3}{5}$, $|-6|$ because $-|8| = -8$, $\frac{4}{9} = 0.444...$, $\frac{3}{5} = 0.6$ and $|-6| = 6$.

77. $\frac{5}{12}$, 0.6, $\frac{2}{3}$, $\frac{19}{25}$, $|-2.6|$ because $\frac{5}{12} = .416416...$, $\frac{2}{3} = .666...$, $\frac{19}{25} = .76$ and $|-2.6| = 2.6$.

79. 4 and -4 since $|4| = |-4| = 4$

For Exercises 81-87, answers will vary. One possible answer is given.

81. Three numbers greater than 4 and less than 6 are $4\frac{1}{2}$, 5, 5.5.

83. Three numbers less than -2 and greater than -6 are -3, -4, -5.

85. Three numbers greater than -3 and greater than 3 are 4, 5, 6.

87. Three numbers greater than $|-2|$ and less than $|-6|$ are 3, 4, 5.

89. a. Between does not include endpoints.

b. Three real numbers between 4 and 6 are 4.1, 5, and $5\frac{1}{2}$.

c. No, 4 is an endpoint.

d. Yes, 5 is greater than 4 and less than 6.

e. True

91.a. The taxes were less than 50 cents all the time

b. In March 2000 the gasoline was greater than \$1.50 for the firs time.

c. Gas was less than \$1.00 from January to March, 1999.

d. March 2000 – May 2001, except for June 2000

93. The result of dividing a number by itself is 1. Thus, the result of dividing a number between 0 and 1 by itself is a number, 1, which is greater than the original number.

95. Yes, 0 since $|0| = 0$ and $-|0| = -(0) = 0$

98. $1\frac{2}{3} - \frac{3}{8}$
$1\frac{2}{3} = \frac{3+2}{3} = \frac{5}{3} = \frac{5}{3} \cdot \frac{8}{8} = \frac{40}{24}$
$\frac{3}{8} = \frac{3}{8} \cdot \frac{3}{3} = \frac{9}{24}$
$1\frac{2}{3} - \frac{3}{8} = \frac{40}{24} - \frac{9}{24} = \frac{31}{24}$ or $1\frac{7}{24}$

99. The set of whole numbers is {0, 1, 2, 3, ...}.

100. The set of counting numbers is {1, 2, 3, 4, ...}.

101. a. 5 is a natural number.

b. 5 and 0 are whole numbers.

c. 5, –2, and 0 are integers.

d. 5, –2, 0, $\frac{1}{3}$, $-\frac{5}{9}$, and 2.3 are rational numbers.

e. $\sqrt{3}$ is an irrational number.

f. 5, –2, 0, $\frac{1}{3}$, $\sqrt{3}$, $-\frac{5}{9}$, and 2.3 are real numbers.

Exercise Set 1.6

1. The 4 basic operations of arithmetic are addition, subtraction, multiplication, and division.

b. One example is 3 and –3.

3. a. No; $-\frac{2}{3} + \frac{3}{2}$ does not equal 0.

b. The opposite of $-\frac{2}{3}$ is $\frac{2}{3}$ because $-\frac{2}{3} + \frac{2}{3} = 0$.

5. The sum of a positive number and a negative number can be either positive or negative. The sum of a positive number and a negative number has the same sign as the number that has larger absolute value.

7. Answers will vary.

9. a. He owed 175, a negative, and then paid a positive amount, 93, toward his debt.

b. $-175 + 93$
The numbers have different signs so find the difference between the absolute values.
$|-175| - |93| = 82$
$|-175|$ is greater so sum is negative.
$-175 + 93 = -82$

c. Answers will vary.

11. Yes, it is correct.

13. The opposite of 9 is –9 since 9 + (–9) = 0.

15. The opposite of –28 is 28 since –28 + 28 = 0.

17. The opposite of 0 is 0 since 0 + 0 = 0.

19. The opposite of $\frac{5}{3}$ is $-\frac{5}{3}$ since $\frac{5}{3} + \left(-\frac{5}{3}\right) = 0$.

21. The opposite of $2\frac{3}{5}$ is $-2\frac{3}{5}$ since
$2\frac{3}{5} + \left(-2\frac{3}{5}\right) = 0$.

23. The opposite of 3.72 is –3.72 since
3.72 + (-3.72) = 0.

25. Numbers have same sign, so add absolute values.
$|5| + |6| = 5 + 6 = 11$
Numbers are positive so sum is positive.
$5 + 6 = 11$

27. Numbers have different signs so find difference between larger and smaller absolute values.
$|4|-|-3|=4-3=1$. $|4|$ is greater than $|-3|$ so the sum is positive.
$4+(-3)=1$

29. Numbers have same sign, so add absolute values.
$|-4|+|-2|=4+2=6$.
Numbers are negative, so sum is negative.
$-4+(-2)=-6$

31. Numbers have different signs, so find difference between absolute values.
$|6|-|-6|=6-6=0$
$6+(-6)=0$

33. Numbers have different signs, so find difference between absolute values.
$|-4|-|4|=4-4=0$
$-4+4=0$

35. Numbers have same sign, so add absolute values.
$|-8|+|-2|=8+2=10$. Numbers are negative, so sum is negative.
$-8+(-2)=-10$

37. Numbers have different signs, so find difference between absolute values.
$|-6|-|6|=6-6=0$
$-6+6=0$

39. Numbers have same sign, so add absolute values.
$|-8|+|-5|=8+5=13$
Numbers are negative, so sum is negative.
$-8+(-5)=-13$

41. $0+0=0$

43. $-6+0=-6$

45. Numbers have different signs, so find difference between larger and smaller absolute values.
$|18|-|-9|=18-9=9$. $|18|$ is greater than $|-9|$ so sum is positive.
$18+(-9)=9$

47. Numbers have same sign, so add absolute values.
$|-33|+|-31|=33+31=64$
Numbers are negative, so sum is negative.
$-33+(-31)=-64$

49. Numbers have same sign, so add absolute values.
$|-42|+|-9|=42+9=51$.
Numbers are negative, so sum is negative.
$-42+(-9)=-51$

51. Numbers have different signs, so find difference between larger and smaller absolute values.
$|-30|-|4|=30-4=26$. $|-30|$ is greater than $|4|$ so sum is negative.
$4+(-30)=-26$

53. Numbers have different signs, so find difference between larger and smaller absolute values.
$|40|-|-35|=40-35=5$. $|40|$ is greater than $|-35|$ so sum is positive.
$-35+40=5$

55. Numbers have different signs, so take difference between larger and smaller absolute values.
$|6.5|-|-4.2|=6.5-4.2=2.3$. $|6.5|$ is greater than $|-4.2|$ so sum is positive.
$-4.2+6.5=2.3$

57. Numbers have same sign, so add absolute values.
$|-9.7|+|-5.4|=9.7+5.4=15.1$. Numbers are negative, so sum is negative..
$-9.7+(-5.4)=-15.1$

59. Numbers have different signs, so find difference between larger and smaller absolute values.
$|-200|-|180|=200-180=20$. $|-200|$ is greater than $|180|$ so sum is negative.
$180+(-200)=-20$

61. Numbers have different signs, so find difference between larger and smaller absolute values.
$|-67|-|28|=67-28=39$. $|-67|$ is greater than $|28|$ so sum is negative.
$-67+28=-39$

63. Numbers have different signs, so find difference between larger and smaller absolute values.
$|184|-|-93|=184-93=91$. $|184|$ is greater than $|-93|$ so sum is positive.
$184+(-93)=91$

65. Numbers have different signs, so find difference between larger and smaller absolute values.
$|-452|-|312|=452-312=140$. $|-452|$ is greater than $|312|$ so sum is negative.
$-452+312=-140$

67. Numbers have same sign, so add absolute values.
$|-26|+|-79|=26+79=105$. Numbers are negative so sum is negative.
$-26+(-79)=-105$

69. Numbers have same sign, so add absolute values.
$|-24.6|+|-13.9|=24.6+13.9=38.5$. Numbers are

negative so sum is negative.
$-24.6 + (-13.9) = -38.5$

71. Numbers have different signs, so find difference between larger and smaller absolute values.
$|110.9| - |106.3| = 110.9 - 106.3 = 4.6$. $|-110.9|$ is greater than $|106.3|$ so sum is negative.
$106.3 + (-110.9) = -4.6$

73. $\frac{3}{5} + \frac{1}{7} = \frac{21}{35} + \frac{5}{35} = \frac{21+5}{35} = \frac{26}{35}$

75. $\frac{5}{12} + \frac{6}{7} = \frac{35}{84} + \frac{72}{84} = \frac{35+72}{84} = \frac{107}{84}$

77. Numbers have different signs, so find difference between larger and smaller absolute values.
$-\frac{8}{11} + \frac{4}{5} = -\frac{40}{55} + \frac{44}{55} = \left|\frac{44}{55}\right| - \left|-\frac{40}{55}\right| = \frac{44}{55} - \frac{40}{55} = \frac{4}{55}$
$\left|\frac{44}{55}\right|$ is greater than $\left|-\frac{40}{55}\right|$ so sum is positive.
$-\frac{8}{11} + \frac{4}{5} = \frac{4}{55}$

79. Numbers have different signs, so find difference between larger and smaller absolute values.
$-\frac{7}{10} + \frac{11}{90} = \left|-\frac{63}{90}\right| - \left|\frac{11}{90}\right| = \frac{63}{90} - \frac{11}{90} = \frac{63-11}{90} = \frac{52}{90} = \frac{26}{45}$
$\left|-\frac{63}{90}\right|$ is greater than $\left|\frac{11}{90}\right|$ so sum is negative.
$-\frac{7}{10} + \frac{11}{90} = -\frac{26}{45}$

81. Numbers have different signs, so find difference between larger and smaller absolute values
$\frac{9}{25} + \left(-\frac{3}{50}\right) = \left|\frac{18}{50}\right| - \left|-\frac{3}{50}\right| = \frac{18}{50} - \frac{3}{50} = \frac{18-3}{50} =$
$\frac{15}{50} = \frac{3}{10}$
$\left|\frac{18}{50}\right|$ is greater than $\left|-\frac{3}{50}\right|$ so sum is positive.
$\frac{9}{25} + \left(-\frac{3}{50}\right) = \frac{3}{10}$

83. Numbers have same sign, so add absolute values.
$-\frac{7}{30} + \left(-\frac{5}{6}\right) = \left|-\frac{7}{30}\right| + \left|-\frac{25}{30}\right| = \frac{7}{30} + \frac{25}{30} = \frac{7+25}{30} =$
$\frac{32}{30} = \frac{16}{15}$
Numbers are negative so sum is negative.
$-\frac{7}{30} + \left(-\frac{5}{6}\right) = -\frac{16}{15}$

85. Numbers have same sign, so add absolute values.
$-\frac{4}{5} + \left(-\frac{5}{75}\right) = \left|-\frac{60}{75}\right| + \left|-\frac{5}{75}\right| = \frac{60}{75} + \frac{5}{75} = \frac{60+5}{75} =$
$\frac{65}{75} = \frac{13}{15}$
Numbers are negative so sum is negative.
$-\frac{4}{5} + \left(-\frac{5}{75}\right) = -\frac{13}{15}$

87. Numbers have different signs, so find difference between larger and smaller absolute values.
$\frac{5}{36} + \left(-\frac{5}{24}\right) = \frac{10}{72} + \left(-\frac{15}{72}\right) = \left|-\frac{15}{72}\right| - \left|\frac{10}{72}\right| =$
$\frac{15}{72} - \frac{10}{72} = \frac{5}{72}$
$\left|-\frac{15}{72}\right|$ is greater than $\left|\frac{10}{72}\right|$ so sum is negative.
$\frac{5}{36} + \left(-\frac{5}{24}\right) = -\frac{5}{72}$

89. Numbers have same sign, so add absolute values.
$-\frac{5}{12} + \left(-\frac{3}{10}\right) = -\frac{25}{60} + \left(-\frac{18}{60}\right) = \left|-\frac{25}{60}\right| + \left|-\frac{18}{60}\right| =$
$\frac{25}{60} + \frac{18}{60} = \frac{43}{60}$
Numbers are negative so sum is negative.
$-\frac{5}{12} + \left(-\frac{3}{10}\right) = -\frac{43}{60}$

91. Numbers have same sign, so add absolute values.
$-\frac{13}{14} + \left(-\frac{7}{42}\right) = -\frac{39}{42} + \left(-\frac{7}{42}\right) = \left|-\frac{39}{42}\right| + \left|-\frac{7}{42}\right| =$
$\frac{39}{42} + \frac{7}{42} = \frac{39+7}{42} = \frac{46}{42} = \frac{23}{21}$
Numbers are negative so sum is negative.
$-\frac{13}{14} + \left(-\frac{7}{42}\right) = -\frac{23}{21}$

93. a. Positive; $|587|$ is greater than $|-197|$ so sum will be positive.

b. $587 + (-197) = 390$

c. Yes; By part a) we expect a positive sum. The magnitude of the sum is the difference between the larger and smaller absolute values.

95. a. Negative; the sum of 2 negative numbers is always negative.

b. $-84 + (-289) = -373$

c. Yes; the sum of 2 negative numbers should be (and is) a larger negative number.

97. a. Negative; $|-947|$ is greater than $|495|$ so sum will be negative.

b. $-947 + 495 = -452$

c. Yes; by part a) we expect a negative sum. Magnitude of the sum is the difference between the larger and smaller absolute values.

99. a. Negative; the sum of 2 negative numbers is always negative.

b. $-496 + (-804) = -1300$

c. Yes; the sum of 2 negative numbers should be (and is) a larger negative number.

101. a. Negative; $|-375|$ is greater than $|263|$ so sum will be negative.

b. $-375 + 263 = -112$

c. Yes; by part a) we expect a negative sum. The magnitude of the sum is the difference between the larger and smaller absolute values.

b. $1127 + (-84) = 1043$

c. Yes; by part a) we expect a positive sum. Magnitude of sum is difference between larger and smaller absolute values.

103. a. Negative; the sum of 2 negative numbers is always negative.

b. $-1833 + (-2047) = -3880$

c. Yes; The sum of 2 negative numbers should be (and is) a larger negative number.

105. a. Positive; $|3124|$ is greater than $|-2013|$ so sum will be positive.

b. $3124 + (-2013) = 1111$

c. Yes; by part a) we expect a positive sum. Magnitude of sum is difference between larger and smaller absolute values.

107. a. Negative; the sum of 2 negative numbers is always negative.

b. $-1025 + (-1025) = -2050$

c. Yes; the sum of 2 negative numbers should be (and is) a larger negative number.

109. True

111. True; the sum of two positive numbers is always positive.

113. False; the sum has the sign of the number with the larger absolute value.

115. Mr. Peter's balance was –\$94. His new balance can be found by adding. $-94 + (-183) = -277$
Mr. Peter owes the bank \$277.

117. Total loss can be represented as $-18 + (-3)$.
$|-18| + |-3| = 18 + 3 = 21$. The total loss in yardage is 21 yards.

119. The depth of the well can be found by adding $-27 + (-34) = -61$. The well is 61 feet deep.

121. Distance from sea level to place on mountain can be represented by $-267 + 198$.
$|-267| - |198| = 267 - 198 = 69$
$|-267| > |198|$ so $-267 + 198 = -69$
Their vertical distance is 69 feet below sea level.

123. a. In 2001, the Postal Service will have a net loss of \$3 billion.

b. 1999: 0,4 billion dollars
2000: - 0.3 billion dollars
2001: -3 billion dollars
$0.4+(-0.3)+(-3)= -2.9$ billion dollars
From 1999 through 2001, the Postal Service had a net loss of \$2.9 billion.

125. $(-4) + (-6) + (-12) = (-10) + (-12) = -22$

127. $29 + (-46) + 37 = (-17) + 37 = 20$

129. $(-12)+(-10)+25+(-3)=(-22)+25+(-3)$
$=3+(-3)$
$=0$

131. $\frac{1}{2}+\left(-\frac{1}{3}\right)+\frac{1}{5}=\left(\frac{3}{6}-\frac{2}{6}\right)+\frac{1}{5}$
$=\frac{1}{6}+\frac{1}{5}$
$=\frac{5}{30}+\frac{6}{30}$
$=\frac{11}{30}$

133. $1+2+3+\cdots+10=(1+10)+(2+9)\cdots+(5+6)$
$=(5)(11)$
$=55$

135. $1\frac{2}{3}=\frac{3+2}{3}=\frac{5}{3}$
$\left(\frac{3}{5}\right)\left(1\frac{2}{3}\right)=\left(\frac{3}{5}\right)\left(\frac{5}{3}\right)=\frac{1}{1}\cdot\frac{1}{1}=1$

136. $3=\frac{3}{1}\cdot\frac{16}{16}=\frac{48}{16}$
$3-\frac{5}{16}=\frac{48}{16}-\frac{5}{16}=\frac{48-5}{16}=\frac{43}{16}$ or $2\frac{11}{16}$

137. $\{1, 2, 3, 4, \ldots\}$

138. $|-3|>2$ since $|-3|=3$

139. $8>|-7|$ since $|-7|=7$

Exercise Set 1.7

1. 2 - 7

3. $*$ − □

5. a. to subtract b from a, add the opposite of b to a.

b. 5 + (-14)

c. 5 + (-14) = -9

7. a. $a-(+b)=a-b$

b. $7-(+9)=7-9$

c. $7-9=7+(-9)=-2$

9. a. $3-(-6)+(-5)=3+6-5$

b. $3+6-5=9-5=4$

11. Yes it is correct.

13. $12-5=12+(-5)=7$

15. $8-9=8+(-9)=-1$

17. $-4-2=-4+(-2)=-6$

19. $-4-(-3)=-4+3=-1$

21. $-4-4=-4+(-4)=-8$

23. $0-7=0+(-7)=-7$

25. $8-8=8+(-8)=0$

27. $-3-1=-3+(-1)=-4$

29. $5-3=5+(-3)=2$

31. $6-(-3)=6+3=9$

33. $0-(-5)=0+5=5$

35. $-9-11=-9+(-11)=-20$

37. $(-4)-(-4)=-4+4=0$

39. $-8-(-12)=-8+12=4$

41. $-6-(-2)=-6+2=-4$

43. $-9-2=-9+(-2)=-11$

45. $-35-(-8)=-35+8=-27$

47. $-90-60=-90+(-60)=-150$

49. $-45-37=-45+(-37)=-82$

51. $70-(-70)=70+70=140$

53. $42.3-49.7=42.3+(-49.7)=-7.4$

55. $-7.85-(-3.92)=-7.85+3.92=-3.93$

57. $-20-9=-20+(-9)=-29$

59. $-8-8=-8+(-8)=-16$

61. $-5-(-3)=-5+3=-2$

63. $9-(-4)=9+4=13$

65. $13-24=13+(-24)=-11$

67. $-6.3-(-12.4)=-6.3+12.4=6.1$

69. $\frac{4}{5}-\frac{5}{6}=\frac{24}{30}-\frac{25}{30}=\frac{24-25}{30}=-\frac{1}{30}$

71. $\frac{8}{15}-\frac{7}{45}=\frac{24}{45}-\frac{7}{45}=\frac{24-7}{45}=\frac{17}{45}$

73. $-\frac{1}{4}-\frac{2}{3}=-\frac{3}{12}-\frac{8}{12}=\frac{-3-8}{12}=-\frac{11}{12}$

75. $-\frac{4}{15}-\frac{3}{20}=-\frac{16}{60}-\frac{9}{60}=\frac{-16-9}{60}=$
$-\frac{25}{60}=-\frac{5}{12}$

77. $\frac{3}{8}-\frac{6}{48}=\frac{18}{48}-\frac{6}{48}=\frac{18-6}{48}=\frac{12}{48}=\frac{1}{4}$

79. $-\frac{7}{12}-\frac{5}{40}=-\frac{70}{120}-\frac{15}{120}=\frac{-70-15}{120}$
$=-\frac{85}{120}=-\frac{17}{24}$

81. $\frac{3}{16}-\left(-\frac{5}{8}\right)=\frac{3}{16}+\frac{10}{16}=\frac{3+10}{16}=\frac{13}{16}$

83. $-\frac{5}{12}-\left(-\frac{3}{8}\right)=-\frac{10}{24}+\frac{9}{24}=\frac{-10+9}{24}=-\frac{1}{24}$

85. $\frac{4}{7}-\frac{7}{9}=\frac{36}{63}-\frac{49}{63}=\frac{36-49}{63}=-\frac{13}{63}$

87. $-\frac{5}{12}-\left(-\frac{3}{10}\right)=-\frac{25}{60}+\frac{18}{60}=\frac{-25+18}{60}=-\frac{7}{60}$

89. a. Positive; $378-279=378+(-279)$
$|378|$ is greater than $|-279|$ so the sum will be positive.

b. $378+(-279)=99$

c. Yes; by part a) we expect a positive sum. The size of the sum is the difference between the absolute values of the 2 numbers.

91. a. Negative; $-482-137=-482+(-137)$
The sum of 2 negative numbers is always negative.

b. $-482+(-137)=-619$

c. Yes; the sum of two negative numbers should be (and is) a larger negative number.

93. a. Positive; $843-(-745)=843+745$.
The sum of 2 positive numbers is always positive.

b. $843+745=1588$

c. Yes; by part a) we expect a positive answer. The size of the sum is the sum of the absolute values of the numbers.

95. a. Positive; $-408-(-604)=-408+604$.
$|604|$ is greater than $|-408|$ so the sum will be positive.

b. $-408+604=196$

c. Yes; by part a) we expect a positive answer. The size of the answer is the difference between the larger and smaller absolute values.

97. a. Negative; $-1024-(-576)=-1024+576$.
$|-1024|$ is greater than $|576|$ so the sum will be negative.

b. $-1024+576=-448$

c. Yes; by part a) we expect a negative answer. The size of the answer is the difference between the larger and the smaller absolute values.

99. a. Positive; $165.7-49.6=165.7+(-49.6)$.
$|165.7|$ is greater than $|-49.6|$ so the sum will be positive.

b. $165.7+(-49.6)=116.1$

c. Yes; by part a) we expect a negative answer. The size of the answer is the difference between the larger and the smaller absolute values.

101. a. Negative; $295-364=295+(-364)$.
Since $|-364|$ is greater than $|295|$ the answer will be negative.

b. $295+(-364)=-69$

c. Yes; by part a) we expect a negative answer. The size of the answer is the difference between the larger and the smaller absolute values.

103. a. Negative; $-1023-647=-1023+(-647)$.
The sum of two negative numbers is always negative.

b. $-1023+(-647)=-1670$

c. Yes; the sum of two negative numbers should be (and is) a larger negative number.

105. a. Zero; $-7.62-(-7.62)=-7.62+7.62$.
The sum of two opposite numbers is always zero.

b. $-7.62+7.62=0$

c. Yes; by part a) we expect zero.

c. Yes; the sum of two negative numbers should be (and is) a larger negative number.

107. $7+5-(+8)=7+5+(-8)=12+(-8)=4$

109. $-6+(-6)+6=-12+6=-6$

111. $-13-(+5)+3=-13+(-5)+3=-18+3=-15$

113. $-9-(-3)+4=-9+3+4=-6+4=-2$

115. $5-(-9)+(-1)=5+9+(-1)=14+(-1)=13$

117. $$\begin{aligned}17+(-8)-(+14)&=17+(-8)+(-14)\\&=9+(-14)\\&=-5\end{aligned}$$

119. $-36-5+9=-36+(-5)+9=-41+9=-32$

121. $-2+7-9=-2+7+(-9)=5+(-9)=-4$

123. $25-19+3=25+(-19)+3=6+3=9$

125. $$\begin{aligned}(-4)+(-6)+5-7&=(-4)+(-6)+5+(-7)\\&=-10+5+(-7)\\&=-5+(-7)\\&=-12\end{aligned}$$

127. $$\begin{aligned}17+(-3)-9-(-7)&=17+(-3)+(-9)+7\\&=14+(-9)+7\\&=5+7\\&=12\end{aligned}$$

129. $$\begin{aligned}-9+(-7)+(-5)-(-3)&=-9+(-7)+(-5)+3\\&=-16+(-5)+3\\&=-21+3\\&=-18\end{aligned}$$

131. a. $300-343=300+(-343)=-43$
They had 43 sweaters on back order.

b. $43+100=143$
They would need to order 143 sweaters.

133. $42-58=42+(-58)=-16$
The bottom of the well is 16 feet below sea level.

135. $44-(-56)=44+56=100$
Thus the temperature dropped 100°F.

137. a. $288+(-12)=276$
In 2002, his score was 276.

b. $-12-(-4)=-12+4=-8$
He had 8 strokes less in 2002 than in 2000.

c. 2000 stroke score: $288+(-4)=284$
2001 stroke score: $288+(-16)=272$
2002 stroke score: $288+(-12)=276$

$$\text{Average}=\frac{284+272+276}{3}=\frac{832}{3}\approx 277.33$$
His average stroke score was 277.33.

139. $$\begin{aligned}&1-2+3-4+5-6+7-8+9-10\\&=(1-2)+(3-4)+(5-6)+(7-8)+(9-10)\\&=(-1)+(-1)+(-1)+(-1)+(-1)\\&=-5\end{aligned}$$

141. a. 7 units

b. $5-(-2)=7$

143. a. $3+2+2+1+1=9$
The ball travels 9 feet vertically.

b. $-3+2+(-2)+1+(-1)=-3$
The net distance is –3 feet.

144. The integers are $\{\ldots,-2,-1,0,1,2,\ldots\}$.

145. The set of rational numbers together with the set of irrational numbers forms the set of real numbers.

146. $|-3|>-5$ since $|-3|=3$

147. $|-6|<|-7|$ since $|-6|=6$ and $|-7|=7$

148. $\frac{4}{5}-\frac{3}{8}=\frac{32}{40}-\frac{15}{40}=\frac{32-15}{40}=\frac{17}{40}$

Exercise Set 1.8

1. Like signs: product is positive. Unlike signs: product is negative.

3. When multiplying 3 or more real numbers, the product is positive if there is an even number of negative numbers and the product is negative if there is an odd number

5. A fraction of the form $\frac{a}{-b}$ is generally written $-\frac{a}{b}$ or $\frac{-a}{b}$.

7. a. With $3-5$ you subtract, but with $3(-5)$ you multiply.

b. $3-5=3+(-5)=-2$
$3(-5)=-15$

9. a. With $x-y$ you subtract, but with $x(-y)$ you multiply.

b. $x-y=5-(-2)=5+2=7$

c. $x(-y)=[-(-2)]=5(2)=10$

d. $-x-y=-5-(-2)=-5+2=-3$

11. The product $(8)(4)(-5)$ is negative since there is an odd number (1) of negatives

13. The product $(-102)(-16)(24)(19)$ is positive since there is an even number of negatives.

15. The product $(-40)(-16)(30)(50)(-13)$ is negative since there is an odd number (3) of negatives.

17. Since the numbers have like signs, the product is positive. $(-5)(-4)=20$

19. Since the number have unlike signs, the product is negative. $6(-3)=-18$

21. Since the numbers have unlike signs, the product is negative. $(-2)(4)=-8$

23. Zero multiplied by any real number equals zero. $0(-5)=0$

25. Since the numbers have like signs, the product is positive. $6(7)=42$

27. Since the numbers have unlike signs, the product is negative. $(7)(-8)=-56$

29. Since the numbers have like signs, the product is positive. $(-5)(-6)=30$

31. Zero multiplied by any real number equals zero. $0(3)(8)=0(8)=0$

33. Since there is one negative number (an odd number), the product will be negative.
$(21)(-1)(4)=(-21)(4)=-84$

35. Since there are three negative numbers (an odd number), the product will be negative.
$-1(-3)(3)(-8)=3(3)(-8)=9(-8)=-72$

37. Since there are two negative numbers (an even number), the product will be positive.
$(-4)(5)(-7)(1)=(-20)(-7)(1)=(140)(1)=140$

39. Zero multiplied by any real number equals zero.
$(-1)(3)(0)(-7)=(-3)(0)(-7)=0(-7)=0$

41. $\left(\frac{-1}{2}\right)\left(\frac{3}{5}\right)=\frac{(-1)(3)}{2\cdot 5}=\frac{-3}{10}=-\frac{3}{10}$

43. $\left(\frac{-5}{9}\right)\left(\frac{-7}{15}\right)=\frac{(-5)(-7)}{9\cdot 15}=\frac{(-1)(-7)}{9\cdot 3}=\frac{7}{27}$

45. $\left(\frac{6}{-3}\right)\left(\frac{4}{-2}\right)=\left(\frac{2}{-1}\right)\left(\frac{2}{-1}\right)=\frac{(2)(2)}{(-1)(-1)}=\frac{4}{1}=4$

47. $\left(\frac{-3}{8}\right)\left(\frac{5}{6}\right)=\left(\frac{-1}{8}\right)\left(\frac{5}{2}\right)=\frac{(-1)(5)}{(8)(2)}=\frac{-5}{16}=-\frac{5}{16}$

49. Since the numbers have like signs, the quotient is positive. $\frac{10}{5}=2$

51. Since the numbers have like signs, the quotient is positive. $-16\div(-4)=\frac{-16}{-4}=4$

53. Since the numbers have like signs, the quotient is positive. $\frac{-36}{-9}=4$

55. Since the numbers have unlike signs, the quotient is negative. $\frac{36}{-2}=-18$

57. Since the numbers have like signs, the quotient is positive. $\frac{-12}{-1}=12$

59. Since the numbers have unlike signs, the quotient is negative. $40/(-4)=\frac{40}{-4}=-10$

61. Since the numbers have unlike signs, the quotient is negative. $\frac{-42}{7}=-6$

63. Since the numbers have unlike signs, the quotient is negative. $\frac{36}{-4} = -9$

65. Since the numbers have like signs, the quotient is positive. $-64 \div (-8) = \frac{-64}{-8} = 8$

67. Zero divided by any nonzero number is zero. $\frac{0}{4} = 0$

69. Since the numbers have unlike signs, the quotient is negative. $\frac{30}{-10} = -3$

71. Since the numbers have unlike signs, the quotient is negative. $\frac{-120}{30} = -4$

73. $$\begin{aligned} \frac{3}{12} \div \left(\frac{-5}{8}\right) &= \frac{3}{12} \cdot \left(\frac{8}{-5}\right) \\ &= \frac{1}{1} \cdot \left(\frac{2}{-5}\right) \\ &= \frac{1 \cdot 2}{1(-5)} \\ &= \frac{2}{-5} \\ &= -\frac{2}{5} \end{aligned}$$

75. $$\frac{-5}{12} \div (-3) = \frac{-5}{12} \cdot \frac{1}{-3} = \frac{-5(1)}{12(-3)} = \frac{-5}{-36} = \frac{5}{36}$$

77. $$\begin{aligned} \frac{-15}{21} \div \left(\frac{-15}{21}\right) &= \frac{-15}{21} \cdot \frac{21}{-15} \\ &= \frac{-1}{1} \cdot \frac{1}{-1} \\ &= \frac{(-1)(1)}{(1)(-1)} \\ &= \frac{-1}{-1} \\ &= 1 \end{aligned}$$

79. $$\begin{aligned} -12 \div \frac{5}{12} &= \frac{-12}{1} \cdot \frac{12}{5} \\ &= \frac{(-12)(12)}{(1)(5)} \\ &= \frac{-144}{5} \\ &= -\frac{144}{5} \end{aligned}$$

81. Since the numbers have unlike signs, the product is negative. $-4(8) = -32$

83. Since the numbers have unlike signs, the quotient is negative. $\frac{100}{-5} = -20$

85. Since the numbers have unlike signs, the product is negative. $-7(2) = -14$

87. Since the numbers have unlike signs, the quotient is negative. $27 \div (-3) = \frac{27}{-3} = -9$

89. Since the numbers have unlike signs, the quotient is negative. $\frac{-100}{5} = -20$

91. Since the numbers have unlike signs, the quotient is negative. $\frac{60}{-60} = -1$

93. Zero divided by any nonzero number is zero. $0 \div 6 = \frac{0}{6} = 0$

95. Any nonzero number divided by zero is undefined. $\frac{5}{0}$ is undefined.

97. Zero divided by any nonzero number is zero. $\frac{0}{1} = 0$

99. Any nonzero number divided by zero is undefined. $\frac{8}{0}$ is undefined.

101. **a.** Since the numbers have unlike signs, the product will be negative.

b. $92(-38) = -3496$

c. Yes; as expected the product is negative.

103. **a.** Since the numbers have unlike signs, the quotient will be negative.

b. $-240 / 15 = \frac{-240}{15} = -16$

c. Yes; as expected the quotient is negative.

105. **a.** Since the numbers have unlike signs, the quotient will be negative.

b. $243 \div (-27) = \frac{243}{-27} = -9$

c. Yes; as expected the quotient is negative.

107. a. Since the numbers have like signs, the product will be positive.

b. $(-49)(-126) = 6174$

c. Yes; as expected the product is positive.

109. a. The quotient will be zero; zero divided by any nonzero number is zero.

b. $\frac{0}{5335} = 0$

c. Yes; as expected the answer is zero.

111. a. Undefined; any nonzero number divided by 0 is undefined.

b. $7.2 \div 0 = \frac{7.2}{0}$ is undefined

c. Yes; as expected the quotient is undefined.

113. a. Since the numbers have like signs, the quotient will be positive.

b. $8 \div 2.5 = \frac{8}{2.5} = 3.2$

c. Yes; as expected the quotient is positive.

115. a. Since there are two negative numbers (an even number), the product will be positive.

b. $(-3.0)(4.2)(-18) = 226.8$

c. Yes; as expected the product is positive.

117. False; the product of two numbers with like signs is a positive number

119. False; the quotient of two numbers with unlike signs is a negative number.

121. True

123. False; zero divided by any nonzero number is zero.

125. False; zero divided by 1 is zero.

127. True

129. $4(-15) = -60$
The total loss was 60 yards.

131. a. $\frac{1}{3}(450) = \frac{450}{3} = 150$
She paid back \$150.

b. $-450 + 150 = -300$
Her new balance is –\$300.

133. $3\left(-1\frac{1}{2}\right) = \frac{3}{1}\left(-\frac{3}{2}\right) = -\frac{9}{2}$ or $-4\frac{1}{2}$
It has lost $4\frac{1}{2}$ points.

135. a. $\frac{\text{change in October}}{\text{change in September}} = \frac{-415{,}000}{200{,}000} \approx 2.075$
The change in October was about 2.075 times larger than in September.

b. $\frac{\text{change in October}}{\text{change in July}} = \frac{-415{,}000}{-50{,}000} \approx 8.3$
The change in October was about 8.3 times larger than in July.

137. a. $220 - 50 = 170$
60% of $170 = 0.6(170) = 102$
75% of $170 = 0.75(170) = 127.5$
Target heart rate is 102 to 128 beats per minute.

b. Answers will vary.

139. $(-2)^3 = (-2)(-2)(-2) = 4(-2) = -8$

141. $1^{100} = 1$

143. The product $(-1)(-2)(-3)(-4)\cdots(-10)$ will be positive because there are an even number (10) of negative numbers.

145. The country will start with D. Most students will select Denmark. They will most likely select kangaroo which leads to orange.

146. $\frac{5}{7} \div \frac{1}{5} = \frac{5}{7} \cdot \frac{5}{1} = \frac{5 \cdot 5}{7 \cdot 1} = \frac{25}{7}$ or $3\frac{4}{7}$

147. $-20 - (-18) = -20 + 18 = -2$

148. $6 - 3 - 4 - 2 = 3 - 4 - 2 = -1 - 2 = -3$

149. $5-(-2)+3-7=5+2+3-7$
$=7+3-7$
$=10-7$
$=3$

150. $-40 \div (-8) = 5$

Exercise Set 1.9

1. In the expression a^b, a is the base and b is the exponent.

3. a. Every number has an understood exponent of 1.

b. In $5x^3y^2z$, 5 has exponent of 1, x has an exponent of 3, y has an exponent of 2, and z has an exponent of 1.

5. a. $x+x+x+x+x=5x$

b. $x \cdot x \cdot x \cdot x \cdot x = x^5$

7. The order of operations are parentheses, exponents, multiplication or division, then addition or subtraction.

9. No; $4+5\times 2=4+10=14$, on a scientific calculator.

11. a. $20 \div 5 - 3 = 4 - 3 = 1$

b. $20 \div (5-3) = 20 \div 2 = 10$

c. The keystrokes in b) are used since the fraction bar is a grouping symbol.

13. b. $[10-(16\div 4)]^2-6^3=[10-4]^2-6^3$
$=6^2-6^3$
$=36-216$
$=-180$

15. b. When $x=5$:
$-4x^2+3x-6=-4(5)^2+3(5)-6$
$=-4(25)+3(5)-6$
$=-100+15-6$
$=-85-6$
$=-91$

17. $5^2=5\cdot 5=25$

19. $1^7=1\cdot 1\cdot 1\cdot 1\cdot 1\cdot 1\cdot 1=1$

21. $-5^2=-(5)(5)=-25$

23. $(-3)^2=(-3)(-3)=9$

25. $(-1)^3=(-1)(-1)(-1)=-1$

27. $(-9)^2=(-9)(-9)=81$

29. $(-6)^2=(-6)(-6)=36$

31. $4^1=4$

33. $(-4)^4=(-4)(-4)(-4)(-4)=256$

35. $-2^4=-(2)(2)(2)(2)=-16$

37. $\left(\frac{3}{4}\right)^2=\frac{3}{4}\cdot\frac{3}{4}=\frac{9}{16}$

39. $\left(-\frac{1}{2}\right)^5=\left(-\frac{1}{2}\right)\left(-\frac{1}{2}\right)\left(-\frac{1}{2}\right)\left(-\frac{1}{2}\right)\left(-\frac{1}{2}\right)=-\frac{1}{32}$

41. $5^2\cdot 3^2=5\cdot 5\cdot 3\cdot 3=225$

43. $4^3\cdot 3^2=4\cdot 4\cdot 4\cdot 3\cdot 3=576$

45. a. Positive; a positive number raised to any power is positive.

b. $7^3=343$

c. Yes; as expected the answer is positive.

47. a. Positive; a positive number raised to any power is positive.

b. $5^4=625$

c. Yes; as expected the answer is positive.

49. a. Negative; a negative number raised to an odd power is negative.

b. $(-3)^5=-243$

c. Yes; as expected the answer is negative.

51. a. Positive; a negative number raised to an even power is positive.

b. $(-5)^4=625$

c. Yes; as expected the answer is positive.

53. a. Positive; a positive number raised to any power is positive.

b. $(4.6)^4 = 447.7456$

c. Yes; as expected the answer is positive.

55. a. Negative; $\left(\frac{7}{8}\right)^2$ is positive therefore, $-\left(\frac{7}{8}\right)^2$ is negative.

b. $-\left(\frac{7}{8}\right)^2 = -0.765625$

c. Yes; as expected the answer is negative.

57. $3 + 2 \cdot 6 = 3 + 12 = 15$

59. $6 - 6 + 8 = 0 + 8 = 8$

61. $1 + 3 \cdot 2^2 = 1 + 3 \cdot 4 = 1 + 12 = 13$

63. $-3^3 + 27 = -27 + 27 = 0$

65. $(4-3)\cdot(5-1)^2 = (1)\cdot(4)^2 = 1 \cdot 16 = 16$

67. $3 \cdot 7 + 4 \cdot 2 = 21 + 8 = 29$

69. $5 - 2(7+5) = 5 - 2(12) = 5 - 24 = -19$

71. $-32 - 5(7-10)^2 = -32 - 5(-3)^2 = -32 - 5(9) = -32 - 45 = -77$

73. $\frac{3}{4} + 2\left(\frac{1}{5}\right)^2 = \frac{3}{4} + 2\left(\frac{1}{25}\right) = \frac{3}{4} + \frac{2}{25} = \frac{75}{100} + \frac{8}{100} = \frac{83}{100}$

75.
$$\begin{aligned} 4^2 - 3 \cdot 4 - 6 &= 16 - 3 \cdot 4 - 6 \\ &= 16 - 12 - 6 \\ &= 4 - 6 \\ &= -2 \end{aligned}$$

77.
$$\begin{aligned} (6 \div 3)^3 + 4^2 \div 8 &= (2)^3 + 4^2 \div 8 \\ &= 8 + 16 \div 8 \\ &= 8 + 2 \\ &= 10 \end{aligned}$$

79.
$$\begin{aligned} [-8(-2+5)]^2 &= [-8(3)]^2 \\ &= (-24)^2 \\ &= 576 \end{aligned}$$

81.
$$\begin{aligned} (3^2 - 1) \div (3+1)^2 &= (9-1) \div (3+1)^2 \\ &= 8 \div (4)^2 \\ &= 8 \div 16 \\ &= \frac{8}{16} \\ &= \frac{1}{2} \end{aligned}$$

83.
$$\begin{aligned} [4 + ((5-2)^2 \div 3)^2]^2 &= [4 + ((3)^2 \div 3)^2]^2 \\ &= [4 + (9 \div 3)^2]^2 \\ &= [4 + (3)^2]^2 \\ &= [4 + 9]^2 \\ &= (13)^2 \\ &= 169 \end{aligned}$$

85.
$$\begin{aligned} &2.5 + 7.56 \div 2.1 + (9.2)^2 \\ &= 2.5 + 7.56 \div 2.1 + 84.64 \\ &= 2.5 + 3.6 + 84.64 \\ &= 6.1 + 84.64 \\ &= 90.74 \end{aligned}$$

87.
$$\begin{aligned} 2[1.55 + 5(3.7)] - 3.35 &= 2[1.55 + 18.5] - 3.35 \\ &= 2(20.05) - 3.35 \\ &= 40.1 - 3.35 \\ &= 36.75 \end{aligned}$$

89.
$$\begin{aligned} \left(\frac{2}{5} + \frac{3}{8}\right) - \frac{3}{20} &= \left(\frac{16}{40} + \frac{15}{40}\right) - \frac{3}{20} \\ &= \frac{31}{40} - \frac{3}{20} \\ &= \frac{31}{40} - \frac{6}{40} \\ &= \frac{25}{40} = \frac{5}{8} \end{aligned}$$

91. $\frac{3}{4} - 4 \cdot \frac{5}{40} = \frac{3}{4} - \frac{4}{1} \cdot \frac{5}{40} = \frac{3}{4} - \frac{4}{8} = \frac{3}{4} - \frac{2}{4} = \frac{1}{4}$

93.
$$\begin{aligned} \frac{4}{5} + \frac{3}{4} \div \frac{1}{2} - \frac{2}{3} &= \frac{4}{5} + \frac{3}{4} \cdot \frac{2}{1} - \frac{2}{3} \\ &= \frac{4}{5} + \frac{3}{2} - \frac{2}{3} \\ &= \frac{24}{30} + \frac{45}{30} - \frac{20}{30} \\ &= \frac{49}{30} \end{aligned}$$

95. $\dfrac{5-[3(6\div 3)-2]}{5^2-4^2\div 2}=\dfrac{5-[3(2)-2]}{25-16\div 2}$
$=\dfrac{5-[6-2]}{25-8}$
$=\dfrac{5-4}{25-8}$
$=\dfrac{1}{17}$

97. $\dfrac{-[4-(6-12)^2]}{[(9\div 3)+4]^2+2^2}=\dfrac{-[4-(-6)^2]}{(3+4)^2+4}$
$=\dfrac{-[4-36]}{7^2+4}$
$=\dfrac{-(-32)}{49+4}$
$=\dfrac{32}{53}$

99. $\{5-2[4-(6\div 2)]^2\}^2=\{5-2[4-3]^2\}^2$
$=\{5-2(1)^2\}^2$
$=\{5-2(1)\}^2$
$=\{5-2\}^2$
$=(3)^2$
$=9$

101. $-\{4-[-3-(2-5)]^2\}=-\{4-[-3-(-3)]^2\}$
$=-\{4-[-3+3]^2\}$
$=-\{4-[0]^2\}$
$=-\{4-0\}$
$=-(4)$
$=-4$

103. Substitute 3 for x

a. $x^2=3^2=3\cdot 3=9$

b. $-x^2=-3^2=-(3)(3)=-9$

c. $(-x)^2=(-3)^2=(-3)(-3)=9$

105. Substitute −4 for x

a. $x^2=(-4)^2=(-4)(-4)=16$

b. $-x^2=-(-4)^2=-(-4)(-4)=-(16)=-16$

c. $(-x)^2=4^2=4\cdot 4=16$

107. Substitute 6 for x

a. $x^2=6^2=6\cdot 6=36$

b. $-x^2=-6^2=-(6\cdot 6)=-36$

c. $(-x)^2=(-6)^2=(-6)(-6)=36$

109. Substitute $-\frac{1}{3}$ for x

a. $x^2=\left(-\frac{1}{3}\right)^2=\left(-\frac{1}{3}\right)\left(-\frac{1}{3}\right)=\frac{1}{9}$

b. $-x^2=-\left(-\frac{1}{3}\right)^2=-\left(-\frac{1}{3}\right)\left(-\frac{1}{3}\right)=-\frac{1}{9}$

c. $(-x)^2=\left(\frac{1}{3}\right)^2=\left(\frac{1}{3}\right)\left(\frac{1}{3}\right)=\frac{1}{9}$

111. Substitute −2 for x in the expression.
$x+6=-2+6=4$

113. Substitute 4 for z in the expression.
$5z-2=5(4)-2=20-2=18$

115. Substitute −3 for a in the expression.
$a^2-6=(-3)^2-6=9-6=3$

117. Substitute −1 for each x in the expression.
$-4x^2-2x+1=-4(-1)^2-2(-1)+1$
$=-4(1)-2(-1)+1$
$=-4+2+1$
$=-2+1$
$=-1$

119. Substitute 2 for each p in the expression.
$3p^2-6p-4=3(2)^2-6(2)-4$
$=3(4)-12-4$
$=12-12-4$
$=0-4$
$=-4$

121. Substitute $\frac{1}{2}$ for each x in the expression.

$$\begin{aligned} -x^2 - 2x + 5 &= -(\tfrac{1}{2})^2 - 2(\tfrac{1}{2}) + 5 \\ &= -\frac{1}{4} - 1 + 5 \\ &= -\frac{1}{4} - \frac{4}{4} + \frac{20}{4} \\ &= -\frac{5}{4} + \frac{20}{4} \\ &= \frac{15}{4} \end{aligned}$$

123. Substitute 5 for each x in the expression.

$$\begin{aligned} 4(3x+1)^2 - 6x &= 4(3(5)+1)^2 - 6(5) \\ &= 4(15+1)^2 - 30 \\ &= 4(16)^2 - 30 \\ &= 4(256) - 30 \\ &= 1024 - 30 \\ &= 994 \end{aligned}$$

125. Substitute 3 for s and 4 for t in the expression.
$-3s + 2t = -3(3) + 2(4) = -9 + 8 = -1$

127. Substitute –2 for r and –3 for s in the expression.
$r^2 - s^2 = (-2)^2 - (-3)^2 = 4 - 9 = -5$

129. Substitute 2 for x and –3 for y in the expression.

$$\begin{aligned} 2(x+2y) + 4x - 3y &= 2[2 + 2(-3)] + 4(2) - 3(-3) \\ &= 2(2 + (-6)) + 8 - (-9) \\ &= 2(-4) + 8 + 9 \\ &= -8 + 8 + 9 \\ &= 0 + 9 \\ &= 9 \end{aligned}$$

131. Substitute 2 for x and –3 for y in the expression.

$$\begin{aligned} 6x^2 + 3xy - y^2 &= 6(2)^2 + 3(2)(-3) - (-3)^2 \\ &= 6(4) + 3(2)(-3) - 9 \\ &= 24 + (-18) - 9 \\ &= 6 - 9 \\ &= -3 \end{aligned}$$

133. $6 \cdot 3$ Multiply 6 by 3
$(6 \cdot 3) - 4$ Subtract 4 from the product
$[(6 \cdot 3) - 4] - 2$ Subtract 2 from the difference
Evaluate:
$[(6 \cdot 3) - 4] - 2 = [18 - 4] - 2 = 14 - 2 = 12$

135. $18 \div 3$ Divide 18 by 3
$(18 \div 3) + 9$ Add 9 to the quotient
$[(18 \div 3) + 9] - 8$ Subtract 8 from the sum
$9\{[(18 \div 3) + 9] - 8\}$ Multiply the difference by 9
Evaluate:

$$\begin{aligned} 9\{[(18 \div 3) + 9]\} - 8 &= 9\{[6 + 9] - 8\} \\ &= 9[15 - 8] \\ &= 9(7) \\ &= 63 \end{aligned}$$

137. $\frac{4}{5} + \frac{3}{7}$ Add $\frac{4}{5}$ to $\frac{3}{7}$
$\left(\frac{4}{5} + \frac{3}{7}\right) \cdot \frac{2}{3}$ Multiply the sum by $\frac{2}{3}$
Evaluate:

$$\begin{aligned} \left(\frac{4}{5} + \frac{3}{7}\right) \cdot \frac{2}{3} &= \left(\frac{28}{35} + \frac{15}{35}\right) \cdot \frac{2}{3} \\ &= \left(\frac{43}{35}\right) \cdot \left(\frac{2}{3}\right) \\ &= \frac{86}{105} \end{aligned}$$

139. $-(x^2) = -x^2$ is true for all real numbers.

141. When $d = 15.99$, $0.07d = 0.07(15.99) \approx 1.12$.
The sales tax is \$1.12.

143. When $d = 15{,}000$,

$$\begin{aligned} d + 0.07d &= 15{,}000 + 0.07(15{,}000) \\ &= 15{,}000 + 1050 \\ &= 16{,}050 \end{aligned}$$

The total cost is \$16,050.

145. a. $2 \div 5^2 = 2 \div 25 = 0.08$

b. $(2 \div 5)^2 = (0.4)^2 = 0.16$

147. When $R = 2$ and $T = 70$,

$$\begin{aligned} &0.2R^2 + 0.003RT + 0.0001T^2 \\ &= 0.02(2)^2 + 0.003(2)(70) + 0.0001(70)^2 \\ &= 0.2(4) + 0.003(2)(70) + 0.0001(4900) \\ &= 0.8 + 0.42 + 0.49 = 1.71 \end{aligned}$$

The growth is 1.71 inches.

149. $12 - (4 - 6) + 10 = 24$

151. a. $2^2 \cdot 2^3 = 2 \cdot 2 \cdot 2 \cdot 2 \cdot 2 = 2^5$

b. $3^3 \cdot 3^3 = 3 \cdot 3 \cdot 3 \cdot 3 \cdot 3 = 3^5$

c. $2^3 \cdot 2^4 = 2 \cdot 2 \cdot 2 \cdot 2 \cdot 2 \cdot 2 \cdot 2 = 2^7$

d. $x^m \cdot x^n = x^{m+n}$

155. a. There are 4 houses with 3 occupants.

b.

Occupants	Number of Houses
1	3
2	5
3	4
4	6
5	2

c. $3(1) + 5(2) + 4(3) + 6(4) + 2(5)$
$= 3 + 10 + 12 + 24 + 10$
$= 13 + 12 + 24 + 10$
$= 25 + 24 + 10$
$= 49 + 10$
$= 59$
There are 59 occupants in all.

d. Number of houses $= 3+5+4+6+2$
$= 8+4+6+2$
$= 12+6+2$
$= 18+2$
$= 20$

$$\text{mean} = \frac{\text{number of occupants}}{\text{number of houses}} = \frac{59}{20} = 2.95$$

There is a mean of 2.95 people per house.

156. $3 = \frac{6}{2} = \frac{1}{2} + \frac{5}{2} = \frac{1}{2} + \frac{20}{8}$
Cost $= \$2.40 + 20(0.20) = \$2.40 + \$4.00 = \$6.40$

157. $(-2)(-4)(6)(-1)(-3) = (8)(6)(-1)(-3)$
$= (48)(-1)(-3)$
$= (-48)(-3)$
$= 144$

158. $\left(\frac{-5}{7}\right) \div \left(\frac{-3}{14}\right) = \left(\frac{-5}{7}\right) \cdot \left(\frac{14}{-3}\right)$
$= \frac{-5}{1} \cdot \frac{2}{-3}$
$= \frac{(-5)(2)}{(1)(-3)}$
$= \frac{10}{3}$ or $3\frac{1}{3}$

Exercise Set 1.10

1. The commutative property of addition states that the sum of two numbers is the same regardless of the order in which they are added. One possible example is $3 + 4 = 4 + 3$.

3. The associative property of addition states that the sum of 3 numbers is the same regardless of the way the numbers are grouped. One possible example is $(2 + 3) + 4 = 2 + (3 + 4)$.

5. a. In $x + (y + z)$ the sum of y and z is added to x whereas in $x(y + z)$, x is multiplied by the sum.

b. When $x = 4$, $y = 5$, and $z = 6$,
$x + (y + z) = 4 + (5 + 6) = 4 + 11 = 15$.

c. When $x = 4$, $y = 5$, and $z = 6$,
$x(y + z) = 4(5 + 6) = 4(11) = 44$.

7. The associative property involves changing parentheses with one operation whereas the distributive property involves distributing a multiplication over an addition.

9. 0

11. a. -6 b. $\frac{1}{6}$

13. a. 3 b. $-\frac{1}{3}$

15. a. $-x$ b. $\frac{1}{x}$

17. a. -1.6 b. $\frac{1}{1.6}$ or 0.625

19. a. $-\frac{1}{5}$ b. 5

21. a. $\frac{3}{5}$ b. $-\frac{5}{3}$

23. Associative property of addition

25. Distributive property

27. Commutative property of multiplication

29. Associative property of multiplication

31. Distributive property

32. Associative property of addition

33. Identity property for addition

35. Inverse property for multiplication

37. $6+x$

39. $(-6\cdot 4)\cdot 2$

41. $x+y$

43. $y\cdot x$

45. $3y+4x$

47. $a+(b+3)$

49. $3x+(4+6)$

51. $(m+n)^3$

53. $4x+4y+12$

55. 0

57. $\frac{5}{2}n$

59. Yes; the order does not affect the outcome so the process is commutative.

61. No; the order affects the outcome, so the process is not commutative.

63. No; the order affects the outcome, so the process is not commutative.

65. Yes; the outcome is not affected by whether you do the first two items first or the last two first, so the process is associative.

67. No; the outcome is affected by whether you do the first two items first or the last two first, so the process is not associative.

69. No; the outcome is affected by whether you do the first two items first or the last two first, so the process is not associative.

71. In $(3+4)+x=x+(3+4)$ the $(3+4)$ is treated as one value.

73. This illustrates the commutative property of addition because the change is 3 + 5 = 5 + 3.

75. No; it illustrates the associative property of addition since the grouping is changed.

77. $2\frac{3}{5}+\frac{2}{3}$

$2\frac{3}{5}=\frac{13}{5}=\frac{3}{3}\cdot\frac{13}{5}=\frac{39}{15}$

$\frac{2}{3}=\frac{2}{3}\cdot\frac{5}{5}=\frac{10}{15}$

$2\frac{3}{5}+\frac{2}{3}=\frac{39}{15}+\frac{10}{15}=\frac{49}{15}$ or $3\frac{4}{15}$

78. $3\frac{5}{8}-2\frac{3}{16}$

$3\frac{5}{8}=\frac{29}{8}=\frac{2}{2}\cdot\frac{29}{8}=\frac{58}{16}$

$2\frac{3}{16}=\frac{35}{16}$

$3\frac{5}{8}-2\frac{3}{16}=\frac{58}{16}-\frac{35}{16}=\frac{23}{16}$ or $1\frac{7}{16}$

79.
$$\begin{aligned}12-24\div 8+4\cdot 3^2 &= 12-24\div 8+4\cdot 9\\ &= 12-3+36\\ &= 9+36\\ &= 45\end{aligned}$$

80. Substitute 2 for x and -3 for y.

$$\begin{aligned}&-4x^2+6xy+3y^2\\ &= -4(2)^2+6(2)(-3)+3(-3)^2\\ &= -4(4)+6(2)(-3)+3(9)\\ &= -16+(-36)+27\\ &= -52+27\\ &= -25\end{aligned}$$

Review Exercises

1.
$$\begin{aligned}60(12)-(162+187+196+95) &= 720-640\\ &= 80\end{aligned}$$

He had 80 hotdogs left over.

2. $1.05[1.05(500.00)]=1.05[525]=551.25$

In 2 years the goods will cost $551.25.

3. Less than; the increase of 20% of the original price is less than the decrease of 20% of the higher price.

4.
$$\begin{aligned}[30+12(25)]-300 &= [30+300]-300\\ &= 330-300\\ &= 30\end{aligned}$$

5. a.
$$\begin{aligned}\text{mean} &= \frac{75+79+86+88+64}{5}\\ &= \frac{392}{5}\\ &= 78.4\end{aligned}$$

The mean grade is 78.4.`

b. 64, 75, 79, 86, 88

The middle number is 79. The median grade is 79.

6. a. mean $= \dfrac{76+79+84+82+79}{5} = \dfrac{400}{5} = 80$
The mean temperature is 80°F.

b. 76, 79, 79, 82, 84
The middle number is 79. The median temperature is 79°F.

7. a. Profit = 18.2% of \$45.79
= (0.182)(45.79)
= 8.33
The drug manufacturer makes \$8.33 profit on the average U.S. prescription.

b. Selling price = Original cost + (Original cost times Markup)
= 60 + 60(0.22)
= 60 + 13.2
= 73.2
The pharmacist will sell the drug for \$73.20.

8. a. U.S. oil reserves – Canadian oil reserves
= 21.8 – 4.7
= 17.1
The U.S. has 17.1 billions of barrels of oil reserves more than Canada.

b. Middle East oil reserves ÷ North American oil reserves = 683.5 ÷ 54.7 ≈ 12.5
The Middle East has about 12.5 times more oil reserves than in North America.

c. 54.8 + 95.2 +17.2 + 59.0 + 74.9 + 683.6 + 44.1 = 1028.8 billion barrels
The world's oil reserves total 1028.8 billion barrels.

9. $\dfrac{3}{5}\cdot\dfrac{5}{6} = \dfrac{1}{1}\cdot\dfrac{1}{2} = \dfrac{1\cdot 1}{1\cdot 2} = \dfrac{1}{2}$

10. $\dfrac{2}{5}\div\dfrac{10}{9} = \dfrac{2}{5}\cdot\dfrac{9}{10} = \dfrac{1}{5}\cdot\dfrac{9}{5} = \dfrac{1\cdot 9}{5\cdot 5} = \dfrac{9}{25}$

11. $\dfrac{5}{12}\div\dfrac{3}{5} = \dfrac{5}{12}\cdot\dfrac{5}{3} = \dfrac{5\cdot 5}{12\cdot 3} = \dfrac{25}{36}$

12. $\dfrac{5}{6}+\dfrac{1}{3} = \dfrac{5}{6}+\dfrac{1}{3}\cdot\dfrac{2}{2} = \dfrac{5}{6}+\dfrac{2}{6} = \dfrac{7}{6}$ or $1\dfrac{1}{6}$

13. $\dfrac{3}{8}-\dfrac{1}{9} = \dfrac{3}{8}\cdot\dfrac{9}{9}-\dfrac{1}{9}\cdot\dfrac{8}{8} = \dfrac{27}{72}-\dfrac{8}{72} = \dfrac{19}{72}$

14. $2\dfrac{1}{3}-1\dfrac{1}{5}$

$2\dfrac{1}{3} = \dfrac{6+1}{3} = \dfrac{7}{3} = \dfrac{7}{3}\cdot\dfrac{5}{5} = \dfrac{35}{15}$

$1\dfrac{1}{5} = \dfrac{5+1}{5} = \dfrac{6}{5} = \dfrac{6}{5}\cdot\dfrac{3}{3} = \dfrac{18}{15}$

$2\dfrac{1}{3}-1\dfrac{1}{5} = \dfrac{35}{15}-\dfrac{18}{15} = \dfrac{17}{15}$ or $1\dfrac{2}{15}$

15. The natural numbers are {1, 2, 3, ...}.

16. The whole numbers are {0, 1, 2, 3, ...}.

17. The integers are {..., –3, –2, –1, 0, 1, 2, ...}.

18. The set of rational numbers is the set of all numbers which can be expressed as the quotient of two integers, denominator not zero.

19. a. 3 and 426 are positive integers.

b. 3, 0, and 426 are whole numbers.

c. 3, –5, –12, 0, and 426 are integers.

d. 3, –5, –12, 0, $\dfrac{1}{2}$, –0.62, 426, and $-3\dfrac{1}{4}$ are rational numbers.

e. $\sqrt{7}$ is an irrational number.

f. 3, –5, –12, 0, $\dfrac{1}{2}$, –0.62, $\sqrt{7}$, 426, and $-3\dfrac{1}{4}$ are real numbers.

20. a. 1 is a natural number.

b. 1 is a whole number.

c. –8 and –9 are negative numbers.

d. –8, –9, and 1 are integers.

e. –2.3, –8, –9, $1\dfrac{1}{2}$, 1, and $-\dfrac{3}{17}$ are rational numbers.

f. –2.3, –8, –9, $1\dfrac{1}{2}$, $\sqrt{2}$, $-\sqrt{2}$, 1, and $-\dfrac{3}{17}$ are real numbers.

21. $-7 < -5$; –7 is to the left of –5 on a number line.

22. $-2.6 > -3.6$; –2.6 is to the right of –3.6 on a number line.

23. $0.50 < 0.509$; 0.50 is to the left 0.509 on a number line.

24. $4.6 > 4.06$; 4.6 is to the right of 4.06 on a number line.

25. $-3.2 < -3.02$; -3.2 is to the left of -3.02 on a number line.

26. $5 > |-3|$ since $|-3|$ equals 3.

27. $-9 < |-7|$ since $|-7|$ equals 7.

28. $|-2.5| = \left|\frac{5}{2}\right|$ since $|-2.5| = \left|\frac{5}{2}\right| = 2.5$.

29. $-9 + (5) = -14$

30. $-6 + 6 = 0$

31. $0 + (-3) = -3$

32. $-10 + 4 = -6$

33. $-8 - (-2) = -8 + 2 = -6$

34. $-2 - (-4) = -2 + 4 = 2$

35. $4 - (-4) = 4 + 4 = 8$

36. $12 - 12 = 12 + (-12) = 0$

37. $7 - (-7) = 7 + 7 = 14$

38. $2 - 7 = 2 + (-7) = -5$

39. $0 - (-4) = 0 + 4 = 4$

40. $-7 - 5 = -7 + (-5) = -12$

41. $\frac{4}{3} - \frac{3}{4} = \frac{16}{12} - \frac{9}{12} = \frac{16-9}{12} = \frac{7}{12}$

42. $\frac{1}{2} + \frac{3}{5} = \frac{5}{10} + \frac{6}{10} = \frac{5+6}{10} = \frac{11}{10}$

43. $\frac{5}{9} - \frac{3}{4} = \frac{20}{36} - \frac{27}{36} = \frac{20-27}{36} = -\frac{7}{36}$

44. $-\frac{5}{7} + \frac{3}{8} = -\frac{40}{56} + \frac{21}{56} = \frac{-40+21}{56} = -\frac{19}{56}$

45. $-\frac{5}{12} - \frac{5}{6} = -\frac{5}{12} - \frac{10}{12} = \frac{-5-10}{12} = -\frac{15}{12} = -\frac{5}{4}$

46. $-\frac{6}{7} + \frac{5}{12} = -\frac{72}{84} + \frac{35}{84} = \frac{-72+35}{84} = -\frac{37}{84}$

47. $\frac{2}{9} - \frac{3}{10} = \frac{20}{90} - \frac{27}{90} = \frac{20-27}{90} = -\frac{7}{90}$

48. $\frac{7}{15} - \left(-\frac{7}{60}\right) = \frac{28}{60} + \frac{7}{60} = \frac{28+7}{60} = \frac{35}{60} = \frac{7}{12}$

49. $9 - 4 + 3 = 5 + 3 = 8$

50. $-5 + 7 - 6 = 2 - 6 = -4$

51. $-5 - 4 - 3 = -9 - 3 = -12$

52. $-2 + (-3) - 2 = -5 - 2 = -7$

53. $7 - (+4) - (-3) = 7 - 4 + 3 = 3 + 3 = 6$

54. $4 - (-2) + 3 = 4 + 2 + 3 = 6 + 3 = 9$

55. Since the numbers have unlike signs, the product is negative; $-3(9) = -27$

56. Since the numbers have like signs, the product is positive; $(-8)(-5) = 40$

57. Since there are an odd number (3) of negatives the product is negative;
$(-4)(-5)(-6) = (20)(-6) = -120$

58. $\left(\frac{3}{5}\right)\left(\frac{-2}{7}\right) = \frac{3(-2)}{5\cdot 7} = \frac{-6}{35} = -\frac{6}{35}$

59. $\left(\frac{10}{11}\right)\left(\frac{3}{-5}\right) = \frac{2}{11}\cdot\frac{3}{-1} = \frac{2\cdot 3}{(11)(-1)} = \frac{6}{-11} = -\frac{6}{11}$

60. $\left(\frac{-5}{8}\right)\left(\frac{-3}{7}\right) = \frac{(-5)(-3)}{8\cdot 7} = \frac{15}{56}$

61. Zero multiplied by any real number is zero.
$0\cdot\frac{4}{9} = 0$

62. Since there are four negative numbers (an even number), the product is positive.
$(-4)(-6)(-2)(-3) = (24)(-2)(-3) = (-48)(-3) = 144$

63. Since the numbers have unlike signs, the quotient is negative; $15 \div (-3) = \frac{15}{-3} = -5$

64. Since the numbers have unlike signs, the quotient is negative; $12 \div (-2) = \frac{12}{-2} = -6$

65. Since the numbers have unlike signs, the quotient is negative; $-20 \div 5 = \frac{-20}{5} = -4$

66. Zero divided by any nonzero number is zero;
$0 \div 4 = \frac{0}{4} = 0$

67. Since the numbers have unlike signs, the quotient is negative; $90 \div (-9) = \frac{90}{-9} = -10$

68. $-4 \div \left(\frac{-4}{9}\right) = \frac{-4}{1} \cdot \frac{9}{-4} = \frac{-1}{1} \cdot \frac{9}{-1} = \frac{-9}{-1} = 9$

69. $\frac{28}{-3} \div \left(\frac{9}{-2}\right) = \left(\frac{28}{-3}\right) \cdot \left(\frac{-2}{9}\right) = \frac{-56}{-27} = \frac{56}{27}$

70. $\frac{14}{3} \div \left(\frac{-6}{5}\right) = \frac{14}{3} \cdot \left(\frac{5}{-6}\right) = \frac{7}{3} \cdot \frac{5}{-3} = \frac{35}{-9} = -\frac{35}{9}$

71. Zero divided by any nonzero number is zero; $0 \div 5 = \frac{0}{5} = 0$

72. Zero divided by any nonzero number is zero; $0 \div (-6) = \frac{0}{-6} = 0$

73. Any real number divided by zero is undefined; $8 \div 0 = \frac{8}{0}$ is undefined.

74. Any real number divided by zero is undefined; $-4 \div 0 = \frac{-4}{0}$ is undefined.

75. Any real number divided by zero is undefined; $\frac{8}{0}$ is undefined

76. Zero divided by any nonzero number is zero; $\frac{0}{-5} = 0$

77. $-5(3-8) = -5(-5) = 25$

78. $2(4-8) = 2(-4) = -8$

79. $(3-6)+4 = -3+4 = 1$

80. $(-4+3)-(2-6) = (-1)-(-4) = -1+4 = 3$

81. $[6+3(-2)]-6 = [6+(-6)]-6 = 0-6 = -6$

82. $(-4-2)(-3) = (-6)(-3) = 18$

83. $[12+(-4)]+(6-8) = 8+(-2) = 6$

84. $9[3+(-4)]+5 = 9(-1)+5 = -9+5 = -4$

85. $-4(-3)+[4 \div (-2)] = (12)+(-2) = 10$

86. $(-3 \cdot 4) \div (-2 \cdot 6) = -12 \div (-12) = 1$

87. $(-3)(-4)+6-3 = 12+6-3 = 18-3 = 15$

88. $[-2(3)+6]-4 = [-6+6]-4 = 0-4 = -4$

89. $7^2 = (7)(7) = 49$

90. $9^3 = (9)(9)(9) = 729$

91. $3^4 = (3)(3)(3)(3) = 81$

92. $(-3)^3 = (-3)(-3)(-3) = -27$

93. $(-1)^9 = (-1)(-1)(-1)(-1)(-1)(-1)(-1)(-1)(-1$
$= -1$

94. $(-2)^5 = (-2)(-2)(-2)(-2)(-2) = -32$

95. $\left(\frac{-4}{5}\right)^2 = \left(\frac{-4}{5}\right)\left(\frac{-4}{5}\right) = \frac{16}{25}$

96. $\left(\frac{2}{5}\right)^3 = \left(\frac{2}{5}\right)\left(\frac{2}{5}\right)\left(\frac{2}{5}\right) = \frac{8}{125}$

97. $xxy = x^2y$

98. $2 \cdot 2 \cdot 3 \cdot 3 \cdot 3xyy = 2^2 \cdot 3^3 xy^2$

99. $5 \cdot 7 \cdot 7 \cdot xxy = 5 \cdot 7^2 x^2 y$

100. $xyxyz = x^2y^2z$

101. $x^2y = xxy$

102. $xz^3 = xzzz$

103. $y^3z = yyyz$

104. $2x^3y^2 = 2xxxyy$

105. $3+5 \cdot 4 = 3+20 = 23$

106. $4 \cdot 6+4 \cdot 2 = 24+8 = 32$

107. $(3-7)^2+6 = (-4)^2+6 = 16+6 = 22$

108. $10-36 \div 4 \cdot 3 = 10-9 \cdot 3 = 10-27 = -17$

109. $6-3^2 \cdot 5 = 6-9 \cdot 5 = 6-45 = -39$

110. $[6-(3\cdot 5)]+5=[6-15]+5=-9+5=-4$

111. $3[9(4^2+3)]\cdot 2=3[9-(16+3)]\cdot 2$
$=3[9-19]\cdot 2$
$=3\cdot(-10)\cdot 2$
$=-30\cdot 2$
$=-60$

112. $(-3^2+4^2)+(3^2\div 3)=(-9+16)+(9\div 3)$
$=(7)+(3)$
$=10$

113. $2^3\div 4+6\cdot 3=8\div 4+6\cdot 3=2+18=20$

114. $(4\div 2)^4+4^2\div 2^2=(2)^4+16\div 4$
$=16+16\div 4$
$=16+4$
$=20$

115. $(8-2^2)^2-4\cdot 3+10=(8-4)^2-4\cdot 3+10$
$=(4)^2-4\cdot 3+10$
$=16-4\cdot 3+10$
$=16-12+10$
$=4+10$
$=14$

116. $4^3\div 4^2-5(2-7)\div 5=64\div 16-5(-5)\div 5$
$=4-(-25)\div 5$
$=4-(-5)$
$=4+5$
$=9$

117. Substitute 5 for x;
$6x-6=6(5)-6=30-6=24$

118. Substitute -5 for x;
$6-4x=6-4(-5)=6-(-20)=6+20=26$

119. Substitute 6 for x;
$2x^2-5x+3=2(6)^2-5(6)+3$
$=2(36)-30+3$
$=72-30+3$
$=42+3$
$=45$

120. Substitute -1 for y;
$5y^2+3y-2=5(-1)^2+3(-1)-2$
$=5(1)-3-2$
$=5-3-2$
$=2-2$
$=0$

121. Substitute 2 for x;
$-x^2+2x-3=-2^2+2(2)-3$
$=-4+4-3$
$=0-3$
$=-3$

122. Substitute -2 for x;
$-x^2+2x-3=-(-2)^2+2(-2)-3$
$=-4+(-4)-3$
$=-8-3$
$=-11$

123. Substitute 1 for x;
$-3x^2-5x+5=-3(1)^2-5(1)+5$
$=-3(1)-5+5$
$=-3-5+5$
$=-8+5$
$=-3$

124. Substitute -3 for x and -2 for y;
$-x^2-8x-12y=-(-3)^2-8(-3)-12(-2)$
$=-9-(-24)+24$
$=-9+24+24$
$=15+24$
$=39$

125. a. $278+(-493)=-215$

b. $|-493|$ is greater than $|278|$ so the sum should be (and is) negative.

126. a. $324-(-29.6)=324+29.6=353.6$

b. The sum of two positive numbers is always positive. As expected, the answer is positive.

127. a. $\dfrac{-17.28}{6}=-2.88$

b. Since the numbers have unlike signs, the quotient is negative, as expected.

128. a. $(-62)(-1.9)=117.8$

b. Since the numbers have like signs, the product is positive, as expected.

129. a. $(-3)^6 = 729$

b. A negative number raised to an even power is positive. As expected, the answer is positive.

130. a. $-(4.2)^3 = -74.088$

b. Since $(4.2)^3$ is positive, $-(4.2)^3$ should be (and is) negative.

131. Associative property of addition

132. Commutative property of multiplication

133. Distributive property

134. Commutative property of multiplication

135. Commutative property of addition

136. Associative property of addition

137. Identity property of addition

138. Inverse property of multiplication

Practice Test

1. a. $2(1.30) + 4.75 + 3(1.10)$
$= 2.60 + 4.75 + 3.30$
$= 7.35 + 3.30$
$= 10.65$
The bill is \$10.65 before tax.

b. $0.07(3.30) \approx 0.23$
The tax on the soda is \$0.23.

c. $10.65 + 0.23 = 10.88$
The total bill is \$10.88.

d. $50 - 10.88 = 39.12$
Her change will be \$39.12.

2. a. The average cost, per employee, for employers in 2002 was about \$4400.

b. \$4400 - \$3000 = \$1400
The difference in average cost, per employee, for employers from 1997 to 2002 was \$1400.

3. a. $\frac{\text{Population}}{\text{Average family size}} = \frac{281 \text{ million}}{2.59}$
$\approx 108.5 \text{ million}$
There were about 108.5 million households in the U.S. in 2000.

b. This means half the population was above and half was below this age.

4. a. 42 is a natural number.

b. 42 and 0 are whole numbers.

c. –6, 42, 0, –7, and –1 are integers.

d. $-6, 42, -3\frac{1}{2}, 0, 6.52, \frac{5}{9}, -7$, and -1 are rational numbers.

e. $\sqrt{5}$ is an irrational number.

f. $-6, 42, -3\frac{1}{2}, 0, 6.52, \sqrt{5}, \frac{5}{9}, -7$, and -1 are real numbers.

5. $-9 > -12$; –9 is to the right of –12 on a number line.

6. $|-3| > |-2|$ since $|-3| = 3$ and $|-2| = 2$.

7. $-7 + (-8) = -15$

8. $-6 - 5 = -6 + (-5) = -11$

9. $15 - 12 - 17 = 3 - 17 = -14$

10. $(-4+6) - 3(-2) = (2) - (-6) = 2 + 6 = 8$

11. $(-4)(-3)(2)(-1) = (12)(2)(-1) = (24)(-1) = -24$

12. $\left(\frac{-2}{9}\right) \div \left(\frac{-7}{8}\right) = \frac{-2}{9} \cdot \frac{8}{-7} = \frac{-16}{-63} = \frac{16}{63}$

13. $\left(-18 \cdot \frac{1}{2}\right) \div 3 = \left(\frac{-18}{1} \cdot \frac{1}{2}\right) \div 3$
$= \left(\frac{-9}{1} \cdot \frac{1}{1}\right) \div 3$
$= -9 \div 3$
$= -3$

14. $-\frac{3}{8} - \frac{4}{7} = -\frac{21}{56} - \frac{32}{56} = \frac{-21-32}{56} = -\frac{53}{56}$

15. $-6(-2-3) \div 5 \cdot 2 = -6(-5) \div 5 \cdot 2$
$= 30 \div 5 \cdot 2$
$= 6 \cdot 2$
$= 12$

16. $\left(\frac{3}{5}\right)^3 = \left(\frac{3}{5}\right)\left(\frac{3}{5}\right)\left(\frac{3}{5}\right) = \frac{27}{125}$

17. $\dfrac{7+9^3 \div 27}{6+4(11-4)} = \dfrac{7+729 \div 27}{6+4(7)}$

$= \dfrac{7+27}{6+28}$

$= \dfrac{34}{34}$

$= 1$

18. $-\{-5[64 \div 2^2 - 3(8-6)]\} = -\{-5[64 \div 4 - 3(2)]\}$

$= -\{-5[64 \div 4 - 6]\}$

$= -\{-5[16-6]\}$

$= -\{-5[10]\}$

$= -\{-50\}$

$= 50$

19. Substitute –4 for x;

$5x^2 - 8 = 5(-4)^2 - 8 = 5(16) - 8 = 80 - 8 = 72$

20. Substitute 3 for x and –2 for y;

$6x - 3y^2 + 4 = 6(3) - 3(-2)^2 + 4$

$= 6(3) - 3(4) + 4$

$= 18 - 12 + 4$

$= 6 + 4$

$= 10$

21. Substitute –2 for each x;

$-x^2 - 6x + 3 = -(-2)^2 - 6(-2) + 3$

$= -4 - (-12) + 3$

$= -4 + 12 + 3$

$= 8 + 3$

$= 11$

22. Substitute 1 for x and –2 for y;

$-x^2 + xy + y^2 = -(1)^2 + (1)(-2) + (-2)^2$

$= -1 + (-2) + 4$

$= -3 + 4$

$= 1$

23. Commutative property of addition

24. Distributive property

25. Associative property of addition

Chapter 2

Exercise Set 2.1

1. a. The terms of an expression are the parts that are added.

b. The terms of $3x-4y-5$ are $3x$, $-4y$, and -5.

c. The terms of $6xy+3x-y-9$ are $6xy$, $3x$, $-y$, and -9.

3. a. The factors of an expression are the parts that are multiplied.

b. In $3x$, 3 and x are factors because they are multiplied together.

c. In $5xy$, 5, x and y are factors because they are multiplied together.

5. a. The numerical part of a term is the numerical coefficient or coefficient of the term.

b. The coefficient of $4x$ is 4.

c. Since $x = 1x$, the coefficient of x is 1.

d. Since $-x = -1x$, the coefficient of $-x$ is -1.

e. Since $\frac{3x}{5} = \frac{3}{5}x$, the coefficient of $\frac{3x}{5}$ is $\frac{3}{5}$.

f. Since $\frac{4}{7}(3t-5) = \frac{12t}{7} - \frac{20}{7} = \frac{12}{7}t - \frac{20}{7}$, the coefficient of $\frac{4}{7}(3t-5)$ is $\frac{12}{7}$.

7. a. The signs of all the terms inside the parentheses are changed when the parentheses are removed.

b. $-(x-8) = -x+8$

b. $+(x-8) = x-8$

9. $5x+3x = 8x$

11. $4x-5x = -x$

13. $y+3+4y = y+4y+3$
$= 5y+3$

15. $-2x+5x = 3x$

17. $2-6x+5 = -6x+2+5$
$= -6x+7$

19. $-2w-3w+5 = -5w+5$

21. $-x+2-x-2 = -x-x+2-2$
$= -2x$

23. $3+6x-3-6x = 6x-6x+3-3 = 0$

25. $5+2x-4x+6 = 2x-4x+5+6$
$= -2x+11$

27. $4r-6-6r-2 = 4r-6r-6-2$
$= -2r-8$

29. $2-3x-2x+y = -3x-2x+y+2$
$= -5x+y+2$

31. $-2x+4x-3 = 2x-3$

33. $b+4+\frac{3}{5} = b+\frac{20}{5}+\frac{3}{5}$
$= b+\frac{23}{5}$

35. $5.1n+6.42-4.3n = 5.1n-4.3n+6.42$
$= 0.8n+6.42$

37. There are no like terms.
$\frac{1}{2}a+3b+1$

39. $2x^2+3y^2+4x+5y^2 = 2x^2+4x+3y^2+5y^2$
$= 2x^2+4x+8y^2$

41. $-x^2+2x^2+y = x^2+y$

43. $2x-7y-5x+2y = 2x-5x-7y+2y$
$= -3x-5y$

45. $4-3n^2+9-2n = -3n^2-2n+4+9$
$= -3n^2-2n+13$

47. $-19.36+40.02x+12.25-18.3x$
$= 40.02x-18.3x-19.36+12.25$
$= 21.72x-7.11$

49. $\frac{3}{5}x-3-\frac{7}{4}x-2=\frac{3}{5}x-\frac{7}{4}x-3-2$
$=\frac{12}{20}x-\frac{35}{20}x-5$
$=-\frac{23}{20}x-5$

51. There are no like terms.
$5w^3+2w^2+w+3$

53. $2z-5z^3-2z^3-z^2=-5z^3-2z^3-z^2+2z$
$=-7z^3-z^2+2z$

55. There are no like terms.
$6x^2-6xy+3y^2$

57. $4a^2-3ab+6ab+b^2=4a^2+3ab+b^2$

59. $2(x+6)=2x+2(6)$
$=2x+12$

61. $5(x+4)=5x+5(4)$
$=5x+20$

63. $-2(x-4)=-2[x+(-4)]$
$=-2x+(-2)(-4)$
$=-2x+8$

65. $-\frac{1}{2}(2x-4)=-\frac{1}{2}[2x+(-4)]$
$=-\frac{1}{2}(2x)+\left(-\frac{1}{2}\right)(-4)$
$=-x+2$

67. $1(-4+x)=1(-4)+1(x)$
$=-4+x$
$=x-4$

69. $\frac{4}{5}(s-5)=\frac{4}{5}s-\frac{4}{5}(5)$
$=\frac{4}{5}s-4$

71. $-0.3(3x+5)=-0.3(3x)+(-0.3)(5)$
$=-0.9x+(-1.5)$
$=-0.9x-1.5$

73. $\frac{1}{3}(3r-12)=\frac{1}{3}(3r)+\frac{1}{3}(-12)$
$=r-4$

75. $0.7(2x+0.5)=0.7(2x)+0.7(0.5)$
$=1.4x+0.35$

77. $-(-x+y)=-1(-x+y)$
$=-1(-x)+(-1)(y)$
$=x+(-y)$
$=x-y$

79. $-(2x+4y-8)=-1[2x+4y+(-8)]$
$=-1(2x)+(-1)(4y)+(-1)(-8)$
$=-2x-4y+8$
$=-2x-4y+8$

81. $1.1(3.1x-5.2y+2.8)$
$=1.1[3.1x+(-5.2y)+2.8]$
$=(1.1)(3.1x)+(1.1)(-5.2y)+(1.1)(2.8)$
$=3.41x+(-5.72y)+3.08$
$=3.41x-5.72y+3.08$

83. $2\left(3x-2y+\frac{1}{4}\right)=2\left[3x+(-2y)+\frac{1}{4}\right]$
$=2(3x)+2(-2y)+2\left(\frac{1}{4}\right)$
$=6x+(-4y)+\frac{1}{2}$
$=6x-4y+\frac{1}{2}$

85. $(x+3y-9)=1[x+3y+(-9)]$
$=1(x)+1(3y)+(1)(-9)$
$=x+3y+(-9)=x+3y-9$

87. $-3(-x+2y+4)=-3(-x)+(-3)(2y)+(-3)(4)$
$=3x+(-6y)+(-12)$
$=3x-6y-12$

89. $3(x-5)-x=3x-15-x$
$=3x-x-15$
$=2x-15$

91. $-2(3-x)+7=-6+2x+7$
$=2x-6+7$
$=2x+1$

93. $6x+2(4x+9)=6x+8x+18$
$=14x+18$

95. $2(x-y)+2x+3=2x-2y+2x+3$
$=2x+2x-2y+3$
$=4x-2y+3$

97. $4(2c-3)-3(c-4) = 8c-12-3c+12$
$= 8c-3c-12+12$
$= 5c$

99. $8x-(x-3) = 8x-x+3$
$= 7x+3$

101. $2(x-3)-(x+3) = 2x-6-x-3$
$= 2x-x-6-3$
$= x-9$

103. $4(x-1)+2(3-x)-4 = 4x-4(1)+2(3)-2x-4$
$= 4x-4+6-2x-4$
$= 4x-2x-4+6-4$
$= 2x-2$

105. $-(3s+4)-(s+2) = -3s-4-s-2$
$= -3s-s-4-2$
$= -4s-6$

107. $-3(x+1)+5x+6 = -3x-3+5x+6$
$= -3x+5x-3+6$
$= 2x+3$

109. $4(m+3)-4m-12 = 4m+4(3)-4m-12$
$= 4m+12-4m-12$
$= 4m-4m+12-12$
$= 0$

111. $0.4+(y+5)+0.6-2 = 0.4+y+5+0.6-2$
$= y+0.4+5+0.6-2$
$= y+4$

113. $4+(3x-4)-5 = 4+3x-4-5$
$= 3x+4-4-5$
$= 3x-5$

115. $4(x+2)-3(x-4)-5$
$= 4x+4(2)-3x-3(-4)-5$
$= 4x+8-3x+12-5$
$= 4x-3x+8+12-5$
$= x+15$

117. $-0.2(6-x)-4(y+0.4)$
$= -0.2(6)-0.2(-x)-4y-4(0.4)$
$= -1.2+0.2x-4y-1.6$
$= 0.2x-4y-1.2-1.6$
$= 0.2x-4y-2.8$

119. $-6x+7y-(3+x)+(x+3)$
$= -6x+7y-3-x+x+3$
$= -6x-x+x+7y-3+3$
$= -6x+7y$

121. $\frac{1}{2}(x+3)+\frac{1}{3}(3x+6) = \frac{1}{2}x+\frac{3}{2}+\frac{3}{3}x+\frac{6}{3}$
$= \frac{1}{2}x+\frac{3}{2}+x+2$
$= \frac{1}{2}x+x+\frac{3}{2}+2$
$= \frac{3}{2}x+\frac{7}{2}$

123. □+⊖+⊖+□+⊖ = 2□+3⊖

125. $x+y+\Delta+\Delta+x+y+y$
$= x+x+y+y+y+\Delta+\Delta$
$= 2x+3y+2\Delta$

127. $1\cdot 12,\ 2\cdot 6,\ 3\cdot 4$
positive factors: 1, 2, 3, 4, 6, 12

129. 3Δ + 5□ − Δ − 3□ = 3Δ − Δ + 5□ − 3□
= 2Δ + 2□

131. $4x^2+5y^2+6(3x^2-5y^2)-4x+3$
$= 4x^2+5y^2+18x^2-30y^2-4x+3$
$= 4x^2+18x^2-4x+5y^2-30y^2+3$
$= 22x^2-25y^2-4x+3$

133. $2x^2-4x+8x^2-3(x+2)-x^2-2$
$= 2x^2-4x+8x^2-3x-6-x^2-2$
$= 2x^2+8x^2-x^2-4x-3x-6-2$
$= 9x^2-7x-8$

135. $|-7| = 7$

136. $-|-16| = -(16) = -16$

137. $-4-3-(-6) = -4-3+6$
$= -7+6$
$= -1$

138. Answers will vary. The answer should include that the order is parentheses, exponents, multiplication and division from left to right, and addition and subtraction from left to right.

139. Substitute −1 for each x in the expression.

$$\begin{aligned}-x^2 + 5x - 6 &= -(-1)^2 + 5(-1) - 6\\ &= -1 + (-5) - 6\\ &= -6 - 6\\ &= -12\end{aligned}$$

Exercise Set 2.2

1. An equation is a statement that shows two algebraic expressions are equal.

3. A solution to an equation may be checked by substituting the value in the equation and determining if it results in a true statement.

5. Equivalent equations are two or more equations with the same solution.

7. Add 4 to both sides of the equation to get the variable by itself.

9. One example is $x + 2 = 1$.

11. Subtraction is defined in terms of addition.

13. Substitute 2 for $x = 2$.

$$\begin{aligned}4x - 3 &= 5\\ 4(2) - 3 &= 5\\ 8 - 3 &= 5\\ 5 &= 5 \quad \text{True}\end{aligned}$$

Since we obtain true statement, 2 is a solution.

15. Substitute −3 for x, $x = -3$.

$$\begin{aligned}2x - 5 &= 5(x + 2)\\ 2(-3) - 5 &= 5[(-3) + 2]\\ -6 - 5 &= 5(-1)\\ -11 &= -5 \quad \text{False}\end{aligned}$$

Since we obtain a false statement, −3 is not a solution.

17. Substitute 0 for p, $p = 0$.

$$\begin{aligned}3p - 4 &= 2(p + 3) - 10\\ 3(0) - 4 &= 2(0 + 3) - 10\\ 0 - 4 &= 2(3) - 10\\ -4 &= 6 - 10\\ -4 &= -4 \quad \text{True}\end{aligned}$$

Since we obtain a true statement, 0 is a solution.

19. Substitute 3.4 for x, $x = 3.4$.

$$\begin{aligned}3(x + 2) - 3(x - 1) &= 9\\ 3(3.4 + 2) - 3(3.4 - 1) &= 9\\ 3(5.4) - 3(2.4) &= 9\\ 16.2 - 7.2 &= 9\\ 9 &= 9 \quad \text{True}\end{aligned}$$

Since we obtain a true statement, 3.4 is a solution.

21. Substitute $\frac{1}{2}$ for x, $x = \frac{1}{2}$.

$$\begin{aligned}4x - 4 &= 2x - 2\\ 4\left(\frac{1}{2}\right) - 4 &= 2\left(\frac{1}{2}\right) - 2\\ 2 - 4 &= 1 - 2\\ -2 &= -1 \quad \text{False}\end{aligned}$$

Since we obtain a false statement, $\frac{1}{2}$ is not a solution.

23. Substitute $\frac{11}{2}$ for x, $x = \frac{11}{2}$.

$$\begin{aligned}3(x + 2) &= 5(x - 1)\\ 3\left(\frac{11}{2} + 2\right) &= 5\left(\frac{11}{2} - 1\right)\\ 3\left(\frac{15}{2}\right) &= 5\left(\frac{9}{2}\right)\\ \frac{45}{2} &= \frac{45}{2} \quad \text{True}\end{aligned}$$

Since we obtain a true statement, $\frac{11}{2}$ is a solution.

25.
$$\begin{aligned}x + 5 &= 8\\ x + 5 - 5 &= 8 - 5\\ x + 0 &= 3\\ x &= 3\end{aligned}$$

Check:
$$\begin{aligned}x + 5 &= 8\\ 3 + 5 &= 8\\ 8 &= 8 \quad \text{True}\end{aligned}$$

27.
$$\begin{aligned}x + 1 &= -6\\ x + 1 - 1 &= -6 - 1\\ x + 0 &= -7\\ x &= -7\end{aligned}$$

Check:
$$\begin{aligned}x + 1 &= -6\\ -7 + 1 &= -6\\ -6 &= -6 \quad \text{True}\end{aligned}$$

29. $x+4=-5$
$x+4-4=-5-4$
$x+0=-9$
$x=-9$
Check: $x+4=-5$
$-9+4=-5$
$-5=-5$ True

31. $x+9=52$
$x+9-9=52-9$
$x+0=43$
$x=43$
Check: $x+9=52$
$43+9=52$
$52=52$ True

33. $-6+w=9$
$-6+6+w=9+6$
$0+w=15$
$w=15$
Check: $-6+w=9$
$-6+15=9$
$9=9$ True

35. $27=x+16$
$27-16=x+16-16$
$11=x+0$
$11=x$
Check: $27=x+16$
$27=11+16$
$27=27$ True

37. $-18=-14+x$
$-18+14=-14+14+x$
$-4=0+x$
$-4=x$
Check: $-18=-14+x$
$-18=-14+(-4)$
$-18=-18$ True

39. $9+x=4$
$9-9+x=4-9$
$0+x=-5$
$x=-5$
Check: $9+x=4$
$9+(-5)=4$
$4=4$ True

41. $4+x=-9$
$4-4+x=-9-4$
$0+x=-13$
$x=-13$
Check: $4+x=-9$
$4+(-13)=-9$
$-9=-9$ True

43. $7+r=-23$

$7+r=-23$
$7-7+r=-23-7$
$0+r=-30$
$r=-30$

Check: $7+r=-23$
$7+(-30)=-23$
$-23=-23$ True

45. $8=8+v$
$8-8=8-8+v$
$0=0+v$
$0=v$
Check: $8=8+v$
$8=8+0$
$8=8$ True

47. $-4=x-3$
$-4+3=x-3+3$
$-1=x+0$
$-1=x$
Check: $-4=x-3$
$-4=-1-3$
$-4=-4$ True

49. $12=16+x$
$12-16=16-16+x$
$-4=0+x$
$-4=x$
Check: $12=16+x$
$12=16+(-4)$
$12=12$ True

51. $15+x=-5$
$15-15+x=-5-15$
$0+x=-20$
$x=-20$
Check: $15+x=-5$
$15+(-20)=-5$
$-5=-5$ True

53. $-10=-10+x$
$-10+10=-10+10+x$
$0=0+x$
$0=x$
Check: $-10=-10+x$
$-10=-10+0$
$-10=-10$ True

55. $5 = x - 12$
$5 + 12 = x - 12 + 12$
$17 = x + 0$
$17 = x$
Check: $5 = x - 12$
$5 = 17 - 12$
$5 = 5$ True

57. $-50 = x - 24$
$-50 + 24 = x - 24 + 24$
$-26 = x + 0$
$-26 = x$
Check: $-50 = x - 24$
$-50 = -26 - 24$
$-50 = -50$ True

59. $43 = 15 + p$
$43 - 15 = 15 - 15 + p$
$28 = 0 + p$
$28 = p$
Check: $43 = 15 + p$
$43 = 15 + 28$
$43 = 43$ True

61. $40.2 + x = -5.9$
$40.2 - 40.2 + x = -5.9 - 40.2$
$0 + x = -46.1$
$x = -46.1$
Check: $40.2 + x = -5.9$
$40.2 + (-46.1) = -5.9$
$-5.9 = -5.9$ True

63. $-37 + x = 9.5$
$-37 + 37 + x = 9.5 + 37$
$0 + x = 46.5$
$x = 46.5$
Check: $-37 + x = 9.5$
$-37 + 46.5 = 9.5$
$9.5 = 9.5$ True

65. $x - 8.77 = -17$
$x - 8.77 + 8.77 = -17 + 8.77$
$x + 0 = -8.23$
$x = -8.23$
Check: $x - 8.77 = -17$
$-8.23 - 8.77 = -17$
$-17 = -17$ True

67. $9.32 = x + 3.75$
$9.32 - 3.75 = x + 3.75 - 3.75$
$5.57 = x + 0$
$5.57 = x$
Check: $9.32 = x + 3.75$
$9.32 = 5.57 + 3.75$
$9.32 = 9.32$ True

69. No; there are no real numbers that can make $x + 1 = x + 2$.

71. $x - \Delta = \square$
$x - \Delta + \Delta = \square + \Delta$
$x = \square + \Delta$

73. ☺ $= \square + \Delta$
☺ $- \Delta = \square + \Delta - \Delta$
☺ $- \Delta = \square$

76. Substitute 4 for each x in the expression.
$3x + 4(x - 3) + 2 = 3(4) + 4(4 - 3) + 2$
$= 12 + 4(1) + 2$
$= 12 + 4 + 2$
$= 16 + 2$
$= 18$

77. Substitute –3 for x in the expression.
$6x - 2(2x + 1) = 6(-3) - 2[2(-3) + 1]$
$= -18 - 2(-6 + 1)$
$= -18 - 2(-5)$
$= -18 + 10$
$= -8$

78. $4x + 3(x - 2) - 5x - 7 = 4x + 3x - 6 - 5x - 7$
$= 4x + 3x - 5x - 6 - 7$
$= 2x - 13$

79. $-(2t + 4) + 3(4t - 5) - 3t = -2t - 4 + 12t - 15 - 3t$
$= -2t + 12t - 3t - 4 - 15$
$= 7t - 19$

Exercise Set 2.3

1. Answers will vary. Answer should include that both sides of an equation can be multiplied by the same nonzero number without changing the solution to the equation.

3. a. $-x = a$
$-1x = a$
$(-1)(-1x) = (-1)a$
$1x = -a$
$x = -a$

b. $-x = 5$
$-1x = 5$
$(-1)(-1x) = (-1)5$
$1x = -5$
$x = -5$

c.
$$-x = -5$$
$$-1x = -5$$
$$(-1)(-1x) = (-1)(-5)$$
$$1x = 5$$
$$x = 5$$

5. Divide by –2 to isolate the variable.

7. Multiply both sides by 3 because $3 \cdot \frac{x}{3} = x$.

9. $4x = 12$
$$\frac{4x}{4} = \frac{12}{4}$$
$$x = 3$$
Check: $4x = 12$
$$4(3) = 12$$
$$12 = 12 \text{ True}$$

11. $\frac{x}{2} = 4$
$$2\left(\frac{x}{2}\right) = 2(4)$$
$$x = 8$$
Check: $\frac{x}{2} = 4$
$$\frac{8}{2} = 4$$
$$4 = 4 \text{ True}$$

13. $-4x = 12$
$$\frac{-4x}{-4} = \frac{12}{-4}$$
$$x = -3$$
Check: $-4x = 12$
$$-4(-3) = 12$$
$$12 = 12 \text{ True}$$

15. $\frac{x}{4} = -2$
$$4\left(\frac{x}{4}\right) = 4(-2)$$
$$x = 4(-2)$$
$$x = -8$$
Check: $\frac{x}{4} = -2$
$$\frac{-8}{4} = -2$$
$$-2 = -2 \text{ True}$$

17. $\frac{x}{5} = 1$
$$5\left(\frac{x}{5}\right) = 5(1)$$
$$x = 5$$
Check: $\frac{x}{5} = 1$
$$\frac{5}{5} = 1$$
$$1 = 1 \text{ True}$$

19. $-27n = 81$
$$\frac{-27n}{-27} = \frac{81}{-27}$$
$$n = -3$$
Check: $-27n = 81$
$$-27(-3) = 81$$
$$81 = 81 \text{ True}$$

21. $-7 = 3r$
$$\frac{-7}{3} = \frac{3r}{3}$$
$$-\frac{7}{3} = r$$
Check: $-7 = 3r$
$$-7 = 3\left(-\frac{7}{3}\right)$$
$$-7 = -7 \text{ True}$$

23.
$$-x = -11$$
$$-1x = -11$$
$$(-1)(-1x) = (-1)(-11)$$
$$1x = 11$$
$$x = 11$$
Check: $-x = -11$
$$-11 = -11 \text{ True}$$

25.
$$10 = -y$$
$$10 = -1y$$
$$(-1)(10) = (-1)(-1y)$$
$$-10 = 1y$$
$$-10 = y$$
Check: $10 = -y$
$$10 = -(-10)$$
$$10 = 10 \text{ True}$$

27. $-\frac{w}{3} = -13$

$\frac{w}{-3} = -13$

$(-3)\left(\frac{w}{-3}\right) = (-3)(-13)$

$w = 39$

Check: $-\frac{w}{3} = -13$

$-\frac{39}{3} = -13$

$-13 = -13$ True

29. $4 = -12x$

$\frac{4}{-12} = \frac{-12x}{-12}$

$-\frac{1}{3} = x$

Check: $4 = -12x$

$4 = -12\left(-\frac{1}{3}\right)$

$4 = 4$ True

31. $-\frac{x}{3} = -2$

$\frac{x}{-3} = -2$

$(-3)\left(\frac{x}{-3}\right) = (-3)(-2)$

$x = 6$

Check: $-\frac{x}{3} = -2$

$-\frac{6}{3} = -2$

$-2 = -2$ True

33. $43t = 26$

$\frac{43t}{43} = \frac{26}{43}$

$t = \frac{26}{43}$

Check: $43t = 26$

$43\left(\frac{26}{43}\right) = 26$

$26 = 26$ True

35. $-4.2x = -8.4$

$\frac{-4.2x}{-4.2} = \frac{-8.4}{-4.2}$

$x = 2$

Check: $-4.2x = -8.4$

$-4.2(2) = -8.4$

$-8.4 = -8.4$ True

37. $7x = -7$

$\frac{7x}{7} = \frac{-7}{7}$

$x = -1$

Check: $7x = -7$

$7(-1) = -7$

$-7 = -7$ True

39. $5x = -\frac{3}{8}$

$\frac{1}{5} \cdot 5x = \left(\frac{1}{5}\right) \cdot \left(-\frac{3}{8}\right)$

$x = \frac{(1) \cdot (-3)}{(5) \cdot (8)}$

$x = -\frac{3}{40}$

Check: $5x = -\frac{3}{8}$

$5\left(-\frac{3}{40}\right) = -\frac{3}{8}$

$-\frac{3}{8} = -\frac{3}{8}$ True

41. $15 = -\frac{x}{4}$

$15 = \frac{x}{-4}$

$(-4)(15) = (-4) \cdot \left(\frac{x}{-4}\right)$

$-60 = x$

Check: $15 = -\frac{x}{4}$

$15 = -\frac{(-60)}{4}$

$15 = 15$ True

43. $-\frac{b}{4} = -60$

$\frac{b}{-4} = -60$

$-4\left(\frac{b}{-4}\right) = (-4)(-60)$

$b = 240$

Check: $-\frac{b}{4} = -60$

$-\frac{240}{4} = -60$

$-60 = -60$ True

45. $\frac{x}{5} = -7$

$5\left(\frac{x}{5}\right) = 5(-7)$

$x = -35$

Check: $\frac{x}{5} = -7$

$\frac{-35}{5} = -7$

$-7 = -7$ True

47. $5 = \frac{x}{4}$

$4 \cdot 5 = 4\left(\frac{x}{4}\right)$

$20 = x$

Check: $5 = \frac{x}{4}$

$5 = \frac{20}{4}$

$5 = 5$ True

49. $\frac{3}{5}d = -30$

$\frac{5}{3} \cdot \frac{3}{5}d = \frac{5}{3}(-30)$

$d = -50$

Check: $\frac{3}{5}d = -30$

$\frac{3}{5}(-50) = -30$

$-30 = -30$ True

51. $\frac{y}{-2} = 0$

$(-2)\left(\frac{y}{-2}\right) = (-2)(0)$

$y = 0$

Check: $\frac{y}{-2} = 0$

$\frac{0}{-2} = 0$

$0 = 0$ True

53. $\frac{-7}{8}w = 0$

$\frac{8}{-7}\left(\frac{-7}{8}w\right) = \frac{8}{-7} \cdot 0$

$w = 0$

Check: $\frac{-7}{8}w = 0$

$\frac{-7}{8}(0) = 0$

$0 = 0$ True

55. $\frac{1}{5}x = 4.5$

$5\left(\frac{1}{5}x\right) = 5(4.5)$

$x = 22.5$

Check: $\frac{1}{5}x = 4.5$

$\frac{1}{5}(22.5) = 4.5$

$4.5 = 4.5$ True

57. $-4 = -\frac{2}{3}z$

$\left(-\frac{3}{2}\right)(-4) = \left(-\frac{3}{2}\right)\left(-\frac{2}{3}\right)z$

$6 = z$

Check: $-4 = -\frac{2}{3}z$

$-4 = -\frac{2}{3} \cdot 6$

$-4 = -4$ True

59. $-1.4x = 28.28$

$\frac{-1.4x}{-1.4} = \frac{28.28}{-1.4}$

$x = -20.2$

Check: $-1.4x = 28.28$

$-1.4(-20.2) = 28.28$

$28.28 = 28.28$ True

61. $-4w = \frac{7}{12}$

$-\frac{1}{4}\cdot -4w = -\frac{1}{4}\cdot\frac{7}{12}$

$w = -\frac{7}{48}$

Check: $-4w = \frac{7}{12}$

$-4\left(-\frac{7}{48}\right) = \frac{7}{12}$

$\frac{28}{48} = \frac{7}{12}$

$\frac{7}{12} = \frac{7}{12}$ True

63. $\frac{2}{3}x = 6$

$\frac{3}{2}\cdot\frac{2}{3}x = \frac{3}{2}\cdot 6$

$x = 9$

Check: $\frac{2}{3}x = 6$

$\frac{2}{3}(9) = 6$

$6 = 6$ True

65. a. In $5 + x = 10$, 5 is added to the variable, whereas in $5x = 10$, 5 is multiplied by the variable.

b. $5x = 10$

$5 + x - 5 = 10 - 5$

$x = 5$

c. $5x = 10$

$\frac{5x}{5} = \frac{10}{5}$

$x = 2$

67. Multiplying by $\frac{3}{2}$ is easier because the equation involves fractions.

$\frac{2}{3}x = 4$

$\left(\frac{3}{2}\right)\left(\frac{2}{3}\right)x = \left(\frac{3}{2}\right)\left(\frac{4}{1}\right)$

$x = \frac{12}{2}$

$x = 6$

69. Multiplying by $\frac{7}{3}$ is easier because the equation involves fractions.

$\frac{3}{7}x = \frac{4}{5}$

$\left(\frac{7}{3}\right)\frac{3}{7}x = \left(\frac{7}{3}\right)\frac{4}{5}$

$x = \frac{28}{15}$

71. a. □

b. Divide both sides of the equation by Δ.

c. ☺ = Δ□

$\frac{☺}{Δ} = \frac{Δ□}{Δ}$

$\frac{☺}{Δ} = □$

73. $-8 - (-4) = -8 + 4$

$= -4$

74. $6 - (-3) - 5 - 4 = 6 + 3 - 5 - 4$

$= 9 - 5 - 4$

$= 4 - 4$

$= 0$

75. $4^2 - 2^3 \cdot 6 \div 3 + 6 = 16 - 8 \cdot 6 \div 3 + 6$

$= 16 - 48 \div 3 + 6$

$= 16 - 16 + 6$

$= 0 + 6$

$= 0$

76. $-(x + 3) - 5(2x - 7) + 6 = -x - 3 - 10x + 35 + 6$

$= -x - 10x - 3 + 35 + 6$

$= -11x + 38$

77. $-48 = x + 9$

$-48 - 9 = x + 9 - 9$

$-57 = x + 0$

$-57 = x$

Exercise Set 2.4

1. No; the variable x is on both sides of the equal sign.

3. $x = \frac{1}{3}$ because $1x = x$.

5. $x = -\frac{1}{2}$ because $-x = -1x$.

7. $x = \frac{3}{5}$ because $-x = -1x$.

9. You solve an equation. An equation that contains a variable is true for certain values of that variable. We solve an equation to find those values.

11. a. Answers will vary.

b. Answers will vary.

13. a. Use the distributive property.
Subtract 8 from both sides of the equation.
Divide both sides of the equation by 6.

b.
$$\begin{aligned} 2(3x+4) &= -4 \\ 6x+8 &= -4 \\ 6x+8-8 &= -4-8 \\ 6x &= -12 \\ \frac{6x}{6} &= -\frac{12}{6} \\ x &= -2 \end{aligned}$$

15.
$$\begin{aligned} 3x+6 &= 12 \\ 3x+6-6 &= 12-6 \\ 3x &= 6 \\ \frac{3x}{3} &= \frac{6}{3} \\ x &= 2 \end{aligned}$$

17.
$$\begin{aligned} -4w-5 &= 11 \\ -4w-5+5 &= 11+5 \\ -4w &= 16 \\ \frac{-4w}{-4} &= \frac{16}{-4} \\ w &= -4 \end{aligned}$$

19.
$$\begin{aligned} 5x-6 &= 19 \\ 5x-6+6 &= 19+6 \\ 5x &= 25 \\ \frac{5x}{5} &= \frac{25}{5} \\ x &= 5 \end{aligned}$$

21.
$$\begin{aligned} 5x-2 &= 10 \\ 5x-2+2 &= 10+2 \\ 5x &= 12 \\ \frac{5x}{5} &= \frac{12}{5} \\ x &= \frac{12}{5} \end{aligned}$$

23.
$$\begin{aligned} -2t+9 &= 21 \\ -2t+9-9 &= 21-9 \\ -2t &= 12 \\ \frac{-2t}{-2} &= \frac{12}{-2} \\ t &= -6 \end{aligned}$$

25.
$$\begin{aligned} 12-x &= 9 \\ 12-12-x &= 9-12 \\ -x &= -3 \\ (-1)(-x) &= (-1)(-3) \\ x &= 3 \end{aligned}$$

27.
$$\begin{aligned} 8+3x &= 19 \\ 8-8+3x &= 19-8 \\ 3x &= 11 \\ \frac{3x}{3} &= \frac{11}{3} \\ x &= \frac{11}{3} \end{aligned}$$

29.
$$\begin{aligned} 16x+5 &= -14 \\ 16x+5-5 &= -14-5 \\ 16x &= -19 \\ \frac{16x}{16} &= \frac{-19}{16} \\ x &= -\frac{19}{16} \end{aligned}$$

31.
$$\begin{aligned} -4.2 &= 3x+25.8 \\ -4.2-25.8 &= 3x+25.8-25.8 \\ -30 &= 3x \\ \frac{-30}{3} &= \frac{3x}{3} \\ -10 &= x \end{aligned}$$

33.
$$\begin{aligned} 7r-16 &= -2 \\ 7r-16+16 &= -2+16 \\ 7r &= 14 \\ \frac{7r}{7} &= \frac{14}{7} \\ r &= 2 \end{aligned}$$

35.
$$\begin{aligned} 60 &= -5s+9 \\ 60-9 &= -5s+9-9 \\ 51 &= -5s \\ \frac{51}{-5} &= \frac{-5s}{-5} \\ -\frac{51}{5} &= s \end{aligned}$$

37.
$$-2x - 7 = -13$$
$$-2x - 7 + 7 = -13 + 7$$
$$-2x = -6$$
$$\frac{-2x}{-2} = \frac{-6}{-2}$$
$$x = 3$$

39.
$$2.3x - 9.34 = 6.3$$
$$2.3x - 9.34 + 9.34 = 6.3 + 9.34$$
$$2.3x = 15.64$$
$$\frac{2.3x}{2.3} = \frac{15.64}{2.3}$$
$$x = 6.8$$

41.
$$x + 0.07x = 16.05$$
$$1.07x = 16.05$$
$$\frac{1.07x}{1.07} = \frac{16.05}{1.07}$$
$$x = 15$$

43.
$$28.8 = x + 1.40x$$
$$28.8 = 2.40x$$
$$\frac{28.8}{2.40} = \frac{2.40x}{2.40}$$
$$12 = x$$

45.
$$\frac{x-4}{6} = 9$$
$$6\left(\frac{x-4}{6}\right) = 6(9)$$
$$x - 4 = 54$$
$$x - 4 + 4 = 54 + 4$$
$$x = 58$$

47.
$$\frac{d+3}{7} = 9$$
$$7\left(\frac{d+3}{7}\right) = 7(9)$$
$$d + 3 = 63$$
$$d + 3 - 3 = 63 - 3$$
$$d = 60$$

49.
$$\frac{1}{3}(t-5) = -6$$
$$\frac{1}{3}t - \frac{5}{3} = -6$$
$$3\left(\frac{1}{3}t - \frac{5}{3}\right) = 3(-6)$$
$$1t - 5 = -18$$
$$1t - 5 + 5 = -18 + 5$$
$$1t = -13$$
$$t = -13$$

51.
$$\frac{3}{4}(x-5) = -12$$
$$\frac{3}{4}x - \frac{15}{4} = -12$$
$$4\left(\frac{3}{4}x - \frac{15}{4}\right) = 4(-12)$$
$$3x - 15 = -48$$
$$3x - 15 + 15 = -48 + 15$$
$$3x = -33$$
$$\frac{3x}{3} = \frac{-33}{3}$$
$$x = -11$$

53.
$$\frac{1}{4} = \frac{z+1}{4}$$
$$4\left(\frac{1}{4}\right) = 4\left(\frac{z+1}{4}\right)$$
$$1 = z + 1$$
$$1 - 1 = z + 1 - 1$$
$$0 = z$$

55.
$$\frac{3}{4} = \frac{4m-5}{6}$$
$$12\left(\frac{3}{4}\right) = 12\left(\frac{4m-5}{6}\right)$$
$$9 = 2(4m - 5)$$
$$9 = 8m - 10$$
$$9 + 10 = 8m - 10 + 10$$
$$19 = 8m$$
$$\frac{19}{8} = \frac{8m}{8}$$
$$\frac{19}{8} = m$$

57.
$$\begin{aligned} 4(n+2) &= 8 \\ 4n+8 &= 8 \\ 4n+8-8 &= 8-8 \\ 4n &= 0 \\ \frac{4n}{4} &= \frac{0}{4} \\ n &= 0 \end{aligned}$$

59.
$$\begin{aligned} 5(3-x) &= 15 \\ 15-5x &= 15 \\ 15-15-5x &= 15-15 \\ -5x &= 0 \\ \frac{-5x}{-5} &= \frac{0}{-5} \\ x &= 0 \end{aligned}$$

61.
$$\begin{aligned} -4 &= -(x+5) \\ -4 &= -x-5 \\ -4+5 &= -x-5+5 \\ 1 &= -x \\ (-1)(1) &= (-1)(-x) \\ -1 &= x \end{aligned}$$

63.
$$\begin{aligned} 12 &= 4(x-3) \\ 12 &= 4x-12 \\ 12+12 &= 4x-12+12 \\ 24 &= 4x \\ \frac{24}{4} &= \frac{4x}{4} \\ 6 &= x \end{aligned}$$

65.
$$\begin{aligned} 22 &= -(3x-4) \\ 22 &= -3x+4 \\ 22-4 &= -3x+4-4 \\ 18 &= -3x \\ \frac{18}{-3} &= \frac{-3x}{-3} \\ -6 &= x \end{aligned}$$

67.
$$\begin{aligned} -3r+4(r+2) &= 11 \\ -3r+4r+8 &= 11 \\ r+8 &= 11 \\ r+8-8 &= 11-8 \\ r &= 3 \end{aligned}$$

69.
$$\begin{aligned} x-3(2x+3) &= 11 \\ x-6x-9 &= 11 \\ -5x-9 &= 11 \\ -5x-9+9 &= 11+9 \\ -5x &= 20 \\ \frac{-5x}{-5} &= \frac{20}{-5} \\ x &= -4 \end{aligned}$$

71.
$$\begin{aligned} 5x+3x-4x-7 &= 9 \\ 4x-7 &= 9 \\ 4x-7+7 &= 9+7 \\ 4x &= 16 \\ \frac{4x}{4} &= \frac{16}{4} \\ x &= 4 \end{aligned}$$

73.
$$\begin{aligned} 0.7(x-3) &= 1.4 \\ 0.7x-2.1 &= 1.4 \\ 0.7x-2.1+2.1 &= 1.4+2.1 \\ 0.7x &= 3.5 \\ \frac{0.7x}{0.7} &= \frac{3.5}{0.7} \\ x &= 5 \end{aligned}$$

75.
$$\begin{aligned} 2.5(4q-3) &= 0.5 \\ 10q-7.5 &= 0.5 \\ 10q-7.5+7.5 &= 0.5+7.5 \\ 10q &= 8 \\ \frac{10q}{10} &= \frac{8}{10} \\ q &= 0.8 \end{aligned}$$

77.
$$\begin{aligned} 3-2(x+3)+2 &= 1 \\ 3-2x-6+2 &= 1 \\ -2x-1 &= 1 \\ -2x-1+1 &= 1+1 \\ -2x &= 2 \\ \frac{-2x}{-2} &= \frac{2}{-2} \\ x &= -1 \end{aligned}$$

79.
$$\begin{aligned} 1+(x+3)+6x &= 6 \\ 1+x+3+6x &= 6 \\ 7x+4 &= 6 \\ 7x+4-4 &= 6-4 \\ 7x &= 2 \\ \frac{7x}{7} &= \frac{2}{7} \\ x &= \frac{2}{7} \end{aligned}$$

81.
$$\begin{aligned} 4.85 - 6.4x + 1.11 &= 22.6 \\ -6.4x + 5.96 &= 22.6 \\ -6.4x + 5.96 - 5.96 &= 22.6 - 5.96 \\ -6.4x &= 16.64 \\ \frac{-6.4x}{-6.4} &= \frac{16.64}{-6.4} \\ x &= -2.6 \end{aligned}$$

83.
$$\begin{aligned} 7 &= 8 - 5(m+3) \\ 7 &= 8 - 5m - 15 \\ 7 &= -5m - 7 \\ 7 + 7 &= -5m - 7 + 7 \\ 14 &= -5m \\ \frac{14}{-5} &= \frac{-5m}{-5} \\ -\frac{14}{5} &= m \end{aligned}$$

85.
$$\begin{aligned} 10 &= \frac{2s+4}{5} \\ 5(10) &= 5\left(\frac{2s+4}{5}\right) \\ 50 &= 2s + 4 \\ 50 - 4 &= 2s + 4 - 4 \\ 46 &= 2s \\ \frac{46}{2} &= \frac{2s}{2} \\ 23 &= s \end{aligned}$$

87.
$$\begin{aligned} x + \frac{2}{3} &= \frac{3}{5} \\ 15\left(x + \frac{2}{3}\right) &= 15\left(\frac{3}{5}\right) \\ 15x + 10 &= 9 \\ 15x + 10 - 10 &= 9 - 10 \\ 15x &= -1 \\ \frac{15x}{15} &= \frac{-1}{15} \\ x &= -\frac{1}{15} \end{aligned}$$

89.
$$\begin{aligned} \frac{t}{4} - t &= \frac{3}{2} \\ 4\left(\frac{t}{4} - t\right) &= 4\left(\frac{3}{2}\right) \\ t - 4t &= 6 \\ -3t &= 6 \\ \frac{-3t}{-3} &= \frac{6}{-3} \\ t &= -2 \end{aligned}$$

91.
$$\begin{aligned} \frac{3}{7} &= \frac{3t}{4} + 1 \\ 28\left(\frac{3}{7}\right) &= 28\left(\frac{3t}{4} + 1\right) \\ 12 &= 21t + 28 \\ 12 - 28 &= 21t + 28 - 28 \\ -16 &= 21t \\ \frac{-16}{21} &= \frac{21t}{21} \\ -\frac{16}{21} &= t \end{aligned}$$

93.
$$\begin{aligned} \frac{1}{2}r + \frac{1}{5}r &= 7 \\ 10\left(\frac{1}{2}r + \frac{1}{5}r\right) &= 10(7) \\ 5r + 2r &= 70 \\ 7r &= 70 \\ \frac{7r}{7} &= \frac{70}{7} \\ r &= 10 \end{aligned}$$

95.
$$\begin{aligned} \frac{x}{3} - \frac{3x}{4} &= \frac{1}{12} \\ 12\left(\frac{x}{3} - \frac{3x}{4}\right) &= 12\left(\frac{1}{12}\right) \\ 4x - 9x &= 1 \\ -5x &= 1 \\ \frac{-5x}{-5} &= \frac{1}{-5} \\ x &= -\frac{1}{5} \end{aligned}$$

97.
$$\frac{1}{2}x + 4 = \frac{1}{6}$$
$$6\left(\frac{1}{2}x + 4\right) = 6\left(\frac{1}{6}\right)$$
$$3x + 24 = 1$$
$$3x + 24 - 24 = 1 - 24$$
$$3x = -23$$
$$\frac{3x}{3} = \frac{-23}{3}$$
$$x = -\frac{23}{3}$$

99.
$$\frac{4}{5}s - \frac{3}{4}s = \frac{1}{10}$$
$$20\left(\frac{4}{5}s - \frac{3}{4}s\right) = 20\left(\frac{1}{10}\right)$$
$$16s - 15s = 2$$
$$s = 2$$

101.
$$\frac{4}{9} = \frac{1}{3}(n - 7)$$
$$\frac{4}{9} = \frac{1}{3}n - \frac{7}{3}$$
$$9\left(\frac{4}{9}\right) = 9\left(\frac{1}{3}n - \frac{7}{3}\right)$$
$$4 = 3n - 21$$
$$4 + 21 = 3n - 21 + 21$$
$$25 = 3n$$
$$\frac{25}{3} = \frac{3n}{3}$$
$$\frac{25}{3} = n$$

103.
$$-\frac{3}{5} = -\frac{1}{9} - \frac{3}{4}x$$
$$180\left(-\frac{3}{5}\right) = 180\left(-\frac{1}{9} - \frac{3}{4}x\right)$$
$$-108 = -20 - 135x$$
$$-108 + 20 = -20 + 20 - 135x$$
$$-88 = -135x$$
$$\frac{-88}{-135} = \frac{-135x}{-135}$$
$$\frac{88}{135} = x$$

105. a. By subtracting first, you will not have to work with fractions.

b.
$$3x + 2 = 11$$
$$3x + 2 - 2 = 11 - 2$$
$$3x = 9$$
$$\frac{3x}{3} = \frac{9}{3}$$
$$x = 3$$

107.
$$3(x-2) - (x+5) - 2(3-2x) = 18$$
$$3x - 6 - x - 5 - 6 + 4x = 18$$
$$3x - x + 4x - 6 - 5 - 6 = 18$$
$$6x - 17 = 18$$
$$6x - 17 + 17 = 18 + 17$$
$$6x = 35$$
$$\frac{6x}{6} = \frac{35}{6}$$
$$x = \frac{35}{6}$$

109.
$$4[3 - 2(x+4)] - (x+3) = 13$$
$$4(3 - 2x - 8) - x - 3 = 13$$
$$4(-2x - 5) - x - 3 = 13$$
$$-8x - 20 - x - 3 = 13$$
$$-9x - 23 = 13$$
$$-9x - 23 + 23 = 13 + 23$$
$$-9x = 36$$
$$\frac{-9x}{-9} = \frac{36}{-9}$$
$$x = -4$$

111. a. Let x = cost of one box of stationery
$3x + 6 = 42$

b.
$$3x + 6 = 42$$
$$3x + 6 - 6 = 42 - 6$$
$$3x = 36$$
$$x = 12$$
A box of stationery costs $12.00.

113.
$$\left[5(2-6) + 3(8 \div 4)^2\right]^2 = \left[5(-4) + 3(2)^2\right]^2$$
$$= [-20 + 3(4)]^2$$
$$= [-20 + 12]^2$$
$$= [-8]^2$$
$$= 64$$

114. Substitute 5 for x in the expression.
$-2x^2 + 3x - 12$
$-2(5)^2 + 3(5) - 12$
$-2(25) + 15 - 12$
$-50 + 15 - 12$
$-35 - 12$
-47

115. To solve an equation, we need to isolate the variable on one side of the equation.

116. To solve the equation, we divide both sides of the equation by –4.

Exercise Set 2.5

1. Answers will vary.

3. a. An identity is an equation that is true for infinitely many values of the variable.

b. The solution is all real numbers.

5. The equation is an identity because both sides of the equation are identical.

7. An equation has no solution if it simplifies to a false statement.

9. a. Use the distributive property.
Subtract $4x$ from both sides of the equation.
Add 30 to both sides of the equation.
Divide both sides of the equation by 2.

b.
$$\begin{aligned} 4(x+3) &= 6(x-5) \\ 4x + 12 &= 6x - 30 \\ 12 &= 2x - 30 \\ 42 &= 2x \\ 21 &= x \text{ or } x = 21 \end{aligned}$$

11.
$$\begin{aligned} 3x &= -2x + 15 \\ 3x + 2x &= -2x + 2x + 15 \\ 5x &= 15 \\ \frac{5x}{5} &= \frac{15}{5} \\ x &= 3 \end{aligned}$$

13.
$$\begin{aligned} -4x + 10 &= 6x \\ -4x + 4x + 10 &= 6x + 4x \\ 10 &= 10x \\ \frac{10}{10} &= \frac{10x}{10} \\ 1 &= x \end{aligned}$$

15.
$$\begin{aligned} 5x + 3 &= 6 \\ 5x + 3 - 3 &= 6 - 3 \\ 5x &= 3 \\ \frac{5x}{5} &= \frac{3}{5} \\ x &= \frac{3}{5} \end{aligned}$$

17.
$$\begin{aligned} 21 - 6p &= 3p - 2p \\ 21 - 6p &= p \\ 21 - 6p + 6p &= p + 6p \\ 21 &= 7p \\ \frac{21}{7} &= \frac{7p}{7} \\ 3 &= p \end{aligned}$$

19.
$$\begin{aligned} 2x - 4 &= 3x - 6 \\ 2x - 2x - 4 &= 3x - 2x - 6 \\ -4 &= x - 6 \\ -4 + 6 &= x - 6 + 6 \\ 2 &= x \end{aligned}$$

21.
$$\begin{aligned} 6 - 2y &= 9 - 8y + 6y \\ 6 - 2y &= 9 - 2y \\ 6 - 2y + 2y &= 9 - 2y + 2y \\ 6 &= 9 \quad \text{False} \end{aligned}$$

Since a false statement is obtained, there is no solution.

23.
$$\begin{aligned} 9 - 0.5x &= 4.5x + 8.50 \\ 9 - 0.5x + 0.5x &= 4.5x + 0.5x + 8.50 \\ 9 &= 5x + 8.50 \\ 9 - 8.50 &= 5x + 8.50 - 8.50 \\ 0.5 &= 5x \\ \frac{0.5}{5} &= \frac{5x}{5} \\ 0.1 &= x \end{aligned}$$

25.
$$\begin{aligned} 0.62x - .065 &= 9.75 - 2.63x \\ 0.62x + 2.63x - 0.65 &= 9.75 - 2.63x + 2.63x \\ 3.25x - 0.65 &= 9.75 \\ 3.25x - 0.65 + 0.65 &= 9.75 + 0.65 \\ 3.25x &= 10.4 \\ \frac{3.25x}{3.25} &= \frac{10.4}{3.25} \\ x &= 3.2 \end{aligned}$$

27.
$$\begin{aligned}5x+3&=2(x+6)\\5x+3&=2x+12\\5x-2x+3&=2x-2x+12\\3x+3&=12\\3x+3-3&=12-3\\3x&=9\\\frac{3x}{3}&=\frac{9}{3}\\x&=3\end{aligned}$$

29.
$$\begin{aligned}x-25&=12x+9+3x\\x-25&=15x+9\\x-x-25&=15x-x+9\\-25&=14x+9\\-25-9&=14x+9-9\\-34&=14x\\\frac{-34}{14}&=\frac{14x}{14}\\-\frac{17}{7}&=x\end{aligned}$$

31.
$$\begin{aligned}2(x-2)&=4x-6-2x\\2x-4&=2x-6\\2x-2x-4&=2x-2x-6\\-4&=-6 \quad \text{False}\end{aligned}$$
Since a false statement is obtained, there is no solution.

33.
$$\begin{aligned}-(w+2)&=-6w+32\\-w-2&=-6w+32\\-w+w-2&=-6w+w+32\\-2&=-5w+32\\-2-32&=-5w+32-32\\-34&=-5w\\\frac{-34}{-5}&=\frac{-5w}{-5}\\\frac{34}{5}&=w\end{aligned}$$

35.
$$\begin{aligned}5-3(2t-5)&=3t+13\\5-6t+15&=3t+13\\-6t+20&=3t+13\\-6t+6t+20&=3t+6t+13\\20&=9t+13\\20-13&=9t+13-13\\7&=9t\\\frac{7}{9}&=\frac{9t}{9}\\\frac{7}{9}&=t\end{aligned}$$

37.
$$\begin{aligned}\frac{a}{5}&=\frac{a-3}{2}\\10\left(\frac{a}{5}\right)&=10\left(\frac{a-3}{2}\right)\\2a&=5(a-3)\\2a&=5a-15\\2a-5a&=5a-5a-15\\-3a&=-15\\\frac{-3a}{-3}&=\frac{-15}{-3}\\a&=5\end{aligned}$$

39.
$$\begin{aligned}6-\frac{x}{4}&=\frac{x}{8}\\8\left(6-\frac{x}{4}\right)&=8\left(\frac{x}{8}\right)\\48-2x&=x\\48-2x+2x&=x+2x\\48&=3x\\\frac{48}{3}&=\frac{3x}{3}\\16&=x\end{aligned}$$

41.
$$\begin{aligned}\frac{5}{2}-\frac{x}{3}&=3x\\6\left(\frac{5}{2}-\frac{x}{3}\right)&=6(3x)\\15-2x&=18x\\15-2x+2x&=18x+2x\\15&=20x\\\frac{15}{20}&=\frac{20x}{20}\\\frac{15}{20}&=x\\\frac{3}{4}&=x\end{aligned}$$

43.
$$\frac{5}{8}+\frac{1}{4}a=\frac{1}{2}a$$
$$8\left(\frac{5}{8}+\frac{1}{4}a\right)=8\left(\frac{1}{2}a\right)$$
$$5+2a=4a$$
$$5+2a-2a=4a-2a$$
$$5=2a$$
$$\frac{5}{2}=\frac{2a}{2}$$
$$\frac{5}{2}=a$$

45.
$$0.1(x+10)=0.3x-4$$
$$0.1x+1=0.3x-4$$
$$0.1x-0.1x+1=0.3x-0.1x-4$$
$$1+4=0.2x-4+4$$
$$5=0.2x$$
$$\frac{5}{0.2}=\frac{0.2x}{0.2}$$
$$25=x$$

47.
$$2(x+4)=4x+3-2x+5$$
$$2x+8=4x+3-2x+5$$
$$2x+8=2x+8$$
Since the left side of the equation is identical to the right side, the equation is true for all values of x. Thus the solution is all real numbers.

49.
$$5(3n+3)=2(5n-4)+6n$$
$$15n+15=10n-8+6n$$
$$15n+15=16n-8$$
$$15n-15n+15=16n-15n-8$$
$$15=n-8$$
$$15+8=n-8+8$$
$$23=n$$

51.
$$-(3-p)=-(2p+3)$$
$$-3+p=-2p-3$$
$$-3+p+2p=-2p+2p-3$$
$$-3+3p=-3$$
$$-3+3+3p=-3+3$$
$$3p=0$$
$$\frac{3p}{3}=\frac{0}{3}$$
$$p=0$$

53.
$$-(x+4)+5=4x+1-5x$$
$$-x-4+5=4x+1-5x$$
$$-x+1=-x+1$$
Since the left side of the equation is identical to the right side, the equation is true for all values of x. Thus the solution is all real numbers.

55.
$$35(2x-1)=7(x+4)+3x$$
$$70x-35=7x+28+3x$$
$$70x-35=10x+28$$
$$70x-10x-35=10x-10x+28$$
$$60x-35=28$$
$$60x-35+35=28+35$$
$$60x=63$$
$$\frac{60x}{60}=\frac{63}{60}$$
$$x=\frac{21}{20}$$

57.
$$0.4(x+0.7)=0.6(x-4.2)$$
$$0.4x+0.28=0.6x-2.52$$
$$0.4x-0.4x+0.28=0.6x-0.4x-2.52$$
$$0.28=0.2x-2.52$$
$$0.28+2.52=0.2x-2.52+2.52$$
$$2.8=0.2x$$
$$\frac{2.8}{0.2}=\frac{0.2x}{0.2}$$
$$14=x$$

59.
$$\frac{3}{5}x+4=\frac{1}{5}x+5$$
$$5\left(\frac{3}{5}x+4\right)=5\left(\frac{1}{5}x+5\right)$$
$$3x+20=x+25$$
$$3x-x+20=x-x+25$$
$$2x+20=25$$
$$2x+20-20=25-20$$
$$2x=5$$
$$\frac{2x}{2}=\frac{5}{2}$$
$$x=\frac{5}{2}$$

61. $\frac{3}{4}(2x-4)=4-2x$

$\frac{6}{4}x-3=4-2x$

$4\left(\frac{6}{4}x-3\right)=4(4-2x)$

$6x-12=16-8x$

$6x+8x-12=16-8x+8x$

$14x-12=16$

$14x-12+12=16+12$

$14x=28$

$\frac{14x}{14}=\frac{28}{14}$

$x=2$

63. $3(x-4)=2(x-8)+5x$

$3x-12=2x-16+5x$

$3x-12=7x-16$

$3x-3x-12=7x-3x-16$

$-12=4x-16$

$-12+16=4x-16+16$

$4=4x$

$\frac{4}{4}=\frac{4x}{4}$

$1=x$

65. $3(x-6)-4(3x+1)=x-22$

$3x-18-12x-4=x-22$

$-9x-22=x-22$

$-9x+9x-22=x+9x-22$

$-22=10x-22$

$-22+22=10x-22+22$

$0=10x$

$\frac{0}{10}=\frac{10x}{10}$

$0=x$

67. $5+2x=6(x+1)-5(x-3)$

$5+2x=6x+6-5x+15$

$5+2x=x+21$

$5+2x-x=x-x+21$

$5+x=21$

$5-5+x=21-5$

$x=16$

69. $7-(-y-5)=2(y+3)-6(y+1)$

$7+y+5=2y+6-6y-6$

$12+y=-4y$

$12+y-y=-4y-y$

$12=-5y$

$\frac{12}{-5}=\frac{-5y}{-5}$

$-\frac{12}{5}=y$

71. $\frac{1}{2}(2d+4)=\frac{1}{3}(4d-4)$

$d+2=\frac{4d}{3}-\frac{4}{3}$

$3(d+2)=3\left(\frac{4d}{3}-\frac{4}{3}\right)$

$3d+6=4d-4$

$3d-3d+6=4d-3d-4$

$6=d-4$

$6+4=d-4+4$

$10=d$

73. $\frac{3(2r-5)}{5}=\frac{3r-6}{4}$

$\frac{6r-15}{5}=\frac{3r-6}{4}$

$20\left(\frac{6r-15}{5}\right)=20\left(\frac{3r-6}{4}\right)$

$4(6r-15)=5(3r-6)$

$24r-60=15r-30$

$24r-15r-60=15r-15r-30$

$9r-60=-30$

$9r-60+60=-30+60$

$9r=30$

$\frac{9r}{9}=\frac{30}{9}$

$r=\frac{30}{9}=\frac{10}{3}$

75.
$$\frac{2}{7}(5x+4)=\frac{1}{2}(3x-4)+1$$
$$\frac{10x}{7}+\frac{8}{7}=\frac{3x}{2}-2+1$$
$$\frac{10x}{7}+\frac{8}{7}=\frac{3x}{2}-1$$
$$14\left(\frac{10x}{7}+\frac{8}{7}\right)=14\left(\frac{3x}{2}-1\right)$$
$$20x+16=21x-14$$
$$20x-20x+16=21x-20x-14$$
$$16=x-14$$
$$16+14=x-14+14$$
$$30=x$$

77.
$$\frac{a-5}{2}=\frac{3a}{4}+\frac{a-25}{6}$$
$$12\left(\frac{a-5}{2}\right)=12\left(\frac{3a}{4}+\frac{a-25}{6}\right)$$
$$6(a-5)=9a+2(a-25)$$
$$6a-30=9a+2a-50$$
$$6a-30=11a-50$$
$$6a-6a-30=11a-6a-50$$
$$-30=5a-50$$
$$-30+50=5a-50+50$$
$$20=5a$$
$$\frac{20}{5}=\frac{5a}{5}$$
$$4=a$$

79. a. One example is $x+x+1=x+2$.

b. It has a single solution.

c. Answers will vary. For equation given in part **a**):
$$x+x+1=x+2$$
$$2x+1=x+2$$
$$2x-x+1=x-x+2$$
$$x+1=2$$
$$x+1-1=2-1$$
$$x=1$$

81. a. One example is $x+x+1=2x+1$.

b. Both sides simplify to the same expression.

c. The solution is all real numbers.

83. a. One example is $x+x+1=2x+2$.

b. It simplifies to a false statement.

c. The solution is that there is no solution.

85.
$$5*-1=4*+5$$
$$5*-4*-1=4*-4*+5$$
$$*-1=5$$
$$*-1+1=5+1$$
$$*=6$$

87. 3☺ − 5 = 2☺ − 5 + ☺

3☺ − 5 = 3☺ − 5

The left side of the equation is identical to the right side. The solution is all real numbers.

89.
$$4-[5-3(x+2)]=x-3$$
$$4-(5-3x-6)=x-3$$
$$4-5+3x+6=x-3$$
$$3x+5=x-3$$
$$3x-x+5=x-x-3$$
$$2x+5=-3$$
$$2x+5-5=-3-5$$
$$2x=-8$$
$$\frac{2x}{2}=\frac{-8}{2}$$
$$x=-4$$

91. a. $|4|=4$

b. $|-7|=7$

c. $|0|=0$

92. $\left(\frac{2}{3}\right)^5\approx 0.131687243$

93. Factors are expressions that are multiplied together; terms are expressions that are added together.

94.
$$2(x-3)+4x-(4-x)=2x-6+4x-4+x$$
$$=7x-10$$

95.
$$2(x-3)+4x-(4-x)=0$$
$$2x-6+4x-4+x=0$$
$$7x-10=0$$
$$7x-10+10=0+10$$
$$7x=10$$
$$\frac{7x}{7}=\frac{10}{7}$$
$$x=\frac{10}{7}$$

96.
$$\begin{aligned}(x+4)-(4x-3)&=16\\ x+4-4x+3&=16\\ -3x+7&=16\\ -3x+7-7&=16-7\\ -3x&=9\\ \frac{-3x}{-3}&=\frac{9}{-3}\\ x&=-3\end{aligned}$$

Exercise Set 2.6

1. A ratio is a quotient of two quantities.

3. The ratio of c to d can be written as c to d, $c{:}d$, and $\frac{c}{d}$.

5. To set up and solve a proportion, we need a give ratio and one of the two parts of a second ratio.

7. No, similar figures have the same shape but not necessarily the same size.

9. Yes; the terms in each ratio are in the same order.

11. No; The terms in each ratio are not in the same order.

13. 6:9 = 2:3

15. 3:6 = 1:2

17. Total grades $= 6+4+9+3+2=24$
Ratio of total grades to D's = 24: 3 = 8: 1

19. 7:4

21. 5:15 = 1:3

23. 3 hours $= 3\times 60 = 180$ minutes
Ratio is $\frac{180}{30}=\frac{6}{1}$ or 6:1.

25. 4 pounds is $4\times 16 = 64$ ounces
Ratio is $\frac{26}{64}=\frac{13}{32}$ or 13:32.

27.
$$\begin{aligned}\text{Gear ratio}&=\frac{\text{number of teeth on driving gear}}{\text{number of teeth on driven gear}}\\ &=\frac{40}{5}=\frac{8}{1}\end{aligned}$$
Gear ratio is 8:1.

29. **a.** 199:140

b. Since $199\div 140\approx 1.42$, $199{:}140\approx 1.42{:}1$.

31. **a.** 1.13:0.38

b. Since $1.13\div 0.38\approx 2.97$, $1.13{:}0.38\approx 2.97{:}1$.

33. **a.** 434:174 or 217:87

b. 374:434 or 187:217

35. **a.** 40:32 or 5:4

b. 15:11

37.
$$\begin{aligned}\frac{3}{x}&=\frac{5}{20}\\ 3\cdot 20&=x\cdot 5\\ 60&=5x\\ \frac{60}{5}&=x\\ 12&=x\end{aligned}$$

39.
$$\begin{aligned}\frac{5}{3}&=\frac{75}{a}\\ 5\cdot a&=3\cdot 75\\ 5a&=225\\ a&=\frac{225}{5}=45\end{aligned}$$

41.
$$\begin{aligned}\frac{90}{x}&=\frac{-9}{10}\\ 90\cdot 10&=x(-9)\\ 900&=-9x\\ \frac{900}{-9}&=x\\ -100&=x\end{aligned}$$

43.
$$\begin{aligned}\frac{15}{45}&=\frac{x}{-6}\\ 45\cdot x&=-6\cdot 15\\ 45x&=-90\\ x&=\frac{-90}{45}=-2\end{aligned}$$

45.
$$\begin{aligned}\frac{3}{z}&=\frac{-1.5}{27}\\ -1.5\cdot z&=3\cdot 27\\ -1.5z&=81\\ z&=\frac{81}{-1.5}=-54\end{aligned}$$

47. $\frac{15}{20}=\frac{x}{8}$
$15\cdot 8=20\cdot x$
$120=20x$
$\frac{120}{20}=x$
$6=x$

49. $\frac{3}{12}=\frac{8}{x}$
$3x=(8)(12)$
$3x=96$
$x=\frac{96}{3}=32$
Thus the side is 32 inches in length.

51. $\frac{4}{7}=\frac{9}{x}$
$4x=(7)(9)$
$4x=63$
$x=\frac{63}{4}=15.75$
Thus the side is 15.75 inches in length.

53. $\frac{16}{12}=\frac{26}{x}$
$16x=(12)(26)$
$16x=312$
$x=\frac{312}{16}=19.5$
Thus the side is 19.5 inches in length.

55. Let x = number of loads one bottle can do.
$\frac{4\text{ fl ounces}}{1\text{ load}}=\frac{100\text{ fl ounces}}{x\text{ loads}}$
$\frac{4}{1}=\frac{100}{x}$
$4x=100$
$x=\frac{100}{4}=25$
One bottle can do 25 loads.

57. Let x = number of miles that can be driven with a full tank.
$\frac{23\text{ miles}}{1\text{ gallon}}=\frac{x}{15.7\text{ gallons}}$
$\frac{23}{1}=\frac{x}{15.7}$
$x=23\cdot 15.7$
$x=361.1$
It can travel 361.1 miles on a full tank.

59. Let x = length of model in feet.
$\frac{1\text{ foot model}}{20\text{ foot train}}=\frac{x\text{ foot model}}{30\text{ foot train}}$
$\frac{1}{20}=\frac{x}{30}$
$20x=30$
$x=\frac{30}{20}=1.5$
The model should be 1.5 feet long.

61. Let x = number of teaspoons needed for sprayer.
$\frac{3\text{ teaspoons}}{1\text{ gallon water}}=\frac{x\text{ teaspoons}}{8\text{ gallons water}}$
$\frac{3}{1}=\frac{x}{8}$
$3\cdot 8=1\cdot x$
$24=x$
Thus 24 teaspoons are needed for the sprayer.

63. Let x = length of beak of blue heron in inches.
$\frac{3.5\text{ inches in photo}}{3.75\text{ feet}}=\frac{0.4\text{ inches in photo}}{x\text{ feet}}$
$\frac{3.5}{3.75}=\frac{0.4}{x}$
$3.5x=3.75\cdot 0.4$
$3.5x=1.5$
$x\approx 0.43$
It's beak is about 0.43 feet long..

65. Let x = length on a map in inches.
$\frac{0.5\text{ inches}}{22\text{ miles}}=\frac{x\text{ inches}}{55\text{ miles}}$
$\frac{0.5}{22}=\frac{x}{55}$
$0.5\cdot 55=22\cdot x$
$27.5=22x$
$1.25=x$
The length on the map will be 1.25 inches.

67. Let x = length of the model bull in feet.
$\frac{2.95\text{ feet metal bull}}{1\text{ feet real bull}}=\frac{28\text{ feet metal bull}}{\text{x feet real bull}}$
$\frac{2.95}{1}=\frac{28}{x}$
$2.95\cdot x=1\cdot 28$
$2.95x=28$
$x\approx 9.49$
The model bull is about 9.49 feet long.

69. Let x = number of milliliters to be given.

$$\frac{1 \text{ milliliter}}{400 \text{ micrograms}} = \frac{x \text{ milliliter}}{220 \text{ micrograms}}$$
$$\frac{1}{400} = \frac{x}{220}$$
$$1 \cdot 220 = 400 \cdot x$$
$$220 = 400x$$
$$\frac{220}{400} = x$$
$$0.55 = x$$

Thus 0.55 milliliter should be given.

71. Let x = time, in minutes, it takes Jason to swim 30 laps.

$$\frac{3 \text{ laps}}{2.3 \text{ minutes}} = \frac{30 \text{ laps}}{x \text{ minutes}}$$
$$\frac{3}{2.3} = \frac{30}{x}$$
$$3 \cdot x = 2.3 \cdot 30$$
$$3x = 69$$
$$x = 23$$

It will take him 23 minutes.

73. Let x = number of children born with Prader-Willi Syndrome.

$$\frac{12{,}000 \text{ births}}{1 \text{ baby with syndrome}} = \frac{4{,}063{,}000 \text{ births}}{x \text{ babies with syndrome}}$$
$$\frac{12{,}000}{1} = \frac{4{,}063{,}000}{x}$$
$$12{,}000x = 4{,}063{,}000$$
$$x = \frac{4{,}063{,}000}{12{,}000} \approx 339$$

Thus, about 339 children were born with Prader-Willi Syndrome.

75.
$$\frac{12 \text{ inches}}{1 \text{ foot}} = \frac{78 \text{ inches}}{x \text{ feet}}$$
$$\frac{12}{1} = \frac{78}{x}$$
$$12x = 78$$
$$x = \frac{78}{12} = 6.5$$

Thus 42 inches equals 6.5 feet.

77.
$$\frac{9 \text{ square feet}}{1 \text{ square yard}} = \frac{26.1 \text{ square feet}}{x \text{ square yards}}$$
$$\frac{9}{1} = \frac{26.1}{x}$$
$$9x = 26.1$$
$$x = \frac{26.1}{9} = 2.9$$

Thus 26.1 square feet equals 2.9 square yards.

79.
$$\frac{2.54 \text{ cm}}{1 \text{ inch}} = \frac{50.8 \text{ cm}}{x \text{ inches}}$$
$$\frac{2.54}{1} = \frac{50.8}{x}$$
$$2.54x = 50.8$$
$$x = \frac{50.8}{2.54} = 20$$

Thus the length of the newborn is 20 inches.

81. Let x = number of home runs needed to be on schedule to break Bond's record.

$$\frac{73 \text{ home runs}}{162 \text{ games}} = \frac{x}{50 \text{ games}}$$
$$\frac{73}{162} = \frac{x}{50}$$
$$162 \cdot x = 73 \cdot 50$$
$$162x = 3650$$
$$x = \frac{3650}{162} \approx 22.53$$

A player would need to hit 23 home runs.

83.
$$\frac{480 \text{ grains}}{408 \text{ dollars}} = \frac{1 \text{ grain}}{\text{x dollars}}$$
$$\frac{480}{408} = \frac{1}{x}$$
$$480x = 408$$
$$x = \frac{408}{480} = 0.85$$

Thus the cost per grain is \$0.85.

85.
$$\frac{3.75 \text{ standard deviations}}{15 \text{ points}} = \frac{1 \text{ standard deviation}}{x \text{ points}}$$
$$\frac{3.75}{15} = \frac{1}{x}$$
$$3.75x = 15$$
$$x = \frac{15}{3.75} = 4$$

Thus 1 standard deviation equals 4 points.

87. The ratio of Mrs. Ruff's low density to high density cholesterol is $\frac{127}{60}$. If we divide 127 by 60 we obtain approximately 2.12. Thus Mrs. Ruff's ratio is approximately equivalent to 2.12:1. Therefore her ratio is less than the desired 4:1 ratio.

89. In $\frac{a}{b}=\frac{c}{d}$, if b and d remain the same while a increases, then c increases because $ad=bc$. If a increases ad increases so bc must increase by increasing c.

91. Let x = number of miles remaining on the life of each tire.
Inches remaining on the life of each tire:
$0.31-0.06=0.25$

$$\frac{0.03 \text{ inches}}{5000 \text{ miles}}=\frac{0.25 \text{ miles}}{x \text{ miles}}$$

$$\frac{0.03}{5000}=\frac{0.25}{x}$$

$$0.03x=5000\cdot 0.25$$

$$0.03x=1250$$

$$x=\frac{1250}{0.03}$$

$$x\approx 41{,}667$$

The tires will last about 41,667 more miles.

93. Let x = number of cubic centimeters of fluid needed.

$$\frac{1}{40}=\frac{x}{25}$$

$$40x=25$$

$$x=\frac{25}{40}$$

$$x=0.625$$

0.625 cubic centimeters of fluid should be drawn up into a syringe.

97. Commutative property of addition

98. Associative property of multiplication

99. Distributive property

100.

$$-(2x+6)=2(3x-6)$$
$$-2x-6=6x-12$$
$$-2x+2x-6=6x+2x-12$$
$$-6=8x-12$$
$$-6+12=8x-12+12$$
$$6=8x$$
$$\frac{6}{8}=\frac{8x}{8}$$
$$\frac{3}{4}=x$$

101.

$$3(4x-3)=6(2x+1)-15$$
$$12x-9=12x+6-15$$
$$12x-9=12x-9$$
$$12x-12x-9=12x-12x-9$$
$$-9=-9 \quad \text{True}$$

Since a true statement is obtained, the solution is all real numbers.

Review Exercises

1. $3(x+4)=3x+3(4)$
$=3x+12$

2. $3(x-2)=3[x+(-2)]$
$=3x+3(-2)$
$=3x+(-6)$
$=3x-6$

3. $-2(x+4)=-2x+(-2)(4)$
$=-2x+(-8)$
$=-2x-8$

4. $-(x+2)=-1(x+2)$
$=(-1)(x)+(-1)(2)$
$=-x+(-2)$
$=-x-2$

5. $-(m+3)=-1(m+3)$
$=(-1)(m)+(-1)(3)$
$=-m-3$

6. $-4(4-x)=-4[4+(-x)]$
$=(-4)(4)+(-4)(-x)$
$=-16+4x$

7. $5(5-p) = 5[5+(-p)]$
$= 5(5) + 5(-p)$
$= 25 + (-5p)$
$= 25 - 5p$

8. $6(4x-5) = 6(4x) - 6(5)$
$= 24x - 30$

9. $-5(5x-5) = -5[5x+(-5)]$
$= -5(5x) + (-5)(-5)$
$= -25x + 25$

10. $4(-x+3) = 4(-x) + 4(3)$
$= -4x + 12$

11. $\frac{1}{2}(2x+4) = \left(\frac{1}{2}\right)(2x) + \left(\frac{1}{2}\right)(4)$
$= x + 2$

12. $-(3+2y) = -1(3+2y)$
$= (-1)(3) + (-1)(2y)$
$= -3 + (-2y)$
$= -3 - 2y$

13. $-(x+2y-z) = -1[x+2y+(-z)]$
$= -1(x) + (-1)(2y) + (-1)(-z)$
$= -x + (-2y) + z$
$= -x - 2y + z$

14. $-3(2a-5b+7) = -3[2a+(-5b)+7]$
$= -3(2a) + (-3)(-5b) + (-3)(7)$
$= -6a + 15b + (-21)$
$= -6a + 15b - 21$

15. $7x - 3x = 4x$

16. $5 - 3y + 3 = -3y + 5 + 3$
$= -3y + 8$

17. $1 + 3x + 2x = 1 + 5x$
$= 5x + 1$

18. $-2x - x + 3y = -3x + 3y$

19. 4m + 2n + 4m + 6n = 4m +4m +2n +6n
= 8m +8n

20. There are no like terms.
$9x + 3y + 2$ cannot be further simplified.

21. $6x - 2x + 3y + 6 = 4x + 3y + 6$

22. $x + 8x - 9x + 3 = 9x - 9x + 3$
$= 3$

23. $-4z - 8z + 3 = -12z + 3$

24. $-2(3a^2-4) + 6a^2 - 8 = -6a^2 + 8 + 6a^2 - 8$
$= -6a^2 + 6a^2 + 8 - 8$
$= 0$

25. $2x + 3(x+4) - 5 = 2x + 3x + 12 - 5$
$= 5x + 7$

26. $4(3-2b) - 2b = 12 - 8b - 2b$
$= -8b - 2x + 12$
$= -10b + 12$

27. $6 - (-x+6) - x = 6 + x - 6 - x$
$= x - x + 6 - 6$
$= 0$

28. $2(2x+5) - 10 - 4 = 4x + 10 - 10 - 4$
$= 4x - 4$

29. $-6(4-3x) - 18 + 4x = -24 + 18x - 18 + 4x$
$= 18x + 4x - 24 - 18$
$= 22x - 42$

30. $4y - 3(x+y) + 6x^2 = 4y - 3x - 3y + 6x^2$
$= 6x^2 - 3x + 4y - 3y$
$= 6x^2 - 3x + y$

31. $\frac{1}{4}d + 2 - \frac{3}{5}d + 5 = \frac{1}{4}d - \frac{3}{5}d + 2 + 5$
$= \frac{5}{20}d - \frac{12}{20}d + 7$
$= -\frac{7}{20}d + 7$

32. $3 - (x-y) + (x-y) = 3 - x + y + x - y$
$= -x + x + y - y + 3$
$= 3$

33. $\frac{5}{6}x - \frac{1}{3}(2x-6) = \frac{5}{6}x - \frac{2}{3}x + 2$
$= \frac{5}{6}x - \frac{4}{6}x + 2$
$= \frac{1}{6}x + 2$

34. $\frac{2}{3}-\frac{1}{4}n-\frac{1}{3}(n+2)=\frac{2}{3}-\frac{1}{4}n-\frac{1}{3}n-\frac{2}{3}$
$=-\frac{1}{4}n-\frac{1}{3}n+\frac{2}{3}-\frac{2}{3}$
$=-\frac{3}{12}n-\frac{4}{12}n+0$
$=-\frac{7}{12}n$

35. $6x=6$
$\frac{6x}{6}=\frac{6}{6}$
$x=1$

36. $x+6=-7$
$x+6-6=-7-6$
$x=-13$

37. $x-4=7$
$x-4+4=7+4$
$x=11$

38. $\frac{x}{3}=-9$
$3\left(\frac{x}{3}\right)=3(-9)$
$x=-27$

39. $2x+4=8$
$2x+4-4=8-4$
$2x=4$
$\frac{2x}{2}=\frac{4}{2}$
$x=2$

40. $14=3+2x$
$14-3=3-3+2x$
$11=2x$
$\frac{11}{2}=\frac{2x}{2}$
$\frac{11}{2}=x$

41. $4c+3=-21$
$4c+3-3=-21-3$
$4c=-24$
$\frac{4c}{4}=\frac{-24}{4}$
$c=-6$

42. $4-2a=10$
$4-4-2a=10-4$
$-2a=6$
$\frac{-2a}{-2}=\frac{6}{-2}$
$a=-3$

43. $-x=-12$
$-1x=-12$
$(-1)(-1x)=(-1)(-12)$
$1x=12$
$x=12$

44. $3(x-2)=6$
$3x-6=6$
$3x-6+6=6+6$
$3x=12$
$\frac{3x}{3}=\frac{12}{3}$
$x=4$

45. $-12=3(2x-8)$
$-12=6x-24$
$-12+24=6x-24+24$
$12=6x$
$\frac{12}{6}=\frac{6x}{6}$
$2=x$

46. $4(6+2x)=0$
$24+8x=0$
$24-24+8x=0-24$
$8x=-24$
$\frac{8x}{8}=\frac{-24}{8}$
$x=-3$

47. $-6n+2n+6=0$
$-4n+6=0$
$-4n+6-6=0-6$
$-4n=-6$
$\frac{-4n}{-4}=\frac{-6}{-4}$
$n=\frac{6}{4}=\frac{3}{2}$

48.
$$-3 = 3w - (4w + 6)$$
$$-3 = 3w - 4w - 6$$
$$-3 = -1w - 6$$
$$-3 + 6 = -1w - 6 + 6$$
$$3 = -1w$$
$$\frac{3}{-1} = \frac{-1w}{-1}$$
$$-3 = w$$

49.
$$6 - (2n + 3) - 4n = 6$$
$$6 - 2n - 3 - 4n = 6$$
$$6 - 3 - 2n - 4n = 6$$
$$3 - 6n = 6$$
$$3 - 3 - 6n = 6 - 3$$
$$-6n = 3$$
$$\frac{-6n}{-6} = \frac{3}{-6}$$
$$n = -\frac{3}{6} = -\frac{1}{2}$$

50.
$$4x + 6 - 7x + 9 = 18$$
$$-3x + 15 = 18$$
$$-3x + 15 - 15 = 18 - 15$$
$$-3x = 3$$
$$\frac{-3x}{-3} = \frac{3}{-3}$$
$$x = -1$$

51.
$$4 + 3(x + 2) = 10$$
$$4 + 3x + 6 = 10$$
$$3x + 10 = 10$$
$$3x + 10 - 10 = 10 - 10$$
$$3x = 0$$
$$\frac{3x}{3} = \frac{0}{3}$$
$$x = 0$$

52.
$$-3 + 3x = -2(x + 1)$$
$$-3 + 3x = -2x - 2$$
$$-3 + 3x + 3 = -2x - 2 + 3$$
$$3x = -2x + 1$$
$$3x + 2x = -2x + 1 + 2x$$
$$5x = 1$$
$$\frac{5x}{5} = \frac{1}{5}$$
$$x = \frac{1}{5}$$

53.
$$8.4r - 6.3 = 6.3 + 2.1r$$
$$8.4r - 2.1r - 6.3 = 6.3 + 2.1r - 2.1r$$
$$6.3r - 6.3 = 6.3$$
$$6.3r - 6.3 + 6.3 = 6.3 + 6.3$$
$$6.3r = 12.6$$
$$\frac{6.3r}{6.3} = \frac{12.6}{6.3}$$
$$r = 2$$

54.
$$19.6 - 21.3t = 80.1 - 9.2t$$
$$19.6 - 21.3t + 21.3t = 80.1 - 9.2t + 21.3t$$
$$19.6 = 80.1 + 12.1t$$
$$19.6 - 80.1 = 80.1 - 80.1 + 12.1t$$
$$-60.5 = 12.1t$$
$$\frac{-60.5}{12.1} = \frac{12.1t}{12.1}$$
$$-5 = t$$

55.
$$0.35(c - 5) = 0.45(c + 4)$$
$$0.35c - 1.75 = 0.45c + 1.8$$
$$0.35c - 0.35c - 1.75 = 0.45c - 0.35c + 1.8$$
$$-1.75 = 0.10c + 1.8$$
$$-1.75 - 1.8 = 0.10c + 1.8 - 1.8$$
$$-3.55 = 0.10c$$
$$\frac{-3.55}{0.10} = \frac{0.10c}{0.10}$$
$$-35.5 = c$$

56.
$$-2.3(x - 8) = 3.7(x + 4)$$
$$-2.3x + 18.4 = 3.7x + 14.8$$
$$-2.3x + 2.3x + 18.4 = 3.7x + 2.3x + 14.8$$
$$18.4 = 6.0x + 14.8$$
$$18.4 - 14.8 = 6.0x + 14.8 - 14.8$$
$$3.6 = 6.0x$$
$$\frac{3.6}{6.0} = \frac{6.0x}{6.0}$$
$$0.6 = x$$

57.
$$\frac{p}{3}+2=\frac{1}{4}$$
$$12\left(\frac{p}{3}+2\right)=12\left(\frac{1}{4}\right)$$
$$4p+24=3$$
$$4p+24-24=3-24$$
$$4p=-21$$
$$\frac{4p}{4}=\frac{-21}{4}$$
$$p=-\frac{21}{4}$$

58.
$$\frac{d}{6}+\frac{1}{7}=2$$
$$42\left(\frac{d}{6}+\frac{1}{7}\right)=42(2)$$
$$7d+6=84$$
$$7d+6-6=84-6$$
$$7d=78$$
$$\frac{7d}{7}=\frac{78}{7}$$
$$d=\frac{78}{7}$$

59.
$$\frac{3}{5}(r-6)=3r$$
$$\frac{3}{5}r-\frac{18}{5}=3r$$
$$5\left(\frac{3}{5}r-\frac{18}{5}\right)=5(3r)$$
$$3r-18=15r$$
$$3r-3r-18=15r-3r$$
$$-18=12r$$
$$\frac{-18}{12}=\frac{12r}{12}$$
$$-\frac{18}{12}=r$$
$$-\frac{3}{2}=r$$

60.
$$\frac{2}{3}w=\frac{1}{7}(w-2)$$
$$\frac{2}{3}w=\frac{1}{7}w-\frac{2}{7}$$
$$21\left(\frac{2}{3}w\right)=21\left(\frac{1}{7}w-\frac{2}{7}\right)$$
$$14w=3w-6$$
$$14w-3w=3w-3w-6$$
$$11w=-6$$
$$\frac{11w}{11}=\frac{-6}{11}$$
$$w=-\frac{6}{11}$$

61.
$$9x-6=-3x+30$$
$$9x+3x-6=-3x+3x+30$$
$$12x-6=30$$
$$12x-6+6=30+6$$
$$12x=36$$
$$\frac{12x}{12}=\frac{36}{12}$$
$$x=3$$

62.
$$-(w+2)=2(3w-6)$$
$$-w-2=6w-12$$
$$-w-2+12=6w-12+12$$
$$-w+10=6w$$
$$-w+w+10=6w+w$$
$$10=7w$$
$$\frac{10}{7}=\frac{7w}{7}$$
$$\frac{10}{7}=w$$

63.
$$2x+6=3x+9-3$$
$$2x+6=3x+6$$
$$2x-2x+6=3x-2x+6$$
$$6=x+6$$
$$6-6=x+6-6$$
$$0=x$$

64. $-5a+3=2a+10$
$-5a+3-10=2a+10-10$
$-5a-7=2a$
$-5a+5a-7=2a+5a$
$-7=7a$
$\frac{-7}{7}=\frac{7a}{7}$
$-1=a$

65. $3x-12x=24-9x$
$-9x=24-9x$
$-9x+9x=24-9x+9x$
$0=24$ False
Since a false statement is obtained, there is no solution.

66. $5p-2=-2(-3p+6)$
$5p-2=6p-12$
$5p-5p-2=6p-5p-12$
$-2=p-12$
$-2+12=p-12+12$
$10=p$

67. $4(2x-3)+4=8x-8$
$8x-12+4=8x-8$
$8x-8=8x-8$
Since the equation is true for all values of x, the solution is all real numbers.

68. $4-c-2(4-3c)=3(c-4)$
$4-c-8+6c=3c-12$
$5c-4=3c-12$
$5c-3c-4=3c-3c-12$
$2c-4=-12$
$2c-4+4=-12+4$
$2c=-8$
$\frac{2c}{2}=\frac{-8}{2}$
$c=-4$

69. $2(x+7)=6x+9-4x$
$2x+14=6x+9-4x$
$2x+14=2x+9$
$2x-2x+14=2x-2x+9$
$14=9$ False
Since a false statement is obtained, there is no solution.

70. $-5(3-4x)=-6+20x-9$
$-15+20x=-6+20x-9$
$-15+20x=-15+20x$
The statement is true for all values of x, thus the solution is all real numbers.

71. $4(x-3)-(x+5)=0$
$4x-12-x-5=0$
$3x-17=0$
$3x-17+17=0+17$
$3x=17$
$\frac{3x}{3}=\frac{17}{3}$
$x=\frac{17}{3}$

72. $-2(4-x)=6(x+2)+3x$
$-8+2x=6x+12+3x$
$-8+2x=9x+12$
$-8-12+2x=9x+12-12$
$-20+2x=9x$
$-20+2x-2x=9x-2x$
$-20=7x$
$\frac{-20}{7}=\frac{7x}{7}$
$-\frac{20}{7}=x$

73. $\frac{x+3}{2}=\frac{x}{2}$
$2(x+3)=2x$
$2x+6=2x$
$2x-2x+6=2x-2x$
$6=0$ False
Since a false statement is obtained, there is no solution.

74. $\frac{x}{6}=\frac{x-4}{2}$
$2\cdot x=6(x-4)$
$2x=6x-24$
$2x-6x=6x-6x-24$
$-4x=-24$
$\frac{-4x}{-4}=\frac{-24}{-4}$
$x=6$

75. $\frac{1}{5}(3s+4)=\frac{1}{3}(2s-8)$

$\frac{3}{5}s+\frac{4}{5}=\frac{2}{3}s-\frac{8}{3}$

$15\left(\frac{3}{5}s+\frac{4}{5}\right)=15\left(\frac{2}{3}s-\frac{8}{3}\right)$

$9s+12=10s-40$

$9s-9s+12=10s-9s-40$

$12=s-40$

$12+40=s-40+40$

$52=s$

76. $\frac{2(2t-4)}{5}=\frac{3t+6}{4}-\frac{3}{2}$

$\frac{4t-8}{5}=\frac{3t+6}{4}-\frac{3}{2}$

$20\left(\frac{4t-8}{5}\right)=20\left(\frac{3t+6}{4}-\frac{3}{2}\right)$

$4(4t-8)=5(3t+6)-30$

$16t-32=15t+30-30$

$16t-32=15t+0$

$16t-16t-32=15t-16t$

$-32=-1t$

$\frac{-32}{-1}=\frac{-1t}{-1}$

$32=t$

77. $\frac{2}{5}(2-x)=\frac{1}{6}(-2x+2)$

$\frac{4}{5}-\frac{2}{5}x=-\frac{2}{6}x+\frac{2}{6}$

$\frac{4}{5}-\frac{2}{5}x=-\frac{1}{3}x+\frac{1}{3}$

$15\left(\frac{4}{5}-\frac{2}{5}x\right)=15\left(-\frac{1}{3}x+\frac{1}{3}\right)$

$12-6x=-5x+5$

$12-6x+6x=-5x+6x+5$

$12=x+5$

$12-5=x+5-5$

$7=x$

78. $\frac{x}{4}+\frac{x}{6}=\frac{1}{2}(x+3)$

$\frac{x}{4}+\frac{x}{6}=\frac{x}{2}+\frac{3}{2}$

$12\left(\frac{x}{4}+\frac{x}{6}\right)=12\left(\frac{x}{2}+\frac{3}{2}\right)$

$3x+2x=6x+18$

$5x=6x+18$

$5x-6x=6x-6x+18$

$-1x=18$

$\frac{-1x}{-1}=\frac{18}{-1}$

$x=-18$

79. 12: 20 = 3: 5

80. 80 ounces $=\frac{80}{16}=5$ pounds
The ratio of 80 ounces to 12 pounds is thus 5:12.

81. 32 ounces $=\frac{32}{16}=2$ pounds
The ratio of 32 ounces to 2 pounds is $\frac{2}{2}=\frac{1}{1}$.
The ratio is 1:1.

82. $\frac{x}{4}=\frac{8}{16}$

$16\cdot x=8\cdot 4$

$16x=32$

$x=\frac{32}{16}=2$

83. $\frac{5}{20}=\frac{x}{80}$

$20\cdot x=80\cdot 5$

$20x=400$

$x=\frac{400}{20}=20$

84. $\frac{3}{x}=\frac{15}{45}$

$3\cdot 45=15\cdot x$

$135=15x$

$\frac{135}{15}=x$

$9=x$

85. $\frac{20}{45}=\frac{15}{x}$

$20\cdot x=15\cdot 45$

$20x=675$

$x=\frac{675}{20}=\frac{135}{4}$

86. $\frac{6}{5}=\frac{-12}{x}$

$6\cdot x=-12\cdot 5$

$6x=-60$

$x=\frac{-60}{6}=-10$

87. $\frac{b}{6}=\frac{8}{-3}$

$-3\cdot b=6\cdot 8$

$-3b=48$

$x=\frac{48}{-3}=-16$

88. $\frac{-4}{9}=\frac{-16}{x}$

$-4\cdot x=-16\cdot 9$

$-4x=-144$

$x=\frac{-144}{-4}=36$

89. $\frac{x}{-15}=\frac{30}{-5}$

$-5\cdot x=-15\cdot 30$

$-5x=-450$

$x=\frac{-450}{-5}=90$

90. $\frac{6}{8}=\frac{30}{x}$

$6\cdot x=8\cdot 30$

$6x=240$

$x=\frac{240}{6}=40$

The length of the side is thus 40 in.

91. $\frac{7}{3.5}=\frac{2}{x}$

$7\cdot x=2\cdot 3.5$

$7x=7$

$x=\frac{7}{7}=1$

The length of the side is thus 1 ft.

92. Let x = time in hours it takes the ship to travel 140 miles

$\frac{40\text{ miles}}{1.8\text{ hours}}=\frac{140\text{ miles}}{x\text{ hours}}$

$\frac{40}{1.8}=\frac{140}{x}$

$40\cdot x=1.8\cdot 140$

$40x=252$

$\frac{40x}{40}=\frac{252}{40}$

$x=6.3$

It will take 6.3 hours to travel 140 miles.

93. Let x = number of calories in a 6-ounce piece of cake.

$\frac{4\text{ ounces}}{160\text{ calories}}=\frac{6\text{ ounces}}{x\text{ calories}}$

$\frac{4}{160}=\frac{6}{x}$

$4\cdot x=160\cdot 6$

$4x=960$

$x=\frac{960}{4}=240$

Thus, a 6-ounce piece of cake has 240 calories.

94. Let x = number of pages that can be copied in 22 minutes.

$\frac{1\text{ minutes}}{20\text{ pages}}=\frac{22\text{ minutes}}{x\text{ pages}}$

$\frac{1}{20}=\frac{22}{x}$

$1\cdot x=22\cdot 20$

$x=440$

440 pages can be copied in 22 minutes.

95. Let x = number of inches representing 380 miles.

$\frac{60\text{ miles}}{1\text{ inch}}=\frac{380\text{ miles}}{x\text{ inches}}$

$\frac{60}{1}=\frac{380}{x}$

$60\cdot x=380\cdot 1$

$60x=380$

$x=\frac{380}{60}=6\frac{1}{3}$

$6\frac{1}{3}$ inches on the map represent 380 miles.

96. Let x = size of actual car in feet

$$\frac{1 \text{ inch}}{1.5 \text{ feet}} = \frac{10.5 \text{ inches}}{x \text{ feet}}$$
$$\frac{1}{1.5} = \frac{10.5}{x}$$
$$1 \cdot x = 1.5 \cdot 10.5$$
$$x = 15.75$$

The size of the actual car is 15.75 ft.

97. Let x = the value of 1 peso in terms of U.S. dollars.

$$\frac{\$1 \text{ U.S.}}{9.165 \text{ pesos}} = \frac{x \text{ dollars}}{1 \text{ peso}}$$
$$\frac{1}{9.165} = \frac{x}{1}$$
$$9.165 \cdot x = 1 \cdot 1$$
$$9.165x = 1$$
$$x = \frac{1}{9.165} \approx 0.109$$

1 peso equals about \$0.109.

98. Let x = number of bottles the machine can fill and cap in 2 minutes.

2 minutes = 120 seconds

$$\frac{50 \text{ seconds}}{80 \text{ bottles}} = \frac{120 \text{ seconds}}{x \text{ bottles}}$$
$$\frac{50}{80} = \frac{120}{x}$$
$$50 \cdot x = 80 \cdot 120$$
$$50x = 9600$$
$$x = \frac{9600}{50} = 192$$

The machine can fill and cap 192 bottles in 2 minutes.

Practice Test

1. $-3(4-2x) = -3[4+(-2x)]$
$= -3(4)+(-3)(-2x)$
$= -12+6x$ or $6x-12$

2. $-(x+3y-4) = -[x+3y+(-4)]$
$= -1[x+3y+(-4)]$
$= (-1)(x)+(-1)(3y)+(-1)(-4)$
$= -x+(-3y)+4$
$= -x-3y+4$

3. $5x-8x+4 = -3x+4$

4. $4+2x-3x+6 = 2x-3x+4+6$
$= -x+10$

5. $-y-x-4x-6 = -x-4x-y-6$
$= -5x-y-6$

6. $a-2b+6a-6b-3 = a+6a-2b-6b-3$
$= 7a-8b-3$

7. $2x^2+3+2(3x-2) = 2x^2+3+6x-4$
$= 2x^2+6x+3-4$
$= 2x^2+6x-1$

8. $5(x-1)-3(y+4)-10x$
$= (5)(x)+(5)(-1)+(-3)(y)+(-3)(4)-10x$
$= 5x-5-3y-12-10x$
$= 5x-10x-3y-5-12$
$= -5x-3y-17$

9.
$$-x+7 = 12$$
$$-x+7-7 = 12-7$$
$$-x = 5$$
$$\frac{-x}{-1} = \frac{5}{-1}$$
$$x = -5$$

10.
$$\frac{1}{3}x = 8$$
$$3\left(\frac{1}{3}x\right) = 3(8)$$
$$x = 24$$

11.
$$4x-9 = 3x$$
$$4x-9+9 = 3x+9$$
$$4x = 3x+9$$
$$4x-3x = 3x-3x+9$$
$$x = 9$$

12.
$$2.4x-3.9 = 3.3$$
$$2.4x-3.9+3.9 = 3.3+3.9$$
$$2.4x = 7.2$$
$$\frac{2.4x}{2.4} = \frac{7.2}{2.4}$$
$$x = 3$$

13. $\frac{5}{6}(x-2)=x-3$

$\frac{5}{6}x-\frac{10}{6}=x-3$

$6\left(\frac{5}{6}x-\frac{10}{6}\right)=6(x-3)$

$5x-10=6x-18$

$5x-5x-10=6x-5x-18$

$-10=x-18$

$-10+18=x-18+18$

$8=x$

14. $6m-(4-2m)=0$

$6m-4+2m=0$

$6m+2m-4=0$

$8m-4=0$

$8m-4+4=0+4$

$8m=4$

$\frac{8m}{8}=\frac{4}{8}$

$m=\frac{4}{8}=\frac{1}{2}$

15. $3w+2(2w-6)=4(3w-3)$

$3w+4w-12=12w-12$

$7w-12=12w-12$

$7w-7w-12=12w-7w-12$

$-12=5w-12$

$-12+12=5w-12+12$

$0=5w$

$\frac{0}{5}=\frac{5w}{5}$

$0=w$

16. $2x-3(-2x+4)=-13+x$

$2x+6x-12=-13+x$

$8x-12=-13+x$

$8x-12+12=-13+12+x$

$8x=-1+x$

$8x-x=-1+x-x$

$7x=-1$

$\frac{7x}{7}=\frac{-1}{7}$

$x=-\frac{1}{7}$

17. $3x-4-x=2(x+5)$

$3x-4-x=2x+10$

$2x-4=2x+10$

$2x-2x-4=2x-2x+10$

$-4=10$ False

Since a false statement is obtained, there is no solution.

18. $-3(2x+3)=-2(3x+1)-7$

$-6x-9=-6x-2-7$

$-6x-9=-6x-9$

Since the equation is true for all values of x, the solution is all real numbers.

19. $\frac{9}{x}=\frac{3}{-15}$

$9(-15)=3x$

$-135=3x$

$\frac{-135}{3}=x$

$-45=x$

20. $\frac{1}{7}(2x-5)=\frac{3}{8}x-\frac{5}{7}$

$\frac{2}{7}x-\frac{5}{7}=\frac{3}{8}x-\frac{5}{7}$

$56\left(\frac{2}{7}x-\frac{5}{7}\right)=56\left(\frac{3}{8}x-\frac{5}{7}\right)$

$16x-40=21x-40$

$16x-16x-40=21x-16x-40$

$-40=5x-40$

$-40+40=5x-40+40$

$0=5x$

$\frac{0}{5}=\frac{5x}{5}$

$0=x$

21. a. An equation that has exactly one solution is a conditional equation.

b. An equation that has no solution is a contradiction.

c. An equation that has all real numbers as its solution is an identity.

22. $\frac{3}{4} = \frac{8}{x}$
$3x = 4 \cdot 8$
$3x = 32$
$x = \frac{32}{3}$
The length of side x is $\frac{32}{3}$ feet or $10\frac{2}{3}$ feet.

23. Let x = number of gallons needed.
$\frac{3 \text{ acres}}{6 \text{ gallons}} = \frac{75 \text{ acres}}{x \text{ gallons}}$
$\frac{3}{6} = \frac{75}{x}$
$3x = 6 \cdot 75$
$3x = 450$
$x = \frac{450}{3} = 150$
150 gallons are needed to treat 75 acres.

24. Let x = number of gallons he needs to sell.
$\frac{\$0.40}{1 \text{ gallon}} = \frac{\$20,000}{x \text{ gallons}}$
$\frac{0.40}{1} = \frac{20,000}{x}$
$0.40x = 20,000$
$x = \frac{20,000}{0.40} = 50,000$
He needs to sell 50,000 gallons.

25. Let x = number of minutes it will take.
$\frac{25 \text{ miles}}{35 \text{ minutes}} = \frac{125 \text{ miles}}{x \text{ minutes}}$
$\frac{25}{35} = \frac{125}{x}$
$25x = 35 \cdot 125$
$25x = 4375$
$x = \frac{4375}{25} = 175$
It would take 175 minutes or 2 hours 55 minutes.

Cumulative Review Test

1. $\frac{52}{15} \cdot \frac{10}{13} = \frac{4}{3} \cdot \frac{2}{1}$
$= \frac{4 \cdot 2}{3 \cdot 1}$
$= \frac{8}{3}$

2. $\frac{5}{24} \div \frac{2}{9} = \frac{5}{24} \cdot \frac{9}{2}$
$= \frac{5}{8} \cdot \frac{3}{2}$
$= \frac{5 \cdot 3}{8 \cdot 2}$
$= \frac{15}{16}$

3. $|-2| > 1$ since $|-2| = 2$ and $2 > 1$.

4. $-5-(-4)+12-8 = -5+4+12-8$
$= -1+12-8$
$= 11-8$
$= 3$

5. $-7-(-6) = -7+6$
$= -1$

6. $20 - 6 \div 3 \cdot 2 = 20 - 2 \cdot 2$
$= 20 - 4$
$= 16$

7. $3\left[6-\left(4-3^2\right)\right]-30 = 3[6-(4-9)]-30$
$= 3[6-(-5)]-30$
$= 3[6+5]-30$
$= 3(11)-30$
$= 33-30$
$= 3$

8. Substitute –2 for each x.
$-2x^2 - 6x + 8 = -2(-2)^2 - 6(-2) + 8$
$= -2(4) - (-12) + 8$
$= -8 + 12 + 8$
$= 4 + 8$
$= 12$

9. Associative property of addition

10. $8x + 2y + 4x - y = 8x + 4x + 2y - y$
$= 12x + y$

11. $9 - \frac{2}{3}x + 16 + \frac{3}{4}x = \frac{3}{4}x - \frac{2}{3}x + 9 + 16$
$= \frac{9}{12}x - \frac{8}{12}x + 25$
$= \frac{1}{12}x + 25$

12. $3x^2+5+4(2x-7)$
$=3x^2+5+4(2x)+4(-7)$
$=3x^2+5+8x-28$
$=3x^2+8x+5-28$
$=3x^2+8x-23$

13.
$$6x+2=10$$
$$6x+2-2=10-2$$
$$6x=8$$
$$\frac{6x}{6}=\frac{8}{6}$$
$$x=\frac{8}{6}=\frac{4}{3}$$

14.
$$\frac{1}{4}x=-11$$
$$4\left(\frac{1}{4}x\right)=4(-11)$$
$$x=-44$$

15.
$$-6x-5x+6=28$$
$$-11x+6=28$$
$$-11x+6-6=28-6$$
$$-11x=22$$
$$\frac{-11x}{-11}=\frac{22}{-11}$$
$$x=-2$$

16.
$$4(x-2)=5(x-1)+3x+2$$
$$4(x)+4(-2)=5(x)+5(-1)+3x+2$$
$$4x-8=5x-5+3x+2$$
$$4x-8=5x+3x-5+2$$
$$4x-8=8x-3$$
$$4x-8x-8=8x-8x-3$$
$$-4x-8=-3$$
$$-4x-8+8=-3+8$$
$$-4x=5$$
$$\frac{-4x}{-4}=\frac{5}{-4}$$
$$x=\frac{-5}{4}$$

17.
$$\frac{3}{4}n-\frac{1}{5}=\frac{2}{3}n$$
$$60\left(\frac{3}{4}n-\frac{1}{5}\right)=60\left(\frac{2}{3}n\right)$$
$$45n-12=40n$$
$$45n-45n-12=40n-45n$$
$$-12=-5n$$
$$\frac{-12}{-5}=\frac{-5n}{-5}$$
$$\frac{12}{5}=n$$

18.
$$\frac{40}{30}=\frac{3}{x}$$
$$40\cdot x=30\cdot 3$$
$$40x=90$$
$$\frac{40x}{40}=\frac{90}{40}$$
$$x=\frac{90}{40}=\frac{9}{4}\text{ or }2.25$$

19. Let x = number of pounds of fertilizer needed.
$$\frac{5000\text{ square feet}}{36\text{ pounds}}=\frac{22{,}000\text{ square feet}}{x\text{ pounds}}$$
$$\frac{5000}{36}=\frac{22{,}000}{x}$$
$$5000x=(36)(22{,}000)$$
$$5000x=792{,}000$$
$$x=\frac{792{,}000}{5000}=158.4$$
158.4 pounds are needed to fertilize a 22,000-square-foot lawn.

20. Let x = amount he earns after 8 hours.
$$\frac{2\text{ hours}}{\$10.50}=\frac{8\text{ hours}}{x\text{ dollars}}$$
$$\frac{2}{10.5}=\frac{8}{x}$$
$$2x=(10.5)(8)$$
$$2x=84$$
$$x=\frac{84}{2}=42$$
He earns \$42 after 8 hours.

Chapter 3

Exercise Set 3.1

1. A formula is an equation used to express a relationship mathematically.

3. The simple interest formula is:
$i = prt$ where i is interest, p is principle, r is the interest rate, and t is time.

5. The diameter of a circle is 2 times its radius.

7. When you multiply a unit by the same unit, you get a square unit.

9. Substitute 6 for s.

$$P = 4s$$
$$= 4(6)$$
$$= 24$$

11. Substitute 7 for s.

$$A = s^2$$
$$A = (7)^2 = 49$$

13. Substitute 8 for l and 5 for w.

$$P = 2l + 2w$$
$$P = 2(8) + 2(5)$$
$$P = 16 + 10$$
$$P = 26$$

15. Substitute 5 for r.

$$A = \pi r^2$$
$$A = \pi(5)^2$$
$$A = 25\pi \approx 78.54$$

17. Substitute 100 for x, 80 for m, and 10 for s.

$$z = \frac{x - m}{s}$$
$$z = \frac{100 - 80}{10}$$
$$z = \frac{20}{10} = 2$$

19. Substitute 60 for V and 12 for B.

$$V = \frac{1}{3}Bh$$
$$60 = \frac{1}{3}(12)h$$
$$60 = 4h$$
$$\frac{60}{4} = \frac{4h}{4}$$
$$15 = h$$

21. Substitute 36 for A and 16 for m.

$$A = \frac{m+n}{2}$$
$$36 = \frac{16+n}{2}$$
$$2(36) = 2\left(\frac{16+n}{2}\right)$$
$$72 = 16 + n$$
$$72 - 16 = n$$
$$56 = n$$

23. Substitute 15 for C.

$$F = \frac{9}{5}C + 32$$
$$F = \frac{9}{5}(15) + 32$$
$$F = 27 + 32 = 59$$

25. Substitute 678.24 for V, and 6 for r.

$$V = \pi r^2 h$$
$$678.24 = \pi(6)^2 h$$
$$678.24 = 36\pi h$$
$$\frac{678.24}{36\pi} = \frac{36\pi h}{36\pi}$$
$$\frac{678.24}{36\pi} = h$$
$$6.00 \approx h$$

27. Substitute 24 for B and 61 for h.

$$B = \frac{703w}{h^2}$$
$$24 = \frac{703w}{(61)^2}$$
$$24(61)^2 = \frac{703w}{61^2}(61)^2$$
$$89,304 = 703w$$
$$\frac{89,304}{703} = \frac{703w}{703}$$
$$127.03 \approx w$$

29. Substitute 160 for C and 0.12 for r.

$$S = C + rC$$
$$S = 160 + (0.12)(160)$$
$$S = 160 + 19.20 = 179.20$$

31. Substitute 6 for b and 4 for h.

$$A = \frac{1}{2}bh$$
$$A = \frac{1}{2}(6)(4)$$
$$A = (3)(4)$$
$$A = 12 \text{ in}^2$$

33. Substitute 4 for r and 9 for h.

$$V = \pi r^2 h$$
$$V = \pi(4)^2(9)$$
$$V = \pi(16)(9)$$
$$V = 144\pi \approx 452.39 \text{ cm}^3$$

35. Substitute 3 for h, 4 for b and 7 for d.

$$A = \frac{1}{2}h(b+d)$$
$$A = \frac{1}{2}(3)(4+7)$$
$$A = \frac{1}{2}(3)(11)$$
$$A = \frac{1}{2}(33)$$
$$A = 16.5 \text{ ft}^2$$

37. $P = 4s$

$$\frac{P}{4} = \frac{4s}{4}$$
$$\frac{P}{4} = s$$

39. $d = rt$

$$\frac{d}{r} = \frac{rt}{r}$$
$$\frac{d}{r} = t$$

41. $V = lwh$

$$\frac{V}{wh} = \frac{lwh}{wh}$$
$$\frac{V}{wh} = l$$

43. $A = \frac{1}{2}bh$

$$2A = 2\left(\frac{1}{2}bh\right)$$
$$2A = bh$$
$$\frac{2A}{h} = \frac{bh}{h}$$
$$\frac{2A}{h} = b$$

45. $P = 2l + 2w$

$$P - 2l = 2l - 2l + 2w$$
$$P - 2l = 2w$$
$$\frac{P-2l}{2} = \frac{2w}{2}$$
$$\frac{P-2l}{2} = w$$

47. $5 - 2t = m$

$$5 - 5 - 2t = m - 5$$
$$-2t = m - 5$$
$$\frac{-2t}{-2} = \frac{m-5}{-2}$$
$$t = -\frac{m-5}{2} = \frac{-m+5}{2}$$

49. $y = mx + b$

$$y - mx = mx - mx + b$$
$$y - mx = b$$

51. $y = mx + b$

$$y - b = mx + b - b$$
$$y - b = mx$$
$$\frac{y-b}{m} = \frac{mx}{m}$$
$$\frac{y-b}{m} = x$$

53. $ax + by = c$

$$ax - ax + by = -ax + c$$
$$by = -ax + c$$
$$\frac{by}{b} = \frac{-ax+c}{b}$$
$$y = \frac{-ax+c}{b}$$

55. $V = \pi r^2 h$

$\frac{V}{\pi r^2} = \frac{\pi r^2 h}{\pi r^2}$

$\frac{V}{\pi r^2} = h$

57. $A = \frac{m+d}{2}$

$2A = 2\left(\frac{m+d}{2}\right)$

$2A = m + d$

$2A - d = m + d - d$

$2A - d = m$

59. $R = \frac{I+3w}{2}$

$2R = 2\left(\frac{I+3w}{2}\right)$

$2R = I + 3w$

$2R - I = I - I + 3w$

$2R - I = 3w$

$\frac{2R-I}{3} = \frac{3w}{3}$

$\frac{2R-I}{3} = w$

61. 3x + y = 5

a. 3x + y − 3x = 5 − 3x

y = 5 − 3x

b. Substitute 2 for x.

y = 5 − 3(2) = 5 − 6 = -1

63. $4x = 6y - 8$

a. $4x + 8 = 6y - 8 + 8$

$4x + 8 = 6y$

$2(2x+4) = 6y$

$\frac{2(2x+4)}{6} = \frac{6y}{6}$

$\frac{2x+4}{3} = y$

b. Substitute 10 for *x*.

$y = \frac{2(10)+4}{3}$

$= \frac{20+4}{3}$

$= \frac{24}{3}$

$= 8$

65. 5y = -12 + 3x

a. $5y = -12 + 3x$

$\frac{5y}{5} = \frac{-12+3x}{5}$

$y = \frac{3x-12}{5}$

b. Substitute 4 for *x*.

$y = \frac{3(4)-12}{5} = \frac{12-12}{5} = \frac{0}{5} = 0$

67. $-3x + 5y = -10$

a. $-3x + 5y + 3x = -10 + 3x$

$5y = -10 + 3x$

$\frac{5y}{5} = \frac{-10+3x}{5}$

$y = \frac{-10+3x}{5}$

b. Substitute 4 for *x*.

$y = \frac{-10+3(4)}{5}$

$= \frac{-10+12}{5}$

$= \frac{2}{5}$

69. 15 − 3x = -6y

a. $15 - 3x = -6y$

$\frac{15-3x}{-6} = \frac{-6y}{-6}$

$\frac{-3x+15}{-6} = y$

$\frac{-3(x-5)}{-6} = y$

$\frac{x-5}{2} = y$

b. Substitute 0 for x.
$$y = \frac{0-5}{2} = \frac{-5}{2}$$

71. $-8 = -x - 2y$

a.
$$x - 8 = x - x - 2y$$
$$x - 8 = -2y$$
$$\frac{x-8}{-2} = \frac{-2y}{-2}$$
$$\frac{-x+8}{2} = y$$

b. Substitute −4 for x.
$$y = \frac{-(-4)+8}{2} = \frac{4+8}{2} = \frac{12}{2} = 6$$

73. a.
$$y + 3 = -\frac{1}{3}(x-4)$$
$$y + 3 = -\frac{1}{3}x + \frac{4}{3}$$
$$y + 3 - 3 = -\frac{1}{3}x + \frac{4}{3} - 3$$
$$y = -\frac{1}{3}x + \frac{4}{3} - \frac{9}{3}$$
$$y = -\frac{1}{3}x - \frac{5}{3} = \frac{-x-5}{3}$$

b. Substitute 6 for x.
$$y = \frac{-(6)-5}{3} = \frac{-6-5}{3} = \frac{-11}{3}$$

75. a.
$$y - \frac{1}{5} = 2\left(x + \frac{1}{3}\right)$$
$$y - \frac{1}{5} = 2x + \frac{2}{3}$$
$$y - \frac{1}{5} + \frac{1}{5} = 2x + \frac{2}{3} + \frac{1}{5}$$
$$y = 2x + \frac{10}{15} + \frac{3}{15}$$
$$y = 2x + \frac{13}{15} = \frac{30x}{15} + \frac{13}{15} = \frac{30x+13}{15}$$

b. Substitute 4 for x.
$$y = \frac{30(4)+13}{15} = \frac{120+13}{15} = \frac{133}{15}$$

77. Substitute 10 for n.
$$d = \frac{1}{2}n^2 - \frac{3}{2}n$$
$$d = \frac{1}{2}(10)^2 - \frac{3}{2}(10)$$
$$= \frac{1}{2}(100) - 15$$
$$= 50 - 15$$
$$= 35$$

79. Substitute 50 for F.
$$C = \frac{5}{9}(F-32)$$
$$C = \frac{5}{9}(50-32)$$
$$= \frac{5}{9}(18)$$
$$= 10$$
The equivalent temperature is 10°C.

81. Substitute 25 for C.
$$F = \frac{9}{5}C + 32$$
$$F = \frac{9}{5}(25) + 32$$
$$= 45 + 32$$
$$= 77$$
The equivalent temperature is 77°F.

83. $P = \frac{KT}{V}$
$$P = \frac{(2)(20)}{1} = \frac{40}{1} = 40$$

85. $P = \frac{KT}{V}$
$$80 = \frac{K(100)}{5}$$
$$80 = 20K$$
$$\frac{80}{20} = \frac{20K}{20}$$
$$4 = K$$

87. $A = s^2$
$$A = (2s)^2 = 4s^2$$
The area is 4 times as large as the original area.

89. Substitute 6 for n.
$$S = n^2 + n$$
$$S = (6)^2 + 6 = 36 + 6 = 42$$

91. $i = prt$
$i = (6000)(0.08)(3) = 1440$
He will pay \$1440 interest.

93. $i = prt$
$450 = p(0.03)(3)$
$450 = 0.09p$
$\frac{450}{0.09} = \frac{0.09p}{0.09}$
$5000 = p$
She placed \$5000 in the savings account.

95. $P = a + b + c$
$P = 5 + 12 + 8 = 25$
The perimeter of the table top is 25 feet.

97. $A = \frac{1}{2}bh$
$A = \frac{1}{2}(36)(31) = 558$
The area is 558 square inches.

99. $A = \pi r^2$
$A = \pi(1.5)^2$
$A = \pi(2.25) \approx 7.07$
The area of the tabletop is about 7.07 square feet.

101. Total area = Area of top triangle + area of bottom triangle
Total Area = $.5b_1 h_1 + .5b_2 h_2$
Total Area = $.5(2)(1) + .5(2)(2)$
Total Area = $1 + 2 = 3$
The area of the kite is 3 square feet.

103. $V = \pi r^2 h$
$= \pi(4)^2(3)$
$= \pi(16)(3)$
$= 48\pi \approx 150.80$
The volume of water in the Jacuzzi is about 150.80 cubic feet.

105. $A = \frac{1}{2}h(b + d)$
$A = \frac{1}{2}(100)(80 + 200)$
$= \frac{1}{2}(100)(280)$
$= (50)(280)$
$= 14,000$
The seating area is 14,000 square feet.

107. The radius is half the diameter, so
$r = \frac{3}{2} = 1.5$ inches.
$V = \frac{1}{3}\pi r^2 h$
$V = \frac{1}{3}\pi(1.5)^2(5)$
$= \frac{1}{3}\pi(2.25)5$
$= 3.75\pi$
≈ 11.78
The volume of the cone is about 11.78 cubic inches.

109. a. $B = \frac{703w}{h^2}$

b. 5 feet 3 inches $= 5(12) + 3$
$= 60 + 3$
$= 63$ inches
$B = \frac{703(135)}{(63)^2} = \frac{94,905}{3969} \approx 23.91$

111. a. $V = lwh$
$V = (3x)(x)(6x - 1)$
$= 3x^2(6x - 1)$
$= 18x^3 - 3x^2$

b. $V = 18x^3 - 3x^2$
$V = 18(7)^3 - 3(7)^2$
$= 6174 - 147$
$= 6027$
Volume is 6027 cm^3.

c. $S = 2lw + 2lh + 2wh$
$S = 2(3x)(x) + 2(3x)(6x - 1) + 2(x)(6x - 1)$
$= 6x^2 + 36x^2 - 6x + 12x^2 - 2x$
$= 54x^2 - 8x$

d. $S = 54x^2 - 8x$
$S = 54(7)^2 - 8(7)$
$= 2646 - 56$
$= 2590$
Surface area is 2590 cm^2.

113. $\left[4\left(12 \div 2^2 - 3\right)^2\right]^2 = \left[4(12 \div 4 - 3)^2\right]^2$
$= \left[4(3-3)^2\right]^2$
$= \left[4(0)^2\right]^2$
$= [0]^2$
$= 0$

114.
$$\begin{aligned} 2(x-4) &= -(3x+9) \\ 2x - 8 &= -3x - 9 \\ 2x + 3x - 8 &= -3x + 3x - 9 \\ 5x - 8 &= -9 \\ 5x - 8 + 8 &= -9 + 8 \\ 5x &= -1 \\ \frac{5x}{5} &= \frac{-1}{5} \\ x &= -\frac{1}{5} \end{aligned}$$

115. $\frac{6}{4} = \frac{3}{2}$ so the ratio of Arabians to Morgans is 3:2.

116. Let x = number of minutes to siphon 13,500 gallons
$$\begin{aligned} \frac{25 \text{ gallons}}{3 \text{ minutes}} &= \frac{13,500 \text{ gallons}}{x \text{ minutes}} \\ \frac{25}{3} &= \frac{13,500}{x} \\ 25 \cdot x &= 3(13,500) \\ 25x &= 40,500 \\ x &= \frac{40,500}{25} = 1620 \end{aligned}$$
It will take 1620 minutes or 27 hours to empty the pool.

Exercise Set 3.2

1. Added to, more than, increased by, and sum indicate the operation of addition.

3. Multiplied by, product of, twice, and three times indicate the operation of multiplication.

5. The cost is increased by 25% of the cost, so the expression needs $0.25c$.

7. $n + 7$

9. $4x$

11. $\frac{x}{2}$

13. $h + 0.8$

15. $p - 0.08$

17. $\frac{1}{10}n - 5$

19. $\frac{8}{9}m + 16,000$

21. $45 + 0.40x$

23. $25x$

25. $16x + y$

27. $n + 0.04n$

29. $p - 0.02p$

31. $220 + 80x$

33. $275x + 25y$

35. Three less than a number

37. One more than four times a number

39. Seven less than six times a number

41. Four times a number, decreased by two

43. Three times a number subtracted from two

45. Twice the difference between a number and one

47. $s + 5$

49. $b - 6$

51. $600 - a$

53. $100 - m$

55. $\frac{2}{3}m - 6$

57. $2r - 673$

59. $2p - 2.7$

61. $3n - 15$

63. $2n - 67,109$

65. $s + 0.20s$

67. $s + 0.15s$

69. $f - 0.12f$

71. $c + 0.07c$

73. $p - 0.50p$

75. a. Let x = first number, then $4x$ = second number.

b. First number + second number = 20
$x + 4x = 20$

77. a. Let x = smaller integer, then $x + 1$ = larger consecutive integer.

b. Smaller + larger = 41
$x + (x + 1) = 41$

79. a. Let x = the number.

b. Twice the number decreased by 8 is 12.
$2x - 8 = 12$

81. a. Let x = the number.

b. One-fifth of the sum of the number and 10 is 150.
$\frac{1}{5}(x + 10) = 150$

83. a. Let s = the distance traveled by the Southern Pacific train.

b. $s + (2s - 4) = 890$

85. a. Let c = the cost of the car.

b. $c + 0.07c = 32{,}600$

87. a. Let c = the cost of the meal.

b. $c + 0.15c = 42.50$

89. a. Let f = the average salary in San Francisco.

b. $1.28f - f = 16{,}762$

91. a. Let s = number of laser vision surgeries in 1999.

b. $s + (s + 0.55) = 2.45$

93. Two more than a number is five.

95. Three times a number, decreased by one, is four more than twice the number.

97. Four times the difference between a number and one is six.

99. Six more than five times a number is the difference between six times the number and one.

101. The sum of a number and the number increased by four is eight.

103. The sum of twice a number and the number increased by three is five.

105. Answers will vary.

107. a. 1 minute = 60 seconds
1 hour = 60 minutes = 3600 seconds
1 day = 24 hours
= 1440 minutes
= 86,400 seconds
$86{,}400d + 3600h + 60m + s$

b. $86{,}400d + 3600h + 60m + s$
$= 86{,}400(4) + 3600(6) + 60(15) + 25$
$= 368{,}125$ seconds

109. $30 = 6t$

111.
$$3[(4 - 16) \div 2] + 5^2 - 3$$
$$3[(-12) \div 2] + 25 - 3$$
$$3(-6) + 25 - 3$$
$$-18 + 25 - 3$$
$$7 - 3$$
$$4$$

112.
$$\frac{3.6}{x} = \frac{10}{7}$$
$$3.6 \cdot 7 = 10 \cdot x$$
$$25.2 = 10x$$
$$\frac{25.2}{10} = \frac{10x}{10}$$
$$2.52 = x$$

113. The units must be the same so change the 4 pounds to ounces. To do this multiply 4 by 16 because 1 pound contains 16 ounces.
4(16)=64
$$\frac{26 \text{ ounces}}{64 \text{ ounces}} = \frac{13}{32} = 13{:}32$$

114. Substitute 40 for P and 5 for w.

$P = 2l + 2w$

$40 = 2l + 2(5)$

$40 = 2l + 10$

$30 = 2l$

$15 = l$

115. $3x - 2y = 6$

$3x - 3x - 2y = -3x + 6$

$-2y = -3x + 6$

$\frac{-2y}{-2} = \frac{-3x+6}{-2}$

$y = \frac{3x-6}{2}$

$y = \frac{3}{2}x - 3$

Substitute 6 for x.

$y = \frac{3(6)-6}{2} = \frac{18-6}{2} = \frac{12}{2} = 6$

Exercise Set 3.3

1. Answers will vary.

3. Let x = smaller integer, then

$x + 1$ = next consecutive integer.

Smaller number + larger number = 85.

$x + (x + 1) = 85$

$2x + 1 = 85$

$2x = 84$

$x = 42$

Smaller number = 42

Larger number = $x + 1 = 42 + 1 = 43$

5. Let x = smaller odd integer, then

$x + 2$ = next consecutive odd integer.

Sum of integers = 104.

$x + (x + 2) = 104$

$2x + 2 = 104$

$2x = 102$

$x = 51$

Smaller integer = 51

Larger integer = 51 + 2 = 53

7. Let x = one number.

Then $2x + 3$ = second number.

First number + second number = 27

$x + (2x + 3) = 27$

$3x + 3 = 27$

$3x = 24$

$x = 8$

First number = 8

Second number = $2x + 3 = 2(8) + 3 = 19$

9. Let x = smaller integer, then

larger integer = $2x - 8$.

Larger integer – smaller integer = 17

$(2x - 8) - x = 17$

$2x - 8 - x = 17$

$x - 8 = 17$

$x = 25$

Smaller number = 25

Larger number = $2x - 8 = 2(25) - 8 = 42$

11. Let x = the life expectancy, in years, for men in the U.S. in 1900, then $2x - 19$ is the life expectancy, in years, for men in the U.S. in 2000.

Life expectancy in 2000 – Life expectancy in 1900 = 27.3

$(2x - 19) - x = 27.3$

$x - 19 = 27.3$

$x = 46.3$

$2x - 19 = 2(46.3) - 19 = 73.6$

The life expectancy of men in the U.S. in 2000 is 73.6 years old.

13. Let x = the number of DVD players sold in 2001, then $2x + 20$ = number of DVD players sold in 2002.

DVD players sold in 2001 + DVD players sold in 2002 = 3260

$x + (2x + 20) = 3260$

$3x + 20 = 3260$

$3x = 3240$

$x = 1080$

$2x + 20 = 2(1080) + 20 = 2180$

In 2002, 2,180 DVD players were sold.

15. Let x = the number of baseball cards given to Richey, then $3x$ = number of baseball cards given to Erin.

Number of cards given to Richey + Number of cards given to Erin = 260

$x + 3x = 260$

$4x = 260$

$4x = 260$

$x = 65$

Grandma gave 65 baseball cards to Richey.

17. Let x = the number of hours it takes to design a horse, then $2x + 1.4$ = the number of hours it takes to attach the gloves to the horse.

Time to design horse + Time to attach gloves to horse = 32.6

$x + (2x + 1.4) = 32.6$

$3x + 1.4 = 32.6$

$3x = 31.2$

$x = 10.4$

$2x + 1.4 = 2(10.4) + 1.4 = 22.3$

It took him 22.2 hours to attach the gloves to the horse.

19. Let x = the number of weeks, then $6x$ = the amount she wishes to add to her collection over x weeks.
Amount started with in collection + Amount added each week to the collection over x weeks = Total number in collection
$624 + 6x = 1000$
$6x = 376$
$x \approx 62.7$
It will take her about 62.7 weeks to get 1000 frogs in her collection.

21. Let x = the time in years, then $1200x$ = the increase in population over x years.
Current population + Increase in population over x years = Future population
$6500 + 1200x = 20{,}600$
$1200x = 14{,}100$
$x = 11.75$
In 11.75 years, the population will reach 20,600.

23. Let x = the number of weeks, then $120x$ = the number of computers shipped after x weeks.
Current supply of computers – Number of computers shipped over x weeks = Future inventory
$3600 - 120x = 2000$
$-120x = -1600$
$x \approx 13.3$
It will take about 13.3 weeks for the computer inventory to drop to 2000.

25. Let x = the number of miles, then $0.30x$ = the cost of driving over x miles.
Daily cost + Mileage cost = Total cost
$50 + 0.30x = 92$
$0.30x = 42$
$x = 140$
Lori can drive a maximum of 140 miles.

27. Let x the number of copies made, then $0.02x$ = the cost to make x number of copies.
Cost of machine + Cost of copies made = Total cost
$2100 + 0.02x = 2462$
$0.02x = 362$
$x = 18{,}100$
In one year, 18,100 copies were made.

29. Let x = the number of minutes, then $x - 500$ = the number of minutes talked over 500 minutes and $0.40(x - 500)$ = the cost to talk after 500 minutes.
Monthly fee + Cost of talk time over 500 minutes = Total cost
$25.95 + 0.40(x - 500) = 61.95$
$25.95 + 0.40x - 200 = 61.95$
$0.40x - 174.05 = 61.95$
$0.40x = 236$
$x = 590$
$x - 500 = 590 - 500 = 90$
Anke used 90 minutes over and above the 500 minutes of free time.

31. Let x = the time before the costs are the same.
Total cost for Kenmore = total cost for Neptune
Price of Kenmore + energy costs = Price of Neptune + energy costs
$362 + 84x = 454 + 38x$
$362 + 46x = 454$
$46x = 92$
$x = 2$
It will take 2 years before the total cost is the same for both machines.

33. Let x = number of years until the salaries are the same.
yearly salary = base salary + (yearly increase) · (number of years)
yearly salary at Data Tech. = yearly salary at Nuteck
$40,000 + 2400x = 49,600 + 800x$
$40,000 + 1600x = 49,600$
$1600x = 9600$
$x = 6$
It will take 6 years for the two salaries to be the same.

35. Let x = the number of pages, then 0.02x = the cost of printing x pages on the HP and $0.03x$ = the cost of printing x pages on the Lexmark.

Cost of the HP printer + Cost of printing x pages on the HP = Cost of the Lexmark printer + Cost of printing x pages on the Lexmark

$149 + 0.02x = 99 + 0.03x$
$50 = 0.01x$
$5000 = x$

For the two printers to have the same cost, 5000 pages would have to be printed.

37. Let x = the number of months, then 980x = the monthly payments in x months with First Union and 910x = the monthly payments in x months with Kensington.
First Union monthly payments = Kensington monthly payments + fees
$980x = 910x + 2000$
$70x = 2000$
$x \approx 28.6$
In about 28.6 months the total cost of both mortgages would be the same.

39. Let x = amount of assets Greg manages for Judy.
(Assets)(percentage) = fee
$(x)(0.01) = 620$
$0.01x = 620$
$x = 62{,}000$
Greg manages \$62,000 in assets for Judy.

41. Let x = the cost of the flight before tax, then $0.07x$ = the sales tax on the flight.
Cost of flight before tax + Sales tax on flight = Total cost
$x + 0.07x = 280$
$1.07x = 280$
$x = 261.68$
The cost of the flight before taxes was \$261.68.

43. Let x = Zhen's present salary, then $x + 0.30x$ = Zhen's salary at his new job.
New salary = 30,200
$x + 0.30x = 30{,}200$
$1.3x = 30{,}200$
$x = 23230.77$
Zhen's present salary is \$23,230.77.

45. Let x = the size of former house in square feet, then $x - 0.18x$ = the size of the new house in square feet.
Size of new house = 2200 square feet
$x - 0.18x = 2200$
$0.82x = 2200$
$x = 2682.93$
The size of the former house was about 2682.93 square feet.

47. Let x = total amount collected at door;
3000 + 3% of admission fees = total amount received.
$3000 + 0.03x = 3750$
$0.03x = 750$
$$x = \frac{7.50}{0.03} = 25{,}000$$
The total amount collected at the door was \$25,000.

49. Let x = average salary before wage cut.
(average salary before cut) – (decrease in salary) = average salary after wage cut
$x - 0.02x = 28{,}600$
$0.98x = 28{,}600$
$x \approx 29{,}183.67$
The average salary before the wage cut was \$29,183.67.

51. Let x = total dollar volume in a week.
Then $0.06x$ = commission.
Salary + commission = 710
$350 + 0.06x = 710$
$0.06x = 360$
$$x = \frac{360}{0.06} = 6000$$
His dollar volume in a week must be \$6000.

53. Let x = the amount of sales in dollars, then $600 + 0.02x$ = Plan 1 salary and $0.10x$ = Plan 2 salary.
Plan 1 salary = Plan 2 salary
$600 + 0.02x = 0.10x$
$600 = 0.08x$
$7500 = x$
Sales of \$7,500 will result in the same salary from both plans.

55. Let x = customer assets in dollar, then $1000 + 0.01x$ = Plan 1 charges and $500 + 0.02x$ = Plan 2 charges.
Plan 1 charges = Plan 2 charges
$1000 + 0.01x = 500 + 0.02x$
$1000 = 500 + 0.01x$
$500 = 0.01x$
$50{,}000 = x$
Customer assets of \$50,000 would result in both plans having the same total cost.

57. Let x = regular membership fee, then amount of reduction = $0.10x$.
regular fee – reduction – 20 = new fee on a Monday
$x - 0.10x - 20 = 250$
$x - 0.10x = 270$
$0.90x = 270$
$$x = \frac{270}{0.90} = 300$$
Regular fee is \$300.

59. Let x = the amount in dollars Phil's daughter receives, then $x + 0.25x$ = amount in dollars Phil's wife receives.
Daughter's share + Wife's share = \$140,000
$x + (x + 0.25x) = 140{,}000$
$2x + 0.25x = 140{,}000$
$2{,}25x = 140{,}000$
$x = 62{,}222.22$
$x + 0.25x = 62{,}222.22 + (0.25)(62{,}222.22)$
$= 77{,}777.78$
Phil's wife will receive \$77,777.78.

61. a. $$\frac{74 + 88 + 76 + x}{4} = 80$$

b. $$\frac{74 + 88 + 76 + x}{4} = 80$$
$74 + 88 + 76 + x = 320$
$238 + x = 320$
$x = 82$

Paul must receive an 82 on his fourth exam.

63. a. Yearly cost with 10% discount $= 600 - 0.10(600) = \$540$. Let $x =$ the number of years it will take for the costs to be equal.
Total cost with driver's ed = Total cost without driver's ed.

$$45 + 540x = 600x$$
$$45 = 60x$$
$$0.75 = x$$

It will take 0.75 years, or 9 months for the costs to be equal.

b. $25 - 18 = 7$ years. The cost with driver's ed is $45 + 540(7) = \$3825$. The cost without driver's ed is $600 \cdot (7) = \$4200$. He will save $4200 - 3825 = \$375$.

65.
$$\frac{1}{4} + \frac{3}{4} \div \frac{1}{2} - \frac{1}{3} = \frac{3}{12} + \frac{9}{12} \div \frac{6}{12} - \frac{4}{12}$$
$$= \frac{3}{12} + \frac{9}{12} \cdot \frac{12}{6} - \frac{4}{12}$$
$$= \frac{3}{12} + \frac{9}{6} - \frac{4}{12}$$
$$= \frac{3}{12} + \frac{18}{12} - \frac{4}{12}$$
$$= \frac{21}{12} - \frac{4}{12}$$
$$= \frac{17}{12}$$

66. Associative property of addition

67. Commutative property of multiplication

68. Distributive property

69. Let $x =$ number of pounds of coleslaw needed.

$$\frac{5 \text{ people}}{\frac{1}{2} \text{ pound coleslaw}} = \frac{560 \text{ people}}{x \text{ pounds coleslaw}}$$
$$\frac{5}{\frac{1}{2}} = \frac{560}{x}$$
$$5x = \left(\frac{1}{2}\right)(560)$$
$$5x = 280$$
$$x = \frac{280}{5} = 56$$

He will need 56 pounds of coleslaw.

70.
$$A = \frac{1}{2}bh$$
$$2A = 2(\frac{1}{2}bh)$$
$$2A = bh$$
$$\frac{2A}{h} = \frac{bh}{h}$$
$$\frac{2A}{h} = b$$

Exercise Set 3.4

1. $A = (2l) \cdot \left(\frac{w}{2}\right) = lw$
The area remains the same

3. $V = 2l \cdot 2w \cdot 2h = 8(lwh)$
The volume is eight times as great.

5. $A = \pi r^2 = \pi(3r)^2 = \pi(9r^2) = 9\pi r^2$
The area is nine times as great.

7. An isosceles triangle is a triangle with 2 equal sides.

9. The sum of the measures of the angles in a triangle is 180°.

11. Let $x =$ the measure of the two equal angles, then $x + 42 =$ the measure of the third angle.
Sum of the three angles = 180°

$$x + x + (x + 42) = 180$$
$$3x + 42 = 180$$
$$3x = 138$$
$$x = 46$$

The two equal angles are each 46°. The third angle is $x + 42° = 46° + 42° = 88°$.

13. Let $x =$ length of each side of the triangle, then $P = x + x + x = 3x$.
Perimeter = 28.5

$$x = \frac{28.5}{3} = 9.5$$

The length of each side is 9.5 inches.

15. Let $x =$ measure of angle B. Then $2x + 21 =$ measure of angle A.
Sum of the 2 angles = 90

$$x + (2x + 21) = 90$$
$$3x + 21 = 90$$
$$3x = 69$$
$$x = 23$$

Measure of angle $A = 2(23) + 21 = 67°$
Measure of angle $B = 23°$

17. Let x = measure of angle A, then
$3x - 8$ = measure of angle B.
Sum of the 2 angles = 180
$$x + (3x - 8) = 180$$
$$4x - 8 = 180$$
$$4x = 188$$
$$x = \frac{188}{4} = 47$$
Measure of angle $A = 47°$
Measure of angle $B = 3(47) - 8 = 141 - 8 = 133°$

19. The two angles have equal measures.
$$2x + 50 = 4x + 12$$
$$38 = 2x$$
$$19 = x$$
$2x + 50 = 2(19) + 50 = 38 + 50 = 88°$
$4x + 12 = 4(19) + 12 = 76 + 12 = 88°$
Each angle measures 88°.

21. Let x = measure of smallest angle. Then second angle $= x + 10$ and third angle $= 2x - 30$.
Sum of the 3 angles = 180
$$x + (x + 10) + (2x - 30) = 180$$
$$4x - 20 = 180$$
$$4x = 200$$
$$x = \frac{200}{4} = 50$$
The first angle is 50°.
The second angle is $50 + 10 = 60°$.
The third angle is $2(50) - 30 = 70°$.

23. Let x = width of rectangle. Then $x + 8$ = length of rectangle.
$$P = 2l + 2w$$
$$48 = 2(x + 8) + 2x$$
$$48 = 2x + 16 + 2x$$
$$48 = 4x + 16$$
$$32 = 4x$$
$$8 = x$$
Width is 8 feet and length is $8 + 8 = 16$ feet.

25. Let x = width of tennis court.
Then $2x + 6$ = length of tennis court.
$$P = 2l + 2w$$
$$228 = 2(2x + 6) + 2x$$
$$228 = 4x + 12 + 2x$$
$$228 = 6x + 12$$
$$216 = 6x$$
$$36 = x$$
The width is 36 feet and the length is $2(36) + 6 = 78$ feet.

27. Let x = measure of each smaller angle.
Then $3x - 20$ = measure of each larger angle.
(measure of the two smaller angles) + (measure of the two larger angles) = 360°
$$x + x + (3x - 20) + (3x - 20) = 360$$
$$8x - 40 = 360$$
$$8x = 400$$
$$x = 50$$
Each smaller angle is 50°. Each larger angle is $3(50) - 20 = 130°$.

29. Let x = measure of the smallest angle.
Then $x + 10$ = measure of the second angle,
$2x + 14$ = measure of third angle, and
$x + 21$ = measure of fourth angle.
Sum of the four angles = 360°
$$x + (x + 10) + (2x + 14) + (x + 21) = 360$$
$$5x + 45 = 360$$
$$5x = 315$$
$$x = 63$$
Thus the angles are 63°, $63 + 10 = 73°$
$2(63) + 14 = 140°$ and $63 + 21 = 84°$.

31. Let x = width of bookcase shelf.
Then $x + 3$ = height of bookcase.
4 shelves + 2 sides = total lumber available.
$$4x + 2(x + 3) = 30$$
$$4x + 2x + 6 = 30$$
$$6x + 6 = 30$$
$$6x = 24$$
$$x = 4$$
The width of each shelf is 4 feet and the height is $4 + 3 = 7$ feet.

33. Let x = length of a shelf.
Then $2x$ = height of bookcase.
4 shelves + 2 sides = total lumber available
$$4x + 2(2x) = 20$$
$$4x + 4x = 20$$
$$8x = 20$$
$$x = \frac{20}{8} = 2.5$$
The width of the bookcase is 2.5 feet. The height of the bookcase is $2(2.5) = 5$ feet.

35. Let x = width of fenced in area.
Then $x + 4$ = length of fenced in area
Five "widths" + one "length" = total fencing
$$5x + (x + 4) = 64$$
$$6x + 4 = 64$$
$$6x = 60$$
$$x = 10$$
Width is 10 feet and length is $10 + 4 = 14$ feet.

37. $ac + ad + bc + bd$

39. $-|-6| < |-4|$ since $-|-6| = -6$ and $|-4| = 4$

40. $|-3| > -|3|$ since $|-3| = 3$ and $-|3| = -3$

41. $-6-(-2)+(-4) = -6+2+(-4)$
$= -4+(-4)$
$= -8$

42. $-6y + x - 3(x-2) + 2y$
$= -6y + x - 3x + 6 + 2y$
$= x - 3x - 6y + 2y + 6$
$= -2x - 4y + 6$

43. $2x + 3y = 9$
$2x - 2x + 3y = -2x + 9$
$3y = -2x + 9$
$\frac{3y}{3} = \frac{-2x+9}{3}$
$y = \frac{-2x+9}{3}$ or $y = -\frac{2}{3}x + 3$
Substitute 3 for *x*.
$y = \frac{-2x+9}{3} = \frac{-6+9}{3} = \frac{3}{3} = 1$

Exercise Set 3.5

1. Rate $= \frac{\text{distance}}{\text{time}} = \frac{150}{3} = 50$. Therefore, her average speed was 50 mph.

3. Thickness = rate · time = (0.2)(12) = 2.4. The door is 2.4 cm thick.

5. Time $= \frac{\text{amount}}{\text{rate}} = \frac{420}{30} = 14$. It will take 14 hours to lay the tile.

7. Rate $= \frac{\text{volume}}{\text{time}} = \frac{1500}{6} = 250$.
Therefore, the flow rate should be 250 cm^3/hr.

9. Time $= \frac{\text{distance}}{\text{rate}} = \frac{5280}{4} = 1320$ seconds
or $\frac{1320}{60} = 22$ minutes
It will take about 22 minutes.

11. Rate $= \frac{\text{distance}}{\text{time}} = \frac{500}{2.635} \approx 189.75$. His average speed was approximately 189.75 mph.

13. Let *t* be the time it takes for Willie and Shanna to be 16.8 miles apart.

Person	Rate	Time	Distance
Willie	3	t	$3t$
Shanna	4	t	$4t$

$3t + 4t = 16.8$
$7t = 16.8$
$t = 2.4$
It will take 2.4 hours.

15. Let *r* be the second rate of the machine.

Machine	Rate	Time	Distance
First	60	7.2	432
Second	r	6.8	$6.8r$

$432 + 6.8r = 908$
$6.8r = 476$
$r = 70$
The second speed the machine was set at was 70 miles per hour.

17. Let *t* be the time they have been walking.

Walker	Rate	Time	Distance
Sadie	220	t	$220t$
Dale	100	t	$100t$

$220t - 100t = 600$
$120t = 600$
$t = 5$
They have been walking for 5 minutes.

19. Let *r* be the speed of the cutter coming from the east (westbound). Then $r + 5$ is the speed of the cutter coming from the west (eastbound).

Cutter	Rate	Time	Distance
Eastbound	$r+5$	3	$3(r+5)$
Westbound	r	3	$3r$

$3(r+5) + 3r = 225$
$3r + 15 + 3r = 225$
$6r = 210$
$r = 35$
The speed of the westbound cutter is 35 mph and the speed of the eastbound cutter is 40 mph.

21. **a.** Distance = rate · time = (2.38)(1.01) ≈ 2.4 miles

b. Distance = rate · time = (21.17)(5.29) ≈ 112.0 miles

c. Distance = rate · time = (8.32)(3.15) ≈ 26.2 miles

d. $2.4 + 112.0 + 26.2 = 140.6$ miles

e. $1.01 + 5.29 + 3.15 = 9.45$ hours

23. Let r be the rate of *Apollo*. Then $r + 4$ is the rate of *Pythagoras*.

Boat	Rate	Time	Distance
Apollo	r	0.7	$0.7r$
Pythagoras	$r+4$	0.7	$0.7(r+4)$

$$0.7r + 0.7(r+4) = 9.8$$
$$0.7r + 0.7r + 2.8 = 9.8$$
$$1.4r = 7$$
$$r = 5$$

The speed of *Apollo* is 5 mph, and the speed of *Pythagoras* is 9 mph.

25. a. Let t be the time it takes for the Coast Guard to catch the bank robber.

Boat	Rate	Time	Distance
Robber	25	$t+\frac{1}{2}$	$25\left(t+\frac{1}{2}\right)$
Coast Guard	35	t	$35t$

$$25\left(t+\frac{1}{2}\right) = 35t$$
$$25t + \frac{25}{2} = 35t$$
$$\frac{25}{2} = 10t$$
$$1.25 = t$$

It will take 1.25 hours for the Coast Guard to catch the bank robber.

b. Distance = rate · time = (35)(1.25) = 43.75 miles.

27. a. Let t be the time it takes for Phil's pass to reach Pete.

Player	Rate	Time	Distance
Pete	25	$t+2$	$25(t+2)$
Phil	50	t	$50t$

$$25(t+2) = 50t$$
$$25t + 50 = 50t$$
$$50 = 25t$$
$$2 = t$$

It will take 2 seconds for Phil's pass to reach Pete.

b. Distance = rate · time = $50 \cdot 2 = 100$ feet

29. Let t = time, in hours, Betty was traveling at 50 mph.

Speed	Rate	Time	Distance
Faster	70 mph	$t-0.5$	$70(t-0.5)$
Slower	50 mph	t	$50t$

a. $50t - 70t + 35 = 5$
$$-20t = -30$$
$$t = 1.5$$

b. Betty traveled for 1.5 hours at 50 mph.

31. Let r be the planned speed of the plane. Then $r + 30$ is the new speed of the plane.

Speed	Rate	Time	Distance
Planned	r	4	$4r$
New	$r+30$	$4-0.2$	$3.8(r+30)$

$$4r = 3.8(r+30)$$
$$4r = 3.8r + 114$$
$$0.2r = 114$$
$$r = 570$$

The plane's planned speed is 570 miles per hour. The plane's new speed is 600 miles per hour.

33. Let r be the rate of clearing the bridge. Then $1.2 + r$ is the rate of clearing the road.

	Rate	Time	Distance
Road	$1.2+r$	20	$20(1.2+r)$
Bridge	r	60	$60r$

$$20(1.2+r) + 60r = 124$$
$$24 + 20r + 60r = 124$$
$$80r = 100$$
$$r = 1.25$$

The crew will clear the bridge at a rate of 1.25 feet/day, and they will clear the road at a rate of 2.45 feet/day.

35. Let x be the amount invested at 5%. Then $9400 - x$ is the amount invested at 7%.

Principal	Rate	Time	Interest
x	5%	1	$0.05x$
$9400-x$	7%	1	$0.07(9400-x)$

$$0.05x + 0.07(9400 - x) = 610$$
$$0.05x + 658 - 0.07x = 610$$
$$-0.02x = -48$$
$$x = 2400$$

They invested $2400 at 5% and $7000 at 7%.

37. Let x be the amount invested at 6%. Then $6000 - x$ is the amount invested at 4%.

Principal	Rate	Time	Interest
x	6%	1	$0.06x$
$6000 - x$	4%	1	$0.04(6000 - x)$

$$0.06x = 0.04(6000 - x)$$
$$0.06x = 240 - 0.04x$$
$$0.10x = 240$$
$$x = 2400$$

She invested $2400 at 6% and $3600 at 4%.

39. Let x be the amount invested at 4%. Then $10{,}000 - x$ is the amount invested at 5%.

Principal	Rate	Time	Interest
x	4%	1	$0.04x$
$10{,}000 - x$	5%	1	$0.05(10{,}000 - x)$

$$0.05(10{,}000 - x) - 0.04x = 320$$
$$500 - 0.05x - 0.04x = 320$$
$$-0.09x = -180$$
$$x = 2000$$

She invested $2,000.00 at 4% and $8,000.00 at 5%.

41. Let t be the time, in months, during which Patricia paid $17.10 per month. Then $12 - t$ is the time during which she paid $18.40 per month.

Rate	Time	Amount
17.10	t	$17.10t$
18.40	$12 - t$	$18.40(12 - t)$

$$17.10t + 18.40(12 - t) = 207.80$$
$$17.10t + 220.80 - 18.40t = 207.80$$
$$-1.30t = -13$$
$$t = 10$$

She paid $17.10 for the first 10 months of the year, and paid $18.40 for the remainder of the year. The rate increase took effect in November.

43. Let x be the number of hours worked at Home Depot ($6.50 per hour). Then $18 - x$ is the number of hours worked at the veterinary clinic ($7.00 per hour).

Rate	Hours	Total
$6.50	x	$6.5x$
$7.00	$18 - x$	$7(18 - x)$

$$6.5x + 7(18 - x) = 122$$
$$6.5x + 126 - 7x = 122$$
$$-0.5x = -4$$
$$x = 8$$

Mihàly worked 8 hours at Home Depot and 10 hours at the clinic.

45. Let t be the number of $1550 computer systems sold. Then $200 - t$ is the number of $1320 computer systems sold.

Rate	Amount sold	Money collected
1550	t	1550
1320	$200 - t$	$1320(200 - t)$

$$1320(200 - t) + 1550t = 282{,}400$$
$$264{,}000 - 1320t + 1550t = 282{,}400$$
$$230t = 18{,}400$$
$$t = 80$$

There were 80 of the $1,550 Computer systems sold.

47. a. Let x be the number of shares of General Electric. Then $5x$ is the number of shares of PepsiCo.

Stock	Price	Shares	Total
GE	$74	x	$74x$
PepsiCo	$35	$5x$	$35 \cdot 5x$

$$74x + 35 \cdot 5x = 8000$$
$$74x + 175x = 8000$$
$$249x = 8000$$
$$x \approx 32.1$$

Since only whole shares can be purchased, he will purchase 32 shares of GE and 160 shares of PepsiCo.

b. Mr. Gilbert spent $32 \cdot 74 + 160 \cdot 35 = \$7968$. He has $8000 – $7968 = $32 left over.

49. Let x be the amount of Family grass seed. Then $10-x$ is the amount of Spot Filler grass seed.

Seed	Price	Amount	Total
Family	\$2.45	x	$2.45x$
Filler	\$2.10	$10-x$	$2.10(10-x)$
Mixture	\$2.20	10	22

$$\begin{aligned} 2.45x + 2.10(10-x) &= 22 \\ 2.45x + 21 - 2.10x &= 22 \\ 0.35x &= 1 \\ x &\approx 2.86 \end{aligned}$$

2.86 pounds of Family grass seed and 7.14 pounds of Spot Filler grass seed should be mixed together.

51. Let x be the number of gallons of regular gasoline.

Gas Type	Cost	Gallons	Total
Regular	\$1.20	x	$1.20x$
Premium Plus	\$1.35	$500 - x$	$1.35(500-x)$
Premium	126	500	1.26(500)

$$\begin{aligned} 1.20x + 1.35(500-x) &= 126(500) \\ 1.20x + 675 - 1.35x &= 630 \\ -.15x &= -45 \\ x &= 300 \end{aligned}$$

He should mix 300 gallons of regular and 200 gallons of premium plus.

53. Let x be the cost per pound of the mixture.

Type	Cost	Pounds	Total
Good & Plenty	\$2.49	3	\$2.49(3)
Sweet Treats	\$2.89	5	\$2.89(5)
Mixture	x	8	$8x$

$$\begin{aligned} 2.49(3) + 2.89(5) &= 8x \\ 7.47 + 14.45 &= 8x \\ 21.92 &= 8x \\ 2.74 &= x \end{aligned}$$

The mixture should sell for \$2.74 per pound.

55. Let x be the percentage of alcohol in the mixture.

Percentage	Liters	Amount of Alcohol
12%	5	0.6
9%	2	0.18
x%	7	$\left(\frac{x}{100}\right)\cdot 7$

$$\begin{aligned} \left(\frac{x}{100}\right)\cdot 7 &= 0.6 + 0.18 \\ 0.07x &= 0.78 \\ x &\approx 11.1 \end{aligned}$$

The alcohol content of the mixture is about 11.1%

57. Let x be the amount of 12% sulfuric acid solution.

Solution	Strength	Liters	Amount
20%	0.20	1	0.20
12%	0.12	x	$0.12x$
Mixture	0.15	$x+1$	$0.15(x+1)$

$$\begin{aligned} 0.20 + 0.12x &= 0.15(x+1) \\ 0.20 + 0.12x &= 0.15x + 0.15 \\ -0.03x &= -0.05 \\ x &= \frac{5}{3} \end{aligned}$$

$1\frac{2}{3}$ liters of 12% sulfuric acid should be used.

59. Let x be the amount of the swimming pool shock treatment to be added to a quart of water.

Product	Percentage	Ounces	Amount
Clorox	5.25%	8	$(0.0525)\cdot 8$
Shock Treatment	10.5%	x	$0.105x$

$$\begin{aligned} (0.0525)\cdot 8 &= 0.105x \\ 0.42 &= 0.105x \\ 4 &= x \end{aligned}$$

Add 4 ounces of the shock treatment to a quart of water.

61. Let x be the percentage of milkfat in whole milk.

Type	Percentage	Gallons	Amount
Whole	$x\%$	4	$\left(\frac{x}{100}\right)\cdot 4$
Low fat	1%	5	$(0.01)\cdot 5$
Reduced fat	2%	9	$(0.02)\cdot 9$

$$\left(\frac{x}{100}\right)\cdot 4+(0.01)\cdot 5=(0.02)\cdot 9$$
$$\frac{4x}{100}+0.05=0.18$$
$$\frac{4x}{100}=0.13$$
$$4x=13$$
$$x=3.25$$

The milkfat content of whole milk is 3.25%.

63. Let x be the percent of orange juice in the new mixture.

Percentage of water without salt	Gallons	Amount
99.1%	50,000	0.991(50,000)
100%	x	$1x$
99.2%	$x + 50{,}000$	$0.992(x + 50{,}000)$

$$0.991(50{,}000) + 1x = 0.992(x + 50{,}000)$$
$$49{,}550 + 1x = 0.992x + 49{,}600$$
$$0.008x = 50$$
$$x = 6{,}250$$

To lower the salt concentration, 6,250 gallons of 0% salt content has to be added.

65. Let x be the number of gallons of Prestone antifreeze added.

Brand	Percentage of antifreeze	Gallons	Amount
Prestone	12%	x	$0.12x$
Xeres	9%	1	(0.09)(1)
Mixture	10%	$x+1$	(0.10)(x+1)

$$0.12x + 0.09 = 0.10(x + 1)$$
$$0.12x + 0.09 = 0.10x + 0.10$$
$$0.02x = 0.01$$
$$x = 0.5$$

Nina added 0.5 gallons of Prestone antifreeze.

67. The time it takes for the transport to make the trip is:

$$\text{Time} = \frac{\text{Distance}}{\text{Rate}} = \frac{1720}{370} \approx 4.65 \text{ hours}$$

The time it takes for the Hornets to make the trip is:

$$\text{Time} = \frac{\text{Distance}}{\text{Rate}} = \frac{1720}{900} \approx 1.91 \text{ hours}$$

It takes the transport $4.65 - 1.91 = 2.74$ hours longer to make the trip. Since it needs to arrive 3 hours before the Hornets, it should leave about $2.74 + 3 = 5.74$ hours before them.

71. a.
$$2\frac{3}{4} \div 1\frac{5}{8} = \frac{11}{4} \div \frac{13}{8}$$
$$= \frac{11}{4} \cdot \frac{8}{13}$$
$$= \frac{22}{13} \text{ or } 1\frac{9}{13}$$

b.
$$2\frac{3}{4} + 1\frac{5}{8} = \frac{11}{4} + \frac{13}{8}$$
$$= \frac{22}{8} + \frac{13}{8}$$
$$= \frac{35}{8} \text{ or } 4\frac{3}{8}$$

72. $6(x-3) = 4x - 18 + 2x$
$6x - 18 = 6x - 18$
All real numbers are solutions.

73.
$$\frac{6}{x} = \frac{72}{9}$$
$$6\cdot 9 = 72x$$
$$54 = 72x$$
$$x = \frac{54}{72} = \frac{3}{4} \text{ or } 0.75$$

74. Let n = smaller integer
Then $n+1$ = next consecutive integer.
Smaller number + larger number = 77

$$n + n + 1 = 77$$
$$2n + 1 = 77$$
$$2n + 1 - 1 = 77 - 1$$
$$2n = 76$$
$$\frac{2n}{2} = \frac{76}{2}$$
$$n = 38$$

The first number is 38 and the second number is 39.

Review Exercises

1. Substitute 6 for r.
$C = 2\pi r$
$C = 2\pi(6)$
$= 12\pi$
≈ 37.70

2. Substitute 4 for l and 5 for w.
$P = 2l + 2w$
$P = 2(4) + 2(5) = 8 + 10 = 18$

3. Substitute 8 for b and 12 for h.
$A = \frac{1}{2}bh$
$A = \frac{1}{2}(8)(12) = 48$

4. Substitute 200 for K and 4 for v.
$K = \frac{1}{2}mv^2$
$200 = \frac{1}{2}m(4)^2$
$200 = \frac{1}{2}m(16)$
$200 = 8m$
$\frac{200}{8} = \frac{8m}{8}$
$25 = m$

5. Substitute 15 for y, 3 for m, and –2 for x.
$y = mx + b$
$15 = (3)(-2) + b$
$15 = -6 + b$
$15 + 6 = -6 + 6 + b$
$21 = b$

6. Substitute 4716.98 for P and 0.06 for i.
$P = \frac{f}{1+i}$
$4716.98 = \frac{f}{1 + 0.06}$
$4716.98 = \frac{f}{1.06}$
$1.06 \times 4716.98 = \frac{f}{1.06} \times 1.06$
$5000 \approx f$

7. a.
$2x = 2y + 4$
$2x - 4 = 2y + 4 - 4$
$2x - 4 = 2y$
$\frac{2x-4}{2} = \frac{2y}{2}$
$\frac{2x}{2} - \frac{4}{2} = y$
$x - 2 = y$

b. Substitute 10 for x.
$10 - 2 = y$
$8 = y$

8. a.
$6x + 3y = -9$
$6x - 6x + 3y = -9 - 6x$
$3y = -9 - 6x$
$\frac{3y}{3} = \frac{-9-6x}{3}$
$y = \frac{-9}{3} - \frac{6x}{3}$
$y = -3 - 2x$

b. Substitute 12 for x.
$y = -3 - 2(12) = -3 - 24 = -27$

9. a. $5x - 2y = 16$
$5x - 5x - 2y = 16 - 5x$
$-2y = 16 - 5x$
$\frac{-2y}{-2} = \frac{16-5x}{-2}$
$y = \frac{16-5x}{-2}$
$y = \frac{5}{2}x - 8$

b. Substitute 2 for x.
$y = \frac{5}{2}(2) - 8 = 5 - 8 = -3$

10. a.
$2x = 3y + 12$
$2x - 12 = 3y + 12 - 12$
$2x - 12 = 3y$
$\frac{2x-12}{3} = \frac{3y}{3}$
$\frac{2x-12}{3} = y$
$y = \frac{2}{3}x - 4$

b. Substitute –6 for x.

$$y = \frac{2}{3}(-6) - 4$$
$$= -4 - 4$$
$$= -8$$

11. $A = lw$

$$\frac{A}{l} = \frac{lw}{l}$$
$$\frac{A}{l} = w$$

12. $A = \frac{1}{2}bh$

$$2A = 2\left(\frac{1}{2}bh\right)$$
$$2A = bh$$
$$\frac{2A}{b} = \frac{bh}{b}$$
$$\frac{2A}{b} = h$$

13. $i = prt$

$$\frac{i}{pr} = \frac{prt}{pr}$$
$$\frac{i}{pr} = t$$

14. $P = 2l + 2w$

$$P - 2l = 2l - 2l + 2w$$
$$P - 2l = 2w$$
$$\frac{P - 2l}{2} = \frac{2w}{2}$$
$$\frac{P - 2l}{2} = w$$

15. $V = \pi r^2 h$

$$\frac{V}{\pi r^2} = \frac{\pi r^2 h}{\pi r^2}$$
$$\frac{V}{\pi r^2} = h$$

16. $V = \frac{1}{3}Bh$

$$3V = 3(\frac{1}{3}Bh)$$
$$3V = Bh$$
$$\frac{3V}{B} = \frac{Bh}{B}$$
$$\frac{3V}{B} = h$$

17. Substitute 600 for p, 0.09 for r, and 2 for t.
$i = prt$
$i = (600)0.09(2) = 108$
Tom will pay \$108 interest.

18. $P = 2l + 2w$
$16 = 2l + 2(2)$
$16 = 2l + 4$
$12 = 2l$
$6 = l$
The length of the rectangle is 6 inches.

19. The sum of a number and the number increased by 5 is 9.

20. The sum of a number and twice the number decreased by 1 is 10.

21. Let x = the smaller number.
Then $x + 8$ = the larger number.
Smaller number + larger number = 74
$x + (x + 8) = 74$
$2x + 8 = 74$
$2x = 66$
$x = 33$
The smaller number is 33 and the larger number is $33 + 8 = 41$.

22. Let x = smaller integer.
Then $x + 1$ = next consecutive integer.
Smaller number + larger number = 237
$x + (x + 1) = 237$
$2x + 1 = 237$
$2x = 236$
$x = 118$
The smaller number is 118 and the larger number is $118 + 1 = 119$.

23. Let x = the smaller integer.
Then $5x + 3$ = the larger integer
Larger number – smaller number = 31
$(5x + 3) - x = 31$
$4x + 3 = 31$
$4x = 28$
$x = 7$

The smaller number is 7 and the larger number is 5(7) + 3 = 38.

24. Let x = cost of car before tax.
Then $0.07x$ = amount of tax.
Cost of car before tax + tax on car
= cost of car after tax

$$x + 0.07x = 23{,}260$$
$$1.07x = 23{,}260$$
$$x = \frac{23{,}260}{1.07} = 21{,}738.32$$

The cost of the car before tax is \$21,738.32.

25. Let x = the number of months, then $20x$ = the increase in production of bagels over x months.
Current production + Increase in production = Future production

$$520 + 20x = 900$$
$$20x = 380$$
$$x = 19$$

It will take 19 months.

26. Let x = weekly dollar sales that would make total salaries from both companies the same.
The commission at present company = $0.03x$ and commission at new company = $0.08x$
Salary + commission for present company = salary + commission for new company

$$500 + 0.03x = 400 + 0.08x$$
$$100 + 0.03x = 0.08x$$
$$100 = 0.05x$$
$$\frac{100}{0.05} = x$$
$$2000 = x$$

Ron's weekly sales would have to be \$2000 for the total salaries from both companies to be the same.

27. Let x = original price of camcorder. Then $0.20x$ = reduction during first week.
Original price – first reduction – second reduction = price during second week

$$x - 0.20x - 25 = 495$$
$$0.8x - 25 = 495$$
$$0.8x = 520$$
$$x = \frac{520}{0.8} = 650$$

The original price of the camcorder was \$650.

28. Let x = the number of months for the total payments for both banks to be the same.
First Federal = Internet Bank

$$900 + 889x = 1200 + 826x$$
$$900 + 63x = 1200$$
$$63x = 300$$
$$x \approx 4.76$$

It will take about 4.76 months.

29. Let x = measure of the smallest angle. Then
$x + 10$ = measure of second angle and
$2x - 10$ = measure of third angle.
Sum of the three angles = 180°

$$x + (x + 10) + (2x - 10) = 180$$
$$4x = 180$$
$$x = 45$$

The angles are 45°, 45 + 10 = 55°, and 2(45) – 10 = 80°.

30. Let x = measure of the smallest angle. Then
$x + 10$ = measure of second angle,
$5x$ = measure of third angle,
$4x + 20$ = measure of the fourth angle.
Sum of the four angles = 360°

$$x + (x + 10) + 5x + (4x + 20) = 360$$
$$11x + 30 = 360$$
$$11x = 330$$
$$x = \frac{330}{11} = 30$$

The angles are 30°, 30 + 10 = 40°.
5(30) = 150°, and 4(30) + 20 = 140°.

31. Let w = width of garden. Then
$w + 4$ = length of garden.

$$P = 2l + 2w$$
$$70 = 2(w + 4) + 2w$$
$$70 = 4w + 8$$
$$62 = 4w$$
$$15.5 = w$$

The width is 15.5 feet and the length is 15.5 + 4 = 19.5 feet.

32. Let x = the width of the room. Then $x + 30$ = the length of the room. The amount of string used is the perimeter of the room plus the wall separating the two rooms.

$$P = 2l + 2w + w$$
$$P = 2l + 3w$$
$$310 = 2(x + 30) + 3x$$
$$310 = 2x + 60 + 3x$$
$$310 = 5x + 60$$
$$250 = 5x$$
$$50 = x$$

The width of the room is 50 feet and the length is 50 + 30 = 80 feet.

33. Let r be the flow rate of the water.
Amount = rate · time
$105 = r \cdot (3.5)$
$\frac{105}{3.5} = \frac{r \cdot (3.5)}{3.5}$
$30 = r$
The flow rate of the water is 30 gallons per hour.

34. Speed $= \frac{\text{distance}}{\text{time}} = \frac{26}{4} = 6.5$ mph

35. Let t be the time it takes for the joggers to be 4 kilometers apart.

Jogger	Rate	Time	Distance
Harold	8	t	$8t$
Susan	6	t	$6t$

$8t - 6t = 4$
$2t = 4$
$t = 2$
It takes the joggers 2 hours to be 4 kilometers apart.

36. Let t be the amount of time it takes for the trains to be 440 miles apart.

Train	Rate	Time	Distance
First	50	t	$50t$
Second	60	t	$60t$

$50t + 60t = 440$
$110t = 440$
$t = 4$
After 4 hours, the two trains will be 440 miles apart.

37. Rate $= \frac{\text{Distance}}{\text{Time}} = \frac{200 \text{ feet}}{22.73 \text{ seconds}} \approx 8.8$ ft / sec
The cars travel at about 8.8 ft/sec.

38. Let x be the amount invested at 8%. Then $12{,}000 - x$ is the amount invested at $7\frac{1}{4}\%$.

Principal	Rate	Time	Interest
x	8%	1	$0.08x$
$12{,}000 - x$	$7\frac{1}{4}\%$	1	$0.0725(12{,}000 - x)$

$0.08x + 0.0725(12{,}000 - x) = 900$
$0.08x + 870 - 0.0725x = 900$
$0.0075x = 30$
$x = 4000$
Tatiana should invest \$4000 at 8% and \$8000 at $7\frac{1}{4}\%$.

39. Let x be the amount invested at 3%. Then $4{,}000 - x$ is the amount invested at 3.5%.

Principal	Rate	Time	Interest
x	3%	1	$0.03x$
$4{,}000 - x$	3.5%	1	$0.035(4{,}000 - x)$

$0.03x + 94.50 = 0.035(4{,}000 - x)$
$0.03x + 94.50 = 140 - .035x$
$0.065x = 45.5$
$x = 700$
Aimee invested \$700 in the 3% account and \$3300 in the 3.5% account.

40. Let x be the number of gallons of pure punch. Then $2 - x$ is the number of liters of the 5% acid solution.

Punch Solution	Strength	Gallons	Amount
98%	0.98	2	0.98(2)
100%	1	x	$1x$
98.5%	0.985	$x + 2$	$0.985(x + 2)$

$1.96 + x = 0.985(x + 2)$
$1.96 + x = 0.985x + 1.97$
$0.015x = 0.01$
$x \approx 0.67$
Marcie should add about 0.67 gallons of pure punch.

41. Let x be the number of small wind chimes sold..

Type	Price	Number Sold	Amount
Small	\$8	x	$8x$
Large	\$20	$30 - x$	$20(30 - x)$

$8x + 20(30 - x) = 492$
$8x + 600 - 20x = 492$
$-12x = -108$
$x = 9$
He sold 9 small and 21 large wind chimes.

42. Let x be the number of liters of the 10% solution. Then $2 - x$ is the number of liters of the 5% acid solution.

Solution	Strength	Liters	Amount
10%	0.10	x	$0.10x$
5%	0.05	$2 - x$	$0.05(2 - x)$
Mixture	0.08	2	0.16

$$0.10x + 0.05(2 - x) = 0.16$$
$$0.10x + 0.10 - 0.05x = 0.16$$
$$0.05x = 0.06$$
$$x = 1.2$$

The chemist should mix 1.2 liters of 10% solution with 0.8 liters of 5% solution.

43. Let x = smaller odd integer. Then $x + 2$ = next consecutive odd integer.
Smaller number + larger number = 208

$$x + (x + 2) = 208$$
$$2x + 2 = 208$$
$$2x = 206$$
$$x = 103$$

The smaller number is 103 and the larger number is $103 + 2 = 105$.

44. Let x = cost of television before tax. Then amount of tax = $0.06x$. Cost of television before tax + tax on television = cost of television after tax.

$$x + 0.06x = 477$$
$$1.06x = 477$$
$$x = \frac{477}{1.06} = 450$$

The cost of the television before tax is \$450.

45. Let x = his dollar sales.
Then $0.05x$ = amount of commission.
Salary + commission = 900

$$300 + 0.05x = 900$$
$$0.05x = 600$$
$$x = \frac{600}{0.05} = 12,000$$

His sales last week were \$12,000.

46. Let x = measure of the smallest angle. Then $x + 8$ = measure of second angle and $2x + 4$ = measure of third angle.
Sum of the three angles = 180°

$$x + (x + 8) + (2x + 4) = 180$$
$$4x + 12 = 180$$
$$4x = 168$$
$$x = 42$$

The angles are 42°, $42 + 8 = 50°$, and $2(42) + 4 = 88°$.

47. Let t = number of years. Then
$25t$ = increase in employees over t years
Present number of employees + increase in employees = future number of employees

$$427 + 25t = 627$$
$$25t = 200$$
$$t = \frac{200}{25} = 8$$

It will take 8 years before they reach 627 employees.

48. Let x = measure of each smaller angle. Then $x + 40$ = measure of each larger angle
(measure of the two smaller angles) +(measure of the two larger angles) = 360°

$$x + x + (x + 40) + (x + 40) = 360$$
$$4x + 80 = 360$$
$$4x = 280$$
$$x = 70$$

Each of the smaller angles is 70° and each of the two larger angles is $70 + 40 = 110°$.

49. a. Let x = number of copies that would result in both centers charging the same. Then charge for copies at Copy King = $0.04x$ and charge for copies at King Kopie = $0.03x$
Monthly fee + charge for copies at Copy King = monthly fee + charge for copies at King Kopie.

$$20 + 0.04x = 25 + 0.03x$$
$$0.04x = 5 + 0.03x$$
$$0.01x = 5$$
$$x = \frac{5}{0.01} = 500$$

500 copies would result in both centers charging the same.

50. a. Let t = time the sisters meet after Chris starts swimming.

Person	Rate	Time	Distance
Chris	60	t	$60t$
Kathy	50	$t + 2$	$50(t + 2)$

$$60t = 50(t+2)$$
$$60t = 50t + 100$$
$$10t = 100$$
$$t = 10$$
The sisters will meet 10 minutes after Chris starts swimming.

b. rate · time = distance
$60 \cdot 10 = 600$
The sisters will be 600 feet from the boat when they meet.

51. Let x be the amount of \$3.50 per pound ground beef. Then $80 - x$ is the amount of \$4.10 per pound of ground beef.

Ground Beef	Price	Amount	Total
\$3.50	3.50	x	$3.50x$
\$4.10	4.10	$80 - x$	$4.10(80 - x)$
Mixture	3.65	80	292

$$3.50x + 4.10(80 - x) = 292$$
$$3.50x + 328 - 4.10x = 292$$
$$-0.60x = -36$$
$$x = 60$$
The butcher mixed 60 lbs of \$3.50 per pound ground beef with 20 lbs of \$4.10 per pound ground beef.

52. Let x = the rate the older brother travels. Then $x + 5$ = the rate the younger brother travels.

Brother	Rate	Time	Distance
Younger	$x+5$	2	$2(x+5)$
Older	x	2	$2x$

Younger brother's distance + older brother's distance = 230 miles.
$$2(x+5) + 2x = 230$$
$$2x + 10 + 2x = 230$$
$$4x + 10 = 230$$
$$4x = 220$$
$$x = 55$$
The older brother travels at 55 miles per hour and the younger brother travels at $55 + 5 = 60$ miles per hour.

53. Let x = the number of liters of 30% solution.

Percent	Liters	Amount
30%	x	$0.30x$
12%	2	$(0.12)(2)$
15%	$x+2$	$0.15(x+2)$

$$0.30x + (0.12)(2) = 0.15(x+2)$$
$$0.30x + 0.24 = 0.15x + 0.30$$
$$0.15x + 0.24 = 0.30$$
$$0.15x = 0.06$$
$$x = 0.4$$
0.4 liters of the 30% acid solution need to be added.

Practice Test

1. Let r = interest rate in decimal form.
$$i = prt$$
$$3240 = (12{,}000)(r)(3)$$
$$3240 = 36{,}000r$$
$$\frac{3240}{36{,}000} = r$$
$$0.09 = r$$
The interest rate is 9%.

2. $P = 2l + 2w$
$P = 2(6) + 2(3) = 12 + 6 = 18$
The perimeter is 18 feet.

3. $A = P + Prt$
$$A = 100 + (100)(0.15)(3)$$
$$= 100 + 45$$
$$= 145$$

4.
$$A = \frac{m+n}{2}$$
$$79 = \frac{73+n}{2}$$
$$2 \times 79 = \left(\frac{73+n}{2}\right) \times 2$$
$$158 = 73 + n$$
$$85 = n$$

5.
$$C = 2\pi r$$
$$50 = 2\pi r$$
$$\frac{50}{2\pi} = \frac{2\pi r}{2\pi}$$
$$7.96 \approx r$$

6. a.
$$4x = 3y + 9$$
$$4x - 9 = 3y + 9 - 9$$
$$4x - 9 = 3y$$
$$\frac{4x-9}{3} = \frac{3y}{3}$$
$$\frac{4x-9}{3} = y$$
$$y = \frac{4}{3}x - 3$$

b. Substitute 12 for x.

$$y = \frac{4}{3}(12) - 3$$
$$= 4(4) - 3$$
$$= 16 - 3$$
$$= 13$$

7. $P = IR$

$$\frac{P}{I} = \frac{IR}{I}$$
$$\frac{P}{I} = R$$

8. $A = \frac{a+b}{3}$

$$3A = 3\left(\frac{a+b}{3}\right)$$
$$3A = a + b$$
$$3A - b = a + b - b$$
$$3A - b = a$$

9. $A = \frac{1}{2}h(b+d)$

$$= \frac{1}{2}(4)[5+9]$$
$$= \frac{1}{2}(4)(14)$$
$$= 2(14)$$
$$= 28$$

The area of the trapezoid is 28 square feet.

10. The area of the skating rink is the area of a rectangle plus the area of two half circles, or one full circle. The radius of the circle is $\frac{30}{2} = 15$ feet.

Area of rectangle $= l \cdot w = 80 \cdot 30 = 2400 \text{ ft}^2$

Area of circle

$= \pi r^2 = \pi(15)^2 = 225\pi \approx 706.86 \text{ ft}^2$

Total area $= 2400 + 706.86 = 3106.86 \text{ ft}^2$

11. $500 - n$

12. $2f + 6000$

13. The sum of a number and the number increased by 4 is 9.

14. Let x = smaller integer.
Then $2x - 10$ = larger integer.
Smaller number + larger number = 158

$$x + (2x - 10) = 158$$
$$3x - 10 = 158$$
$$3x = 168$$
$$x = 56$$

The smaller number is 56 and the larger number is $2(56) - 10 = 102$

15. Let x = smallest integer.
Then $x + 1$ is the consecutive integer.
Sum of the two integers = 43

$$x + (x + 1) = 43$$
$$2x + 1 = 43$$
$$2x = 42$$
$$x = 21$$

The integers are 21 and 21 + 1 = 22.

16. Let c = the cost of the furniture before tax
Tax amount = (cost) · (tax rate) = $0.06c$
Total cost = cost before tax + tax amount

$$2650 = c + 0.06c$$
$$2640 = 1.06c$$
$$\frac{2650}{1.06} = \frac{1.06c}{1.06}$$
$$2500 = c$$

The cost of the furniture before tax was $2500.

17. Let x = price of most expensive meal he can order.
Then $0.15x$ = tip and $0.07x$ = tax
Price of meal + tip + tax = 40

$$x + 0.15x + 0.07x = 40$$
$$1.22x = 40$$
$$x = \frac{40}{1.22} \approx 32.79$$

The price of the most expensive meal he can order is $32.79.

18. Let x = the amount of money Peter invested.
Then $2x$ = the amount of money Julie invested.
Then $2x$ = the amount of profit Julie receives.
Peter's profit + Julie's profit
= Total profit

$$x + 2x = 120{,}000$$
$$3x = 120{,}000$$
$$x = 40{,}000$$

Peter will receive $40,000 and Julie receives 2($40,000) = $80,000.

19. Let x = the number of times the plow is needed for the costs to be equal.
Elizabeth's charge = $80 + 5x$
Jan charge = $50 + 10x$
The charges are equal when:
$80 + 5x = 50 + 10x$

$$30 + 5x = 10x$$
$$30 = 5x$$
$$6 = x$$

The snow would need to be plowed 6 times for the costs to be the same.

20. Let x = number of months for the total cost of both mortgages to be equal.
Then $980x$ = monthly payments for x months at Bank of Washington,
and $1025x$ = monthly payments for x months at First Trust.
Bank of Washington = First Trust

$$980x + 1500 = 1025x$$
$$1500 = 45x$$
$$33.3 \approx x$$

It would take about 33.3 months for the costs to be the same.

21. Let x = length of smallest side.
Then $x + 15$ = length of second side and
$2x$ = length of third side.
Sum of the three sides = perimeter

$$x + (x + 15) + 2x = 75$$
$$4x + 15 = 75$$
$$4x = 60$$
$$x = 15$$

The three sides are 15 inches,
15 + 15 = 30 inches, and 2(15) = 30 inches.

22. Let w = width of flag
Then $2w - 4$ = length of flag.
$2l + 2w$ = perimeter

$$2(2w - 4) + 2w = 28$$
$$4w - 8 + 2w = 28$$
$$6w - 8 = 28$$
$$6w = 36$$
$$w = 6$$

The width is 6 feet and the length is 8 feet.

23. Let x = the rate Harlene digs.
Then $x + 0.2$ is the rate Ellis digs.

Name	Rate	Time	Distance
Harlene	x	84	$84x$
Ellis	$x + 0.2$	84	$84(x + 0.2)$

Distance Harlene digs + distance Ellis digs
= total length of trench

$$84x + 84(x + 0.2) = 67.2$$
$$84x + 84x + 16.8 = 67.2$$
$$168x = 50.4$$
$$x = 0.3$$

Harlene digs at 0.3 feet per minute and Ellis digs at 0.3 + 2 = 0.5 feet per minute.

24. Let x be the number of pounds of Jelly Belly candy. Then $3 - x$ is the number of pounds of Kit candy.

Type	Cost	Pounds	Total
Jelly Belly	\$2.20	x	\$2.20(x)
Kits	\$2.75	$3 - x$	\$2.75($3 - x$)
Mixture	\$2.40	3	\$2.40(3)

$$2.20x + 2.75(3 - x) = 2.40(3)$$
$$2.2x + 8.25 - 2.75x = 7.2$$
$$-0.55x + 8.25 = 7.2$$
$$-0.55x = -1.05$$
$$x \approx 1.91$$

The mixture should contain about 1.91 pounds of Jelly Belly candy and about 1.09 pounds of Kits candy.

25. Let x = amount of 20% salt solution to be added.

Percent	Liters	Amount
20%	x	$0.20x$
40%	60	(0.40)(60)
35%	$x + 60$	$0.35(x + 60)$

$$0.20x + (0.40)(60) = 0.35(x + 60)$$
$$0.20x + 24 = 0.35x + 21$$
$$-0.15x + 24 = 21$$
$$-0.15x = -3$$
$$x = 20$$

20 liters of 20% solution must be added.

Cumulative Review Test

1. 40% of \$40,000 per year $= (0.40)(40{,}000)$
$= 16{,}000$
Emily receives \$16,000 in social security.

2. **a.** 42.3 million – 35.4 million = 6.9 million

b. $\dfrac{42.3}{35.4} \approx 1.19$ times greater

3. **a.** $\dfrac{5 + 6 + 8 + 12 + 5}{5} = 7.2$
The mean level was 7.2 parts per million.

b. Carbon dioxide levels in order: 5, 5, 6, 8, 12
The median is 6 parts per million.

4. $\frac{5}{12} \div \frac{3}{4} = \frac{5}{12} \cdot \frac{4}{3} = \frac{20}{36} = \frac{5}{9}$

5. $\frac{2}{3} - \frac{3}{8} = \frac{2 \cdot 8}{3 \cdot 8} - \frac{3 \cdot 3}{8 \cdot 3} = \frac{16}{24} - \frac{9}{24} = \frac{7}{24}$

$\frac{2}{3}$ inch is $\frac{7}{24}$ inch greater than $\frac{3}{8}$ inch.

6. a. $\{1, 2, 3, 4, \ldots\}$

b. $\{0, 1, 2, 3, \ldots\}$

c. A rational number is a quotient of two integers, denominator not 0.

7. a. $|-9| = 9$

b. $|-5| = 5$ and $|-3| = 3$. Since $5 > 3$, $|-5| > |-3|$.

8. $$\begin{aligned} 2 - 6^2 \div 2 \cdot 2 &= 2 - 36 \div 2 \cdot 2 \\ &= 2 - 18 \cdot 2 \\ &= 2 - 36 \\ &= -34 \end{aligned}$$

9. $$\begin{aligned} 4(2x-3) - 2(3x+5) - 6 &= 8x - 12 - 6x - 10 - 6 \\ &= 2x - 28 \end{aligned}$$

10. $$\begin{aligned} 4x - 6 &= x + 12 \\ 4x - 6 - x &= x - x + 12 \\ 3x - 6 &= 12 \\ 3x - 6 + 6 &= 12 + 6 \\ 3x &= 18 \\ \frac{3x}{3} &= \frac{18}{3} \\ x &= 6 \end{aligned}$$

11. $$\begin{aligned} 6r &= 2(r+3) - (r+5) \\ 6r &= 2r + 6 - r - 5 \\ 6r &= r + 1 \\ 6r - r &= r - r + 1 \\ 5r &= 1 \\ \frac{5r}{5} &= \frac{1}{5} \\ r &= \frac{1}{5} \end{aligned}$$

12. $$\begin{aligned} 2(x+5) &= 3(2x-4) - 4x \\ 2x + 10 &= 6x - 12 - 4x \\ 2x + 10 &= 2x - 12 \\ 2x - 2x + 10 &= 2x - 2x - 12 \\ 10 &= -12 \end{aligned}$$

The equation has no solution.

13. $$\begin{aligned} \frac{4.8}{x} &= -\frac{3}{5} \\ 4.8(5) &= -3(x) \\ 24 &= -3x \\ \frac{24}{-3} &= \frac{-3x}{-3} \\ -8 &= x \end{aligned}$$

14. $$\begin{aligned} \frac{50 \text{ miles}}{2 \text{ gallons}} &= \frac{225 \text{ miles}}{x \text{ gallons}} \\ \frac{50}{2} &= \frac{225}{x} \\ 50x &= 450 \\ x &= 9 \end{aligned}$$

It will need 9 gallons.

15. Substitute 6 for r.

$A = \pi r^2$

$A = \pi(6)^2 = 36\pi \approx 113.10$

16. a. $$\begin{aligned} 4x + 8y &= 16 \\ 4x - 4x + 8y &= 16 - 4x \\ 8y &= 16 - 4x \\ \frac{8y}{8} &= \frac{16 - 4x}{8} \\ y &= \frac{16}{8} - \frac{4x}{8} \\ y &= 2 - \frac{1}{2}x \\ y &= -\frac{1}{2}x + 2 \end{aligned}$$

b. Substitute −4 for x.

$y = -\frac{1}{2}(-4) + 2 = 2 + 2 = 4$

17. $P = 2l + 2w$

$P - 2l = 2l - 2l + 2w$

$P - 2l = 2w$

$\frac{P-2l}{2} = \frac{2w}{2}$

$\frac{P-2l}{2} = w$

18. Let x = smaller number.
Then $2x + 11$ = larger number.
smaller number + larger number = 29

$$x + (2x + 11) = 29$$
$$3x + 11 = 29$$
$$3x = 18$$
$$x = 6$$

The smaller number is 6 and the larger number is $2(6) + 11 = 23$.

19. Let x = number of minutes for the two plans to have the same cost.
Cost for Plan A = $19.95 + 0.35x$
Cost for Plan B = $29.95 + 0.10x$
The costs will be equal when the cost for Plan A = cost for Plan B.

$$19.95 + 0.35x = 29.95 + 0.10x$$
$$0.35x = 10 + 0.10x$$
$$0.25x = 10$$
$$x = 40$$

Lori would need to talk 40 minutes in a month for the plans to have the same cost.

20. Let x = smallest angle. Then $x + 5$ = second angle, $x + 50$ = third angle, and $4x + 25$ = fourth angle. Sum of the angle measures = 360°

$$x + (x + 5) + (x + 50) + (4x + 25) = 360$$
$$7x + 80 = 360$$
$$7x = 280$$
$$x = 40$$

The angle measures are 40°, 40 + 5 = 45°, 40 + 50 = 90°, and 4(40) + 25 = 185°.

Chapter 4

Exercise Set 4.1

1. The x-coordinate is always listed first.

3. a. The horizontal axis is the x-axis.

b. The vertical axis is the y-axis.

5. Axis is singular, while axes is plural.

7. The graph of a linear equation is an illustration of the set of points whose coordinates satisfy the equation.

9. a. Two points are needed to graph a linear equation.

b. It is a good idea to use three or more points when graphing a linear equation to catch errors.

11. $ax + by = c$

13.

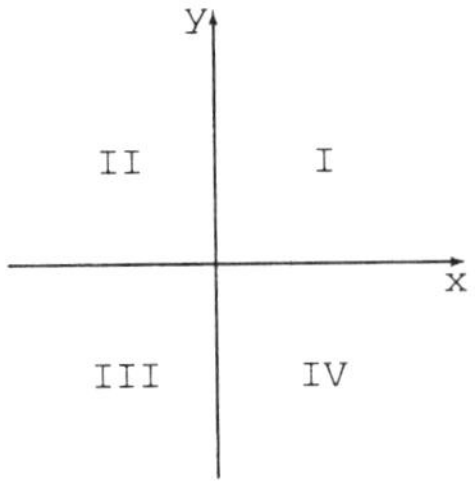

15. II

17. IV

19. II

21. III

23. III

25. II

27. $A(3, 1)$; $B(-3, 0)$; $C(1, -3)$; $D(-2, -3)$; $E(0, 3)$; $F\left(\frac{3}{2}, -1\right)$

29.

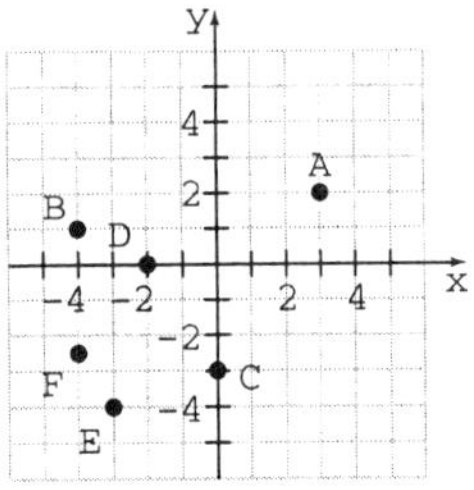

31.

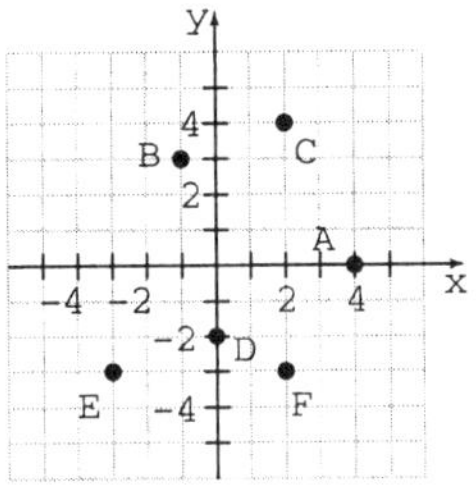

33.

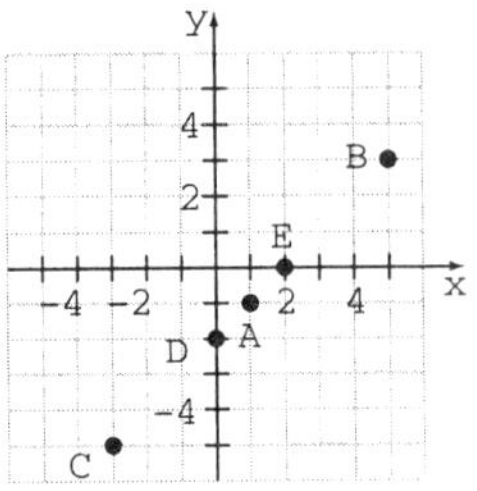

The points are collinear.

35.

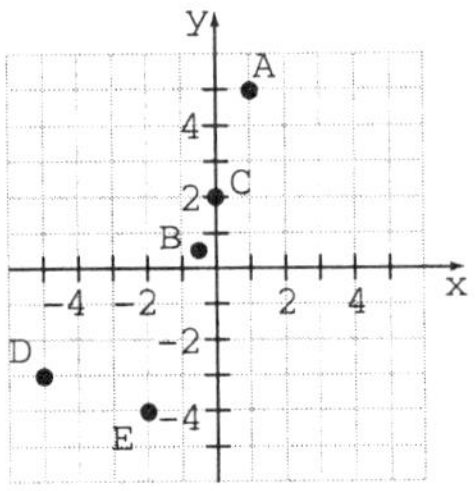

The points are not collinear since $(-5, -3)$ is not on the line.

37. a.

$y = x + 2$		$y = x + 2$	
$4 = 2 + 2$		$0 = -2 + 2$	
$4 = 4$	True	$0 = 0$	True

$y = x + 2$		$y = x + 2$	
$3 = 2 + 2$		$2 = 0 + 2$	
$3 = 4$	False	$2 = 2$	True

Point c) does not satisfy the equation.

b.

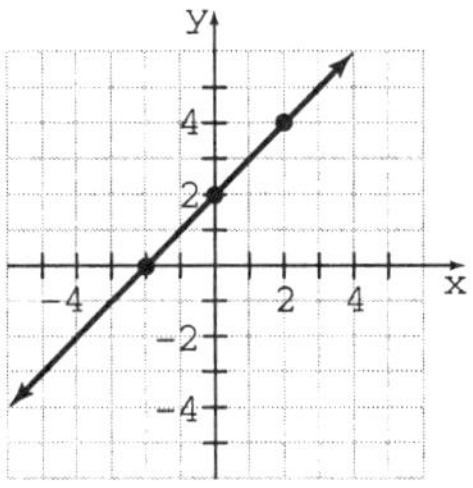

39. a.

$$3x - 2y = 6 \qquad 3x - 2y = 6$$
$$3(4) - 2(0) = 6 \qquad 3(2) - 2(0) = 6$$
$$12 = 6 \quad \text{False} \qquad 6 = 6 \quad \text{True}$$
$$3x - 2y = 6 \qquad 3x - 2y = 6$$
$$3\left(\frac{2}{3}\right) - 2(-2) = 6 \qquad 3\left(\frac{4}{3}\right) - 2(-1) = 6$$
$$2 + 4 = 6 \qquad 4 + 2 = 6$$
$$6 = 6 \quad \text{True} \qquad 6 = 6 \quad \text{True}$$

Point a) does not satisfy the equation.

b.

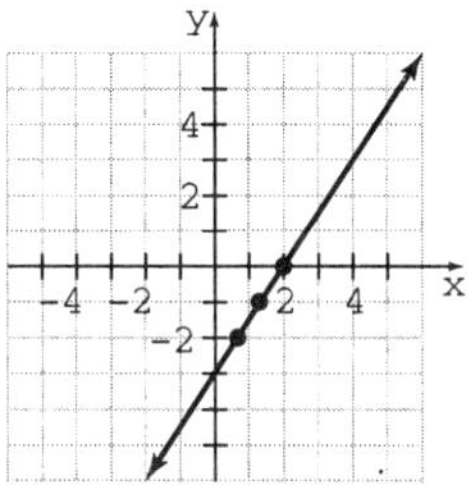

41. a.

$$\frac{1}{2}x + 4y = 4 \qquad \frac{1}{2}x + 4y = 4$$
$$\frac{1}{2}(2) + 4(-1) = 4 \qquad \frac{1}{2}(2) + 4\left(\frac{3}{4}\right) = 4$$
$$1 - 4 = 4 \qquad 1 + 3 = 4$$
$$-3 = 4 \quad \text{False} \qquad 4 = 4 \quad \text{True}$$

$$\frac{1}{2}x + 4y = 4 \qquad \frac{1}{2}x + 4y = 4$$
$$\frac{1}{2}(0) + 4(1) = 4 \qquad \frac{1}{2}(-4) + 4\left(\frac{3}{2}\right) = 4$$
$$0 + 4 = 4 \qquad -2 + 6 = 4$$
$$4 = 4 \quad \text{True} \qquad 4 = 4 \quad \text{True}$$

Point a) does not satisfy the equation.

b.

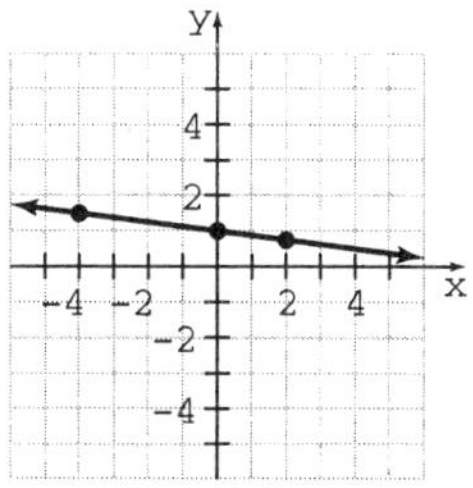

43. $y = 3x - 4$
$y = 3(2) - 4$
$y = 6 - 4$
$y = 2$

45. $y = 3x - 4$
$y = 3(0) - 4$
$y = 0 - 4$
$y = -4$

47.
$$\begin{aligned} 2x + 3y &= 12 \\ 2(3) + 3y &= 12 \\ 6 + 3y &= 12 \\ 3y &= 6 \\ y &= 2 \end{aligned}$$

49.
$$\begin{aligned} 2x + 3y &= 12 \\ 2\left(\frac{1}{2}\right) + 3y &= 12 \\ 1 + 3y &= 12 \\ 3y &= 11 \\ y &= \frac{11}{3} \end{aligned}$$

51. The value of y is 0 when a straight line crosses the x-axis, because any point on the x-axis is neither above or below the origin.

53 **a.** Latitude: 16°N Longitude: 56°W
b. Latitude: 29°N Longitude: 90.5°W
c. Latitude: 26°N Longitude: 80.5°W
d. Answers will vary.

57. The general form of a linear equation in one variable is $ax + b = c$.

58. It is a linear equation that has only one solution.

59.
$$\begin{aligned} -2(-3x + 5) + 6 &= 4(x - 2) \\ -2(-3x) + (-2)(5) + 6 &= 4(x) + 4(-2) \\ 6x - 10 + 6 &= 4x - 8 \\ 6x - 4 &= 4x - 8 \\ 6x - 4x - 4 &= 4x - 4x - 8 \\ 2x - 4 &= -8 \\ 2x - 4 + 4 &= -8 + 4 \\ 2x &= -4 \\ \frac{2x}{2} &= \frac{-4}{2} \\ x &= -2 \end{aligned}$$

60. $C = 2\pi r$ $\qquad A = \pi r^2$
$= 2\pi(3)$ $\qquad = \pi(3)^2$
$= 6\pi$ $\qquad = 9\pi$
≈ 18.84 inches $\qquad \approx 28.27$ in.^{2}

61.
$$\begin{aligned} 2x - 5y &= 6 \\ 2x - 2x - 5y &= -2x + 6 \\ -5y &= -2x + 6 \\ \frac{-5y}{-5} &= \frac{-2x}{-5} + \frac{6}{-5} \\ y &= \frac{2}{5}x - \frac{6}{5} \end{aligned}$$

Exercise Set 4.2

1. To find the x-intercept, substitute 0 for y and find the corresponding value of x. To find the y-intercept, substitute 0 for x and find the corresponding value of y.

3. The graph of $y = b$ is a horizontal line.

5. You may not be able to read exact answers from a graph.

7. Yes. The equation goes through the origin because the point (0, 0) satisfies the equation.

9.
$$\begin{aligned} 3x + y &= 9 \\ 3(3) + y &= 9 \\ 9 + y &= 9 \\ y &= 0 \end{aligned}$$

11.
$$\begin{aligned} 3x + y &= 9 \\ 3x + (-6) &= 9 \\ 3x - 6 &= 9 \\ 3x &= 15 \\ x &= 5 \end{aligned}$$

12.
$$\begin{aligned} 3x + y &= 9 \\ 3x + (-3) &= 9 \\ 3x - 3 &= 9 \\ 3x &= 12 \\ x &= 4 \end{aligned}$$

13. $3x + y = 9$
$3x + 0 = 9$
$3x = 9$
$x = 3$

15. $3x - 2y = 8$
$3 \cdot 4 - 2y = 8$
$12 - 2y = 8$
$-2y = -4$
$y = 2$

17. $3x - 2y = 8$
$3x - 2(0) = 8$
$3x = 8$
$x = \dfrac{8}{3}$

19. $3x - 2y = 8$
$3(-4) - 2y = 8$
$-12 - 2y = 8$
$-2y = 20$
$y = -10$

21. An equation of the form $x = c$ is a vertical line with x-intercept at $(c, 0)$.

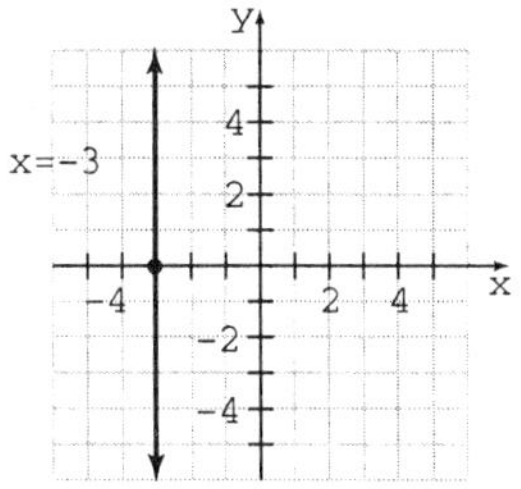

23. An equation of the form $y = c$ is a horizontal line with y-intercept at $(0, c)$.

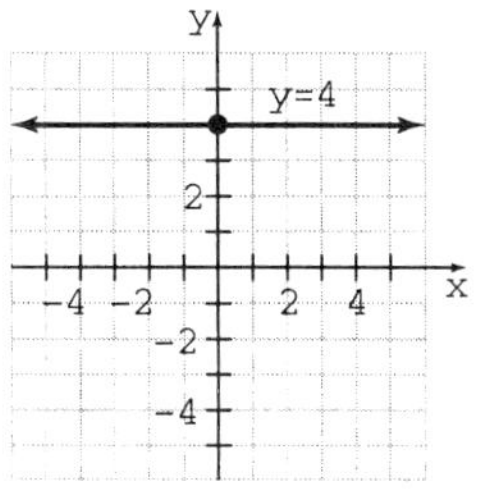

25. Let $x = 0$, $y = 3(0) - 1 = -1$, $(0, -1)$
Let $x = 1$, $y = 3(1) - 1 = 2$, $(1, 2)$
Let $x = 2$, $y = 3(2) - 1 = 5$, $(2, 5)$

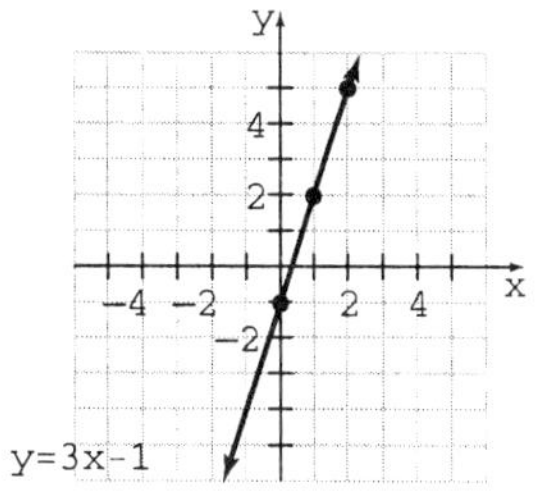

27. Let $x = -1$, $y = 6(-1) + 2 = -4$, $(-1, -4)$
Let $x = 0$, $y = 6(0) + 2 = 2$, $(0, 2)$
Let $x = 1$, $y = 6(1) + 2 = 8$, $(1, 8)$, $(1, 8)$

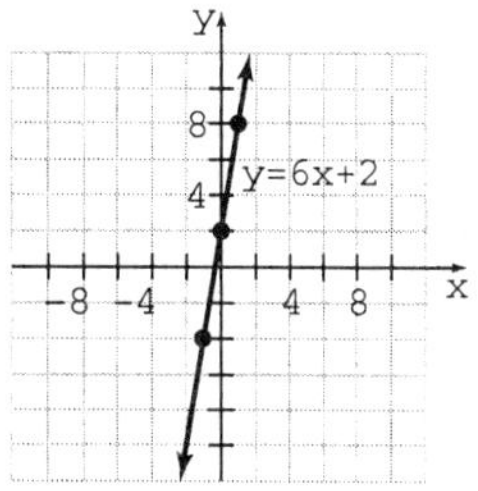

29. Let $x = 0$, $y = -\dfrac{1}{2}(0) + 3 = 3$, $(0, 3)$

Let $x = 2$, $y = -\dfrac{1}{2}(2) + 3 = 2$, $(2, 2)$

Let $x = 4$, $y = -\dfrac{1}{2}(4) + 3 = 1$, $(4, 1)$

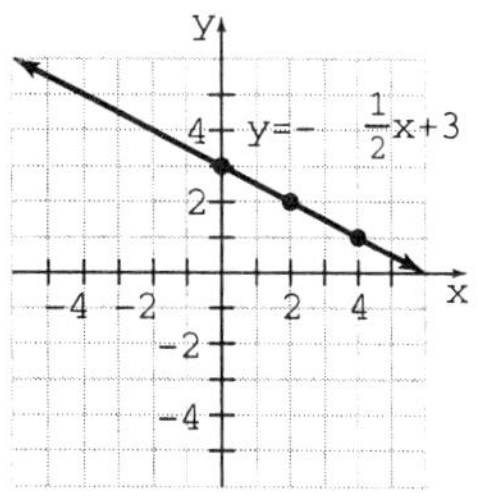

31. $2x - 4y = 4$
$-4y = -2x + 4$
$y = \dfrac{1}{2}x - 1$

Let $x = 0$, $y = \dfrac{1}{2}(0) - 1 = -1$, $(0, -1)$

Let $x = 2$, $y = \dfrac{1}{2}(2) - 1 = 0$, $(2, 0)$

Let $x = 4$, $y = \frac{1}{2}(4) - 1 = 1$, $(4, 1)$

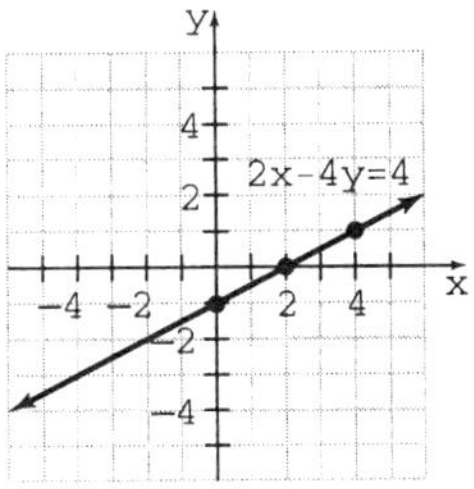

33. $5x - 2y = 8$
$-2y = -5x + 8$
$y = \frac{5}{2}x - 4$

Let $x = 0$, $y = \frac{5}{2}(0) - 4 = -4$, $(0, -4)$

Let $x = 2$, $y = \frac{5}{2}(2) - 4 = 1$, $(2, 1)$

Let $x = 4$, $y = \frac{5}{2}(4) - 4 = 6$, $(4, 6)$

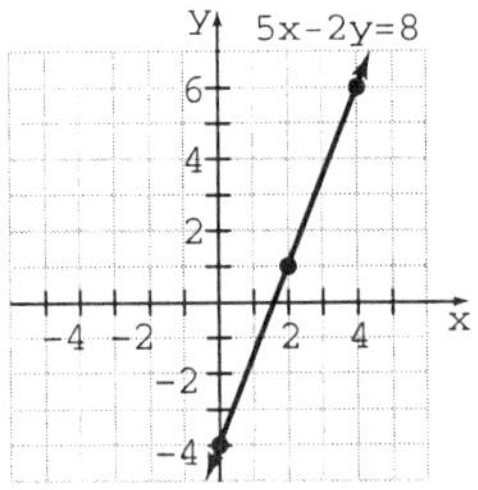

35. $6x + 5y = 30$
$5y = -6x + 30$
$y = -\frac{6}{5}x + 6$

Let $x = 0$, $y = -\frac{6}{5}(0) + 6 = 6$, $(0, 6)$

Let $x = 5$, $y = -\frac{6}{5}(5) + 6 = 0$, $(5, 0)$

Let $x = 10$, $y = -\frac{6}{5}(10) + 6 = -6$, $(10, -6)$

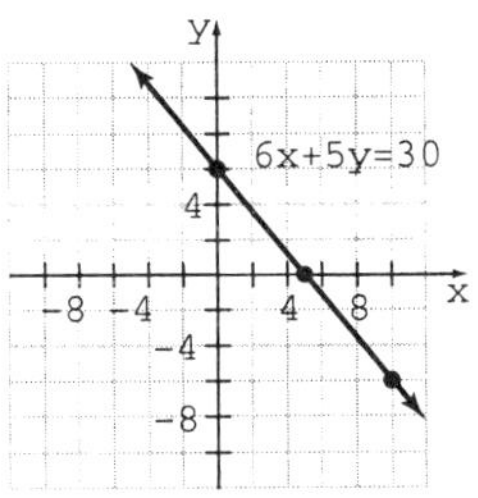

37. $-4x + 5y = 0$
$5y = 4x$
$y = \frac{4}{5}x$

Let $x = -5$, $y = \frac{4}{5}(-5) = -4$, $(-5, -4)$

Let $x = 0$, $y = \frac{4}{5}(0) = 0$, $(0, 0)$

Let $x = 5$, $y = \frac{4}{5}(5) = 4$, $(5, 4)$

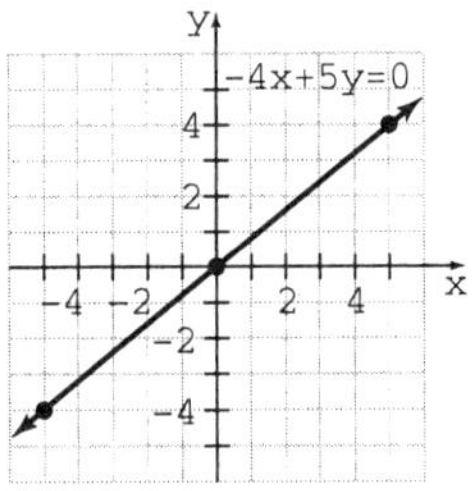

39. Let $x = 0$, $y = -20(0) + 60 = 60$, $(0, 60)$
Let $x = 2$, $y = -20(2) + 60 = 20$, $(2, 20)$
Let $x = 4$, $y = -20(4) + 60 = -20$, $(4, -20)$

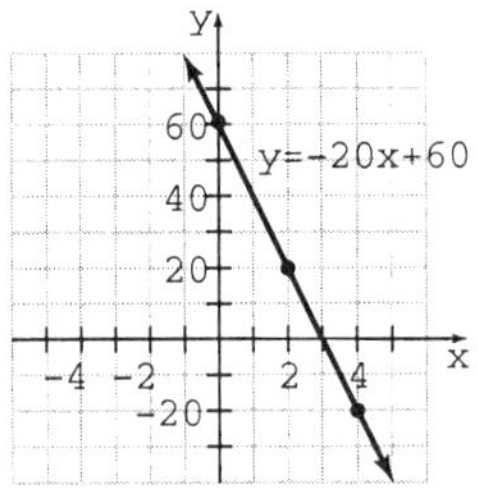

41. Let $x = -3$, $y = \frac{4}{3}(-3) = -4$, $(-3, -4)$

Let $x = 0$, $y = \frac{4}{3}(0) = 0$, $(0, 0)$

Let $x = 3$, $y = \frac{4}{3}(3) = 4$, $(3, 4)$

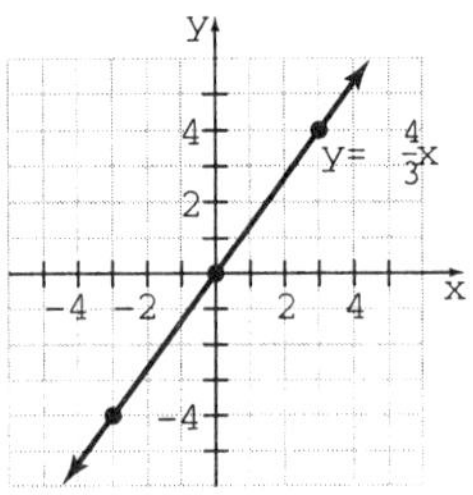

43. Let $x = 0$, $y = \frac{1}{2}(0) + 4 = 4$, $(0, 4)$

Let $x = 2$, $y = \frac{1}{2}(2) + 4 = 5$, $(2, 5)$

Let $x = 4$, $y = \frac{1}{2}(4) + 4 = 6$, $(4, 6)$

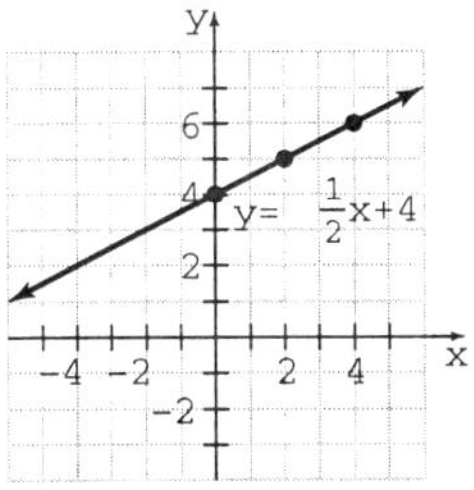

45. Let $x = 0$ Let $y = 0$

$y = 3x + 3$ $y = 3x + 3$

$y = 3(0) + 3$ $0 = 3x + 3$

$y = 3$ $-3x = 3$

$x = -1$

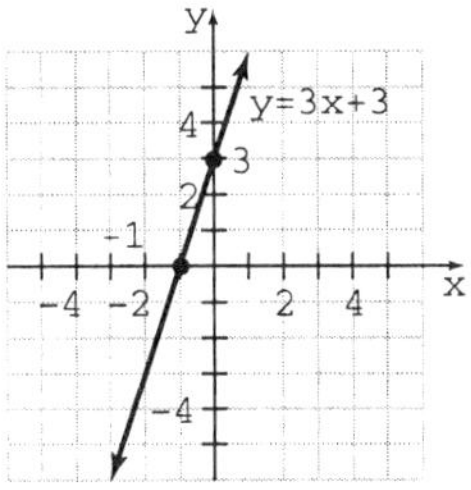

47. Let $x = 0$ Let $y = 0$

$y = -4x + 2$ $y = -4x + 2$

$y = -4(0) + 2$ $0 = -4x + 2$

$y = 2$ $-2 = -4x$

$x = \frac{1}{2}$

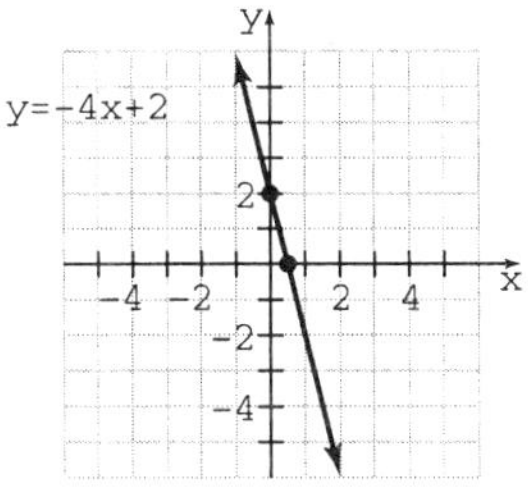

49. Let $x = 0$ Let $y = 0$

$y = -5x + 4$ $y = -5x + 4$

$y = -5(0) + 4$ $0 = -5x + 4$

$y = 4$ $-4 = -5x$

$x = \frac{4}{5}$

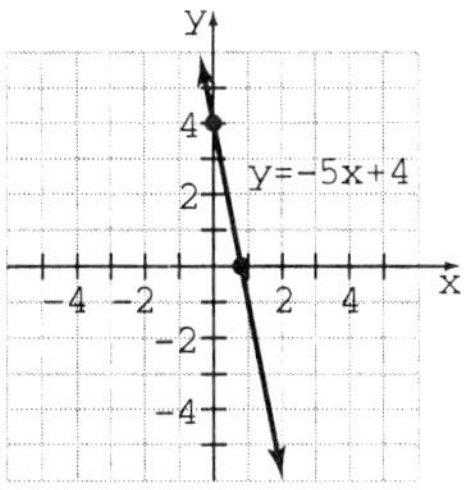

51. $4y + 6x = 24$

$4y = -6x + 24$

$y = -\frac{3}{2}x + 6$

Let $x = 0$ Let $y = 0$

$y = -\frac{3}{2}(0) + 6$ $0 = -\frac{3}{2}x + 6$

$y = 6$ $\frac{3}{2}x = 6$

$x = 4$

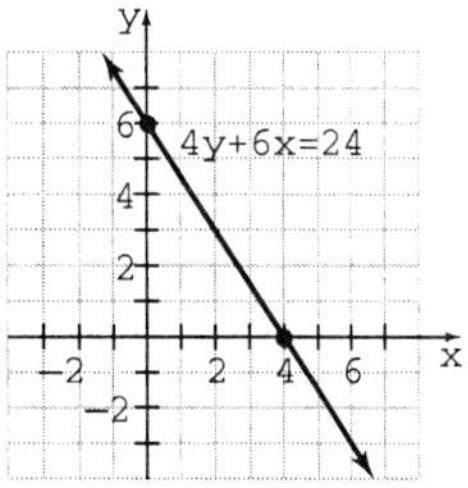

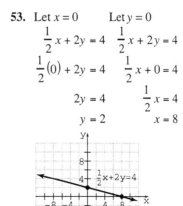

53. Let $x = 0$ Let $y = 0$

$\frac{1}{2}x + 2y = 4$ $\frac{1}{2}x + 2y = 4$

$\frac{1}{2}(0) + 2y = 4$ $\frac{1}{2}x + 0 = 4$

$2y = 4$ $\frac{1}{2}x = 4$

$y = 2$ $x = 8$

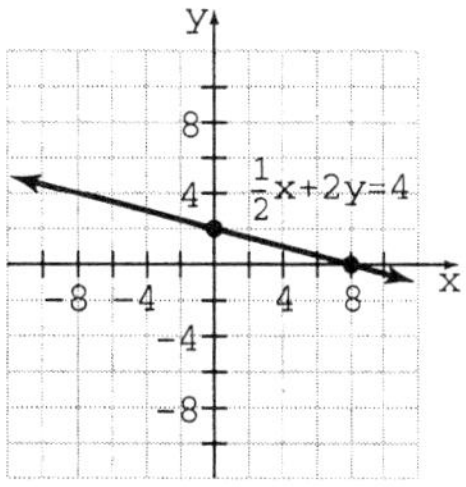

55. Let $x = 0$

$$6x - 12y = 24$$
$$6(0) - 12y = 24$$
$$-12y = 24$$
$$y = -2$$

Let $y = 0$

$$6x - 12y = 24$$
$$6x - 12(0) = 24$$
$$6x = 24$$
$$x = 4$$

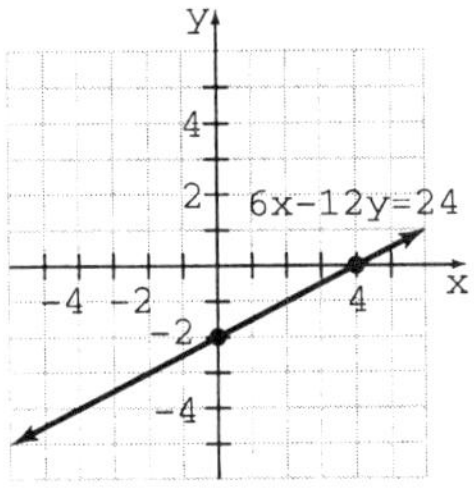

57. Let $x = 0$

$$8y = 6x - 12$$
$$8y = 6(0) - 12$$
$$8y = -12$$
$$y = -\frac{3}{2}$$

Let $y = 0$

$$8y = 6x - 12$$
$$8(0) = 6x - 12$$
$$0 = 6x - 12$$
$$-6x = -12$$
$$x = 2$$

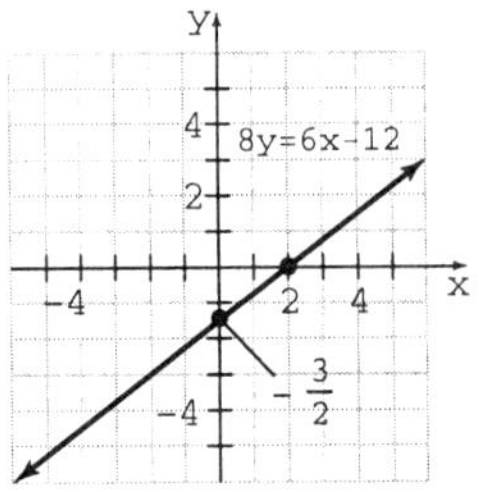

59. Let $x = 0$

$$-20y - 30x = 40$$
$$-20y - 30(0) = 40$$
$$-20y = 40$$
$$y = -2$$

Let $y = 0$

$$-20y - 30x = 40$$
$$-20(0) - 30x = 40$$
$$-30x = 40$$
$$x = -\frac{4}{3}$$

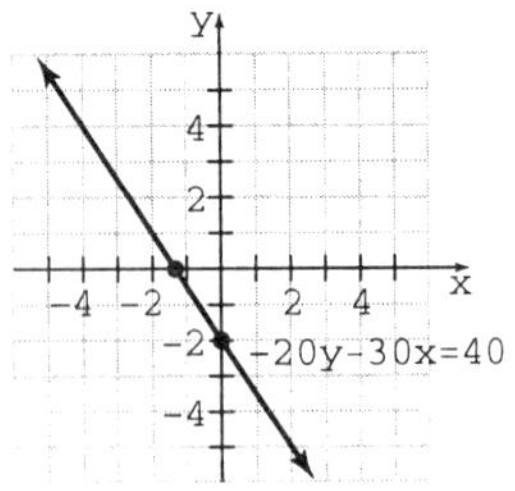

61. Let $x = 0$

$$\frac{1}{3}x + \frac{1}{4}y = 12$$
$$\frac{1}{3}(0) + \frac{1}{4}y = 12$$
$$\frac{1}{4}y = 12$$
$$y = 48$$

Let $y = 0$

$$\frac{1}{3}x + \frac{1}{4}y = 12$$
$$\frac{1}{3}x + \frac{1}{4}(0) = 12$$
$$\frac{1}{3}x = 12$$
$$x = 36$$

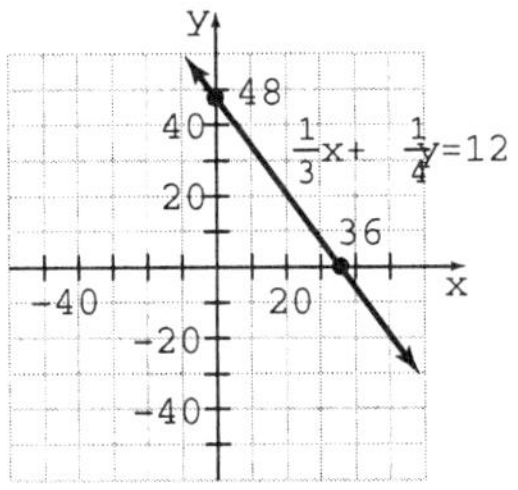

63. Let $x = 0$

$$\frac{1}{2}x = \frac{2}{5}y - 80$$
$$\frac{1}{2}(0) = \frac{2}{5}y - 80$$
$$0 = \frac{2}{5}y - 80$$
$$-\frac{2}{5}y = -80$$
$$y = 200$$

Let $y = 0$

$$\frac{1}{2}x = \frac{2}{5}y - 80$$
$$\frac{1}{2}x = \frac{2}{5}(0) - 80$$
$$\frac{1}{2}x = -80$$
$$x = -160$$

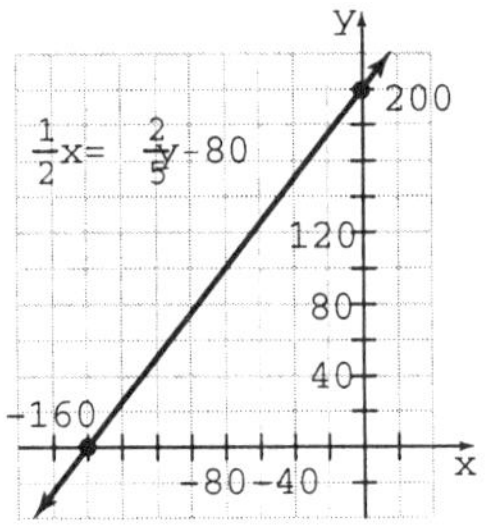

65. $x = -2$

67. $y = 6$

69.

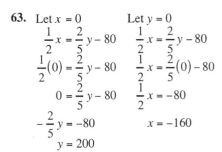

$$ax + 4y = 8$$
$$a(2) + 4(0) = 8$$
$$2a + 0 = 8$$
$$2a = 8$$
$$a = 4$$

71.

$$3x + by = 10$$
$$3(0) + b(5) = 10$$
$$0 + 5b = 10$$
$$5b = 10$$
$$b = 2$$

73. Yes. For each 15 minutes of time, the number of calories burned increases by 200 calories.

75. a. $C = 2n + 30$

b.

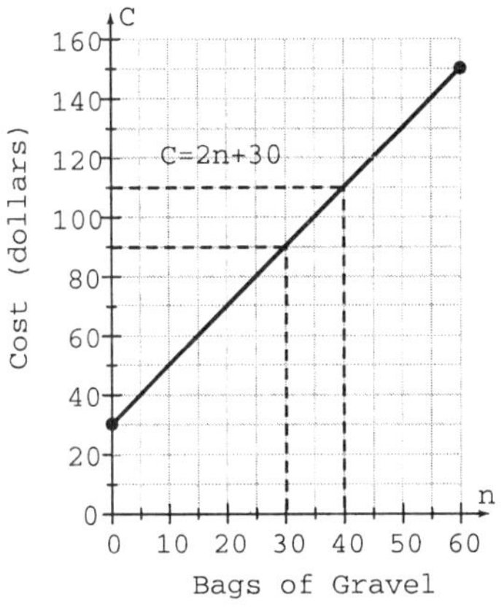

c. \$90

d. 40 bags

77. a. $C = m + 40$

b.

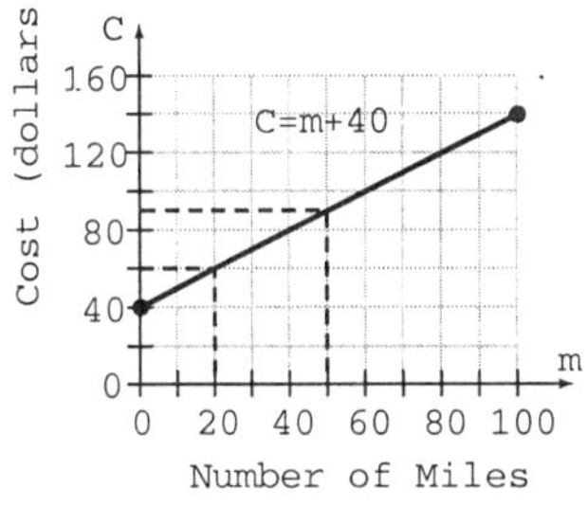

c. \$90

d. 20 miles

79. a.

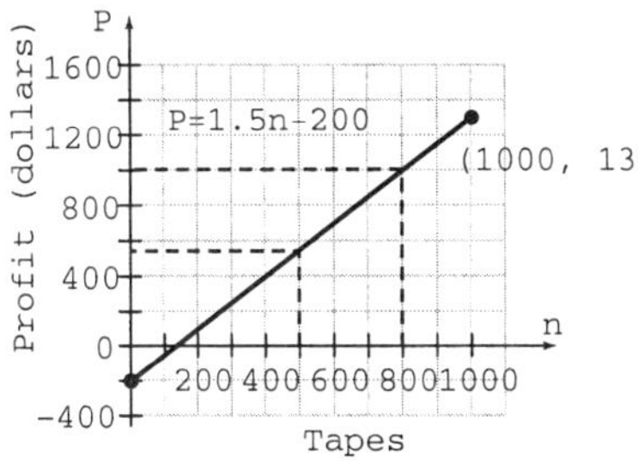

b. \$550

c. 800 tapes

81. Since each shaded area multiplied by the corresponding intercept must equal 20, the coefficients are 5 and 4 respectively.

83. Since the first shaded area multiplied by the x-intercept must equal -12, the coefficient of x is 6. Since the opposite of the second shaded area multiplied by the y-intercept must equal -12, the coefficient of y is 4.

88.
$$\begin{aligned} 2[6-(4-5)] \div 2 - 5^2 &= 2[6-(-1)] \div 2 - 25 \\ &= 2[7] \div 2 - 25 \\ &= 14 \div 2 - 25 \\ &= 7 - 25 \\ &= -18 \end{aligned}$$

89.
$$\begin{aligned} \frac{-3^2 \cdot 4 \div 2}{\sqrt{9} - 2^2} &= \frac{-9 \cdot 4 \div 2}{3-4} \\ &= \frac{-36 \div 2}{-1} \\ &= \frac{-18}{-1} \\ &= 18 \end{aligned}$$

90.
$$\frac{8 \text{ ounces}}{3 \text{ gallons}} = \frac{x \text{ ounces}}{2.5 \text{ gallons}}$$
$$\frac{8}{3} = \frac{x}{2.5}$$
$$20 = 3x$$
$$6.67 \approx x$$
You should use 6.67 ounces of cleaner.

91. Let x = smaller integer. Then $3x + 1$ = larger integer.
$$\begin{aligned} x + (3x + 1) &= 37 \\ 4x + 1 &= 37 \\ 4x &= 36 \\ x &= 9 \end{aligned}$$
The smaller integer is 9.
The larger is $3(9) + 1 = 27 + 1 = 28$.

Exercise Set 4.3

1. The slope of a line is the ratio of the vertical change to the horizontal change between any two points on the line.

3. A line with a positive slope rises from left to right.

5. Lines that rise from the left to right have a positive slope. Lines that fall from left to right have a negative slope.

7. No, since we cannot divide by 0, the slope is undefined.

9. Their slopes are the same.

11. $m = \frac{5-1}{7-5} = \frac{4}{2} = 2$

13. $m = \frac{-2-0}{5-9} = \frac{-2}{-4} = \frac{1}{2}$

15. $m = \frac{\frac{1}{2}-\frac{1}{2}}{-3-3} = \frac{0}{-6} = 0$

17. $m = \frac{6-6}{-2-(-7)} = \frac{0}{5} = 0$

19. $m = \frac{-2-4}{6-6} = \frac{-6}{0}$ undefined

21. $m = \frac{3-0}{-2-6} = \frac{3}{-8} = -\frac{3}{8}$

23. $m = \frac{1-\frac{3}{2}}{-\frac{3}{4}-0} = \frac{-\frac{1}{2}}{-\frac{3}{4}} = \frac{-1}{2}\cdot\frac{4}{-3} = \frac{-4}{-6} = \frac{2}{3}$

25. $m = \frac{6}{3} = 2$

27. $m = \frac{6}{-3} = -2$

29. $m = \frac{4}{-7} = -\frac{4}{7}$

31. $m = \frac{7}{4}$

33. $m = \frac{0}{3} = 0$

35. Vertical line, slope is undefined.

37. Horizontal line, slope is 0.

39.

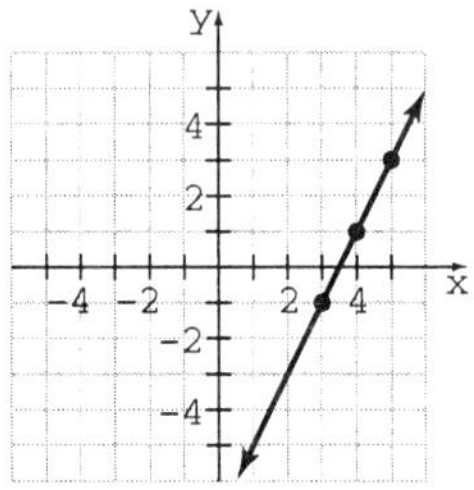

41.

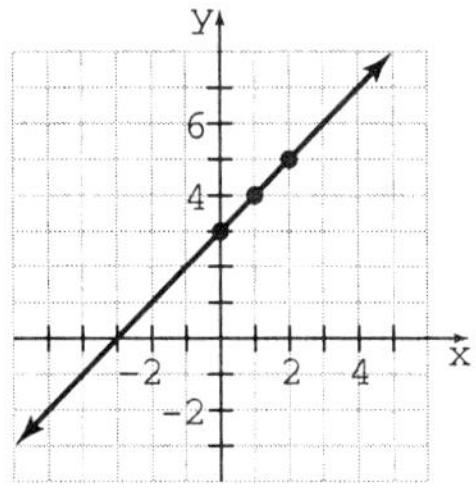

43.

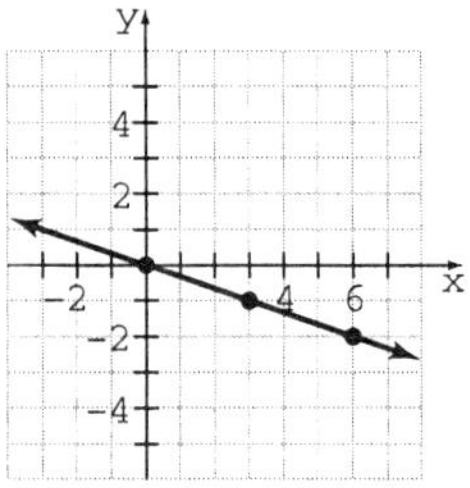

45.

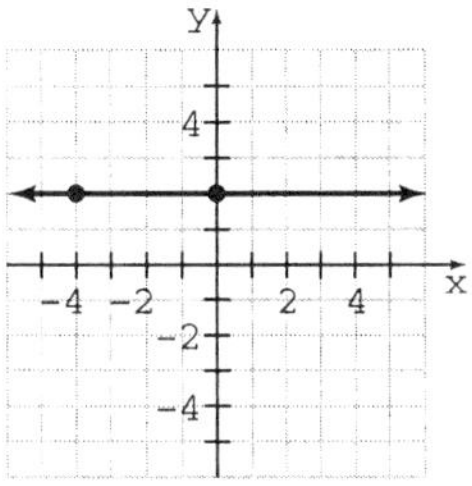

47.

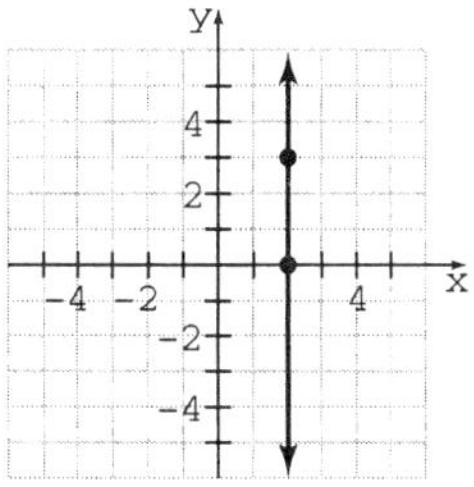

49. The lines are parallel because the slopes are the same.

51. The lines are perpendicular because the slopes are negative reciprocals.

53. The lines are perpendicular because the slopes are negative reciprocals.

55. The lines are neither parallel or perpendicular.

57. The lines are neither parallel or perpendicular.

59. The lines are parallel because the slopes are the same.

61. The lines are parallel because the slopes are the same.

63. The lines are perpendicular because if the slope is 0, the line is horizontal and if the slope is undefined the line is vertical.

65. Its slope would be 2.

67. Its slope would be $\frac{1}{4}$.

69. The first graph appears to pass through the points (–1, 0) and (0, 6). It's slope is $m = \frac{6-0}{0-(-1)} = \frac{6}{1} = 6$. The second graph appears to pass through the points (–4, 0) and (0, 6). It's slope is $m = \frac{6-0}{0-(-4)} = \frac{6}{4} = \frac{6}{4} = \frac{3}{2}$. The first graph has the greater slope.

71. a. $m = \frac{45-22}{1957-1961} = \frac{23}{-4} = -\frac{23}{4}$

b. $m = \frac{51-7}{1989-1985} = \frac{44}{4} = 11$

73. $m = \frac{-2-6}{3-1} = \frac{-8}{2} = -4$

A line parallel to the given line would have a slope of -4.

75. $m = \frac{5-(-3)}{2-1} = \frac{5+3}{1} = 8$

A line perpendicular to the given line would have a slope of $\frac{1}{8}$.

77.

$$m = \frac{-\frac{7}{2}-\left(-\frac{3}{8}\right)}{-\frac{4}{9}-\frac{1}{2}} = \frac{-\frac{28}{8}+\frac{3}{8}}{-\frac{8}{18}-\frac{9}{18}} = \frac{-\frac{25}{8}}{-\frac{17}{18}} = \left(-\frac{25}{8}\right)\left(-\frac{18}{17}\right) = \frac{(-25)(-9)}{(4)(17)} = \frac{225}{68}$$

79. a.

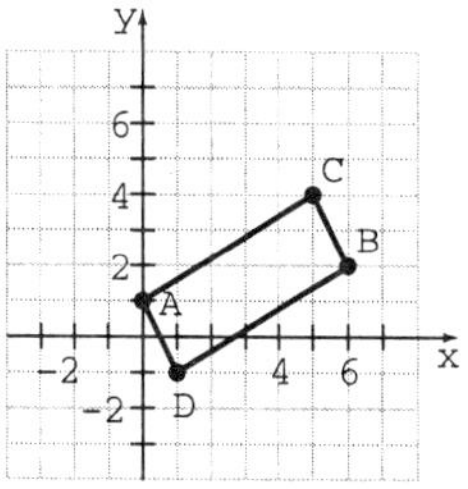

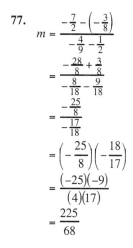

b. $AC;\ m = \frac{4-1}{5-0} = \frac{3}{5}$

$CB;\ m = \frac{4-2}{5-6} = \frac{2}{-1} = -2$

$$DB;\ m = \frac{2-(-1)}{6-1} = \frac{3}{5}$$

$$AD;\ m = \frac{-1-1}{1-0} = \frac{-2}{1} = -2$$

c. Yes; opposite sides are parallel.

83. $4x^2 + 3x + \frac{x}{2} = 4(0)^2 + 3(0) + \frac{0}{2}$

$$= 0 + 0 + 0$$
$$= 0$$

84. a.
$$-x = -\frac{3}{2}$$
$$(-1)(-x) = (-1)\left(-\frac{3}{2}\right)$$
$$x = \frac{3}{2}$$

b.
$$5x = 0$$
$$\frac{5x}{5} = \frac{0}{5}$$
$$x = 0$$

85.
$$\frac{2}{3}x + \frac{1}{7}x = 4$$
$$21\left(\frac{2}{3}x + \frac{1}{7}x\right) = 21(4)$$
$$21\left(\frac{2}{3}x\right) + 21\left(\frac{1}{7}x\right) = 84$$
$$14x + 3x = 84$$
$$17x = 84$$
$$\frac{17x}{17} = \frac{84}{17}$$
$$x = \frac{84}{17}$$

86.
$$d = a + b + c$$
$$d - a = a - a + b + c$$
$$d - a = b + c$$
$$d - a - b = b - b + c$$
$$d - a - b = c$$

87. $5x - 3y = 15$

$x = 0$	$y = 0$
$5(0) - 3y = 15$	$5x - 3(0) = 15$
$-3y = 15$	$5x = 15$
$y = -5$	$x = 5$
$(0, -5)$	$(5, 0)$

Exercise Set 4.4

1. $y = mx + b$

3. $y = 3x - 5$

5. Compare their slopes: If slopes are the same and their y-intercepts are different, the lines are parallel.

7. $y - y_1 = m(x - x_1)$

9. $m = 4$; y-intercept: $(0, -6)$

11. $m = \frac{4}{3}$; y-intercept: $(0, -5)$

13. $m = 1$; y-intercept: $(0, -3)$

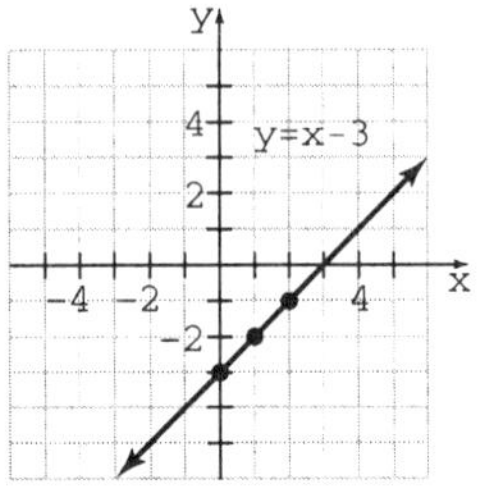

15. $m = 3$; y-intercept: $(0, 2)$

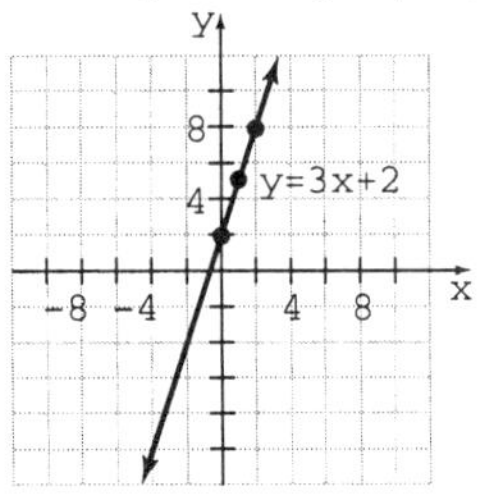

17. $m = -4$; y-intercept: $(0, 0)$

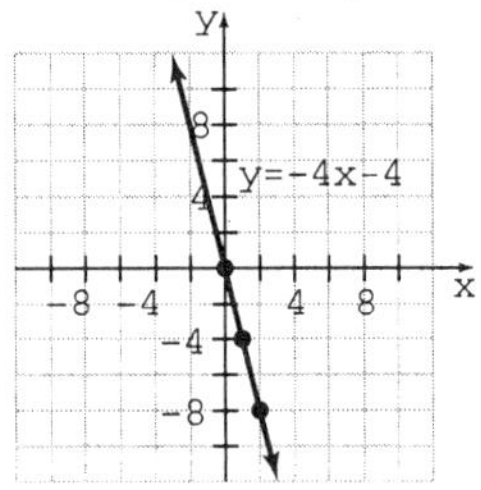

19. $-2x + y = -3$

$y = 2x - 3$

$m = 2$; y-intercept: $(0, -3)$

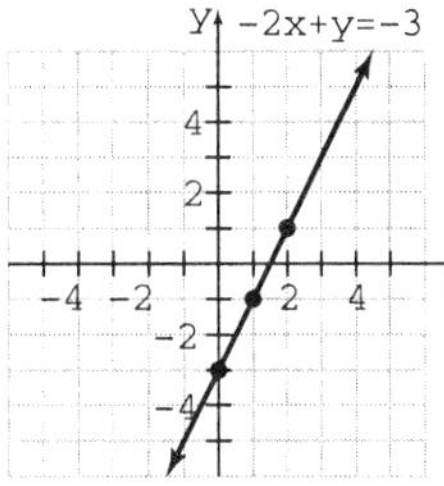

21. $5x - 2y = 10$

$-2y = -5x + 10$

$y = \frac{5}{2}x - 5$

$m = \frac{5}{2}$; y-intercept: $(0, -5)$

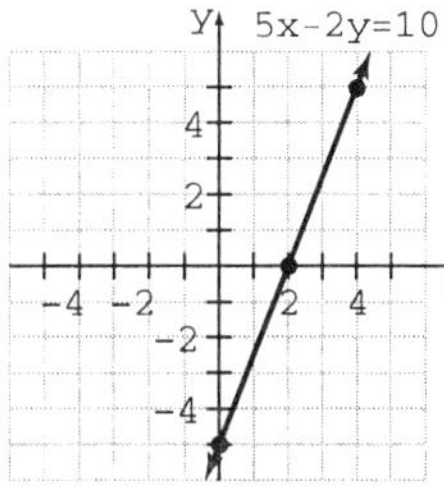

23. $6x + 12y = 18$

$12y = -6x + 18$

$y = -\frac{1}{2}x + \frac{3}{2}$

$m = -\frac{1}{2}$; y-intercept: $\left(0, \frac{3}{2}\right)$

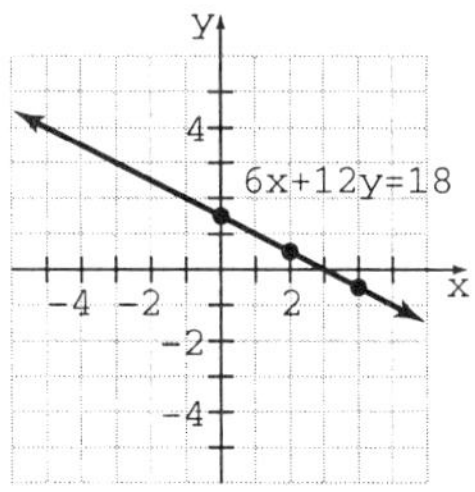

25. $-6x + 2y - 8 = 0$

$2y = 6x + 8$

$y = 3x + 4$

$m = 3$; y-intercept: $(0, 4)$

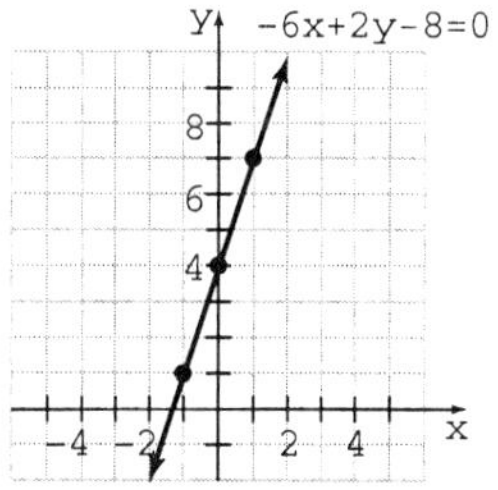

27. $3x = 2y - 4$

$-2y = -3x - 4$

$y = \frac{3}{2}x + 2$

$m = \frac{3}{2}$; y-intercept: $(0, 2)$

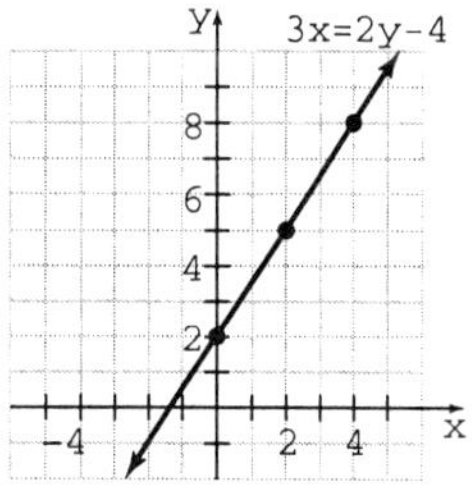

29. $m = \frac{4}{4} = 1$, $b = -2$

$y = x - 2$

31. $m = \frac{-2}{6} = -\frac{1}{3}$, $b = 2$

$y = -\frac{1}{3}x + 2$

33. $m = \frac{10}{30} = \frac{1}{3}$, $b = 5$

$y = \frac{1}{3}x + 5$

35. $m = \frac{-3}{1} = -3$, $b = 4$

$y = -3x + 4$

37. Since the slopes of the lines are the same and y-intercepts are different, the lines are parallel.

39. $4x + 2y = 9$ $\quad$ $4x = 8y + 4$

$2y = -4x + 9$ $\quad$ $-8y = -4x + 4$

$y = -2x + \frac{9}{2}$ $\quad$ $y = \frac{1}{2}x - \frac{1}{2}$

Since the slopes of the lines are opposite reciprocals, the lines are perpendicular.

41. $3x + 5y = 9$ $\quad$ $6x = -10y + 9$

$5y = -3x + 9$ $\quad$ $10y = -6x + 9$

$y = -\frac{3}{5}x + \frac{9}{5}$ $\quad$ $y = \frac{-6}{10}x + \frac{9}{10}$

$y = -\frac{3}{5}x + \frac{9}{10}$

Since the slopes of the lines are the same and y-intercepts are different, the lines are parallel.

43. $y = \frac{1}{2}x - 4$ $\quad$ $2y = 6x + 9$

$y = 3x + \frac{9}{2}$

Since the slopes of the lines are not equal and are not opposite reciprocals, the lines are neither parallel nor perpendicular.

45. $5y = 2x + 3$ $\quad$ $-10x = 4y + 8$

$y = \frac{2}{5}x + \frac{3}{5}$ $\quad$ $-4y = 10x + 8$

$y = -\frac{5}{2}x - 2$

Since the slopes of the lines are opposite reciprocals, the lines are perpendicular.

47. $3x + 7y = 8$ $\quad$ or $\quad$ $7x + 3y = 8$

$7y = -3x + 8$ $\quad$ $3y = -7x + 8$

$y = -\frac{3}{7}x + \frac{8}{7}$ $\quad$ $y = -\frac{7}{3}x + \frac{8}{3}$

Since the slopes of the lines are not equal and are not opposite reciprocals, the lines are neither parallel nor perpendicular.

49. $y - 2 = 3(x - 0)$

$y = 3x + 2$

51. $y - 5 = -3[x - (-4)]$

$y - 5 = -3(x + 4)$

$y - 5 = -3x - 12$

$y = -3x - 7$

53. $y - (-5) = \frac{1}{2}[x - (-1)]$

$y + 5 = \frac{1}{2}(x + 1)$

$y + 5 = \frac{1}{2}x + \frac{1}{2}$

$y = \frac{1}{2}x - \frac{9}{2}$

55. $y - 6 = \frac{2}{5}(x - 0)$

$y - 6 = \frac{2}{5}x$

$y = \frac{2}{5}x + 6$

57. $m = \frac{4 - (-2)}{-2 - (-4)} = \frac{6}{2} = 3$

$y - (-2) = 3[x - (-4)]$

$y + 2 = 3(x + 4)$

$y + 2 = 3x + 12$

$y = 3x + 10$

59. $m = \frac{-9 - 9}{6 - (-6)} = \frac{-18}{12} = -\frac{3}{2}$

$y - 9 = -\frac{3}{2}(x - (-6))$

$y - 9 = -\frac{3}{2}(x + 6)$

$y - 9 = -\frac{3}{2}x - 9$

$y = -\frac{3}{2}x$

61. $m = \frac{-2 - 3}{0 - 10} = \frac{-5}{-10} = \frac{1}{2}$

$y - 3 = \frac{1}{2}(x - 10)$

$y - 3 = \frac{1}{2}x - 5$

$y = \frac{1}{2}x - 2$

63. $y - (-4.5) = 6.3(x - 0)$

$y + 4.5 = 6.3x$

$y = 6.3x - 4.5$

65. a. $y = 5x + 60$

b. $y = 5(30) + 60 = 150 + 60 = \$210$

67. a. Use the slope-intercept form.

b. Use the point-slope form.

c. Use the point-slope form but first find the slope.

69. a. No. The equations will look different because different points are used.

b. $y-(-4)=2[x-(-5)]$
$y+4=2(x+5)$

c. $y-10=2(x-2)$

d. $y+4=2(x+5)$
$y+4=2x+10$
$y=2x+6$

e. $y-10=2(x-2)$
$y-10=2x-4$
$y=2x+6$

f. Yes

71. a. Use the points (0,0) and (200,293) and substitute into the slope formula.
$m=\frac{293-0}{200-0}=\frac{293}{200}\approx 1.465$

b. Using the point-slope formula, the equation of the line is:
$y-y_1=m(x-x_1)$
$f-0=1.465(m-0)$
$f=1.465m$

c. $f=1.465m$
$f=1.465(130.81)$
$f\approx 191.64$
The speed was 191.64 feet per second.

d. It is about 150 feet per second.

e. It is about 55 miles per hour.

73. First, find the slope of the line $2x+y=6$.
$2x+y=6$
$y=-2x+6$
$m=-2$
Use m = -2 and b = 4 in the slope-intercept equation.
$y=-2x+4$

75. $3x-4y=6$
$4y=3x-6$
$y=\frac{3}{4}x-\frac{3}{2}$
$m=\frac{3}{4}$
$y-(-1)=\frac{3}{4}[x-(-4)]$
$y+1=\frac{3}{4}(x+4)$
$y+1=\frac{3}{4}x+3$
$y=\frac{3}{4}x+2$

78. $|-9|>|-6|$ because $9>6$.

79. True; the product of two numbers with like signs is a positive number.

80. True; the sum of two negative numbers is a negative number.

81. False; the sum has the sign of the number with the larger absolute value.

82. False; the quotient of two numbers with like signs is a positive number.

83. $4^3=4\cdot 4\cdot 4$
$=16\cdot 4$
$=64$

84. $i=prt$
$\frac{i}{pt}=\frac{prt}{pt}$
$\frac{i}{pt}=r$

Review Exercises

1.

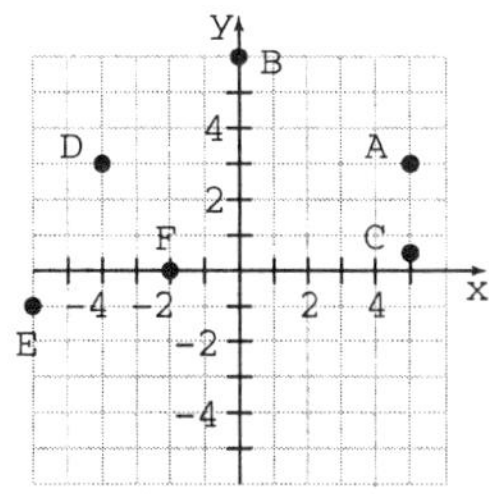

2.

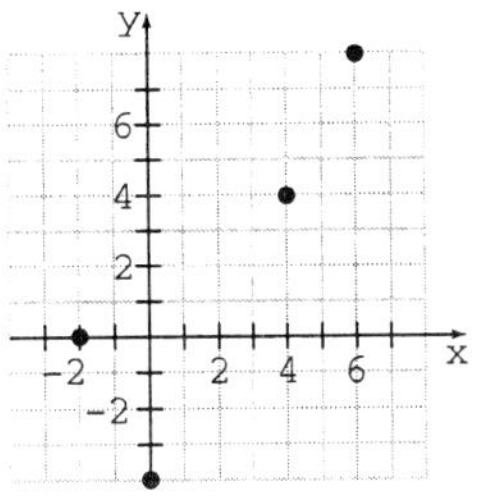

The points are not collinear.

3. a.
$$\begin{aligned} 2x + 3y &= 9 \\ 2\cdot 5 + 3\cdot\left(-\tfrac{1}{3}\right) &= 9 \\ 10 - 1 &= 9 \\ 9 &= 9 \quad \text{True} \end{aligned}$$

b.
$$\begin{aligned} 2x + 3y &= 9 \\ 2\cdot 0 + 3\cdot 3 &= 9 \\ 9 &= 9 \quad \text{True} \end{aligned}$$

4. a.
$$\begin{aligned} 3x - 2y &= 8 \\ 3\cdot 4 - 2y &= 8 \\ 12 - 2y &= 8 \\ -2y &= -4 \\ y &= 2 \end{aligned}$$

b.
$$\begin{aligned} 3x - 2y &= 8 \\ 3\cdot 0 - 2y &= 8 \\ -2y &= 8 \\ y &= -4 \end{aligned}$$

c.
$$\begin{aligned} 3x - 2y &= 8 \\ 3x - 2\cdot 4 &= 8 \\ 3x - 8 &= 8 \\ 3x &= 16 \\ x &= \frac{16}{3} \end{aligned}$$

d.
$$\begin{aligned} 3x - 2y &= 8 \\ 3x - 2\cdot 0 &= 8 \\ 3x &= 8 \\ x &= \frac{8}{3} \end{aligned}$$

5. $y = 4$ is a horizontal line with y-intercept = (0, 4).

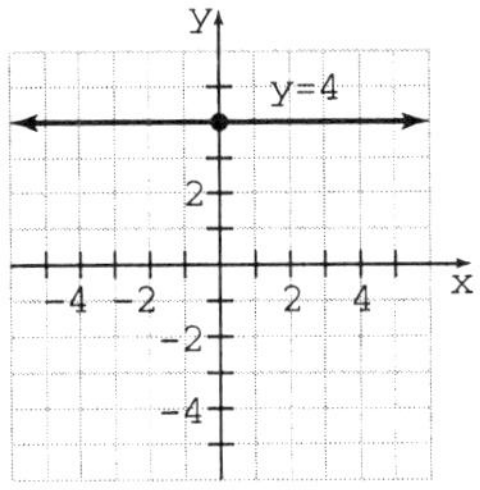

6. $x = 2$ is a vertical line with x-intercept = (2, 0).

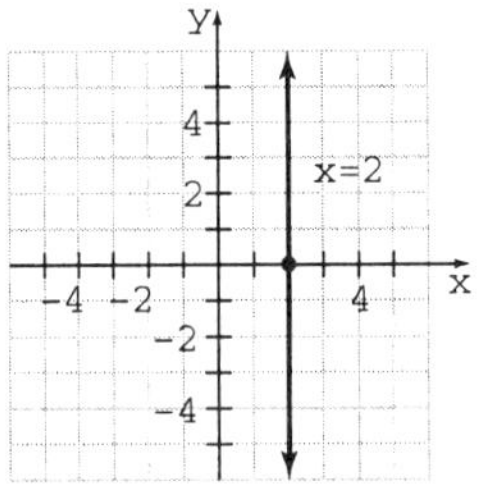

7. Let $x = -1$, $y = 3(-1) = -3$, $(-1, -3)$
Let $x = 0$, $y = 3\cdot 0 = 0$, $(0, 0)$
Let $x = 1$, $y = 3\cdot 1 = 3$, $(1, 3)$

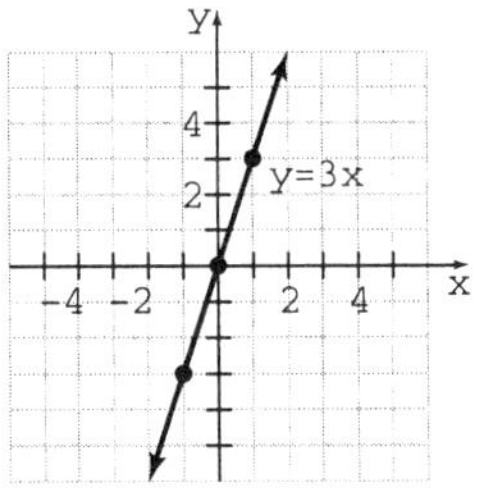

8. Let $x = 0$, $y = 2\cdot 0 - 1 = -1$, $(0, -1)$
Let $x = 1$, $y = 2\cdot 1 - 1 = 1$, $(1, 1)$
Let $x = 2$, $y = 2\cdot 2 - 1 = 3$, $(2, 3)$

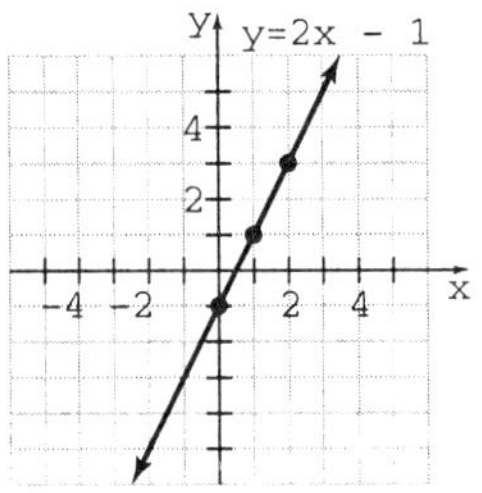

9. Let $x = 0$, $y = -2\cdot 0 + 5 = 5$, $(0, 5)$
Let $x = 1$, $y = -2\cdot 1 + 5 = 3$, $(1, 3)$
Let $x = 2$, $y = -2\cdot 2 + 5 = 1$, $(2, 1)$

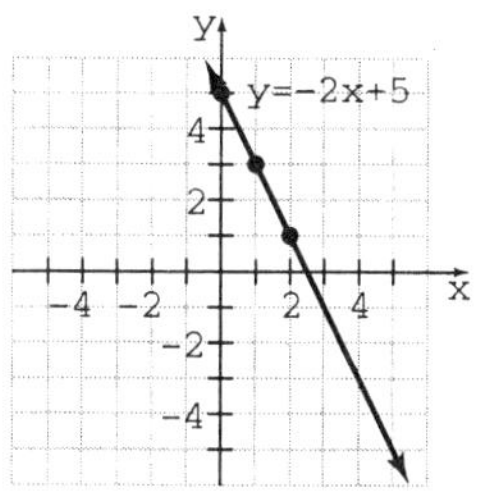

10. Let $x = 0$, $y = -\frac{1}{2}(0) + 4 = 4$, $(0, 4)$

Let $x = 2$, $y = -\frac{1}{2}(2) + 4 = 3$, $(2, 3)$

Let $x = 4$, $y = -\frac{1}{2}(4) + 4 = 2$, $(4, 2)$

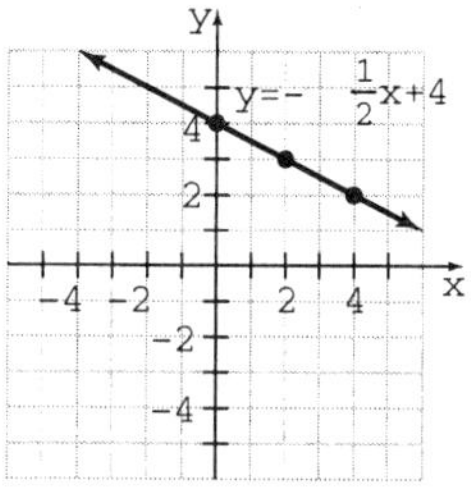

11. Let $x = 0$ Let $y = 0$

$$\begin{aligned} -2x + 3y &= 6 & -2x + 3y &= 6 \\ -2\cdot 0 + 3y &= 6 & -2x + 3\cdot 0 &= 6 \\ 3y &= 6 & -2x &= 6 \\ y &= 2 & x &= -3 \end{aligned}$$

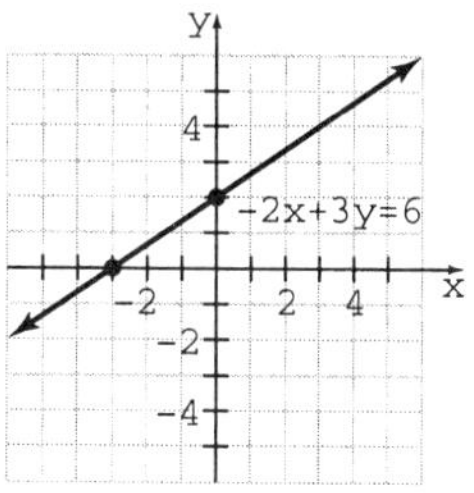

12. Let $x = 0$ Let $y = 0$

$$\begin{aligned} -5x - 2y &= 10 & -5x - 2y &= 10 \\ -5\cdot 0 - 2y &= 10 & -5x - 2\cdot 0 &= 10 \\ -2y &= 10 & -5x &= 10 \\ y &= -5 & x &= -2 \end{aligned}$$

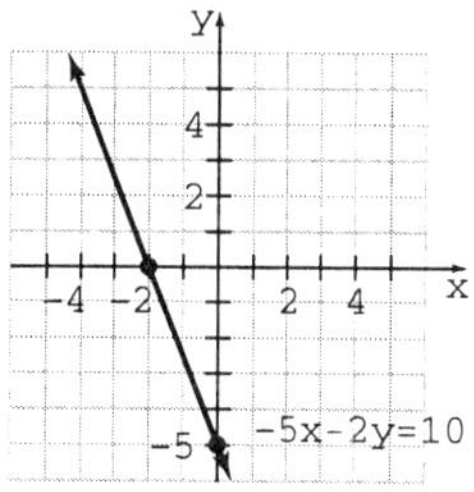

13. Let $x = 0$ Let $y = 0$

$$\begin{aligned} 25x + 50y &= 100 & 25x + 50y &= 100 \\ 25\cdot 0 + 50y &= 100 & 25x + 50\cdot 0 &= 100 \\ 50y &= 100 & 25x &= 100 \\ y &= 2 & x &= 4 \end{aligned}$$

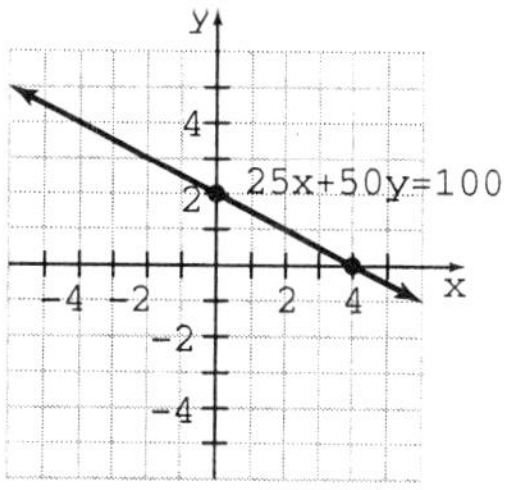

14. Let $x = 0$ Let $y = 0$

$$\begin{aligned} \frac{2}{3}x &= \frac{1}{4}y + 20 & \frac{2}{3}x &= \frac{1}{4}y + 20 \\ \frac{2}{3}\cdot 0 &= \frac{1}{4}y + 20 & \frac{2}{3}x &= \frac{1}{4}\cdot 0 + 20 \\ 0 &= \frac{1}{4}y + 20 & \frac{2}{3}x &= 20 \\ -\frac{1}{4}y &= 20 & x &= 30 \\ y &= -80 & & \end{aligned}$$

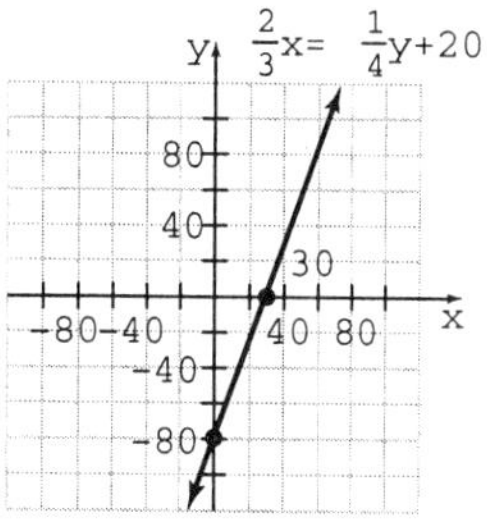

15. $m = \frac{5-(-4)}{-2-3}$

$= \frac{9}{-5}$

$= -\frac{9}{5}$

16. $m = \dfrac{-3-(-2)}{8-(-4)}$
$= \dfrac{-1}{12}$
$= -\dfrac{1}{12}$

17. $m = \dfrac{3-(-1)}{-4-(-2)}$
$= \dfrac{4}{-2}$
$= -2$

18. The slope of a horizontal line is 0.

19. The slope of a vertical line is undefined.

20. The slope of a straight line is the ratio of the vertical change to the horizontal change between any two points on the line.

21. $m = \dfrac{-6}{3}$
$= -2$

22. $m = \dfrac{2}{8}$
$= \dfrac{1}{4}$

23. Neither. For the lines to be parallel the slopes have to be the same. For the lines to be perpendicular the slopes have to be opposite reciprocals.

24. Perpendicular, because the slopes are opposite reciprocals.

25. a. $m = \dfrac{201-145}{1995-1993}$
$= \dfrac{56}{2}$
$= 28$

b. $m = \dfrac{415-268}{1996-1999}$
$= \dfrac{147}{-3}$
$= -49$

26. $6x + 7y = 14$
$7y = -6x + 14$
$y = -\dfrac{6}{7}x + \dfrac{14}{7}$
$y = -\dfrac{6}{7}x + 2$
$m = -\dfrac{6}{7}, b = 2$

The slope is $-\dfrac{6}{7}$; the y-intercept is (0, 2).

27. $2x + 5 = 0$
$2x = -5$
$x = -\dfrac{5}{2}$

This is a vertical line, so the slope is undefined and there is no y-intercept.

28. $3y + 9 = 0$
$3y = -9$
$y = -3$

This is a horizontal line, so the slope is 0 and the y-intercept is (0, –3).

29. $m = \dfrac{3}{1} = 3, b = -3$
$y = 3x - 3$

30. $m = \dfrac{-2}{4} = -\dfrac{1}{2}, b = 2$
$y = -\dfrac{1}{2}x + 2$

31. $y = 2x - 6 \quad 6y = 12x + 6$
$y = 2x + 1$

Since the slopes are the same and the y-intercepts are different, the lines are parallel.

32. $2x - 3y = 9 \qquad 3x + 2y = 6$
$-3y = -2x + 9 \qquad +2y = -3x + 6$
$y = \dfrac{2}{3}x - 3 \qquad y = -\dfrac{3}{2}x + 3$

Since the slopes are opposite reciprocals, the lines are perpendicular.

33. $y - 4 = 3(x - 2)$
$y - 4 = 3x - 6$
$y = 3x - 2$

34. $y-2=-\frac{2}{3}(x-3)$
$y-2=-\frac{2}{3}x+2$
$y=-\frac{2}{3}x+4$

35. $y-2=0(x-4)$
$y-2=0$
$y=2$

36. Lines with undefined slopes are vertical and have the form $x = c$ where c is the value of x for any point on the line.
$x = 4$

37. $m=\frac{-4-3}{0-(-2)}=\frac{-7}{2}=-\frac{7}{2}$
$y-3=-\frac{7}{2}[x-(-2)]$
$y-3=-\frac{7}{2}(x+2)$
$y-3=-\frac{7}{2}x-7$
$y=-\frac{7}{2}x-4$

38. $m=\frac{3-(-2)}{-4-(-4)}=\frac{5}{0}=$ undefined
Lines with undefined slopes are vertical and have the form $x = c$ where c is the value of x for any point on the line.
$x = -4$

Practice Test

1. A graph is an illustration of the set of points that satisfy an equation.

2. a. IV

b. III

3. A (0, –2), B (–2, 5), C (5, 0), D (–4, –1), E (3, –4)

4. a. $ax + by = c$

b. $y = mx + b$

c. $y-y_1=m(x-x_1)$

5. a. $3y=5x-9$
$3\cdot 2=5\cdot 4-9$
$6=20-9$
$6=11$ False

b. $3y=5x-9$
$3\cdot 0=5\cdot\frac{9}{5}-9$
$0=9-9$
$0=0$ True

c. $3y=5x-9$
$3(-6)=5(-2)-9$
$-18=-10-9$
$-18=-19$ False

d. $3y=5x-9$
$3\cdot(-3)=5\cdot 0-9$
$-9=-9$ True

$\left(\frac{9}{5}, 0\right)$ and (0, –3) satisfy the equation.

6. To find y, substitute $\frac{1}{3}$ for x in the equation $3x - 4y = 9$.
$3x-4y=9$
$3\left(\frac{1}{3}\right)-4y=9$
$1-4y=9$
$1-1-4y=9-1$
$-4y=8$
$\frac{-4y}{-4}=\frac{8}{-4}$
$y=-2$

7. $m=\frac{3-(-5)}{-4-2}=\frac{8}{-6}=-\frac{4}{3}$

8. $m=\frac{-8-7}{3-3}=\frac{-15}{0}=$ undefined

9. $4x-9y=15$
$-9y=-4x+15$
$y=\frac{4}{9}x-\frac{5}{3}$
$m=\frac{4}{9}, b=-\frac{5}{3}$
The slope is $\frac{4}{9}$; the y-intercept is $\left(0, -\frac{5}{3}\right)$

10. $4x - 9y = 15$

$-9y = -4x + 15$

$y = \frac{4}{9}x - \frac{5}{3}$

$m = \frac{4}{9},\ b = -\frac{5}{3}$

The slope is $\frac{4}{9}$; the y-intercept is $\left(0,\ -\frac{5}{3}\right)$.

11. $m = \frac{-1}{1} = -1,\ b = -1$

$y = -x - 1$

12. $y = -4$

13. $x = -3$ is a vertical line with x-intercept = (–3, 0).

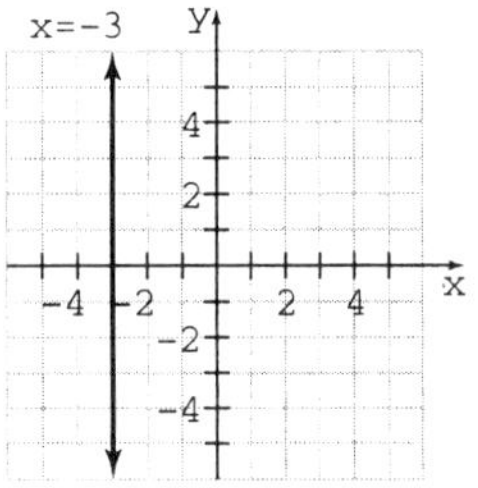

14. $y = 2$ is a horizontal line with y-intercept = (0, 2).

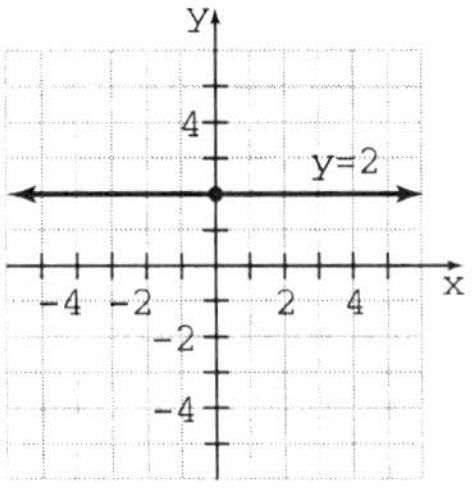

15. Let $x = 0,\ y = 3 \cdot 0 - 2 = -2$, (0, –2)
Let $x = 1,\ y = 3 \cdot 1 - 2 = 1$, (1, 1)
Let $x = 2,\ y = 3 \cdot 2 - 2 = 4$, (2, 4)

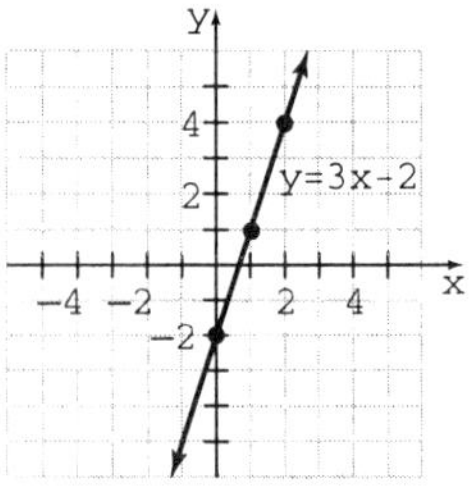

16. a. $2x - 4y = 8$

$-4y = -2x + 8$

$y = \frac{1}{2}x - 2$

b. Let $x = 0,\ y = \frac{1}{2}(0) - 2 = -2$, (0, –2)

Let $x = 2,\ y = \frac{1}{2}(2) - 2 = -1$, (2, –1)

Let $x = 4,\ y = \frac{1}{2}(4) - 2 = 0$, (4, 0)

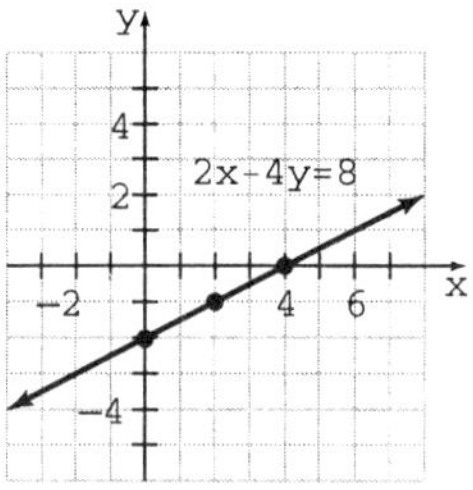

17. To find the x-intercept, set $y = 0$ and solve for x.

$4x - 2y = 8$

$4x - 2(0) = 8$

$4x - 0 = 8$

$4x = 8$

$\frac{4x}{4} = \frac{8}{4}$

$x = 2$

The x-intercept is (2, 0).

18. $3x + 5y = 15$

Let $x = 0$: $3(0) + 5y = 15$, $5y = 15$, $y = 3$

Let $y = 0$: $3x + 5(0) = 15$, $3x = 15$, $x = 5$

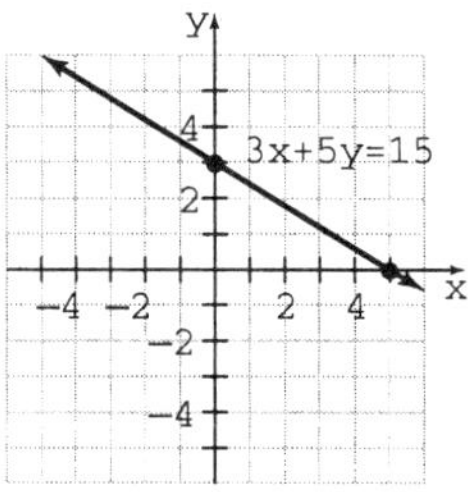

19. $m = \frac{2}{3}$ and $b = -7$

$y = mx + b$

$y = \frac{2}{3}x - 7$

20. $y-(-5)=4(x-2)$

$y+5=4x-8$

$y=4x-13$

21. $m=\dfrac{2-(-1)}{-4-3}=\dfrac{3}{-7}=-\dfrac{3}{7}$

$y-(-1)=-\dfrac{3}{7}(x-3)$

$y+1=-\dfrac{3}{7}x+\dfrac{9}{7}$

$y=-\dfrac{3}{7}x+\dfrac{2}{7}$

22. $2y=3x-6 \qquad y-\dfrac{3}{2}x=-5$

$y=\dfrac{3}{2}x-3 \qquad y=\dfrac{3}{2}x-5$

The lines are parallel since they have the same slope but different y-intercepts.

23. slope = 3, y intercept is (0, –4).

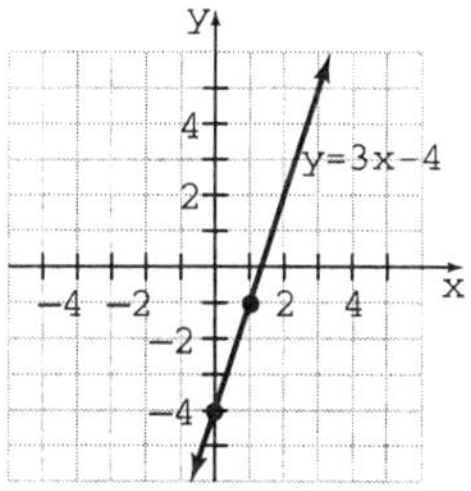

24. $4x-2y=6$

$-2y=-4x+6$

$y=2x-3$

Slope = 2, y intercept is (0, –3).

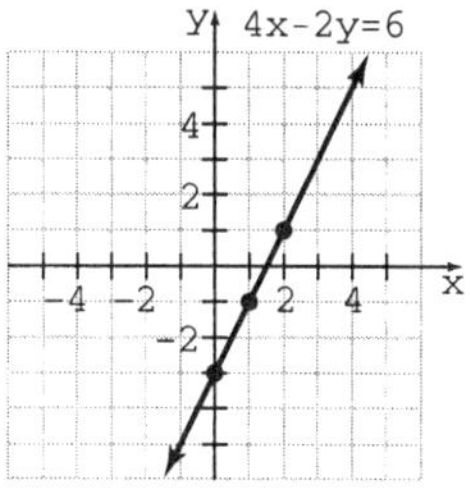

25. a.

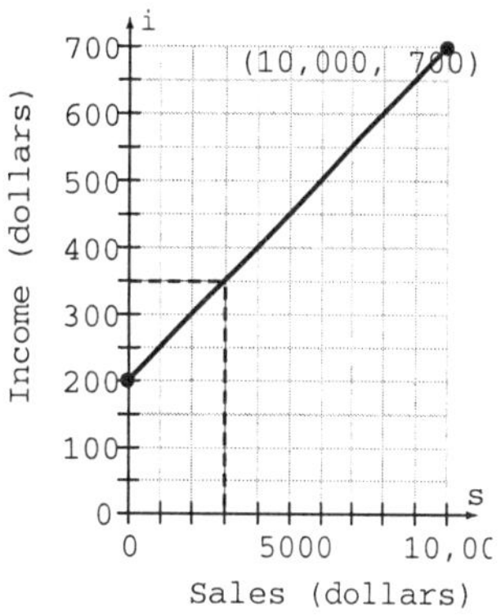

b. \$350

Cumulative Review Test

1. a. $\{1, 2, 3, \ldots\}$

b. $\{0, 1, 2, 3, \ldots\}$

2. $4-9-(-10)+13=4-9+10+13$

$=-5+10+13$

$=5+13$

$=18$

3. $(8\div 4)^3+9^2\div 3=2^3+9^2\div 3$

$=8+81\div 3$

$=8+27$

$=35$

4. $(10\div 5\cdot 5+5-5)^2$

$(2\cdot 5+5-5)^2$

$(10+5-5)^2$

$(15-5)^2$

10^2

100

5. a. Distributive Property

b. Commutative Property of Addition

6. $2x+5=3(x-5)$

$2x+5=3x-15$

$-x+5=-15$

$-x=-20$

$x=20$

7. $3(x-2)-(x+4)=2x-10$

$3x-6-x-4=2x-10$

$2x-10=2x-10$

$0=0$

All real numbers are solutions.

8. $\frac{2}{20}=\frac{x}{200}$

$2(200)=20(x)$

$400=20x$

$\frac{400}{20}=\frac{20x}{20}$

$20=x$

9. $\frac{1 \text{ gallon}}{825 \text{ square feet}}=\frac{x \text{ gallons}}{5775 \text{ square feet}}$

$\frac{1}{825}=\frac{x}{5775}$

$825x=5775$

$\frac{825x}{825}=\frac{5775}{825}$

$x=7$

It will take 7 gallons of paint.

10. $v=lwh$

$\frac{v}{lh}=\frac{lwh}{lh}$

$\frac{v}{lh}=w$

11. Let n = the number

$11+2n=19$

$11-11+2n=19-11$

$2n=8$

$\frac{2n}{2}=\frac{8}{2}$

$n=4$

The number is 4.

12. Let n = the first integer. Then $n + 2$ is the next even integer. The sum of the two consecutive even integers is $n + (n + 2) = 2n + 2$

$2[2n+2]=20$

$2(2n)+2(2)=20$

$4n+4=20$

$4n+4-4=20-4$

$4n=16$

$\frac{4n}{4}=\frac{16}{4}$

$n=4$

The first number is 4 and the second number is $n+2=4+2=6$.

13. Let x = number of hours until the runners are 28 miles apart.

Runner	Rate	Time	Distance
First	6 mph	x	$6x$
Second	8 mph	x	$8x$

(Distance run by first runner) + (Distance run by second runner) = 28 miles

$6x+8x=28$

$14x=28$

$x=2$

It will take 2 hours.

14. Answers will vary.

15. Let $x=0$, $y=f(0)=3\cdot 0-5=-5$, $(0, -5)$

Let $x=1$, $y=f(1)=3\cdot 1-5=-2$, $(1, -2)$

Let $x=2$, $y=f(2)=3\cdot 2-5=1$, $(2, 1)$

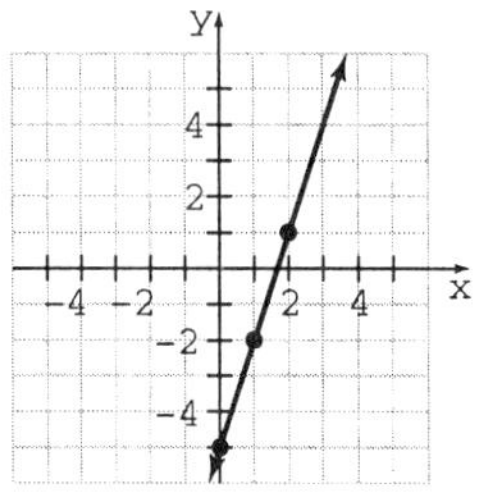

16. $6x - 3y = -12$

Let $x = 0$ Let $y = 0$

$6(0) - 3y = -12$ $6x - 3(0) = -12$

$-3y = -12$ $6x = -12$

$y = 4$ $x = -2$

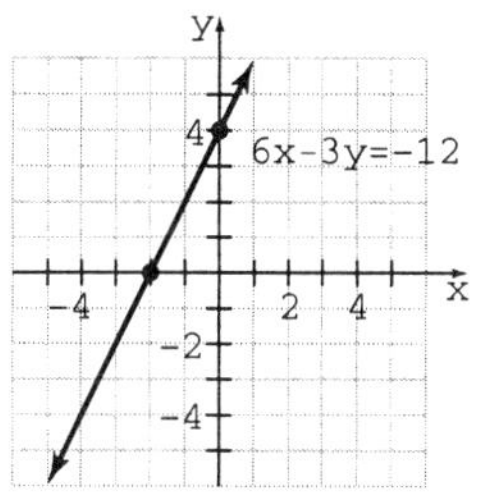

17. $m = \dfrac{-5-(-6)}{6-5} = \dfrac{-5+6}{1} = 1$

18. Parallel lines have the same slope. The slope will be –7.

19. Slope $= \dfrac{2}{3}$, y intercept is $(0, -3)$

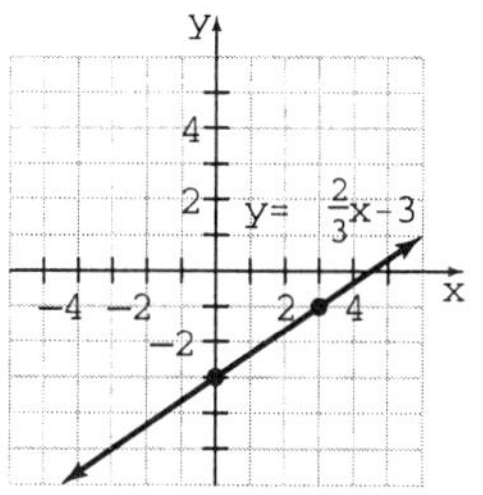

20. $y - 2 = 3(x - 5)$

Chapter 5

Exercise Set 5.1

1. In the expression c^r, c is the base, r is the exponent.

3. a. $\dfrac{x^m}{x^n} = x^{m-n}, x \neq 0$

   b. Answers will vary.

5. a. $\left(x^m\right)^n = x^{m \cdot n}$

   b. Answers will vary.

7. $x^0 \neq 1$ when $x = 0$.

9. $x^5 \cdot x^4 = x^{5+4} = x^9$

11. $z^4 \cdot z = z^{4+1} = z^5$

13. $3^2 \cdot 3^3 = 3^{2+3} = 3^5 = 243$

15. $y^3 \cdot y^2 = y^{3+2} = y^5$

17. $z^3 \cdot z^5 = z^{3+5} = z^8$

19. $y^6 \cdot y = y^6 \cdot y^1 = y^{6+1} = y^7$

21. $\dfrac{6^2}{6} = \dfrac{6^2}{6^1} = 6^{2-1} = 6^1 = 6$

23. $\dfrac{x^{10}}{x^3} = x^{10-3} = x^7$

25. $\dfrac{3^5}{3^2} = 3^{5-2} = 3^3 = 27$

27. $\dfrac{y^4}{y^6} = \dfrac{y^4}{y^4 \cdot y^2} = \dfrac{1}{1 \cdot y^2} = \dfrac{1}{y^2}$

29. $\dfrac{c^4}{c^4} = c^{4-4} = c^0 = 1$

31. $\dfrac{a^3}{a^7} = \dfrac{a^3}{a^3 \cdot a^4} = \dfrac{1}{1 \cdot a^4} = \dfrac{1}{a^4}$

33. $x^0 = 1$

35. $3x^0 = 3 \cdot 1 = 3$

37. $4(5d)^0 = 4(5^0 d^0) = 4(1 \cdot 1) = 4 \cdot 1 = 4$

39. $$\begin{aligned} -3(-4x)^0 &= -3(-4)^0 \cdot x^0 \\ &= -3(1 \cdot 1) \\ &= -3(1) \\ &= -3 \end{aligned}$$

41. $5x^3yz^0 = 5x^3y(1) = 5x^3y$

43. $-5r(st)^0 = -5rs^0t^0 = -5r \cdot 1 \cdot 1 = -5r$

45. $\left(x^4\right)^2 = x^{4 \cdot 2} = x^8$

47. $\left(x^5\right)^5 = x^{5 \cdot 5} = x^{25}$

49. $\left(x^3\right)^1 = x^{3 \cdot 1} = x^3$

51. $\left(x^4\right)^3 = x^{4 \cdot 3} = x^{12}$

53. $\left(n^6\right)^3 = n^{6 \cdot 3} = n^{18}$

55. $(1.3x)^2 = (1.3)^2 x^2 = 1.69x^2$

57. $\left(-3x^3\right)^3 = (-3)^3 x^{3 \cdot 3} = (-27)x^9$

59. $\left(3a^2b^4\right)^3 = 3^3 \cdot a^{2 \cdot 3} b^{4 \cdot 3} = 27a^6b^{12}$

61. $\left(\dfrac{x}{3}\right)^2 = \dfrac{x^2}{3^2} = \dfrac{x^2}{9}$

63. $\left(\dfrac{y}{x}\right)^4 = \dfrac{y^4}{x^4}$

65. $\left(\dfrac{6}{x}\right)^3 = \dfrac{6^3}{x^3} = \dfrac{216}{x^3}$

67. $\left(\dfrac{3x}{y}\right)^3 = \dfrac{3^3x^3}{y^3} = \dfrac{27x^3}{y^3}$

69. $\left(\dfrac{4p}{5}\right)^2 = \dfrac{4^2p^2}{5^2} = \dfrac{16p^2}{25}$

71. $\left(\frac{2y^3}{x}\right)^4 = \frac{2^4 y^{3\cdot 4}}{x^4} = \frac{16y^{12}}{x^4}$

73. $\frac{x^6 y}{xy^3} = \frac{x \cdot x^5 \cdot y}{x \cdot y \cdot y^2} = \frac{x^5}{y^2}$

75. $\frac{10x^3y^8}{2xy^{10}} = \frac{2\cdot 5 \cdot x \cdot x^2 \cdot y^8}{2 \cdot x \cdot y^8 \cdot y^2} = \frac{5x^2}{y^2}$

77. $\frac{3ab}{27a^3b^4} = \frac{3\cdot a \cdot b}{3\cdot 9 \cdot a \cdot a^2 \cdot b \cdot b^3} = \frac{1}{9a^2b^3}$

79. $\frac{35x^4y^9}{15x^9y^{12}} = \frac{5\cdot 7 \cdot x^4 \cdot y^9}{5\cdot 3 \cdot x^4 \cdot x^5 \cdot y^9 \cdot y^3} = \frac{7}{3x^5y^3}$

81. $\frac{-36xy^7z}{12x^4y^5z} = -\frac{3\cdot 12 \cdot x \cdot y^5 \cdot y^2 \cdot z}{12 \cdot x \cdot x^3 \cdot y^5 \cdot z} = -\frac{3y^2}{x^3}$

83. $-\frac{6x^2y^7z}{3x^5y^9z^6} = -\frac{2\cdot 3 \cdot x^2 \cdot y^7 \cdot z}{3\cdot x^2 \cdot x^3 \cdot y^7 \cdot y^2 \cdot z \cdot z^5}$
$= -\frac{2}{x^3y^2z^5}$

85. $\left(\frac{10x^4}{5x^6}\right)^3 = \left(\frac{10}{5}\cdot\frac{x^4}{x^6}\right)^3$
$= \left(\frac{2}{x^2}\right)^3$
$= \frac{2^3}{x^{2\cdot 3}}$
$= \frac{8}{x^6}$

87. $\left(\frac{6y^6}{2y^3}\right)^3 = \left(\frac{6}{2}\cdot\frac{y^6}{y^3}\right)^3$
$= \left(3y^3\right)^3$
$= 3^3y^{3\cdot 3}$
$= 27y^9$

89. $\left(\frac{9a^2b^4}{3a^2b^9}\right)^0 = 1$

91. $\left(\frac{x^4y^3}{x^2y^5}\right)^2 = \left(\frac{x^4}{x^2}\cdot\frac{y^3}{y^5}\right)^2$
$= \left(\frac{x^2}{y^2}\right)^2$
$= \frac{x^{2\cdot 2}}{y^{2\cdot 2}}$
$= \frac{x^4}{y^4}$

93. $\left(\frac{9y^2z^7}{18y^9z}\right)^4 = \left(\frac{9}{18}\cdot\frac{y^2}{y^9}\cdot\frac{z^7}{z}\right)^4$
$= \left(\frac{z^6}{2y^7}\right)^4$
$= \frac{z^{6\cdot 4}}{2^4y^{7\cdot 4}}$
$= \frac{z^{24}}{16y^{28}}$

95. $\left(\frac{4xy^5}{y}\right)^3 = \left(4x\cdot\frac{y^5}{y}\right)^3$
$= (4xy^4)^3$
$= 4^3x^3y^{4\cdot 3}$
$= 64x^3y^{12}$

97. $\left(5xy^4\right)^2 = 5^2x^2y^{4\cdot 2} = 25x^2y^8$

99. $\left(2x^4y\right)\left(-y^5\right) = -2x^4\left(y^{5+1}\right) = -2x^4y^6$

101. $(-2xy)(3xy) = (-2\cdot 3)\left(x^{1+1}\right)\left(y^{1+1}\right)$
$= -6x^2y^2$

103. $\left(5x^2y\right)\left(3xy^5\right) = (5\cdot 3)\left(x^{2+1}\right)\left(y^{1+5}\right) = 15x^3y^6$

105. $\left(-3p^2q\right)^2\left(-pq\right) = [(-3)^2p^{2\cdot 2}q^2](-p^2q)$
$= (9p^4q^2)(-pq)$
$= (9\cdot -1)(p^{4+2})(q^{2+1})$
$= -9p^6q^3$

107. $\left(5r^3s^2\right)^2\left(5r^3s^4\right)^0 = (5^2\cdot r^{3\cdot 2}s^{2\cdot 2})(1) = 25r^6s^4$

109. $(-x)^2 = (-x)(-x) = x^2$

111. $\left(\frac{x^5y^5}{xy^5}\right)^3 = \left(\frac{x^5}{x}\cdot\frac{y^5}{y^5}\right)^3$
$= \left(x^4\cdot 1\right)^3$
$= \left(x^4\right)^3$
$= x^{4\cdot 3} = x^{12}$

113. $\left(2.5x^3\right)^2 = 2.5^2\cdot x^{3\cdot 2} = 6.25x^6$

115. $\frac{x^7y^2}{xy^6} = \frac{x^7}{x}\cdot\frac{y^2}{y^6} = \frac{x^6}{y^4}$

117. $\left(\frac{-m^4}{n^3}\right)^3 = \frac{(-1)^3m^{4\cdot 3}}{n^{3\cdot 3}} = -\frac{m^{12}}{n^9}$

119. $\left(-6x^3y^2\right)^3 = (-6)^3x^{3\cdot 3}y^{2\cdot 3} = -216x^9y^6$

121. $\left(-2x^4y^2z\right)^3 = (-2)^3x^{4\cdot 3}y^{2\cdot 3}z^{1\cdot 3} = -8x^{12}y^6z^3$

123. $\left(9r^4s^5\right)^3 = 9^3\cdot r^{4\cdot 3}\cdot s^{5\cdot 3} = 729r^{12}s^{15}$

125. $\left(4x^2y\right)\left(3xy^2\right)^3 = \left(4x^2y\right)\left(3^3x^3y^{2\cdot 3}\right)$
$= \left(4x^2y\right)\left(27x^3y^6\right)$
$= 4\cdot 27x^{2+3}y^{1+6}$
$= 108x^5y^7$

127. $\left(7.3x^2y^4\right)^2 = 7.3^2x^{2\cdot 2}y^{4\cdot 2} = 53.29x^4y^8$

129. $\left(x^7y^5\right)\left(xy^2\right)^4 = \left(x^7y^5\right)\left(x^{1\cdot 4}y^{2\cdot 4}\right)$
$= \left(x^7y^5\right)\left(x^4y^8\right)$
$= x^{7+4}y^{5+8}$
$= x^{11}y^{13}$

131. $\left(\frac{-x^4z^7}{x^2z^5}\right)^4 = \left(-1\cdot\frac{x^4}{x^2}\cdot\frac{z^7}{z^5}\right)^4$
$= (-x^2z^2)^4$
$= (-1)^4x^{2\cdot 4}z^{2\cdot 4}$
$= x^8z^8$

133. $\frac{x+y}{x}$ cannot be simplified.

135. $\frac{y^2+3}{y}$ cannot be simplified.

137. $\frac{6yz^4}{yz^2} = 6y^{1-1}\cdot z^{4-2} = 6z^2$

139. $\frac{x}{x+1}$ cannot be simplified.

141. $x^2y = 4^2\cdot 2 = 16\cdot 2 = 32$

143. $(xy)^0 = (2\cdot 4)^0 = 8^0 = 1$

145. The sign will be negative because a negative number with an odd number for an exponent will be negative. This is because $(-1)^m = -1$ when m is odd.

147. Area = Length × Width = $7x\cdot x = 7x^2$

149. Area = (Area of Shape 1) + (Area of Shape 2)
$= x\cdot 3x + x\cdot 2y + xy + xy$
$= 3x^2 + 2xy + 2xy$
$= 3x^2 + 4xy$

151. $\left(\frac{3x^4y^5}{6x^6y^8}\right)^3\left(\frac{9x^7y^8}{3x^3y^5}\right)^2 = \left(\frac{3}{6}\cdot\frac{x^4}{x^6}\cdot\frac{y^5}{y^8}\right)^3\left(\frac{9}{3}\cdot\frac{x^7}{x^3}\cdot\frac{y^8}{y^5}\right)^2$

$= \left(\frac{1}{2x^2y^3}\right)^3\left(\frac{3x^4y^3}{1}\right)^2$

$= \frac{1^3}{2^3x^{2\cdot3}y^{3\cdot3}}\cdot\frac{3^2x^{4\cdot2}y^{3\cdot2}}{1^2}$

$= \frac{9x^8y^6}{8x^6y^9}$

$= \frac{9x^2}{8y^3}$

154. $3^4 \div 3^3 - (5-8) + 7 = 81 \div 27 - (-3) + 7$

$= 3 - (-3) + 7$

$= 6 + 7$

$= 13$

155. $-4(x-3) + 5x - 2 = -4x + 12 + 5x - 2$

$= -4x + 5x + 12 - 2$

$= x + 10$

156. $2(x+4) - 3 = 5x + 4 - 3x + 1$

$2x + 8 - 3 = 2x + 5$

$2x + 5 = 2x + 5$

All real numbers are solutions to this equation.

157. a. $P = 2l + 2w$

$26 = 2(x+5) + 2(x)$

$26 = 2x + 10 + 2x$

$26 = 4x + 10$

$16 = 4x$

$4 = x$

$x + 5 = 4 + 5 = 9$

The sides are 4 and 9 inches long.

b. $P = 2l + 2w$

$P - 2l = 2l + 2w - 2l$

$P - 2l = 2w$

$\frac{P-2l}{2} = w$

158. Slope $= \frac{y_2 - y_1}{x_2 - x_1}$

Let $(x_1, y_1) = (-6, 2)$ and $(x_2, y_2) = (-3, 4)$.

$m = \frac{y_2 - y_1}{x_2 - x_1}$

$= \frac{4-2}{-3-(-6)}$

$= \frac{2}{-3+6}$

$= \frac{2}{3}$

Exercise Set 6.2

1. Answers will vary.

3. No, it is not simplified because of the negative exponent.

$x^5y^{-3} = \frac{x^5}{y^3}$

5. The given simplification is not correct since $\left(y^4\right)^{-3} = y^{4\cdot(-3)} = y^{-12} = \frac{1}{y^{12}}$.

7. a. The numerator has one term, x^5y^2.

b. The factors of the numerator are x^5 and y^2.

9. The sign of the exponent changes when a factor is moved from the numerator to the denominator of a fraction.

11. $x^{-6} = \frac{1}{x^6}$

13. $5^{-1} = \frac{1}{5}$

15. $\frac{1}{x^{-3}} = x^3$

17. $\frac{1}{x^{-1}} = x^1 = x$

19. $\frac{1}{6^{-2}} = 6^2 = 36$

21. $\left(x^{-2}\right)^3 = x^{-2\cdot 3} = x^{-6} = \frac{1}{x^6}$

23. $\left(y^{-5}\right)^4 = y^{-5\cdot 4} = y^{-20} = \frac{1}{y^{20}}$

25. $\left(x^4\right)^{-2} = x^{4(-2)} = x^{-8} = \frac{1}{x^8}$

27. $\left(2^{-2}\right)^{-3} = 2^{(-2)(-3)} = 2^6 = 64$

29. $y^4 \cdot y^{-2} = y^{4+(-2)} = y^2$

31. $x^7 \cdot x^{-5} = x^{7-5} = x^2$

33. $3^{-2} \cdot 3^4 = 3^{-2+4} = 3^2 = 9$

35. $\frac{r^5}{r^6} = r^{5-6} = r^{-1} = \frac{1}{r}$

37. $\frac{p^0}{p^{-3}} = p^{0-(-3)} = p^3$

39. $\frac{x^{-7}}{x^{-3}} = x^{-7-(-3)} = x^{-4} = \frac{1}{x^4}$

41. $\frac{3^2}{3^{-1}} = 3^{2-(-1)} = 3^3 = 27$

43. $3^{-3} = \frac{1}{3^3} = \frac{1}{27}$

45. $\frac{1}{z^{-9}} = z^9$

47. $\left(p^{-4}\right)^{-6} = p^{-4(-6)} = p^{24}$

49. $\left(y^{-2}\right)^{-3} = y^{(-2)(-3)} = y^6$

51. $x^3 \cdot x^{-7} = x^{3-7} = x^{-4} = \frac{1}{x^4}$

53. $x^{-8} \cdot x^{-7} = x^{-8-7} = x^{-15} = \frac{1}{x^{15}}$

55. $-4^{-2} = -\frac{1}{4^2} = -\frac{1}{16}$

57. $-(-4)^{-2} = -\frac{1}{(-4)^2} = -\frac{1}{16}$

59. $(-4)^{-3} = \frac{1}{(-4)^3} = -\frac{1}{64}$

61. $(-6)^{-2} = \frac{1}{(-6)^2} = \frac{1}{36}$

63. $\frac{x^{-5}}{x^5} = x^{-5-5} = x^{-10} = \frac{1}{x^{10}}$

65. $\frac{n^{-5}}{n^{-7}} = n^{-5-(-7)} = n^2$

67. $\frac{2^{-3}}{2^{-3}} = 2^{-3-(-3)} = 2^0 = 1$

69. $\left(2^{-1} + 3^{-1}\right)^0 = 1$

71. $\frac{2}{2^{-5}} = 2^{1-(-5)} = 2^6 = 64$

73. $\left(x^{-4}\right)^{-2} = x^{(-4)(-2)} = x^8$

75. $\left(x^0\right)^{-2} = (1)^{-2} = 1$

77. $2^{-3} \cdot 2 = 2^{-3+1} = 2^{-2} = \frac{1}{2^2} = \frac{1}{4}$

79. $6^{-4} \cdot 6^2 = 6^{-4+2} = 6^{-2} = \frac{1}{6^2} = \frac{1}{36}$

81. $\frac{x^{-1}}{x^{-4}} = x^{-1-(-4)} = x^3$

83. $\left(4^2\right)^{-1} = 4^{2\cdot -1} = 4^{-2} = \frac{1}{4^2} = \frac{1}{16}$

85. $\frac{5}{5^{-2}} = 5^{1-(-2)} = 5^3 = 125$

87. $\frac{3^{-4}}{3^{-2}} = 3^{-4-(-2)} = 3^{-2} = \frac{1}{3^2} = \frac{1}{9}$

89. $\frac{7^{-1}}{7^{-1}} = 7^{-1-(-1)} = 7^0 = 1$

91. $\left(6x^2\right)^{-2} = 6^{-2}x^{2(-2)} = 6^{-2}x^{-4} = \frac{1}{6^2x^4} = \frac{1}{36x^4}$

93. $3x^{-2}y^2 = 3\cdot\frac{1}{x^2}\cdot y^2 = \frac{3y^2}{x^2}$

95. $\left(\frac{1}{2}\right)^{-2} = \left(\frac{2}{1}\right)^2 = 2^2 = 4$

96. $\left(\frac{3}{5}\right)^{-2} = \left(\frac{5}{3}\right)^2 = \frac{5^2}{3^2} = \frac{25}{9}$

97. $\left(\frac{5}{4}\right)^{-3} = \left(\frac{4}{5}\right)^3 = \frac{4^3}{5^3} = \frac{64}{125}$

98. $\left(\frac{3}{5}\right)^{-3} = \left(\frac{5}{3}\right)^3 = \frac{5^3}{3^3} = \frac{125}{27}$

99. $\left(\frac{x^2}{y}\right)^{-2} = \left(\frac{y}{x^2}\right)^2 = \frac{y^2}{x^{2\cdot 2}} = \frac{y^2}{x^4}$

101. $-\left(\frac{r^4}{s}\right)^{-4} = -\left(\frac{s}{r^4}\right)^4 = -\frac{s^4}{r^{4\cdot 4}} = -\frac{s^4}{r^{16}}$

103. $7a^{-3}b^{-4} = 7\cdot\frac{1}{a^3}\cdot\frac{1}{b^4} = \frac{7}{a^3b^4}$

105. $\left(x^5y^{-3}\right)^{-3} = x^{5(-3)}y^{(-3)(-3)} = x^{-15}y^9 = \frac{y^9}{x^{15}}$

107. $\left(4y^{-2}\right)\left(5y^{-3}\right) = 4\cdot 5\cdot y^{-2}\cdot y^{-3} = 20y^{-5} = \frac{20}{y^5}$

109. $4x^4\left(-2x^{-4}\right) = 4\cdot(-2)\cdot x^4\cdot x^{-4} = -8x^0 = -8$

111. $\left(4x^2y\right)\left(3x^3y^{-1}\right) = 4\cdot 3\cdot x^2\cdot x^3\cdot y\cdot y^{-1} = 12x^5$

113. $\left(5y^2\right)\left(4y^{-3}z^5\right) = 5\cdot 4\cdot y^2\cdot y^{-3}\cdot z^5$

$= 20y^{-1}z^5$

$= \frac{20z^5}{y}$

115. $\frac{12c^9}{4c^4} = \frac{12}{4}\cdot c^{9-4} = 3c^5$

117. $\frac{36x^{-4}}{9x^{-2}} = \frac{36}{4}\cdot\frac{x^{-4}}{x^{-2}} = 4\cdot\frac{1}{x^2} = \frac{4}{x^2}$

119. $\frac{3x^4y^{-2}}{6y^3} = \frac{3}{6}\cdot x^4\cdot\frac{y^{-2}}{y^3} = \frac{1}{2}\cdot x^4\cdot\frac{1}{y^5} = \frac{x^4}{2y^5}$

121. $\frac{32x^4y^{-2}}{4x^{-2}y^0} = \left(\frac{32}{4}\right)x^{4-(-2)}y^{-2-0}$

$= 8x^6y^{-2}$

$= \frac{8x^6}{y^2}$

123. $\left(\frac{2x^2y^{-3}}{z}\right)^{-4} = \left(\frac{z}{2x^2y^{-3}}\right)^4$

$= \frac{z^4}{2^4x^{2\cdot 4}y^{-3\cdot 4}}$

$= \frac{z^4}{16x^8y^{-12}}$

$= \frac{y^{12}z^4}{16x^8}$

125. $\left(\frac{2r^{-5}s^9}{t^{12}}\right)^{-4} = \left(\frac{t^{12}}{2r^{-5}s^9}\right)^4$

$= \frac{t^{12\cdot 4}}{2^4r^{-5\cdot 4}s^{9\cdot 4}}$

$= \frac{t^{48}}{16r^{-20}s^{36}}$

$= \frac{r^{20}t^{48}}{16s^{36}}$

127. $\left(\frac{x^3y^{-4}z}{y^{-2}}\right)^{-6} = \left(\frac{y^{-2}}{x^3y^{-4}z}\right)^6$

$= \frac{y^{-2\cdot 6}}{x^{3\cdot 6}y^{-4\cdot 6}z^6}$

$= \frac{y^{-12}}{x^{18}y^{-24}z^6}$

$= \frac{y^{-12-(-24)}}{x^{18}z^6}$

$= \frac{y^{12}}{x^{18}z^6}$

129. $\left(\frac{a^2b^{-2}}{3a^4}\right)^3 = \left(\frac{a^{2-4}b^{-2}}{3}\right)^3$
$= \left(\frac{a^{-2}b^{-2}}{3}\right)^3$
$= \frac{a^{-2\cdot 3}b^{-2\cdot 3}}{3^3}$
$= \frac{a^{-6}b^{-6}}{27}$
$= \frac{1}{27a^6b^6}$

131. a. Yes, $a^{-1}b^{-1} = \frac{1}{a}\cdot\frac{1}{b} = \frac{1}{ab}$.

b. No, $a^{-1} + b^{-1} = \frac{1}{a} + \frac{1}{b} \neq \frac{1}{a+b}$.

133. $4^2 + 4^{-2} = 16 + \frac{1}{4^2} = 16 + \frac{1}{16} = 16\frac{1}{16}$

135. $2^2 + 2^{-2} = 4 + \frac{1}{2^2} = 4 + \frac{1}{4} = 4\frac{1}{4}$

137. $5^0 - 3^{-1} = 1 - \frac{1}{3^1} = 1 - \frac{1}{3} = \frac{2}{3}$

139. $2\cdot 4^{-1} + 4\cdot 3^{-1} = 2\left(\frac{1}{4^1}\right) + 4\left(\frac{1}{3^1}\right)$
$= 2\left(\frac{1}{4}\right) + 4\left(\frac{1}{3}\right)$
$= \frac{2}{4} + \frac{4}{3}$
$= \frac{6}{12} + \frac{16}{12}$
$= \frac{22}{12}$
$= \frac{11}{6}$

141. $2\cdot 4^{-1} - 3^{-1} = 2\cdot\frac{1}{4} - \frac{1}{3}$
$= \frac{2\cdot 1}{4} - \frac{1}{3}$
$= \frac{2}{4} - \frac{1}{3}$
$= \frac{6}{12} - \frac{4}{12}$
$= \frac{2}{12} = \frac{1}{6}$

143. $3\cdot 5^0 - 5\cdot 3^{-2} = 3\cdot 1 - 5\cdot\frac{1}{3^2}$
$= 3 - 5\cdot\frac{1}{9}$
$= 3 - \frac{5}{9}$
$= \frac{27}{9} - \frac{5}{9}$
$= \frac{22}{9}$

145. The missing number is –2 since
$3^{-2} = \frac{1}{3^2} = \frac{1}{9}$.

147. The missing number is –2 since $\frac{1}{3^{-2}} = 3^2 = 9$.

149. The missing number is –2 since
$(x^{-2})^{-2} = x^{(-2)(-2)} = x^4$.

151. The missing number is –3 since
$(x^4)^{-3} = x^{4(-3)} = x^{-12} = \frac{1}{x^{12}}$ and
$(y^{-3})^{-3} = y^{(-3)(-3)} = y^9$.

155. Let x = the number of miles.
$\frac{3\text{ miles}}{48\text{ minutes}} = \frac{x}{80\text{ minutes}}$
$\frac{3}{48} = \frac{x}{80}$
$48x = 240$
$x = 5$
It will sail 5 miles.

156. Substitute 5 for r and 12 for h.

$$\begin{aligned} V &= \pi r^2 h \\ &= \pi(5)^2(12) \\ &= \pi(25)(12) \\ &\approx 942.48 \end{aligned}$$

The volume is about 942.48 cubic inches.

157. Let x = the larger integer.
Then $37 - x$ = the smaller integer.

$$\begin{aligned} x &= 3(37 - x) + 1 \\ x &= 111 - 3x + 1 \\ 4x &= 112 \\ x &= 28 \end{aligned}$$

The numbers are 28 and $37 - 28 = 9$.

158. Let p = price of the item before the Increase.

$$\begin{aligned} p + .20p &= 150 \\ 1.20p &= 150 \\ p &= 125 \end{aligned}$$

The item cost $125 before the increase.

159. Let n = the first integer, then $n + 1$ is the next integer.

$$\begin{aligned} n + (n + 1) &= 75 \\ 2n + 1 &= 75 \\ 2n &= 74 \\ n &= 37 \end{aligned}$$

The integers are 37 and 38.

160. $$\begin{aligned} \left(6xy^5\right)\left(3x^2y^4\right) &= 6 \cdot 3 \cdot x \cdot x^2 \cdot y^5 \cdot y^4 \\ &= 18x^{1+2}y^{5+4} \\ &= 18x^3y^9 \end{aligned}$$

Exercise Set 5.3

1. A number in scientific notation is written as a number greater than or equal to 1 and less than 10 that is multiplied by some power of 10.

b. 42,100 in scientific notation is 4.21×10^4.

3. a. Answers will vary.

b. 0.00568 in scientific notation is 5.68×10^{-3}

5. You will move the decimal point 5 places to the left.

7. The exponent will be negative when the number is less than 1.

9. The exponent will be negative since $0.00734 < 1$.

11. $0.000001 = 1 \times 10^{-6}$

13. $350,000 = 3.5 \times 10^5$

15. $450 = 4.5 \times 10^2$

17. $0.053 = 5.3 \times 10^{-2}$

19. $19,000 = 1.9 \times 10^4$

21. $0.00000186 = 1.86 \times 10^{-6}$

23. $0.00000914 = 9.14 \times 10^{-6}$

25. $220,300 = 2.203 \times 10^5$

27. $.005104 = 5.104 \times 10^{3}$

29. $4.3 \times 10^4 = 43,000$

31. $5.43 \times 10^{-3} = .00543$

33. $2.13 \times 10^{-5} = 0.0000213$

35. $6.25 \times 10^5 = 625,000$

37. $9 \times 10^6 = 9,000,000$

39. $5.35 \times 10^2 = 535$

41. $6.201 \times 10^{-4} = 0.0006201$

43. $1 \times 10^4 = 10,000$

45. 8 micrometers = $8 \times 10^{-6} = 0.000008$ meters.

47. 125 gigawatts = $125 \times 10^9 = 125,000,000,000$ watts.

49. 15.3 km = $15.3 \times 10^3 = 15,300$ meters.

51. 15 micrograms = $15 \times 10^{-6} = 0.000015$ grams.

53. $$\begin{aligned} \left(2 \times 10^2\right)\left(3 \times 10^5\right) &= (2 \times 3)\left(10^2 \times 10^5\right) \\ &= 6 \times 10^7 \\ &= 60,000,000 \end{aligned}$$

55. $$\begin{aligned} \left(2.7 \times 10^{-6}\right)\left(9 \times 10^4\right) &= (2.7 \times 9)\left(10^{-6} \times 10^4\right) \\ &= 24.3 \times 10^{-2} \\ &= 0.243 \end{aligned}$$

57. $\left(1.3\times10^{-8}\right)\left(1.74\times10^{6}\right)=(1.3\times1.74)\left(10^{-8}\times10^{6}\right)$
$=2.262\times10^{-2}$
$=0.02262$

59. $\dfrac{8.4\times10^{6}}{2\times10^{3}}=\left(\dfrac{8.4}{2}\right)\left(\dfrac{10^{6}}{10^{3}}\right)=4.2\times10^{3}=4,200$

61. $\dfrac{7.5\times10^{6}}{3\times10^{3}}=\left(\dfrac{7.5}{3}\right)\left(\dfrac{10^{6}}{10^{3}}\right)=2.5\times10^{3}=2500$

63. $\dfrac{4\times10^{2}}{8\times10^{5}}=\left(\dfrac{4}{8}\right)\left(\dfrac{10^{2}}{10^{5}}\right)=.5\times10^{-3}=0.0005$

65. $(700,000)(6,000,000)=\left(7\times10^{5}\right)\left(6\times10^{6}\right)$
$=(7\times6)\left(10^{5}\times10^{6}\right)$
$=42\times10^{11}$
$=4.2\times10^{12}$

67. $(500,000)(25,000)=\left(5\times10^{5}\right)\left(2.5\times10^{4}\right)$
$=(5\times2.5)\left(10^{5}\times10^{4}\right)$
$=12.5\times10^{9}$
$=1.25\times10^{10}$

69. $\dfrac{1,400,000}{700}=\dfrac{1.4\times10^{6}}{7\times10^{2}}$
$=\left(\dfrac{1.4}{7}\right)\left(\dfrac{10^{6}}{10^{2}}\right)$
$=0.2\times10^{4}$
$=2\times10^{3}$

71. $\dfrac{0.00035}{0.000002}=\dfrac{3.5\times10^{-4}}{2.0\times10^{-6}}$
$=\left(\dfrac{3.5}{2.0}\right)\left(\dfrac{10^{-4}}{10^{-6}}\right)$
$=1.75\times10^{2}$

73. $8.3\times10^{-4}, 3.2\times10^{-1}, 4.6, 4.8\times10^{5}$

75. a. $\left(6.20\times10^{9}\right)-\left(2.81\times10^{8}\right)=\left(62.0\times10^{8}\right)-\left(2.81\times10^{8}\right)$
$=(62.0-2.81)\times10^{8}$
$=59.19\times10$
$=5,919,000,000$

The people that live outside the U.S. total about 5,919,000,000.

b. $\dfrac{6.20\times10^{9}}{2.81\times10^{8}}=\left(\dfrac{6.20}{2.81}\right)\left(\dfrac{10^{9}}{10^{8}}\right)\approx2.21\times10\approx22.1$

The world is about 22.1 times greater than the U.S. population.

77. $\text{Minimum volume} = \left(100{,}000 \text{ ft}^3/\text{sec}\right)(60 \text{ sec}/\text{min})(60 \text{ min}/\text{hr})(24 \text{ hrs})$

$$= \left(1\times10^5\right)\left(6\times10^1\right)\left(6\times10^1\right)\left(2.4\times10^1\right)\text{ft}^3$$
$$= (1\times6\times6\times2.4)\left(10^5\times10^1\times10^1\times10^1\right)\text{ft}^3$$
$$= 86.4\times10^8\text{ft}^3$$
$$= 8{,}640{,}000{,}000 \text{ ft}^3$$

79. $\left(2\times10^{-6}\right)\times\left(8\times10^{12}\right) = (2\times8)\left(10^{-6}\times10^{12}\right)$

$$= 16\times10^6$$
$$= 1.6\times10^7$$

It would take 1.6×10^7 seconds.

81. a. $18 \text{ billion} = 18{,}000{,}000{,}000$

$$= 1.8\times10^{10}$$

b. Distance to moon $= 2.38\times10^5$ miles
7 round trips is 14 one-way lengths.

$$14\left(2.38\times10^5\right) = 33.32\times10^5$$
$$= 3.332\times10^6$$

The length of the diapers placed end to end is 3.332×10^6 or 3,332,000 miles.

b. $\dfrac{4.5\times10^7}{6.9\times10^6} = \left(\dfrac{4.5}{6.9}\right)\left(\dfrac{10^7}{10^6}\right) \approx .65\times10^1 \approx 6.5$

In 2002, the amount spent on incarceration was 6.5 times greater than in 1980.

83. a. $\left(1.05\times10^5\right)-\left(2.23\times10^4\right) = \left(1.05\times10^5\right)-\left(.223\times10^5\right)$

$$= (1.05-.223)\times10^5$$
$$= .827\times10^5$$
$$= 82{,}700$$

The starting salary for an umpire was \$82,700 more than for a referee.

b. $\dfrac{1.05\times10^5}{2.23\times10^4} = \left(\dfrac{1.05}{2.23}\right)\left(\dfrac{10^5}{10^4}\right) = .471\times10^1 = 4.71$

The starting salary for an umpire was 4.71 times greater than for a referee.

85. $\dfrac{9.3\times10^7}{1.86\times10^5} = \left(\dfrac{9.3}{1.86}\right)\left(\dfrac{10^7}{10^5}\right) = 5\times10^2 = 500$ seconds or 8.33 minutes

It takes 500 seconds for light from the sun to reach Earth.

87. a. $2\times\left(5.92\times10^9\right) = 11.84\times10^9$

$$= 1.184\times10^{10}$$

The world's population in 2052 will be about 1.184×10^{10}

b. $(53 \text{ years})(365 \text{ days/year}) = 19{,}345 \text{ days} = 1.9345 \times 10^4$ days
Increase of 5.92×10^9 people
$\frac{5.92 \times 10^9}{1.9345 \times 10^4} \approx 3.06 \times 10^5 = 306{,}000$
About 306,000 people are added per day.

89. a. $27{,}000{,}000 = 2.7 \times 10^7$ is the worldwide production of digital cameras in 2002.

b. 22% of 10,342,000
$(2.2 \times 10^{-1})(1.0342 \times 10^7) = (2.2 \times 1.0342)(10^{-1} \times 10^7) = 2.27524 \times 10^6 = 2{,}275{,}240$
Olympus produced 2,275,240 cameras in 2000.

91. a. In decimal form, 6.02×10^{23} would contain 24 digits.

b.
$$\frac{6.02 \times 10^{23}}{12} = \frac{6.02 \times 10^{23}}{1.2 \times 10^1}$$
$$= \left(\frac{6.02}{1.2}\right)\left(\frac{10^{23}}{10^1}\right)$$
$$\approx 5.02 \times 10^{22}$$

93. Answers will vary.

95. 1 nanosecond = 1×10^{-9} and 1 millisecond = 1×10^{-3}

$$\frac{1 \times 10^{-3}}{1 \times 10^{-9}} = 1 \times 10^{-3-(-9)} = 1 \times 10^6 = 1{,}000{,}000 \text{ times smaller}$$

98.
$$4x^2 + 3x + \frac{x}{2} = 4 \cdot 0^2 + 3 \cdot 0 + \frac{0}{2}$$
$$= 0 + 0 + 0$$
$$= 0$$

99. a. If $x = \frac{3}{2}$, $-x = -\frac{3}{2}$.

b. If $5x = 0$, then $x = 0$.

100.
$$2x - 3(x-2) = x + 2$$
$$2x - 3x + 6 = x + 2$$
$$-x + 6 = x + 2$$
$$4 = 2x$$
$$2 = x$$

101.
$$\left(-\frac{2x^5y^7}{8x^8y^3}\right)^3 = \left(\frac{-2}{8} \cdot \frac{x^5}{x^8} \cdot \frac{y^7}{y^3}\right)^3$$
$$= \left(\frac{-1}{4} \cdot \frac{1}{x^3} \cdot y^4\right)^3$$
$$= \left(-\frac{y^4}{4x^3}\right)^3$$
$$= \frac{(-1)^3 y^{4 \cdot 3}}{4^3 x^{3 \cdot 3}}$$
$$= -\frac{y^{12}}{64x^9}$$

Exercise Set 5.4

1. A polynomial is an expression containing the sum of a finite number of terms of the form ax^n where a is a real number and n is a whole number.

3. a. The exponent on the variable is the degree of the term.

b. The degree of the polynomial is the same as the degree of the highest degree term in the polynomial.

5. Add the exponents on the variable.

7. $(3x+2)-(4x-6)=3x+2-4x+6=-x+8$

9. Because the exponent on the variable in a constant term is 0.

11. a. Answers will vary.

b. $4x^3+5x-7$ will be rewritten as $4x^3+0x^2+5x-7$

13. No, it contains a fractional exponent.

15. Fifth

17. Fourth

19. Seventh

21. Third

23. Tenth

25. Twelfth

27. Binomial

29. Monomial

31. Binomial

33. Monomial

35. Not a polynomial

37. Polynomial

39. Trinomial

41. Polynomial

43. $5x+4$, first

45. x^2-2x-4, second

47. $3x^2+x-8$, second

49. Already in descending order, first

51. $2t^2$, second

53. $4x^3-3x^2+x-4$, third

55. $-2x^4+3x^2+5x-6$, fourth

57. $$\begin{aligned}(5x+4)+(x-5)&=5x+4+x-5\\&=5x+x+4-5\\&=6x-1\end{aligned}$$

59. $$\begin{aligned}(-4x+8)+(2x+3)&=-4x+8+2x+3\\&=-4x+2x+8+3\\&=-2x+11\end{aligned}$$

61. $$\begin{aligned}(t+7)+(-3t-8)&=t-3t+7-8\\&=-2t-1\end{aligned}$$

63. $$\begin{aligned}(x^2+2.6x-3)+(4x+3.8)&=x^2+2.6x-3+4x+3.8\\&=x^2+2.6x+4x-3+3.8\\&=x^2+6.6x+0.8\end{aligned}$$

65. $$\begin{aligned}(4m-3)+(5m^2-4m+7)&=4m-3+5m^2-4m+7\\&=5m^2+4m-4m-3+7\\&=5m^2+4\end{aligned}$$

67. $$\begin{aligned}(2x^2-3x+5)+(-x^2+6x-8)&=2x^2-3x+5-x^2+6x-8\\&=2x^2-x^2+6x-3x+5-8\\&=x^2+3x-3\end{aligned}$$

69. $\left(-x^2-4x+8\right)+\left(5x-2x^2+\frac{1}{2}\right)=-x^2-4x+8+5x-2x^2+\frac{1}{2}$
$=-x^2-2x^2-4x+5x+8+\frac{1}{2}$
$=-3x^2+x+\frac{17}{2}$

71. $\left(8x^2+4\right)+\left(-2.6x^2-5x-2.3\right)=8x^2+4-2.6x^2-5x-2.3$
$=8x^2-2.6x^2-5x+4-2.3$
$=5.4x^2-5x+1.7$

73. $\left(-7x^3-3x^2+4\right)+\left(4x+5x^3-7\right)=-7x^3-3x^2+4+4x+5x^3-7$
$=-7x^3+5x^3-3x^2+4x+4-7$
$=-2x^3-3x^2+4x-3$

75. $\left(8x^2+2xy+4\right)+\left(-x^2-3xy-8\right)=8x^2+2xy+4-x^2-3xy-8$
$=8x^2-x^2+2xy-3xy+4-8$
$=7x^2-xy-4$

77. $\left(2x^2y+2x-3\right)+\left(3x^2y-5x+5\right)=2x^2y+2x-3+3x^2y-5x+5$
$=2x^2y+3x^2y+2x-5x-3+5$
$=5x^2y-3x+2$

79.
$$\begin{array}{r} 3x-6 \\ \underline{4x+5} \\ 7x-1 \end{array}$$

81.
$$\begin{array}{r} 4y^2-2y+4 \\ \underline{3y^2\qquad\ +1} \\ 7y^2-2y+5 \end{array}$$

83.
$$\begin{array}{r} -x^2-3x+3 \\ \underline{5x^2+5x-7} \\ 4x^2+2x-4 \end{array}$$

85.
$$\begin{array}{r} 2x^3+3x^2+6x\ \ -9 \\ \underline{-4x^2\qquad\quad +7} \\ 2x^3-\ x^2+6x\ \ -2 \end{array}$$

87.
$$\begin{array}{r} 4n^3-5n^2+n-6 \\ \underline{-n^3-6n^2-2n+8} \\ 3n^3-11n^2-n+2 \end{array}$$

89. $(4x-4)-(2x+2)=4x-4-2x-2$
$=4x-2x-4-2$
$=2x-6$

91. $(-2x-3)-(-5x-7)=-2x-3+5x+7$
$=-2x+5x-3+7$
$=3x+4$

93. $(-r+5)-(2r+5)=-r+5-2r-5$
$=-r-2r+5-5$
$=-3r$

95. $\left(9x^2+7x-5\right)-\left(3x^2+3.5\right)$
$=9x^2+7x-5-3x^2-3.5$
$=9x^2-3x^2+7x-5-3.5$
$=6x^2+7x-8.5$

97. $\left(5x^2-x-1\right)-\left(-3x^2-2x-5\right)$
$=5x^2-x-1+3x^2+2x+5$
$=5x^2+3x^2-x+2x-1+5$
$=8x^2+x+4$

99. $\left(-6m^2-2m\right)-\left(3m^2-7m+6\right)$
$=-6m^2-3m^2-2m+7m-6$
$=-9m^2+5m-6$

101. $\left(8x^3-2x^2-4x+5\right)-\left(5x^2+8\right)=8x^3-2x^2-4x+5-5x^2-8$
$=8x^3-2x^2-5x^2-4x-8+5$
$=8x^3-7x^2-4x-3$

103. $\left(2x^3-4x^2+5x-7\right)-\left(3x+\frac{3}{5}x^2-5\right)=2x^3-4x^2+5x-7-3x-\frac{3}{5}x^2+5$
$=2x^3-4x^2-\frac{3}{5}x^2+5x-3x-7+5$
$=2x^3-\frac{23}{5}x^2+2x-2$

105. $(8x+2)-(5x+4)=8x+2-5x-4$
$=8x-5x+2-4$
$=3x-2$

107. $\left(2x^2-4x+8\right)-(5x-6)=2x^2-4x+8-5x+6$
$=2x^2-4x-5x+8+6$
$=2x^2-9x+14$

109. $\left(3x^3+5x^2+9x-7\right)-\left(4x^3-6x^2\right)=3x^3+5x^2+9x-7-4x^3+6x^2$
$=3x^3-4x^3+5x^2+6x^2+9x-7$
$=-x^3+11x^2+9x-7$

111.
$$\begin{array}{r}6x+5\\ \underline{-(3x-3)}\end{array}\quad\text{or}\quad\begin{array}{r}6x+5\\ \underline{-3x+3}\\ 3x+8\end{array}$$

113.
$$\begin{array}{r}-6d+8\\ \underline{-(-3d-4)}\end{array}\quad\text{or}\quad\begin{array}{r}-6d+8\\ \underline{3d+4}\\ -3d+12\end{array}$$

115.
$$\begin{array}{r}7x^2-3x-4\\ \underline{-(6x^2\quad\;\; -1)}\end{array}\quad\text{or}\quad\begin{array}{r}7x^2-3x-4\\ \underline{-6x^2+0x+1}\\ x^2-3x-3\end{array}$$

117.
$$\begin{array}{r}m-6\\ \underline{-\left(-5m^2+6m\right)}\end{array}\quad\text{or}\quad\begin{array}{r}m-6\\ \underline{5m^2-6m}\\ 5m^2-5m-6\end{array}$$

119.
$$\begin{array}{r}4x^3-6x^2+7x-9\\ \underline{-\left(x^2+6x-7\right)}\end{array}\quad\text{or}\quad\begin{array}{r}4x^3-6x^2+7x-9\\ \underline{-x^2-6x+7}\\ 4x^3-7x^2+x-2\end{array}$$

121. Answers will vary.

123. Answers will vary.

125. Sometimes

127. Sometimes

129. Answers will vary; one example is: x^5+x^4+x

131. No, all three terms must have degree 5 or 0.

133. $a^2+2ab+b^2$

135. $x^2+xz+yz$

137. $(3x^2-6x+3)-(2x^2-x-6)-(x^2+7x-9) = 3x^2-6x+3-2x^2+x+6-x^2-7x+9$
$= (3x^2-2x^2-x^2)+(-6x+x-7x)+(3+6+9)$
$= -12x+18$

139. $4(x^2+2x-3)-6(2-4x-x^2)-2x(x+2) = 4x^2+8x-12-12+24x+6x^2-2x^2-4x$
$= (4x^2+6x^2-2x^2)+(8x+24x-4x)+(-12-12)$
$= 8x^2+28x-24$

141. $\frac{5}{9}=\frac{2.5}{x}$
$5\cdot x = 9\cdot 2.5$
$5x = 22.5$
$\frac{5x}{5}=\frac{22.5}{5}$
$x = 4.5$

142. $n-5$

143. $3y = 9$
$\frac{3y}{3}=\frac{9}{3}$
$y = 3$

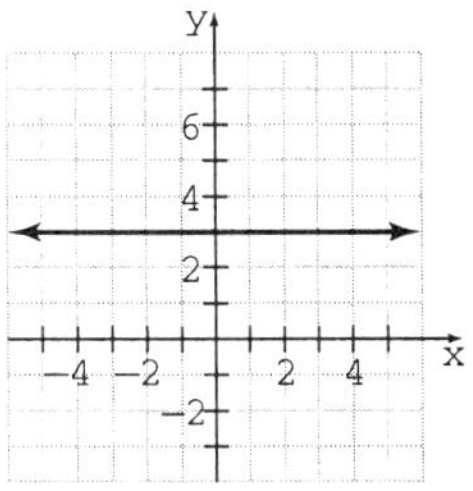

144.

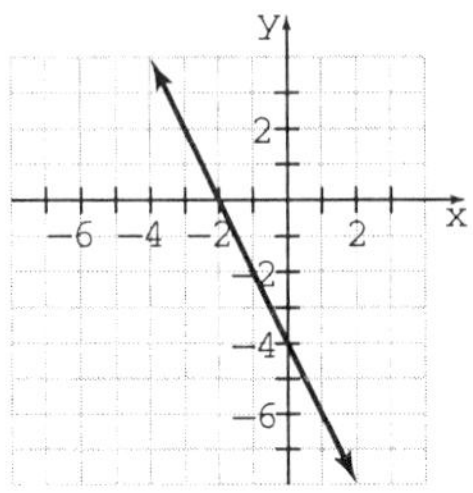

145. $\left(\frac{4x^3y^5}{12x^7y^4}\right)^3 = \left(\frac{4}{12}\cdot\frac{x^3}{x^7}\cdot\frac{y^5}{y^4}\right)^3 = \left(\frac{1}{3}\cdot x^{3-7}\cdot y^{5-4}\right)^3$
$= \left(\frac{1}{3}\cdot x^{-4}\cdot y\right)^3 = \left(\frac{y}{3x^4}\right)^3$
$= \frac{y^3}{3^3x^{4\cdot 3}}$
$= \frac{y^3}{27x^{12}}$

146. 0.035

Exercise Set 5.5

1. The distributive property is used when multiplying a monomial by a polynomial.

3. First, Outer, Inner, Last

5. Yes, FOIL is simply a way to remember the procedure.

7. $(a+b)^2 = a^2+2ab+b^2$
$(a-b)^2 = a^2-2ab+b^2$

9. No, $(x+5)^2 = x^2+10x+25$

11. Answers will vary.

13. Answers will vary.

15. $x^3\cdot 2xy = 2x^{3+1}y = 2x^4y$

17. $5x^3y^5(4x^2y) = (5\cdot 4)x^{3+2}y^{5+1} = 20x^5y^6$

19. $4x^4y^6(-7x^2y^9) = -28x^{4+2}y^{6+9}$
$= -28x^6y^{15}$

21. $9xy^6 \cdot 6x^5y^8 = 9 \cdot 6x^{1+5}y^{6+8}$
$= 54x^6y^{14}$

23. $\left(6x^2y\right)\left(\frac{1}{2}x^4\right) = 6 \cdot \frac{1}{2}x^{2+4}y$
$= 3x^6y$

25. $\left(3.3x^4\right)\left(1.8x^4y^3\right) = (3.3 \cdot 1.8)x^{4+4}y^3$
$= 5.94x^8y^3$

27. $5(x+4) = 5 \cdot x + 5(4) = 5x + 20$

29. $-3x(2x-2) = -3x(2x) - 3x(-2) = -6x^2 + 6x$

31. $-2(8y+5) = (-2)8y + (-2)(5)$
$= -16y - 10$

33. $-2x\left(x^2 - 2x + 5\right) = (-2x)\left(x^2\right) + (-2x)(-2x) + (-2x)(5)$
$= -2x^3 + 4x^2 - 10x$

35. $5x\left(-4x^2 + 6x - 4\right) = 5x\left(-4x^2\right) + 5x(6x) + 5x(-4)$
$= -20x^3 + 30x^2 - 20x$

37. $0.5x^2\left(x^3 - 6x^2 - 1\right) = 0.5x^2(x^3) + 0.5x^2(-6x^2) + 0.5x^2(-1)$
$= 0.5x^5 - 3x^4 - 0.5x^2$

39. $0.3x(2xy + 5x - 6y) = (0.3x)(2xy) + (0.3x)(5x) + (0.3x)(-6y)$
$= 0.6x^2y + 1.5x^2 - 1.8xy$

41. $\left(x^2 - 4y^3 - 3\right)y^4 = x^2 \cdot y^4 + \left(-4y^3\right)y^4 + (-3)y^4$
$= x^2y^4 - 4y^{3+4} - 3y^4$
$= x^2y^4 - 4y^7 - 3y^4$

43. $(x+3)(x+4) = x \cdot x + x \cdot 4 + 3 \cdot x + 3 \cdot 4$
$= x^2 + 4x + 3x + 12$
$= x^2 + 7x + 12$

45. $(2x+5)(3x-6) = (2x)(3x) + 2x(-6) + 5 \cdot 3x + 5(-6)$
$= 6x^2 - 12x + 15x - 30$
$= 6x^2 + 3x - 30$

47. $(2x-4)(2x+4) = (2x)(2x) + (2x)(4) + (-4)(2x) + (-4)(4)$
$= 4x^2 + 8x - 8x - 16$
$= 4x^2 - 16$

49. $(5-3x)(6+2x) = 5 \cdot 6 + 5(2x) + (-3x)(6) + (-3x)(2x)$
$= 30 + 10x - 18x - 6x^2$
$= 30 - 8x - 6x^2$
$= -6x^2 - 8x + 30$

51. $(6x-1)(-2x+5) = 6x(-2x)+(6x)5+(-1)(-2x)+(-1)5$
$= -12x^2 + 30x + 2x - 5$
$= -12x^2 + 32x - 5$

53. $(x-2)(4x-2) = 4x \cdot x - 2 \cdot x - 2 \cdot 4x + (-2)(-2)$
$= 4x^2 - 2x - 8x + 4$
$= 4x^2 - 10x + 4$

55. $(3k-6)(4k-2) = (3k)(4k)+(3k)(-2)+(-6)(4k)+(-6)(-2)$
$= 12k^2 - 6k - 24k + 12$
$= 12k^2 - 30k + 12$

57. $(x-2)(x+2) = x \cdot x + x \cdot 2 + (-2) \cdot x + (-2) \cdot 2$
$= x^2 + 2x - 2x - 4$
$= x^2 - 4$

59. $(2x-3)(2x-3) = 2x \cdot 2x - 3 \cdot 2x - 3 \cdot 2x + (-3)(-3)$
$= 4x^2 - 6x - 6x + 9$
$= 4x^2 - 12x + 9$

61. $(6z-4)(7-z) = 6z \cdot 7 - 6z \cdot z - 4 \cdot 7 - 4(-z)$
$= 42z - 6z^2 - 28 + 4z$
$= -6z^2 + 46z - 28$

63. $(2x+3)(4-2x) = (2x)4 + (2x)(-2x) + 3 \cdot 4 + 3(-2x)$
$= 8x - 4x^2 + 12 - 6x$
$= -4x^2 + 2x + 12$

65. $(x+y)(x-y) = x \cdot x + x(-y) + y \cdot x + y \cdot y$
$= x^2 - xy + xy - y^2$
$= x^2 - y^2$

67. $(2x-3y)(3x+2y) = (2x)(3x)+(2x)(2y)+(-3y)(3x)+(-3y)(2y)$
$= 6x^2 + 4xy - 9xy - 6y^2$
$= 6x^2 - 5xy - 6y^2$

69. $(3x+y)(2+2x) = 3x \cdot 2 + 2x \cdot 3x + 2 \cdot y + 2x \cdot y$
$= 6x + 6x^2 + 2y + 2xy$
$= 6x^2 + 6x + 2xy + 2y$

71. $(x+0.6)(x+0.3) = x \cdot x + x(0.3) + (0.6)x + (0.6)(0.3)$
$= x^2 + 0.3x + 0.6x + 0.18$
$= x^2 + 0.9x + 0.18$

73. $(2y-4)\left(\frac{1}{2}x-1\right) = 2y\cdot\left(\frac{1}{2}x\right) - 1\cdot 2y - 4\left(\frac{1}{2}x\right) + (-4)(-1)$
$= xy - 2y - 2x + 4$
$= xy - 2x - 2y + 4$

75. $(x+6)(x-6) = x^2 - 6^2$
$= x^2 - 36$

77. $(3x-3)(3x+3) = (3x)^2 - 3^2 = 9x - 9$

79. $(x+y)^2 = (x)^2 + 2(x)(y) + (y)^2$
$= x^2 + 2xy + y^2$

81. $(x-0.2)^2 = (x)^2 - 2(x)(0.2) + (0.2)^2$
$= x^2 - 0.4x + 0.04$

83. $(4x+5)(4x+5) = (4x)^2 + 2(4x)(5) + (5)^2 = 16x^2 + 40x + 25$

85. $(0.4x+y)^2 = (0.4x)^2 + 2(0.4x)(y) + (y)^2$
$= 0.16x^2 + 0.8xy + y^2$

87. $(5a-7b)(5a+7b) = (5a)^2 - (7b)^2 = 25a^2 - 49b^2$

89. $(-2x+6)(-2x-6) = (-2x)^2 - 6^2 = 4x^2 - 36$

91. $(7a+2)^2 = (7a)^2 + 2(7a)(2) + 2^2 = 49a^2 + 28a + 4$

93. $(x+4)(3x^2+4x-1) = x(3x^2+4x-1) + 4(3x^2+4x-1)$
$= 3x^3 + 4x^2 - x + 12x^2 + 16x - 4$
$= 3x^3 + 16x^2 + 15x - 4$

95. $(3x+2)(4x^2-x+5) = (3x)(4x^2-x+5) + 2(4x^2-x+5)$
$= 12x^3 - 3x^2 + 15x + 8x^2 - 2x + 10$
$= 12x^3 + 5x^2 + 13x + 10$

97. $(-2x^2-4x+1)(7x-3) = -2x^2(7x-3) - 4x(7x-3) + 1(7x-3)$
$= -14x^3 + 6x^2 - 28x^2 + 12x + 7x - 3$
$= -14x^3 - 22x^2 + 19x - 3$

99. $(-3a+5)(2a^2+4a-3) = -3a(2a^2+4a-3) + 5(2a^2+4a-3)$
$= -6a^3 - 12a^2 + 9a + 10a^2 + 20a - 15$
$= -6a^3 - 2a^2 + 29a - 15$

101. $\left(3x^2-2x+4\right)\left(2x^2+3x+1\right)=3x^2\left(2x^2+3x+1\right)-2x\left(2x^2+3x+1\right)+4\left(2x^2+3x+1\right)$
$=6x^4+9x^3+3x^2-4x^3-6x^2-2x+8x^2+12x+4$
$=6x^4+5x^3+5x^2+10x+4$

103. $\left(x^2-x+3\right)\left(x^2-2x\right)=x^2\left(x^2-2x\right)-x\left(x^2-2x\right)+3\left(x^2-2x\right)$
$=x^4-2x^3-x^3+2x^2+3x^2-6x$
$=x^4-3x^3+5x^2-6x$

105. $(a+b)\left(a^2-ab+b^2\right)=a\left(a^2-ab+b^2\right)+b\left(a^2-ab+b^2\right)$
$=a^3-a^2b+ab^2+a^2b-ab^2+b^3$
$=a^3+b^3$

107. $(x+2)^3=(x+2)(x+2)^2$
$=(x+2)\left(x^2+4x+4\right)$
$=x\left(x^2+4x+4\right)+2\left(x^2+4x+4\right)$
$=x^3+4x^2+4x+2x^2+8x+8$
$=x^3+6x^2+12x+8$

109. $(3a-5)^3=(3a-5)(3a-5)^2$
$=(3a-5)\left(9a^2-30a+25\right)$
$=3a\left(9a^2-30a+25\right)-5\left(9a^2-30a+25\right)$
$=27a^3-90a^2+75a-45a^2+150a-125$
$=27a^3-135a^2+225a-125$

111. No, it will always be a binomial.

113. No, it could have 2 or 4 terms.

115. The missing exponents are 6, 3, and 1 since
$3x^2(2x^6-5x^3+3x^1)=3x^2(2x^6)-3x^2(5x^3)+3x^2(3x^1)$
$=6x^8-15x^5+9x^3$

117. **a.** $A=(x+2)(2x+1)$
$=x(2x)+x\cdot 1+2\cdot 2x+2\cdot 1$
$=2x^2+x+4x+2$
$=2x^2+5x+2$

b. If $x=4$, $A=2\cdot 4^2+5\cdot 4+2=54$.
The area is 54 square feet.

c. For the rectangle to be a square, all sides must have the same length. Thus, $x+2=2x+1$.

$$x+2=2x+1$$
$$2=x+1$$
$$1=x$$

The rectangle is a square when $x = 1$ foot.

119. a. $a+b$

b. $a+b$

c. Yes

d. $(a+b)^2$

e. Area of small square = $a \cdot a$
Area of larger square = $b \cdot b$
Area of each rectangle = $a \cdot b$

$$(a+b)^2 = a \cdot a + b \cdot b + 2(a \cdot b)$$
$$= a^2 + 2ab + b^2$$

121. $(2x^3 - 6x^2 + 5x - 3)(3x^3 - 6x + 4)$

$= 2x^3(3x^3 - 6x + 4) - 6x^2(3x^3 - 6x + 4) + 5x(3x^3 - 6x + 4) - 3(3x^3 - 6x + 4)$

$= 2x^3(3x^3) + 2x^3(-6x) + 2x^3(4) - 6x^2(3x^3) - 6x^2(-6x) - 6x^2(4) + 5x(3x^3) + 5x(-6x)$
$\quad + 5x(4) - 3(3x^3) - 3(-6x) - 3(4)$

$= 6x^6 - 12x^4 + 8x^3 - 18x^5 + 36x^3 - 24x^2 + 15x^4 - 30x^2 + 20x - 9x^3 + 18x - 12$

$= 6x^6 - 18x^5 + 3x^4 + 35x^3 - 54x^2 + 38x - 12$

123.

$$3(x-4) = 2(x-8) + 5x$$
$$3x - 12 = 2x - 16 + 5x$$
$$3x - 12 = 7x - 16$$
$$3x - 7x - 12 = 7x - 7x - 16$$
$$-4x - 12 = -16$$
$$-4x - 12 + 12 = -16 + 12$$
$$-4x = -4$$
$$\frac{-4x}{-4} = \frac{-4}{-4}$$
$$x = 1$$

124. Let x equal the maximum distance. Then

$$2 + 1.5(x-1) = 20$$
$$2 + 1.5x - 1.5 = 20$$
$$1.5x + 0.5 = 20$$
$$1.5x = 19.5$$
$$x = \frac{19.5}{1.5} = 13$$

Bill can go a maximum distance of 13 miles.

125. $\left(\frac{3xy^4}{6y^6}\right)^4 = \left(\frac{3}{6}\cdot x\cdot\frac{y^4}{y^6}\right)^4$

$= \left(\frac{1}{2}\cdot x\cdot\frac{1}{y^2}\right)^4$

$= \left(\frac{x}{2y^2}\right)^4$

$= \frac{x^4}{2^4 y^{2\cdot 4}}$

$= \frac{x^4}{16y^8}$

126. a. $-6^3 = -\left(6^3\right) = -216$

b. $6^{-3} = \frac{1}{6^3} = \frac{1}{216}$

127. $\left(-x^2 - 6x + 5\right) - \left(5x^2 - 4x - 3\right) = -x^2 - 6x + 5 - 5x^2 + 4x + 3$

$= -x^2 - 5x^2 - 6x + 4x + 5 + 3$

$= -6x^2 - 2x + 8$

Exercise Set 5.6

1. To divide a polynomial by a monomial, divide each term in the polynomial by the monomial.

3. $\frac{y+5}{y} = \frac{y}{y} + \frac{5}{y} = 1 + \frac{5}{y}$

5. Terms should be listed in descending order.

7. $\frac{x^3 - 14x + 15}{x-3} = \frac{x^3 + 0x^2 - 14x + 15}{x-3}$

9. $(x+5)(x-3) - 2 = x^2 + 2x - 15 - 2$

$= x^2 + 2x - 17$

11. $\frac{x^2 + x - 20}{x-4} = x + 5$ or $\frac{x^2 + x - 20}{x+5} = x - 4$

13. $\frac{2x^2 + 5x + 3}{2x+3} = x + 1$ or $\frac{2x^2 + 5x + 3}{x+1} = 2x + 3$

15. $\frac{4x^2 - 9}{2x+3} = 2x - 3$ or $\frac{4x^2 - 9}{2x-3} = 2x + 3$

17. $\frac{3x+6}{3} = \frac{3x}{3} + \frac{6}{3} = x + 2$

19. $\frac{4n+10}{2} = \frac{4n}{2} + \frac{10}{2} = 2n + 5$

21. $\frac{3x+8}{2} = \frac{3x}{2} + \frac{8}{2} = \frac{3}{2}x + 4$

23. $\frac{-6x+4}{2} = \frac{-6x}{2} + \frac{4}{2} = -3x + 2$

25. $\frac{-9x-3}{-3} = 3x + 1$

27. $\frac{2x+16}{4} = \frac{2x}{4} + \frac{16}{4}$

$= \frac{x}{2} + 4$

29. $\frac{4-10w}{-4} = \frac{(-1)(4-10w)}{(-1)(-4)}$

$= \frac{-4+10w}{4}$

$= -\frac{4}{4} + \frac{10w}{4}$

$= -1 + \frac{5}{2}w$

31. $(3x^2 + 6x - 9) \div 3x^2 = \frac{3x^2 + 6x - 9}{3x^2}$

$= \frac{3x^2}{3x^2} + \frac{6x}{3x^2} + \frac{-9}{3x^2}$

$= 1 + \frac{2}{x} - \frac{3}{x^2}$

33. $\dfrac{-4x^5+6x+8}{2x^2}=\dfrac{-4x^5}{2x^2}+\dfrac{6x}{2x^2}+\dfrac{8}{2x^2}$
$=-2x^3+\dfrac{3}{x}+\dfrac{4}{x^2}$

35. $(x^5+3x^4-3)\div x^3=\dfrac{x^5+3x^4-3}{x^3}$
$=\dfrac{x^5}{x^3}+\dfrac{3x^4}{x^3}+\dfrac{-3}{x^3}$
$=x^2+3x-\dfrac{3}{x^3}$

37. $\dfrac{6x^5-4x^4+12x^3-5x^2}{2x^3}=\dfrac{6x^5}{2x^3}-\dfrac{4x^4}{2x^3}+\dfrac{12x^3}{2x^3}-\dfrac{5x^2}{2x^3}$
$=3x^2-2x+6-\dfrac{5}{2x}$

39. $\dfrac{8k^3+6k^2-8}{-4k}=\dfrac{(-1)(8k^3+6k^2-8)}{(-1)(-4k)}$
$=\dfrac{-8k^3-6k^2+8}{4k}$
$=\dfrac{-8k^3}{4k}-\dfrac{6k^2}{4k}+\dfrac{8}{4k}$
$=-2k^2-\dfrac{3}{2}k+\dfrac{2}{k}$

41. $\dfrac{12x^5+3x^4-10x^2-9}{-3x^2}$
$=\dfrac{12x^5}{-3x^2}+\dfrac{3x^4}{-3x^2}-\dfrac{10x^2}{-3x^2}-\dfrac{9}{-3x^2}$
$=-4x^3-x^2+\dfrac{10}{3}+\dfrac{3}{x^2}$

43.
$$\begin{array}{r|l} & x+3 \\ \hline x+1 & x^2+4x+3 \\ & \underline{x^2+x} \\ & \quad 3x+3 \\ & \quad \underline{3x+3} \\ & \qquad 0 \end{array}$$
$\dfrac{x^2+4x+3}{x+1}=x+3$

45.
$$\begin{array}{r|l} & 2x+3 \\ \hline x-6 & 2x^2-9x-18 \\ & \underline{2x^2-12x} \\ & \quad 3x-18 \\ & \quad \underline{3x-18} \\ & \qquad 0 \end{array}$$
$\dfrac{2x^2-9x-18}{x-6}=2x+3$

47.
$$\begin{array}{r|l} & 2x+4 \\ \hline 3x+2 & 6x^2+16x+8 \\ & \underline{6x^2+4x} \\ & \quad 12x+8 \\ & \quad \underline{12x+8} \\ & \qquad 0 \end{array}$$
$\dfrac{6x^2+16x+8}{3x+2}=2x+4$

49. $\dfrac{x^2-16}{-4+x}=\dfrac{x^2+0x-16}{x-4}$
$$\begin{array}{r|l} & x+4 \\ \hline x-4 & x^2+0x-16 \\ & \underline{x^2-4x} \\ & \quad 4x-16 \\ & \quad \underline{4x-16} \\ & \qquad 0 \end{array}$$
$\dfrac{x^2-16}{-4+x}=x+4$

51.
$$\begin{array}{r|l} & x+5 \\ \hline 2x-3 & 2x^2+7x-18 \\ & \underline{2x^2-3x} \\ & \quad 10x-18 \\ & \quad \underline{10x-15} \\ & \qquad -3 \end{array}$$
$(2x^2+7x-18)\div(2x-3)=x+5-\dfrac{3}{2x-3}$

53. $(4a^2 - 25) \div (2a - 5) = (4a^2 + 0a - 25) \div (2a - 5)$

$$\begin{array}{r} 2a + 5 \\ 2a - 5 \overline{)\ 4a^2 + 0a - 25} \\ \underline{4a^2 - 10a} \\ 10a - 25 \\ \underline{10a - 25} \\ 0 \end{array}$$

$$\frac{4a^2 - 25}{2a - 5} = 2a + 5$$

55. $\dfrac{6x + 8x^2 - 25}{4x + 9} = \dfrac{8x^2 + 6x - 25}{4x + 9}$

$$\begin{array}{r} 2x - 3 \\ 4x + 9 \overline{)\ 8x^2 + 6x - 25} \\ \underline{8x^2 + 18x} \\ -12x - 25 \\ \underline{-12x - 27} \\ 2 \end{array}$$

$$\frac{6x + 8x^2 - 25}{4x + 9} = 2x - 3 + \frac{2}{4x + 9}$$

57. $\dfrac{6x + 8x^2 - 12}{2x + 3} = \dfrac{8x^2 + 6x - 12}{2x + 3}$

$$\begin{array}{r} 4x - 3 \\ 2x + 3 \overline{)\ 8x^2 + 6x - 12} \\ \underline{8x^2 + 12x} \\ -6x - 12 \\ \underline{-6x - 9} \\ -3 \end{array}$$

$$\frac{6x + 8x^2 - 12}{2x + 3} = 4x - 3 - \frac{3}{2x + 3}$$

59.

$$\begin{array}{r} 3x^2 - 5 \\ x + 6 \overline{)\ 3x^3 + 18x^2 - 5x - 30} \\ \underline{3x^3 + 18x^2} \\ -5x - 30 \\ \underline{-5x - 30} \\ 0 \end{array}$$

$$\frac{3x^3 + 18x^2 - 5x - 30}{x + 6} = 3x^2 - 5$$

61. $\dfrac{2x^3 - 4x^2 + 12}{x - 2} = \dfrac{2x^3 - 4x^2 + 0x + 12}{x - 2}$

$$\begin{array}{r} 2x \\ x - 2 \overline{)\ 2x^3 - 4x^2 + 0x + 12} \\ \underline{2x^3 - 4x^2} \\ 12 \end{array}$$

$$\frac{2x^3 - 4x^2 + 12}{x - 2} = 2x^2 + \frac{12}{x - 2}$$

63. $(w^3 - 8) \div (w - 3) = (w^3 + 0w^2 + 0w - 8) \div (w - 3)$

$$\begin{array}{r} w^2 + 3w + 9 \\ w - 3 \overline{)\ w^3 + 0w^2 + 0w - 8} \\ \underline{w^3 - 3w^2} \\ 3w^2 + 0w \\ \underline{3w^2 - 9w} \\ 9w - 8 \\ \underline{9w - 27} \\ 19 \end{array}$$

$$(w^3 - 8) \div (w - 3) = w^2 + 3w + 9 + \frac{19}{w - 3}$$

65. $\dfrac{x^3 - 27}{x - 3} = \dfrac{x^3 + 0x^2 + 0x - 27}{x - 3}$

$$\begin{array}{r} x^2 + 3x + 9 \\ x - 3 \overline{)\ x^3 + 0x^2 + 0x - 27} \\ \underline{x^3 - 3x^2} \\ 3x^2 + 0x \\ \underline{3x^2 - 9x} \\ 9x - 27 \\ \underline{9x - 27} \\ 0 \end{array}$$

$$\frac{x^3 - 27}{x - 3} = x^2 + 3x + 9$$

67. $\dfrac{4x^3-5x}{2x-1}=\dfrac{4x^3+0x^2-5x+0}{2x-1}$

$$\begin{array}{r}
2x^2+x-2 \\
2x-1\,\big)\,\overline{4x^3+0x^2-5x+0} \\
\underline{4x^3-2x^2} \\
2x^2-5x \\
\underline{2x^2-x} \\
-4x+0 \\
\underline{4x+2} \\
-2
\end{array}$$

$\dfrac{4x^3-5x}{2x-1}=2x^2+x-2-\dfrac{2}{2x-1}$

69.
$$\begin{array}{r}
-m^2-7m-5 \\
m-1\,\big)\,\overline{-m^3-6m^2+2m-3} \\
\underline{-m^3+m^2} \\
-7m^2+2m \\
\underline{-7m^2+7m} \\
-5m-3 \\
\underline{-5m+5} \\
-8
\end{array}$$

$\dfrac{-m^3-6m^2+2m-3}{m-1}=-m^2-7m-5-\dfrac{8}{m-1}$

71. $\dfrac{9n^3-6n+4}{3n-3}=\dfrac{9n^3+0n^2-6n+4}{3n-3}$

$$\begin{array}{r}
3n^2+3n+1 \\
3n-3\,\big)\,\overline{9n^3+0n^2-6n+4} \\
\underline{9n^3-9n^2} \\
9n^2-6n \\
\underline{9n^2-9n} \\
3n+4 \\
\underline{3n-3} \\
7
\end{array}$$

$\dfrac{9n^3-6n+4}{3n-3}=3n^2+3n+1+\dfrac{7}{3n-3}$

73.
$$\begin{array}{r|rrr}
-1 & 1 & 7 & 6 \\
 & & -1 & -6 \\
\hline
 & 1 & 6 & 0
\end{array}$$

Thus, $\dfrac{x^2+7x+6}{x+1}=x+6$

75.
$$\begin{array}{r|rrr}
-2 & 1 & 5 & 6 \\
 & & -2 & -6 \\
\hline
 & 1 & 3 & 0
\end{array}$$

Thus, $\dfrac{x^2+5x+6}{x+2}=x+3$

77.
$$\begin{array}{r|rrr}
4 & 1 & -11 & 28 \\
 & & 4 & -28 \\
\hline
 & 1 & -7 & 0
\end{array}$$

Thus, $\dfrac{x^2-11x+28}{x-4}=x-7$

79.
$$\begin{array}{r|rrr}
3 & 1 & 5 & -12 \\
 & & 3 & 24 \\
\hline
 & 1 & 8 & 12
\end{array}$$

Thus, $\dfrac{x^2+5x-12}{x-3}=x+8+\dfrac{12}{x-3}$

81.
$$\begin{array}{r|rrr}
4 & 3 & -7 & -10 \\
 & & 12 & 20 \\
\hline
 & 3 & 5 & 10
\end{array}$$

Thus, $\dfrac{3x^2-7x-10}{x-4}=3x+5+\dfrac{10}{x-4}$

83.
$$\begin{array}{r|rrrr}
1 & 4 & -3 & 2 & 0 \\
 & & 4 & 1 & 3 \\
\hline
 & 4 & 1 & 3 & 3
\end{array}$$

Thus, $\dfrac{4x^3-3x^2+2x}{x-1}=4x^2+x+3+\dfrac{3}{x-1}$

85.
$$\begin{array}{r|rrrr}
-3 & 3 & 7 & -4 & 18 \\
 & & -9 & 6 & -6 \\
\hline
 & 3 & -2 & 2 & 12
\end{array}$$

Thus, $\dfrac{3c^3+7c^2-4c+18}{c+3}=3c^2-2c+2+\dfrac{12}{c+3}$

87.
$$\begin{array}{r|rrrrr}
1 & 1 & 0 & 0 & 0 & -1 \\
 & & 1 & 1 & 1 & 1 \\
\hline
 & 1 & 1 & 1 & 1 & 0
\end{array}$$

Thus, $\dfrac{y^4-1}{y-1}=y^3+y^2+y+1$

89.

–4	1	0	0	0	16
		–4	16	–64	256
	1	–4	16	–64	272

Thus, $\frac{x^4+16}{x+4} = x^3 - 4x^2 + 16x - 64 + \frac{272}{x+4}$

91.

–1	1	1	0	0	0	–10
		–1	0	0	0	0
	1	0	0	0	0	–10

Thus, $\frac{x^5+x^4-10}{x+1} = x^4 - \frac{10}{x+1}$

93.

–1	1	4	0	0	0	–12
		–1	–3	3	–3	3
	1	3	–3	3	–3	–9

Thus,
$\frac{b^5+4b^4-10}{b+1} = b^4 + 3b^3 - 3b^2 + 3b - 3 - \frac{9}{b+1}$

95.

$\frac{1}{3}$	3	2	–4	1
		1	1	–1
	3	3	–3	0

Thus, $\frac{3x^3+2x^2-4x+1}{x-\frac{1}{3}} = 3x^2 + 3x - 3$

97.

$\frac{1}{2}$	2	–1	2	–3	1
		1	0	1	–1
	2	0	2	–2	0

Thus, $\frac{2x^4-x^3+2x^2-3x+1}{x-\frac{1}{2}} = 2x^3 + 2x - 2$

99. To find the remainder, use synthetic division:

2	4	–5	4
		8	6
	4	3	10

The remainder is 10.

101. To find the remainder, use synthetic division:

2	1	–2	4	–8
		2	0	8
	1	0	4	0

The remainder is 0 which means that $x - 2$ is a factor.

103. To find the remainder, use synthetic division.

$\frac{1}{2}$	–2	–6	2	–4
		–1	$-\frac{7}{2}$	$-\frac{3}{4}$
	–2	–7	$-\frac{3}{2}$	$-\frac{19}{4}$ or –4.75

The remainder is $-\frac{19}{4}$ or –4.75.

105. No, $\frac{2x+1}{x^2} = \frac{2x}{x^2} + \frac{1}{x^2} = \frac{2}{x} + \frac{1}{x^2}$

107. $(x+4)(2x+3)+4 = 2x^2 + 3x + 8x + 12 + 4$
$= 2x^2 + 11x + 16$

109. First Degree

111. It has to be $4x$ since that is what must be multiplied with $4x^3$ in the quotient to get $16x^4$ in the dividend.

113. When dividing by $2x^2$, each exponent will decrease by two. So, the shaded areas must be 5, 3, 2, 1, respectively.

115. $\dfrac{4x^3-4x+6}{2x+3}=\dfrac{4x^3+0x^2-4x+6}{2x+3}$

$$\begin{array}{r}
2x^2-3x+\frac{5}{2} \\
2x+3\overline{\big)\,4x^3+0x^2-4x+6} \\
\underline{4x^3+6x^2} \qquad\qquad \\
-6x^2-4x \qquad \\
\underline{-6x^2-9x} \qquad \\
5x+6 \\
\underline{5x+\frac{15}{2}} \\
-\frac{3}{2}
\end{array}$$

$\dfrac{4x^3-4x+6}{2x+3}=2x^2-3x+\dfrac{5}{2}-\dfrac{3}{2(2x+3)}$

117.
$$\begin{array}{r}
-3x+3 \\
-x-3\overline{\big)\,3x^2+6x-10} \\
\underline{3x^2+9x} \qquad \\
-3x-10 \\
\underline{-3x-\ 9} \\
-1
\end{array}$$

$\dfrac{3x^2+6x-10}{-x-3}=-3x+3+\dfrac{1}{x+3}$

120. a. 2 is a natural number.

b. 2 and 0 are whole numbers.

c. $2, -5, 0, \frac{2}{5}, -6.3$, and $-\frac{23}{34}$ are rational numbers.

d. $\sqrt{7}$ and $\sqrt{3}$ are irrational numbers.

e. All of the numbers are real numbers.

121. a. $\dfrac{0}{1}=0$

b. $\dfrac{1}{0}$ is undefined

122. Evaluate expressions in parentheses first, then exponents, followed by multiplications and divisions from left to right, and finally additions and subtractions from left to right.

123.
$$\begin{aligned}
2(x+3)+2x &= x+4 \\
2x+6+2x &= x+4 \\
4x+6 &= x+4 \\
4x &= x-2 \\
3x &= -2 \\
x &= -\frac{2}{3}
\end{aligned}$$

124.

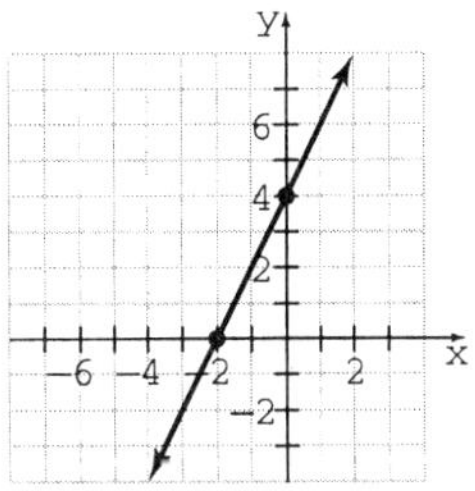

125. $\dfrac{x^7}{x^{-3}}=x^{7-(-3)}=x^{7+3}=x^{10}$

Review Exercises

1. $x^5\cdot x^2=x^{5+2}=x^7$

2. $x^2\cdot x^4=x^{2+4}=x^6$

3. $3^2\cdot 3^3=3^{2+3}=3^5=243$

4. $2^4\cdot 2=2^{4+1}=2^5=32$

5. $\dfrac{x^4}{x}=x^{4-1}=x^3$

6. $\dfrac{a^5}{a^5}=a^{5-5}=a^0=1$

7. $\dfrac{5^5}{5^3}=5^{5-3}=5^2=25$

8. $\dfrac{2^5}{2}=2^{5-1}=2^4=16$

9. $\dfrac{x^6}{x^8}=\dfrac{1}{x^{8-6}}=\dfrac{1}{x^2}$

10. $\dfrac{y^4}{y}=y^{4-1}=y^3$

11. $x^0=1$

12. $4x^0=4\cdot 1=4$

13. $(3x)^0 = 1$

14. $6^0 = 1$

15. $(5x)^2 = 5^2x^2 = 25x^2$

16. $(3a)^3 = 3^3a^3 = 27a^3$

17. $(6s)^3 = 6^3s^3 = 216s^3$

18. $(-3x)^3 = (-3)^3x^3 = -27x^3$

19. $\left(2x^2\right)^4 = 2^4x^{2\cdot 4} = 16x^8$

20. $\left(-x^4\right)^6 = (-1)^6x^{4\cdot 6} = x^{24}$

21. $\left(-m^4\right)^5 = (-1)^5m^{4\cdot 5} = -m^{20}$

22. $\left(\frac{2x^3}{y}\right)^2 = \frac{2^2x^{3\cdot 2}}{y^2} = \frac{4x^6}{y^2}$

23.
$$\left(\frac{5y^2}{2b}\right)^2 = \frac{5^2y^{2\cdot 2}}{2^2b^2}$$
$$= \frac{25y^4}{4b^2}$$

24.
$$6x^2 \cdot 4x^3 = 6\cdot 4x^{2+3}$$
$$= 24x^5$$

25.
$$\frac{16x^2y}{4xy^2} = \frac{16}{4}\cdot\frac{x^2}{x}\cdot\frac{y}{y^2}$$
$$= 4x\frac{1}{y}$$
$$= \frac{4x}{y}$$

26.
$$2x(3xy^3)^2 = 2x(3^2x^2y^{3\cdot 2})$$
$$= 2x(9x^2y^6)$$
$$= 2\cdot 9x^{1+2}y^6$$
$$= 18x^3y^6$$

27.
$$\left(\frac{9x^2y}{3xy}\right)^2 = \left(\frac{9}{3}\cdot\frac{x^2}{x}\cdot\frac{y}{y}\right)^2$$
$$= (3x)^2$$
$$= 3^2x^2$$
$$= 9x^2$$

28.
$$(2x^2y)^3(3xy^4) = (2^3x^{2\cdot 3}y^3)(3xy^4)$$
$$= (8x^6y^3)(3xy^4)$$
$$= 8\cdot 3x^{6+1}y^{3+4}$$
$$= 24x^7y^7$$

29.
$$4x^2y^3\left(2x^3y^4\right)^2 = 4x^2y^3\left(2^2x^{3\cdot 2}y^{4\cdot 2}\right)$$
$$= 4x^2y^3\left(4x^6y^8\right)$$
$$= 4\cdot 4x^{2+6}y^{3+8}$$
$$= 16x^8y^{11}$$

30. $3c^2(2c^4d^3) = 3\cdot 2\cdot c^{2+4}\cdot d^3 = 6c^6d^3$

31.
$$\left(\frac{8x^4y^3}{2xy^5}\right)^2 = \left(\frac{8}{2}\cdot\frac{x^4}{x}\cdot\frac{y^3}{y^5}\right)^2$$
$$= \left(4x^3\cdot\frac{1}{y^2}\right)^2$$
$$= \left(\frac{4x^3}{y^2}\right)^2$$
$$= \frac{4^2x^{3\cdot 2}}{y^{2\cdot 2}}$$
$$= \frac{16x^6}{y^4}$$

32.
$$\left(\frac{21x^4y^3}{7y^2}\right)^3 = \left(\frac{21}{7}\cdot x^4\cdot\frac{y^3}{y^2}\right)^3 = \left(3x^4y\right)^3$$
$$= 3^3x^{4\cdot 3}y^3$$
$$= 27x^{12}y^3$$

33. $x^{-4} = \frac{1}{x^4}$

34. $3^{-3} = \frac{1}{3^3} = \frac{1}{27}$

35. $5^{-2} = \frac{1}{5^2} = \frac{1}{25}$

36. $\frac{1}{z^{-2}} = z^2$

37. $\frac{1}{x^{-7}} = x^7$

38. $\frac{1}{3^{-2}} = 3^2 = 9$

39. $y^5 \cdot y^{-8} = y^{5-8} = y^{-3} = \frac{1}{y^3}$

40. $x^{-2} \cdot x^{-3} = x^{-2-3} = x^{-5} = \frac{1}{x^5}$

41. $p^{-6} \cdot p^4 = p^{-6+4} = p^{-2} = \frac{1}{p^2}$

42. $a^{-2} \cdot a^{-3} = a^{-2+(-3)} = a^{-5} = \frac{1}{a^5}$

43. $\frac{x^3}{x^{-3}} = x^{3-(-3)} = x^6$

44. $\frac{x^5}{x^{-2}} = x^{5-(-2)} = x^7$

45. $\frac{x^{-3}}{x^3} = \frac{1}{x^{3+3}} = \frac{1}{x^6}$

46. $(3x^4)^{-2} = 3^{-2}x^{4(-2)}$

$= 3^{-2}x^{-8}$

$= \frac{1}{3^2x^8}$

$= \frac{1}{9x^8}$

47. $(4x^{-3}y)^{-3} = 4^{-3}x^{(-3)(-3)}y^{-3}$

$= 4^{-3}x^9y^{-3}$

$= \frac{x^9}{4^3y^3}$

$= \frac{x^9}{64y^3}$

48. $(-2m^{-3}n)^2 = (-2)^2m^{-3\cdot 2}n^{1\cdot 2}$

$= 4m^{-6}n^2$

$= \frac{4n^2}{m^6}$

49. $6y^{-2} \cdot 2y^4 = 6 \cdot 2y^{-2+4} = 12y^2$

50. $(5y^{-3}z)^3 = 5^3y^{(-3)3}z^3$

$= 125y^{-9}z^3$

$= \frac{125z^3}{y^9}$

51. $(4x^{-2}y^3)^{-2} = 4^{-2}x^{(-2)(-2)}y^{3(-2)}$

$= 4^{-2}x^4y^{-6}$

$= \frac{x^4}{4^2y^6}$

$= \frac{x^4}{16y^6}$

52. $2x(3x^{-2}) = 2 \cdot 3x^{1-2} = 6x^{-1} = \frac{6}{x}$

53. $(5x^{-2}y)(2x^4y) = 5 \cdot 2x^{-2+4}y^{1+1} = 10x^2y^2$

54. $4x^5(6x^{-7}y^2) = 4 \cdot 6x^{5-7}y^2$

$= 24x^{-2}y^2$

$= \frac{24y^2}{x^2}$

55. $4y^{-2}(3x^2y) = 4 \cdot 3x^2y^{-2+1}$

$= 12x^2y^{-1}$

$= \frac{12x^2}{y}$

56. $\frac{6xy^4}{2xy^{-1}} = \frac{6}{2} \cdot \frac{x}{x} \cdot \frac{y^4}{y^{-1}} = 3y^5$

57. $\frac{12x^{-2}y^3}{3xy^2} = \frac{12}{3} \cdot \frac{x^{-2}}{x} \cdot \frac{y^3}{y^2}$

$= 4 \cdot \frac{1}{x^3} \cdot y$

$= \frac{4y}{x^3}$

58. $\frac{49x^2y^{-3}}{7x^{-3}y} = \frac{49}{7} \cdot \frac{x^2}{x^{-3}} \cdot \frac{y^{-3}}{y}$

$= \left(\frac{49}{7}\right)x^{2-(-3)}y^{-3-1}$

$= 7x^5y^{-4}$

$= \frac{7x^5}{y^4}$

59. $\frac{36x^4y^7}{9x^5y^{-3}} = \frac{36}{9} \cdot \frac{x^4}{x^5} \cdot \frac{y^7}{y^{-3}}$

$= 4 \cdot \frac{1}{x} \cdot y^{10}$

$= \frac{4y^{10}}{x}$

60. $\frac{4x^8y^{-2}}{8x^7y^3} = \frac{4}{8} \cdot \frac{x^8}{x^7} \cdot \frac{y^{-2}}{y^3}$

$= \frac{1}{2} \cdot x \cdot \frac{1}{y^5}$

$= \frac{x}{2y^5}$

61. $1{,}720{,}000 = 1.72 \times 10^6$

62. $0.153 = 1.53 \times 10^{-1}$

63. $0.00763 = 7.63 \times 10^{-3}$

64. $47{,}000 = 4.7 \times 10^4$

65. $4{,}820 = 4.82 \times 10^3$

66. $0.000314 = 3.14 \times 10^{-4}$

67. $8.4 \times 10^{-3} = 0.0084$

68. $6.52 \times 10^{-4} = 0.000652$

69. $9.7 \times 10^5 = 970{,}000$

70. $4.38 \times 10^{-6} = 0.00000438$

71. $3.14 \times 10^{-5} = 0.0000314$

72. $1.103 \times 10^7 = 11{,}030{,}000$

73. 6 gigameters $= 6 \times 10^9 = 6{,}000{,}000{,}000$ meters

74. 92 milliliters $= 92 \times 10^{-3} = 0.092$ liters

75. 19.2 kilograms $= 19.2 \times 10^3 = 19{,}200$ grams

76. 12.8 micrograms $= 12.8 \times 10^{-6}$

$= 0.0000128$ grams

77. $(2.5 \times 10^2)(3.4 \times 10^{-4}) = (2.5 \times 3.4)(10^2 \times 10^{-4})$

$= 8.5 \times 10^{-2}$

$= 0.085$

78. $(4.2 \times 10^{-3})(3 \times 10^5) = (4.2 \times 3)(10^{-3} \times 10^5)$

$= 12.6 \times 10^2$

$= 1260$

79. $(3.5 \times 10^{-2})(7.0 \times 10^3) = (3.5 \times 7.0)(10^{-2} \times 10^3)$

$= 24.5 \times 10^1$

$= 245$

80. $\frac{7.94 \times 10^6}{2 \times 10^{-2}} = \left(\frac{7.94}{2}\right)\left(\frac{10^6}{10^{-2}}\right)$

$= 3.97 \times 10^8$

$= 397{,}000{,}000$

81. $\frac{6.5 \times 10^4}{2.0 \times 10^6} = \left(\frac{6.5}{2.0}\right)\left(\frac{10^4}{10^6}\right)$

$= 3.25 \times 10^{-2}$

$= 0.0325$

82. $\frac{15 \times 10^{-3}}{5 \times 10^2} = \left(\frac{15}{5}\right)\left(\frac{10^{-3}}{10^2}\right)$

$= 3 \times 10^{-5}$

$= 0.00003$

83. $(14{,}000)(260{,}000) = (1.4 \times 10^4)(2.6 \times 10^5)$

$= (1.4 \times 2.6)(10^4 \times 10^5)$

$= 3.64 \times 10^9$

84. $(12{,}500)(400{,}000) = (1.25 \times 10^4)(4.0 \times 10^5)$

$= (1.25 \times 4.0) \times (10^4 \times 10^5)$

$= 5.0 \times 10^9$

85. $(0.00053)(40,000) = (5.3\times 10^{-4})(4\times 10^{4})$
$= (5.3\times 4)(10^{-4}\times 10^{4})$
$= 21.2\times 10^{0}$
$= 2.12\times 10^{1}$

86. $\dfrac{250}{500,000} = \dfrac{2.5\times 10^{2}}{5.0\times 10^{5}}$
$= \left(\dfrac{2.5}{5.0}\right)\left(\dfrac{10^{2}}{10^{5}}\right)$
$= 0.5\times 10^{-3}$
$= 5.0\times 10^{-4}$

87. $\dfrac{0.000068}{0.02} = \dfrac{6.8\times 10^{-5}}{2\times 10^{-2}}$
$= \left(\dfrac{6.8}{2}\right)\left(\dfrac{10^{-5}}{10^{-2}}\right)$
$= 3.4\times 10^{-3}$

88. $\dfrac{850,000}{0.025} = \dfrac{8.5\times 10^{5}}{2.5\times 10^{-2}}$
$= \left(\dfrac{8.50}{2.50}\right)\left(\dfrac{10^{5}}{10^{-2}}\right)$
$= 3.40\times 10^{7}$

89. $\dfrac{6.4\times 10^{6}}{1.28\times 10^{2}} = \left(\dfrac{6.4}{1.28}\right)\left(\dfrac{10^{6}}{10^{2}}\right)$
$= 5\times 10^{4}$
$= 50,000$ gallons
The milk tank holds 50,000 gallons.

90. **a.** $(1.38\times 10^{10}) - (8.54\times 10^{9}) = (13.8\times 10^{9}) - (8.54\times 10^{9})$
$= (13.8 - 8.54)\times 10^{9}$
$= 5.26\times 10^{9}$
$= 5,260,000,000$
There was \$5,260,000, 000 more in circulation of the \$10 bills than in \$5 bills.

b. $\dfrac{1.38\times 10^{10}}{8.54\times 10^{9}} = \left(\dfrac{1.38}{8.54}\right)\left(\dfrac{10^{10}}{10^{9}}\right)$
$= .162\times 10^{1}$
$= 1.62$
The amount of \$10 bills in circulation is 1.62 times greater than in \$5 bills.

91. Not a polynomial

92. Monomial, zero degree

93. $x^2 + 3x - 4$, trinomial, second degree

94. $4x^2 - x - 3$, trinomial, second degree

95. $13x^3 - 4$, Binomial, third degree

96. Not a polynomial

97. $-4x^2 + x$, binomial, second degree

98. Not a polynomial

99. $2x^3 + 4x^2 - 3x - 7$, polynomial, third degree

100. $(x-5)+(2x+4) = x-5+2x+4$
$= x+2x-5+4$
$= 3x-1$

101. $(2d-3)+(5d+7) = 2d-3+5d+7$
$= 2d+5d-3+7$
$= 7d+4$

102. $(-x-10)+(-2x+5) = -x-10-2x+5$
$= -x-2x-10+5$
$= -3x-5$

103. $(-x^2+6x-7)+(-2x^2+4x-8) = -x^2+6x-7-2x^2+4x-8$
$= -x^2-2x^2+6x+4x-7-8$
$= -3x^2+10x-15$

104. $(-m^2+5m-8)+(6m^2-5m-2)$
$= -m^2+6m^2+5m-5m-8-2$
$= 5m^2-10$

105. $(6.2p-4.3)+(1.9p+7.1)$
$= 6.2p+1.9p-4.3+7.1$
$= 8.1p+2.8$

106. $(-4x+8)-(-2x+6) = -4x+8+2x-6$
$= -4x+2x+8-6$
$= -2x+2$

107. $(4x^2-9x)-(3x+15) = 4x^2-9x-3x-15$
$= 4x^2-12x-15$

108. $(5a^2-6a-9)-(2a^2-a+12)$
$= 5a^2-6a-9-2a^2+a-12$
$= 5a^2-2a^2-6a+a-9-12$
$= 3a^2-5a-21$

109. $(-2x^2+8x-7)-(3x^2+12) = -2x^2+8x-7-3x^2-12$
$= -2x^2-3x^2+8x-7-12$
$= -5x^2+8x-19$

110. $(x^2+7x-3)-(x^2+3x-5) = x^2+7x-3-x^2-3x+5$
$= x^2-x^2+7x-3x-3+5$
$= 4x+2$

111. $\frac{1}{7}x(21x+21) = \frac{1}{7}x(21x)+\frac{1}{7}x(21)$
$= \frac{21}{7}x^2+\frac{21}{7}x$
$= 3x^2+3x$

112. $-3x(5x+4) = -3x \cdot 5x + (-3x)4$
$= -15x^2 - 12x$

113. $3x(2x^2 - 4x + 7) = 3x(2x^2) + 3x(-4x) + 3x(7)$
$= 6x^3 - 12x^2 + 21x$

114. $-c(2c^2 - 3c + 5) = (-c)(2c^2) + (-c)(-3c) + (-c)(5)$
$= -2c^3 + 3c^2 - 5c$

115. $-4z(-3z^2 - 2z - 8) = (-4z)(-3z^2) + (-4z)(-2z) + (-4z)(-8)$
$= 12z^3 + 8z^2 + 32z$

116. $(x+4)(x+5) = x \cdot x + x \cdot 5 + 4 \cdot x + 4 \cdot 5$
$= x^2 + 5x + 4x + 20$
$= x^2 + 9x + 20$

117. $(3x+6)(-4x+1) = 3x(-4x) + 3x(1) + 6(-4x) + 6(1)$
$= -12x^2 + 3x - 24x + 6$
$= -12x^2 - 21x + 6$

118. $(-2x+6)^2 = (-2x)^2 + 2(-2x)(6) + (6)^2$
$= 4x^2 - 24x + 36$

119. $(6-2x)(2+3x) = 6 \cdot 2 + 6 \cdot 3x + (-2x)(2) + (-2x)(3x)$
$= 12 + 18x - 4x - 6x^2$
$= 12 + 14x - 6x^2$
$= -6x^2 + 14x + 12$

120. $(r+5)(r-5) = (r)^2 - (5)^2$
$= r^2 - 25$

121. $(3x+1)(x^2 + 2x + 4) = 3x(x^2 + 2x + 4) + 1(x^2 + 2x + 4)$
$= 3x^3 + 6x^2 + 12x + x^2 + 2x + 4$
$= 3x^3 + 7x^2 + 14x + 4$

122. $(x-1)(3x^2 + 4x - 6) = x(3x^2 + 4x - 6) - 1(3x^2 + 4x - 6)$
$= 3x^3 + 4x^2 - 6x - 3x^2 - 4x + 6$
$= 3x^3 + x^2 - 10x + 6$

123. $(-4x+2)(3x^2 - x + 7) = -4x(3x^2 - x + 7) + 2(3x^2 - x + 7)$
$= -12x^3 + 4x^2 - 28x + 6x^2 - 2x + 14$
$= -12x^3 + 10x^2 - 30x + 14$

124. $\dfrac{2x+4}{2}=\dfrac{2x}{2}+\dfrac{4}{2}$
$=x+2$

125. $\dfrac{10x+12}{2}=\dfrac{10x}{2}+\dfrac{12}{2}$
$=5x+6$

126. $\dfrac{8x^2+4x}{x}=\dfrac{8x^2}{x}+\dfrac{4x}{x}$
$=8x+4$

127. $\dfrac{6x^2+9x-4}{3}=\dfrac{6x^2}{3}+\dfrac{9x}{3}-\dfrac{4}{3}$
$=2x^2+3x-\dfrac{4}{3}$

128. $\dfrac{6w^2-5w+3}{3w}=\dfrac{6w^2}{3w}-\dfrac{5w}{3w}-\dfrac{3}{3w}$
$=2w-\dfrac{5}{3}-\dfrac{1}{w}$

129. $\dfrac{8x^5-4x^4+3x^2-2}{2x}=\dfrac{8x^5}{2x}-\dfrac{4x^4}{2x}+\dfrac{3x^2}{2x}-\dfrac{2}{2x}$
$=4x^4-2x^3+\dfrac{3}{2}x-\dfrac{1}{x}$

130. $\dfrac{8m-4}{-2}=\dfrac{(-1)(8m-4)}{(-1)(-2)}$
$=\dfrac{-8m+4}{2}$
$=\dfrac{-8m}{2}+\dfrac{4}{2}$
$=-4m+2$

131. $\dfrac{5x^2-6x+15}{3x}=\dfrac{5x^2}{3x}-\dfrac{6x}{3x}+\dfrac{15}{3x}$
$=\dfrac{5x}{3}x-2+\dfrac{5}{x}$

132. $\dfrac{5x^3+10x+2}{2x^2}=\dfrac{5x^3}{2x^2}+\dfrac{10x}{2x^2}+\dfrac{2}{2x^2}$
$=\dfrac{5x}{2}+\dfrac{5}{x}+\dfrac{1}{x^2}$

133.
$$\begin{array}{r|l} & x+4 \\ \hline x-3 & x^2+x-12 \\ & \underline{x^2-3x} \\ & \quad 4x-12 \\ & \quad \underline{4x-12} \\ & \qquad 0 \end{array}$$
$\dfrac{x^2+x-12}{x-3}=x+4$

134.
$$\begin{array}{r|l} & n+3 \\ \hline 6n+1 & 6n^2+19n+3 \\ & \underline{6n^2+\ \ n} \\ & \quad 18n+3 \\ & \quad \underline{18n+3} \\ & \qquad 0 \end{array}$$
$\dfrac{6n^2+19n+3}{6n+1}=n+3$

135.
$$\begin{array}{r|l} & 5x-2 \\ \hline x+6 & 5x^2+28x-10 \\ & \underline{5x^2+30x} \\ & \quad -2x-10 \\ & \quad \underline{-2x-12} \\ & \qquad 2 \end{array}$$
$\dfrac{5x^2+28x-10}{x+6}=5x-2+\dfrac{2}{x+6}$

136.
$$\begin{array}{r|l} & 2x^2+3x-4 \\ \hline 2x+3 & 4x^3+12x^2+x-12 \\ & \underline{4x^3+6x^2} \\ & \quad 6x^2+x \\ & \quad \underline{6x^2+9x} \\ & \qquad -8x-12 \\ & \qquad \underline{-8x-12} \\ & \qquad\quad 0 \end{array}$$
$\dfrac{4x^3+12x^2+x-12}{2x+3}=2x^2+3x-4$

137.
$$\begin{array}{r} 2x-3 \\ 2x-3\overline{\big)\,4x^2-12x+9} \\ \underline{4x^2-\;6x} \\ -6x+9 \\ \underline{-6x+9} \\ 0 \end{array}$$

$$\frac{4x^2-12x+9}{2x-3}=2x-3$$

138.

3	3	−2	0	10
		9	21	63
	3	7	21	73

Thus, $\frac{3x^3-2x^2+10}{x-3}=3x^2+7x+21+\frac{73}{x-3}$

138.

−1	2	0	−10	0	1	−1
		−2	2	8	−8	7
	2	−2	−8	8	−7	6

Thus, $\frac{2y^5-10y^3+y-1}{y+1}$

$=2y^4-2y^3-8y^2+8y-7+\frac{6}{y+1}$

140.

2	1	0	0	0	0	−20
		2	4	8	16	32
	1	2	4	8	16	12

Thus,

$\frac{x^5-20}{x-2}=x^4+2x^3+4x^2+8x+16+\frac{12}{x-2}$

141.

$\frac{1}{2}$	2	1	5	−3
		1	1	3
	2	2	6	0

Thus, $\frac{2x^3+x^2+5x-3}{x-\frac{1}{2}}=2x^2+2x+6$

142. To find the remainder, use synthetic division:

3	1	−4	11
		3	−3
	1	−1	8

Thus, the remainder is 8.

143. To find the remainder, use synthetic division:

−4	2	−6	3	0
		−8	56	−236
	2	−14	59	−236

Thus, the remainder is −236.

144. To find the remainder, use synthetic division:

$\frac{1}{3}$	3	0	0	−6
		1	$\frac{1}{3}$	$\frac{1}{9}$
	3	1	$\frac{1}{3}$	$-\frac{53}{9}$ or $-5.\overline{8}$

Thus, the remainder is $-\frac{53}{9}$ or $-5.\overline{8}$.

145. To find the remainder, use synthetic division:

−2	2	0	−6	0	−8
		−4	8	−4	8
	2	−4	2	−4	0

Since the remainder is 0, then $x+2$ is a factor.

Practice Test

1. $5x^4\cdot 3x^2=5\cdot 3x^{4+2}$
$=15x^6$

2. $(3xy^2)^3=3^3x^3y^{2\cdot 3}$
$=27x^3y^6$

3. $\frac{12d^5}{4d}=\frac{12}{4}d^{5-1}$
$=3d^4$

4. $\left(\frac{3x^2y}{6xy^3}\right)^3=\left(\frac{3}{6}\cdot\frac{x^2}{x}\cdot\frac{y}{y^3}\right)^3$
$=\left(\frac{1}{2}\cdot x\cdot\frac{1}{y^2}\right)^3$
$=\left(\frac{x}{2y^2}\right)^3$
$=\frac{x^3}{2^3y^{2\cdot 3}}$
$=\frac{x^3}{8y^6}$

5. $(2x^3y^{-2})^{-2}=2^{-2}x^{3(-2)}y^{(-2)(-2)}$
$=2^{-2}x^{-6}y^4$
$=\frac{y^4}{2^2x^6}$
$=\frac{y^4}{4x^6}$

6. $\frac{30x^6y^2}{45x^{-1}y} = \frac{30}{45}x^{6-(-1)}y^{2-1}$
$= \frac{2}{3}x^7y$
$= \frac{2x^7y}{3}$

7. $(4x^0)(3x^2)^0 = (4 \cdot 1) \cdot 1$
$= 4$

8. $(175{,}000)(30{,}000) = (1.75 \times 10^5)(3.0 \times 10^4)$
$= (1.75 \times 3.0)(10^5 \times 10^4)$
$= 5.25 \times 10^9$

9. $\frac{0.0008}{4000} = \frac{8.0 \times 10^{-4}}{4.0 \times 10^3}$
$= \left(\frac{8.0}{4.0}\right)\left(\frac{10^{-4}}{10^3}\right)$
$= 2.0 \times 10^{-7}$

10. $4x$ is a monomial

11. $3b + 2$, binomial

12. $x^{-2} + 4$, not a polynomial

13. $-5 + 6x^3 - 2x^2 + 5x = 6x^3 - 2x^2 + 5x - 5$, third degree

14. $(6x - 4) + (2x^2 - 5x - 3) = 6x - 4 + 2x^2 - 5x - 3$
$= 2x^2 + 6x - 5x - 4 - 3$
$= 2x^2 + x - 7$

15. $(x^2 - 4x + 7) - (3x^2 - 8x + 7)$
$= x^2 - 4x + 7 - 3x^2 + 8x - 7$
$= x^2 - 3x^2 - 4x + 8x + 7 - 7$
$= -2x^2 + 4x$

16. $(4x^2 - 5) - (x^2 + x - 8) = 4x^2 - 5 - x^2 - x + 8$
$= 4x^2 - x^2 - x - 5 + 8$
$= 3x^2 - x + 3$

17. $-5d(-3d + 8) = -5d(-3d) - 5d(8)$
$= 15d^2 - 40d$

18. $(4x + 7)(2x - 3)$
$= (4x)(2x) + (4x)(-3) + 7(2x) + 7(-3)$
$= 8x^2 - 12x + 14x - 21$
$= 8x^2 + 2x - 21$

19. $(3x - 5)(2x^2 + 4x - 5)$
$= 3x(2x^2 + 4x - 5) - 5(2x^2 + 4x - 5)$
$= 3x \cdot 2x^2 + 3x \cdot 4x - 3x \cdot 5 - 5 \cdot 2x^2 - 5 \cdot 4x - 5(-5)$
$= 6x^3 + 12x^2 - 15x - 10x^2 - 20x + 25$
$= 6x^3 + 2x^2 - 35x + 25$

20. $\frac{-12x^2 - 6x + 5}{-3x} = \frac{(-1)(-12x^2 - 6x + 5)}{(-1)(-3x)}$
$= \frac{12x^2 + 6x - 5}{3x}$
$= \frac{12x^2}{3x} + \frac{6x}{3x} - \frac{5}{3x}$
$= 4x + 2 - \frac{5}{3x}$

21.
$$\begin{array}{r} 4x+5 \\ 2x-3\overline{)\,8x^2-2x-15} \\ \underline{8x^2-12x} \\ 10x-15 \\ \underline{10x-15} \\ 0 \end{array}$$

$$\frac{8x^2-2x-15}{2x-3}=4x+5$$

22.
$$\begin{array}{r} 3x-2 \\ 4x+5\overline{)\,12x^2+7x-12} \\ \underline{12x^2+15x} \\ -8x-12 \\ \underline{-8x-10} \\ -2 \end{array}$$

$$\frac{12x^2+7x-12}{4x+5}=3x-2-\frac{2}{4x+5}$$

23.

5	3	−12	0	−60	4
		15	15	75	75
	3	3	15	15	79

Thus,
$$\frac{3x^4-12x^3-60x+4}{x-5}=3x^3+3x^2+15x+15+\frac{79}{x-5}$$

24.

−3	2	−6	−5	4
		−6	36	−93
	2	−12	31	−89

Thus, the remainder is −89.

25. a. $5730=5.73\times10^3$

b. $\frac{4.46\times10^9}{5.73\times10^3}=\left(\frac{4.46}{5.73}\right)\left(\frac{10^9}{10^3}\right)$
$\approx0.778\times10^6$
$\approx7.78\times10^5$

Cumulative Review Test

1. $12+8\div2^2+3=12+8\div4+3$
$=12+2+3$
$=17$

2. Substitute −2 for x.
$-4x^2+x-7=-4(-2)^2+(-2)-7$
$=-4(4)-2-7$
$=-16-2-7$
$=-18-7$
$=-25$

3. $7-(2x-3)+2x-8(1-x)$
$=7-2x+3+2x-8+8x$
$=7+3-8-2x+2x+8x$
$=2+8x$
$=8x+2$

4. $\frac{5}{8}=\frac{5t}{6}+2$
$24\left(\frac{5}{8}\right)=24\left(\frac{5t}{6}+2\right)$
$15=20t+48$
$15-48=20t+48-48$
$-33=20t$
$\frac{-33}{20}=\frac{20t}{20}$
$-\frac{33}{20}=t$

5. $3x+5=4(x-2)$
$3x+5=4x-8$
$13=x$

6. $3(x+2)+3x-5=4x+1$
$3x+6+3x-5=4x+1$
$6x+1=4x+1$
$2x=0$
$x=0$

7. $3x-2=y-7$
$y-7=3x-2$
$y=3x-2+7$
$y=3x+5$

8. Let (1,3) be (x_1, y_1) and (5,1) be (x_2, y_2).

$$m = \frac{y_2 - y_1}{x_2 - x_1} = \frac{1-3}{5-1} = \frac{-2}{4} = -\frac{1}{2}$$

9. Put each equation in slope-intercept form and determine their slopes.

$$\begin{aligned} 3x - 5y &= 7 \\ 3x - 3x - 5y &= -3x + 7 \\ -5y &= -3x + 7 \\ \frac{-5y}{-5} &= \frac{-3x}{-5} + \frac{7}{-5} \\ y &= \frac{3}{5}x - \frac{7}{5} \\ m &= \frac{3}{5} \end{aligned}$$

$$\begin{aligned} 5y + 3x &= 2 \\ 5y + 3x - 3x &= -3x + 2 \\ 5y &= -3x + 2 \\ \frac{5y}{5} &= \frac{-3x}{5} + \frac{2}{5} \\ y &= -\frac{3}{5}x + \frac{2}{5} \\ m &= -\frac{3}{5} \end{aligned}$$

Since the slopes are different, the lines are not parallel.

10.
$$\begin{aligned} \left(2x^4y^3\right)^3\left(5x^2y\right) &= \left(2^3x^{4\cdot 3}y^{3\cdot 3}\right)\left(5x^2y\right) \\ &= \left(8x^{12}y^9\right)\left(5x^2y\right) \\ &= 8 \cdot 5x^{12+2}y^{9+1} \\ &= 40x^{14}y^{10} \end{aligned}$$

11. $-5x + 2 - 7x^2 = -7x^2 - 5x + 2$
second degree

12.
$$\begin{aligned} \left(x^2 + 4x - 3\right) + \left(2x^2 + 5x + 1\right) &= x^2 + 4x - 3 + 2x^2 + 5x + 1 \\ &= x^2 + 2x^2 + 4x + 5x - 3 + 1 \\ &= 3x^2 + 9x - 2 \end{aligned}$$

13.
$$\begin{aligned} \left(6a^2 + 3a + 2\right) - \left(a^2 - 3a - 3\right) &= 6a^2 + 3a + 2 - a^2 + 3a + 3 \\ &= 6a^2 - a^2 + 3a + 3a + 2 + 3 \\ &= 5a^2 + 6a + 5 \end{aligned}$$

14.
$$\begin{aligned} (3y - 5)(2y + 3) &= 3y \cdot 2y + 3 \cdot 3y - 5 \cdot 2y - 5 \cdot 3 \\ &= 6y^2 + 9y - 10y - 15 \\ &= 6y^2 - y - 15 \end{aligned}$$

15.
$$\begin{aligned} (2x - 1)\left(3x^2 - 5x + 2\right) &= 2x\left(3x^2 - 5x + 2\right) - 1\left(3x^2 - 5x + 2\right) \\ &= 6x^3 - 10x^2 + 4x - 3x^2 + 5x - 2 \\ &= 6x^3 - 13x^2 + 9x - 2 \end{aligned}$$

16. $\frac{10d^2+12d-8}{4d} = \frac{10d^2}{4d} + \frac{12d}{4d} - \frac{8}{4d}$

$= \frac{5}{2}d + 3 - \frac{2}{d}$

17.
$$\begin{array}{r|l} & 2x+5 \\ \hline 3x-2 & 6x^2+11x-10 \\ & \underline{6x^2-4x} \\ & \quad 15x-10 \\ & \quad \underline{15x-10} \\ & \qquad 0 \end{array}$$

$\frac{6x^2+11x-10}{3x-2} = 2x+5$

18. $\frac{x}{8} = \frac{1.25}{3}$

$3x = 8(1.25)$

$3x = 10$

$x = \frac{10}{3} \approx 3.33$

Eight cans of soup cost \$3.33.

19. Let x = the width of the rectangle.
Then $3x - 2$ = the length of the rectangle.

$P = 2l + 2w$

$28 = 2(3x-2) + 2x$

$28 = 6x - 4 + 2x$

$32 = 8x$

$4 = x$

The width of the rectangle is 4 feet and the length is $3(4) - 2 = 10$ feet.

20. Let b = Bob's average speed.
Then $b + 7$ = Nick's average speed.
$d = r \cdot t$. Both Bob and Nick drove for 0.5 hour and the total distance they covered was 60 miles.

$0.5b + 0.5(b+7) = 60$

$0.5b + 0.5b + 3.5 = 60$

$b = 56.5$

Bob's average speed was 56.5 miles per hour and Nick's average speed was $56.5 + 7 = 63.5$ miles per hour.

Chapter 6

Exercise Set 6.1

1. A prime number is an integer greater than 1 that has exactly two factors, itself and 1.

3. To factor an expression means to write the expression as the product of factors.

5. The greatest common factor of two or more numbers is the greatest number that divides into all the numbers.

7. A factoring problem may be checked by multiplying the factors.

9. $$\begin{aligned} 56 &= 8\cdot 7 \\ &= 2\cdot 4\cdot 7 \\ &= 2\cdot 2\cdot 2\cdot 7 \\ &= 2^3\cdot 7 \end{aligned}$$

11. $$\begin{aligned} 90 &= 9\cdot 10 \\ &= 3\cdot 3\cdot 2\cdot 5 \\ &= 2\cdot 3^2\cdot 5 \end{aligned}$$

13. $$\begin{aligned} 196 &= 4\cdot 49 \\ &= 2\cdot 2\cdot 7\cdot 7 \\ &= 2^2\cdot 7^2 \end{aligned}$$

15. $20 = 2^2\cdot 5$, $24 = 2^3\cdot 3$, so the greatest common factor is 2^2 or 4.

17. $70 = 2\cdot 5\cdot 7$, $98 = 2\cdot 7^2$, so the greatest common factor is $2\cdot 7$ or 14.

19. $80 = 2^4\cdot 5$, $126 = 2\cdot 3^2\cdot 7$ so the greatest common factor is 2.

21. The greatest common factor is x.

23. The greatest common factor is $3x$.

25. The greatest common factor is 1.

27. The greatest common factor is mn.

29. The greatest common factor is x^3y^5.

31. The greatest common factor is 5.

33. The greatest common factor is x^2y^2.

35. The greatest common factor is x.

37. The greatest common factor is $x + 3$.

39. The greatest common factor is $2x - 3$.

41. The greatest common factor is $3w + 5$.

43. The greatest common factor is $x - 4$.

45. The greatest common factor is $x - 1$.

47. The greatest common factor is $x + 3$.

49. The greatest common factor is 4.
$$\begin{aligned} 4x - 8 &= 4\cdot x - 4\cdot 2 \\ &= 4(x-2) \end{aligned}$$

51. The greatest common factor is 5.
$$\begin{aligned} 15x - 5 &= 5\cdot 3x - 5\cdot 1 \\ &= 5(3x-1) \end{aligned}$$

53. The greatest common factor is 6.
$$\begin{aligned} 6p + 12 &= 6\cdot p + 6\cdot 2 \\ &= 6(p+2) \end{aligned}$$

55. The greatest common factor is $3x$.
$$\begin{aligned} 9x^2 - 12x &= 3x\cdot 3x - 3x\cdot 4 \\ &= 3x(3x-4) \end{aligned}$$

57. The greatest common factor is $2p$.
$$\begin{aligned} 26p^2 - 8p &= 2p\cdot 13p - 2p\cdot 4 \\ &= 2p(13p-4) \end{aligned}$$

59. The greatest common factor is $3x^2$.
$$\begin{aligned} 3x^5 - 12x^2 &= 3x^2\cdot x^3 - 3x^2\cdot 4 \\ &= 3x^2(x^3-4) \end{aligned}$$

61. The greatest common factor is $12x^8$.
$$\begin{aligned} 36x^{12} + 24x^8 &= 12x^8\cdot 3x^4 + 12x^8\cdot 2 \\ &= 12x^8(3x^4+2) \end{aligned}$$

63. The greatest common factor is $9y^3$.
$$\begin{aligned} 27y^{15} - 9y^3 &= 9y^3\cdot 3y^{12} - 9y^3\cdot 1 \\ &= 9y^3(3y^{12}-1) \end{aligned}$$

65. The greatest common factor is x.
$$\begin{aligned} x + 3xy^2 &= x\cdot 1 + x\cdot 3y^2 \\ &= x(1+3y^2) \end{aligned}$$

67. The greatest common factor is a^2.
$$7a^4 + 3a^2 = a^2 \cdot 7a^2 + a^2 \cdot 3 = a^2(7a^2 + 3)$$

69. The greatest common factor is $4xy$.
$$16xy^2z + 4x^3y = 4xy \cdot 4yz + 4xy \cdot x^2 = 4xy(4yz + x^2)$$

71. The greatest common factor is $16mn^2$.
$$48m^4n^2 - 16mn^2 = 16mn^2 \cdot 3m^3 - 16mn^2 \cdot 1 = 16mn^2(3m^3 - 1)$$

73. The greatest common factor is $25x^2yz$.
$$25x^2yz^3 + 25x^3yz = 25x^2yz \cdot z^2 + 25x^2yz \cdot x = 25x^2yz(z^2 + x)$$

75. The greatest common factor is y^2z^3.
$$13y^5z^3 - 11xy^2z^5 = y^2z^3 \cdot 13y^3 - y^2z^3 \cdot 11xz^2 = y^2z^3(13y^3 - 11xz^2)$$

77. The greatest common factor is 4.
$$8c^2 - 4c - 32 = 4 \cdot 2c^2 - 4 \cdot c - 4 \cdot 8 = 4(2c^2 - c - 8)$$

79. The greatest common factor is 3.
$$9x^2 + 18x + 3 = 3 \cdot 3x^2 + 3 \cdot 6x + 3 \cdot 1 = 3(3x^2 + 6x + 1)$$

81. The greatest common factor is $4x$.
$$4x^3 - 8x^2 + 12x = 4x \cdot x^2 - 4x \cdot 2x + 4x \cdot 3 = 4x(x^2 - 2x + 3)$$

83. The greatest common factor is 5.
$$35x^2 - 15y + 10 = 5 \cdot 7x^2 - 5 \cdot 3y + 5 \cdot 2 = 5(7x^2 - 3y + 2)$$

85. The greatest common factor is 3.
$$15p^2 - 6p + 9 = 3 \cdot 5p^2 - 3 \cdot 2p + 3 \cdot 3 = 3(5p^2 - 2p + 3)$$

87. The greatest common factor is $3a$.
$$9a^4 - 6a^3 + 3ab = 3a \cdot 3a^3 - 3a \cdot 2a^2 + 3a \cdot b = 3a(3a^3 - 2a^2 + b)$$

89. The greatest common factor is xy.
$$8x^2y + 12xy^2 + 5xy = xy \cdot 8x + xy \cdot 12y + xy \cdot 5 = xy(8x + 12y + 5)$$

91. The greatest common factor is $x + 4$.
$$x(x + 4) + 3(x + 4) = (x + 4)(x + 3)$$

93. The greatest common factor is $a - 2$.
$$3b(a - 2) - 4(a - 2) = (a - 2)(3b - 4)$$

95. The greatest common factor is $2x + 1$.
$$4x(2x + 1) + 1(2x + 1) = (2x + 1)(4x + 1)$$

97. The greatest common factor is $2x + 1$.
$$5x(2x + 1) + 2x + 1 = 5x(2x + 1) + 1(2x + 1) = (2x + 1)(5x + 1)$$

99. The greatest common factor is $2z + 3$.
$$4z(2z + 3) - 3(2z + 3) = (2z + 3)(4z - 3)$$

101. $3\bigstar + 6 = 3\bigstar \cdot + 3 \cdot 2 = 3(\bigstar + 2)$

103. $$35\Delta^3 - 7\Delta^2 + 14\Delta = 7\Delta \cdot 5\Delta^2 - 7\Delta \cdot \Delta + 7\Delta \cdot 2 = 7\Delta(5\Delta^2 - \Delta + 2)$$

105. The greatest common factor is $2(x-3)$.

$$4x^2(x-3)^3 - 6x(x-3)^2 + 4(x-3) = 2(x-3)\cdot 2x^2(x-3)^2 - 2(x-3)\cdot 3x(x-3) - 2(x-3)\cdot 2$$
$$= 2(x-3)\left[2x^2(x-3)^2 - 3x(x-3) + 2\right]$$

107. First factor $x^{1/3}$ from terms.

$$x^{7/3} + 5x^{4/3} + 2x^{1/3} = x^{1/3}\left(x^2 + 5x + 2\right)$$

109. $x^2 + 2x + 3x + 6 = x\cdot x + x\cdot 2 + 3\cdot x + 3\cdot 2$
$= x(x+2) + 3(x+2)$
$= (x+2)(x+3)$

110. $2x - (x-5) + 4(3-x) = 2x - x + 5 + 12 - 4x$
$= x - 4x + 17$
$= -3x + 17$

111. $4 + 3(x-8) = x - 4(x+2)$
$4 + 3x - 24 = x - 4x - 8$
$3x - 20 = -3x - 8$
$6x = 12$
$x = 2$

112. $4x - 5y = 20$
$-5y = -4x + 20$
$y = -\dfrac{-4x+20}{-5}$
$y = \dfrac{4}{5}x - 4$

113. $V = \dfrac{1}{3}\pi r^2 h$
$= \dfrac{1}{3}\pi(4)^2(12)$
$= \dfrac{1}{3}\pi(16)(12)$
$= 64\pi \text{ in.}^3$ or 201.06 in.^3

114. Let x = smaller number, then $2x - 1$ = the larger number.
$x + 2x - 1 = 41$
$3x - 1 = 41$
$3x = 42$
$x = 14$
$2x - 1 = 2(14) - 1 = 28 - 1 = 27$
The numbers are 14 and 27.

115. $\left(\dfrac{3x^2y^3}{2x^5y^2}\right)^2 = \left(\dfrac{3y}{2x^3}\right)^2 = \dfrac{(3y)^2}{(2x^3)^2} = \dfrac{3^2y^2}{2^2(x^3)^2} = \dfrac{9y^2}{4x^6}$

Exercise Set 6.2

1. The first step in any factoring by grouping problem is to factor out a common factor, if one exists.

3. If you multiply $(x-2)(x+4)$ by the FOIL method, you get the polynomial $x^2 + 4x - 2x - 8$

5. Answers will vary.

7. $x^2 + 3x + 2x + 6 = x(x+3) + 2(x+3)$
$= (x+3)(x+2)$

9. $x^2 + 5x + 4x + 20 = x(x+5) + 4(x+5)$
$= (x+5)(x+4)$

11. $x^2 + 2x + 5x + 10 = x(x+2) + 5(x+2)$
$= (x+2)(x+5)$

13. $x^2 + 3x - 5x - 15 = x(x+3) - 5(x+3)$
$= (x+3)(x-5)$

15. $4b^2 - 10b + 10b - 25 = 2b(2b-5) + 5(2b-5)$
$= (2b-5)(2b+5)$

17. $3x^2 + 9x + x + 3 = 3x(x+3) + 1(x+3)$
$= (x+3)(3x+1)$

19. $6x^2 + 3x - 2x - 1 = 3x(2x+1) - 1(2x+1)$
$= (2x+1)(3x-1)$

21. $8x^2 + 32x + x + 4 = 8x(x+4) + 1(x+4)$
$= (x+4)(8x+1)$

23. $12t^2 - 8t - 3t + 2 = 4t(3t - 2) - 1(3t - 2)$
$= (3t - 2)(4t - 1)$

25. $2x^2 - 4x - 3x + 6 = 2x(x - 2) - 3(x - 2)$
$= (x - 2)(2x - 3)$

27. $6p^2 + 15p - 4p - 10 = 3p(2p + 5) - 2(2p + 5)$
$= (2p + 5)(3p - 2)$

29. $x^2 + 2xy - 3xy - 6y^2 = x(x + 2y) - 3y(x + 2y)$
$= (x + 2y)(x - 3y)$

31. $3x^2 + 2xy - 9xy - 6y^2 = x(3x + 2y) - 3y(3x + 2y)$
$= (3x + 2y)(x - 3y)$

33. $10x^2 - 12xy - 25xy + 30y^2 = 2x(5x - 6y) - 5y(5x - 6y)$
$= (5x - 6y)(2x - 5y)$

35. $x^2 + bx + ax + ab = x(x + b) + a(x + b)$
$= (x + b)(x + a)$

37. $xy + 5x - 3y - 15 = x(y + 5) - 3(y + 5)$
$= (y + 5)(x - 3)$

39. $a^2 + 3a + ab + 3b = a(a + 3) + b(a + 3)$
$= (a + 3)(a + b)$

41. $xy - x + 5y - 5 = x(y - 1) + 5(y - 1)$
$= (x + 5)(y - 1)$

43. $12 + 8y - 3x - 2xy = 4(3 + 2y) - x(3 + 2y)$
$= (4 - x)(3 + 2y)$

45. $z^3 + 3z^2 + z + 3 = z^2(z + 3) + 1(z + 3)$
$= (z^2 + 1)(z + 3)$

47. $x^3 + 4x^2 - 3x - 12 = x^2(x + 4) - 3(x + 4)$
$= (x^2 - 3)(x + 4)$

49. $2x^2 - 12x + 8x - 48$
$= 2 \cdot x^2 - 2 \cdot 6x + 2 \cdot 4x - 2 \cdot 24$
$= 2(x^2 - 6x + 4x - 24)$
$= 2[x(x - 6) + 4(x - 6)]$
$= 2(x - 6)(x + 4)$

51. $4x^2 + 8x + 8x + 16 = 4 \cdot x^2 + 4 \cdot 2x + 4 \cdot 2x + 4 \cdot 4$
$= 4(x^2 + 2x + 2x + 4)$
$= 4[x(x + 2) + 2(x + 2)]$
$= 4(x + 2)(x + 2)$
$= 4(x + 2)^2$

53. $6x^3 + 9x^2 - 2x^2 - 3x$
$= x \cdot 6x^2 + x \cdot 9x - x \cdot 2x - x \cdot 3$
$= x(6x^2 + 9x - 2x - 3)$
$= x[3x(2x + 3) - 1(2x + 3)]$
$= x(2x + 3)(3x - 1)$

55. $x^3 + 3x^2y - 2x^2y - 6xy^2$
$= x \cdot x^2 + x \cdot 3xy - x \cdot 2xy - x \cdot 6y^2$
$= x(x^2 + 3xy - 2xy - 6y^2)$
$= x[x(x + 3y) - 2y(x + 3y)]$
$= x(x + 3y)(x - 2y)$

57. $5x + 3y + xy + 15 = xy + 5x + 3y + 15$
$= x(y + 5) + 3(y + 5)$
$= (y + 5)(x + 3)$

59. $6x + 5y + xy + 30 = 6x + xy + 5y + 30$
$= x(6 + y) + 5(y + 6)$
$= (x + 5)(y + 6)$

61. $ax + by + ay + bx = ax + ay + bx + by$
$= a(x + y) + b(x + y)$
$= (a + b)(x + y)$

63. $cd - 12 - 4d + 3c = cd - 4d + 3c - 12$
$= d(c - 4) + 3(c - 4)$
$= (d + 3)(c - 4)$

65. $ac - bd - ad + bc = ac - ad + bc - bd$
$= a(c - d) + b(c - d)$
$= (a + b)(c - d)$

67. Not *any* arrangement of the terms of a polynomial is factorable by grouping. $xy + 2x + 5y + 10$ is factorable but $xy + 10 + 2x + 5y$ is not factorable in this arrangement.

69. $\odot^2 + 3\odot - 5\odot - 15 = \odot(\odot + 3) - 5(\odot + 3)$
$= (\odot + 3)(\odot - 5)$

71. a. $3x^2 + 10x + 8 = 3x^2 + 6x + 4x + 8$

b. $3x^2 + 6x + 4x + 8 = 3x(x + 2) + 4(x + 2)$
$= (x + 2)(3x + 4)$

73. a. $2x^2 - 11x + 15 = 2x^2 - 6x - 5x + 15$

b. $2x^2 - 6x - 5x + 15 = 2x(x - 3) - 5(x - 3)$
$= (x - 3)(2x - 5)$

75. a. $4x^2 - 17x - 15 = 4x^2 - 20x + 3x - 15$

b. $4x^2 - 20x + 3x - 15 = 4x(x - 5) + 3(x - 5)$
$= (4x + 3)(x - 5)$

77. $\star\odot + 3\star + 2\odot + 6 = \star(\odot + 3) + 2(\odot + 3)$
$= (\odot + 3)(\star + 2)$

79.
$$\begin{aligned} 5 - 3(2x - 7) &= 4(x + 5) - 6 \\ 5 - 6x + 21 &= 4x + 20 - 6 \\ -6x + 26 &= 4x + 14 \\ 12 &= 10x \\ \frac{12}{10} &= x \\ \frac{5}{6} &= x \end{aligned}$$

80. Let w = the number of pounds of chocolate wafers
p = the number of pounds of peppermint candies
$$w + p = 50$$
$$6.25w + 2.50p = 4.75(50)$$
or
$$w + p = 50$$
$$6.25w + 2.50p = 237.50$$

Multiply the first equation by -2.5 and then add.
$-2.5[w + p = 50]$
gives
$$\begin{aligned} -2.5w - 2.5p &= -125 \\ 6.25w + 2.5p &= 237.5 \\ \hline 3.75w \quad &= 112.5 \\ w &= 30 \end{aligned}$$
$$\begin{aligned} w + p &= 50 \\ 30 + p &= 50 \\ p &= 20 \end{aligned}$$
They should mix 30 pounds of chocolate wafers with 20 pounds of peppermint hard candies.

81. $\dfrac{15x^3 - 6x^2 - 9x + 5}{3x} = \dfrac{15x^3}{3x} - \dfrac{6x^2}{3x} - \dfrac{9x}{3x} + \dfrac{5}{3x}$
$= 5x^2 - 2x - 3 + \dfrac{5}{3x}$

82.
$$\begin{array}{r} x + 3 \\ x - 3 \overline{\smash{\big)} x^2 \qquad - 9} \\ \underline{x^2 - 3x} \qquad \\ 3x - 9 \\ \underline{3x - 9} \\ 0 \end{array}$$

$$\frac{x^2 - 9}{x - 3} = x + 3$$

Exercise Set 6.3

1. Since 8000 is positive, both signs will be the same. Since 180 is positive, both signs will be positive.

3. Since –8000 is negative, one sign will be positive, the other will be negative.

5. Since 8000 is positive, both signs will be the same. Since –240 is negative, both signs will be negative.

7. The trinomial $x^2 + 4xy - 12y^2$ is obtained by multiplying the factors using the FOIL method.

9. The trinomial $4a^2 - 4b^2$ is obtained by multiplying all the factors and combing like terms.

11. The answer is not fully factored. A 2 can be factored from $(2x - 4)$.

13. To determine the factors when factoring a trinomial of the form $x^2 + bx + c$. First, find two numbers whose product is c, and whose sum is b. The factors are (x + first number) and (x + second number).

15. $x^2 - 7x + 10 = (x-5)(x-2)$

17. $x^2 + 6x + 8 = (x+4)(x+2)$

19. $x^2 + 7x + 12 = (x+4)(x+3)$

21. $x^2 + 4x - 6$ is prime.

23. $y^2 - 13y + 12 = (y-12)(y-1)$

25. $a^2 - 2a - 8 = (a-4)(x+2)$

27. $r^2 - 2r - 15 = (r-5)(r+3)$

29. $b^2 - 11b + 18 = (b-9)(b-2)$

31. $x^2 - 8x - 15$ is prime.

33. $a^2 + 12a + 11 = (a+1)(a+11)$

35. $x^2 - 7x - 30 = (x-10)(x+3)$

37. $x^2 + 4x + 4 = (x+2)(x+2)$
$= (x+2)^2$

39. $p^2 + 6p + 9 = (p+3)(p+3)$
$= (p+3)^2$

41. $p^2 - 12p + 36 = (p-6)(p-6)$
$= (p-6)^2$

43. $w^2 - 18w + 45 = (w-15)(w-3)$

45. $x^2 + 10x - 39 = (x+13)(x-3)$

47. $x^2 - x - 20 = (x-5)(x+4)$

49. $y^2 + 9y + 14 = (y+7)(y+2)$

51. $x^2 + 12x - 64 = (x+16)(x-4)$

53. $s^2 + 14s - 24$ is prime.

55. $x^2 - 20x + 64 = (x-16)(x-4)$

57. $b^2 - 18b + 65 = (b-5)(b-13)$

59. $x^2 + 2 + 3x = x^2 + 3x + 2 = (x+2)(x+1)$

61. $7w - 18 + w^2 = w^2 + 7w - 18 = (w+9)(w-2)$

63. $x^2 - 8xy + 15y^2 = (x-3y)(x-5y)$

65. $m^2 - 6mn + 9n^2 = (m-3n)(m-3n)$
$= (m-3n)^2$

67. $x^2 + 8xy + 15y^2 = (x+3y)(x+5y)$

69. $m^2 - 5mn - 24n^2 = (m+3n)(m-8n)$

71. $6x^2 - 30x + 24 = 6(x^2 - 5x + 4)$
$= 6(x-4)(x-1)$

73. $5x^2 + 20x + 15 = 5(x^2 + 4x + 3)$
$= 5(x+1)(x+3)$

75. $2x^2 - 14x + 24 = 2(x^2 - 7x + 12)$
$= 2(x-4)(x-3)$

77. $b^3 - 7b^2 + 10b = b(b^2 - 7b + 10)$
$= b(b-5)(b-2)$

79. $3z^3 - 21z^2 - 54z = 3z(z^2 - 7z - 18)$
$= 3z(z - 9)(z + 2)$

81. $x^3 + 8x^2 + 16x = x(x^2 + 8x + 16)$
$= x(x + 4)(x + 4)$
$= x(x + 4)^2$

83. $4a^2 - 24ab + 32b^2 = 4(a^2 - 6b + 8b^2)$
$= 4(a - 4b)(a - 2b)$

85. $r^2s + 7rs^2 + 12s^3 = s(r^2 + 7rs + 12s^2)$
$= s(r + 3s)(r + 4s)$

87. $x^4 - 4x^3 - 21x^2 = x^2(x^2 - 4x - 21)$
$= x^2(x - 7)(x + 3)$

89.

Sign of Coefficient of x-term	Sign of Constant of Trinomial	Signs of Constant Terms in the Binomial Factors
–	+	both negative
–	–	one positive and one negative
+	–	one positive and one negative
+	+	both positive

91. $x^2 + 5x + 4 = (x + 1)(x + 4)$

93. $x^2 + 12x + 32 = (x + 8)(x + 4)$

95. $x^2 + 0.6x + 0.08 = (x + 0.4)(x + 0.2)$

97. $x^2 + \frac{2}{5}x + \frac{1}{25} = \left(x + \frac{1}{5}\right)\left(x + \frac{1}{5}\right)$
$= \left(x + \frac{1}{5}\right)^2$

99. $x^2 + 5x - 300 = (x + 20)(x - 15)$

101.
$$4(2x - 4) = 5x + 11$$
$$8x - 16 = 5x + 11$$
$$8x - 5x - 16 = 5x - 5x + 11$$
$$3x - 16 = 11$$
$$3x - 16 + 16 = 11 + 16$$
$$3x = 27$$
$$x = 9$$

102. Let x be the percent of acid in the mixture.

Solution	Strength	Liters	Amount
18%	0.18	4	0.72
26%	0.26	1	0.26
Mixture	$\frac{x}{100}$	5	$\frac{5x}{100}$

$$0.72 + 0.26 = \frac{5x}{100}$$
$$0.98 = \frac{x}{20}$$
$$19.6 = x$$

The mixture is a 19.6% acid solution.

103.

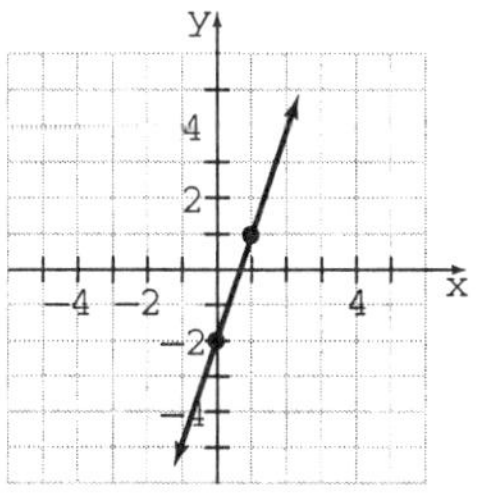

104.
$$\begin{aligned}(2x^2 + 5x - 6)(x - 2) &= 2x^2(x - 2) + 5x(x - 2) - 6(x - 2)\\ &= 2x^3 - 4x^2 + 5x^2 - 10x - 6x + 12\\ &= 2x^3 + x^2 - 16x + 12\end{aligned}$$

105.
$$\begin{array}{r|l} & 3x + 2 \\ x - 4 & 3x^2 - 10x - 10 \\ & \underline{3x^2 - 12x} \\ & \quad 2x - 10 \\ & \quad \underline{2x - 8} \\ & \qquad -2 \end{array}$$

$$\frac{3x^2 - 10x - 10}{x - 4} = 3x + 2 - \frac{2}{x - 4}$$

106.
$$\begin{aligned}3x^2 + 5x - 6x - 10 &= x(3x + 5) - 2(3x + 5)\\ &= (3x + 5)(x - 2)\end{aligned}$$

Exercise Set 6.4

1. Factoring trinomials is the reverse process of multiplying binomials.

3. When factoring` a trinomial of the form $ax^2 + bx + c$, the product of the constants in the binomial factors must equal the constant, c, of the trinomial.

5. $2x^2 + 11x + 5 = (2x + 1)(x + 5)$

7. $3x^2 + 14x + 8 = (3x + 2)(x + 4)$

9. $5x^2 - 9x - 2 = (5x + 1)(x - 2)$

11. $3r^2 + 13r - 10 = (3r - 2)(r + 5)$

13.
$$\begin{aligned}4z^2 - 12z + 9 &= (2z - 3)(2z - 3)\\ &= (2z - 3)^2\end{aligned}$$

15. $5y^2 - y - 4 = (5y + 4)(y - 1)$

17. $5a^2 - 12a + 6$ is prime.

19. $6z^2 + z - 12 = (2z + 3)(3z - 4)$

21. $3x^2 + 11x + 4$ is prime.

23. $5y^2 - 16y + 3 = (5y - 1)(y - 3)$

25. $7x^2 + 43x + 6 = (7x + 1)(x + 6)$

27. $7x^2 - 8x + 1 = (7x - 1)(x - 1)$

29. $5b^2 - 23b + 12 = (5b - 3)(b - 4)$

31. $5z^2 - 6z - 8 = (5z + 4)(z - 2)$

33. $4y^2 + 5y - 6 = (4y - 3)(y + 2)$

35. $10x^2 - 27x + 5 = (5x - 1)(2x - 5)$

37. $10d^2 - 7d - 12 = (5d + 4)(2d - 3)$

39. $$\begin{aligned} 6x^2 - 22x - 8 &= 2 \cdot 3x^2 - 2 \cdot 11x - 2 \cdot 4 \\ &= 2(3x^2 - 11x - 4) \\ &= 2(3x + 1)(x - 4) \end{aligned}$$

41. $$\begin{aligned} 10t + 3 + 7t^2 &= 7t^2 + 10t + 3 \\ &= (7t + 3)(t + 1) \end{aligned}$$

43. $$\begin{aligned} 6x^2 + 16x + 10 &= 2 \cdot 3x^2 + 2 \cdot 8x + 2 \cdot 5 \\ &= 2(3x^2 + 8x + 5) \\ &= 2(3x + 5)(x + 1) \end{aligned}$$

45. $$\begin{aligned} 6x^3 - 5x^2 - 4x &= x \cdot 6x^2 - x \cdot 5x - x \cdot 4 \\ &= x(6x^2 - 5x - 4) \\ &= x(2x + 1)(3x - 4) \end{aligned}$$

47. $$\begin{aligned} 12x^3 + 28x^2 + 8x &= 4x \cdot 3x^2 + 4x \cdot 7x + 4x \cdot 2 \\ &= 4x(3x^2 + 7x + 2) \\ &= 4x(3x + 1)(x + 2) \end{aligned}$$

49. $$\begin{aligned} 4x^3 - 2x^2 - 12x &= 2x \cdot 2x^2 - 2x \cdot x - 2x \cdot 6 \\ &= 2x(2x^2 - x - 6) \\ &= 2x(2x + 3)(x - 2) \end{aligned}$$

51. $$\begin{aligned} 36z^2 + 6z - 6 &= 6 \cdot 6z^2 + 6 \cdot z - 6 \cdot 1 \\ &= 6(6z^2 + z - 1) \\ &= 6(3z - 1)(2z + 1) \end{aligned}$$

53. $$\begin{aligned} 72 + 3r^2 - 30r &= 3r^2 - 30r + 72 \\ &= 3 \cdot r^2 - 3 \cdot 10r + 3 \cdot 24 \\ &= 3(r^2 - 10r + 24) \\ &= 3(r - 4)(r - 6) \end{aligned}$$

55. $2x^2 + 5xy + 2y^2 = (2x + y)(x + 2y)$

57. $2x^2 - 7xy + 3y^2 = (2x - y)(x - 3y)$

59. $$\begin{aligned} 12x^2 + 10xy - 8y^2 &= 2 \cdot 6x^2 + 2 \cdot 5xy - 2 \cdot 4y^2 \\ &= 2(6x^2 + 5xy - 4y^2) \\ &= 2(2x - y)(3x + 4y) \end{aligned}$$

61. $$\begin{aligned} 6x^2 - 9xy - 27y^2 &= 3 \cdot 2x^2 - 3 \cdot 3xy - 3 \cdot 9y^2 \\ &= 3(2x^2 - 3xy - 9y^2) \\ &= 3(x - 3y)(2x + 3y) \end{aligned}$$

63. $6m^2 - mn - 2n^2 = (3m - 2n)(2m + n)$

65. $$\begin{aligned} 8x^3 + 10x^2y + 3xy^2 &= x \cdot 8x^2 + x \cdot 10xy + x \cdot 3y^2 \\ &= x(8x^2 + 10xy + 3y^2) \\ &= x(4x + 3y)(2x + y) \end{aligned}$$

67. $$\begin{aligned} 4x^4 + 8x^3y + 3x^2y^2 &= x^2 \cdot 4x^2 + x^2 \cdot 8xy + x^2 \cdot 3y^2 \\ &= x^2(4x^2 + 8xy + 3y^2) \\ &= x^2(2x + y)(2x + 3y) \end{aligned}$$

69. $3x^2 - 20x - 7$. This polynomial was obtained by multiplying the factors.

71. $10x^2 + 35x + 15$. This polynomial was obtained by multiplying the factors.

73. $2x^4 - x^3 - 3x^2$. This polynomial was obtained by multiplying the factors.

75. a. The second factor can be found by dividing the trinomial by the binomial.

b. $$\begin{array}{r} 6x+11 \\ 3x+10\overline{)\,18x^2+93x+110} \\ \underline{18x^2+60x} \\ 33x+110 \\ \underline{33x+110} \\ 0 \end{array}$$

The other factor is $6x + 11$.

77. $18x^2 + 9x - 20 = (6x - 5)(3x + 4)$

79. $15x^2 - 124x + 160 = (5x - 8)(3x - 20)$

81. $$\begin{aligned} 72x^2 - 180x - 200 &= 4 \cdot 18x^2 - 4 \cdot 45x - 4 \cdot 50 \\ &= 4(18x^2 - 45x - 50) \\ &= 4(6x + 5)(3x - 10) \end{aligned}$$

83. The other factor is $2x + 45$. The product of the three first terms must equal $6x^3$, and the product of the constants must equal $2250x$.

85. $$\begin{aligned} &-x^2 - 4(y + 3) + 2y^2 \\ &= -(-3)^2 - 4(-5 + 3) + 2(-5)^2 \\ &= -9 - 4(-2) + 2(25) \\ &= -9 + 8 + 50 \\ &= 49 \end{aligned}$$

86. $\dfrac{500}{3.82} \approx 130.89$

His average speed was about 130.89 miles per hour.

87. $$\begin{aligned} 36x^4y^3 - 12xy^2 + 24x^5y^6 &= 12xy^2 \cdot 3x^3y - 12xy^2 \cdot 1 + 12xy^2 \cdot 2x^4y^4 \\ &= 12xy^2(3x^3y - 1 + 2x^4y^4) \end{aligned}$$

88. $x^2 - 15x + 54 = (x - 9)(x - 6)$

Exercise Set 6.5

1. a. $a^2 - b^2 = (a + b)(a - b)$

b. Answers will vary.

3. a. $a^3 - b^3 = (a - b)(a^2 + ab + b^2)$

b. Answers will vary.

5. No, there is no special formula for factoring the sum of two squares.

7. $x^2 + 9$ is prime.

9. $4a^2 + 16 = 4(a^2 + 4)$

11. $16m^2 + 36n^2 = 4(4m^2 + 9n^2)$

13. $y^2 - 25 = y^2 - 5^2 = (y + 5)(y - 5)$

15. $$\begin{aligned} z^2 - 81 &= z^2 - 9^2 \\ &= (z + 9)(z - 9) \end{aligned}$$

17. $$\begin{aligned} x^2 - 49 &= x^2 - 7^2 \\ &= (x + 7)(x - 7) \end{aligned}$$

19. $x^2 - y^2 = (x + y)(x - y)$

21. $$\begin{aligned} 9y^2 - 25z^2 &= (3y)^2 - (5z)^2 \\ &= (3y + 5z)(3y - 5z) \end{aligned}$$

23. $$\begin{aligned} 64a^2 - 36b^2 &= 4(16a^2 - 9b^2) \\ &= 4[(4a)^2 - (3b)^2] \\ &= 4(4a + 3b)(4a - 3b) \end{aligned}$$

25. $$\begin{aligned} 49x^2 - 36 &= (7x)^2 - 6^2 \\ &= (7x + 6)(7x - 6) \end{aligned}$$

27. $$\begin{aligned} z^4 - 81x^2 &= (z^2)^2 - (9x)^2 \\ &= (z^2 + 9x)(z^2 - 9x) \end{aligned}$$

29. $$\begin{aligned} 9x^4 - 81y^2 &= 9(x^4 - 9y^2) \\ &= 9[(x^2)^2 - (3y)^2] \\ &= 9(x^2 + 3y)(x^2 - 3y) \end{aligned}$$

31. $$\begin{aligned} 36m^4 - 49n^2 &= (6m^2)^2 - (7n)^2 \\ &= (6m^2 + 7n)(6m^2 - 7n) \end{aligned}$$

33. $10x^2 - 160 = 10(x^2 - 16)$
$= 10(x^2 - 4^2)$
$= 10(x+4)(x-4)$

35. $16x^2 - 100y^4 = 4(4x^2 - 25y^4)$
$= 4\left[(2x)^2 - (5y^2)^2\right]$
$= 4(2x + 5y^2)(2x - 5y^2)$

37. $x^3 + y^3 = (x+y)(x^2 - xy + y^2)$

39. $a^3 - b^3 = (a-b)(a^2 + ab + b^2)$

41. $x^3 + 8 = x^3 + 2^3$
$= (x+2)(x^2 - 2x + 4)$

43. $x^3 - 27 = x^3 - 3^3$
$= (x-3)(x^2 + 3x + 9)$

45. $a^3 + 1 = a^3 + 1^3$
$= (a+1)(a^2 - a + 1)$

47. $27x^3 - 1 = (3x)^3 - 1^3$
$= (3x-1)(9x^2 + 3x + 1)$

49. $27a^3 - 125 = (3a)^3 - 5^3$
$= (3a-5)(9a^2 + 15a + 25)$

51. $27 - 8y^3 = 3^3 - (2y)^3$
$= (3-2y)(9 + 6y + 4y^2)$

53. $64m^3 + 27n^3 = (4m)^3 + (3n)^3$
$= (4m + 3n)(16m^2 - 12mn + 9n^2)$

55. $8a^3 - 27b^3 = (2a)^3 - (3b)^3$
$= (2a - 3b)(4a^2 + 6ab + 9b^2)$

57. $2x^2 + 8x + 8 = 2(x^2 + 4x + 4)$
$= 2(x+2)^2$

59. $a^2b - 25b = b(a^2 - 25)$
$= b(a^2 - 5^2)$
$= b(a+5)(a-5)$

61. $3c^2 - 18c + 27 = 3(c^2 - 6c + 9)$
$= 3(c-3)^2$

63. $5x^2 - 10x - 15 = 5(x^2 - 2x - 3)$
$= 5(x-3)(x+1)$

65. $3xy - 6x + 9y - 18 = 3(xy - 2x + 3y - 6)$
$= 3[x(y-2) + 3(y-2)]$
$= 3(x+3)(y-2)$

67. $2x^2 - 50 = 2(x^2 - 25)$
$= 2(x^2 - 5^2)$
$= 2(x+5)(x-5)$

69. $3x^2y - 27y = 3y(x^2 - 9)$
$= 3y(x^2 - 3^2)$
$= 3y(x+3)(x-3)$

71. $3x^3y^2 + 3y^2 = 3y^2(x^3 + 1)$
$= 3y^2(x^3 + 1^3)$
$= 3y^2(x+1)(x^2 - x + 1)$

73. $2x^3 - 16 = 2(x^3 - 8)$
$= 2(x^3 - 2^3)$
$= 2(x-2)(x^2 + 2x + 4)$

75. $18x^2 - 50 = 2(9x^2 - 25)$
$= 2((3x)^2 - 5^2)$
$= 2(3x+5)(3x-5)$

77. $12x^2 + 36x + 27 = 3(4x^2 + 12x + 9)$
$= 3(2x+3)(2x+3)$
$= 3(2x+3)^2$

79. $6x^2 - 4x + 24x - 16 = 2(3x^2 - 2x + 12x - 8)$
$= 2[x(3x - 2) + 4(3x - 2)]$
$= 2(3x - 2)(x + 4)$

81. $2rs^2 - 10rs - 48r = 2r(s^2 - 5s - 24)$
$= 2r(s + 3)(s - 8)$

83. $4x^2 + 5x - 6 = (x + 2)(4x - 3)$

85. $25b^2 - 100 = 25(b^2 - 4)$
$= 25(b^2 - 2^2)$
$= 25(b + 2)(b - 2)$

87. $a^5b^2 - 4a^3b^4 = a^3b^2(a^2 - 4b^2)$
$= a^3b^2[a^2 - (2b)^2]$
$= a^3b^2(a + 2b)(a - 2b)$

89. $3x^4 - 18x^3 + 27x^2 = 3x^2(x^2 - 6x + 9)$
$= 3x^2(x - 3)(x - 3)$
$= 3x^2(x - 3)^2$

91. $x^3 + 25x = x(x^2 + 25)$

93. $y^4 - 16 = (y^2)^2 - 4^2$
$= (y^2 + 4)(y^2 - 4)$
$= (y^2 + 4)(y^2 - 2^2)$
$= (y^2 + 4)(y + 2)(y - 2)$

95. $36a^2 - 15ab - 6b^2 = 3(12a^2 - 5ab - 2b^2)$
$= 3(3a - 2b)(4a + b)$

97. $2ab - 3b + 4a - 6 = b(2a - 3) + 2(2a - 3)$
$= (2a - 3)(b + 2)$

99. $9 - 9y^4 = 9(1 - y^4)$
$= 9[1^2 - (y^2)^2]$
$= 9(1 + y^2)(1 - y^2)$
$= 9(1 + y^2)(1^2 - y^2)$
$= 9(1 + y^2)(1 + y)(1 - y)$

101. ◆✹ + 2◆ + ☺✹ + 2☺
= ◆(✹ + 2) + ☺(✹ + 2)
= (◆ + ☺)(✹ + 2)

103. 4◆^{2}✹ − 6◆✹ − 20✹◆ + 30✹
= 2✹(2◆^{2} − 3◆ − 10◆ + 15)
= 2✹(2◆^{2} − 13◆ + 15)
= 2✹(2◆ − 3)(◆ − 5)

105. $x^6 - 27y^9 = (x^2)^3 - (3y^3)^3$
$= (x^2 - 3y^3)(x^4 + 3x^2y^3 + 9y^6)$

107. $x^2 + 10x + 25 - y^2 + 4y - 4$
$= (x + 5)^2 - (y^2 - 4y + 4)$
$= (x + 5)^2 - (y - 2)^2$
$= [(x + 5) + (y - 2)][(x + 5) - (y - 2)]$
$= (x + y + 3)(x - y + 7)$

109. $7x - 2(x + 6) = 2x - 5$
$7x - 2x - 12 = 2x - 5$
$5x - 12 = 2x - 5$
$5x - 2x - 12 = 2x - 2x - 5$
$3x - 12 = -5$
$3x - 12 + 12 = -5 + 12$
$3x = 7$
$\frac{3x}{3} = \frac{7}{3}$
$x = \frac{7}{3}$

110. Substitute 36 for A, 6 for b, and 12 for d.

$$A = \frac{1}{2}h(b+d)$$
$$36 = \frac{1}{2}h(6+12)$$
$$36 = \frac{1}{2}h(18)$$
$$36 = 9h$$
$$4 = h$$

The height is 4 inches.

111. $x + (5 - 2x) = 2$
The sum of a number, and 5 decreased by twice the number is 2.

112.
$$\left(\frac{4x^4y}{6xy^5}\right)^3 = \left(\frac{4}{6}\cdot\frac{x^4}{x}\cdot\frac{y}{y^5}\right)^3 = \left(\frac{2}{3}\cdot x^3\cdot\frac{1}{y^4}\right)^3 = \left(\frac{2x^3}{3y^4}\right)^3 = \frac{2^3x^{3\cdot 3}}{3^3y^{4\cdot 3}} = \frac{8x^9}{27y^{12}}$$

113.
$$x^{-2}x^{-3} = x^{-2-3} = x^{-5} = \frac{1}{x^5}$$

Exercise Set 6.6

1. Answers will vary.

3. The standard form of a quadratic equation is $ax^2 + bx + c = 0$.

5. a. The zero-factor property may only be used when one side of the equation is equal to 0.

b.
$$(x+1)(x-2) = 4$$
$$x^2 - 2x + x - 2 = 4$$
$$x^2 - x - 6 = 0$$
$$(x-3)(x+2) = 0$$
$x - 3 = 0$ or $x + 2 = 0$
$x = 3$ $\qquad x = -2$

7. $x(x+2) = 0$
$x = 0$ or $x + 2 = 0$
$x = 0$ $\qquad x = -2$

9. $7x(x-8) = 0$
$x = 0$ or $x - 8 = 0$
$\qquad x = 8$

11. $(2x+5)(x-3) = 0$
$2x + 5 = 0$ or $x - 3 = 0$
$2x = -5$ $\qquad x = 3$
$x = -\frac{5}{2}$

13.
$$x^2 - 9 = 0$$
$$(x+3)(x-3) = 0$$
$x + 3 = 0$ or $x - 3 = 0$
$x = -3$ $\qquad x = 3$

15. $x^2 - 12x = 0$
$x(x-12) = 0$
$x = 0$ or $x - 12 = 0$
$\qquad x = 12$

17. $9x^2 + 27x = 0$
$9x(x+3) = 0$
$x = 0$ or $x + 3 = 0$
$\qquad x = -3$

19.
$$x^2 - 8x + 16 = 0$$
$$(x-4)(x-4) = 0$$
$$x - 4 = 0$$
$$x = 4$$

21.
$$x^2 + 12x = -20$$
$$x^2 + 12x + 20 = 0$$
$$(x+10)(x+2) = 0$$
$x + 10 = 0$ or $x + 2 = 0$
$x = -10$ $\qquad x = -2$

23. $z^2 + 3z = 18$
$z^2 + 3z - 18 = 0$
$(z+6)(z-3) = 0$
$z+6 = 0$ or $z-3 = 0$
$z = -6$ $z = 3$

25. $4a^2 - 4a - 48 = 0$
$4(a^2 - a - 12) = 0$
$4(a-4)(a+3) = 0$
$a-4 = 0$ or $a+3 = 0$
$a = 4$ $a = -3$

27. $23p - 24 = -p^2$
$p^2 + 23p - 24 = 0$
$(p+24)(p-1) = 0$
$p+24 = 0$ or $p-1 = 0$
$p = -24$ $p = 1$

29. $33w + 90 = -3w^2$
$3w^2 + 33w + 90 = 0$
$3(w^2 + 11w + 30) = 0$
$3(w+5)(w+6) = 0$
$w+5 = 0$ or $w+6 = 0$
$w = -5$ $w = -6$

31. $-2x - 15 = -x^2$
$x^2 - 2x - 15 = 0$
$(x-5)(x+3) = 0$
$x-5 = 0$ or $x+3 = 0$
$x = 5$ $x = -3$

33. $-x^2 + 29x + 30 = 0$
$x^2 - 29x - 30 = 0$
$(x-30)(x+1) = 0$
$x-30 = 0$ or $x+1 = 0$
$x = 30$ $x = -1$

35. $12 = 3n^2 + 16n$
$3n^2 + 16n - 12 = 0$
$(n+6)(3n-2) = 0$
$n+6 = 0$ or $3n-2 = 0$
$n = 6$ $3n = 2$
$n = \frac{2}{3}$

37. $9p^2 = -21p - 6$
$9p^2 + 21p + 6 = 0$
$3(3p^2 + 7p + 2) = 0$
$3(3p+1)(p+2) = 0$
$3p+1 = 0$ or $p+2 = 0$
$3p = -1$ $p = -2$
$p = -\frac{1}{3}$

39. $3r^2 + 13r = 10$
$3r^2 + 13r - 10 = 0$
$(3r-2)(r+5) = 0$
$3r-2 = 0$ or $r+5 = 0$
$3r = 2$ $r = -5$
$r = \frac{2}{3}$

41. $4x^2 + 4x - 48 = 0$
$4(x^2 + x - 12) = 0$
$4(x+4)(x-3) = 0$
$x+4 = 0$ or $x-3 = 0$
$x = -4$ $x = 3$

43. $8x^2 + 2x = 3$
$8x^2 + 2x - 3 = 0$
$(4x+3)(2x-1) = 0$
$4x+3 = 0$ or $2x-1 = 0$
$4x = -3$ $2x = 1$
$x = -\frac{3}{4}$ $x = \frac{1}{2}$

45. $2n^2 + 36 = -18n$
$2n^2 + 18n + 36 = 0$
$2(n^2 + 9n + 18) = 0$
$2(n+6)(n+3) = 0$
$n+6 = 0$ or $n+3 = 0$
$n = -6$ $n = -3$

47. $2x^2 = 50x$
$2x^2 - 50x = 0$
$2x(x-25) = 0$
$2x = 0$ or $x-25 = 0$
$x = 0$ $x = 25$

49.
$$x^2 = 100$$
$$x^2 - 100 = 0$$
$$x^2 - 10^2 = 0$$
$$(x+10)(x-10) = 0$$
$$x + 10 = 0 \quad \text{or} \quad x - 10 = 0$$
$$x = -10 \qquad x = 10$$

51. $(x-2)(x-1) = 12$
$$x^2 - 3x + 2 = 12$$
$$x^2 - 3x - 10 = 0$$
$$(x+2)(x-5) = 0$$
$$x + 2 = 0 \quad \text{or} \quad x - 5 = 0$$
$$x = -2 \qquad x = 5$$

53. $(x-1)(2x-5) = 9$
$$2x^2 - 7x + 5 = 9$$
$$2x^2 - 7x - 4 = 0$$
$$(2x+1)(x-4) = 0$$
$$2x + 1 = 0 \quad \text{or} \quad x - 4 = 0$$
$$x = -\tfrac{1}{2} \qquad x = 4$$

55.
$$x(x+5) = 6$$
$$x^2 + 5x = 6$$
$$x^2 + 5x - 6 = 0$$
$$(x+6)(x-1) = 0$$
$$x + 6 = 0 \quad \text{or} \quad x - 1 = 0$$
$$x = -6 \qquad x = 1$$

57. The solutions are 4 and –2, so the factors are $x - 4$ and $x + 2$.
$$(x-4)(x+2) = x^2 - 2x - 8$$
The equation is $x^2 - 2x - 8 = 0$.

59. The solutions are 6 and 0, so the factors are $x - 0$ and $x - 6$.
$$(x-0)(x-6) = x(x-6)$$
$$= x^2 - 6x$$
The equation is $x^2 - 6x = 0$.

61. **a.** The solutions are
$$x = \frac{1}{2} \quad \text{or} \quad x = -\frac{1}{3}$$
$$2x = 1 \qquad 3x = -1$$
$$2x - 1 = 0 \qquad 3x + 1 = 0$$
Thus the factors are:
$(2x-1)$ and $(3x+1)$.

b.
The equation is $(2x-1)(3x+1) = 6x^2 - x - 1 = 0$

63. $(x-3)(x-2) = (x+5)(2x-3) + 21$
$$x^2 - 5x + 6 = 2x^2 + 7x - 15 + 21$$
$$x^2 - 5x + 6 = 2x^2 + 7x + 6$$
$$0 = x^2 + 12x$$
$$0 = x(x+12)$$
$$x = 0 \quad \text{or} \quad x + 12 = 0$$
$$x = -12$$

65. $x(x-3)(x+2) = 0$
$$x = 0 \quad \text{or} \quad x - 3 = 0 \quad \text{or} \quad x + 2 = 0$$
$$x = 3 \qquad x = -2$$

67. $\frac{3}{5} - \frac{4}{7} = \frac{21}{35} - \frac{20}{35} = \frac{21-20}{35} = \frac{1}{35}$

68. **a.** Identity
b. Contradiction

69. Let x be the number of people admitted in 60 minutes.
$$\frac{160 \text{ people}}{13 \text{ minutes}} = \frac{x}{60 \text{ minutes}}$$
$$13x = 160(60)$$
$$13x = 9600$$
$$x \approx 738$$
About 738 people were admitted in 60 minutes.

70.
$$\left(\frac{2x^4y^5}{x^5y^7}\right)^3 = \left(2 \cdot x^{4-5} \cdot y^{5-7}\right)^3$$
$$= \left(2x^{-1}y^{-2}\right)^3$$
$$= \left(\frac{2}{xy^2}\right)^3$$
$$= \frac{2^3}{x^3y^{2\cdot 3}}$$
$$= \frac{8}{x^3y^6}$$

71. monomial

72. binomial

73. not a polynomial

74. trinomial

Exercise Set 6.7

1. A right triangle is a triangle with a 90° angle.

3. The Pythagorean Theorem is $a^2 + b^2 = c^2$.

5.
$$a^2 + b^2 = c^2$$
$$?^2 + 4^2 = 5^2$$
$$?^2 + 16 = 25$$
$$?^2 = 9$$
$$? = 3$$

7.
$$a^2 + b^2 = c^2$$
$$8^2 + 15^2 = ?^2$$
$$64 + 225 = ?^2$$
$$289 = ?^2$$
$$17 = ?$$

9.
$$a^2 + b^2 = c^2$$
$$3^2 + ?^2 = 5^2$$
$$9 + ?^2 = 25$$
$$?^2 = 16$$
$$? = 4$$

11.
$$a^2 + b^2 = c^2$$
$$15^2 + 36^2 = ?^2$$
$$225 + 1296 = ?^2$$
$$1521 = ?^2$$
$$39 = ?$$

13. Let x be the smaller of the two positive integers. Then $x + 4$ is the other integer.
$$x(x+4) = 117$$
$$x^2 + 4x + 117 = 0$$
$$(x-9)(x+13) = 0$$
$$x - 9 = 0 \text{ or } x + 13 = 0$$
$$x = 9 \qquad x = -13$$
Since x must be positive, the two integers are 9 and $9 + 4 = 13$.

15. Let x = first positive number. Then $2x + 2$ is the other number.
$$x(2x+2) = 84$$
$$2x^2 + 2x = 84$$
$$2x^2 + 2x - 84 = 0$$
$$2(x^2 + x - 42) = 0$$
$$2(x+7)(x-6) = 0$$
$$x + 7 = 0 \quad \text{or} \quad x - 6 = 0$$
$$x = -7 \qquad x = 6$$
The numbers have to be positive. Thus the numbers are 6 and $2x + 2 = 14$.

17. Let x be the smaller of the two consecutive positive odd integers. Then $x + 2$ is the other integer.
$$x(x+2) = 63$$
$$x^2 + 2x - 63 = 0$$
$$(x+9)(x-7) = 0$$
$$x - 9 = 0 \quad \text{or} \quad x - 7 = 0$$
$$x = -9 \qquad x = 7$$
Since x must be positive, the two integers are 7 and $7 + 2 = 9$.

19. Let w = width. Then length = $4w$.
$$A = lw$$
$$36 = (4w)(w)$$
$$36 = 4w^2$$
$$0 = 4w^2 - 36$$
$$0 = 4(w^2 - 9)$$
$$0 = 4(w+3)(w-3)$$
$$w + 3 = 0 \quad \text{or} \quad w - 3 = 0$$
$$w = -3 \qquad w = 3$$
Since dimensions must be positive, the width is 3 feet and the length is $4(3) = 12$ feet.

21. Let w = width of the garden, l = length of the garden.
$$w = \frac{2}{3}l$$
$$lw = 150$$
$$l\left(\frac{2}{3}l\right) = 150$$
$$2l^2 = 450$$
$$2l^2 - 450 = 0$$
$$2(l^2 - 225) = 0$$
$$2(l^2 - 15^2) = 0$$
$$2(l+15)(l-15) = 0$$

$l + 15 = 0$ or $l - 15 = 0$

$l = -15$ $\quad$ $l = 15$

Since dimensions must be positive, the length is 15 feet and the width is $\frac{2}{3}(15) = 10$ feet.

23. Let a = length of a side of the original square. Then $a + 4$ is the length of the new side.

$(a+4)(a+4) = 49$

$a^2 + 8a + 16 = 49$

$a^2 + 8a - 33 = 0$

$(a-3)(a+11) = 0$

$a - 3 = 0$ or $a + 11 = 0$

$a = 3$ or $a = -11$

Since length must be positive, the original square had sides of length 3 meters.

25. $d = 16t^2$

$256 = 16t^2$

$\frac{256}{16} = t^2$

$16 = t^2$

$0 = t^2 - 16$

$0 = (t+4)(t-4)$

$t + 4 = 0$ or $t - 4 = 0$

$t = -4$ $\quad$ $t = 4$

Since time must be positive, it would take the egg 4 seconds to hit the ground.

27. $a^2 + b^2 = c^2$

$7^2 + 24^2 = 25^2$

$49 + 576 = 625$

$625 = 625$ True

Since these values are true for the Pythagorean Theorem, a right triangle can exist.

29. $a^2 + b^2 = c^2$

$9^2 + 40^2 = 41^2$

$81 + 1600 = 1681$

$1681 = 1681$ True

Since these values are true for the Pythagorean Theorem, a right triangle can exist.

31. $a^2 + b^2 = c^2$

$30^2 + ?^2 = 34^2$

$900 + ?^2 = 1156$

$?^2 = 256$

$? = 16$ feet

33. $a^2 + b^2 = c^2$

$?^2 + 24^2 = 26^2$

$?^2 + 576 = 676$

$?^2 = 100$

$? = 10$ inches

35. Let x be the length of one leg of the triangle. Then $x + 2$ is the length of the other leg.

$a^2 + b^2 = c^2$

$x^2 + (x+2)^2 = 10^2$

$x^2 + x^2 + 4x + 4 = 100$

$2x^2 + 4x + 4 = 100$

$2x^2 + 4x - 96 = 0$

$2(x^2 + 2x - 48) = 0$

$2(x+8)(x-6) = 0$

$x + 8 = 0$ or $x - 6 = 0$

$x = -8$ $\quad$ $x = 6$

Since length is positive, the lengths are 6 ft, 8ft and 10ft.

37. Let d be the distance traveled by Bob. Then $3d - 3$ is the distance traveled by Alice and $2d + 3$ is the distance between Bob and Alice.

$a^2 + b^2 = c^2$

$d^2 + (3d-3)^2 = (2d+3)^2$

$d^2 + 9d^2 - 18d + 9 = 4d^2 + 12d + 9$

$10d^2 - 18d + 9 = 4d^2 + 12d + 9$

$6d^2 - 30d = 0$

$6d(d-5) = 0$

$6d = 0$ or $d - 5 = 0$

$d = 0$ $\quad$ $d = 5$

Since distance is positive, the distance between Bob and Alice is $2d + 3 = 2(5) + 3 = 13$ miles.

39. Let w be the width of the rectangle. Then $3w + 3$ is the length and $3w + 4$ is the length of the diagonal.

$$a^2 + b^2 = c^2$$
$$w^2 + (3w+3)^2 = (3w+4)^2$$
$$w^2 + 9w^2 + 18w + 9 = 9w^2 + 24w + 16$$
$$10w^2 + 18w + 9 = 9w^2 + 24w + 16$$
$$w^2 - 6w - 7 = 0$$
$$(w-7)(w+1) = 0$$
$$w - 7 = 0 \quad \text{or} \quad w + 1 = 0$$
$$w = 7 \qquad\qquad w = -1$$

Since length is positive, the dimensions of the garden are 7 ft. for the width and $3w + 3 = 3(7) + 3 = 24$ ft. for the length.

41.
$$P = x^2 - 15x - 50$$
$$x^2 - 15x - 50 = 400$$
$$x^2 - 15 - 450 = 0$$
$$(x-30)(x+15) = 0$$
$$x - 30 = 0 \quad \text{or} \quad x + 15 = 0$$
$$x = 30 \qquad\qquad x = -15$$

Since x must be positive, she must sell 30 videos for a profit of $400.

43. a.
$$n^2 + n = 20$$
$$n^2 + n - 20 = 0$$
$$(n-4)(n+5) = 0$$
$$n - 4 = 0 \quad \text{or} \quad n + 5 = 0$$
$$n = 4 \qquad\qquad n = -5$$

Since n must be positive, $n = 4$.

b.
$$n^2 + n = 90$$
$$n^2 + n - 90 = 0$$
$$(n-9)(n+10) = 0$$
$$n - 9 = 0 \quad \text{or} \quad n + 10 = 0$$
$$n = 9 \qquad\qquad n = -10$$

Since n must be positive, $n = 9$.

45. Before area can be determined, the length of the rectangle must be found first. Let x be the length of the rectangle.

$$a^2 + b^2 = c^2$$
$$x^2 + 18^2 = 30^2$$
$$x^2 + 324 = 900$$
$$x^2 = 576$$
$$x = 24$$

Area of a rectangle can be found by multiplying the length and width.

$$A = lw$$
$$= 24 \cdot 18$$
$$= 432$$

The area of the rectangle is 432 square feet.

47.
$$x^3 + 3x^2 - 10x = 0$$
$$x(x^2 + 3x - 10) = 0$$
$$x(x+5)(x-2) = 0$$
$$x = 0 \quad \text{or} \quad x + 5 = 0 \quad \text{or} \quad x - 2 = 0$$
$$x = -5 \qquad\qquad x = 2$$

49. The numbers are the constants in the binomials that are factors of $x^2 + 3x - 40$.

$x + 3x - 40 = (x-5)(x+8)$

The numbers are –5 and 8.

53. "Five less than twice a number" is $2x - 5$.

54.
$$(3x+2) - (x^2 - 4x + 6) = 3x + 2 - x^2 + 4x - 6$$
$$= -x^2 + 3x + 4x + 2 - 6$$
$$= -x^2 + 7x - 4$$

55.
$$(3x^2 + 2x - 4)(2x - 1) = 3x^2(2x-1) + 2x(2x-1) - 4(2x-1)$$
$$= 6x^3 - 3x^2 + 4x^2 - 2x - 8x + 4$$
$$= 6x^3 + x^2 - 10x + 4$$

56.
$$\begin{array}{r} 2x-3 \\ 3x-5\overline{)\ 6x^2-19x+15} \\ \underline{6x^2-10x} \\ -9x+15 \\ \underline{-9x+15} \\ 0 \end{array}$$

$$\frac{6x^2-19x+15}{3x-5}=2x-3$$

57. $\frac{6x^2-19x+15}{3x-5}=\frac{(3x-5)(2x-3)}{3x-5}$
$=2x-3$

Review Exercises

1. The greatest common factor is y^3.
2. The greatest common factor is $3p$.
3. The greatest common factor is $5a^2$.
4. The greatest common factor is $5x^2y^2$.
5. The greatest common factor is 1.
6. The greatest common factor is 1.
7. The greatest common factor is $x-5$.
8. The greatest common factor is $x+5$.
9. $4x-12=4(x-3)$
10. $35x-5=5(7x-1)$
11. $24y^2-4y=4y(6y-1)$
12. $55p^3-20p^2=5p^2(11p-4)$
13. $60a^2b-36ab^2=12ab(5a-3b)$
14. $6xy-12x^2y=6xy(1-2x)$
15. $20x^3y^2+8x^9y^3-16x^5y^2$
$=4x^3y^2(5+2x^6y-4x^2)$
16. $24x^2-13y^2+6xy$ is prime.
17. $14a^2b-7b-a^3$ is prime.
18. $x(5x+3)-2(5x+3)=(5x+3)(x-2)$
19. $5x(x+2)-2(x+2)=(x+2)(5x-2)$
20. $2x(4x-3)+4x-3=2x(4x-3)+1(4x-3)$
$=(4x-3)(2x+1)$
21. $x^2+6x+2x+12=x(x+6)+2(x+6)$
$=(x+6)(x+2)$
22. $x^2-5x+4x-20=x(x-5)+4(x-5)$
$=(x-5)(x+4)$
23. $y^2-9y-9y+81=y(y-9)-9(y-9)$
$=(y-9)(y-9)$
$=(y-9)^2$
24. $4a^2-4ab-a+b=4a(a-b)-1(a-b)$
$=(a-b)(4a-1)$
25. $3xy+3x+2y+2=3x(y+1)+2(y+1)$
$=(y+1)(3x+2)$
26. $x^2+3x-2xy-6y=x(x+3)-2y(x+3)$
$=(x+3)(x-2y)$
27. $2x^2+12x-x-6=2x(x+6)-1(x+6)$
$=(x+6)(2x-1)$
28. $5x^2-xy+20xy-4y^2=x(5x-y)+4y(5x-y)$
$=(5x-y)(x+4y)$
29. $4x^2+12xy-5xy-15y^2=4x(x+3y)-5y(x+3y)$
$=(x+3y)(4x-5y)$
30. $6a^2-10ab-3ab+5b^2=2a(3a-5b)-b(3a-5b)$
$=(3a-5b)(2a-b)$
31. $ab-a+b-1=a(b-1)+1(b-1)$
$=(b-1)(a+1)$

32. $3x^2 - 9xy + 2xy - 6y^2 = 3x(x - 3y) + 2y(x - 3y)$
$= (x - 3y)(3x + 2y)$

33. $7a^2 + 14ab - ab - 2b^2 = 7a(a + 2b) - b(a + 2b)$
$= (a + 2b)(7a - b)$

34. $6x^2 + 9x - 2x - 3 = 3x(2x + 3) - 1(2x + 3)$
$= (2x + 3)(3x - 1)$

35. $x^2 - x - 6 = (x + 2)(x - 3)$

36. $x^2 + 4x - 15$ is prime.

37. $x^2 - 13x + 42 = (x - 6)(x - 7)$

38. $b^2 + b - 20 = (b - 4)(b + 5)$

39. $n^2 + 3n - 40 = (n + 8)(n - 5)$

40. $x^2 - 15x + 56 = (x - 8)(x - 7)$

41. $c^2 - 10c - 20$ is prime.

42. $x^2 + 11x - 24$ is prime.

43. $x^3 - 17x^2 + 72x = x(x^2 - 17x + 72)$
$= x(x - 9)(x - 8)$

44. $x^3 - 3x^2 - 40x = x(x^2 - 3x - 40)$
$= x(x - 8)(x + 5)$

45. $x^2 - 2xy - 15y^2 = (x - 5y)(x + 3y)$

46. $4x^3 + 32x^2y + 60xy^2 = 4x(x^2 + 8xy + 15y^2)$
$= 4x(x + 3y)(x + 5y)$

47. $2x^2 - x - 15 = (2x + 5)(x - 3)$

48. $3x^2 - 13x + 4 = (3x - 1)(x - 4)$

49. $4x^2 - 9x + 5 = (4x - 5)(x - 1)$

50. $5m^2 - 14m + 8 = (5m - 4)(m - 2)$

51. $9x^2 + 3x - 2 = (3x - 1)(3x + 2)$

52. $5x^2 - 32x + 12 = (5x - 2)(x - 6)$

53. $2t^2 + 14t + 9$ is prime.

54. $6s^2 + 13s + 5 = (2s + 1)(3s + 5)$

55. $5x^2 + 37x - 24 = (5x - 3)(x + 8)$

56. $6x^2 + 11x - 10 = (3x - 2)(2x + 5)$

57. $12x^2 + 2x - 4 = 2(6x^2 + x - 2)$
$= 2(3x + 2)(2x - 1)$

58. $9x^2 - 6x + 1 = (3x - 1)(3x - 1)$
$= (3x - 1)^2$

59. $9x^3 - 12x^2 + 4x = x(9x^2 - 12x + 4)$
$= x(3x - 2)(3x - 2)$
$= x(3x - 2)^2$

60. $18x^3 + 12x^2 - 16x = 2x(9x^2 + 6x - 8)$
$= 2x(3x + 4)(3x - 2)$

61. $16a^2 - 22ab - 3b^2 = (8a + b)(2a - 3b)$

62. $4a^2 - 16ab + 15b^2 = (2a - 3b)(2a - 5b)$

63. $x^2 - 36 = x^2 - 6^2$
$= (x + 6)(x - 6)$

64. $x^2 - 100 = x^2 - 10^2$
$= (x + 10)(x - 10)$

65. $4x^2 - 16 = 4(x^2 - 4)$
$= 4(x^2 - 2^2)$
$= 4(x + 2)(x - 2)$

66. $81x^2 - 9y^2 = 9(9x^2 - y^2)$
$= 9[(3x)^2 - y^2]$
$= 9(3x + y)(3x - y)$

67. $81 - a^2 = 9^2 - a^2$
$= (9 + a)(9 - a)$

68. $64 - x^2 = 8^2 - x^2$
$= (8 + x)(8 - x)$

69. $16x^4 - 49y^2 = (4x^2)^2 - (7y)^2$
$= (4x^2 + 7y)(4x^2 - 7y)$

70. $100x^4 - 121y^4 = (10x^2)^2 - (11y^2)^2$
$= (10x^2 + 11y^2)(10x^2 - 11y^2)$

71. $x^3 - y^3 = (x - y)(x^2 + xy + y^2)$

72. $x^3 + y^3 = (x + y)(x^2 - xy + y^2)$

73. $x^3 - 1 = (x - 1)(x^2 + x + 1)$

74. $x^3 + 8 = x^3 + 2^3$
$= (x + 2)(x^2 - 2x + 4)$

75. $a^3 + 27 = a^3 + 3^3$
$= (a + 3)(a^2 - 3a + 9)$

76. $b^3 - 64 = b^3 - 4^3$
$= (b - 4)(b^2 + 4b + 16)$

77. $125a^3 + b^3 = (5a)^3 + b^3$
$= (5a + b)(25a^2 - 5ab + b^2)$

78. $27 - 8y^3 = 3^3 - (2y)^3$
$= (3 - 2y)(9 + 6y + 4y^2)$

79. $27x^4 - 75y^2 = 3(9x^4 - 25y^2)$
$= 3\left[(3x^2)^2 - (5y)^2\right]$
$= 3(3x^2 + 5y)(3x^2 - 5y)$

80. $3x^3 - 192y^3 = 3(x^3 - 64y^3)$
$= 3\left[x^3 - (4y)^3\right]$
$= 3(x - 4y)(x^2 + 4xy + 16y^2)$

81. $x^2 - 14x + 48 = (x - 6)(x - 8)$

82. $3x^2 - 18x + 27 = 3(x^2 - 6x + 9)$
$= 3(x - 3)^2$

83. $4a^2 - 64 = 4(a^2 - 16)$
$= 4(a^2 - 4^2)$
$= 4(a + 4)(a - 4)$

84. $4y^2 - 36 = 4(y^2 - 9)$
$= 4(y^2 - 3^2)$
$= 4(y + 3)(y - 3)$

85. $8x^2 + 16x - 24 = 8(x^2 + 2x - 3)$
$= 8(x + 3)(x - 1)$

86. $x^2 - 6x - 27 = (x - 9)(x + 3)$

87. $9x^2 - 6x + 1 = (3x - 1)(3x - 1)$
$= (3x - 1)^2$

88. $4x^2 + 7x - 2 = (4x - 1)(x + 2)$

89. $6x^3 - 6 = 6(x^3 - 1)$
$= 6(x^3 - 1^3)$
$= 6(x - 1)(x^2 + x + 1)$

90. $x^3y - 27y = y(x^3 - 27)$
$= y(x^3 - 3^3)$
$= y(x - 3)(x^2 + 3x + 9)$

91. $a^2b - 2ab - 15b = b(a^2 - 2a - 15)$
$= b(a + 3)(a - 5)$

92. $6x^3 + 30x^2 + 9x^2 + 45x = 3x(2x^2 + 10x + 3x + 15)$
$= 3x[2x(x + 5) + 3(x + 5)]$
$= 3x(2x + 3)(x + 5)$

93. $x^2 - 4xy + 3y^2 = (x - 3y)(x - y)$

94. $3m^2 + 2mn - 8n^2 = (3m - 4n)(m + 2n)$

95. $4x^2 - 20xy + 25y^2 = (2x - 5y)(2x - 5y)$
$= (2x - 5y)^2$

96. $25a^2 - 49b^2 = (5a)^2 - (7b)^2$
$= (5a + 7b)(5a - 7b)$

97. $xy - 7x + 2y - 14 = x(y - 7) + 2(y - 7)$
$= (x + 2)(y - 7)$

98. $16y^5 - 25y^7 = y^5(16 - 25y^2)$
$= y^5[4^2 - (5y)^2]$
$= y^5(4 + 5y)(4 - 5y)$

99. $4x^3 + 18x^2y + 20xy^2 = 2x(2x^2 + 9xy + 10y^2)$
$= 2x(2x + 5y)(x + 2y)$

100. $6x^2 + 5xy - 21y^2 = (2x - 3y)(3x + 7y)$

101. $16x^4 - 8x^3 - 3x^2 = x^2(16x^2 - 8x - 3)$
$= x^2(4x + 1)(4x - 3)$

102. $a^4 - 1 = (a^2)^2 - 1^2$
$= (a^2 + 1)(a^2 - 1)$
$= (a^2 + 1)(a + 1)(a - 1)$

103. $x(x - 5) = 0$
$x = 0$ or $x - 5 = 0$
$x = 5$

104. $(a - 2)(a + 6) = 0$
$a - 2 = 0$ or $a + 6 = 0$
$a = 2$ $a = -6$

105. $(x + 5)(4x - 3) = 0$
$x + 5 = 0$ or $4x - 3 = 0$
$x = -5$ $4x = 3$
$x = \frac{3}{4}$

106. $x^2 - 3x = 0$
$x(x - 3) = 0$
$x = 0$ or $x - 3 = 0$
$x = 3$

107. $5x^2 + 20x = 0$
$5x(x + 4) = 0$
$x = 0$ or $x + 4 = 0$
$x = -4$

108. $6x^2 + 18x = 0$
$6x(x + 3) = 0$
$x = 0$ or $x + 3 = 0$
$x = -3$

109. $r^2 + 9r + 18 = 0$
$(r + 3)(r + 6) = 0$
$r + 3 = 0$ or $r + 6 = 0$
$r = -3$ $r = -6$

110. $x^2 - 12 = -x$
$x^2 + x - 12 = 0$
$(x + 4)(x - 3) = 0$
$x + 4 = 0$ or $x - 3 = 0$
$x = -4$ $x = 3$

111. $x^2 - 3x = -2$
$x^2 - 3x + 2 = 0$
$(x - 1)(x - 2) = 0$
$x - 1 = 0$ or $x - 2 = 0$
$x = 1$ $x = 2$

112. $15x + 12 = -3x^2$
$3x^2 + 15x + 12 = 0$
$3(x^2 + 5x + 4) = 0$
$3(x + 1)(x + 4) = 0$
$x + 1 = 0$ or $x + 4 = 0$
$x = -1$ $x = -4$

113. $x^2 - 6x + 8 = 0$
$(x - 4)(x - 2) = 0$
$x - 4 = 0$ or $x - 2 = 0$
$x = 4$ $x = 2$

114. $3p^2 + 6p = 45$
$3p^2 + 6p - 45 = 0$
$3(p^2 + 2p - 15) = 0$
$3(x - 3)(x + 5) = 0$
$x - 3 = 0$ or $x + 5 = 0$
$x = 3$ $x = -5$

115.
$$8x^2 - 3 = -10x$$
$$8x^2 + 10x - 3 = 0$$
$$(4x - 1)(2x + 3) = 0$$
$$4x - 1 = 0 \quad \text{or} \quad 2x + 3 = 0$$
$$4x = 1 \qquad 2x = -3$$
$$x = \frac{1}{4} \qquad x = -\frac{3}{2}$$

116.
$$2x^2 + 15x = 8$$
$$2x^2 + 15x - 8 = 0$$
$$(2x - 1)(x + 8) = 0$$
$$2x - 1 = 0 \quad \text{or} \quad x + 8 = 0$$
$$2x = 1 \qquad x = -8$$
$$x = \frac{1}{2}$$

117.
$$4x^2 - 16 = 0$$
$$4(x^2 - 4) = 0$$
$$4(x^2 - 2^2) = 0$$
$$4(x + 2)(x - 2) = 0$$
$$x + 2 = 0 \quad \text{or} \quad x - 2 = 0$$
$$x = -2 \qquad x = 2$$

118.
$$49x^2 - 100 = 0$$
$$(7x)^2 - 10^2 = 0$$
$$(7x + 10)(7x - 10) = 0$$
$$7x + 10 = 0 \quad \text{or} \quad 7x - 10 = 0$$
$$7x = -10 \qquad 7x = 10$$
$$x = -\frac{10}{7} \qquad x = \frac{10}{7}$$

119.
$$8x^2 - 14x + 3 = 0$$
$$(2x - 3)(4x - 1) = 0$$
$$2x - 3 = 0 \quad \text{or} \quad 4x - 1 = 0$$
$$2x = 3 \qquad 4x = 1$$
$$x = \frac{3}{2} \qquad x = \frac{1}{4}$$

120.
$$-48x = -12x^2 - 45$$
$$12x^2 - 48x + 45 = 0$$
$$3(4x^2 - 16x + 15) = 0$$
$$3(2x - 3)(2x - 5) = 0$$
$$2x - 3 = 0 \quad \text{or} \quad 2x - 5 = 0$$
$$2x = 3 \qquad 2x = 5$$
$$x = \frac{3}{2} \qquad x = \frac{5}{2}$$

121. $a^2 + b^2 = c^2$

122. hypotenuse

123. $a^2 + b^2 = c^2$
$$?^2 + 5^2 = 13^2$$
$$?^2 + 25 = 169$$
$$?^2 = 144$$
$$? = 12$$

124. $a^2 + b^2 = c^2$
$$6^2 + 8^2 = ?^2$$
$$36 + 64 = ?^2$$
$$100 = ?^2$$
$$10 = ?$$

125. Let x be the smaller integer. The larger is $x + 2$.
$$x(x + 2) = 48$$
$$x^2 + 2x = 48$$
$$x^2 + 2x - 48 = 0$$
$$(x + 8)(x - 6) = 0$$
$$x + 8 = 0 \quad \text{or} \quad x - 6 = 0$$
$$x = -8 \qquad x = 6$$
Since the integers must be positive, they are 6 and 8.

126. Let x be the smaller integer. Then the larger is $2x + 6$.
$$x(2x + 6) = 56$$
$$2x^2 + 6x = 56$$
$$2x^2 + 6x - 56 = 0$$
$$2(x^2 + 3x - 28) = 0$$
$$(x + 7)(x - 4) = 0$$
$$x + 7 = 0 \quad \text{or} \quad x - 4 = 0$$
$$x = -7 \qquad x = 4$$
Since the integers must be positive, they are 4 and 14.

127. Let w be the width of the rectangle. Then the length is $w + 2$.
$$w(w + 2) = 63$$
$$w^2 + 2w = 63$$
$$w^2 + 2w - 63 = 0$$
$$(w + 9)(w - 7) = 0$$
$$w + 9 = 0 \quad \text{or} \quad w - 7 = 0$$
$$w = -9 \qquad w = 7$$
Since the width must be positive, it is 7 feet, and the length is 9 feet.

128. Let x be the length of a side of the original square. Then $x-4$ is the length of a side of the smaller square.

$$(x-4)^2 = 25$$
$$x^2 - 8x + 16 = 25$$
$$x^2 - 8x + 16 - 25 = 0$$
$$x^2 - 8x - 9 = 0$$
$$(x-9)(x+1) = 0$$
$$x-9=0 \quad \text{or} \quad x+1=0$$
$$x = 9 \qquad x = -1$$

Since lengths must be positive, the length of a side of the original square is 9 inches.

129. Let x be the length of one leg of the triangle. Then $x+7$ is the length of the other leg and $x+9$ is the length of the hypotenuse.

$$a^2 + b^2 = c^2$$
$$x^2 + (x+7)^2 = (x+9)^2$$
$$x^2 + x^2 + 14x + 49 = x^2 + 18x + 81$$
$$2x^2 + 14x + 49 = x^2 + 18x + 81$$
$$x^2 - 4x - 32 = 0$$
$$(x-8)(x+4) = 0$$
$$x-8=0 \quad \text{or} \quad x+4=0$$
$$x = 8 \qquad x = -4$$

Since lengths must be positive, the lengths of the three sides are 8ft, 15ft, and 17ft.

130. Let w be the width of the pool. Then the length is $w+2$ and the diagonal is $w+4$.

$$a^2 + b^2 = c^2$$
$$w^2 + (w+2)^2 = (w+4)^2$$
$$w^2 + w^2 + 4x + 4 = w^2 + 8x + 16$$
$$2w^2 + 4x + 4 = w^2 + 8x + 16$$
$$w^2 - 4w - 12 = 0$$
$$(w-6)(w+2) = 0$$
$$w-6=0 \quad \text{or} \quad w+2=0$$
$$w = 6 \qquad w = -2$$

Since lengths must be positive, the diagonal is $w+4 = 6+4 = 10$ ft.

131. $d = 16t^2$

$$16 = 16t^2$$
$$1 = t^2$$
$$1 = t$$

It will take 1 second for the apple to hit the ground.

132. $C = x^2 - 79x + 20$

$$100 = x^2 - 79x + 20$$
$$0 = x^2 - 79x - 80$$
$$0 = (x-80)(x+1)$$
$$x-80=0 \quad \text{or} \quad x+1=0$$
$$x = 80 \qquad x = -1$$

Since only a positive number of dozens of cookies can be made, the association can make 80 dozen cookies.

Practice Test

1. The greatest common factor is $3y^3$.

2. The greatest common factor is $3xy^2$.

3. $5x^2y^3 - 15x^5y^2 = 5x^2y^2(y - 3x^3)$

4. $8a^3b - 12a^2b^2 + 28a^2b = 4a^2b(2a - 3b + 7)$

5. $5x^2 - 15x + 2x - 6 = 5x(x-3) + 2(x-3)$
$= (x-3)(5x+2)$

6. $a^2 - 4ab - 5ab + 20b^2 = a(a-4b) - 5b(a-4b)$
$= (a-4b)(a-5b)$

7. $r^2 + 5r - 24 = (r+8)(r-3)$

8. $25a^2 - 5ab - 6b^2 = (5a-3b)(5a+2b)$

9. $4x^2 - 16x - 48 = 4(x^2 - 4x - 12)$
$= 4(x+2)(x-6)$

10. $2x^3 - 3x^2 + x = x(2x^2 - 3x + 1)$
$= x(2x-1)(x-1)$

11. $12x^2 - xy - 6y^2 = (3x+2y)(4x-3y)$

12. $x^2 - 9y^2 = x^2 - (3y)^2$
$= (x+3y)(x-3y)$

13. $x^3 + 27 = x^3 + 3^3$
$= (x+3)(x^2 - 3x + 9)$

14. $(5x-3)(x-1)=0$
$5x-3=0$ or $x-1=0$
$5x=3$ $\quad x=1$
$x=\frac{3}{5}$

15. $x^2-6x=0$
$x(x-6)=0$
$x=0$ or $x-6=0$
$x=6$

16. $x^2=64$
$x^2-64=0$
$x^2-8^2=0$
$(x+8)(x-8)=0$
$x+8=0$ or $x-8=0$
$x=-8$ $\quad x=8$

17. $x^2-14x+49=0$
$(x-7)^2=0$
$x-7=0$
$x=7$

18. $x^2+6=-5x$
$x^2+5x+6=0$
$(x+2)(x+3)=0$
$x-2=0$ or $x+3=0$
$x=-2$ $\quad x=-3$

19. $x^2-7x+12=0$
$(x-3)(x-4)=0$
$x-3=0$ or $x-4=0$
$x=3$ $\quad x=4$

20. Use Pythagorean Theorem.
$a^2+b^2=c^2$
$?^2+10^2=26^2$
$?^2+100=676$
$?^2=576$
$?=24$ in.

21. Let x be the length of one leg. Then $2x-2$ is the length of the other leg and $2x+2$ is the length of the hypotenuse.
$x^2+(2x-2)^2=(2x+2)^2$
$x^2+4x^2-8x+4=4x^2+8x+4$
$5x^2-8x+4=4x^2+8x+4$
$x^2-16x=0$
$x(x-16)=0$
$x=0$ or $x-16=0$
$x=16$
Since length has to be positive, the hypotenuse is $2x+2=2(16)+2=32+2=34$ ft.

22. Let x be the smaller of the two integers. Then $2x+1$ is the larger.
$x(2x+1)=36$
$2x^2+x-36=0$
$(x-4)(2x+9)=0$
$x-4=0$ or $2x+9=0$
$x=4$ $\quad 2x=-9$
$x=-\frac{9}{2}$
Since x must be positive and an integer, the smaller integer is 4 and the larger is $2\cdot 4+1=9$.

23. Let x be the smaller of the two consecutive odd integers. Then $x+2$ is the larger.
$x(x+2)=99$
$x^2+2x-99=0$
$(x-9)(x+11)=0$
$x-9=0$ or $x+11=0$
$x=9$ $\quad x=-11$
Since x must be positive, then the smaller integer is 9 and the larger is 11.

24. Let w be the width of the rectangle. Then the length is $w+2$.
$w(w+2)=24$
$w^2+2w=24$
$w^2+2w-24=0$
$(w+6)(w-4)=0$
$w+6=0$ or $w-4=0$
$w=-6$ $\quad w=4$
Since the width is positive, it is 4 meters, and the length is 6 meters.

25. $d = 16t^2$

$$1600 = 16t^2$$
$$16t^2 - 1600 = 0$$
$$16(t^2 - 100) = 0$$
$$16(t^2 - 10^2) = 0$$
$$16(t + 10)(t - 10) = 0$$
$$t + 10 = 0 \quad \text{or} \quad t - 10 = 0$$
$$t = -10 \qquad t = 10$$

Since time must be positive, then it would take the object 10 seconds to fall 1600 feet to the ground..

Cumulative Review Test

1. $4 - 5(2x + 4x^2 - 21) = 4 - 5[2(-3) + 4(-3)^2 - 21]$
$= 4 - 5[-6 + 4(9) - 21]$
$= 4 - 5(-6 + 36 - 21)$
$= 4 - 5(9)$
$= 4 - 45$
$= -41$

2. $5x^2 - 3y + 7(2 + y^2 - 4x) = 5(3)^2 - 3(-2) + 7[2 + (-2)^2 - 4(3)]$
$= 5(9) + 6 + 7(2 + 4 - 12)$
$= 45 + 6 + 7(-6)$
$= 51 - 42$
$= 9$

3. Let x = the cost of the room before tax.
$x + (.15x) = 103.50$
$1.15x = 103.50$
$x = 90$
The hotel room costs \$90 before taxes.

4. a. 7 is a natural number

b. $-6,\ -0.2,\ \frac{3}{5},\ 7,\ 0,\ -\frac{5}{9}$, and 1.34 are rational numbers.

c. $\sqrt{7}$ and $-\sqrt{2}$ are irrational numbers.

d. All of the numbers are real numbers.

5. $|-4|$ is greater than $-|2|$ since $-|2| = -(2) = -2$.

6. $4x - 2 = 4(x - 7) + 2x$
$4x - 2 = 4x - 28 + 2x$
$4x - 2 = 6x - 28$
$26 = 2x$
$13 = x$

7. $\frac{5}{12} = \frac{8}{x}$
$5x = 8(12)$
$5x = 96$
$x = \frac{96}{5} = 19.2$

8. Let x = the number of gallons of paint needed.

$$\frac{1 \text{ gallon}}{825 \text{ sq. ft.}} = \frac{x \text{ gallons}}{5775 \text{ sq. ft.}}$$

$$\frac{1}{825} = \frac{x}{5775}$$

$$825x = 5775$$

$$\frac{825x}{825} = \frac{5775}{825}$$

$$x = 7$$

The paint needed to cover the house is 7 gallons.

9. $5x - 2y = 6$

$$-2y = -5x + 6$$

$$y = \frac{-5x + 6}{-2}$$

$$y = \frac{5}{2}x - 3$$

10. Let t = the number of hours that Brooke has been skiing. Then Bob has been skiing for $\left(t + \frac{1}{4}\right)$ hours.

	Rate	Time	Distance
Brooke	8 kph	t	$8t$
Bob	4 kph	$t + \frac{1}{4}$	$4\left(t + \frac{1}{4}\right)$

Brooke catches Bob when they have both gone the same distance.

$$8t = 4\left(t + \frac{1}{4}\right)$$

$$8t = 4t + 1$$

$$4t = 1$$

$$t = \frac{1}{4}$$

It will take Brooke $\frac{1}{4}$ hour to catch Bob.

11. Let x be the amount of 10% acid solution needed.

$$0.10x + 0.04(3) = 0.08(x + 3)$$

$$0.10x + 0.12 = 0.08x + 0.24$$

$$0.02x + 0.12 = 0.24$$

$$0.02x = 0.12$$

$$x = 6$$

Six liters of the 10% solution is needed.

12. $m = -\frac{3}{5}$ $b = 1$

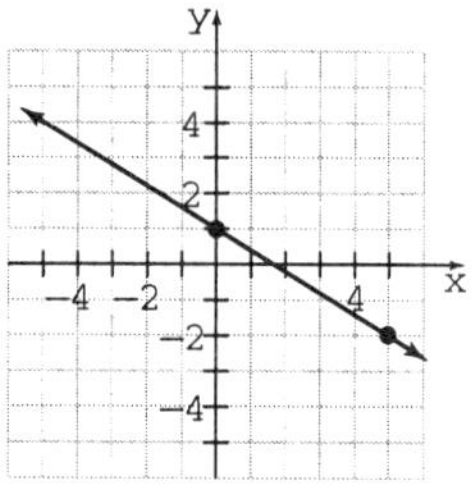

13. The slope must be found first.

Let (3, 7) be (x_1, y_1) and (-2, 4) be (x_2, y_2).

$$m = \frac{y_2 - y_1}{x_2 - x_1}$$

$$m = \frac{4 - 7}{-2 - 3}$$

$$m = \frac{-3}{-5}$$

$$m = \frac{3}{5}$$

Now choose one of the points and the point-slope formula.

$$y - y_1 = m(x - x_1)$$

$$y - 7 = \frac{3}{5}(x - 3)$$

$$y - 7 = \frac{3}{5}x - \frac{9}{5}$$

$$y - 7 + 7 = \frac{3}{5}x - \frac{9}{5} + 7$$

$$y = \frac{3}{5}x - \frac{9}{5} + \frac{35}{5}$$

$$y = \frac{3}{5}x + \frac{26}{5}$$

14.
$$(2x^{-3})^{-2}(4x^{-3}y^2)^3 = 2^{-2}x^{-3(-2)}4^3x^{-3(3)}y^{2(3)}$$
$$= 2^{-2}x^6 4^3 x^{-9}y^6$$
$$= 2^{-2} \cdot 4^3 x^{-3}y^6$$
$$= \frac{4^3y^6}{2^2x^3}$$
$$= \frac{64y^6}{4x^3}$$
$$= \frac{16y^6}{x^3}$$

15. $(x^3 - x^2 + 6x - 5) - (4x^3 - 3x^2 + 7)$
$= x^3 - x^2 + 6x - 5 - 4x^3 + 3x^2 - 7$
$= x^3 - 4x^3 - x^2 + 3x^2 + 6x - 5 - 7$
$= -3x^3 + 2x^2 + 6x - 12$

16. $(3x - 2)(x^2 + 5x - 6)$
$= 3x(x^2) - 2(x^2) + 3x(5x) - 2(5x) + 3x(-6) - 2(-6)$
$= 3x^3 - 2x^2 + 15x^2 - 10x - 18x + 12$
$= 3x^3 + 13x^2 - 28x + 12$

17.
$$\begin{array}{r} x - 5 \\ x+3\overline{\smash{)}\,x^2 - 2x + 6} \\ \underline{x^2 + 3x} \\ -5x + 6 \\ \underline{-5x - 15} \\ 21 \end{array}$$

$$\frac{x^2 - 2x + 6}{x + 3} = x - 5 + \frac{21}{x + 3}$$

18. $ab + 3b - 6a - 18 = b(a + 3) - 6(a + 3)$
$= (a + 3)(b - 6)$

19. $x^2 - 2x - 63 = (x - 9)(x + 7)$

20. $5x^3 - 125x = 5x(x^2 - 25)$
$= 5x(x^2 - 5^2)$
$= 5x(x + 5)(x - 5)$

Chapter 7

Exercise Set 7.1

1. Answers will vary.

3. The value of the variable does not make the denominator equal to 0.

5. There is no factor common to both the numerator and denominator of $\frac{2+3x}{4}$.

7. The denominator cannot be 0.

9. $x-2=0$
$x \neq 2$

11. $$-\frac{x+5}{5-x} = \frac{x+5}{-(5-x)} = \frac{x+5}{-5+x} = \frac{x+5}{x-5}$$
$\frac{x+5}{x-5} \neq -1$
No

13. The expression is defined for all real numbers except $x = 0$.

15. $4n-12=0$
$4n = 12$
$n = 3$
The expression is defined for all real numbers except $n = 3$.

17. $x^2-4=0$
$(x-2)(x+2)=0$
The expression is defined for all real numbers except $x = 2$, $x = -2$.

19. $2x^2-9x+9=0$
$(2x-3)(x-3)=0$
The expression is defined for all real numbers except $x = \frac{3}{2}$, $x = 3$.

21. All real numbers because $x^2 + 16 \neq 0$.

23. $4p^2-25=0$
$4p^2 = 25$
$p^2 = \frac{25}{4}$
$p = \pm\frac{5}{2}$
The expression is defined for all real numbers except $p = \pm\frac{5}{2}$.

25. $$\frac{7x^3y}{21x^2y^5} = \frac{7}{21}\cdot x^{3-2}\cdot y^{1-5} = \frac{1}{3}xy^{-4} = \frac{x}{3y^4}$$

27. $$\frac{\left(2a^4b^5\right)^3}{2a^{12}b^{20}} = \frac{2^3\cdot a^{4(3)}\cdot b^{5(3)}}{2a^{12}b^{20}} = \frac{8a^{12}b^{15}}{2a^{12}b^{20}} = 4a^{12-12}b^{15-20} = 4a^0b^{-5} = \frac{4}{b^5}$$

29. $$\frac{x}{x+xy} = \frac{x}{x(1+y)} = \frac{1}{1+y}$$

31. $$\frac{5x+15}{x+3} = \frac{5(x+3)}{x+3} = 5$$

33. $$\frac{x^3+6x^2+3x}{2x} = \frac{x\left(x^2+6x+3\right)}{2x} = \frac{x^2+6x+3}{2}$$

35. $$\frac{r^2-r-2}{r+1} = \frac{(r-2)(r+1)}{r+1} = r-2$$

37. $\frac{x^2+2x}{x^2+4x+4}=\frac{x(x+2)}{(x+2)^2}$
$=\frac{x}{x+2}$

39. $\frac{k^2-6k+9}{k^2-9}=\frac{(k-3)(k-3)}{(k-3)(k+3)}$
$=\frac{k-3}{k+3}$

41. $\frac{x^2-2x-3}{x^2-x-6}=\frac{(x+1)(x-3)}{(x+2)(x-3)}$
$=\frac{x+1}{x+2}$

43. $\frac{2x-3}{3-2x}=\frac{2x-3}{-(2x-3)}$
$=-1$

45. $\frac{x^2-2x-8}{4-x}=\frac{(x-4)(x+2)}{-(x-4)}$
$=-(x+2)$

47. $\frac{x^2+3x-18}{-2x^2+6x}=\frac{(x+6)(x-3)}{-2x(x-3)}$
$=-\frac{x+6}{2x}$

49. $\frac{2x^2+5x-3}{1-2x}=\frac{(2x-1)(x+3)}{-(2x-1)}$
$=-(x+3)$

51. $\frac{m-2}{4m^2-13m+10}=\frac{m-2}{(4m-5)(m-2)}=\frac{1}{4m-5}$

53. $\frac{x^2-25}{(x+5)^2}=\frac{(x-5)(x+5)}{(x+5)^2}$
$=\frac{x-5}{x+5}$

55. $\frac{6x^2-13x+6}{3x-2}=\frac{(3x-2)(2x-3)}{3x-2}$
$=2x-3$

57. $\frac{x^2-3x+4x-12}{x-3}=\frac{x(x-3)+4(x-3)}{x-3}$
$=\frac{(x+4)(x-3)}{x-3}$
$=x+4$

59. $\frac{2x^2-8x+3x-12}{2x^2+8x+3x+12}=\frac{2x(x-4)+3(x-4)}{2x(x+4)+3(x+4)}$
$=\frac{(x-4)(2x+3)}{(x+4)(2x+3)}$
$=\frac{x-4}{x+4}$

61. $\frac{a^3-8}{a-2}=\frac{(a-2)(a^2+2a+4)}{a-2}$
$=a^2+2a+4$

63. $\frac{9s^2-16t^2}{3s-4t}=\frac{(3s+4t)(3s-4t)}{3s-4t}=3s+4t$

65. $\frac{4x+6y}{2x^2+xy-3y^2}=\frac{2(2x+3y)}{(2x+3y)(x-y)}=\frac{2}{x-y}$

67. $\frac{3☺}{12}=\frac{3☺}{3\cdot 4}=\frac{☺}{4}$

69. $\frac{7\Delta}{14\Delta+21}=\frac{7\Delta}{7(2\Delta+3)}=\frac{\Delta}{2\Delta+3}$

71. $\frac{3\Delta-2}{2-3\Delta}=\frac{-(2-3\Delta)}{(2-3\Delta)}=-1$

73. $x^2-x-6=(x-3)(x+2)$
Denominator $=x+2$

75. $(x+3)(x+4)=x^2+7x+12$
Numerator $=x^2+7x+12$

77. a. $\frac{x+3}{x^2-2x+3x-6}=\frac{x+3}{x(x-2)+3(x-2)}$
$=\frac{x+3}{(x+3)(x-2)}$

$x\neq -3, x\neq 2$

b. $\frac{x+3}{(x+3)(x-2)}=\frac{1}{x-2}$

79. a. $\frac{x+5}{2x^3+7x^2-15x} = \frac{x+5}{x(2x^2+7x-15)}$

$= \frac{x+5}{x(2x-3)(x+5)}$

$x \neq 0,\ x \neq \frac{3}{2},\ x \neq -5$

b. $\frac{x+5}{x(2x-3)(x+5)} = \frac{1}{x(2x-3)}$

81. $\frac{\frac{1}{5}x^5 - \frac{2}{3}x^4}{\frac{1}{5}x^5 - \frac{2}{3}x^4} = 1$

84. $z = \frac{x-y}{2}$

$2z = x - y$

$2z - x = -y$

$y = x - 2z$

85. Let x = measure of the smallest angle. Then the second angle = $x + 30$ and third angle = $3x + 10$.
angle 1 + angle 2 + angle 3 = 180°

$x + (x+30) + (3x+10) = 180$

$5x + 40 = 180$

$5x = 140$

$x = 28$

$x + 30 = 28 + 30 = 58$

$3x + 10 = 3(28) + 10 = 84 + 10 = 94$. The three angles are 28°, 58°, and 94°.

86. $\left(\frac{4x^2y^2}{9x^4y^3}\right)^2 = \left(\frac{4}{9x^2y}\right)^2$

$= \frac{16}{81x^4y^2}$

87. $3x^2 - 4x - 8 - (-3x^2 + 6x + 9)$

$= 3x^2 - 4x - 8 + 3x^2 - 6x - 9$

$= 6x^2 - 10x - 17$

88. $3a^2 - 30a + 72 = 3(a^2 - 10a + 24)$

$= 3(a-4)(a-6)$

89. $a^2 + b^2 = c^2$

$5^2 + 12^2 = c^2$

$25 + 144 = c^2$

$169 = c^2$

$\sqrt{169} = \sqrt{c^2}$

$13 = c$

The hypotenuse is 13 inches long.

Exercise Set 7.2

1. Answers will vary.

3. $\frac{x+3}{x-4} \cdot \frac{\square}{x+3} = x + 2$

Numerator must be $(x+2)(x-4) = x^2 - 2x - 8$

5. $\frac{x-4}{x+5} \cdot \frac{x+5}{\square} = \frac{1}{x+3}$

Denominator must be
$(x+3)(x-4) = x^2 - x - 12$

7. $\frac{5x}{4y} \cdot \frac{y^2}{10} = \frac{5x}{4y} \cdot \frac{y^2}{5 \cdot 2}$

$= \frac{xy}{8}$

9. $\frac{16x^2}{y^4} \cdot \frac{5x^2}{y^2} = \frac{80x^4}{y^6}$

11. $\frac{6x^5y^3}{5z^3} \cdot \frac{6x^4}{5yz^4} = \frac{36x^9y^2}{25z^7}$

13. $\frac{3x-2}{3x+2} \cdot \frac{4x-1}{1-4x} = \frac{3x-2}{3x+2} \cdot \frac{4x-1}{-(4x-1)}$

$= \frac{-3x+2}{3x+2}$

15. $\frac{x^2+7x+12}{x+4} \cdot \frac{1}{x+3} = \frac{(x+4)(x+3)}{(x+4)(x+3)}$

$= 1$

17. $\frac{a}{a^2-b^2} \cdot \frac{a+b}{a^2+ab} = \frac{a(a+b)}{(a+b)(a-b) \cdot a(a+b)}$

$= \frac{1}{(a-b)(a+b)}$

$= \frac{1}{a^2-b^2}$

19. $\dfrac{6x^2-14x-12}{6x+4}\cdot\dfrac{x+3}{2x^2-2x-12}=\dfrac{2(3x^2-7x-6)}{2(3x+2)}\cdot\dfrac{x+3}{2(x^2-x-6)}$
$=\dfrac{2(3x+2)(x-3)(x+3)}{2(3x+2)\cdot 2(x-3)(x+2)}$
$=\dfrac{x+3}{2(x+2)}$

21. $\dfrac{3x^2-13x-10}{x^2-2x-15}\cdot\dfrac{x^2+x-2}{3x^2-x-2}=\dfrac{(3x+2)(x-5)}{(x+3)(x-5)}\cdot\dfrac{(x+2)(x-1)}{(3x+2)(x-1)}$
$=\dfrac{x+2}{x+3}$

23. $\dfrac{x+3}{x-3}\cdot\dfrac{x^3-27}{x^2+3x+9}=\dfrac{(x+3)(x-3)(x^2+3x+9)}{(x-3)(x^2+3x+9)}$
$=x+3$

25. $\dfrac{9x^3}{y^2}\div\dfrac{3x}{y^3}=\dfrac{9x^3}{y^2}\cdot\dfrac{y^3}{3x}$
$=\dfrac{3\cdot 3x^3}{y^2}\cdot\dfrac{y^3}{3x}$
$=3x^2y$

27. $\dfrac{15xy^2}{4z}\div\dfrac{5x^2y^2}{12z^2}=\dfrac{3\cdot 5xy^2}{4z}\cdot\dfrac{12z^2}{5x^2y^2}$
$=\dfrac{3\cdot 5xy^2}{4z}\cdot\dfrac{3\cdot 4z^2}{5x^2y^2}$
$=\dfrac{9z}{x}$

29. $\dfrac{5xy}{7ab^2}\div\dfrac{6xy}{7}=\dfrac{5xy}{7ab^2}\cdot\dfrac{7}{6xy}$
$=\dfrac{5}{6ab^2}$

31. $\dfrac{10r+5}{r}\div\dfrac{2r+1}{r^2}=\dfrac{5(2r+1)}{r}\cdot\dfrac{r^2}{2r+1}$
$=5r$

33. $\dfrac{x^2+5x-14}{x}\div\dfrac{x-2}{x}=\dfrac{(x-2)(x+7)}{x}\cdot\dfrac{x}{(x-2)}$
$=x+7$

35. $\dfrac{x^2-12x+32}{x^2-6x-16}\div\dfrac{x^2-x-12}{x^2-5x-24}=\dfrac{x^2-12x+32}{x^2-6x-16}\cdot\dfrac{x^2-5x-24}{x^2-x-12}$
$=\dfrac{(x-8)(x-4)}{(x-8)(x+2)}\cdot\dfrac{(x-8)(x+3)}{(x-4)(x+3)}$
$=\dfrac{x-8}{x+2}$

37. $\dfrac{2x^2+9x+4}{x^2+7x+12}\div\dfrac{2x^2-x-1}{(x+3)^2}=\dfrac{(2x+1)(x+4)}{(x+3)(x+4)}\cdot\dfrac{(x+3)^2}{(2x+1)(x-1)}$
$=\dfrac{x+3}{x-1}$

39. $\dfrac{x^2-y^2}{x^2-2xy+y^2} \div \dfrac{x+y}{y-x} = \dfrac{(x-y)(x+y)}{(x-y)^2} \cdot \dfrac{-(x-y)}{x+y}$
$= -1$

41. $\dfrac{5x^2-4x-1}{5x^2+6x+1} \div \dfrac{x^2-5x+4}{x^2+2x+1} = \dfrac{(5x+1)(x-1)}{(5x+1)(x+1)} \cdot \dfrac{(x+1)(x+1)}{(x-4)(x-1)}$
$= \dfrac{x+1}{x-4}$

43. $\dfrac{9x}{6y^2} \cdot \dfrac{24x^2y^4}{9x} = 4x^2y^2$

45. $\dfrac{63a^2b^3}{16c^3} \cdot \dfrac{4c^4}{9a^3b^5} = \dfrac{4 \cdot 7 \cdot 9a^2b^3c^4}{9 \cdot 4 \cdot 4a^3b^5c^3} = \dfrac{7c}{4ab^2}$

47. $\dfrac{-xy}{a} \div \dfrac{-2ax}{6y} = \dfrac{-xy}{a} \cdot \dfrac{6y}{-2ax}$
$= \dfrac{3y^2}{a^2}$

49. $\dfrac{100m^6}{21x^5y^7} \cdot \dfrac{14x^{12}y^5}{25m^5} = \dfrac{5 \cdot 5 \cdot 4m^6}{3 \cdot 7x^5y^7} \cdot \dfrac{2 \cdot 7x^{12}y^5}{5 \cdot 5m^5}$
$= \dfrac{8mx^7}{3y^2}$

51. $(3x+5) \cdot \dfrac{1}{6x+10} = (3x+5) \cdot \dfrac{1}{2(3x+5)}$
$= \dfrac{1}{2}$

53. $\dfrac{1}{4x^2y^2} \div \dfrac{1}{28x^3y} = \dfrac{1}{4x^2y^2} \cdot \dfrac{4 \cdot 7x^3y}{1}$
$= \dfrac{7x}{y}$

55. $\dfrac{(4m)^2}{8n^3} \div \dfrac{m^6n^8}{4} = \dfrac{16m^2}{8n^3} \cdot \dfrac{4}{m^6n^8}$
$= \dfrac{8 \cdot 2m^2}{8n^3} \cdot \dfrac{4}{m^6n^8}$
$= \dfrac{8}{m^4n^{11}}$

57. $\dfrac{r^2+5r+6}{r^2+9r+18} \cdot \dfrac{r^2+4r-12}{r^2-5r+6} = \dfrac{(r+2)(r+3)}{(r+6)(r+3)} \cdot \dfrac{(r+6)(r-2)}{(r-3)(r-2)} = \dfrac{r+2}{r-3}$

59. $\dfrac{x^2-10x+24}{x^2-8x+12} \div \dfrac{x^2-7x+12}{x^2-6x+8} = \dfrac{x^2-10x+24}{x^2-8x+12} \cdot \dfrac{x^2-6x+8}{x^2-7x+12}$
$= \dfrac{(x-6)(x-4)}{(x-6)(x-2)} \cdot \dfrac{(x-4)(x-2)}{(x-4)(x-3)}$
$= \dfrac{x-4}{x-3}$

61. $\dfrac{3z^2-4z-4}{z^2-4} \cdot \dfrac{2z^2+5z+2}{2z^2-3z-2} = \dfrac{(3z+2)(z-2)}{(z+2)(z-2)} \cdot \dfrac{(2z+1)(z+2)}{(2z+1)(z-2)} = \dfrac{3z+2}{z-2}$

63. $\dfrac{2x^2-19x+24}{x^2-12x+32} \div \dfrac{2x^2+x-6}{x^2+7x+10} = \dfrac{2x^2-19x+24}{x^2-12x+32} \cdot \dfrac{x^2+7x+10}{2x^2+x-6} = \dfrac{(2x-3)(x-8)}{(x-4)(x-8)} \cdot \dfrac{(x+5)(x+2)}{(2x-3)(x+2)} = \dfrac{x+5}{x-4}$

65. $\dfrac{4n^2-9}{9n^2-1} \cdot \dfrac{3n^2-2n-1}{2n^2-5n+3} = \dfrac{(2n+3)(2n-3)}{(3n+1)(3n-1)} \cdot \dfrac{(3n+1)(n-1)}{(2n-3)(n-1)} = \dfrac{2n+3}{3n-1}$

67. $\dfrac{6\Delta^2}{12} \cdot \dfrac{12}{36\Delta^5} = \dfrac{6\Delta^2}{12} \cdot \dfrac{12}{6 \cdot 6\Delta^5} = \dfrac{1}{6\Delta^3}$

69. $\dfrac{\Delta-\text{☺}}{9\Delta-9\text{☺}} \div \dfrac{\Delta^2-\text{☺}^2}{\Delta^2+2\Delta\text{☺}+\text{☺}^2}$

$= \dfrac{\Delta-\text{☺}}{9(\Delta-\text{☺})} \cdot \dfrac{(\Delta+\text{☺})^2}{(\Delta+\text{☺})(\Delta-\text{☺})} = \dfrac{\Delta+\text{☺}}{9(\Delta-\text{☺})}$

71. $(x+2)(x+1) = x^2+x+2x+2$

$= x^2+3x+2$

Numerator is x^2+3x+2.

73. $(x-5)(x+2) = x^2+2x-5x-10$

$= x^2-3x-10$

Numerator is $x^2-3x-10$.

75. $(x^2-4)(x-1) \div (x+2)$

$(x-2)(x+2)(x-1) \cdot \dfrac{1}{x+2} = x^2-x-2x+2$

$= x^2-3x+2$

Numerator is x^2-3x+2.

77. $\left(\dfrac{x+2}{x^2-4x-12} \cdot \dfrac{x^2-9x+18}{x-2}\right) \div \dfrac{x^2+5x+6}{x^2-4} = \dfrac{x+2}{(x-6)(x+2)} \cdot \dfrac{(x-3)(x-6)}{x-2} \cdot \dfrac{(x+2)(x-2)}{(x+2)(x+3)}$

$= \dfrac{x-3}{x+3}$

79. $\left(\dfrac{x^2+4x+3}{x^2-6x-16}\right) \div \left(\dfrac{x^2+5x+6}{x^2-9x+8} \cdot \dfrac{x^2-1}{x^2+4x+4}\right) = \dfrac{x^2+4x+3^2}{x^2-6x-16} \cdot \dfrac{x^2-9x+8}{x^2+5x+6} \cdot \dfrac{x^2+4x+4}{x^2-1}$

$= \dfrac{(x+1)(x+3)}{(x-8)(x+2)} \cdot \dfrac{(x-8)(x-1)}{(x+2)(x+3)} \cdot \dfrac{(x+2)^2}{(x-1)(x+1)}$

$= 1$

81. $\dfrac{(x-3)(x-2)}{(x+4)(x-5)} \cdot \dfrac{(x+4)(x-1)}{(x-3)(x-1)} = \dfrac{x-2}{x-5}$

The numerator is $(x-3)(x-2) = x^2-5x+6$.

The denominator is $(x+4)(x-5) = x^2-x-20$.

84. Let x = the time it takes the tug boat to reach the barge.
Then $x + 2$ = the time it takes the tug boat to return to the dock.

	Rate	Time	Distance
Trip Out	15	x	$15(x)$
Return Trip	5	$x+2$	$5(x+2)$

Distance to barge = Distance back to dock

$15(x) = 5(x+2)$

$15x = 5x + 10$

$10x = 10$

$x = 1$

It took 1 hour for the tug boat to reach the barge.

85. $\left(4x^3y^2z^4\right)\left(5xy^3z^7\right) = 4\cdot 5\cdot x^3xy^2y^3z^4z^7$
$= 20x^4y^5z^{11}$

86.
$$\begin{array}{r} 2x^2+x-2 \\ 2x-1\overline{\smash{)}\,4x^3+0x^2-5x+0} \\ \underline{4x^3-2x^2} \\ 2x^2-5x \\ \underline{2x^2-\;x} \\ -4x+0 \\ \underline{-4x+2} \\ -2 \end{array}$$

$$\frac{4x^3-5x}{2x-1} = 2x^2+x-2-\frac{2}{2x-1}$$

87. $3x^2-9x-30 = 3\left(x^2-3x-10\right)$
$= 3(x-5)(x+2)$

88. $3x^2-9x-30=0$
$3\left(x^2-3x-10\right)=0$
$3(x-5)(x+2)=0$
$x-5=0$ or $x+2=0$
$x=5$ $x=-2$

Exercise Set 7.3

1. Answers will vary.

3. Answers will vary.

5. $\dfrac{5}{x+6}-\dfrac{2}{x}$
The only factor (other than 1) of the first denominator is $x+6$. The only factor (other than 1) of the second denominator is x. The LCD is therefore $x(x+6)$.

7. $\dfrac{2}{x+3}+\dfrac{1}{x}+\dfrac{1}{3}$
The only factor (other than 1) of the first denominator is $x+3$. The only factor (other than 1) of the second denominator is x. The only factor (other than 1) of the third denominator is 3. The LCD is therefore $3x(x+3)$.

9. a. The negative sign in $-(2x-7)$ was not distributed.

b.
$$\frac{4x-3}{5x+4}-\frac{2x-7}{5x+4} = \frac{4x-3-(2x-7)}{5x+4}$$
$$= \frac{4x-3-2x+7}{5x+4}$$
$$\neq \frac{4x-3-2x-7}{5x+4}$$

11. a. The negative sign in $-\left(3x^2-4x+5\right)$ was not distributed.

b.
$$\frac{6x-2}{x^2-4x+3}-\frac{3x^2-4x+5}{x^2-4x+3} = \frac{6x-2-\left(3x^2-4x+5\right)}{x^2-4x+3}$$
$$= \frac{6x-2-3x^2+4x-5}{x^2-4x+3}$$
$$\neq \frac{6x-2-3x^2-4x+5}{x^2-4x+3}$$

13. $\dfrac{x-2}{7}+\dfrac{2x}{7} = \dfrac{x-2+2x}{7}$
$= \dfrac{3x-2}{7}$

15. $\dfrac{3r+2}{4}-\dfrac{3}{4} = \dfrac{3r+2-3}{4}$
$= \dfrac{3r-1}{4}$

17. $\frac{2}{x}+\frac{x+4}{x}=\frac{2+x+4}{x}$
$=\frac{x+6}{x}$

19. $\frac{n-5}{n}-\frac{n+7}{n}=\frac{n-5-n-7}{n}$
$=-\frac{12}{n}$

21. $\frac{x}{x-1}+\frac{4x+7}{x-1}=\frac{x+4x+7}{x-1}$
$=\frac{5x+7}{x-1}$

23. $\frac{4t+7}{5t^2}-\frac{3t+4}{5t^2}=\frac{4t+7-(3t+4)}{5t^2}$
$=\frac{4t+7-3t-4}{5t^2}$
$=\frac{t+3}{5t^2}$

25. $\frac{5x+4}{x^2-x-12}+\frac{-4x-1}{x^2-x-12}=\frac{5x+4-4x-1}{x^2-x-12}$
$=\frac{x+3}{(x+3)(x-4)}$
$=\frac{1}{x-4}$

27. $\frac{x+4}{3x+2}-\frac{x+4}{3x+2}=\frac{x+4-(x+4)}{3x+2}$
$=0$

29. $\frac{2p-5}{p-5}-\frac{p+5}{p-5}=\frac{2p-5-(p+5)}{p-5}$
$=\frac{2p-5-p-5}{p-5}$
$=\frac{p-10}{p-5}$

31. $\frac{x^2+4x+3}{x+2}-\frac{5x+9}{x+2}=\frac{x^2+4x+3-(5x+9)}{x+2}$
$=\frac{x^2+4x+3-5x-9}{x+2}$
$=\frac{x^2-x-6}{x+2}$
$=\frac{(x-3)(x+2)}{x+2}$
$=x-3$

33. $\frac{3x+11}{2x+10}-\frac{2(x+3)}{2x+10}=\frac{3x+11-2(x+3)}{2x+10}$
$=\frac{3x+11-2x-6}{2x+10}$
$=\frac{x+5}{2(x+5)}$
$=\frac{1}{2}$

35. $\frac{b^2-2b-3}{b^2-b-6}+\frac{b-3}{b^2-b-6}=\frac{b^2-2b-3+b-3}{b^2-b-6}$
$=\frac{b^2-b-6}{b^2-b-6}$
$=1$

37. $\frac{t-3}{t+3}-\frac{-3t-15}{t+3}=\frac{t-3-(-3t-15)}{t+3}$
$=\frac{t-3+3t+15}{t+3}$
$=\frac{4t+12}{t+3}$
$=\frac{4(t+3)}{t+3}$
$=4$

39. $\frac{x^2+2x}{(x+6)(x-3)}-\frac{15}{(x+6)(x-3)}=\frac{x^2+2x-15}{(x+6)(x-3)}$
$=\frac{(x+5)(x-3)}{(x+6)(x-3)}$
$=\frac{x+5}{x+6}$

41. $\frac{3x^2-7x}{4x^2-8x}+\frac{x}{4x^2-8x}=\frac{3x^2-7x+x}{4x^2-8x}$
$=\frac{3x^2-6x}{4x^2-8x}$
$=\frac{3x(x-2)}{4x(x-2)}$
$=\frac{3}{4}$

43. $$\frac{3x^2-4x+4}{3x^2+7x+2}-\frac{10x+9}{3x^2+7x+2}=\frac{3x^2-4x+4-(10x+9)}{3x^2+7x+2}$$
$$=\frac{3x^2-4x+4-10x-9}{3x^2+7x+2}$$
$$=\frac{3x^2-14x-5}{3x^2+7x+2}$$
$$=\frac{(3x+1)(x-5)}{(3x+1)(x+2)}$$
$$=\frac{x-5}{x+2}$$

45. $$\frac{x^2+3x-6}{x^2-5x+4}-\frac{-2x^2+4x-4}{x^2-5x+4}=\frac{x^2+3x-6-(-2x^2+4x-4)}{x^2-5x+4}$$
$$=\frac{x^2+3x-6+2x^2-4x+4}{x^2-5x+4}$$
$$=\frac{3x^2-x-2}{x^2-5x+4}$$
$$=\frac{(3x+2)(x-1)}{(x-4)(x-1)}$$
$$=\frac{3x+2}{x-4}$$

47. $$\frac{5x^2+40x+8}{x^2-64}+\frac{x^2+9x}{x^2-64}=\frac{5x^2+40x+8+x^2+9x}{x^2-64}$$
$$=\frac{6x^2+49x+8}{x^2-64}$$
$$=\frac{(6x+1)(x+8)}{(x-8)(x+8)}$$
$$=\frac{6x+1}{x-8}$$

49. $\frac{x}{5}+\frac{x+4}{5}$
Least common denominator = 5

51. $\frac{1}{n}+\frac{1}{5n}$
Least common denominator = $5n$

53. $\frac{3}{5x}+\frac{7}{4}$
Least common denominator $= 5x\cdot 4 = 20x$

55. $\frac{6}{p}+\frac{3}{p^3}$
Least common denominator = p^3

57. $\frac{m+3}{3m-4}+m=\frac{m+3}{3m-4}+\frac{m}{1}$
Least common denominator = $3m - 4$

59. $\frac{x}{2x+3}+\frac{4}{x^2}$
Least common denominator $= x^2(2x+3)$

61. $\frac{x+1}{12x^2y}-\frac{7}{9x^3}=\frac{x+1}{3\cdot 4x^2y}-\frac{7}{3^2x^3}$
Least common denominator
$= 4\cdot 3^2\cdot x^3\cdot y = 36x^3y$

63. $\frac{4}{2r^4s^5} - \frac{5}{9r^3s^7} = \frac{4}{2r^4s^5} - \frac{5}{3\cdot 3r^3s^7}$

Least common denominator =

$2\cdot 3\cdot 3r^4s^7 = 18r^4s^7$

65. $\frac{x^2-7}{24x} - \frac{x+3}{9(x+5)} = \frac{x^2-7}{3\cdot 8x} - \frac{x+3}{3^2(x+5)}$

Least common denominator

$= 8\cdot 3^2 x(x+5) = 72x(x+5)$

67. $\frac{5x-2}{x^2+x} - \frac{x^2}{x} = \frac{5x-2}{x(x+1)} - \frac{x^2}{x}$

Least common denominator $= x(x+1)$

69. $\frac{n}{4n-1} + \frac{n-2}{1-4n} = \frac{n}{4n-1} + \frac{(-1)(n-2)}{(-1)(1-4n)}$

$= \frac{n}{4n-1} + \frac{-n+2}{4n-1}$

Least common denominator $= 4n-1$

71. $\frac{6}{4k-5r} - \frac{5}{-4k+5r} = \frac{6}{4k-5r} - \frac{(-1)5}{(-1)(-4k+5r)}$

$= \frac{6}{4k-5r} - \frac{-5}{4k-5r}$

Least common denominator $= 4k - 5r$

73. $\frac{5}{2q^2+2q} - \frac{5}{3q} = \frac{5}{2q(q+1)} - \frac{5}{3q}$

Least common denominator $= 2\cdot 3q(q+1)$

$= 6q(q+1)$

75. $\frac{21}{24x^2y} + \frac{x+4}{15xy^3} = \frac{21}{3\cdot 8x^2y} + \frac{x+4}{3\cdot 5xy^3}$

Least common denominator

$= 8\cdot 3\cdot 5x^2y^3 = 120x^2y^3$

77. $\frac{3}{3x+12} + \frac{3x+6}{2x+4} = \frac{3}{3(x+4)} + \frac{3x+6}{2(x+2)}$

Least common denominator

$= 3\cdot 2(x+4)(x+2) = 6(x+4)(x+2)$

79. $\frac{9x+4}{x+6} - \frac{3x-6}{x+5}$

Least common denominator $= (x+6)(x+5)$

81. $\frac{x-2}{x^2-5x-24} + \frac{3}{x^2+11x+24}$

$= \frac{x-2}{(x-8)(x+3)} + \frac{3}{(x+8)(x+3)}$

Least common denominator

$= (x-8)(x+3)(x+8)$

83. $\frac{7}{(a-4)^2} - \frac{a+2}{a^2-7a+12}$

$= \frac{7}{(a-4)^2} - \frac{a+2}{(a-4)(a-3)}$

Least common denominator

$= (a-4)^2(a-3)$

85. $\frac{2x}{x^2+6x+5} - \frac{5x^2}{x^2+4x+3}$

$= \frac{2x}{(x+5)(x+1)} - \frac{5x^2}{(x+3)(x+1)}$

Least common denominator

$= (x+5)(x+1)(x+3)$

87. $\frac{3x-5}{x^2-6x+9} + \frac{3}{x-3}$

$= \frac{3x+5}{(x-3)^2} + \frac{3}{x-3}$

Least common denominator $= (x-3)^2$

89. $\frac{8x^2}{x^2-7x+6} + x - 3$

$= \frac{8x^2}{(x-6)(x-1)} + \frac{x-3}{1}$

Least common denominator $= (x-6)(x-1)$

91. $\frac{t-1}{3t^2+10t-8} - \frac{6}{3t^2+11t-4}$

$= \frac{t-1}{(3t-2)(t+4)} - \frac{6}{(3t-1)(t+4)}$

Least common denominator

$= (3t-2)(t+4)(3t-1)$

93. $\frac{2x-3}{4x^2+4x+1} + \frac{x^2-4}{8x^2+10x+3}$

$= \frac{2x-3}{(2x+1)^2} + \frac{x^2-4}{(2x+1)(4x+3)}$

Least common denominator

$= (2x+1)^2(4x+3)$

95. $x^2 - 6x + 3 + \square = 2x^2 - 5x - 6$

$\square = 2x^2 - 5x - 6 - (x^2 - 6x + 3)$

$= 2x^2 - 5x - 6 - x^2 + 6x - 3$

$= x^2 + x - 9$

Sum of numerators must be $2x^2 - 5x - 6$

97. $-x^2 - 4x + 3 + \square = 5x - 7$

$\square = 5x - 7 - (-x^2 - 4x + 3)$

$= 5x - 7 + x^2 + 4x - 3$

$= x^2 + 9x - 10$

Sum of numerator must be $5x - 7$

99. $\frac{3}{☺} + \frac{4}{5☺}$

Least common denominator = 5☺

101. $\frac{8}{\Delta^2 - 9} - \frac{2}{\Delta + 3} = \frac{8}{(\Delta + 3)(\Delta - 3)} - \frac{2}{\Delta + 3}$

Least common denominator $= (\Delta + 3)(\Delta - 3)$

103. $\frac{4x - 1}{x^2 - 25} - \frac{3x^2 - 8}{x^2 - 25} + \frac{8x - 3}{x^2 - 25} = \frac{4x - 1 - (3x^2 - 8) + (8x - 3)}{x^2 - 25}$

$= \frac{4x - 1 - 3x^2 + 8 + 8x - 3}{x^2 - 25}$

$= \frac{-3x^2 + 12x + 4}{x^2 - 25}$

105. $\frac{7}{6x^5y^9} - \frac{9}{2x^3y} + \frac{4}{5x^{12}y^2} = \frac{7}{2 \cdot 3x^5y^9} - \frac{9}{2x^3y} + \frac{4}{5x^{12}y^2}$

Least common denominator $= 2 \cdot 3 \cdot 5x^{12}y^9 = 30x^{12}y^9$

107. $\frac{4}{x^2 - x - 12} + \frac{3}{x^2 - 6x + 8} + \frac{5}{x^2 + x - 6} = \frac{4}{(x - 4)(x + 3)} + \frac{3}{(x - 4)(x - 2)} + \frac{5}{(x + 3)(x - 2)}$

Least common denominator $= (x - 4)(x + 3)(x - 2)$

109. $4\frac{3}{5} - 2\frac{5}{9} = \frac{23}{5} - \frac{23}{9}$

$= \frac{207}{45} - \frac{115}{45}$

$= \frac{92}{45}$

$= 2\frac{2}{45}$

110. $6x + 4 = -(x + 2) - 3x + 4$

$6x + 4 = -x - 2 - 3x + 4$

$6x + 4 = -4x + 2$

$6x + 4x = 2 - 4$

$10x = -2$

$x = \frac{-2}{10} = -\frac{1}{5}$

111. $\frac{6}{128} = \frac{x}{48}$
$128x = 6 \cdot 48$
$128x = 288$
$x = \frac{288}{128}$
$x = \frac{9}{4} = 2\frac{1}{4}$
You should use 2.25 ounces of concentrate.

112. Let h = the number of hours played.
The cost under Plan 1 is $C = 125 + 2.5h$ while the cost under Plan 2 is $C = 300$.

Set the two costs equal
$125 + 2.5h = 300$
$2.5h = 175$
$h = 70$
If Malcolm plays 70 hours in a year, the cost of the two plans is equal.

113. Use the two points given and find the slope between the two points. This is the slope of the line.
$m = \frac{y_2 - y_1}{x_2 - x_1}$
$m = \frac{5-(-2)}{4-(-3)}$
$m = \frac{7}{7} = 1$
The slope of the line is 1.

114. $\frac{4.2 \times 10^8}{2.1 \times 10^{-3}} = 2.0 \times 10^{8-(-3)} = 2.0 \times 10^{11}$

115. $2x^2 - 3 = x$
$2x^2 - x - 3 = 0$
$(2x-3)(x+1) = 0$
$2x - 3 = 0$ or $x + 1 = 0$
$2x = 3 \qquad x = -1$
$x = \frac{3}{2}$

Exercise Set 7.4

1. For each fraction, divide the LCD by the denominator.

3. a. Answers will vary.

b. $\frac{x}{x^2 - x - 6} + \frac{3}{x^2 - 4}$
$= \frac{x}{(x-3)(x+2)} + \frac{3}{(x-2)(x+2)}$
$= \frac{x(x-2)}{(x-3)(x+2)(x-2)} + \frac{3(x-3)}{(x-2)(x+2)(x-3)}$
$= \frac{x^2 - 2x + 3x - 9}{(x-3)(x+2)(x-2)}$
$= \frac{x^2 + x - 9}{(x-3)(x+2)(x-2)}$

5. a. $\frac{y}{4z} + \frac{5}{6z^2}$
$4z = 2 \cdot 2 \cdot z$
$6z^2 = 2 \cdot 3 \cdot z^2$
Least common denominator
$2 \cdot 2 \cdot 3 \cdot z^2 = 12z^2$

b. $\frac{y}{4z} + \frac{5}{6z^2} = \frac{y}{4z} \cdot \frac{3z}{3z} + \frac{5}{6z^2} \cdot \frac{2}{2}$
$= \frac{3yz}{12z^2} + \frac{10}{12z^2}$
$= \frac{3yz + 10}{12z^2}$

c. Yes. After factoring out the common factors, the reduced form would be the same.

7. $\frac{1}{4x} + \frac{3}{x} = \frac{1}{4x} + \frac{3 \cdot 4}{4x}$
$= \frac{1+12}{4x}$
$= \frac{13}{4x}$

9. $\frac{5}{x^2} + \frac{3}{2x} = \frac{5 \cdot 2}{2x^2} + \frac{3x}{2x^2}$
$= \frac{3x + 10}{2x^2}$

11. $3 + \frac{5}{x} = \frac{3x}{x} + \frac{5}{x} = \frac{3x+5}{x}$

13. $\frac{2}{x^2} + \frac{3}{5x} = \frac{2 \cdot 5}{5x^2} + \frac{3x}{5x^2} = \frac{3x + 10}{5x^2}$

15. $\frac{7}{4x^2y}+\frac{3}{5xy^2}=\frac{7\cdot 5y}{4x^2y\cdot 5y}+\frac{3\cdot 4x}{5xy^2\cdot 4x}$
$=\frac{35y}{20x^2y^2}+\frac{12x}{20x^2y^2}$
$=\frac{35y+12x}{20x^2y^2}$

17. $3y+\frac{x}{y}=\frac{3y\cdot y}{y}+\frac{x}{y}=\frac{3y^2+x}{y}$

19. $\frac{3a-1}{2a}+\frac{2}{3a}=\frac{(3a-1)3}{2a\cdot 3}+\frac{2\cdot 2}{3a\cdot 2}$
$=\frac{9a-3+4}{6a}$
$=\frac{9a+1}{6a}$

21. $\frac{4x}{y}+\frac{2y}{xy}=\frac{4x\cdot x}{xy}+\frac{2y}{xy}$
$=\frac{4x^2+2y}{xy}$

23. $\frac{4}{b}-\frac{4}{5a^2}=\frac{4\cdot 5a^2}{b\cdot 5a^2}-\frac{4\cdot b}{5a^2\cdot b}$
$=\frac{20a^2-4b}{5a^2b}$

25. $\frac{4}{x}+\frac{7}{x-3}=\frac{4\cdot(x-3)}{x\cdot(x-3)}+\frac{7\cdot x}{(x-3)\cdot x}$
$=\frac{4x-12+7x}{x(x-3)}$
$=\frac{11x-12}{x(x-3)}$

27. $\frac{9}{p+3}+\frac{2}{p}=\frac{9p}{p(p+3)}+\frac{2(p+3)}{p(p+3)}$
$=\frac{9p+2p+6}{p(p+3)}$
$=\frac{11p+6}{p(p+3)}$

29. $\frac{5}{6d}-\frac{d}{3d+5}=\frac{5\cdot(3d+5)}{6d\cdot(3d+5)}-\frac{d\cdot(6d)}{(3d+5)\cdot(6d)}$
$=\frac{15d+25-6d^2}{6d(3d+5)}$
$=\frac{-6d^2+15d+25}{6d(3d+5)}$

31. $\frac{4}{p-3}+\frac{2}{3-p}=\frac{4}{p-3}-\frac{2}{p-3}$
$=\frac{2}{p-3}$

33. $\frac{9}{x+7}-\frac{5}{-x-7}=\frac{9}{x+7}+\frac{5}{x+7}$
$=\frac{14}{x+7}$

35. $\frac{6}{a-2}+\frac{a}{2a-4}=\frac{6\cdot 2}{(a-2)\cdot 2}+\frac{a}{2(a-2)}$
$=\frac{a+12}{2(a-2)}$

37. $\frac{x+5}{x-5}-\frac{x-5}{x+5}=\frac{(x+5)^2}{(x-5)(x+5)}-\frac{(x-5)^2}{(x-5)(x+5)}$
$=\frac{x^2+10x+25-(x^2-10x+25)}{(x-5)(x+5)}$
$=\frac{x^2+10x+25-x^2+10x-25}{(x-5)(x+5)}$
$=\frac{20x}{(x-5)(x+5)}$

39. $\frac{5}{6n+3}-\frac{4}{n}=\frac{5}{3(2n+1)}-\frac{4}{n}$
$=\frac{5\cdot n}{3(2n+1)\cdot n}-\frac{4\cdot 3(2n+1)}{n\cdot 3(2n+1)}$
$=\frac{5n-24n-12}{3n(2n+1)}$
$=\frac{-19n-12}{3n(2n+1)}$

41. $\frac{3}{2w+10}+\frac{5}{w+2}=\frac{3}{2(w+5)}+\frac{5}{w+2}$
$=\frac{3\cdot(w+2)}{2(w+5)\cdot(w+2)}+\frac{5\cdot 2(w+5)}{(w+2)\cdot 2(w+5)}$
$=\frac{3w+6+10w+50}{2(w+5)(w+2)}$
$=\frac{13w+56}{2(w+5)(w+2)}$

43. $\dfrac{z}{z^2-16}+\dfrac{4}{z+4}=\dfrac{z}{(z+4)(z-4)}+\dfrac{4\cdot(z-4)}{(z+4)\cdot(z-4)}$

$=\dfrac{z+4z-16}{(z+4)(z-4)}$

$=\dfrac{5z-16}{(z+4)(z-4)}$

45. $\dfrac{x+2}{x^2-4}-\dfrac{2}{x+2}=\dfrac{x+2}{(x+2)(x-2)}-\dfrac{2(x-2)}{(x+2)(x-2)}$

$=\dfrac{x+2-(2x-4)}{(x+2)(x-2)}$

$=\dfrac{x+2-2x+4}{(x+2)(x-2)}$

$=\dfrac{-x+6}{(x+2)(x-2)}$

47. $\dfrac{3r+2}{r^2-10r+24}-\dfrac{2}{r-6}=\dfrac{3r+2}{(r-6)(r-4)}-\dfrac{2\cdot(r-4)}{(r-6)\cdot(r-4)}$

$=\dfrac{3r+2-(2r-4)}{(r-6)(r-4)}$

$=\dfrac{3r+2-2r+8}{(r-6)(r-4)}$

$=\dfrac{r+10}{(r-6)(r-4)}$

49. $\dfrac{x^2}{x^2+2x-8}-\dfrac{x-4}{x+4}=\dfrac{x^2}{(x+4)(x-2)}-\dfrac{(x-4)(x-2)}{(x+4)(x-2)}$

$=\dfrac{x^2-\left(x^2-6x+8\right)}{(x+4)(x-2)}$

$=\dfrac{x^2-x^2+6x-8}{(x+4)(x-2)}$

$=\dfrac{6x-8}{(x+4)(x-2)}$

51. $\dfrac{x-3}{x^2+10x+25}+\dfrac{x-3}{x+5}=\dfrac{x-3}{(x+5)^2}+\dfrac{(x-3)(x+5)}{(x+5)^2}$

$=\dfrac{x-3+x^2+2x-15}{(x+5)^2}$

$=\dfrac{x^2+3x-18}{(x+5)^2}$

53. $\dfrac{5}{a^2-9a+8}-\dfrac{3}{a^2-6a-16}=\dfrac{5}{(a-8)(a-1)}-\dfrac{3}{(a-8)(a+2)}$

$=\dfrac{5(a+2)}{(a-8)(a-1)(a+2)}-\dfrac{3(a-1)}{(a-8)(a-1)(a+2)}$

$=\dfrac{5a+10-(3a-3)}{(a-8)(a-1)(a+2)}$

$=\dfrac{5a+10-3a+3}{(a-8)(a-1)(a+2)}$

$=\dfrac{2a+13}{(a-8)(a-1)(a+2)}$

55. $\frac{2}{x^2+6x+9}+\frac{3}{x^2+x-6}=\frac{2}{(x+3)(x+3)}+\frac{3}{(x+3)(x-2)}$

$=\frac{2(x-2)}{(x+3)(x+3)(x-2)}+\frac{3(x+3)}{(x+3)(x+3)(x-2)}$

$=\frac{2x-4+3x+9}{(x+3)(x+3)(x-2)}$

$=\frac{5x+5}{(x+3)^2(x-2)}$

57. $\frac{x}{2x^2+7x+3}-\frac{3}{3x^2+7x-6}=\frac{x}{(2x+1)(x+3)}-\frac{3}{(3x-2)(x+3)}$

$=\frac{x(3x-2)}{(2x+1)(3x-2)(x+3)}-\frac{3(2x+1)}{(2x+1)(3x-2)(x+3)}$

$=\frac{3x^2-2x-(6x+3)}{(2x+1)(3x-2)(x+3)}$

$=\frac{3x^2-2x-6x-3}{(2x+1)(3x-2)(x+3)}$

$=\frac{3x^2-8x-3}{(2x+1)(3x-2)(x+3)}$

59. $\frac{x}{4x^2+11x+6}-\frac{2}{8x^2+2x-3}=\frac{x}{(4x+3)(x+2)}-\frac{2}{(4x+3)(2x-1)}$

$=\frac{x(2x-1)}{(4x+3)(x+2)(2x-1)}-\frac{2(x+2)}{(4x+3)(x+2)(2x-1)}$

$=\frac{2x^2-x-(2x+4)}{(4x+3)(x+2)(2x-1)}$

$=\frac{2x^2-x-2x-4}{(4x+3)(x+2)(2x-1)}$

$=\frac{2x^2-3x-4}{(4x+3)(x+2)(2x-1)}$

61. $\frac{3w+12}{w^2+w-12}-\frac{2}{w-3}=\frac{3(w+4)}{(w-3)(w+4)}-\frac{2}{w-3}$

$=\frac{3}{w-3}-\frac{2}{w-3}$

$=\frac{3-2}{w-3}$

$=\frac{1}{w-3}$

63. $$\frac{3r}{2r^2-10r+12}+\frac{3}{r-2}=\frac{3r}{2(r^2-5r+6)}+\frac{3}{r-2}$$
$$=\frac{3r}{2(r-3)(r-2)}+\frac{3}{r-2}$$
$$=\frac{3r}{2(r-3)(r-2)}+\frac{3\cdot 2(r-3)}{(r-2)\cdot 2(r-3)}$$
$$=\frac{3r+6r-18}{2(r-3)(r-2)}$$
$$=\frac{9r-18}{2(r-3)(r-2)}$$
$$=\frac{9(r-2)}{2(r-3)(r-2)}$$
$$=\frac{9}{2(r-3)}$$

65. $\frac{2}{x}+6$ is defined for all real numbers except $x = 0$.

67. $\frac{5}{x-4}+\frac{7}{x+6}$ is defined for all real numbers except $x = 4$ and $x = -6$.

69. $$\frac{3}{\Delta-2}-\frac{1}{2-\Delta}=\frac{3}{\Delta-2}+\frac{1}{\Delta-2}$$
$$=\frac{3+1}{\Delta-2}$$
$$=\frac{4}{\Delta-2}$$

71. $\frac{5}{a+b}+\frac{3}{a}$

$a + b = 0$ when $a = -b$. The expression is defined for all real numbers except $a = 0$ and $a = -b$.

73. $$\frac{x}{x^2-9}+\frac{3x}{x+3}+\frac{3x^2-8x}{9-x^2}=\frac{x}{(x+3)(x-3)}+\frac{3x(x-3)}{(x+3)(x-3)}-\frac{3x^2-8x}{(x+3)(x-3)}$$
$$=\frac{x+3x^2-9x-(3x^2-8x)}{(x+3)(x-3)}$$
$$=\frac{x+3x^2-9x-3x^2+8x}{(x+3)(x-3)}$$
$$=\frac{0}{(x+3)(x-3)}$$
$$=0$$

75. $$\frac{x+6}{4-x^2}-\frac{x+3}{x+2}+\frac{x-3}{2-x}=\frac{x+6}{(2-x)(2+x)}-\frac{(x+3)(2-x)}{(2+x)(2-x)}+\frac{(x-3)(2+x)}{(2-x)(2+x)}$$
$$=\frac{x+6-(-x^2-x+6)+(x^2-x-6)}{(2-x)(2+x)}$$
$$=\frac{x+6+x^2+x-6+x^2-x-6}{(2-x)(2+x)}$$
$$=\frac{2x^2+x-6}{(2-x)(2+x)}$$
$$=\frac{(2x-3)(x+2)}{(2-x)(2+x)}$$
$$=\frac{2x-3}{2-x}$$

77. $$\frac{2}{x^2-x-6}+\frac{3}{x^2-2x-3}+\frac{1}{x^2+3x+2}=\frac{2}{(x+2)(x-3)}+\frac{3}{(x-3)(x+1)}+\frac{1}{(x+2)(x+1)}$$
$$=\frac{2(x+1)}{(x+2)(x-3)(x+1)}+\frac{3(x+2)}{(x+2)(x-3)(x+1)}+\frac{x-3}{(x+2)(x-3)(x+1)}$$
$$=\frac{2x+2+3x+6+x-3}{(x+2)(x-3)(x+1)}$$
$$=\frac{6x+5}{(x+2)(x-3)(x+1)}$$

80. Let x = the number of hours it takes the train to travel 42 miles.
$$\frac{22 \text{ miles}}{0.8 \text{ hours}}=\frac{42 \text{ miles}}{x \text{ hours}}$$
$$22x=0.8(42)$$
$$22x=33.6$$
$$x\approx 1.53$$
It takes about 1.53 hours for the train to travel 42 miles.

81. $$m=\frac{-2-6}{3-(-4)}=\frac{-8}{7}$$

82.
$$\begin{array}{r|l} & 4x-3 \\ 2x+3 & 8x^2+6x-13 \\ & \underline{8x^2+12x} \\ & \quad -6x-13 \\ & \quad \underline{-6x-9} \\ & \qquad -4 \end{array}$$

$$(8x^2+6x-13)\div(2x+3)=4x-3-\frac{4}{2x+3}$$

83. $$\frac{x^2+xy-6y^2}{x^2-xy-2y^2}\cdot\frac{y^2-x^2}{x^2+2xy-3y^2}=\frac{(x+3y)(x-2y)}{(x+y)(x-2y)}\cdot\frac{-1(x-y)(x+y)}{(x+3y)(x-y)}$$
$$=-1$$

Exercise Set 7.5

1. A complex fraction is a fraction whose numerator or denominator (or both) contains a fraction.

3. a. $\dfrac{\frac{x+3}{4}}{\frac{7}{x^2+5x+6}}$

   Numerator: $\dfrac{x+3}{4}$

   Denominator: $\dfrac{7}{x^2+5x+6}$

   b. $\dfrac{\frac{1}{2y}+x}{\frac{3}{y}+x}$

   Numerator, $\dfrac{1}{2y}+x$

   Denominator, $\dfrac{3}{y}+x$

5. $\dfrac{4+\frac{2}{3}}{2+\frac{1}{3}} = \dfrac{\left(4+\frac{2}{3}\right)3}{\left(2+\frac{1}{3}\right)3}$
$= \dfrac{12+2}{6+1}$
$= \dfrac{14}{7}$
$= 2$

7. $\dfrac{2+\frac{3}{8}}{1+\frac{1}{3}} = \dfrac{\left(2+\frac{3}{8}\right)8\cdot 3}{\left(1+\frac{1}{3}\right)8\cdot 3}$
$= \dfrac{2\cdot 8\cdot 3+3\cdot 3}{8\cdot 3+8}$
$= \dfrac{48+9}{24+8}$
$= \dfrac{57}{32}$

9. $\dfrac{\frac{2}{7}-\frac{1}{4}}{6-\frac{2}{3}} = \dfrac{\left(\frac{2}{7}-\frac{1}{4}\right)3\cdot 4\cdot 7}{\left(6-\frac{2}{3}\right)3\cdot 4\cdot 7}$
$= \dfrac{2\cdot 3\cdot 4-1\cdot 3\cdot 7}{6\cdot 3\cdot 4\cdot 7-2\cdot 4\cdot 7}$
$= \dfrac{24-21}{504-56}$
$= \dfrac{3}{448}$

11. $\dfrac{\frac{xy^2}{9}}{\frac{3}{x^2}} = \dfrac{xy^2}{9}\cdot\dfrac{x^2}{3}$
$= \dfrac{x^3y^2}{27}$

13. $\dfrac{\frac{6a^2b}{5}}{\frac{9ac^2}{b^2}} = \dfrac{6a^2b}{5}\cdot\dfrac{b^2}{9ac^2}$
$= \dfrac{6a^2b^3}{45ac^2}$
$= \dfrac{2ab^3}{15c^2}$

15. $\dfrac{a-\frac{a}{b}}{\frac{1+a}{b}} = \dfrac{\left(a-\frac{a}{b}\right)b}{\left(\frac{1+a}{b}\right)b}$
$= \dfrac{ab-a}{1+a}$

17. $\dfrac{\frac{9}{x}+\frac{3}{x^2}}{3+\frac{1}{x}} = \dfrac{\left(\frac{9}{x}+\frac{3}{x^2}\right)x^2}{\left(3+\frac{1}{x}\right)x^2}$
$= \dfrac{9x+3}{3x^2+x}$
$= \dfrac{3(3x+1)}{x(3x+1)}$
$= \dfrac{3}{x}$

19. $\dfrac{5-\frac{1}{x}}{4-\frac{1}{x}} = \dfrac{\left(5-\frac{1}{x}\right)x}{\left(4-\frac{1}{x}\right)x}$
$= \dfrac{5x-1}{4x-1}$

21. $\dfrac{\frac{m}{n}-\frac{n}{m}}{\frac{m+n}{n}} = \dfrac{\left(\frac{m}{n}-\frac{n}{m}\right)mn}{\left(\frac{m+n}{n}\right)mn}$
$= \dfrac{m^2-n^2}{m(m+n)}$
$= \dfrac{(m+n)(m-n)}{m(m+n)}$
$= \dfrac{m-n}{m}$

23. $\dfrac{\frac{a^2}{b}-b}{\frac{b^2}{a}-a}=\dfrac{\left(\frac{a^2}{b}-b\right)ab}{\left(\frac{b^2}{a}-a\right)ab}$

$=\dfrac{\left(a^2-b^2\right)a}{\left(b^2-a^2\right)b}$

$=-\dfrac{a}{b}$

25. $\dfrac{5-\frac{a}{b}}{\frac{a}{b}-5}=\dfrac{\left(5-\frac{a}{b}\right)b}{\left(\frac{a}{b}-5\right)b}$

$=\dfrac{5b-a}{a-5b}$

$=\dfrac{-1(-5b+a)}{a-5b}$

$=\dfrac{-1(a-5b)}{(a-5b)}$

$=-1$

27. $\dfrac{\frac{a^2-b^2}{a}}{\frac{a+b}{a^3}}=\dfrac{a^2-b^2}{a}\cdot\dfrac{a^3}{a+b}$

$=\dfrac{(a+b)(a-b)}{a}\cdot\dfrac{a^3}{a+b}$

$=a^2(a-b)$

29. $\dfrac{\frac{1}{a}-\frac{1}{b}}{\frac{1}{ab}}=\dfrac{\left(\frac{1}{a}-\frac{1}{b}\right)ab}{\left(\frac{1}{ab}\right)ab}$

$=\dfrac{b-a}{1}$

$=b-a$

31. $\dfrac{\frac{a}{b}+\frac{1}{a}}{\frac{b}{a}+\frac{1}{a}}=\dfrac{\left(\frac{a}{b}+\frac{1}{a}\right)ab}{\left(\frac{b}{a}+\frac{1}{a}\right)ab}$

$=\dfrac{a^2+b}{b^2+b}$

$=\dfrac{a^2+b}{b(b+1)}$

33. $\dfrac{\frac{1}{xy}}{\frac{1}{x}-\frac{1}{y}}=\dfrac{\left(\frac{1}{xy}\right)xy}{\left(\frac{1}{x}-\frac{1}{y}\right)xy}$

$=\dfrac{1}{y-x}$

35. $\dfrac{\frac{3}{a}+\frac{3}{a^2}}{\frac{3}{b}+\frac{3}{b^2}}=\dfrac{\left(\frac{3}{a}+\frac{3}{a^2}\right)a^2b^2}{\left(\frac{3}{b}+\frac{3}{b^2}\right)a^2b^2}$

$=\dfrac{3ab^2+3b^2}{3a^2b+3a^2}$

$=\dfrac{3b^2(a+1)}{3a^2(b+1)}$

$=\dfrac{ab^2+b^2}{a^2(b+1)}$

37. a. Answers will vary.

b. c. $\dfrac{5+\frac{3}{5}}{\frac{1}{8}-4}=\dfrac{\frac{25}{5}+\frac{3}{5}}{\frac{1}{8}-\frac{32}{8}}$

$=\dfrac{\frac{28}{5}}{-\frac{31}{8}}$

$=\dfrac{28}{5}\cdot\left(-\dfrac{8}{31}\right)$

$=-\dfrac{224}{155}$

$\dfrac{5+\frac{3}{5}}{\frac{1}{8}-4}=\dfrac{\left(5+\frac{3}{5}\right)5\cdot 8}{\left(\frac{1}{8}-4\right)5\cdot 8}$

$=\dfrac{5\cdot 5\cdot 8+3\cdot 8}{5-4\cdot 5\cdot 8}$

$=\dfrac{200+24}{5-160}$

$=-\dfrac{224}{155}$

39.a. Answers will vary.

b. c. $$\frac{\frac{x-y}{x+y}+\frac{3}{x+y}}{2-\frac{7}{x+y}}=\frac{\frac{x-y+3}{x+y}}{\frac{2(x+y)}{x+y}-\frac{7}{x+y}}$$

$$=\frac{\frac{x-y+3}{x+y}}{\frac{2x+2y-7}{x+y}}$$

$$=\frac{x-y+3}{x+y}\cdot\frac{x+y}{2x+2y-7}$$

$$=\frac{x-y+3}{2x+2y-7}$$

$$\frac{\frac{x-y}{x+y}+\frac{3}{x+y}}{2-\frac{7}{x+y}}=\frac{\frac{x-y+3}{x+y}}{2-\frac{7}{x+y}}$$

$$=\frac{\left(\frac{x-y+3}{x+y}\right)(x+y)}{\left(2-\frac{7}{x+y}\right)(x+y)}$$

$$=\frac{x-y+3}{2(x+y)-7}$$

$$=\frac{x-y+3}{2x+2y-7}$$

41. a. $$\frac{\frac{5}{12x}}{\frac{8}{x^2}-\frac{4}{3x}}$$

b. $$\frac{\frac{5}{12x}}{\frac{8}{x^2}-\frac{4}{3x}}=\frac{\frac{5}{12x}}{\frac{24-4x}{3x^2}}=\frac{5}{12x}\cdot\frac{3x^2}{(24-4x)}$$

$$=\frac{5x}{4(24-4x)}$$

$$=\frac{5x}{96-16x}$$

43. $$\frac{x^{-1}+y^{-1}}{2}=\frac{\frac{1}{x}+\frac{1}{y}}{2}$$

$$=\frac{\left(\frac{1}{x}+\frac{1}{y}\right)xy}{2xy}$$

$$=\frac{y+x}{2xy}$$

45. $$\frac{x^{-1}+y^{-1}}{x^{-1}y^{-1}}=\frac{\frac{1}{x}+\frac{1}{y}}{\frac{1}{xy}}$$

$$=\frac{\left(\frac{1}{x}+\frac{1}{y}\right)xy}{\left(\frac{1}{xy}\right)xy}$$

$$=\frac{y+x}{1}$$

$$=x+y$$

47. a. $$E=\frac{\frac{1}{2}\left(\frac{2}{3}\right)}{\frac{2}{3}+\frac{1}{2}}$$

$$=\frac{\frac{2}{6}}{\frac{4+3}{6}}$$

$$=\frac{2}{6}\cdot\frac{6}{7}$$

$$=\frac{2}{7}$$

b. $$E=\frac{\frac{1}{2}\left(\frac{4}{5}\right)}{\frac{4}{5}+\frac{1}{2}}$$

$$=\frac{\frac{4}{10}}{\frac{8+5}{10}}$$

$$=\frac{4}{10}\cdot\frac{10}{13}$$

$$=\frac{4}{13}$$

49. $$\frac{\frac{a}{b}+b-\frac{1}{a}}{\frac{a}{b^2}-\frac{b}{a}+\frac{1}{a^2}}=\frac{\frac{a^2+b^2a-b}{ba}}{\frac{a^3-ab^3+b^2}{a^2b^2}}$$

$$=\frac{a^2+b^2a-b}{ba}\cdot\frac{a^2b^2}{a^3-ab^3+b^2}$$

$$=\frac{(a^2+b^2a-b)ab}{a^3-ab^3+b^2}$$

$$=\frac{a^3b+a^2b^3-ab^2}{a^3-ab^3+b^2}$$

51. $2x-8(5-x)=9x-3(x+2)$

$2x-40+8x=9x-3x-6$

$10x-40=6x-6$

$4x=34$

$x=\frac{34}{4}=\frac{17}{2}$

52. A polynomial is a sum of terms of the form ax^n where a is a real number and n is a whole number.

53. $x^2 - 13x + 42 = (x-6)(x-7)$

54.
$$\frac{x}{3x^2+17x-6} - \frac{2}{x^2+3x-18} = \frac{x}{(3x-1)(x+6)} - \frac{2}{(x+6)(x-3)}$$
$$= \frac{x(x-3)}{(3x+1)(x+6)(x-3)} - \frac{2(3x-1)}{(3x-1)(x+6)(x-3)}$$
$$= \frac{x^2-3x-(6x-2)}{(3x-1)(x+6)(x-3)}$$
$$= \frac{x^2-3x-6x+2}{(3x-1)(x+6)(x-3)}$$
$$= \frac{x^2-9x+2}{(3x-1)(x+6)(x-3)}$$

Exercise Set 7.6

1. a. Answers will vary.

b.
$$\frac{1}{x-1} - \frac{1}{x+1} = \frac{3x}{x^2-1}$$
$$\frac{1}{x-1} - \frac{1}{x+1} = \frac{3x}{(x-1)(x+1)}$$

Multiply both sides of the equation by the least common denominator, $(x-1)(x+1)$.

$$(x-1)(x+1)\left(\frac{1}{x-1} - \frac{1}{x+1}\right) = \left(\frac{3x}{(x-1)(x+1)}\right)(x-1)(x+1)$$
$$(x-1)(x+1)\left(\frac{1}{x-1}\right) - (x-1)(x+1)\left(\frac{1}{x+1}\right) = 3x$$
$$x+1-(x-1) = 3x$$
$$x+1-x+1 = 3x$$
$$2 = 3x$$
$$\frac{2}{3} = x$$

3. a. The problem on the left is an expression to be simplified while the problem on the right is an equation to be solved.

b. Left: Write the fractions with the LCD, $12(x-1)$, then combine numerators.
Right: Multiply both sides of the equation by the LCD, $12(x-1)$, then solve.

c. Left:
$$\frac{x}{3} - \frac{x}{4} + \frac{1}{x-1} = \frac{x\cdot 4(x-1)}{3\cdot 4(x-1)} - \frac{x\cdot 3(x-1)}{4\cdot 3(x-1)} + \frac{1\cdot 3\cdot 4}{(x-1)3\cdot 4}$$
$$= \frac{4x(x-1) - 3x(x-1) + 12}{3\cdot 4(x-1)}$$
$$= \frac{4x^2 - 4x - 3x^2 + 3x + 12}{12(x-1)}$$
$$= \frac{x^2 - x + 12}{12(x-1)}$$

Right:

$$\frac{x}{3}-\frac{x}{4}=\frac{1}{x-1}$$
$$12(x-1)\left(\frac{x}{3}-\frac{x}{4}\right)=\left(\frac{1}{x-1}\right)12(x-1)$$
$$12(x-1)\left(\frac{x}{3}\right)-12(x-1)\left(\frac{x}{4}\right)=12$$
$$4x(x-1)-3x(x-1)=12$$
$$4x^2-4x-3x^2+3x=12$$
$$x^2-x-12=0$$
$$(x-4)(x+3)=0$$
$$x-4=0 \quad \text{or} \quad x+3=0$$
$$x=4 \qquad\qquad x=-3$$

5. You must check for extraneous when there is a variable in the denominator.

7. 2 cannot be a solution because it makes the denominator zero in the first term.

9. No, because there are no variables in the denominator.

11. Yes, because there is a variable in the denominator.

13. $$\frac{x}{3}-\frac{x}{4}=1$$
$$12\left(\frac{x}{3}-\frac{x}{4}\right)=12(1)$$
$$4x-3x=12$$
$$x=12$$

15. $$\frac{r}{3}=\frac{r}{4}+\frac{1}{6}$$
$$12\left(\frac{r}{3}\right)=12\left(\frac{r}{4}+\frac{1}{6}\right)$$
$$4r=3r+2$$
$$r=2$$

17. $$\frac{z}{2}+3=\frac{z}{5}$$
$$10\left(\frac{z}{2}+3\right)=10\left(\frac{z}{5}\right)$$
$$5z+30=2z$$
$$3z=-30$$
$$z=-10$$

19. $$\frac{z}{6}+\frac{1}{3}=\frac{z}{5}-\frac{2}{3}$$
$$30\left(\frac{z}{6}+\frac{1}{3}\right)=30\left(\frac{z}{5}-\frac{2}{3}\right)$$
$$5z+10=6z-20$$
$$-z=-30$$
$$z=30$$

21. $$d+1=\frac{3}{2}d-1$$
$$2(d+1)=2\left(\frac{3}{2}d-1\right)$$
$$2d+2=3d-2$$
$$-d=-4$$
$$d=4$$

23. $$k+\frac{1}{6}=2k-4$$
$$6\left(k+\frac{1}{6}\right)=6(2k-4)$$
$$6k+1=12k-24$$
$$-6k=-25$$
$$k=\frac{25}{6}$$

25. $$\frac{(n+6)}{3}=\frac{2(n-8)}{4}$$
$$4(n+6)=3\cdot 2(n-8)$$
$$4n+24=6n-48$$
$$-2n=-72$$
$$n=36$$

27. $\frac{x-5}{15}=\frac{3}{5}-\frac{x-4}{10}$

$30\left(\frac{x-5}{15}\right)=30\left(\frac{3}{5}-\frac{x-4}{10}\right)$

$2(x-5)=6(3)-3(x-4)$

$2x-10=18-3x+12$

$2x-10=30-3x$

$5x=40$

$x=8$

29. $\frac{-p+1}{4}+\frac{13}{20}=\frac{p}{5}-\frac{p-1}{2}$

$20\left(\frac{-p+1}{4}+\frac{13}{20}\right)=20\left(\frac{p}{5}-\frac{p-1}{2}\right)$

$5(-p+1)+13=4p-10(p-1)$

$-5p+5+13=4p-10p+10$

$-5p+18=-6p+10$

$p=-8$

31. $\frac{d-3}{4}+\frac{1}{5}=\frac{2d+1}{3}-\frac{32}{15}$

$60\left(\frac{d-3}{4}+\frac{1}{5}\right)=60\left(\frac{2d+1}{3}-\frac{32}{15}\right)$

$15(d-3)+12=20(2d+1)-4(32)$

$15d-45+12=40d+20-128$

$15d-33=40d-108$

$-25d=-75$

$d=3$

33. $2+\frac{3}{x}=\frac{11}{4}$

$4x\left(2+\frac{3}{x}\right)=4x\left(\frac{11}{4}\right)$

$8x+12=44$

$8x=32$

$x=4$

Check: $2+\frac{3}{4}=\frac{11}{4}$

$\frac{8}{4}+\frac{3}{4}=\frac{11}{4}$

$\frac{11}{4}=\frac{11}{4}$ True

35. $6-\frac{3}{x}=\frac{9}{2}$

$2x\left(6-\frac{3}{x}\right)=2x\left(\frac{9}{2}\right)$

$12x-6=9x$

$3x=6$

$x=2$

Check: $6-\frac{3}{2}=\frac{9}{2}$

$\frac{12}{2}-\frac{3}{2}=\frac{9}{2}$

$\frac{9}{2}=\frac{9}{2}$ True

37. $\frac{4}{n}-\frac{3}{2n}=\frac{1}{2}$

$2n\left(\frac{4}{n}-\frac{3}{2n}\right)=2n\left(\frac{1}{2}\right)$

$8-3=n$

$5=n$

Check: $\frac{4}{5}-\frac{3}{10}=\frac{1}{2}$

$\frac{8}{10}-\frac{3}{10}=\frac{1}{2}$

$\frac{5}{10}=\frac{1}{2}$

$\frac{1}{2}=\frac{1}{2}$ True

39. $\frac{x-1}{x-5}=\frac{4}{x-5}$

$(x-5)\left(\frac{x-1}{x-5}\right)=\left(\frac{4}{x-5}\right)(x-5)$

$x-1=4$

$x=5$

Check: $\frac{x-1}{x-5}=\frac{4}{x-5}$

$\frac{5-1}{5-5}=\frac{4}{5-5}$

$\frac{4}{0}=\frac{4}{0}$

Since $\frac{4}{0}$ is not a real number, 5 is an extraneous solution. This equation has no solution.

41. $\frac{3-5y}{4}=\frac{2-4y}{3}$

$3(3-5y)=4(2-4y)$

$9-15y=8-16y$

$y=-1$

Check: $\frac{3-5y}{4}=\frac{2-4y}{3}$

$\frac{3-5(-1)}{4}=\frac{2-4(-1)}{3}$

$\frac{3+5}{4}=\frac{2+4}{3}$

$\frac{8}{4}=\frac{6}{3}$ True

43. $\frac{4}{y-3}=\frac{6}{y+3}$

$4(y+3)=6(y-3)$

$4y+12=6y-18$

$30=2y$

$15=y$

Check: $\frac{4}{y-3}=\frac{6}{y+3}$

$\frac{4}{15-3}=\frac{6}{15+3}$

$\frac{4}{12}=\frac{6}{18}$

$\frac{1}{3}=\frac{1}{3}$ True

45. $\frac{2x-3}{x-4}=\frac{5}{x-4}$

$(x-4)\left(\frac{2x-3}{x-4}\right)=\left(\frac{5}{x-4}\right)(x-4)$

$2x-3=5$

$2x=8$

$x=4$

Check: $\frac{2x-3}{x-4}=\frac{5}{x-4}$

$\frac{2(4)-3}{4-4}=\frac{5}{4-4}$

$\frac{5}{0}=\frac{5}{0}$

Since $\frac{5}{0}$ is not a real number, 4 is an extraneous solution. This equation has no solution.

47. $\frac{x^2}{x-4}=\frac{16}{x-4}$

$(x-4)\left(\frac{x^2}{x-4}\right)=(x-4)\left(\frac{16}{x-4}\right)$

$x^2=16$

$x^2-16=0$

$(x+4)(x-4)=0$

$x+4=0$ or $x-4=0$

$x=-4$ $\qquad$ $x=4$

Check: $\frac{(-4)^2}{-4-4}=\frac{16}{-4-4}$ $\qquad$ $\frac{(4)^2}{4-4}=\frac{16}{4-4}$

$\frac{16}{-8}=\frac{16}{-8}$ $\qquad$ $\frac{16}{0}=\frac{16}{0}$

$-2=-2$ True

Since $\frac{16}{0}$ is not a real number, 4 is an extraneous solution. The solution is $x=-4$.

49. $\frac{n+10}{n+2}=\frac{n+4}{n-3}$

$(n+10)(n-3)=(n+2)(n+4)$

$n^2+7n-30=n^2+6n+8$

$n=38$

Check: $\frac{38+10}{38+2}=\frac{38+4}{38-3}$

$\frac{48}{40}=\frac{42}{35}$

$\frac{6}{5}=\frac{6}{5}$ True

51. $\frac{1}{3r} = \frac{r}{8r+3}$

$1(8r+3) = 3r^2$

$0 = 3r^2 - 8r - 3$

$0 = (3r+1)(r-3)$

$3r+1=0$ or $r-3=0$

$3r = -1$ $\quad r = 3$

$r = -\frac{1}{3}$

Check: $\frac{1}{3\left(-\frac{1}{3}\right)} = \frac{\frac{-1}{3}}{8\left(\frac{-1}{3}\right)+3}$

$\frac{1}{-1} = \frac{\frac{-1}{3}}{\frac{-8}{3}+\frac{9}{3}}$

$-1 = \frac{\frac{-1}{3}}{\frac{1}{3}}$

$-1 = -1$ True

$\frac{1}{3(3)} = \frac{3}{8(3)+3}$

$\frac{1}{9} = \frac{3}{27}$

$\frac{1}{9} = \frac{1}{9}$ True

53. $\frac{k}{k+2} = \frac{3}{k-2}$

$k(k-2) = 3(k+2)$

$k^2 - 2k = 3k + 6$

$k^2 - 5k - 6 = 0$

$(k-6)(k+1) = 0$

$k-6=0$ or $k+1=0$

$k = 6$ $\quad k = -1$

Check: $\frac{6}{6+2} = \frac{3}{6-2}$

$\frac{6}{8} = \frac{3}{4}$

$\frac{3}{4} = \frac{3}{4}$ True

$\frac{-1}{-1+2} = \frac{3}{-1-2}$

$\frac{-1}{1} = \frac{3}{-3}$

$-1 = -1$ True

55. $\frac{3}{r} + r = \frac{19}{r}$

$r\left(\frac{3}{r} + r\right) = r\left(\frac{19}{r}\right)$

$3 + r^2 = 19$

$r^2 - 16 = 0$

$(r+4)(r-4) = 0$

$r+4=0$ or $r-4=0$

$r = -4$ $\quad r = 4$

Check: $\frac{3}{4} + 4 = \frac{19}{4}$

$\frac{3}{4} + \frac{16}{4} = \frac{19}{4}$

$\frac{19}{4} = \frac{19}{4}$ True

$\frac{3}{-4} - 4 = \frac{19}{-4}$

$\frac{3}{-4} - \frac{16}{4} = \frac{19}{-4}$

$-\frac{19}{4} = -\frac{19}{4}$ True

57. $x + \frac{20}{x} = -9$

$x\left(x + \frac{20}{x}\right) = -9x$

$x^2 + 20 = -9x$

$x^2 + 9x + 20 = 0$

$(x+4)(x+5) = 0$

$x+4=0$ or $x+5=0$

$x = -4$ $\quad x = -5$

Check $x = -4$: $x + \frac{20}{x} = -9$

$-4 + \frac{20}{-4} = -9$

$-4 + (-5) = -9$ True

Check $x = -5$: $x + \frac{20}{x} = -9$

$-5 + \frac{20}{-5} = -9$

$5 + (-4) = -9$ True

59. $$\frac{3y-2}{y+1} = 4 - \frac{y+2}{y-1}$$
$$(y+1)(y-1)\left(\frac{3y-2}{y+1}\right) = \left(4 - \frac{y+2}{y-1}\right)(y+1)(y-1)$$
$$(y-1)(3y-2) = 4(y+1)(y-1) - \left(\frac{y+2}{y-1}\right)(y+1)(y-1)$$
$$3y^2 - 5y + 2 = 4(y^2 - 1) - (y+2)(y+1)$$
$$3y^2 - 5y + 2 = 4y^2 - 4 - (y^2 + 3y + 2)$$
$$3y^2 - 5y + 2 = 3y^2 - 3y - 6$$
$$-5y + 2 = -3y - 6$$
$$8 = 2y$$
$$4 = y$$

Check: $$\frac{3y-2}{y+1} = 4 - \frac{y+2}{y-1}$$
$$\frac{3(4)-2}{4+1} = 4 - \frac{4+2}{4-1}$$
$$\frac{12-2}{5} = 4 - \frac{6}{3}$$
$$\frac{10}{5} = 4 - 2$$
$$2 = 2 \text{ True}$$

Check: $$\frac{2b}{b+1} = 2 - \frac{5}{2b}$$
$$\frac{2(-5)}{-5+1} = 2 - \frac{5}{2(-5)}$$
$$\frac{-10}{-4} = 2 - \frac{5}{-10}$$
$$\frac{5}{2} = 2 + \frac{1}{2}$$
$$\frac{5}{2} = \frac{5}{2} \text{ True}$$

61. $$\frac{1}{x+3} + \frac{1}{x-3} = \frac{-5}{x^2-9}$$
$$\frac{1}{x+3} + \frac{1}{x-3} = \frac{-5}{(x-3)(x+3)}$$
$$(x-3)(x+3)\left[\frac{1}{x+3} + \frac{1}{x-3}\right] = \left[\frac{-5}{(x-3)(x+3)}\right](x-3)(x+3)$$
$$(x-3)(x+3)\left(\frac{1}{x+3}\right) + (x-3)(x+3)\left(\frac{1}{x-3}\right) = -5$$
$$x - 3 + x + 3 = -5$$
$$2x = -5$$
$$x = -\frac{5}{2}$$

Check: $\frac{1}{x+3}+\frac{1}{x-3}=\frac{-5}{x^2-9}$

$$\frac{1}{-\frac{5}{2}+3}+\frac{1}{-\frac{5}{2}-3}=\frac{-5}{\left(-\frac{5}{2}\right)^2-9}$$

$$\frac{1}{\frac{1}{2}}-\frac{1}{-\frac{11}{2}}=\frac{-5}{\frac{25}{4}-9}$$

$$2-\frac{2}{11}=\frac{-5}{-\frac{11}{4}}$$

$$\frac{20}{11}=\frac{20}{11} \text{ True}$$

63. $\frac{x}{x-3}+\frac{3}{2}=\frac{3}{x-3}$

$$2(x-3)\left(\frac{x}{x-3}+\frac{3}{2}\right)=2(x-3)\left(\frac{3}{x-3}\right)$$

$$2x+3(x-3)=2(3)$$

$$2x+3x-9=6$$

$$5x=15$$

$$x=3$$

Check: $\frac{x}{x-3}+\frac{3}{2}=\frac{3}{x-3}$

$$\frac{3}{3-3}+\frac{3}{2}=\frac{3}{3-3}$$

$$\frac{3}{0}+\frac{3}{2}=\frac{3}{0}$$

Since $\frac{3}{0}$ is not a real number, 3 is an extraneous solution. This equation has no solution.

65. $\frac{3}{x-5}-\frac{4}{x+5}=\frac{11}{x^2-25}$

$$\frac{3}{x-5}-\frac{4}{x+5}=\frac{11}{(x-5)(x+5)}$$

$$(x-5)(x+5)\left[\frac{3}{x-5}-\frac{4}{x+5}\right]=\left[\frac{11}{(x-5)(x+5)}\right](x-5)(x+5)$$

$$(x-5)(x+5)\left(\frac{3}{x-5}\right)-(x-5)(x+5)\left(\frac{4}{x+5}\right)=11$$

$$3(x+5)-4(x-5)=11$$

$$3x+15-4x+20=11$$

$$-x+35=11$$

$$24=x$$

$$\text{Check: } \frac{3}{x-5}-\frac{4}{x+5}=\frac{11}{x^2-25}$$
$$\frac{3}{24-5}-\frac{4}{24+5}=\frac{11}{24^2-25}$$
$$\frac{3}{19}-\frac{4}{29}=\frac{11}{576-25}$$
$$\frac{87}{551}-\frac{76}{551}=\frac{11}{551}$$
$$\frac{11}{551}=\frac{11}{551} \quad \text{True}$$

67.

$$\frac{y}{2y+2}+\frac{2y-16}{4y+4}=\frac{y-3}{y+1}$$
$$\frac{y}{2(y+1)}+\frac{2y-16}{4(y+1)}=\frac{y-3}{y+1}$$
$$4(y+1)\left[\frac{y}{2(y+1)}+\frac{2y-16}{4(y+1)}\right]=\left[\frac{y-3}{y+1}\right]4(y+1)$$
$$4(y+1)\left[\frac{y}{2(y+1)}\right]+4(y+1)\left[\frac{2y-16}{4(y+1)}\right]=4(y-3)$$
$$2y+2y-16=4y-12$$
$$4y-16=4y-12$$
$$-16=-12 \qquad \text{False}$$

Since this is a false statement, the equation has no solution.

69.

$$\frac{1}{y-1}+\frac{1}{2}=\frac{2}{y^2-1}$$
$$\frac{1}{y-1}+\frac{1}{2}=\frac{2}{(y+1)(y-1)}$$
$$2(y+1)(y-1)\left[\frac{1}{y-1}+\frac{1}{2}\right]=2(y+1)(y-1)\left[\frac{2}{(y+1)(y-1)}\right]$$
$$2(y+1)(y-1)\left(\frac{1}{y-1}\right)+2(y+1)(y-1)\left(\frac{1}{2}\right)=2(2)$$
$$2(y+1)+(y+1)(y-1)=4$$
$$2y+2+y^2-1=4$$
$$y^2+2y-3=0$$
$$(y+3)(y-1)=0$$

$$y+3=0 \quad \text{or} \quad y-1=0$$
$$y=-3 \qquad\qquad y=1$$

Check: $y = -3$

$$\frac{1}{y-1}+\frac{1}{2}=\frac{2}{y^2-1}$$
$$\frac{1}{-3-1}+\frac{1}{2}=\frac{2}{(-3)^2-1}$$
$$\frac{1}{-4}+\frac{1}{2}=\frac{2}{8}$$
$$\frac{1}{4}=\frac{1}{4} \quad \text{True}$$

Check: $y = 1$

$$\frac{1}{y-1}+\frac{1}{2}=\frac{2}{y^2-1}$$
$$\frac{1}{1-1}+\frac{1}{2}=\frac{2}{1^2-1}$$
$$\frac{1}{0}+\frac{1}{2}=\frac{2}{0}$$

The solution to the equation is –3. Since $\frac{1}{0}$ and $\frac{2}{0}$ are not real numbers, 1 is an extraneous solution.

71.
$$\frac{2t}{4t+4}+\frac{t}{2t+2}=\frac{2t-3}{t+1}$$
$$\frac{2t}{4(t+1)}+\frac{t}{2(t+1)}=\frac{2t-3}{t+1}$$
$$4(t+1)\left[\frac{2t}{4(t+1)}+\frac{t}{2(t+1)}\right]=4(t+1)\left(\frac{2t-3}{t+1}\right)$$
$$2t+2t=4(2t-3)$$
$$4t=8t-12$$
$$-4t=-12$$
$$t=3$$

Check:
$$\frac{2(3)}{4(3)+4}+\frac{3}{2(3)+2}=\frac{2(3)-3}{3+1}$$
$$\frac{6}{12+4}+\frac{3}{6+2}=\frac{6-3}{4}$$
$$\frac{6}{16}+\frac{3}{8}=\frac{3}{4}$$
$$\frac{3}{8}+\frac{3}{8}=\frac{3}{4}$$
$$\frac{6}{8}=\frac{3}{4}$$
$$\frac{3}{4}=\frac{3}{4} \quad \text{True}$$

73. The solution is 5. Since $3 = x-2$, $x = 5$.

75. The solution is 0.
Since $x + x = 0$, $x = 0$.

77. x can be any real number.
$x-2+x-2=2x-4$.

79.
$$\frac{1}{p}+\frac{1}{q}=\frac{1}{f}$$
$$\frac{1}{30}+\frac{1}{q}=\frac{1}{10}$$
$$\frac{1}{q}=\frac{1}{10}-\frac{1}{30}$$
$$\frac{1}{q}=\frac{2}{30}$$
$$2q=30$$
$$q=15$$

The image will appear 15 cm from the mirror.

81.
$$\frac{x-4}{x^2-2x}=\frac{-4}{x^2-4}$$
$$\frac{x-4}{x(x-2)}=\frac{-4}{(x+2)(x-2)}$$
$$x(x+2)(x-2)\left(\frac{x-4}{x(x-2)}\right)=x(x+2)(x-2)\left(\frac{-4}{(x+2)(x-2)}\right)$$
$$(x+2)(x-4)=-4x$$
$$x^2-2x-8=-4x$$
$$x^2+2x-8=0$$
$$(x+4)(x-2)=0$$
$$x+4=0 \text{ or } x-2=0$$
$$x=-4 \qquad x=2$$

Since $\frac{-4}{0}$ and $\frac{-2}{0}$ is not a real number, 2 is an extraneous solution.
The solution to the equation is –4.

83. No, it is impossible for both sides of the equation to be equal.

85. Let x be the number of minutes of internet access over 5 hours.
Plan 1: $7.95+0.15x$
Plan 2: 19.95
$$7.95+0.15x=19.95$$
$$0.15x=12$$
$$x=80$$
80 minutes = $\frac{80}{60}$ hours = $1\frac{1}{3}$ hours
Jake would have to use the internet more than $5+1\frac{1}{3}=6\frac{1}{3}$ hours.

86. Let x = measure of larger angle
y = measure of smaller angle
$$x+y=180$$
$$\frac{1}{2}x-30=y$$
Substitute $\frac{1}{2}x-30$ for y in the first equation.
$$x+\frac{1}{2}x-30=180$$
$$\frac{3}{2}x=210$$
$$x=140$$
$$y=\frac{1}{2}x-30$$
$$y=\frac{1}{2}(140)-30$$
$$y=70-30$$
$$y=40$$
The angles measure 40° and 140°.

87. $\frac{600 \text{ gallons}}{8 \text{ gallons / minute}}=75$ minutes

88. $(3.4\times10^{-5})(2\times10^{7})=(3.4)(2)\times(10^{-5})(10^{7})$
$=6.8\times10^{-5+7}$
$=6.8\times10^{2}$

89. A linear equation is an equation that can be written in the form $ax+b=c$ where a, b, and c are real numbers and $a\neq0$. An example would be $5x+8=19$. A quadratic equation has the form $ax^2+bx+c=0$ where a, b, and c are real numbers and $a\neq0$. An example would be $4x^2+7x+2=0$.

Exercise Set 7.7

1. Some examples are:
$A=\frac{1}{2}bh$, $A=\frac{1}{2}h(b_1+b_2)$, $V=\frac{1}{3}\pi r^2h$, and $V=\frac{4}{3}\pi r^3$

3. It represents 1 complete task.

5. Let w = width, then

$\frac{2}{3}w + 4 =$ length

area = width · length

$$90 = w\left(\frac{2}{3}w + 4\right)$$
$$90 = \frac{2w^2}{3} + 4w$$
$$3(90) = 3\left(\frac{2w^2}{3} + 4w\right)$$
$$270 = 2w^2 + 12w$$
$$2w^2 + 12w - 270 = 0$$
$$w^2 + 6w - 135 = 0$$
$$(w+15)(w-9) = 0$$
$$w + 15 = 0 \quad \text{or} \quad w - 9 = 0$$
$$w = -15 \qquad w = 9$$

Since the width cannot be negative, $w = 9$

$$l = \frac{2}{3}w + 4$$
$$l = \frac{2}{3}(9) + 4$$
$$l = 6 + 4 = 10$$

The length = 10 inches and the width = 9 inches.

7. Let x = height, then $x + 5$ = base

area $= \frac{1}{2} \cdot$ height · base

$$42 = \frac{1}{2}x(x+5)$$
$$2(42) = 2\left[\frac{1}{2}x(x+5)\right]$$
$$84 = x(x+5)$$
$$84 = x^2 + 5x$$
$$0 = x^2 + 5x - 84$$
$$0 = (x-7)(x+12)$$
$$x - 7 = 0 \quad \text{or} \quad x + 12 = 0$$
$$x = 7 \qquad x = -12$$

Since the height cannot be negative, $x = 7$.

base $= x + 5 = 7 + 15 = 12$

The base is 12 cm and the height is 7 cm.

9. Let b = the base of the triangle, then $\left(\frac{1}{2}b - 1\right)$ is the height.

$$A = \frac{1}{2}bh$$
$$12 = \frac{1}{2}b\left(\frac{1}{2}b - 1\right)$$
$$12 = \frac{1}{4}b^2 - \frac{1}{2}b$$
$$4(12) = 4\left(\frac{1}{4}b^2 - \frac{1}{2}b\right)$$
$$48 = b^2 - 2b$$
$$0 = b^2 - 2b - 48$$
$$0 = (b-8)(b+6)$$
$$b - 8 = 0 \quad \text{or} \quad b + 6 = 0$$
$$b = 8 \quad \text{or} \quad b = -6$$

Since a length cannot be negative, the base is 8 feet.

11. Let one number be x, then the other number is $10x$.

$$\frac{1}{x} - \frac{1}{10x} = 3$$
$$10x\left(\frac{1}{x} - \frac{1}{10x}\right) = 10x(3)$$
$$10 - 1 = 30x$$
$$9 = 30x$$
$$\frac{9}{30} = x$$
$$\frac{3}{10} = x$$
$$3 = 10x$$

The numbers are $\frac{3}{10}$ and 3.

13. Let x = amount by which the numerator was increased.

$$\frac{3+x}{4} = \frac{5}{2}$$
$$4\left(\frac{3+x}{4}\right) = \left(\frac{5}{2}\right)4$$
$$3 + x = 10$$
$$x = 7$$

The numerator was increased by 7.

15. Let r = speed of Creole Queen paddle boat.

$$t = \frac{d}{r}$$

Time upstream = time downstream

$$\frac{4}{r-2}=\frac{6}{r+2}$$
$$4(r+2)=6(r-2)$$
$$4r+8=6r-12$$
$$20=2r$$
$$10=r$$

The boat's speed in still water is 10 mph.

17. Let d = distance and $t=\frac{d}{r}$.

Time going + Time returning = $\frac{5}{2}$.

$$\frac{d}{15}+\frac{d}{15}=\frac{5}{2}$$
$$\frac{2d}{15}=\frac{5}{2}$$
$$4d=75$$
$$d=\frac{75}{4}=18.75$$

The trolley traveled 18.75 miles in one direction.

19. Let r be the speed of the propeller plane, then $4r$ is the speed of the jet.

$$\frac{d}{r}=t$$

time by jet + time by propeller plane = 6 hr

$$\frac{1600}{4r}+\frac{500}{r}=6$$
$$\frac{400}{r}+\frac{300}{r}=6$$
$$\frac{900}{r}=6$$
$$r=\frac{900}{6}=150$$

The speed of the propeller plane is 150 mph and the speed of the jet is 600 mph.

21. Let d = distance traveled by car, then $200-d$ = distance traveled by train.

$$t=\frac{d}{r}$$

time traveled by car + time traveled by train = total time

$$\frac{d}{40}+\frac{200-d}{120}=2.2$$
$$120\left(\frac{d}{40}+\frac{200-d}{120}\right)=2.2(120)$$
$$3d+200-d=264$$
$$2d=64$$
$$d=32$$
$$200-32=168$$

She travels 32 miles by car and 168 miles by train.

23. Let d = the distance flown with the wind, then $2800-d$ = the time flown against the wind.

time with wind $=\frac{d}{600}$

time against wind $=\frac{2800-d}{500}$

$$\frac{d}{600}+\frac{2800-d}{500}=5$$
$$3000\left(\frac{d}{600}+\frac{2800-d}{500}\right)=5(3000)$$
$$5d+16{,}800-6d=15{,}000$$
$$d=1800$$

time with wind $=\frac{1800}{600}=3$

time against wind $=\frac{1000}{500}=2$

It flew 3 hours at 600 mph and 2 hours at 500 mph.

25. time at 30 ft/s $=\frac{d}{30}$

time at 20 ft/s $=\frac{d}{20}$

$$\frac{d}{20}=\frac{d}{30}+15$$
$$60\left(\frac{d}{20}\right)=\left(\frac{d}{30}+15\right)60$$
$$3d=2d+900$$
$$d=900$$

The boat traveled 900 feet in one direction.

27. Felicia's rate $=\frac{1}{6}$

Reynaldo's rate $=\frac{1}{8}$

$$\frac{t}{6}+\frac{t}{8}=1$$
$$48\left(\frac{t}{6}+\frac{t}{8}\right)=48(1)$$
$$8t+6t=48$$
$$14t=48$$
$$t=3\frac{3}{7}$$

It will take them $3\frac{3}{7}$ hours.

29. input rate $= \frac{1}{2}$

output rate $= \frac{1}{3}$

$$\frac{t}{2} - \frac{t}{3} = 1$$
$$6\left(\frac{t}{2} - \frac{t}{3}\right) = 6(1)$$
$$3t - 2t = 6$$
$$t = 6$$

It will take 6 hours.

31. Input rate for small hose $= \frac{1}{5}$

Input rat for larger hose $= \frac{1}{t}$

In 2 hours, the smaller hose does $\frac{2}{5}$ of the filling and the larger hose does $\frac{2}{t}$.

$$\frac{2}{5} + \frac{2}{t} = 1$$
$$5t\left(\frac{2}{5} + \frac{2}{t}\right) = 5t$$
$$2t + 10 = 5t$$
$$10 = 3t$$
$$t = \frac{10}{3} = 3\frac{1}{3}$$

It would take the larger hose $3\frac{1}{3}$ hours

33. Rate for first backhoe $= \frac{1}{12}$

Rate for second backhoe $= \frac{1}{15}$

Work done by first $= \frac{1}{12} \cdot 5 = \frac{5}{12}$

Work done by second $= \frac{1}{15} \cdot t = \frac{t}{15}$

$$\frac{5}{12} + \frac{t}{15} = 1$$
$$60\left(\frac{5}{12} + \frac{t}{15}\right) = 1 \cdot 60$$
$$25 + 4t = 60$$
$$4t = 35$$
$$t = \frac{35}{4} = 8\frac{3}{4}$$

It takes the smaller backhoe $8\frac{3}{4}$ days to finish the trench.

35. Ken's rate $= \frac{1}{4}$

Bettina's rate $= \frac{1}{6}$

$$\frac{t}{6} + \frac{t+3}{4} = 1$$
$$12\left(\frac{t}{6} + \frac{t+3}{4}\right) = 12$$
$$2t + 3(t + 3) = 12$$
$$2t + 3t + 9 = 12$$
$$5t = 3$$
$$t = \frac{3}{5}$$

It will take them $\frac{3}{5}$ hour or 36 minutes longer.

37. Rate of first skimmer $= \frac{1}{60}$

Rate of second skimmer $= \frac{1}{50}$

Rate of transfer $= \frac{1}{30}$

$$\frac{t}{60} + \frac{t}{50} - \frac{t}{30} = 1$$
$$300\left(\frac{t}{60} + \frac{t}{50} - \frac{t}{30}\right) = 300$$
$$5t + 6t - 10t = 300$$
$$t = 300$$

It will take 300 hours to fill the tank.

39. Let x be the number.

$$\frac{4}{x} + 5x = 12$$
$$x\left(\frac{4}{x} + 5x\right) = 12x$$
$$4 + 5x^2 = 12x$$
$$5x^2 - 12x + 4 = 0$$
$$(5x - 2)(x - 2) = 0$$
$$5x - 2 = 0 \quad \text{or} \quad x - 2 = 0$$
$$x = \frac{2}{5} \qquad x = 2$$

The numbers are 2 or $\frac{2}{5}$

41. Ed's rate $= \frac{1}{8}$

Samantha's rate $= \frac{1}{4}$

$$\frac{p}{4} - 1 = \frac{p}{8}$$
$$8\left(\frac{p}{4} - 1\right) = 8\left(\frac{p}{8}\right)$$
$$2p - 8 = p$$
$$p = 8$$

Each must pick 8 pints.

43.
$$\frac{1}{2}(x+3) - (2x+6) = \frac{1}{2}x + \frac{3}{2} - 2x - 6$$
$$= \frac{x}{2} - \frac{4x}{2} + \frac{3}{2} - \frac{12}{2}$$
$$= -\frac{3x}{2} - \frac{9}{2}$$

44.

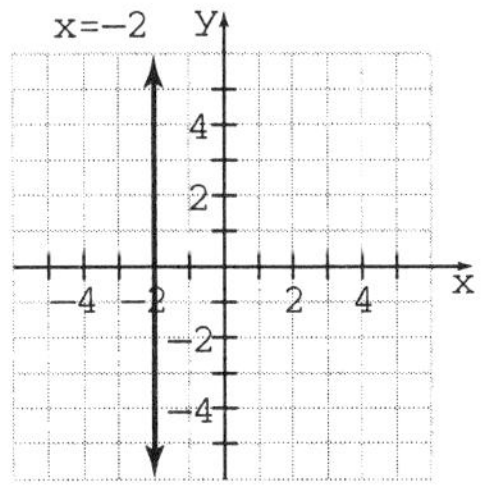

45.
$$\frac{x^2 - 14x + 48}{x^2 - 5x - 24} \div \frac{2x^2 - 13x + 6}{2x^2 + 5x - 3} = \frac{x^2 - 14x + 48}{x^2 - 5x - 24} \cdot \frac{2x^2 + 5x - 3}{2x^2 - 13x + 6}$$
$$= \frac{(x-6)(x-8)}{(x+3)(x-8)} \cdot \frac{(2x-1)(x+3)}{(2x-1)(x-6)}$$
$$= 1$$

46.
$$\frac{x}{6x^2 - x - 15} - \frac{5}{9x^2 - 12x - 5} = \frac{x}{(2x+3)(3x-5)} - \frac{5}{(3x+1)(3x-5)}$$
$$= \frac{x(3x+1)}{(2x+3)(3x-5)(3x+1)} - \frac{5(2x+3)}{(2x+3)(3x-5)(3x+1)}$$
$$= \frac{3x^2 + x - (10x + 15)}{(2x+3)(3x-5)(3x+1)}$$
$$= \frac{3x^2 + x - 10x - 15}{(2x+3)(3x-5)(3x+1)}$$
$$= \frac{3x^2 - 9x - 15}{(2x+3)(3x-5)(3x+1)}$$

Exercise Set 7.8

1. a. As one quantity increases, the other increases.

b. Answers will vary.

c. Answers will vary.

3. One quantity varies as a product of two or more quantities.

5. a. *y* decreases

b. inverse variation

b. direct variation

7. Direct

9. Inverse

11. Direct

13. Direct

15. Direct

17. Inverse

19. Direct

21. Inverse

23. Inverse

25. a. $x = ky$

b. Substitute 12 for y and 6 for k.
$x = 6(12)$
$x = 72$

27. a. $y = kR$

b. Substitute 180 for R and 1.7 for k.
$y = 1.7(180)$
$y = 306$

29. a. $R = \dfrac{k}{W}$

b. Substitute 160 for W and 8 for k.
$R = \dfrac{8}{160}$
$R = \dfrac{1}{20} = 0.05$

31. a. $A = \dfrac{kB}{C}$

b. Substitute 12 for B, 4 for C, and 3 for k.
$A = \dfrac{3(12)}{4} = \dfrac{36}{4} = 9$

33. a. $x = ky$

b. To find k, substitute 12 for x and 3 for y.
$12 = k(3)$
$\dfrac{12}{3} = k$
$4 = k$
Thus, $x = 4y$.
Now substitute 5 for y.
$x = 4(5) = 20$

35. a. $y = kR^2$

b. To find k, substitute 5 for y and 5 for R.
$5 = k(5)^2$
$5 = k(25)$
$\dfrac{5}{25} = k$
$\dfrac{1}{5} = k$
Thus $y = \dfrac{1}{5}R^2$.
Now substitute 10 for R.
$y = \dfrac{1}{5}(10)^2 = \dfrac{1}{5}(100) = 20$

37. a. $C = \dfrac{k}{J}$

b. To find k, substitute 7 for C and 0.7 for J.
$7 = \dfrac{k}{0.7}$
$7(0.7) = k$
$4.9 = k$
Thus $C = \dfrac{4.9}{J}$.
Now substitute 12 for J.
$C = \dfrac{4.9}{12} \approx 0.41$

39. a. $F = \dfrac{kM_1M_2}{d}$

b. To find k, substitute 5 for M_1, 10 for M_2, 0.2 for d, and 20 for F.
$20 = \dfrac{k(5)(10)}{0.2}$
$20 = k(250)$
$\dfrac{20}{250} = k$
$0.08 = k$
Thus $F = \dfrac{0.08M_1M_2}{d}$.
Now substitute 10 for M_1, 20 for M_2, and 0.4 for d.
$F = \dfrac{0.08(10)(20)}{0.4} = \dfrac{16}{0.4} = 40$

41. $a = kb$
$k(2b) = 2(kb) = 2a$
If b is doubled, a is doubled.

43. $y = \dfrac{k}{x}$
$\dfrac{k}{2x} = \dfrac{1}{2}\left(\dfrac{k}{x}\right) = \dfrac{1}{2}y$
If x is doubled, y is halved.

45. $F = \frac{km_1 m_2}{d^2}$

$\frac{k(2m_1)m_2}{d^2} = \frac{2km_1m_2}{d^2} = 2 \cdot \frac{km_1m_2}{d^2} = 2F$

If m_1 is doubled, F is doubled.

47. $F = \frac{km_1 m_2}{d^2}$

$\frac{k(2m_1)\left(\frac{1}{2}m_2\right)}{d^2} = \frac{2 \cdot \frac{1}{2} km_1m_2}{d^2} = \frac{1 \cdot km_1m_2}{d^2} = F$

If m_1 is doubled and m_2 is halved, F is unchanged.

49. $F = \frac{km_1 m_2}{d^2}$

$\frac{k\left(\frac{1}{2}m_1\right)(4m_2)}{d^2} = \frac{\frac{1}{2} \cdot 4 km_1m_2}{d^2} =$

$\frac{2 \cdot km_1m_2}{d^2} = 2 \cdot \frac{km_1m_2}{d^2} = 2F$

If m_1 is halved and m_2 is quadrupled, F is doubled.

51. Notice that as x gets bigger, y gets smaller. This suggests that the variation is inverse rather than direct. Therefore use the equation $y = \frac{k}{x}$. To determine the value of k, choose one of the ordered pairs and substitute the values into the equation $y = \frac{k}{x}$ and solve for k. We'll use the ordered pair (5, 1).

$y = \frac{k}{x}$.

$1 = \frac{k}{5} \Rightarrow k = 5$

53. The equation is $p = kl$ To find k substitute 150 for l and 2542.50 for p.

$2542.50 = k(150)$

$k = \frac{2542.50}{150}$

$k = 16.95$

Thus $p = 16.95l$.

Now substitute 520 for l.

$p = 16.95(520)$

$p = 8814$

The profit would be \$8814.

55. The equation is $d = kw$. To find k, substitute 2376 for d and 132 for w.

$2376 = k132$

$k = \frac{2376}{132}$

$k = 18$

Thus $d = 18w$. Now substitute 172 for w.

$d = 18(172)$

$d = 3096$

The recommended dosage for Bill is 3096 mg.

57. The equation is $t = \frac{k}{n}$. To find k, substitute 8 for t and 5 for n.

$8 = \frac{k}{5}$

$k = 8(5)$

$k = 40$

Thus $t = \frac{40}{n}$.

Now substitute 4 for n.

$t = \frac{40}{4}$

$t = 10$

It will take 4 bricklayers 10 hours.

59. The equation is $V = \frac{k}{P}$. To find k, substitute 800 for V and 200 for P.

$800 = \frac{k}{200}$

$800(200) = k$

$160{,}000 = k$

Thus $V = \frac{160{,}000}{P}$.

Now substitute 25 for P.

$V = \frac{160{,}000}{25} = 6400$

The volume is 6400 cc.

61. The equation is $t = \frac{k}{s}$. To find k, substitute 6 for s and 2.6 for t.

$2.6 = \frac{k}{6}$
$k = 6(2.6)$
$k = 15.6$
Thus $t = \frac{15.6}{s}$.
Now substitute 5 for s.
$t = \frac{15.6}{5}$
$t = 3.12$
Leif will take 3.12 hours.

63. The equation is $t = \frac{k}{s}$. To find k, substitute 122 for s and 0.21 for t.
$0.21 = \frac{k}{122}$
$k = 0.21(122)$
$k = 25.62$
Thus $t = \frac{25.62}{s}$.
Now substitute 80 for s.
$t = \frac{25.62}{80} \approx 0.32$
It takes about 0.32 seconds for the ball to hit the ground in the service box.

65. The equation is $d = ks^2$. To find k, substitute 40 for s and 60 for d.
$d = ks^2$
$60 = k(40)^2$
$60 = 1600k$
$k = \frac{60}{1600} = 0.0375$
Thus $d = 0.0375s^2$. Now substitute 56 for s.
$d = 0.0375s^2$
$d = 0.0375(56)^2$
$d = 117.6$
The stopping distance is 117.6 feet.

67. The equation is $P = krm$. To find k, substitute 50,000 for m and 0.07 for r and 332.5 for P.
$332.5 = k(0.07)(50{,}000)$
$332.5 = 3500k$
$k = \frac{332.5}{3500} = 0.095$
Thus $P = 0.095rm$.
Now substitute 66,000 for m and 0.07 for r.
$P = 0.095(0.07)(66{,}000)$
$P = 438.9$
The payment is $438.90.

69. The equation is $R = \frac{kA}{P}$. To find k, substitute 400 for A, 2 for P, and 4600 for R.
$4600 = \frac{k(400)}{2}$
$4600 = 200k$
$k = \frac{4600}{200} = 23$
Thus $R = \frac{23A}{P}$. Now substitute 500 for A and 2.50 for P.
$R = \frac{23(500)}{2.50}$
$R = \frac{11{,}500}{2.50}$
$R = 4600$ tapes
They would rent 4600 tapes per week.

71. The equation is $R = \frac{kL}{A}$. To find k, substitute 0.2 for R, 200 for L, and 0.05 for A.
$0.02 = \frac{k(200)}{0.05}$
$0.2 = k(4000)$
$\frac{0.2}{4000} = k$
$0.00005 = k$
Thus $R = \frac{0.00005L}{A}$. Now substitute 5000 for L and 0.01 for A.
$R = \frac{0.00005(5000)}{0.01}$
$R = \frac{0.25}{0.01}$
$R = 25$
The resistance is 25 ohms.

73. The equation is $W = \frac{kTA\sqrt{F}}{R}$
To find k, substitute 78 for T, 5.6 for R, 4 for F, 1000 for A, and 68 for W.
$68 = \frac{k(78)(1000)\sqrt{4}}{5.6}$
$68 = \frac{156{,}000k}{5.6}$
$\frac{(68)(5.6)}{156{,}000} = k$
$0.002441 \approx k$
Thus $W = \frac{0.002441TA\sqrt{E}}{R}$
Now, substitute 78 for T, 5.6 for R, 6 for E, and 1500 for A.
$W = \frac{0.002441(78)(1500)\sqrt{6}}{5.6} \approx 124.92$
The water bill is about $124.92.

75. a. The equation is $F = \dfrac{km_1m_2}{d^2}$.

b. If m_1 becomes $2m_1$, m_2 becomes $3m_2$ and d becomes $\frac{d}{2}$, then

$$F = \frac{k(2m_1)(3m_2)}{\left(\frac{d}{2}\right)^2} = \frac{6km_1m_2}{\frac{d^2}{4}} = 24\frac{km_1m_2}{d^2}$$

The new force is 24 times the original force.

77.

$$\begin{array}{r} 2x-3 \\ 4x+9\overline{)8x^2+6x-25} \\ \underline{8x^2+18x} \\ -12x-25 \\ \underline{-12x-27} \\ 2 \end{array}$$

$$\frac{8x^2+6x-25}{4x+9} = 2x-3+\frac{2}{4x+9}$$

78. $y(z-2)+3(z-2)$

$(z-2)(y+3)$

79.

$$3x^2-24=-6x$$
$$3x^2+6x-24=0$$
$$3(x^2+2x-8)=0$$
$$3(x+4)(x-2)=0$$

$x+4=0$ or $x-2=0$

$x=-4$ $\quad x=2$

80. $\dfrac{x+3}{x-3}\cdot\dfrac{x^3-27}{x^2+3x+9} = \dfrac{x+3}{x-3}\cdot\dfrac{(x-3)(x^2+3x+9)}{x^2+3x+9}$

$= x+3$

Review Exercises

1. $\dfrac{5}{2x-12}$

$2x-12=0$

$2(x-6)=0$

$x-6=0$

The expression is defined for all real numbers except $x=6$.

2. $\dfrac{2}{x^2-8x+15}$

$x^2-8x+15=0$

$(x-3)(x-5)=0$

$x-3=0, x-5=0$

The expression is defined for all real numbers except $x=3$, $x=5$.

3. $\dfrac{2}{5x^2+4x-1}$

$5x^2+4x-1=0$

$(5x-1)(x+1)=0$

$5x-1=0, x+1=0$

The expression is defined for all real numbers except $x=\frac{1}{5}$, $x=-1$.

4. $\dfrac{y}{xy-3y} = \dfrac{y}{y(x-3)}$

$= \dfrac{1}{x-3}$

5. $\dfrac{x^3+4x^2+12x}{x} = \dfrac{x(x^2+4x+12)}{x}$

$= x^2+4x+12$

6. $\dfrac{9x^2+6xy}{3x} = \dfrac{3x(3x+2y)}{3x}$

$= 3x+2y$

7. $\dfrac{x^2+2x-8}{x-2} = \dfrac{(x-2)(x+4)}{x-2}$

$= x+4$

8. $\dfrac{a^2-36}{a-6} = \dfrac{(a-6)(a+6)}{a-6}$

$= a+6$

9. $\dfrac{-2x^2+7x+4}{x-4}=\dfrac{-1(2x^2+7x+4)}{(x-4)}$
$=\dfrac{-1(2x+1)(x-4)}{(x-4)}$
$=-(2x+1)$

10. $\dfrac{b^2-8b+15}{b^2-3b-10}=\dfrac{(b-5)(b-3)}{(b-5)(b+2)}$
$=\dfrac{b-3}{b+2}$

11. $\dfrac{4x^2-11x-3}{4x^2-7x-2}=\dfrac{(4x+1)(x-3)}{(4x+1)(x-2)}$
$=\dfrac{x-3}{x-2}$

12. $\dfrac{2x^2-21x+40}{4x^2-4x-15}=\dfrac{(x-8)(2x-5)}{(2x+3)(2x-5)}$
$=\dfrac{x-8}{2x+3}$

13. $\dfrac{5a^2}{6b}\cdot\dfrac{2}{4a^2b}=\dfrac{5\cdot 2}{6\cdot 4}\cdot\dfrac{a^2}{a^2}\cdot\dfrac{1}{b\cdot b}$
$=\dfrac{10}{24b^2}$
$=\dfrac{5}{12b^2}$

14. $\dfrac{15x^2y^3}{3z}\cdot\dfrac{6z^3}{5xy^3}=\dfrac{15\cdot 6}{15}\left(\dfrac{x^2}{x}\right)\left(\dfrac{y^3}{y^3}\right)\left(\dfrac{z^3}{z}\right)$
$=6xz^2$

15. $\dfrac{40a^3b^4}{7c^3}\cdot\dfrac{14c^5}{5a^5b}=\dfrac{40}{5}\cdot\dfrac{14}{7}\cdot\dfrac{a^3}{a^5}\dfrac{b^4}{b}\dfrac{c^5}{c^3}$
$=16\dfrac{1}{a^2}b^3c^2$
$=\dfrac{16b^3c^2}{a^2}$

16. $\dfrac{1}{x-4}\cdot\dfrac{4-x}{3}=\dfrac{1}{x-4}\cdot\dfrac{-1(x-4)}{3}$
$=-\dfrac{1}{3}$

17. $\dfrac{-m+4}{15m}\cdot\dfrac{10m}{m-4}=\dfrac{-1(m-4)}{3\cdot 5m}\cdot\dfrac{2\cdot 5m}{m-4}$
$=\dfrac{-2}{3}$

18. $\dfrac{a-2}{a+3}\cdot\dfrac{a^2+4a+3}{a^2-a-2}=\dfrac{(a-2)(a+3)(a+1)}{(a+3)(a-2)(a+1)}$
$=1$

19. $\dfrac{16x^6}{y^2}\div\dfrac{x^4}{4y}=\dfrac{16x^6}{y^2}\cdot\dfrac{4y}{x^4}$
$=\dfrac{64x^6y}{x^4y^2}$
$=\dfrac{64x^2}{y}$

20. $\dfrac{8xy^2}{z}\div\dfrac{x^4y^2}{4z^2}=\dfrac{8xy^2}{z}\cdot\dfrac{4z^2}{x^4y^2}$
$=\dfrac{32z}{x^3}$

21. $\dfrac{5a+5b}{a^2}\div\dfrac{a^2-b^2}{a^2}=\dfrac{5(a+b)}{a^2}\cdot\dfrac{a^2}{(a+b)(a-b)}$
$=\dfrac{5}{a-b}$

22. $\dfrac{1}{a^2+8a+15}\div\dfrac{3}{a+5}=\dfrac{1}{(a+5)(a+3)}\cdot\dfrac{a+5}{3}$
$=\dfrac{1}{3(a+3)}$

23. $(t+8)\div\dfrac{t^2+5t-24}{t-3}=(t+8)\cdot\dfrac{t-3}{(t-3)(t+8)}$
$=1$

24. $\dfrac{x^2+xy-2y^2}{4y}\div\dfrac{x+2y}{12y^2}=\dfrac{(x-y)(x+2y)}{4y}\cdot\dfrac{12y^2}{x+2y}$
$=3y(x-y)$

25. $\dfrac{n}{n+5}-\dfrac{5}{n+5}=\dfrac{n-5}{n+5}$

26. $\dfrac{3x}{x+7}+\dfrac{21}{x+7}=\dfrac{3x+21}{x+7}$
$=\dfrac{3(x+7)}{x+7}$
$=3$

27. $\frac{9x-4}{x+8}+\frac{76}{x+8}=\frac{9x-4+76}{x+8}$
$=\frac{9x+72}{x+8}$
$=\frac{9(x+8)}{x+8}$
$=9$

28. $\frac{7x-3}{x^2+7x-30}-\frac{3x+9}{x^2+7x-30}=\frac{7x-3-(3x+9)}{x^2+7x-30}$
$=\frac{7x-3-3x-9}{x^2+7x-30}$
$=\frac{4x-12}{x^2+7x-30}$
$=\frac{4(x-3)}{(x+10)(x-3)}$
$=\frac{4}{x+10}$

29. $\frac{5h^2+12h-1}{h+5}-\frac{h^2-5h+14}{h+5}=\frac{5h^2+12h-1-(h^2-5h+14)}{h+5}$
$=\frac{5h^2+12h-1-h^2+5h-14}{h+5}$
$=\frac{4h^2+17h-15}{h+5}$
$\frac{(4h-3)(h+5)}{h+5}$
$=4h-3$

30. $\frac{6x^2-4x}{2x-3}-\frac{-3x+12}{2x-3}=\frac{6x^2+4x-(-3x+12)}{2x-3}$
$=\frac{6x^2-4x+3x-12}{2x-3}$
$=\frac{6x^2-x-12}{2x-3}$
$=\frac{(3x+4)(2x-3)}{2x-3}$
$=3x+4$

31. $\frac{a}{10}+\frac{4a}{3}$
Least common denominator = 2(5)(3) = 30

32. $\frac{8}{x+3}+\frac{4}{x+3}$
Least common denominator = $x+3$

33. $\frac{5}{4xy^3}-\frac{7}{10x^2y}$
Least common denominator = $20x^2y^3$

34. $\frac{6}{x+1}-\frac{3x}{x}$
Least common denominator = $x(x+1)$

35. $\frac{4}{n-5}+\frac{2n-3}{n-4}$
Least common denominator = $(n-5)(n-4)$

36. $\frac{7x-12}{x^2+x}-\frac{4}{x+1}=\frac{7x-12}{x(x+1)}-\frac{4}{x+1}$
Least common denominator = $x(x+1)$

37. $\frac{2r-9}{r-s}-\frac{6}{r^2-s^2}=\frac{2r-9}{r-s}-\frac{6}{(r+s)(r-s)}$
Least common denominator = $(r+s)(r-s)$

38. $\frac{4x^2}{x-7}+8x^2=\frac{4x^2}{x-y}+\frac{8x^2}{1}$
Least common denominator = $x-7$

39. $\frac{19x-5}{x^2+2x-35}+\frac{3x-2}{x^2+9x+14}$
$=\frac{19x-5}{(x+7)(x-5)}+\frac{3x-2}{(x+7)(x+2)}$
Least common denominator
$=(x+7)(x-5)(x+2)$

40. $$\frac{4}{3y^2}+\frac{y}{2y}=\frac{4}{3y^2}+\frac{1}{2}$$
$$=\frac{4}{3y^2}\cdot\frac{2}{2}+\frac{1}{2}\cdot\frac{3y^2}{3y^2}$$
$$=\frac{8}{6y^2}+\frac{3y^2}{6y^2}$$
$$=\frac{8+3y^2}{6y^2}$$
$$=\frac{3y^2+8}{6y^2}$$

41. $$\frac{2x}{xy}+\frac{1}{5x}=\frac{2x}{xy}\cdot\frac{5}{5}+\frac{1}{5x}\cdot\frac{y}{y}$$
$$=\frac{10x}{5xy}+\frac{y}{5xy}$$
$$=\frac{10x+y}{5xy}$$

42. $$\frac{5x}{3xy}-\frac{4}{x^2}=\frac{5x}{3xy}\cdot\frac{x}{x}-\frac{4}{x^2}\cdot\frac{3y}{3y}$$
$$=\frac{5x^2}{3x^2y}-\frac{12y}{3x^2y}$$
$$=\frac{5x^2-12y}{3x^2y}$$

43. $$6-\frac{2}{x+2}=6\left(\frac{x+2}{x+2}\right)-\frac{2}{x+2}$$
$$=\frac{6x+12-2}{x+2}$$
$$=\frac{6x+10}{x+2}$$

44. $$\frac{x-y}{y}-\frac{x+y}{x}=\frac{x-y}{y}\cdot\frac{x}{x}-\frac{x+y}{x}\cdot\frac{y}{y}$$
$$=\frac{x(x-y)}{xy}-\frac{y(x+y)}{xy}$$
$$=\frac{x^2-xy-xy-y^2}{xy}$$
$$=\frac{x^2-2xy-y^2}{xy}$$

45. $$\frac{7}{x+4}+\frac{4}{x}=\frac{7}{x+4}\cdot\frac{x}{x}+\frac{4}{x}\cdot\frac{x+4}{x+4}$$
$$=\frac{7x}{x(x+4)}+\frac{4(x+4)}{x(x+4)}$$
$$=\frac{7x+4x+16}{x(x+4)}$$
$$=\frac{11x+16}{x(x+4)}$$

46. $$\frac{2}{3x}-\frac{3}{3x-6}=\frac{2}{3x}-\frac{3}{3(x-2)}$$
$$=\frac{2}{3x}\cdot\frac{x-2}{x-2}-\frac{3}{3(x-2)}\cdot\frac{x}{x}$$
$$=\frac{2(x-2)}{3x(x-2)}-\frac{3x}{3x(x-2)}$$
$$=\frac{2x-4-3x}{3x(x-2)}$$
$$=\frac{-x-4}{3x(x-2)}$$

47. $$\frac{3}{z+5}+\frac{7}{(z+5)^2}=\frac{3}{z+5}\cdot\frac{z+5}{z+5}+\frac{7}{(z+5)^2}$$
$$=\frac{3(z+5)}{(z+5)^2}+\frac{7}{(z+5)^2}$$
$$=\frac{3z+15+7}{(z+5)^2}$$
$$=\frac{3z+22}{(z+5)^2}$$

48.
$$\frac{x+2}{x^2-x-6}+\frac{x-3}{x^2-8x+15}=\frac{x+2}{(x-3)(x+2)}+\frac{x-3}{(x-3)(x-5)}$$
$$=\frac{1}{x-3}+\frac{1}{x-5}$$
$$=\frac{1}{x-3}\cdot\frac{x-5}{x-5}+\frac{1}{x-5}\cdot\frac{x-3}{x-3}$$
$$=\frac{x-5}{(x-3)(x-5)}+\frac{x-3}{(x-5)(x-3)}$$
$$=\frac{x-5+x-3}{(x-5)(x-3)}$$
$$=\frac{2x-8}{(x-5)(x-3)}$$

49.
$$\frac{x+4}{x+6}-\frac{x-5}{x+2}=\frac{(x+4)(x+2)}{(x+6)(x+2)}-\frac{(x-5)(x+6)}{(x+2)(x+6)}$$
$$=\frac{x^2+6x+8-(x^2+x-30)}{(x+6)(x+2)}$$
$$=\frac{x^2+6x+8-x^2+x+30}{(x+6)(x+2)}$$
$$=\frac{5x+38}{(x+6)(x+2)}$$

50.
$$3+\frac{x}{x-4}=\frac{3(x-4)}{x-4}+\frac{x}{x-4}$$
$$=\frac{3x-12+x}{x-4}$$
$$=\frac{4x-12}{x-4}$$

51.
$$\frac{a+2}{b}\div\frac{a-2}{4b^2}=\frac{a+2}{b}\cdot\frac{4b^2}{a-2}$$
$$=\frac{4b(a+2)}{a-2}$$
$$=\frac{4ab+8b}{a-2}$$

52.
$$\frac{x+3}{x^2-9}+\frac{2}{x+3}=\frac{x+3}{(x-3)(x+3)}+\frac{2}{x+3}$$
$$=\frac{x+3}{(x-3)(x+3)}+\frac{2(x-3)}{(x-3)(x+3)}$$
$$=\frac{x+3+2x-6}{(x-3)(x+3)}$$
$$=\frac{3x-3}{(x-3)(x+3)}$$

53.
$$\frac{5p+10q}{p^2q}\cdot\frac{p^4}{p+2q}=\frac{5(p+2q)p^4}{(p+2q)p^2q}$$
$$=\frac{5p^2}{q}$$

54. $\dfrac{4}{(x+2)(x-3)}-\dfrac{4}{(x-2)(x+2)}=\dfrac{4(x-2)}{(x+2)(x-3)(x-2)}-\dfrac{4(x-3)}{(x+2)(x-3)(x-2)}$

$=\dfrac{4(x-2)-4(x-3)}{(x+2)(x-3)(x-2)}$

$=\dfrac{4x-8-4x+12}{(x+2)(x-3)(x-2)}$

$=\dfrac{4}{(x+2)(x-3)(x-2)}$

55. $\dfrac{x+7}{x^2+9x+14}-\dfrac{x-10}{x^2-49}=\dfrac{x+7}{(x+7)(x+2)}-\dfrac{x-10}{(x+7)(x-7)}$

$=\dfrac{x+7}{(x+7)(x+2)}\cdot\dfrac{(x-7)}{(x-7)}-\dfrac{x-10}{(x+7)(x-7)}\cdot\dfrac{(x+2)}{(x+2)}$

$=\dfrac{x^2-49-(x^2-8x-20)}{(x+7)(x-7)(x+2)}$

$=\dfrac{8x-29}{(x+7)(x-7)(x+2)}$

56. $\dfrac{x-y}{x+y}\cdot\dfrac{xy+x^2}{x^2-y^2}=\dfrac{x-y}{x+y}\cdot\dfrac{x(y+x)}{(x+y)(x-y)}$

$=\dfrac{x}{x+y}$

57. $\dfrac{3x^2-27y^2}{20}\div\dfrac{(x-3y)^2}{4}=\dfrac{3(x^2-9y^2)}{20}\cdot\dfrac{4}{(x-3y)^2}$

$=\dfrac{3(x-3y)(x+3y)}{4\cdot 5}\cdot\dfrac{4}{(x-3y)^2}$

$=\dfrac{3(x+3y)}{5(x-3y)}$

58. $\dfrac{a^2-9a+20}{a-4}\cdot\dfrac{a^2-8a+15}{a^2-10a+25}=\dfrac{(a-4)(a-5)}{a-4}\cdot\dfrac{(a-5)(a-3)}{(a-5)^2}$

$=a-3$

59. $\dfrac{a}{a^2-1}-\dfrac{2}{3a^2-2a-5}=\dfrac{a}{(a-1)(a+1)}-\dfrac{2}{(3a-5)(a+1)}$

$=\dfrac{a(3a-5)}{(a-1)(a+1)(3a-5)}-\dfrac{2(a-1)}{(a-1)(a+1)(3a-5)}$

$=\dfrac{a(3a-5)-2(a-1)}{(a-1)(a+1)(3a-5)}$

$=\dfrac{3a^2-5a-2a+2}{(a-1)(a+1)(3a-5)}$

$=\dfrac{3a^2-7a+2}{(a-1)(a+1)(3a-5)}$

60. $\frac{2x^2+6x-20}{x^2-2x} \div \frac{x^2+7x+10}{4x^2-16} = \frac{2(x^2+3x-10)}{x(x-2)} \cdot \frac{4(x^2-4)}{(x+5)(x+2)}$
$= \frac{2(x+5)(x-2)}{x(x-2)} \cdot \frac{4(x+2)(x-2)}{(x+5)(x+2)}$
$= \frac{8(x-2)}{x}$
$= \frac{8x-16}{x}$

61. $\frac{3+\frac{2}{3}}{\frac{3}{5}} = \frac{15\left(3+\frac{2}{3}\right)}{15\left(\frac{3}{5}\right)}$
$= \frac{45+10}{9}$
$= \frac{55}{9}$

62. $\frac{1+\frac{5}{8}}{4-\frac{9}{16}} = \frac{16\left(1+\frac{5}{8}\right)}{16\left(4-\frac{9}{16}\right)}$
$= \frac{16+10}{64-9}$
$= \frac{26}{55}$

63. $\frac{\frac{12ab}{9c}}{\frac{4a}{c^2}} = \frac{12ab}{9c} \cdot \frac{c^2}{4a}$
$= \frac{bc}{3}$

64. $\frac{36x^4y^2}{\frac{9xy^5}{4z^2}} = \frac{36x^4y^2 \cdot 4z^2}{\frac{9xy^5}{4z^2} \cdot 4z^2}$
$= \frac{144x^4y^2z^2}{9xy^5}$
$= \frac{16x^3z^2}{y^3}$

65. $\frac{a-\frac{a}{b}}{\frac{1+a}{b}} = \frac{\left(a-\frac{a}{b}\right)b}{\left(\frac{1+a}{b}\right)b} = \frac{ab-a}{1+a}$

66. $\frac{r^2+\frac{1}{s}}{s^2} = \frac{\left(r^2+\frac{1}{s}\right)s}{\left(s^2\right)s}$
$= \frac{r^2s+1}{s^3}$

67. $\frac{\frac{4}{x}+\frac{2}{x^2}}{6-\frac{1}{x}} = \frac{\left(\frac{4}{x}+\frac{2}{x^2}\right)x^2}{\left(6-\frac{1}{x}\right)x^2}$
$= \frac{4x+2}{6x^2-x}$
$= \frac{4x+2}{x(6x-1)}$

68. $\frac{\frac{x}{x+y}}{\frac{x^2}{2x+2y}} = \frac{x}{x+y} \cdot \frac{2x+2y}{x^2}$
$= \frac{1}{x+y} \cdot \frac{2(x+y)}{x}$
$= \frac{2}{x}$

69. $\frac{\frac{3}{x}}{\frac{3}{x^2}} = \frac{3}{x} \cdot \frac{x^2}{3} = x$

70. $\frac{\frac{1}{a}+2}{\frac{1}{a}+\frac{1}{a}} = \frac{\frac{1}{a}+2}{\frac{2}{a}}$
$= \frac{\left(\frac{1}{a}+2\right)a}{\left(\frac{2}{a}\right)a}$
$= \frac{1+2a}{2}$

71. $\dfrac{\frac{1}{x^2}-\frac{1}{x}}{\frac{1}{x^2}+\frac{1}{x}} = \dfrac{x^2\left(\frac{1}{x^2}-\frac{1}{x}\right)}{x^2\left(\frac{1}{x^2}+\frac{1}{x}\right)}$

$= \dfrac{1-x}{1+x}$

$= \dfrac{-x+1}{x+1}$

72. $\dfrac{\frac{3x}{y}-x}{\frac{y}{x}-1} = \dfrac{\left(\frac{3x}{y}-x\right)xy}{\left(\frac{y}{x}-1\right)xy}$

$= \dfrac{3x^2-x^2y}{y^2-xy}$

$= \dfrac{3x^2-x^2y}{y(y-x)}$

73. $\dfrac{5}{9} = \dfrac{5}{x+3}$

$5(x+3) = 5(9)$

$5x+15 = 45$

$5x = 30$

$x = 6$

74. $\dfrac{x}{6} = \dfrac{x-4}{2}$

$2x = 6(x-4)$

$2x = 6x-24$

$24 = 4x$

$6 = x$

75. $\dfrac{n}{5}+12 = \dfrac{n}{2}$

$10\left(\dfrac{n}{5}+12\right) = 10\left(\dfrac{n}{2}\right)$

$2n+120 = 5n$

$120 = 3n$

$40 = n$

76. $\dfrac{3}{x}-\dfrac{1}{6} = \dfrac{1}{x}$

$6x\left(\dfrac{3}{x}-\dfrac{1}{6}\right) = 6x\left(\dfrac{1}{x}\right)$

$18-x = 6$

$-x = -12$

$x = 12$

77. $\dfrac{-4}{d} = \dfrac{3}{2}+\dfrac{4-d}{d}$

$2d\left(\dfrac{-4}{d}\right) = 2d\left(\dfrac{3}{2}+\dfrac{4-d}{d}\right)$

$-8 = 3d+8-2d$

$-8 = d+8$

$-16 = d$

78. $\dfrac{1}{x-7}+\dfrac{1}{x+7} = \dfrac{1}{x^2-49}$

$\dfrac{1}{x-7}+\dfrac{1}{x+7} = \dfrac{1}{(x-7)(x+7)}$

$(x-7)(x+7)\left[\dfrac{1}{x-7}+\dfrac{1}{x+7}\right] = \left[\dfrac{1}{(x-7)(x+7)}\right](x-7)(x+7)$

$x+7+x-7 = 1$

$2x = 1$

$x = \dfrac{1}{2}$

79.
$$\frac{x-3}{x-2}+\frac{x+1}{x+3}=\frac{2x^2+x+1}{x^2+x-6}$$
$$\frac{x-3}{x-2}+\frac{x+1}{x+3}=\frac{2x^2+x+1}{(x-2)(x+3)}$$
$$(x-2)(x+3)\left(\frac{x-3}{x-2}+\frac{x+1}{x+3}\right)=(x-2)(x+3)\left[\frac{2x^2+x+1}{(x-2)(x+3)}\right]$$
$$(x+3)(x-3)+(x-2)(x+1)=2x^2+x+1$$
$$x^2-9+x^2-x-2=2x^2+x+1$$
$$2x^2-x-11=2x^2+x+1$$
$$-12=2x$$
$$-6=x$$

80.
$$\frac{a}{a^2-64}+\frac{4}{a+8}=\frac{3}{a-8}$$
$$\frac{a}{(a+8)(a-8)}+\frac{4}{a+8}=\frac{3}{a-8}$$
$$(a+8)(a-8)\left[\frac{a}{(a+8)(a-8)}+\frac{4}{a+8}\right]=(a+8)(a-8)\left(\frac{3}{a-8}\right)$$
$$a+4(a-8)=3(a+8)$$
$$a+4a-32=3a+24$$
$$5a-32=3a+24$$
$$2a=56$$
$$a=28$$

81.
$$\frac{d}{d-2}-2=\frac{2}{d-2}$$
$$(d-2)\left(\frac{d}{d-2}-2\right)=(d-2)\left(\frac{2}{d-2}\right)$$
$$d-2(d-2)=2$$
$$d-2d+4=2$$
$$-d+4=2$$
$$-d=-2$$
$$d=2$$

Since $\frac{2}{0}$ is not a real number, there is no solution.

82. John and Amy rate $=\frac{1}{6}$

Paul and Cindy rate $=\frac{1}{5}$

$$\frac{t}{6}+\frac{t}{5}=1$$
$$30\left(\frac{t}{6}+\frac{t}{5}\right)=30(1)$$
$$5t+6t=30$$
$$11t=30$$
$$t=\frac{30}{11}=2\frac{8}{11}$$

It will take the 4 people $2\frac{8}{11}$ hours.

83. $\frac{3}{4}$-inch hose's rate $=\frac{1}{7}$

$\frac{5}{16}$-inch hose's rate $=\frac{1}{12}$

$$\frac{t}{7}-\frac{t}{12}=1$$
$$\frac{12t-7t}{84}=1$$
$$5t=84$$
$$t=\frac{84}{5}=16\frac{4}{5}$$

It will take $16\frac{4}{5}$ hours to fill the pool.

84. Let x be one number then, $5x$ is the other number.

$$\frac{1}{x}+\frac{1}{5x}=6$$
$$5x\left(\frac{1}{x}+\frac{1}{5x}\right)=5x(6)$$
$$5+1=30x$$
$$6=30x$$
$$\frac{6}{30}=x$$
$$\frac{1}{5}=x$$
$$5x=5\left(\frac{1}{5}\right)=1$$

The numbers are $\frac{1}{5}$ and 1.

85. Let x = Robert's speed, then $3.5+x$ = Tran's speed

$$t=\frac{d}{r}$$

Robert's time = Tran's time

$$\frac{3}{x}=\frac{8}{3.5+x}$$
$$3(3.5+x)=8x$$
$$10.5+3x=8x$$
$$10.5=5x$$
$$2.1=x$$
$$x+3.5=2.1+3.5=5.6$$

Robert's speed is 2.1 mph and Tran's speed is 5.6 mph.

86. The equation is $x=ky^2$. To find k, substitute 45 for x and 3 for y.

$$45=k(3)^2$$
$$45=9k$$
$$\frac{45}{9}=k$$
$$5=k$$

Thus $x=5y^2$. Now substitute 2 for y.

$$A=5(2)^2=5(4)=20$$

87. The equation is $W=\frac{kL^2}{A}$. To find k, substitute 4 for W, 2 for L, and 10 for A.

$$4=\frac{k(2)^2}{10}$$
$$4=\frac{4k}{10}$$
$$40=4k$$
$$10=k$$

Thus $W=\frac{10L^2}{A}$. Now substitute 5 for L and 20 for A.

$$W=\frac{10\cdot(5)^2}{20}=\frac{250}{20}=\frac{25}{2}$$

88. The equation is $z=\frac{kxy}{r^2}$. To find k, substitute 12 for z, 20 for x, 8 for y, and 8 for r.

$$12=\frac{k(20)(8)}{(8)^2}$$
$$12=\frac{160k}{64}$$
$$12(64)=160k$$
$$768=160k$$
$$\frac{768}{160}=k$$
$$k=4.8$$

Thus $z=\frac{4.8xy}{r^2}$. Now substitute 10 for x, 80 for y, and 3 for r.

$$z=\frac{4.8(10)(80)}{3^2}\approx 426.7$$

89. Let s represent the surcharge and let E represent the energy used in kilowatt-hours. The equation is

$s=kE$.

To find k, substitute 7.20 for s and 3600 for E.

$$s=kE$$
$$7.20=k(3600)\ \Rightarrow k=\frac{7.20}{3600}=0.002$$

Thus $s=0.002E$. Now substitute 4200 for E.

$$s=0.002E$$
$$s=0.002(4200)$$
$$s=8.40$$

The surcharge is \$8.40.

90. The equation is $d=kt^2$. To find k, substitute 16 for d and 1 for t.

$$16=k(1)^2$$
$$16=k(1)$$
$$16=k$$

Thus $d=16t^2$. Now substitute 10 for t.

$$d=16(10)^2=16(100)=1600$$

An object will fall 1600 feet.

91. The equation is $A = kr^2$. To find k, substitute 78.5 for A and 5 for r.

$$78.5 = k(5)^2$$
$$78.5 = k(25)$$
$$\frac{78.5}{25} = k$$
$$3.14 = k$$

Thus $A = 3.14r^2$. Now substitute 8 for r.

$$A = 3.14(8)^2$$
$$A = 3.14(64)$$
$$A = 200.96$$

The area is 200.96 square units.

92. The equation is $t = \frac{k}{w}$ where t is time and w is water temperature. To find k, substitute 1.7 for t and 70 for w.

$$1.7 = \frac{k}{70}$$
$$(1.7)(70) = k$$
$$119 = k$$

Thus $t = \frac{119}{w}$. Now substitute 50 for w.

$$t = \frac{119}{50} = 2.38$$

It takes the ice cube 2.38 minutes to melt.

Practice Test

1. $\frac{-6+x}{x-6} = \frac{x-6}{x-6} = 1$

2. $\frac{x^3-1}{x^2-1} = \frac{(x-1)(x^2+x+1)}{(x-1)(x+1)}$

$= \frac{x^2+x+1}{x+1}$

3. $\frac{15x^2y^3}{4z^2} \cdot \frac{8xz^3}{5xy^4} = \frac{3 \cdot 5x^2y^3}{4z^2} \cdot \frac{2 \cdot 4xz^3}{5xy^4}$

$= \frac{6x^3y^3z^3}{xy^4z^2}$

$= \frac{6x^2z}{y}$

4. $\frac{a^2-9a+14}{a-2} \cdot \frac{a^2-4a-21}{(a-7)^2} = \frac{(a-7)(a-2)}{a-2} \cdot \frac{(a-7)(a+3)}{(a-7)^2}$

$= a+3$

5. $\frac{x^2-x-6}{x^2-9} \cdot \frac{x^2-6x+9}{x^2+4x+4} = \frac{(x-3)(x+2)}{(x-3)(x+3)} \cdot \frac{(x-3)(x-3)}{(x+2)(x+2)}$

$= \frac{(x-3)^2}{(x+3)(x+2)}$

$= \frac{x^2-6x+9}{(x+3)(x+2)}$

6. $\frac{x^2-1}{x+2} \cdot \frac{2x+4}{2-2x^2} = \frac{(x-1)(x+1)}{x+2} \cdot \frac{2(x+2)}{-2(x-1)(x+1)}$

$= -1$

7. $\frac{x^2-4y^2}{3x+12y} \div \frac{x+2y}{x+4y} = \frac{(x-2y)(x+2y)}{3(x+4y)} \cdot \frac{x+4y}{x+2y}$

$= \frac{x-2y}{3}$

8. $\dfrac{15}{y^2+2y-15} \div \dfrac{3}{y-3} = \dfrac{15}{(y-3)(y+5)} \cdot \dfrac{y-3}{3}$
$= \dfrac{5}{y+5}$

9. $\dfrac{m^2+3m-18}{m-3} \div \dfrac{m^2-8m+15}{3-m} = \dfrac{(m+6)(m-3)}{m-3} \cdot \dfrac{-1(m-3)}{(m-5)(m-3)}$
$= \dfrac{-(m+6)}{m-5}$
$= -\dfrac{m+6}{m-5}$

10. $\dfrac{4x+3}{2y} + \dfrac{2x-5}{2y} = \dfrac{4x+3+2x-5}{2y}$
$= \dfrac{6x-2}{2y}$
$= \dfrac{2(3x-1)}{2y}$
$= \dfrac{3x-1}{y}$

11. $\dfrac{7x^2-4}{x+3} - \dfrac{6x+7}{x+3} = \dfrac{7x^2-4-(6x+7)}{x+3}$
$= \dfrac{7x^2-4-6x-7}{x+3}$
$= \dfrac{7x^2-6x-11}{x+3}$

12. $\dfrac{4}{xy} - \dfrac{3}{xy^3} = \dfrac{4}{xy} \cdot \dfrac{y^2}{y^2} - \dfrac{3}{xy^3}$
$= \dfrac{4y^2}{xy^3} - \dfrac{3}{xy^3}$
$= \dfrac{4y^2-3}{xy^3}$

13. $4 - \dfrac{5z}{z-5} = 4\left(\dfrac{z-5}{z-5}\right) - \dfrac{5z}{z-5}$
$= \dfrac{4z-20}{z-5} - \dfrac{5z}{z-5}$
$= \dfrac{4z-20-5z}{z-5}$
$= \dfrac{-z-20}{z-5}$
$= \dfrac{-1(z+20)}{z-5}$
$= -\dfrac{z+20}{z-5}$

14. $$\begin{aligned}\frac{x-5}{x^2-16}-\frac{x-2}{x^2+2x-8}&=\frac{x-5}{(x-4)(x+4)}-\frac{x-2}{(x-2)(x+4)}\\&=\frac{x-5}{(x-4)(x+4)}-\frac{1}{x+4}\\&=\frac{x-5}{(x-4)(x+4)}-\frac{1}{x+4}\cdot\frac{x-4}{x-4}\\&=\frac{x-5-(x-4)}{(x-4)(x+4)}\\&=\frac{x-5-x+4}{(x-4)(x+4)}\\&=\frac{-1}{(x-4)(x+4)}\end{aligned}$$

15. $$\begin{aligned}\frac{5+\frac{1}{2}}{3-\frac{1}{5}}&=\frac{10\left(5+\frac{1}{2}\right)}{10\left(3-\frac{1}{5}\right)}\\&=\frac{50+5}{30-2}\\&=\frac{55}{28}\end{aligned}$$

16. $$\begin{aligned}\frac{x+\frac{x}{y}}{\frac{1}{x}}&=\left(x+\frac{x}{y}\right)\frac{x}{1}\\&=\left(\frac{xy+x}{y}\right)\left(\frac{x}{1}\right)\\&=\frac{yx^2+x^2}{y}\end{aligned}$$

17. $$\begin{aligned}\frac{2+\frac{3}{x}}{\frac{2}{x}-5}&=\frac{x\left(2+\frac{3}{x}\right)}{x\left(\frac{2}{x}-5\right)}\\&=\frac{2x+3}{2-5x}\end{aligned}$$

18. $$\begin{aligned}6+\frac{2}{x}&=7\\x\left(6+\frac{2}{x}\right)&=7x\\6x+2&=7x\\2&=x\end{aligned}$$

19. $$\begin{aligned}\frac{2x}{3}-\frac{x}{4}&=x+1\\12\left(\frac{2x}{3}-\frac{x}{4}\right)&=12(x+1)\\8x-3x&=12x+12\\5x&=12x+12\\-7x&=12\\x&=-\frac{12}{7}\end{aligned}$$

20. $$\begin{aligned}\frac{x}{x-8}+\frac{6}{x-2}&=\frac{x^2}{x^2-10x+16}\\\frac{x}{x-8}+\frac{6}{x-2}&=\frac{x^2}{(x-8)(x-2)}\\(x-8)(x-2)\left(\frac{x}{x-8}+\frac{6}{x-2}\right)&=(x-8)(x-2)\left[\frac{x^2}{(x-8)(x-2)}\right]\\x(x-2)+6(x-8)&=x^2\\x^2-2x+6x-48&=x^2\\4x-48&=0\\4x&=48\\x&=12\end{aligned}$$

21.
$$\frac{t}{8}+\frac{t}{5}=1$$
$$\frac{5t+8t}{40}=1$$
$$13t=40$$
$$t=\frac{40}{13}=3\frac{1}{13}$$
It will take them $3\frac{1}{13}$ hours to level one acre together.

22. Let x be the number.
$$x+\frac{1}{x}=2$$
$$x\left(x+\frac{1}{x}\right)=x(2)$$
$$x^2+1=2x$$
$$x^2-2x+1=0$$
$$(x-1)(x-1)=0$$
$$x-1=0$$
$$x=1$$
The number is 1.

23. Let d = the distance she rollerblades, then $12-d$ is the distance she bicycles.
$$t=\frac{d}{r}$$
$$\frac{d}{4}+\frac{12-d}{10}=1.5$$
$$20\left(\frac{d}{4}+\frac{12-d}{10}\right)=20(1.5)$$
$$5d+2(12-d)=30$$
$$5d+24-2d=30$$
$$3d=6$$
$$d=2$$
She rollerblades for 2 miles.

24.
$$w=\frac{k}{f}$$
$$4.3=\frac{k}{263}$$
$$1130.9=k$$

Now substitute 1000 for f and 1130.9 for k.
$$w=\frac{k}{f}$$
$$w=\frac{1130.9}{1000}$$
$$w=1.1309$$
The wavelength would be about 1.13 feet.

Cumulative Review Test

1.
$$3x^2-5xy^2+3=3(-4)^2-5(-4)(-2)^2+3$$
$$=3(16)-5(-4)(4)+3$$
$$=48+80+3$$
$$=131$$

2.
$$[7-[3(8\div 4)]^2+9\cdot 4]^2=[7-[3(2)]^2+9\cdot 4]^2$$
$$=[7-(6)^2+9\cdot 4]^2$$
$$=[7-36+9\cdot 4]^2$$
$$=[7-36+36]^2$$
$$=[7]^2$$
$$=49$$

3.
$$5z+4=-3(z-7)$$
$$5z+4=-3z+21$$
$$5z+3z=21-4$$
$$8z=17$$
$$z=\frac{17}{8}$$

4.
$$\frac{12 \text{ inches}}{1 \text{ foot}}=\frac{162 \text{ inches}}{x \text{ feet}}$$
$$\frac{12}{1}=\frac{162}{x}$$
$$12x=162$$
$$x=13.5 \text{ feet}$$

5. Let $(-7, 8)$ be (x_1, y_1) and $(3, 8)$ be (x_2, y_2).
$$m=\frac{y_2-y_1}{x_2-x_1}$$
$$m=\frac{8-8}{3-(-7)}$$
$$m=\frac{0}{10}=0$$
The slope is 0.

6. To put in slope-intercept form, solve for y.
$$3x-4y=12$$
$$3x-3x-4y=-3x+12$$
$$-4y=-3x+12$$
$$\frac{-4y}{-4}=\frac{-3x}{-4}+\frac{12}{-4}$$
$$y=\frac{3}{4}x-3$$

7. $\left(6x^2-3x-5\right)-\left(-2x^2-8x-9\right)=6x^2-3x-5+2x^2+8x+9$
$$=6x^2+2x^2-3x+8x-5+9$$
$$=8x^2+5x+4$$

8. $\left(3n^2-4n+3\right)(2n-5)=3n^2(2n-5)-4n(2n-5)+3(2n-5)$
$$=6n^3-15n^2-8n^2+20n+6n-15$$
$$=6n^3-23n^2+26n-15$$

9. $\dfrac{4x-34}{8}=\dfrac{4x}{8}-\dfrac{34}{8}=\dfrac{1}{2}x-\dfrac{17}{4}$

10. $6a^2-6a-5a+5=6a(a-1)-5(a-1)$
$$=(6a-5)(a-1)$$

11. $13x^2+26x-39=13\left(x^2+2x-3\right)$
$$=13(x+3)(x-1)$$

12.
$$2x^2=11x-12$$
$$2x^2-11x+12=0$$
$$(x-4)(2x-3)=0$$
$x-4=0$ or $2x-3=0$

$x=4$ $\qquad x=\frac{3}{2}$

13. $\dfrac{x^2-9}{x^2-x-6}\cdot\dfrac{x^2-2x-8}{2x^2-7x-4}=\dfrac{(x-3)(x+3)}{(x-3)(x+2)}\cdot\dfrac{(x-4)(x+2)}{(2x+1)(x-4)}$
$$=\frac{x+3}{2x+1}$$

14. $\dfrac{r}{r+2}-\dfrac{6}{r-5}=\dfrac{r}{r+2}\cdot\dfrac{r-5}{r-5}-\dfrac{6}{r-5}\cdot\dfrac{r+2}{r+2}$
$$=\frac{r(r-5)}{(r+2)(r-5)}-\frac{6(r+2)}{(r+2)(r-5)}$$
$$=\frac{r^2-5r-(6r+12)}{(r+2)(r-5)}$$
$$=\frac{r^2-5r-6r-12}{(r+2)(r-5)}$$
$$=\frac{r^2-11r-12}{(r+2)(r-5)}$$

15. $\frac{4}{x^2-3x-10}+\frac{2}{x^2+5x+6}=\frac{4}{(x-5)(x+2)}+\frac{2}{(x+2)(x+3)}$

$=\frac{4(x+3)}{(x-5)(x+2)(x+3)}+\frac{2(x-5)}{(x-5)(x+2)(x+3)}$

$=\frac{4(x+3)+2(x-5)}{(x-5)(x+2)(x+3)}$

$=\frac{4x+12+2x-10}{(x-5)(x+2)(x+3)}$

$=\frac{6x+2}{(x-5)(x+2)(x+3)}$

16. $\frac{x}{9}-\frac{x}{6}=\frac{1}{12}$

$36\left(\frac{x}{9}-\frac{x}{6}\right)=36\left(\frac{1}{12}\right)$

$4x-6x=3$

$-2x=3$

$x=-\frac{3}{2}$

17. $\frac{7}{x+3}+\frac{5}{x+2}=\frac{5}{x^2+5x+6}$

$\frac{7}{x+3}+\frac{5}{x+2}=\frac{5}{(x+3)(x+2)}$

$(x+3)(x+2)\left(\frac{7}{x+3}+\frac{5}{x+2}\right)=(x+3)(x+2)\left[\frac{5}{(x+3)(x+2)}\right]$

$7(x+2)+5(x+3)=5$

$7x+14+5x+15=5$

$12x+29=5$

$12x=-24$

$x=-2$

Check:

$\frac{7}{x+3}+\frac{5}{x+2}=\frac{5}{x^2+5x+6}$

$\frac{7}{-2+3}+\frac{5}{-2+2}=\frac{5}{(-2)^2+5(-2)+6}$

$\frac{7}{1}+\frac{5}{0}=\frac{5}{0}$

Since $\frac{5}{0}$ is not a real number, there is no solution.

18. Let x = the total medical bills. The cost under plan 1 is $0.10x$, while the cost under plan 2 is $100+0.05x$.

$0.10x=100+0.05x$

$0.05x=100$

$x=2000$

The cost under both plans is the same for $2000 in total medical bills.

19. Let x = pounds of sunflower seed and
y = pounds of premixed assorted seed mix

$$x + y = 50$$
$$0.50x + 0.15y = 14.50$$

Solve the first equation for y.

$$y = 50 - x$$

Substitute $50 - x$ for y in the second equation.

$$0.50x + 0.15(50 - x) = 14.50$$
$$0.50x + 7.5 - 0.15x = 14.50$$
$$0.35x = 7$$
$$x = 20$$
$$y = 50 - x = 50 - 20 = 30$$

He will have to use 20 pounds of sunflower seed and 30 pounds of premixed assorted seed mix.

20. Let d = distance on first leg, then the distance on the second leg is $12.75 - d$.

$$t = \frac{d}{r}$$

Time for first leg + time for second leg = total time

$$\frac{d}{6.5} + \frac{12.75 - d}{9.5} = 1.5$$
$$(9.5)(6.5)\left[\frac{d}{6.5} + \frac{12.75 - d}{9.5}\right] = (9.5)(6.5)(1.5)$$
$$9.5d + 6.5(12.75 - d) = 92.625$$
$$9.5d + 82.875 - 6.5d = 92.625$$
$$3d = 9.75$$
$$d = 3.25$$
$$12.75 - d = 12.75 - 3.25 = 9.5$$

The distance traveled during the first leg of the race was 3.25 miles and the distance traveled in the second leg of the race was 9.5 miles.

Chapter 8

Exercise Set 8.1

1. No, a nonlinear equation does no have to have a *y*-intercept. An example is $y = 1/x$.

3. When graphing $y = \frac{1}{x}$, we cannot substitute 0 for x.
$\frac{1}{0}$ is undefined.

5.

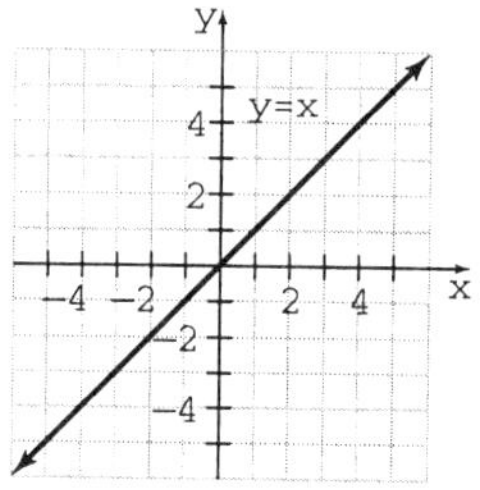

7.

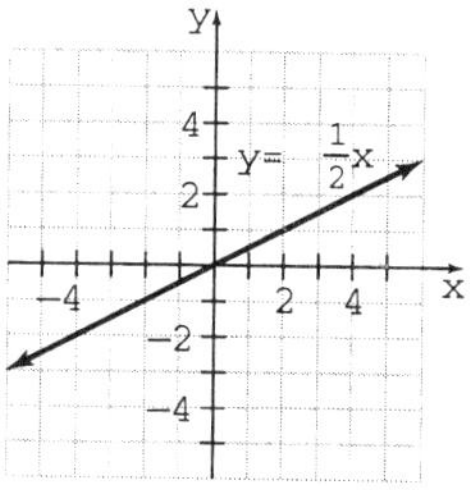

9.

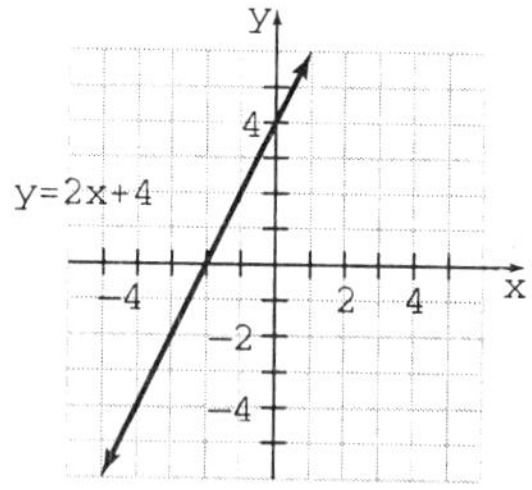

11.

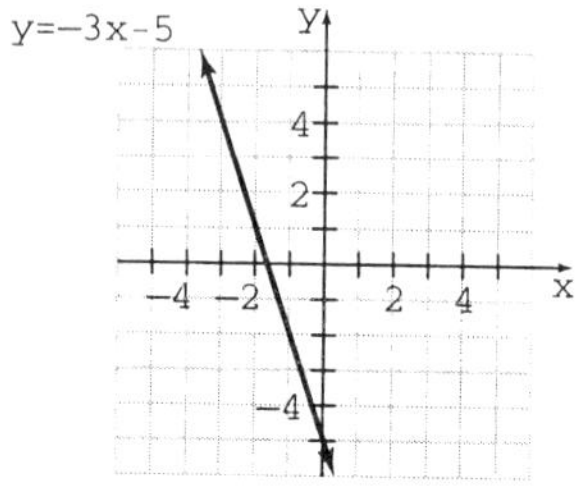

13.

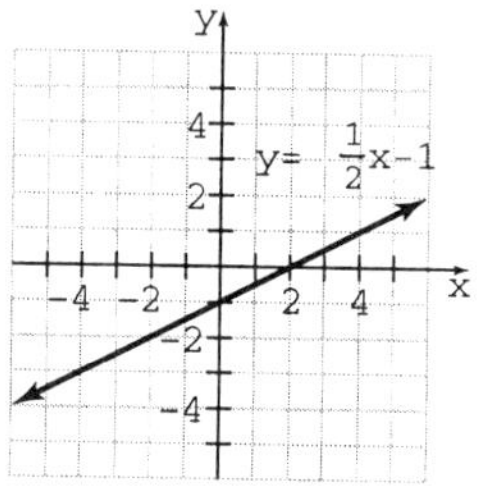

15.

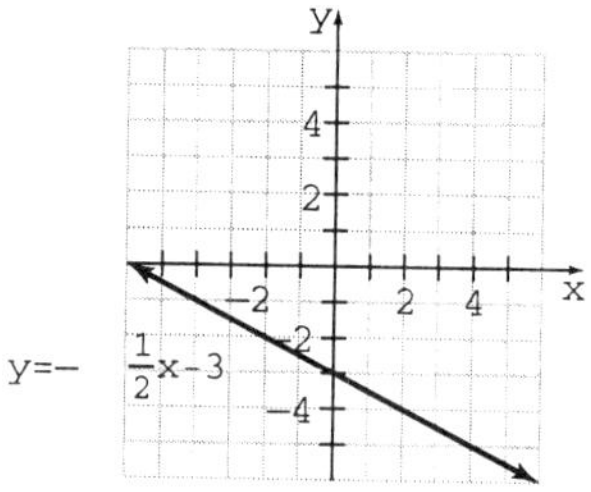

17.

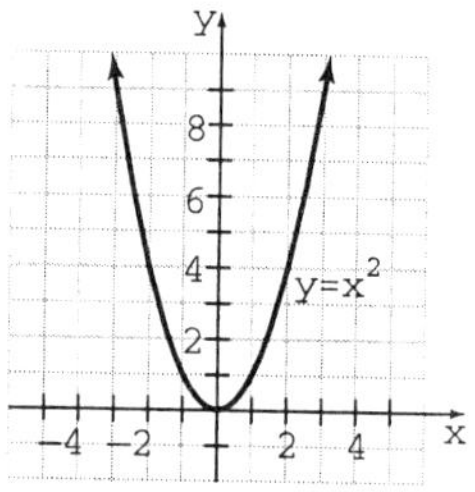

19.

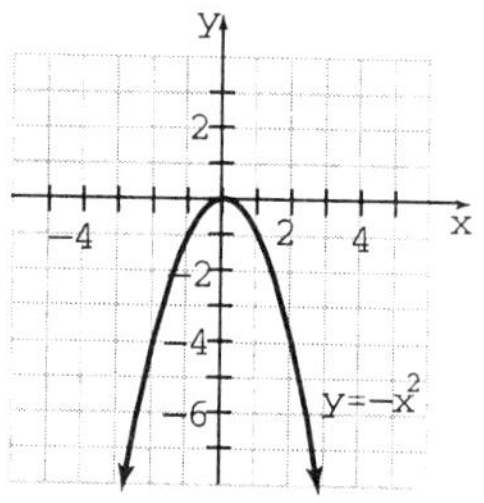

21.

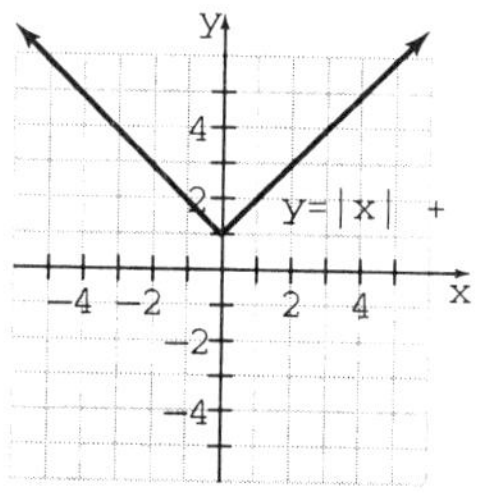

23.

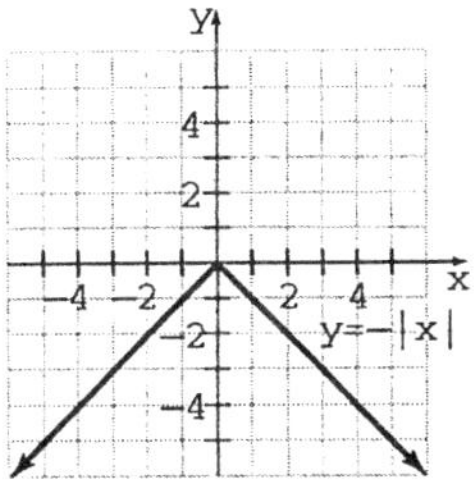

25.

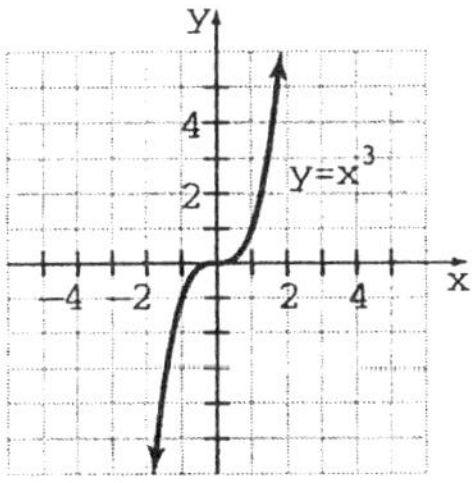

27.

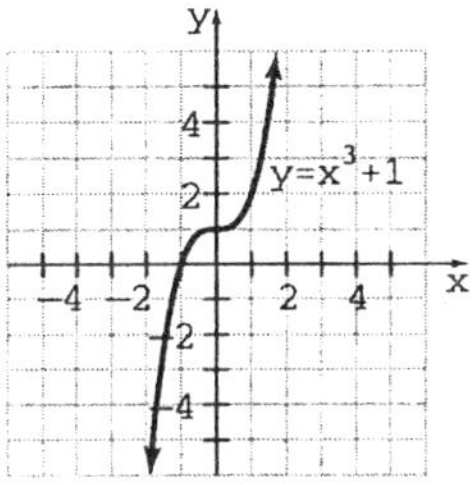

29.

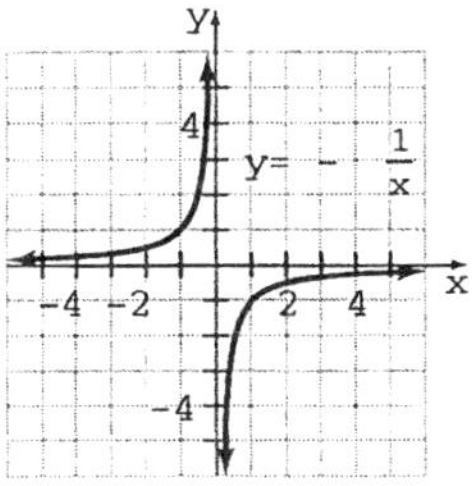

31.

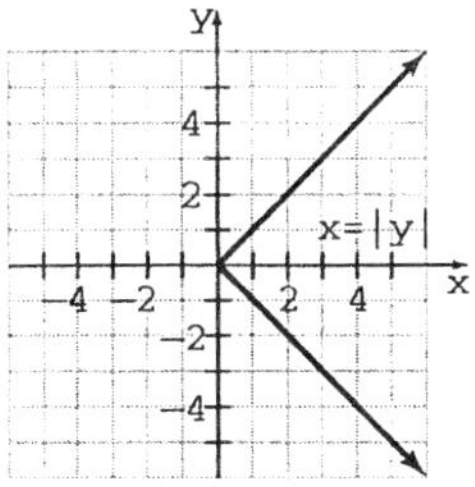

33.

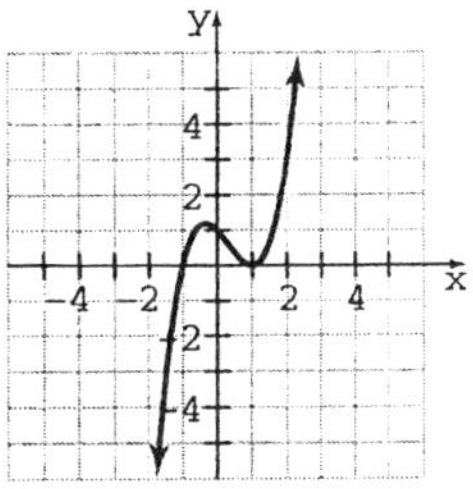

35.

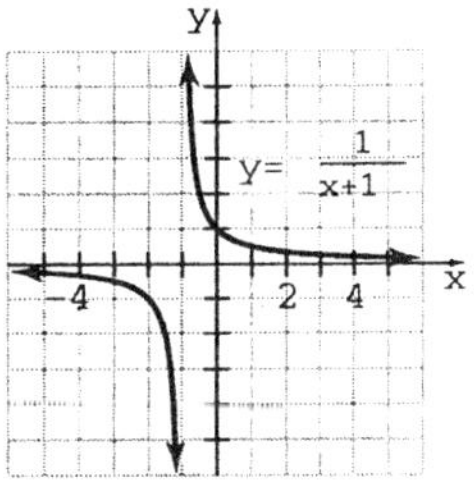

37.

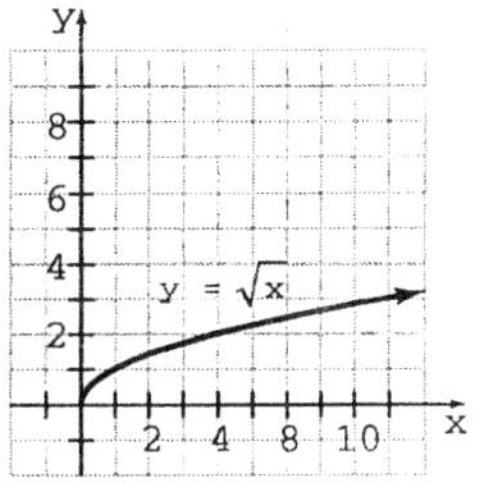

39.

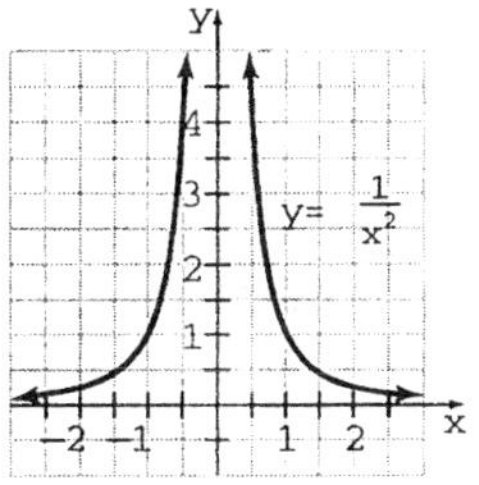

41.

$$y = \frac{x^2}{x+1}$$

$$\left(\frac{1}{12}\right) \stackrel{?}{=} \frac{\left(\frac{1}{3}\right)^2}{\left(\frac{1}{3}\right)+1}$$

$$\frac{1}{12} \stackrel{?}{=} \frac{\frac{1}{9}}{\frac{4}{3}}$$

$$\frac{1}{12} \stackrel{?}{=} \frac{1}{9} \cdot \frac{3}{4}$$

$$\frac{1}{12} \stackrel{?}{=} \frac{1}{12} \text{ true}$$

Yes, $\left(\frac{1}{3}, \frac{1}{12}\right)$ is on the graph of the equation $y = \frac{x^2}{x+1}$.

43. a.

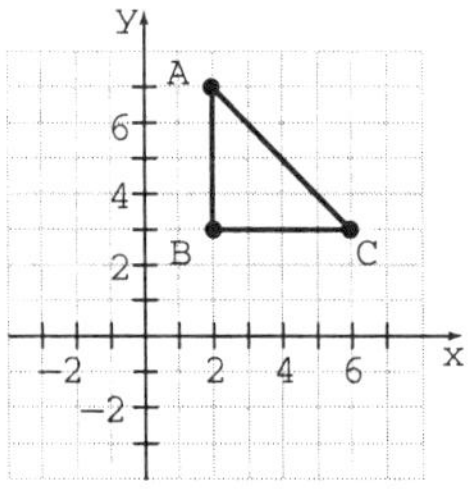

b.

$$\text{area} = \frac{1}{2}bh = \frac{1}{2}(4)(4) = 8$$

The area is 8 square units.

45. a. The estimated shipments for 1999 is 115 million.

b. The estimated shipments for 2003 is 190 million.

c. Estimated shipments of personal computers are greater than 140 million units for the years 2001, 2002, and 2003.

d. Yes, the increase in worldwide shipments of personal computers from 1999 to 2003 appears to be approximately linear. Worldwide shipments of personal computers appear to be increasing at approximately 19 million units per year from 1999 to 2003.

46. a. The area devoted to genetically modified crops in developing nations in 1999 is about 7 million hectares.

b. The area devoted to genetically modified crops in industrial nations in 1999 is about 33 million hectares.

c. The area worldwide devoted to genetically modified crops was less than 20 million hectares for the years 1995, 1996, and 1997.

d. The area worldwide devoted to genetically modified crops was greater than 35 million hectares for the years 1999, and 2000.

47.

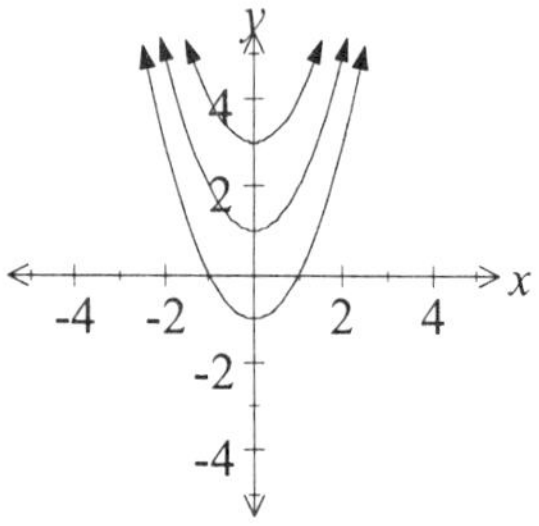

a. Each graph crosses the y-axis at the point corresponding to the constant term in the graph's equation.

b. Yes, all the equations seem to have the same shape.

49.

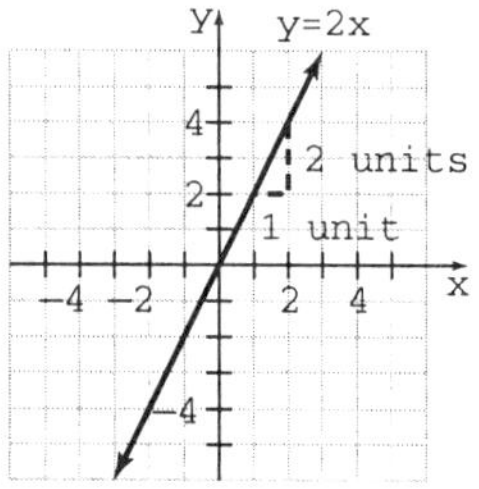

For each unit change in x, y changes 2 units. Therefore, the rate of change of y with respect to x is 2.

51.

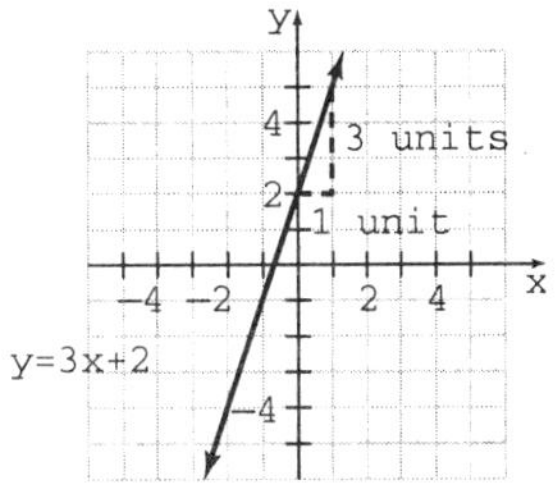

For each unit change in *x*, *y* changes 3 units. Therefore, the rate of change of *y* with respect to *x* is 3.

53. Starting at (3, –6):
For a unit change, *x* changes from 3 to $3 + 1 = 4$.
At the same time, *y* changes from –6 to $-6 + 4 = -2$
So, (4, –2) is a solution to the equation starting at (4, –2):
For a unit change, *x* changes from 4 to $4 + 1 = 5$.
At the same time, *y* changes from –2 to $-2 + 4 = 2$.
So, (5, 2) is a solution to the equation.
Answers may vary. One possible answer is the points (4, –2) and (5, 2).

55. c

57. a

59. d

61. b

63. b

65. d

67. b

69. d

71.

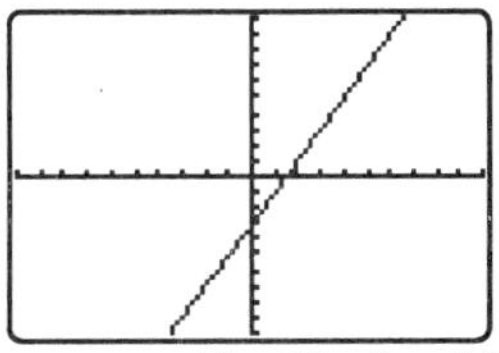

-10, 10, 1, -10

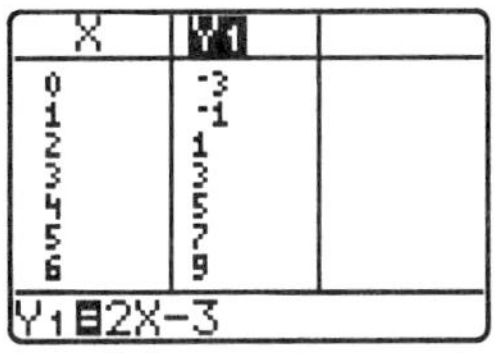

73.

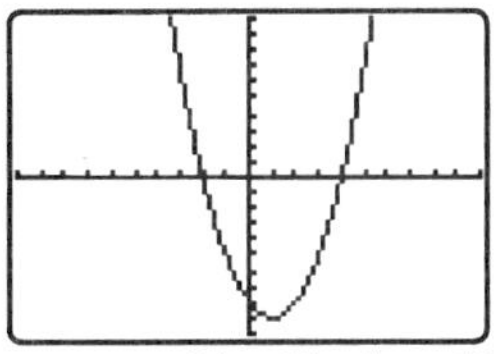

-10, 10, 1, -10

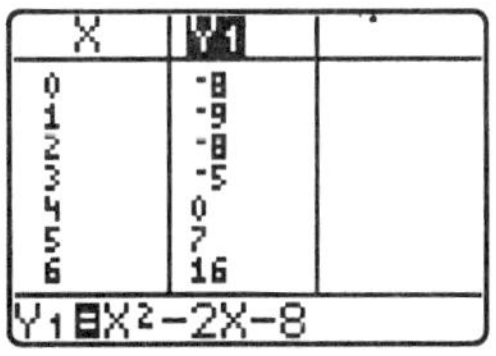

75.

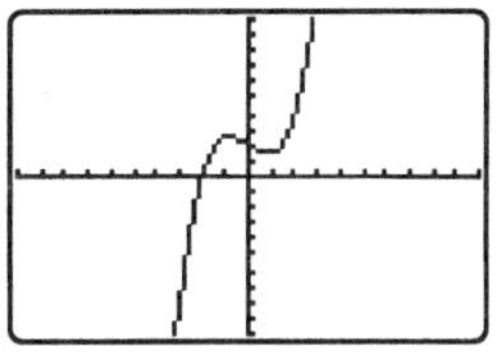

-10, 10, 1, -20

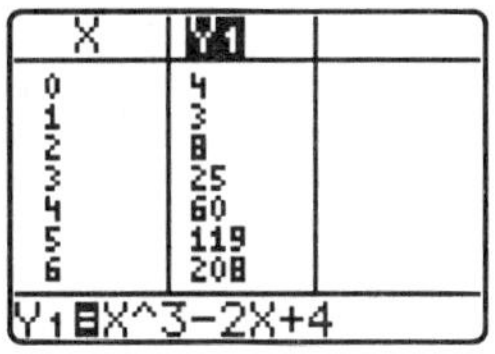

77.

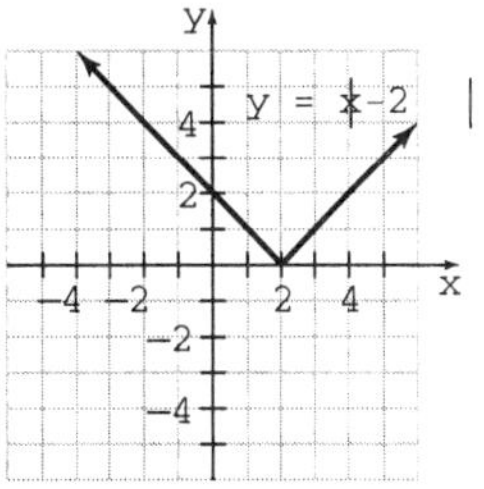

81. $x^2 - 4x + 5$

$(-3)^2 - 4(-3) + 5$

$9 + 12 + 5$

$21 + 5$

26

82. $m = \dfrac{y_2 - y_1}{x_2 - x_1}$
$= \dfrac{7-3}{6-(-5)}$
$= \dfrac{4}{11}$

83.
$$\frac{2x}{x^2-4}+\frac{1}{x-2}=\frac{2}{x+2}$$
$$(x+2)(x-2)\left[\frac{2x}{x^2-4}+\frac{1}{x-2}\right]=(x+2)(x-2)\left[\frac{2}{x+2}\right]$$
$$2x+1(x+2)=(x-2)2$$
$$2x+x+2=2x-4$$
$$x+2=-4$$
$$x=-6$$

84.
$$\frac{4x}{x^2+6x+9}-\frac{2x}{x+3}=\frac{x+1}{x+3}$$
$$(x+3)(x+3)\left[\frac{4x}{x^2+6x+9}-\frac{2x}{x+3}\right]=(x+3)(x+3)\left[\frac{x+1}{x+3}\right]$$
$$4x-2x(x+3)=(x+3)(x+1)$$
$$4x-2x^2-6x=x^2+4x+3$$
$$-2x-2x^2=x^2+4x+3$$
$$0=3x^2+6x+3$$
$$0=3(x^2+2x+1)$$
$$0=3(x+1)^2$$
$$(x+1)^2=0$$
$$\sqrt{(x+1)^2}=\sqrt{0}$$
$$x+1=0$$
$$x=-1$$

Exercise Set 8.2

1. A function is a correspondence between a first set of elements, the domain, and a second set of elements, the range, such that each element of the domain corresponds to exactly one element in the range.

3. No, all relations are not functions. A relation can have two ordered pairs with the same first element but a function cannot.

5. {5, 6, 7, 8}

7. If each vertical line drawn through any part of the graph intersects the graph in at most one point, the graph represents a function.

9. The range is the set of values for the dependent variable.

11. Domain: R or $(-\infty, \infty)$
There are no restrictions on values of x that can be used.
Range: R or $(-\infty, \infty)$
All values of y are represented in the function.

13. If y depends on x, then y is the dependent variable.

15. $f(x)$ is read "f of x."

17. **a.** Yes, the relation is a function.
b. Domain: {3, 5, 10}, Range: {6, 10, 20}

19. **a.** Yes, the relation is a function.
b. Domain: {Cameron, Tyrone, Vishnu}, Range:{1, 2}

21. **a.** No, the relation is not a function.
b. Domain: {1990, 2001, 2002}, Range: {20, 34, 37}

23. **a.** A function
b. Domain: {1, 2, 3, 4, 5}, Range: {1, 2, 3, 4, 5}

25. **a.** A function
b. Domain: {1, 2, 3, 4, 5, 7}, Range: {–1, 0, 2, 4, 5}

27. **a.** Not a function
b. Domain: {1, 2, 3}, Range: {1, 2, 4, 5, 6}

29. **a.** Not a function
b. Domain: {0, 1, 2}, Range: {–7, –1, 2, 3}

31. $A = \{0\}$

33. $C = \{18, 20\}$

35. $E = \{0, 1, 2\}$

37. $H = \{0, 7, 14, 21, 28, \ldots\}$

39. $J = \{1, 2, 3, 4, \ldots\}$, or $J = N$

41. a. Set A is the set of all x such that x is a natural number less than 8.

b. $A = \{1, 2, 3, 4, 5, , 6, 7\}$

43. $\{x|x \geq 0\}$

45. $\{z|z \leq 3\}$

47. $\{p|-4 \leq p < 3\}$

49. $\{q|q > -2 \text{ and } q \in N\}$

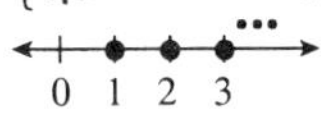

51. $\{r|r \leq \pi \text{ and } r \in W\}$

53. $\{x|x \geq 2\}$

55. $\{x|x < 5 \text{ and } x \in I\}$ or $\{x|x \leq 4 \text{ and } x \in I\}$

57. $\{x|-3 < x \leq 5\}$

59. $\{x|-2.5 \leq x < 4.2\}$

61. $\{x|-3 \leq x \leq 1 \text{ and } x \in I\}$

63. a. A function

b. Domain: R, Range: R

c. $x = 2$

65. a. Not a function

b. Domain: $\{x \mid 0 \leq x \leq 2\}$,
Range: $\{y \mid -3 \leq y \leq 3\}$

c. ≈ 1.5

67. a. A function

b. Domain: R, Range: $\{y|y \geq 0\}$

c. $x = 1$ or $x = 3$

69. a. A function

b. Domain: $\{-1, 0, 1, 2, 3\}$,
Range: $\{-1, 0, 1, 2, 3\}$

c. $x = 2$

71. a. Not a function

b. Domain: $\{x|x \geq 2\}$,
Range: R

c. $x = 3$

73. a. A function

b. Domain: $\{x|-2 \leq x \leq 2\}$,
Range: $\{y|-1 \leq y \leq 2\}$

c. $x = -2$ or $x = 2$

75. a. $f(2) = -2(2) + 5 = -4 + 5 = 1$

b. $f(-3) = -2(-3) + 5 = 6 + 5 = 11$

77. a. $h(0) = (0)^2 - (0) - 6 = -6$

b. $h(-1) = (-1)^2 - (-1) - 6 = 1 + 1 - 6 = -4$

79. a.
$$\begin{aligned} r(1) &= -(1)^3 - 2(1)^2 + (1) + 4 \\ &= -1 - 2 + 1 + 4 = 2 \end{aligned}$$

b.
$$\begin{aligned} r(1) &= -(-2)^3 - 2(-2)^2 + (-2) + 4 \\ &= -(-8) - 2(4) + (-2) + 4 \\ &= 8 - 8 - 2 + 4 \\ &= 2 \end{aligned}$$

81. a.
$$\begin{aligned} h(6) &= |5 - 2(6)| \\ &= |5 - 12| \\ &= |-7| \\ &= 7 \end{aligned}$$

b.
$$\begin{aligned} h\left(\frac{5}{2}\right) &= \left|5 - 2\left(\frac{5}{2}\right)\right| \\ &= |5 - 5| \\ &= 0 \end{aligned}$$

83. a. $s(-2) = \sqrt{(-2)+2}$
$= \sqrt{0}$
$= 0$

b. $s(7) = \sqrt{(7)+2}$
$= \sqrt{9}$
$= 3$

85. a. $g(0) = \dfrac{(0)^3 - 2}{(0) - 2}$
$= \dfrac{-2}{-2}$
$= 1$

b. $g(2) = \dfrac{(2)^3 - 2}{(2) - 2}$
$= \dfrac{8-2}{0}$ undefined

87. a. $A(2) = 6(2) = 12$
The area is 12 square feet.

b. $A(4.5) = 6(4.5) = 27$
The area is 27 square feet.

89. a. $A(r) = \pi r^2$

b. $A(10) = \pi(10)^2 = 100\pi \approx 314.2$
The area is about 314.2 square yards.

91. a. $C(F) = \frac{5}{9}(F - 32)$

b. $C(-40) = \frac{5}{9}(-40 - 32) = \frac{5}{9}(-72) = -40$
The Celsius temperature that corresponds to –40°F is –40°C.

93. a. $T(3) = -0.03(3)^2 + 1.5(3) + 14$
$= -0.27 + 4.5 + 14$
$= 18.23$
The temperature is 18.23°C.

b. $T(12) = -0.03(12)^2 + 1.5(12) + 14$
$= -4.32 + 18 + 14$
$= 27.68$
The temperature is 27.68°C.

95. a. $T(4) = -0.02(4)^2 - 0.34(4) + 80$
$= -0.32 - 1.36 + 80$
$= 78.32$
The temperature is 78.32°.

b. $T(12) = -0.02(12)^2 - 0.34(12) + 80$
$= -2.88 - 4.08 + 80$
$= 73.04$
The temperature is 73.04°.

97. a. $T(6) = \frac{1}{3}(6)^3 + \frac{1}{2}(6)^2 + \frac{1}{6}(6)$
$= 72 + 18 + 1$
$= 91$
91 oranges

b. $T(8) = \frac{1}{3}(8)^3 + \frac{1}{2}(8)^2 + \frac{1}{6}(8)$
$= \frac{512}{3} + 32 + \frac{4}{3}$
204 oranges

99. Answers will vary. One possible interpretation: The person warms up slowly, possibly by walking for 5 minutes, then begins jogging slowly over a period of 5 minutes. For the next 15 minutes, the person jogs at a steady pace. For the next 5 minutes, he walks slowly and his heart rate decreases to his normal resting heart rate. The rate stays the same for the next 5 minutes.

101. Answers will vary. One possible interpretation: The man walks on level ground, about 30 feet above sea level, for 5 minutes. For the next 5 minutes he walks uphill to 45 feet above sea level. For 5 minutes he walks on level ground then walks quickly downhill for 3 minutes to an elevation of 20 feet above sea level. For 7 minutes he walks on level ground. Then he walks quickly uphill.

103. Answers may vary. One possible interpretation: A woman drives in stop-and-go traffic for 5 minutes. Then she drives on the highway for 15 minutes, gets off onto a country road for a few minutes, stops for a couple of minutes, and returns to stop-and-go traffic.

105. a. Yes, it passes the vertical line test.

b. The independent variable is the year.

c. $f(2000)$ is about \$115 billion.

d. percent increase $= \dfrac{115-80}{80} \times 100$

$= .4375 \times 100$

$= 43.75$

The percent increase from 1997 through 2000 was about 43.75%.

107. **a.** Yes, it passes the vertical line test.

b. Yes, it passes the vertical line test.

c. No, the graph does not appear to be a straight line.

d. Yes, the graph does seem to be a straight line.

e. $f(t) = \$80$ billion for $t = 1999$.

f. $g(t) = \$18$ billion for $t = 2000$.

109. **a.**

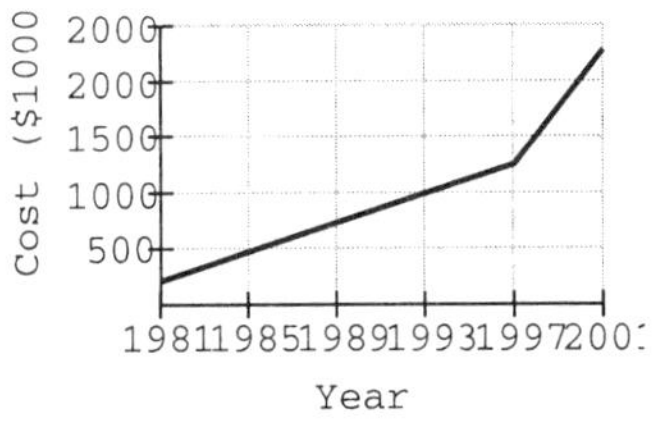

b. No, the points don't lie on a straight line.

c. The cost of a 30-second commercial in 2000 was about $2,000,000.

111. **a.**

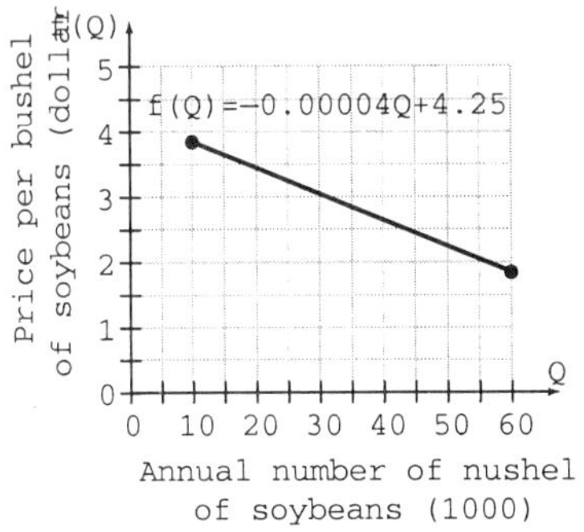

b. $f(40{,}000) = -0.00004(40{,}000) + 4.25$

$= -1.6 + 4.25$

$= 2.65$

The cost of a bushel of soybeans if 40,000 bushels are produced is approximately $2.65 per bushel.

114.

$$
\begin{aligned}
3x - 2 &= \frac{1}{3}(3x - 3) \\
3x - 2 &= \frac{1}{3}(3x) - \frac{1}{3}(3) \\
3x - 2 &= x - 1 \\
3x - x &= -1 + 2 \\
2x &= 1 \\
x &= \frac{1}{2}
\end{aligned}
$$

115.

$$
\begin{aligned}
E &= a_1 p_1 + a_2 p_2 + a_3 p_3 \\
E - a_1 p_1 - a_3 p_3 &= a_2 p_2 \\
p_2 &= \frac{E - a_1 p_1 - a_3 p_3}{a_2}
\end{aligned}
$$

116.

$$
\begin{aligned}
3x + 6y &= 9 \\
6y &= -3x + 9 \\
\frac{6y}{6} &= \frac{-3x}{6} + \frac{9}{6} \\
y &= -\frac{1}{2}x + \frac{3}{2}
\end{aligned}
$$

$m = -\frac{1}{2}$ $\quad y$ – intercept $(0, \frac{3}{2})$

117.

$$
\begin{aligned}
&\frac{3x^2 - 16x - 12}{3x^2 - 10x - 8} \div \frac{x^2 - 7x + 6}{3x^2 - 11x - 4} \\
&= \frac{3x^2 - 16x - 12}{3x^2 - 10x - 8} \cdot \frac{3x^2 - 11x - 4}{x^2 - 7x + 6} \\
&= \frac{(3x+2)(x-6)}{(3x+2)(x-4)} \cdot \frac{(3x+1)(x-4)}{(x-6)(x-1)} \\
&= \frac{3x+1}{x-1}
\end{aligned}
$$

Exercise Set 8.3

1. The graph of a linear function is a straight line.

3. To find the x-intercept, set $y = 0$ and solve for x. To find the y-intercept, set $x = 0$ and solve for y.

5. To solve an equation in one variable, graph both sides of the equation. The solution is the x-coordinate of the intersection.

7. $f(x) = 3x - 2$

x	$f(x)$
−1	−5
0	−2
2	4

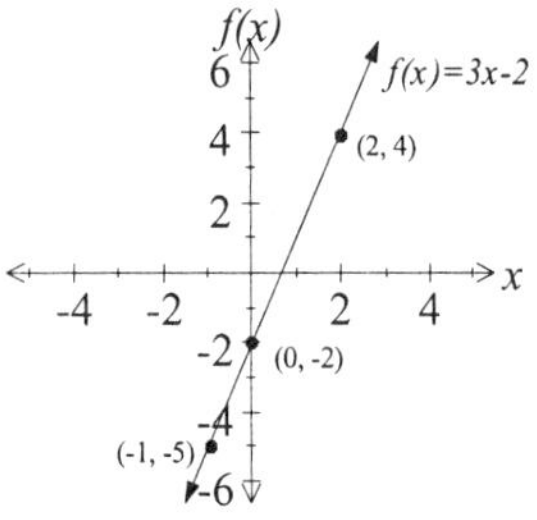

9. $g(x) = -\frac{2}{3}x + 4$

x	$g(x)$
−3	6
0	4
3	2

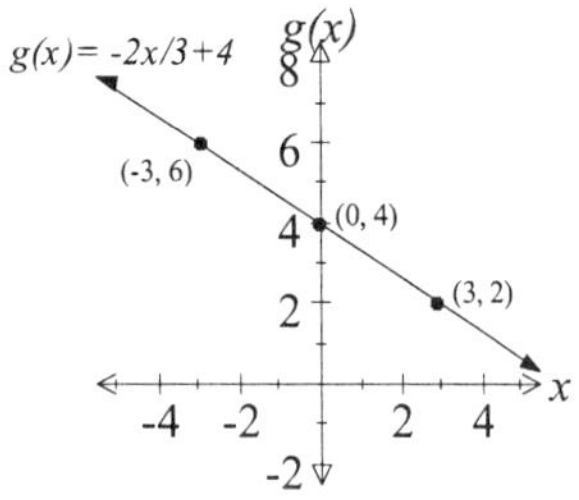

11. $f(x) = \frac{3}{4}x + 1$

x	$f(x)$
−4	−2
0	1
4	4

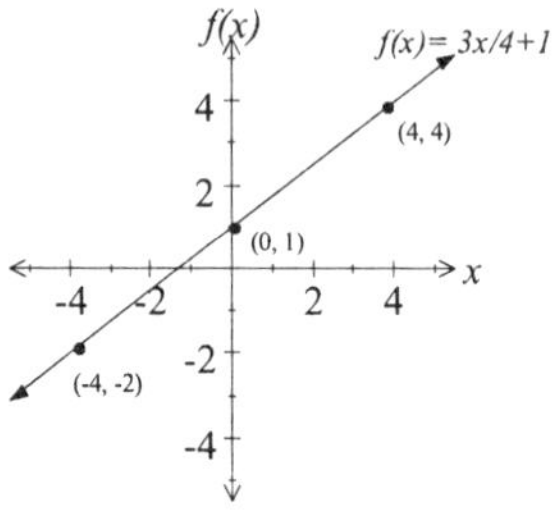

13. $f(x) = 3x - 6$
For the y-intercept, set $x = 0$ and solve for $f(x)$.

$$\begin{aligned} f(x) &= 3x - 6 \\ &= 3(0) - 6 \\ &= 0 - 6 \\ &= -6 \end{aligned}$$

The y-intercept is at (0, −6).
For the x-intercept, set $f(x) = 0$ and solve for x:

$$\begin{aligned} f(x) &= 3x - 6 \\ 0 &= 3x - 6 \\ 6 &= 3x \\ 2 &= x \end{aligned}$$

The x-intercept is at (2, 0).

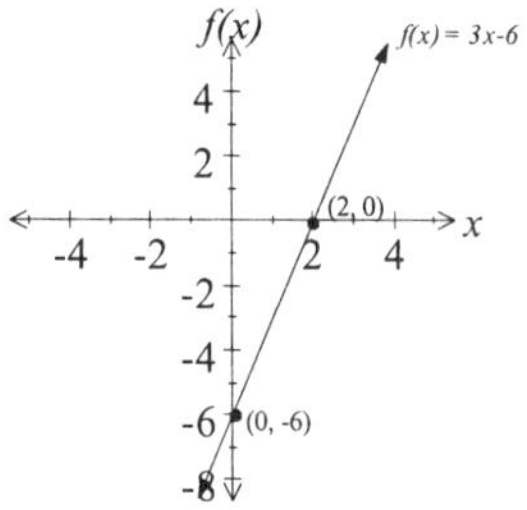

15. $f(x) = 2x + 3$
For the y-intercept, set $x = 0$ and solve for $f(x)$.

$$\begin{aligned} f(x) &= 2x + 3 \\ &= 2(0) + 3 \\ &= 0 + 3 \\ &= 3 \end{aligned}$$

The y-intercept is at (0, 3).
For the x-intercept, set $f(x) = 0$ and solve for x:

$$\begin{aligned} f(x) &= 2x + 3 \\ 0 &= 2x + 3 \\ -3 &= 2x \\ -\frac{3}{2} &= x \end{aligned}$$

The x-intercept is at $\left(-\frac{3}{2},\ 0\right)$.

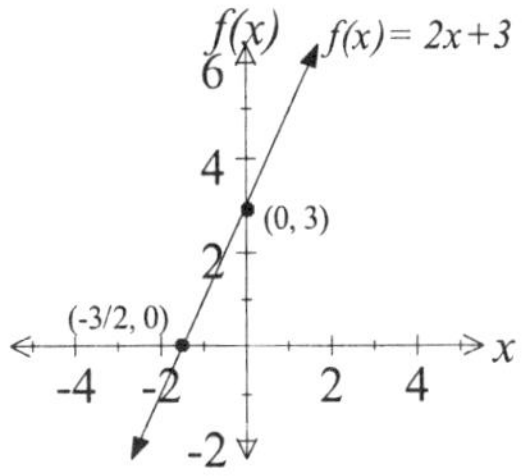

17. $g(x) = 4x - 8$

For the y-intercept, set $x = 0$ and solve for $g(x)$.

$$\begin{aligned} g(x) &= 4x - 8 \\ &= 4(0) - 8 \\ &= 0 - 8 \\ &= -8 \end{aligned}$$

The y-intercept is at $(0, -8)$.

For the x-intercept, set $g(x) = 0$ and solve for x:

$$\begin{aligned} g(x) &= 4x - 8 \\ 0 &= 4x - 8 \\ 8 &= 4x \\ 2 &= x \end{aligned}$$

The x-intercept is at $(2, 0)$.

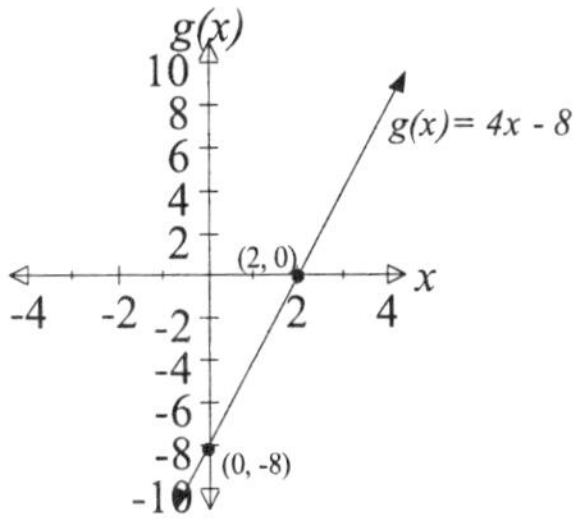

19. $s(x) = \frac{4}{3}x + 3$

For the y-intercept, set $x = 0$ and solve for $s(x)$:

$$\begin{aligned} s(x) &= \frac{4}{3}(0) + 3 \\ &= 0 + 3 \\ &= 3 \end{aligned}$$

The y-intercept is at $(0, 3)$.

For the x-intercept, set $s(x) = 0$ and solve for x:

$$\begin{aligned} 0 &= \frac{4}{3}x + 3 \\ -3 &= \frac{4}{3}x \\ 3(-3) &= 3\left(\frac{4}{3}x\right) \\ -9 &= 4x \\ -\frac{9}{4} &= x \end{aligned}$$

The x-intercept is at $\left(-\frac{9}{4},\ 0\right)$.

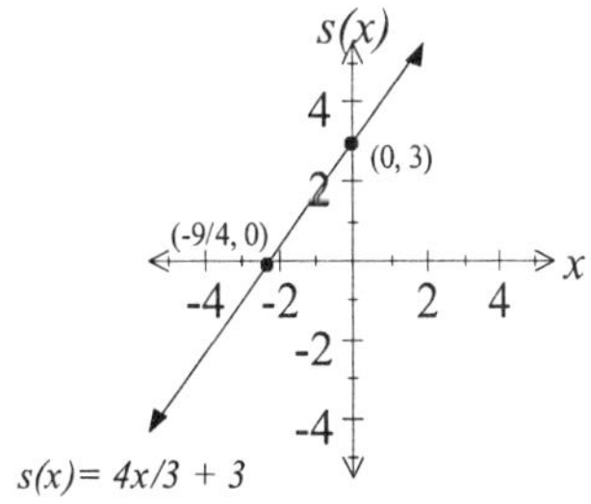

21. $h(x) = -\frac{6}{5}x + 2$

For the y-intercept, set $x = 0$ and solve for $h(x)$:

$$\begin{aligned} h(x) &= -\frac{6}{5}x + 2 \\ &= -\frac{6}{5}(0) + 2 \\ &= 0 + 2 \\ &= 2 \end{aligned}$$

The y-intercept is at $(0, 2)$.

For the x-intercept, set $h(x) = 0$ and solve for x:

$$\begin{aligned} h(x) &= -\frac{6}{5}x + 2 \\ 0 &= -\frac{6}{5}x + 2 \\ -2 &= -\frac{6}{5}x \\ 5(-2) &= 5\left(-\frac{6}{5}x\right) \\ -10 &= -6x \\ \frac{5}{3} &= x \end{aligned}$$

The x-intercept is at $\left(\frac{5}{3}, 0\right)$

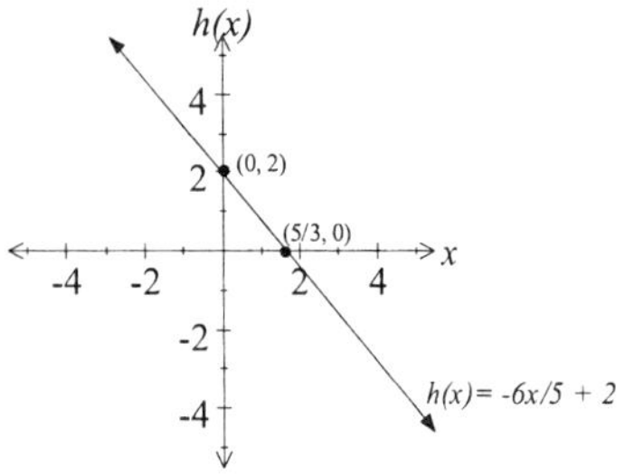

23. $g(x) = -\frac{1}{2}x + 2$

For the y-intercept, set $x = 0$ and solve for $g(x)$:

$$g(x) = -\frac{1}{2}(0) + 2$$
$$= 0 + 2$$
$$= 2$$

The y-intercept is at (0, 2).

For the x-intercept, set $g(x) = 0$ and solve for x:

$$g(x) = -\frac{1}{2}x + 2$$
$$0 = -\frac{1}{2}x + 2$$
$$-2 = -\frac{1}{2}x$$
$$-2(-2) = -2\left(-\frac{1}{2}x\right)$$
$$4 = x$$

The x-intercept is at (4, 0).

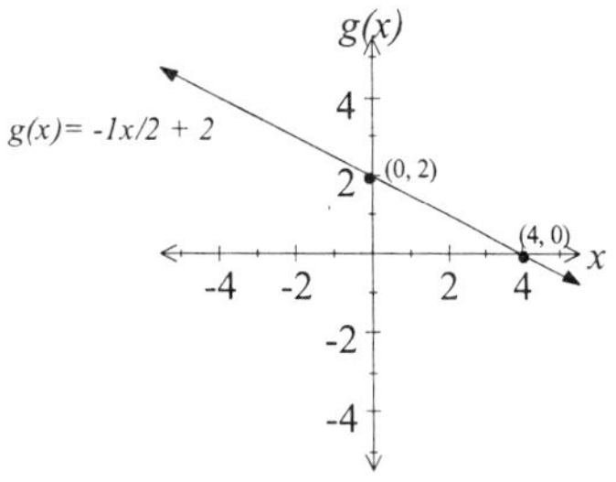

25. $w(x) = \frac{1}{3}x - 2$

For the y-intercept, set $x = 0$ and solve for $w(x)$:

$$w(x) = \frac{1}{3}x - 2$$
$$= \frac{1}{3}(0) - 2$$
$$= 0 - 2$$
$$= -2$$

The y-intercept is at (0, –2).

For the x-intercept, set $w(x) = 0$ and solve for x:

$$w(x) = \frac{1}{3}x - 2$$
$$0 = \frac{1}{3}x - 2$$
$$2 = \frac{1}{3}x$$
$$3(2) = 3\left(\frac{1}{3}x\right)$$
$$6 = x$$

The x-intercept is at (6, 0).

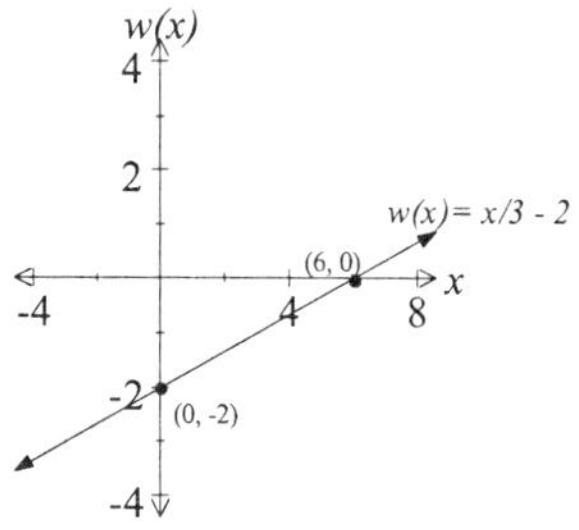

27. $s(x) = -\frac{4}{3}x + 48$

For the y-intercept, set $x = 0$ and solve for $s(x)$:

$$s(x) = -\frac{4}{3}x + 48$$
$$= -\frac{4}{3}(0) + 48$$
$$= 0 + 48$$
$$= 48$$

The y-intercept is at (0, 48).

For the x-intercept, set $s(x) = 0$ and solve for x:

$$s(x) = -\frac{4}{3}x + 48$$
$$0 = -\frac{4}{3}x + 48$$
$$-48 = -\frac{4}{3}x$$
$$-3(-48) = -3\left(-\frac{4}{3}x\right)$$
$$144 = 4x$$
$$36 = x$$

The x-intercept is at (36, 0).

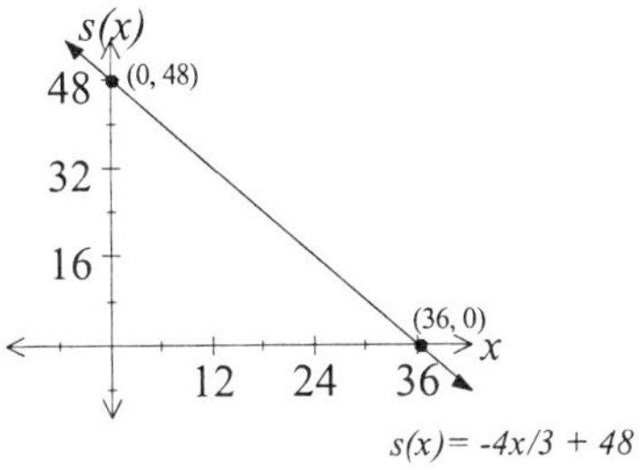

29. The equation is $d = 30t$. To graph, plot a few points.

t	Calculation	d
0	30(0)	0
1	30(1)	30
4	30(4)	120

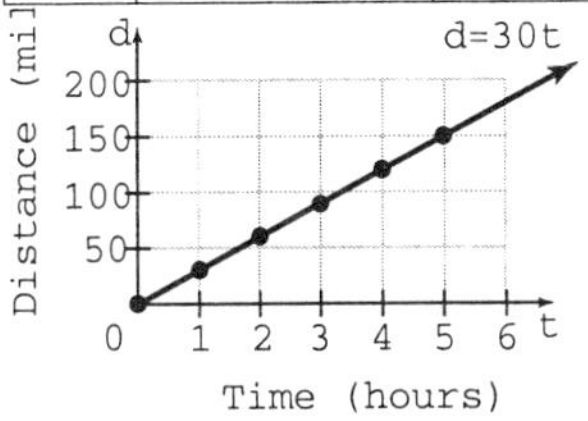

31. a. $p = 60x - 80{,}000$. To graph, plot a few points.

x	Calculation	p
0	60(0) – 80,000	–80,000
2500	60(2500) – 80,000	70,000
5000	60(5000) – 80,000	220,000

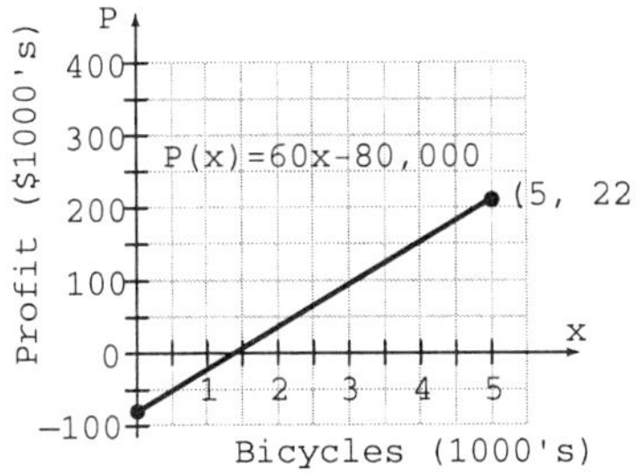

b. To break even, the profit would be zero. That is, set $p = 0$ and solve for x:

$0 = 60x - 80{,}000$

$-60x = -80{,}000$

$$x = \frac{-80{,}000}{-60} \approx 1333$$

The company must sell about 1,300 bicycles to break even.

c. To earn a profit of \$150,000, set $p = 150{,}000$ and solve for x:

$150{,}000 = 60x - 80{,}000$

$230{,}000 = 60x$

$\frac{230{,}000}{60} = x$ or $x = 3833$

The company must sell about 3,800 bicycles to make a \$150,000 profit.

33. a. $s(x) = 500 + 0.15x$

b. To graph, plot a few points.

x	Calculation	s
0	$500 + 0.15(0)$	500
1000	$500 + 0.15(1000)$	650
5000	$500 + 0.15(5000)$	1250

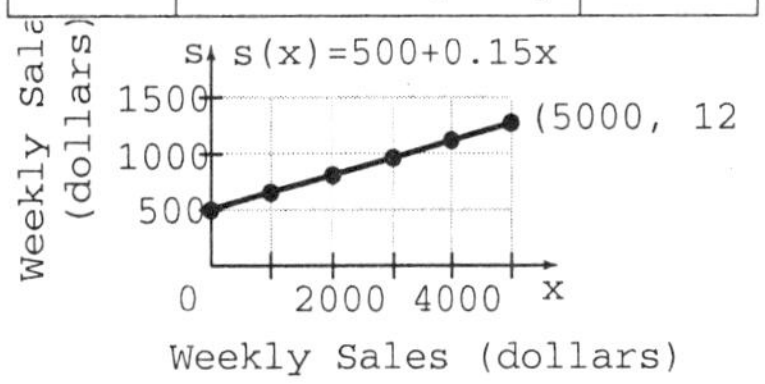

c. For weekly sales of \$2500,

$s(2500) = 500 + 0.15(2500)$

$= 500 + 375$

$= 875$

Her salary is \$875.

d. For a salary of \$1025, set $s = 1025$ and solve for x.

$1025 = 500 + 0.15x$

$525 = 0.15x$

$3500 = x$

Her weekly sales are \$3500.

35. a. There is only one y-value for each x-value.

b. The independent variable is length. The dependent variable is weight.

c. Yes, the graph of weight versus length is approximately linear.

d. The weight of the average girl who is 85 centimeters long is 11.5 kilograms.

e. The average length of a girl with a weight of 7 kilograms is 65 centimeters.

f. For a girl 95 centimeters long, the weights 12.0–15.5 kilograms are considered normal.

g. As the lengths increase, the normal range of weights increases. Yes, this is expected: as the girl grows, it is reasonable that her weight would increase with her length.

37. The x- and y-intercepts of a graph will be the same when the graph goes through the origin.

39. Answers may vary. One possible answer is, $f(x) = 4$ is a function whose graph has no x-intercept but has a y-intercept of (0, 4).

41. The x- and y-intercepts will both be 0.

43. a.

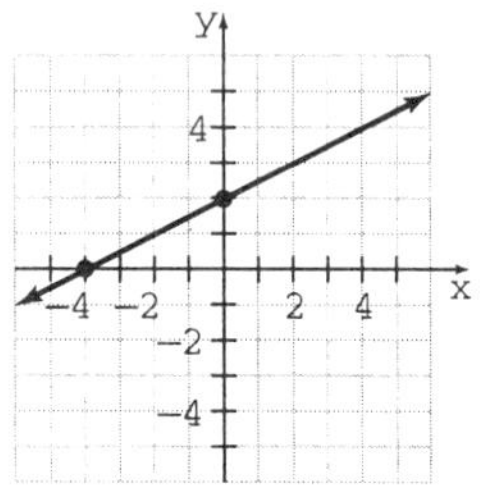

b. vertical change = 2 – 0 = 2

c. horizontal change = 0 – (–4) = 4

d. $\dfrac{\text{vertical change}}{\text{horizontal change}} = \dfrac{2}{4} = \dfrac{1}{2}$

The ratio represents the slope of the line.

45. Graph $f(x) = 3x + 2$ and $g(x) = 2x + 3$, and find the intersection.

The solution is $x = 1$.

47. Graph $f(x) = 0.3(x + 5)$ and $g(x) = -0.6(x + 2)$, and find the intersection.

The solution is $x = -3$.

49. The x-intercept is (–3.2, 0). The y-intercept is (0, 6.4).

51. To use the graphing calculator, we must rewrite the equation in the form $y = f(x)$.

$$-4x - 3.2y = 8$$
$$-3.2y = 4x + 8$$
$$y = -\frac{1}{3.2}(4x + 8)$$

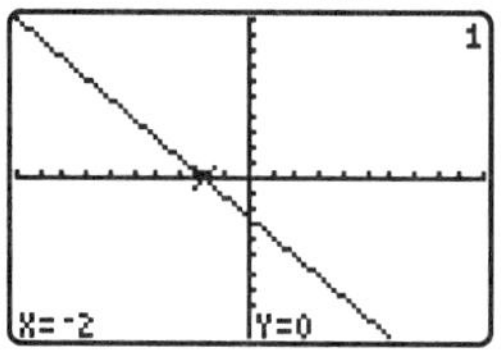

−10, 10, 1, −10

The x-intercept is (–2, 0).
The y-intercept is (0, –2.5).

53.
$$\begin{array}{r} 7x^2 - 3x - 100 \\ \underline{-4x^2 - 9x + 12} \\ 3x^2 - 12x - 88 \end{array}$$

54.
$$3x^2 - 12x - 96$$
$$3(x^2 - 4x - 32)$$
$$3(x + 4)(x - 8)$$

55. $\dfrac{x-2}{x^2-4} = \dfrac{x-2}{(x+2)(x-2)} = \dfrac{1}{x+2}$

56.
$$x\left(x + \frac{24}{x}\right) = x(10)$$
$$x^2 + 24 = 10x$$
$$x^2 - 10x + 24 = 0$$
$$(x - 6)(x - 4) = 0$$
$$x - 6 = 0 \qquad x - 4 = 0$$
$$x = 6 \qquad x = 4$$

The solutions are 4 and 6.

Exercise Set 8.4

1. When the slope is given as a rate of change it means the change in y for a unit change in x.

3. a. If a graph is translated up 3 units, it is lifted or moved up 3 units.

b. If the y-intercept is (0, – 4) and the graph is translated up 5 units, the new y-intercept will be at $y = -4 + 5 = 1$. The new y-intercept is (0, 1).

5. If their slopes are negative reciprocals or if one line is vertical and the other is horizontal, the lines are perpendicular.

7. The slope is negative and y decreases 6 units when x increases 2 units. Thus, $m = -\frac{6}{2} = -3$.
The line crosses the y-axis at 0 so $b = 0$. Hence, $m = -3$ and $b = 0$ and the equation of the line is $y = -3x + 0$ or $y = -3x$.

9. The slope is undefined since the change in x is 0. The equation of this vertical line is $x = -2$.

11. The slope is negative and y decreases 1 unit when x increases 3 units. Thus, $m = -\frac{1}{3}$. The line crosses the y-axis at 2 so $b = 2$. Hence, $m = -\frac{1}{3}$ and $b = 2$ and the equation of the line is $y = -\frac{1}{3}x + 2$.

13. The slope is negative and y decreases 15 units when x increases 10 units. Thus, $m = -\frac{15}{10} = -\frac{3}{2}$. The line crosses the y-axis at 15 so $b = 15$. Hence, $m = -\frac{3}{2}$ and $b = 15$ and the equation of the line is $y = -\frac{3}{2}x + 15$.

15.

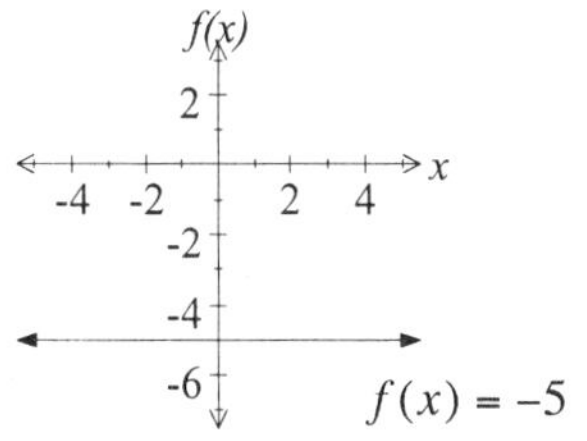

17.

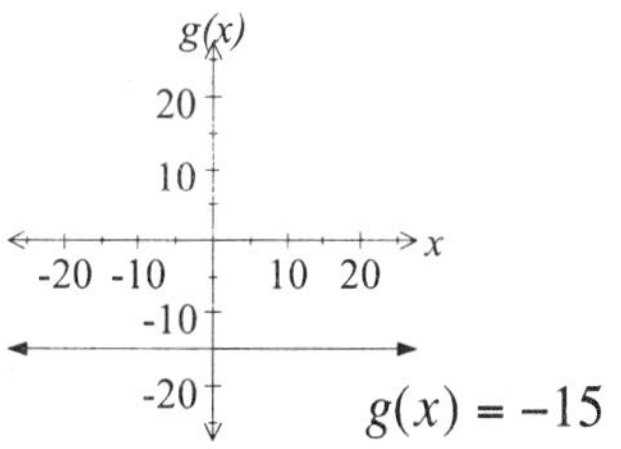

19. $m_1 = \frac{2-0}{0-2} = \frac{2}{-2} = -1$
$m_2 = \frac{5-0}{0-5} = \frac{5}{-5} = -1$
Since their slopes are equal, l_1 and l_2 are parallel.

21. $m_1 = \frac{7-1}{5-1} = \frac{6}{4} = \frac{3}{2}$
$m_2 = \frac{4-(-1)}{1-(-1)} = \frac{5}{2}$
Since their slopes are different and since the product of their slopes is not -1, l, and l_2 are neither parallel nor perpendicular.

23. $m_1 = \frac{-2-2}{-1-3} = \frac{-4}{-4} = 1$
$m_2 = \frac{-1-0}{3-2} = \frac{-1}{1} = -1$
Since the product of their slopes is -1, the lines are perpendicular.

25. $y = \frac{1}{5}x + 1$, so $m_1 = \frac{1}{5}$
$y = -5x + 2$, so $m_2 = -5$
Since the product of their slopes is -1, the lines are perpendicular.

27.

$4x + 2y = 8$	$8x = 4 - 4y$
$2y = -4x + 8$	$4y = -8x + 4$
$y = -2x + 4$	$y = -2x + 1$
$m_1 = -2$	$m_2 = -2$

Since their slopes are equal, the lines are parallel.

29.

$4x + 2y = 6$	$-x + 4y = 4$
$2y = -4x + 6$	$4y = x + 4$
$y = -2x + 3$	$y = \frac{1}{4}x + 1$
$m_1 = -2$	$m_2 = \frac{1}{4}$

Since their slopes are different and since the product of their slopes is not -1, the lines are neither parallel nor perpendicular.

31.

$y = \frac{1}{2}x - 6$	$-3y = 6x + 9$
	$y = -2x - 3$
$m_1 = \frac{1}{2}$	$m_2 = -2$

Since the product of their slopes is -1, the lines are perpendicular.

33. $y = \frac{1}{2}x + 3$

$m_1 = \frac{1}{2}$

$-2x + 4y = 8$

$4y = 2x + 8$

$y = \frac{1}{2}x + 2$

$m_2 = \frac{1}{2}$

Since the slopes are equal, the lines are parallel.

35. $x - 3y = -9$

$-3y = -x - 9$

$y = \frac{1}{3}x + 3$

$m_1 = \frac{1}{3}$

$y = 3x + 6$

$m_2 = 3$

Since their slopes are different and since the product of their slopes is not -1, the lines are neither parallel nor perpendicular.

37. The slope of the given line, $y = 2x + 4$, is $m_1 = 2$. So $m_2 = 2$. Now use the point-slope form with $m = 2$ and $(x_1, y_1) = (2, 5)$ to obtain the slope-intercept form.

$y - y_1 = m(x - x_1)$

$y - 5 = 2(x - 2)$

$y - 5 = 2x - 4$

$y = 2x + 1$

39. Find the slope of the given line.

$2x - 5y = 7$

$-5y = -2x + 7$

$y = \frac{2}{5}x - \frac{7}{5}$

$m_1 = \frac{2}{5}$, so $m_2 = \frac{2}{5}$ Now use the point-slope form with $m = \frac{2}{5}$ and $(x_1, y_1) = (-3, -5)$ to obtain the standard form.

$y - y_1 = m(x - x_1)$

$y - (-5) = \frac{2}{5}(x - (-3))$

$y + 5 = \frac{2}{5}(x + 3)$

$y + 5 = \frac{2}{5}x + \frac{6}{5}$

$5(y + 5) = 5\left(\frac{2}{5}x + \frac{6}{5}\right)$

$5y + 25 = 2x + 6$

$-2x + 5y = -19$

$2x - 5y = 19$

41. Find the slope of the line with the given intercepts.

$m = \frac{5 - 0}{0 - 3} = -\frac{5}{3}$

With $m = -\frac{5}{3}$ and y-intercept (0, 5) the slope-intercept form of the equation is $y = -\frac{5}{3}x + 5$.

43. The slope of the given line $f(x) = \frac{1}{3}x + 1$ is $m_1 = \frac{1}{3}$. So $m_2 = -3$. Now use the point-slope form with $m = -3$ and $(x_1, y_1) = (5, -1)$ to obtain the function notation.

$y - y_1 = m(x - x_1)$

$y - (-1) = -3(x - 5)$

$y + 1 = -3x + 15$

$y = -3x + 14$

$f(x) = -3x + 14$

45. Find the slope of the line with the given intercepts.

$m_1 = \frac{-3 - 0}{0 - 2} = \frac{3}{2}$

So m_2 is the negative reciprocal, or

$m_2 = -\frac{1}{m_1} = -\frac{1}{\frac{3}{2}} = -\frac{2}{3}$.

Now use the point-slope form with $m = -\frac{2}{3}$ and $(x_1, y_1) = (6, 2)$ and obtain the slope-intercept form.

$y - y_1 = m(x - x_1)$

$y - 2 = -\frac{2}{3}(x - 6)$

$y - 2 = -\frac{2}{3}x + 4$

$y = -\frac{2}{3}x + 6$

47. $m = \frac{2-4}{-4-6} = \frac{-2}{-10} = \frac{1}{5}$

Thus, for a unit change in *x*, *y* changes $\frac{1}{5}$ or 0.2 unit.

49. a, b.

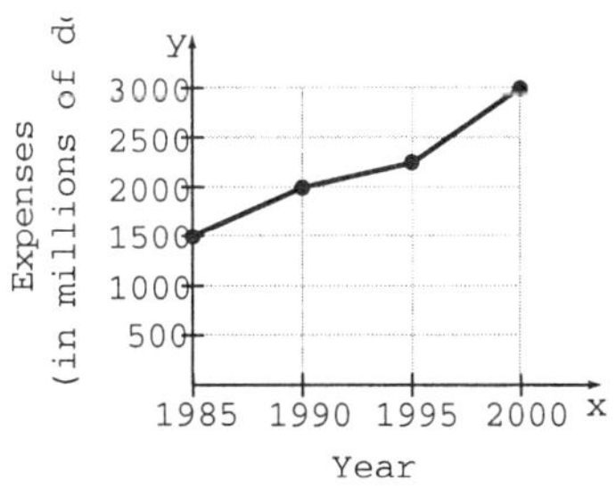

c. From 1985 to 1990,

$m = \frac{2012-1600}{1990-1985} = \frac{412}{5} \approx 82.4$

From 1990 to 1995,

$m = \frac{2257-2012}{1995-1990} = \frac{245}{5} = 49$

From 1995 - 2000,

$m = \frac{2876-2257}{2000-1995} = \frac{619}{5} = 123.8$

d. The greatest average rate of change occurred during the period 1995 to 2000, because the largest slope corresponds to these years.

51. a. If *x* is the number of years after age 20, two points on the graph are (0, 200) and (50, 150). The slope is

$m = \frac{150-200}{50-0} = \frac{-50}{50} = -1$ and the

y-intercept is (0, 200), so $b = 200$. Thus, the equation for the line is $h(x) = -1 \cdot x + 200$, or

$h(x) = -x + 200$.

b. For a 34-year-old man, $x = 34 - 20 = 14$. Therefore,

$h(x) = -x + 200$

$h(14) = -14 + 200 = 186$

186 beats per minute is the maximum recommended heart rate.

53. a. Note that $t = 0$ represents 1996 and that $t = 4$ represents 2000. Therefore, two ordered pairs of the function are (0, 36.5) and (4, 31.1). The slope of the line is

$m = \frac{31.1-36.5}{2000-1996} = \frac{-5.4}{4} = -1.35$ Since

the slope is $m = -1.35$ and the *y*-intercept is (0, 36.5), the linear function is

$N(t) = -1.35t + 36.5$.

b. Note that $t = 2$ represents 1998.

$N(2) = -1.35(2) + 36.5$

$= -2.7 + 36.5$

$= 33.8$

The number of people below the poverty threshold in 1998 was about 33.8 million.

c. Note that $t = 9$ represents 2005.

$N(9) = -1.35(9) + 36.5$

$= -12.15 + 36.5$

$= 24.35$

The number of people below the poverty threshold in 2005 will be about 24.35 million.

d. Set the function equal to 25 and solve for *t*.

$-1.35t + 36.5 = 25$

$-1.35t = -11.5$

$t = \frac{-11.5}{-1.35}$

$t \approx 8.5$

The number of people below the poverty threshold will reach 25 million sometime during the eighth year which is the year 2004.

55. a. Note that $t = 0$ represents 1975 and that $t = 25$ represents 2000. Therefore, two ordered pairs of the function are (0, 7156) and (25, 5890). The slope of the line is

$m = \frac{5890-7156}{2000-1975} = \frac{-1266}{25} = -50.64$

Since the slope is $m = -50.64$ and the *y*-intercept is (0, 7156), the linear function is

$n(t) = -50.64t + 7156$.

b. Note that $t = 20$ represents 1995.

$n(20) = -50.64(20) + 7156$

$= -1012.8 + 7156$

$= 6143.2$

There were about 6,143 hospitals in the United States in 1995.

c. Note that $t = 30$ represents 2005.

$n(30) = -50.64(30) + 7156$

$= -1519.2 + 7156$

$= 5636.8$

There will be about 5,637 hospitals in the United States in 2005.

d. Set the function equal to 5000 and solve for t.

$-50.64t + 7156 = 5000$

$-50.64t = -2156$

$t = \dfrac{-2156}{-50.64}$

$t \approx 42.6$

The number of hospitals in the United States will drop to 5000 during the forty-second year after 1975 which is the year 2017.

57. a. Note that $t = 0$ represents 1995 and that $t = 5$ represents 2000. Therefore, two ordered pairs of the function are (0, \$110,500) and (5, \$139,000). The slope of the line is

$m = \dfrac{139000 - 110500}{2000 - 1995} = \dfrac{28500}{5} = 5700$

Since the slope is $m = 5700$ and the y-intercept is (0, 110500), the linear function is $P(t) = 5700t + 110500$.

b. Note that $t = 2$ represents 1997.

$P(2) = 5,700(2) + 110,500$

$= 11,400 + 110,500$

$= 121,900$

The median home sale price in 1997 was about \$121,900.

c. Note that $t = 15$ represents 2010.

$P(15) = 5,700(15) + 110,500$

$= 85500 + 110,500$

$= 196,000$

The median home sale price in 2010 will be about \$196,000.

d. Set the function equal to 200,000 and solve for t.

$5700t + 110,500 = 200,000$

$5700t = 89,500$

$t = \dfrac{89,500}{5,700}$

$t \approx 15.7$

The median home sale price will reach \$200,000 during the fifteen year after 1995 which is the year 2010.

59. a. To find the function, use the points (2.5, 210) and (6, 370) to determine the slope.

$m = \dfrac{370 - 210}{6 - 2.5} = \dfrac{160}{3.5} \approx 45.7$

Now use the point-slope form with $m = 45.7$ and $(s_1, C_1) = (2.5, 210)$

$C - C_1 = m(s - s_1)$

$C - 210 = 45.7(s - 2.5)$

$C - 210 = 45.7s - 114.25$

$C = 45.7s + 95.75$

$C(s) = 45.7s + 95.8$

b. For a speed of 5 miles per hour:

$C(s) = 45.7s + 95.8$

$C(5) = 45.7(5) + 95.8$

$= 228.5 + 95.8$

$= 324.3$

The average person will burn about 324.3 calories.

61. a. To find the function, use the points (200, 50) and (300, 30) to determine the slope.

$m = \dfrac{30 - 50}{300 - 200} = \dfrac{-20}{100} = -0.20$

Now use the point-slope form with $m = -0.20$ and $(p_1, d_1) = (200,50)$

$d - d_1 = m(p - p_1)$

$d - 50 = -0.20(p - 200)$

$d - 50 = -0.20p + 40$

$d = -0.20p + 90$

$d(p) = -0.20p + 90$

b. For a price of \$260:

$d(p) = -0.20p + 90$

$d(260) = -0.20(260) + 90$

$= -52 + 90$

$= 38$

The demand will be 38 DVD players.

c. Set the function equal to 45 and solve for p.

$d(p) = -0.20p + 90$

$45 = -0.20p + 90$

$0.20p = 45$

$p = 225$

In order to have a demand of 45 DVD players, the price should be \$225.

63. a. To find the function, use the points (2.00, 130) and (4.00, 320) to determine the slope.

$$m = \frac{320-130}{4.00-2.00} = \frac{190}{2.00} = 95$$

Now use the point-slope form with $m = 95$ and $(p_1, s_1) = (2.00, 130)$

$$s - s_1 = m(p - p_1)$$
$$s - 130 = 95(p - 2.00)$$
$$s - 130 = 95p - 190$$
$$s = 95p - 60$$
$$s(p) = 95p - 60$$

b. For a price of \$2.80:

$$s(p) = 95p - 60$$
$$s(2.80) = 95(2.80) - 60$$
$$= 266 - 60$$
$$= 206$$

The supply will be 206 kites.

c. Set the function equal to 225 and solve for p.

$$s(p) = 95p - 60$$
$$225 = 95p - 60$$
$$285 = 95p$$
$$p = 3$$

In order to have a supply of 225 kites, the price should be \$3.00.

65. a. To find the function, use the points (45, 40) and (90, 25) to determine the slope.

$$m = \frac{25-40}{90-45} = \frac{-15}{45} = -\frac{1}{3}$$

Now use the point-slope form with $m = -\frac{1}{3}$ and $(s_1, m_1) = (45, 40)$

$$m - m_1 = m(s - s_1)$$
$$m - 40 = -\frac{1}{3}(s - 45)$$
$$m - 40 = -\frac{1}{3}s + 15$$
$$m = -\frac{1}{3}s + 55$$
$$m(s) = -\frac{1}{3}s + 55$$

b. For a speed of 60 mph:

$$m(s) = -\frac{1}{3}s + 55$$
$$m(60) = -\frac{1}{3}(60) + 55$$
$$= -20 + 55$$
$$= 35$$

The car's gas mileage will be 35 mpg.

c. Set the function equal to 30 and solve for s.

$$m(s) = -\frac{1}{3}s + 55$$
$$30 = -\frac{1}{3}s + 55$$
$$-25 = -\frac{1}{3}s$$
$$s = 75$$

In order to have a gas mileage of 30 mpg, the speed should be 75 mph.

67. a. To find the function, use the points (10, 3477) and (20, 4168) to determine the slope.

$$m = \frac{4168-3477}{20-10} = \frac{691}{10} = 69.1$$

Now use the point-slope form with $m = 69.1$ and $(s_1, p_1) = (10, 3477)$

$$p - p_1 = m(s - s_1)$$
$$p - 3477 = 69.1(s - 10)$$
$$p - 3477 = 69.1s - 691$$
$$p = 69.1s + 2786$$
$$p(s) = 69.1s + 2786$$

b. For 18 years of service:

$$p(s) = 69.1s + 2786$$
$$p(18) = 69.1(18) + 2786$$
$$= 1243.8 + 2786$$
$$= 4029.8$$

The monthly salary will be \$4,029.80.

c. Set the function equal to 4000 and solve for s.

$$p(s) = 69.1s + 2786$$
$$4000 = 69.1s + 2786$$
$$1214 = 69.1s$$
$$s \approx 17.6$$

In order to have a monthly salary of \$4,000, one would need about 18 years of service.

69. a. To find the function, use the points (50, 36.0) and (70, 18.7) to determine the slope.

$$m = \frac{18.7 - 36.0}{70 - 50} = \frac{-17.3}{20} = -0.865$$

Now use the point-slope form with $m = -0.865$ and $(a_1, y_1) = (50,36.0)$

$$y - y_1 = m(a - a_1)$$
$$y - 36.0 = -0.865(a - 50)$$
$$y - 36.0 = -0.865a + 43.25$$
$$y = -0.865a + 79.25$$
$$y(a) = -0.865a + 79.25$$

b. For a person who is currently 37 years old:

$$y(a) = -0.865a + 79.25$$
$$y(37) = -0.865(37) + 79.25$$
$$= -32.005 + 79.25$$
$$= 47.245$$

The additional life expectancy will be about 47.2 years.

c. Set the function equal to 25 and solve for s.

$$y(a) = -0.865a + 79.25$$
$$25 = -0.865a + 79.25$$
$$-54.25 = -0.865a$$
$$a \approx 62.7$$

In order to have an additional life expectancy of 25 years, one would need to be currently about 62.7 years old.

71. a. To find the function, use the points (18, 14) and (36, 17.4) to determine the slope.

$$m = \frac{17.4 - 14}{36 - 18} = \frac{3.4}{18} \approx 0.189$$

Now use the point-slope form with $m = 0.189$ and $(a_1, w_1) = (18, 14)$

$$w - w_1 = m(a - a_1)$$
$$w - 14 = 0.189(a - 18)$$
$$w - 14 = 0.189a - 3.402$$
$$w = 0.189a + 10.598$$
$$w(a) = 0.189a + 10.6$$

b. For a 22-month-old boy who is in the 95th percentile for weight:

$$w(a) = 0.189a + 10.6$$
$$w(22) = 0.189(22) + 10.6$$
$$= 4.158 + 10.6$$
$$= 14.758$$

The boy will weigh about 14.158 kg.

73. The y-intercept of the translated graph has a y-value 17 more than the y-intercept of the original graph. $-13 + 17 = 4$ Thus the y-intercept of the translated graph is (0, 4).

75. The y-intercept of the translated graph has a y-value 9 less than the y-intercept of the original graph. $6 - 9 = -3$ Thus the y-intercept of the translated graph is (0, –3).

77. a. The slope is 3 and the y -intercept is 1. Thus, the equation is $y = 3x + 1$.

b. The green line is a vertical translation down 6 units. The y-intercept of the translated graph has a y-value 6 less than the y-intercept of the original graph. $1 - 6 = -5$ Thus the y-intercept of the translated graph is (0, –5). The equation is $y = 3x - 5$.

79. a. The slope is 1, the same as the original function.

b. The y-intercept is (0, 2). The y-value is 3 units greater than the y-value on the original function.

c. The equation is $y = x + 2$.

81. Solve for y to get the slope.

$$3x - 2y = 6$$
$$-2y = -3x + 6$$
$$y = \frac{3}{2}x - 3$$
$$\text{slope is } \frac{3}{2}$$

The y-intercept is (0, –7). It is 4 units less than the y-value on the original function. The equation is $y = \frac{3}{2}x - 7$.

83. The y-intercept of $y = 3x + 6$ is 6; on the screen, the y-intercept is not 6. The y-intercept is wrong.

85. The slope of $y = \frac{1}{2}x + 4$ is $\frac{1}{2}$; on the screen, the slope is not $\frac{1}{2}$. The slope is wrong.

87. There are 91 steps and the total vertical distance is 1292.2 in. Therefore, the average height of a step is $\frac{1292.2}{91} = 14.2$ inches.
If the slope is 2.21875 and the average height, or "rise", is 14.2 inches., the average width, or "run" is found as follows:

$$\text{slope} = \frac{\text{rise}}{\text{run}}$$
$$m = \frac{\text{height}}{\text{width}}$$
$$2.21875 = \frac{14.2}{\text{width}}$$
$$\text{width} = \frac{14.2}{2.21875} = 6.4$$

The average width is 6.4 inches.

91.
$$\frac{-6^2 - 16 \div 2 \div |-4|}{5 - 3 \cdot 2 - 4 \div 2^2} = \frac{-36 - 16 \div 2 \div 4}{5 - 6 - 4 \div 4}$$
$$= \frac{-36 - 8 \div 4}{5 - 6 - 1}$$
$$= \frac{-36 - 2}{-1 - 1}$$
$$= \frac{-38}{-2}$$
$$= 19$$

92.
$$2.6x - (-1.4x + 3.4) = 6.2$$
$$2.6x + 1.4x - 3.4 = 6.2$$
$$2.6x + 1.4x = 6.2 + 3.4$$
$$4.0x = 9.6$$
$$x = \frac{9.6}{4.0}$$
$$x = 2.4$$

93. Multiply both sides by LCM, 60.
$$\frac{3}{4}x + \frac{1}{5} = \frac{2}{3}(x - 2)$$
$$60\left(\frac{3}{4}x + \frac{1}{5}\right) = 60 \cdot \frac{2}{3}(x - 2)$$
$$15 \cdot 3x + 12 = 20 \cdot 2(x - 2)$$
$$45x + 12 = 40x - 80$$
$$45x - 40x = -12 - 80$$
$$5x = -92$$
$$x = -\frac{92}{5}$$

94. a. A relation is any set of ordered pairs.

b. A function is a correspondence where each member of the domain corresponds to exactly one member in the range.

c. Answers will vary.

95. D:{3, 4, 5, 6}, R:{−2, −1, 2, 3}

Exercise Set 8.5

1. Yes, $f(x) + g(x) = (f + g)(x)$ for all values of x. This is how addition of functions is defined.

3. $f(x)/g(x) = (f/g)(x)$ provided $g(x) \neq 0$. This is because division by zero is undefined.

5. No, $(f - g)(x) \neq (g - f)(x)$ for all values of x since subtraction is not commutative. For example, if $f(x) = x^2 + 1$ and $g(x) = x$, then
$$(f - g)(x) = f(x) - g(x)$$
$$= (x^2 + 1) - (x)$$
$$= x^2 - x + 1$$
$$(g - f)(x) = g(x) \quad f(x)$$
$$= (x) - (x^2 + 1)$$
$$= -x^2 + x - 1$$
So $(f - g)(x) \neq (g - f)(x)$.

7. a. $(f + g)(-2) = f(-2) + g(-2) = -3 + 5 = 2$

b. $(f - g)(-2) = f(-2) - g(-2) = -3 - 5 = -8$

c. $(f \cdot g)(-2) = f(-2) \cdot g(-2) = (-3) \cdot (5) = -15$

d. $(f/g)(-2) = f(-2)/g(-2) = \frac{-3}{5} = -\frac{3}{5}$

9. a.
$$(f + g)(x) = f(x) + g(x)$$
$$= (x + 1) + (x^2 + x)$$
$$= x^2 + 2x + 1$$

b. $(f + g)(a) = a^2 + 2a + 1$

c.
$$(f + g)(2) = (2)^2 + 2(2) + 1$$
$$= 4 + 4 + 1$$
$$= 9$$

13. a.
$$(f + g)(x) = f(x) + g(x)$$
$$= (4x^3 - 3x^2 - x) + (3x^2 + 4)$$
$$= 4x^3 - x + 4$$

b. $(f + g)(a) = 4a^3 - a + 4$

c.
$$(f + g)(2) = 4(2)^3 - (2) + 4$$
$$= 32 - 2 + 4$$
$$= 34$$

15. $f(3)=(3)^2-4=5$
$g(3)=-5(3)+3=-12$
$f(3)+g(3)=5+(-12)=-7$

17. $f(-2)=(-2)^2-4=0$
$g(-2)=-5(-2)+3=13$
$f(-2)-g(-2)=0-13=-13$

19. $f(3)=3^2-4=5$
$g(3)=-5(3)+3=-12$
$f(3)\cdot g(3)=5(-12)=-60$

21. $f(\frac{3}{5})=\left(\frac{3}{5}\right)^2-4=-\frac{91}{25}$
$g\left(\frac{3}{5}\right)=-5\left(\frac{3}{5}\right)+3=0$
$f\left(\frac{3}{5}\right)/g\left(\frac{3}{5}\right)$ is undefined since $g\left(\frac{3}{5}\right)=0$

23. $f(-3)=(-3)^2-4=5$
$g(-3)=-5(-3)+3=18$
$g(-3)-f(-3)=18-5=13$

25. $f(0)=0^2-4=-4$
$g(0)=-5(0)+3=3$
$g(0)/f(0)=3/-4=-\frac{3}{4}$

27. $(f+g)(x)=f(x)+g(x)$
$=(2x^2-x)+(x-6)$
$=2x^2-6$

29. $(f+g)(0)=2(0)^2-6$
$=0-6$
$=-6$

31. $(f-g)(-3)=f(-3)-g(-3)$
$=\left(2\cdot(-3)^2-(-3)\right)-((-3)-6)$
$=21-(-9)$
$=30$

33. $(f\cdot g)(0)=f(0)\cdot g(0)$
$f(0)=2(0)^2-(0)=0$
$g(0)=(0)-6=-6$
$f(0)\cdot g(0)=0\cdot(-6)=0$

35. $(f/g)(-1)=f(-1)/g(-1)$
$f(-1)=2(-1)^2-(-1)=3$
$g(-1)=(-1)-6=-7$
$f(-1)/g(-1)=3/(-7)=-\frac{3}{7}$

37. $(g/f)(5)=g(5)/f(5)$
$f(5)=2(5)^2-5=45$
$g(5)=5-6=-1$
$g(5)/f(5)=(-1)/45=-\frac{1}{45}$

39. $(g-f)(x)=g(x)-f(x)$
$=(x-6)-(2x^2-x)$
$=-2x^2+2x-6$

41. $(f+g)(0)=f(0)+g(0)=2+1=3$

43. $(f\cdot g)(2)=f(2)\cdot g(2)=4\cdot(-1)=-4$

45. $(g-f)(-1)=g(-1)-f(-1)=2-1=1$

47. $(g/f)(4)=g(4)/f(4)=1/0$, undefined

49. $(f+g)(3)=f(3)+g(3)=1+3=4$

51. $(f\cdot g)(1)=f(1)\cdot g(1)=0\cdot 1=0$

53. $(f/g)(4)=f(4)/g(4)=3/1=-3$

55. $(g/f)(2)=g(2)/f(2)=2/(-1)=-2$

57. **a.** The total expenditures ,*T*, is the sum of the private expenditures, *r*, and the public expenditures, *u*.

b. The total amount of expenditures increased the least during 1970 - 1980.

c. The total amount of expenditures increased the most during 1990 - 2000.

59. **a.** There were about 6,200,000 people who were aged receiving Medicare hospital insurance in 2000.

b. There were about 1,000,000 people who were disabled receiving Medicare hospital insurance in 2000.

c. There was a total of about 7,200,000 people receiving Medicare hospital insurance in 2000.

61. If $(f+g)(a)=0$, then, $f(a)$ and $g(a)$ must either be opposites or both be equal to 0.

63. If $(f-g)(a)=0$, then $f(a)=g(a)$.

65. If $(f/g)(a) < 0$, then $f(a)$ and $g(a)$ must have opposite signs.

67. 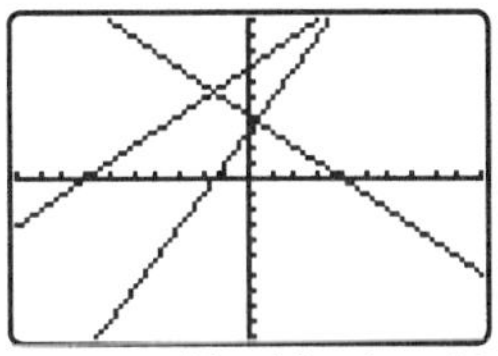

-10, 10, 1, -10

69. 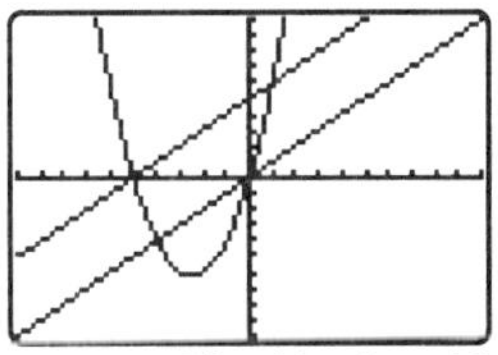

-10, 10, 1, -10

72.
$$A = \frac{1}{2}bh$$
$$2 \cdot A = 2 \cdot \frac{1}{2}bh$$
$$2A = bh$$
$$\frac{2A}{b} = \frac{bh}{b}$$
$$\frac{2A}{b} = h \text{ or } h = \frac{2A}{b}$$

73. Let the pre-tax cost of the washing machine be x.
$$x + 0.06x = 477$$
$$1.06x = 477$$
$$x = \frac{477}{1.06}$$
$$x = 450$$
The pre-tax cost of the washing machine was \$450.

74. Set $y = 0$ to find the x-intercept.
$$3x - 4(0) = 12$$
$$3x = 12$$
$$x = 4$$
The x-intercept is (4, 0). Set $x = 0$ to find the y-intercept.
$$3(0) - 4y = 12$$
$$-4y = 12$$
$$y = -3$$
The y-intercept is (0, –3).

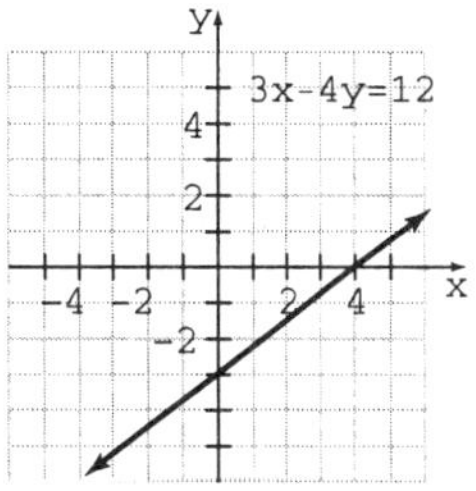

75.

x	y
–3	1
–2	0
–1	–1
0	–2
1	–1
2	0
3	1

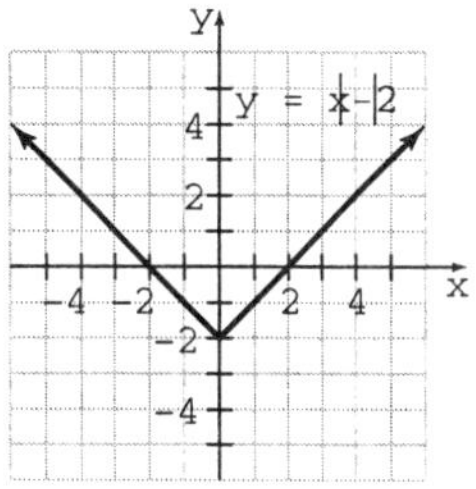

Review Exercises

1. $y = \frac{1}{2}x$

x	y
–2	$y = 0.5(-2) = -1$
0	$y = 0.5(0) = 0$
2	$y = 0.5(2) = 1$

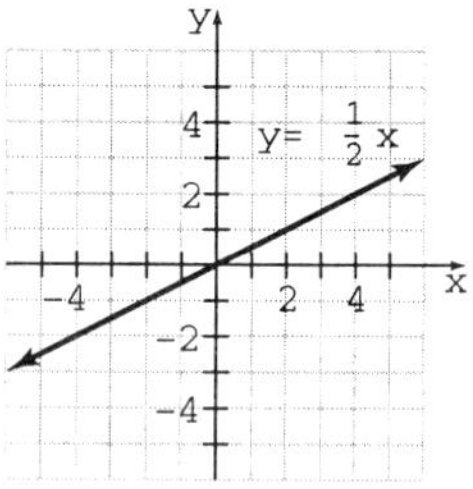

2. $y = -2x - 1$

x	y
0	$y = -2(0) - 1 = -1$
1	$y = -2(1) - 1 = -3$
2	$y = -2(2) - 1 = -4$

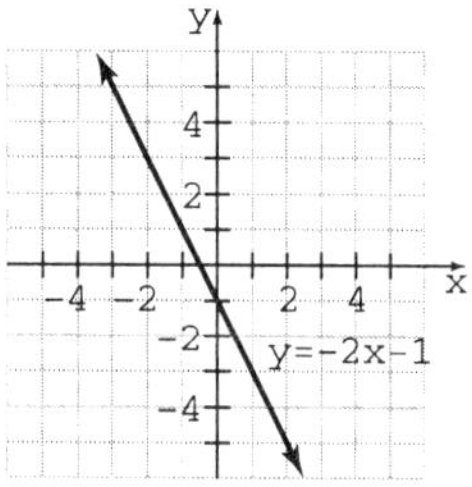

3. $y = \frac{1}{2}x + 3$

x	y
0	$y = \frac{1}{2}(0) + 3 = 3$
–2	$y = \frac{1}{2}(-2) + 3 = 2$
–4	$y = \frac{1}{2}(-4) + 3 = 1$

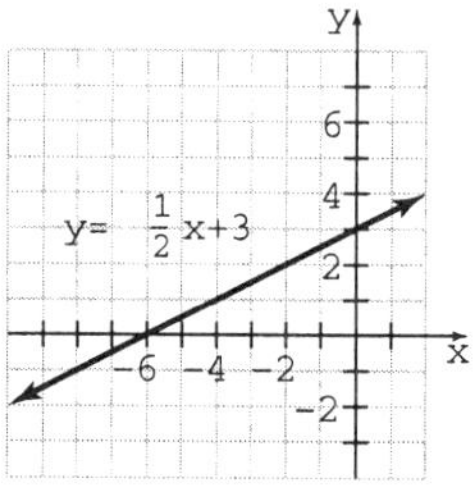

4. $y = -\frac{3}{2}x + 1$

x	y
–2	$y = -\frac{3}{2}(-2) + 1 = 4$
0	$y = -\frac{3}{2}(0) + 1 = 1$
2	$y = -\frac{3}{2}(2) + 1 = -2$

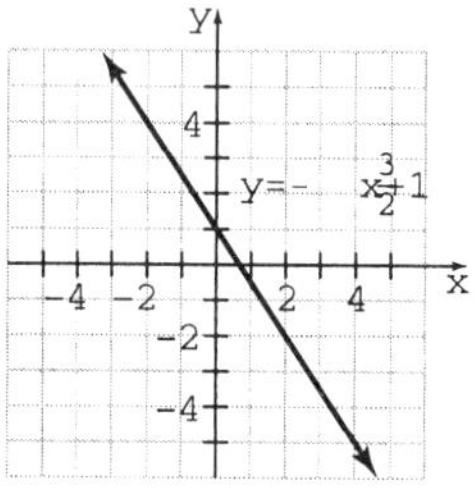

5. $y = x^2$

x	y
–3	$y = (-3)^2 = 9$
–1	$y = (-1) =^2 -1$
0	$y = 0^2 = 0$
2	$y = 2^2 = 4$

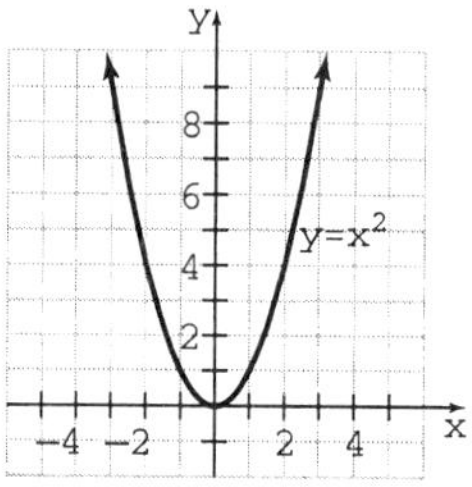

6. $y = x^2 - 1$

x	y
−3	$y = (-3)^2 - 1 = 8$
−1	$y = (-1)^2 - 1 = 0$
0	$y = 0^2 - 1 = 0$
1	$y = 1^2 - 1 = 0$
2	$y = 2^2 - 1 = 3$

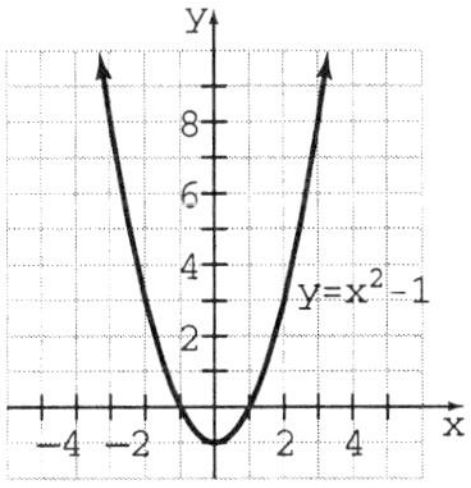

7. $y = |x|$

x	y
−4	$y = \lvert -4 \rvert = 4$
−1	$y = \lvert -1 \rvert = 1$
0	$y = \lvert 0 \rvert = 0$
2	$y = \lvert 2 \rvert = 2$

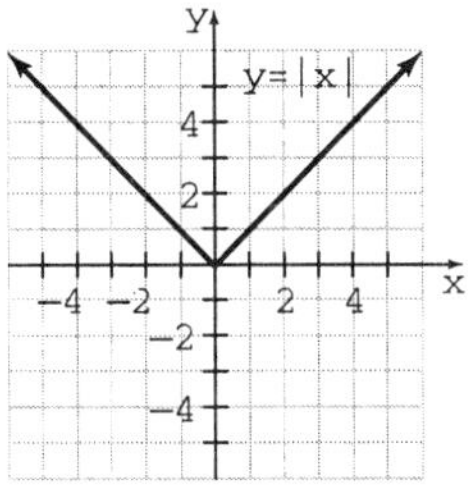

8. $y = |x| - 1$

x	y
−4	$y = \lvert -4 \rvert - 1 = 3$
−1	$y = \lvert -1 \rvert - 1 = 0$
0	$y = \lvert 0 \rvert - 1 = -1$
2	$y = \lvert 2 \rvert - 1 = 1$

x	y
−4	$y = \lvert -4 \rvert - 1 = 3$
−1	$y = \lvert -1 \rvert - 1 = 0$
0	$y = \lvert 0 \rvert - 1 = -1$
2	$y = \lvert 2 \rvert - 1 = 1$

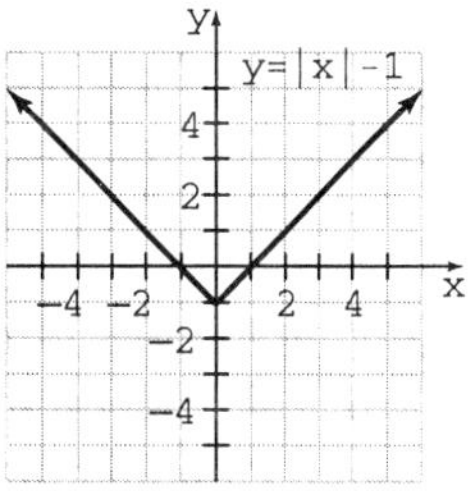

9. $y = x^3$

x	y
−2	$y = (-2)^3 = -8$
−1	$y = (-1)^3 = -1$
0	$y = 0^3 = 0$
1	$y = 1^3 = 1$
2	$y = 2^3 = 8$

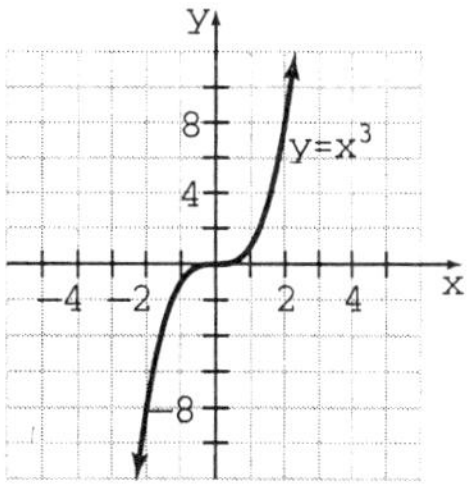

10. $y = x^3 + 4$

x	y
-2	$y = (-2)^3 + 4 = -4$
-1	$y = (-1)^3 + 4 = 3$
0	$y = 0^3 + 4 = 4$
1	$y = 1^3 + 4 = 5$

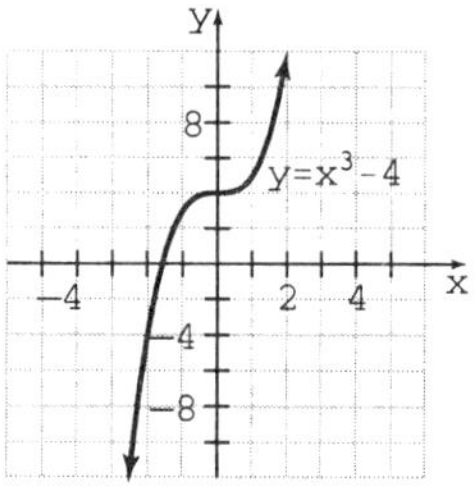

11. A function is a correspondence where each member of the domain corresponds to exactly one member of the range.

12. No, every relation is not a function. {(4, 2), (4, –2)} is a relation but not a function. Yes, every function is a relation because it is a set of ordered pairs.

13. Yes, each member of the domain corresponds to exactly one member of the range.

14. No, the domain element 2 corresponds to more than one member of the range (5 and –5).

15. {3, 4, 5, 6 }

16. {6, 7, 8, ...}

17. $(4, \infty)$

18. $(-\infty, -2]$

19. $(-1.5, 2.7]$

20. a. No, the relation is not a function.

b. Domain: $\{x | -1 \le x \le 1\}$
Range: $\{y | -1 \le y \le 1\}$

21. a. No, the relation is not a function.

b. Domain: $\{x | -2 \le x \le 2\}$
Range: $\{y | -1 \le y \le 1\}$

22. a. Yes, the relation is a function.

b. Domain: R
Range: $\{y | y \le 0\}$

23. a. Yes, the relation is a function.

b. Domain: R
Range: R

24. $f(x) = -x^2 + 3x - 5$

a. $f(2) = -(2)^2 + 3(2) - 5 = -4 + 6 - 5 = -3$

b. $f(h) = -h^2 + 3h - 5$

25. $g(t) = 2t^3 - 3t^2 + 1$

a. $g(-1) = 2(-1)^3 - 3(-1)^2 + 1$
$= -2 - 3 + 1$
$= -4$

b. $g(a) = 2a^3 - 3a^2 + 1$

26. Answers will vary. One possible interpretation: Car speeds up to 50 mph. Stays at 50 mph for about 11 minutes. Speeds up to about 68 mph. Stays at that speed for 5 minutes. Stops quickly. Stays stopped for 5 minutes. In stop and go traffic for 5 minutes.

27. $N(x) = 40x - 0.2x^2$

a. $N(20) = 40(20) - 0.2(20)^2$
$= 800 - 80$
$= 720$
720 baskets of apples are produced by 20 trees.

b. $N(50) = 40(50) - 0.2(50)^2$
$= 2000 - 500$
$= 1500$
1500 baskets of apples are produced by 50 trees.

28. $h(t) = -16t^2 + 100$

a. $h(1) = -16(1)^2 + 100 = 84$
After 1 second, the height of the ball is 84 feet.

b. $h(2) = -16(2)^2 + 100 = 36$
After 2 seconds, the height of the ball is 36 feet.

29.

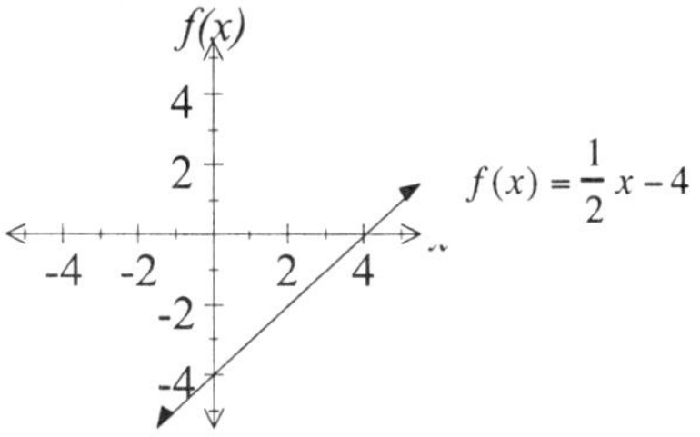

30.

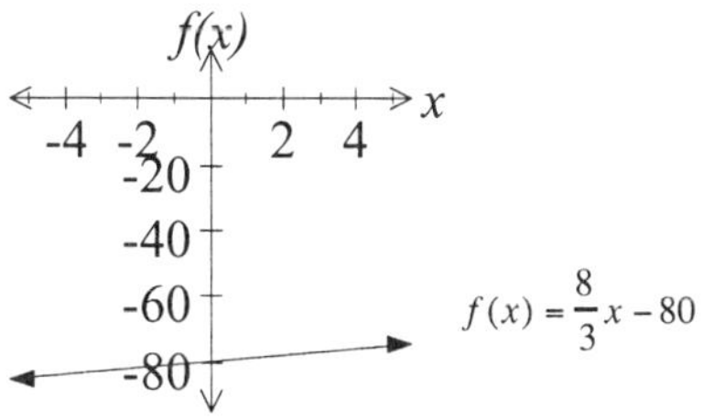

31. $f(x) = 4$ is a horizontal line 4 units above the x-axis.

x	y
−2	4
0	4
2	4

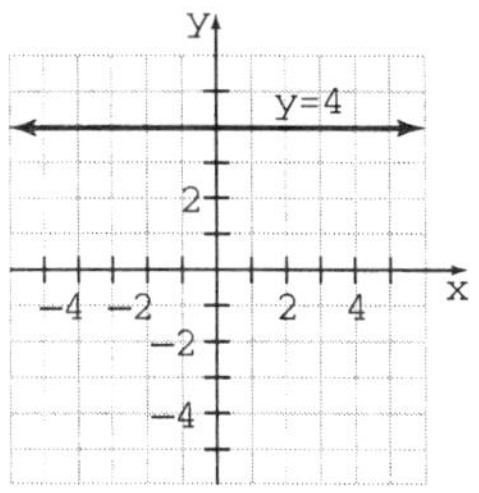

32. **a.** $p = 0.1x - 5000$

x	y
0	$p = 0.1(0) - 5000 = -5000$
50,000	$p = 0.1(50{,}000) - 5000 = 0$
100,000	$p = 0.1(100{,}000) - 5000 = 5000$

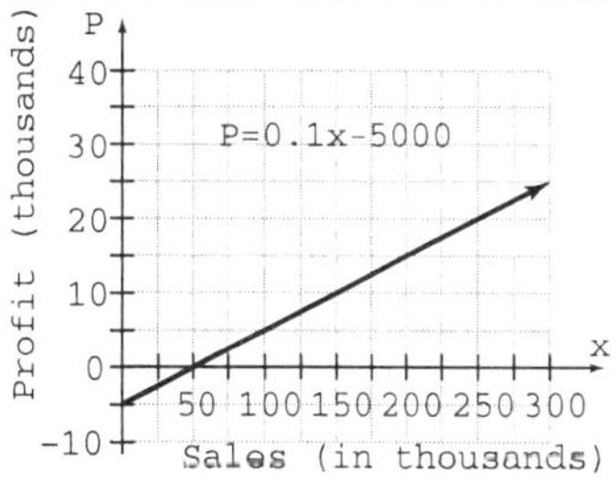

b. Approximately 50,000 bagels are sold when the company breaks even.

c. Approximately 250,000 bagels are sold for \$20,000 profit.

33. The principle is \$12,000 and the time is one year. Use the decimal form of the interest rate.
$I = 12{,}000r$

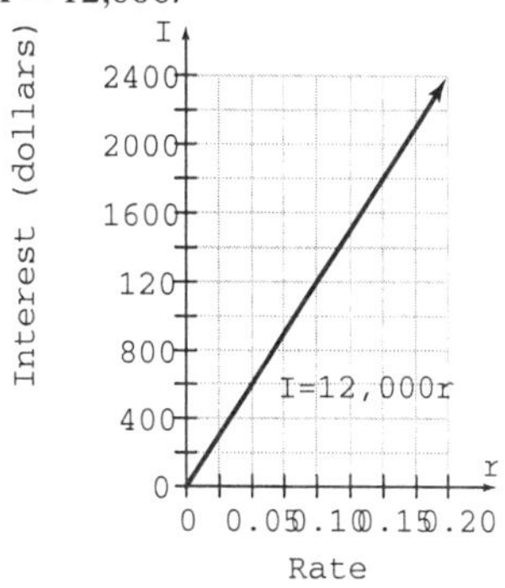

34. This is a horizontal line ($m = 0$) having a y-intercept of 3.
The equation is $y = 3$.

35. This is a vertical line (m is undefined) having an x-intercept of 2. The equation is $x = 2$.

36. The slope is negative and y changes 2 units when x changes 4 units. Thus, $m = -\frac{2}{4} = -\frac{1}{2}$. Hence,

$m = -\frac{1}{2}$ and $b = 2$.

$y = -\frac{1}{2}x + 2$

37. a. The slope of the translated graph is the same as the slope of the original graph: $m = -2$.

b. The y-intercept of the translated graph is the 4 less than the y-intercept of the original graph: $3 - 4 = -1$. The y-intercept is $(0, -1)$.

c. The equation of the translated graph is $y = -2x - 1$

38. The y-intercept of the translated graph is 8 more than the y-intercept if the original graph: $-8 + 8 = 0$. The y-intercept is $(0, 0)$.

39. a.

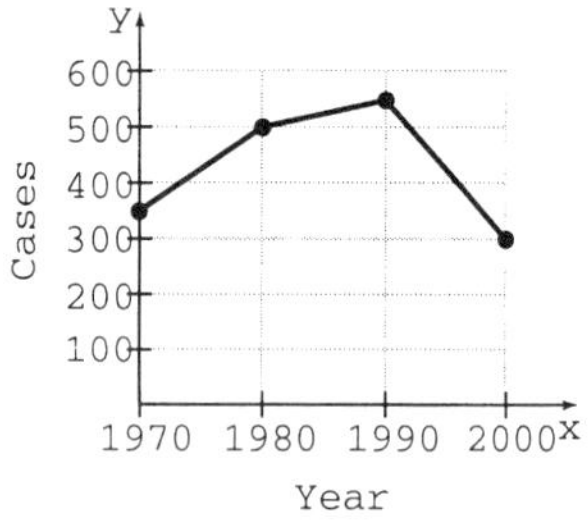

b. 1970 to 1980:

$$m_1 = \frac{510 - 346}{1980 - 1970} = \frac{164}{10} = 16.4$$

1980 to 1990:

$$m_2 = \frac{552 - 510}{1990 - 1980} = \frac{42}{10} = 4.2$$

1990 to 2000:

$$m_3 = \frac{317 - 552}{2000 - 1990} = \frac{-235}{10} = -23.5$$

c. The number of reported cases of typhoid fever increased the most during the 10-year period of 1970 – 1980.

40. First, find the slope.

$$m = \frac{98.2 - 35.6}{2070 - 1980} = \frac{62.6}{90} \approx 0.7$$

Let t be the number of years since 1980. Use the point-slope form with $m = 0.7$ and $(t_1, n_1) = (0, 35.6)$.

$$n - m_1 = m(t - t_1)$$
$$n - 35.6 = 0.7(t - 0)$$
$$n - 35.6 = 0.7t$$
$$n = 0.7t + 35.6$$
$$n(t) = 0.7t + 35.6$$

41. Write each equation in slope-intercept form by solving for y.

$$2x - 3y = 10$$
$$-3y = -2x + 10$$
$$\frac{-3y}{-3} = \frac{-2x + 10}{-3}$$
$$y = \frac{2}{3}x - \frac{10}{3}$$

$$y = \frac{2}{3}x - 4$$

Since $m = \frac{2}{3}$ for both lines, the lines are parallel.

42. Write each equation in slope-intercept form by solving for y.

$$2x - 3y = 9$$
$$-3y = -2x + 9$$
$$y = \frac{-2x + 9}{-3}$$
$$y = \frac{2}{3}x - 3$$

$$-3x - 2y = 6$$
$$-2y = 3x + 6$$
$$y = \frac{3x + 6}{-2}$$
$$y = -\frac{3}{2}x - 3$$

Since the slopes are $\frac{2}{3}$ and $-\frac{3}{2}$ which are negative reciprocals, the lines are perpendicular.

43. Write each equation in slope-intercept form by solving for y.

$$4x - 2y = 10$$
$$-2y = -4x + 10$$
$$y = \frac{-4x + 10}{-2}$$
$$y = 2x - 5$$

$$-2x + 4y = -8$$
$$4y = 2x - 8$$
$$y = \frac{2x - 8}{4}$$
$$y = \frac{1}{2}x - 2$$

Since the slopes are 2 and $\frac{1}{2}$ which are neither equal nor negative reciprocals, the lines are neither parallel nor perpendicular.

44. Use the point-slope form with $m = \frac{1}{2}$ and $(x_1, y_1) = (4, 5)$.

$$y - 5 = \frac{1}{2}(x - 4)$$
$$y - 5 = \frac{1}{2}x - 2$$
$$y = \frac{1}{2}x + 3$$

45. First, find the slope: $m = \frac{-4-1}{2-(-3)} = \frac{-5}{5} = -1$.

Now, use the point-slope form with $m = -1$ and $(x_1, y_1) = (-3, 1)$.

$y - 1 = -1(x - (-3))$

$y - 1 = -1(x + 3)$

$y - 1 = -x - 3$

$y = -x - 2$

46. The slope of the line $y = -\frac{2}{3}x + 1$ is $m = -\frac{2}{3}$. Since the new line is parallel to this line, its slope is also $m = -\frac{2}{3}$. Use the slope-intercept form with $m = -\frac{2}{3}$ and y-intercept of (0, 4).

$y = -\frac{2}{3}x + 4$

47. To find the slope of the line $5x - 2y = 7$, solve for y.

$-2y = -5x + 7$

$y = \frac{-5x + 7}{-2}$

$y = \frac{5}{2}x - \frac{7}{2}$

The slope of this line is $\frac{5}{2}$, and since the new line is parallel to this line, its slope is also $\frac{5}{2}$.

Use the point-slope form with $m = \frac{5}{2}$ and $(x_1, y_1) = (2, 3)$.

$y - 3 = \frac{5}{2}(x - 2)$

$y - 3 = \frac{5}{2}x - 5$

$y = \frac{5}{2}x - 2$

48. The slope of the line $y = \frac{3}{5}x + 5$ is $\frac{3}{5}$. Since the new line is perpendicular to this line, its slope is $-\frac{5}{3}$. Use the point-slope form with $m = -\frac{5}{3}$ and $(x_1, y_1) = (-3, 1)$.

$y - 1 = -\frac{5}{3}[x - (-3)]$

$y - 1 = -\frac{5}{3}(x + 3)$

$y - 1 = -\frac{5}{3}x - 5$

$y = -\frac{5}{3}x - 4$

49. To find the slope of the line $4x - 2y = 8$, solve for y.

$-2y = -4x + 8$

$y = \frac{-4x + 8}{-2}$

$y = 2x - 4$

The slope of this line is 2. Since the new line is perpendicular to this line, its slope is $-\frac{1}{2}$. Use the point-slope form with $m = -\frac{1}{2}$ and $(x_1, y_1) = (4, 2)$.

$y - 2 = -\frac{1}{2}(x - 4)$

$y - 2 = -\frac{1}{2}x + 2$

$y = -\frac{1}{2}x + 4$

50. $m_1 = \frac{3-(-3)}{4-0} = \frac{3+3}{4-0} = \frac{6}{4} = \frac{3}{2}$

$m_2 = \frac{-1-(-2)}{1-2} = \frac{-1+2}{1-2} = \frac{1}{-1} = -1$

Since the slopes are neither the same nor negative reciprocals, the lines are neither parallel nor perpendicular.

51. $m_1 = \frac{2-3}{3-2} = \frac{-1}{1} = -1$

$m_2 = \frac{1-4}{4-1} = \frac{-3}{3} = -1$

Since the slopes are the same, the lines are parallel.

52. $m_1 = \frac{0-3}{4-1} = \frac{-3}{3} = -1$

$m_2 = \frac{2-3}{5-6} = \frac{-1}{-1} = 1$

Since the slopes are negative reciprocals, the lines are perpendicular.

53. $m_1 = \dfrac{5-3}{-3-2} = \dfrac{2}{-5} = -\dfrac{2}{5}$

$m_2 = \dfrac{-2-2}{-4-(-1)} = \dfrac{-2-2}{-4+1} = \dfrac{-4}{-3} = \dfrac{4}{3}$

Since the slopes are neither the same nor negative reciprocals, the lines are neither parallel nor perpendicular.

54. a. First, find the slope of the linear function using the points (35, 10.76) and (50, 19.91)

$m = \dfrac{19.91-10.76}{50-35} = \dfrac{9.15}{15} = 0.61$

Use the point-slope form with $m = 0.61$ and $(a_1, r_1) = (35, 10.76)$

$r - r_1 = m(a - a_1)$

$r - 10.76 = 0.61(a - 35)$

$r - 10.76 = 0.61a - 21.35$

$r = 0.61a - 10.59$

$r(a) = 0.61a - 10.59$

b. For a 42-year-old man:

$r(a) = 0.61a - 10.59$

$r(42) = 0.61(42) - 10.59$

$= 15.03$

The monthly rate is about \$15.03.

55. a. First, find the slope of the linear function using the points (30, 489) and (50, 525)

$m = \dfrac{525-489}{50-30} = \dfrac{36}{20} = 1.8$

Use the point-slope form with $m = 1.8$ and $(r_1, C_1) = (30, 489)$

$C - C_1 = m(r - r_1)$

$C - 489 = 1.8(r - 30)$

$C - 489 = 1.8r - 54$

$C = 1.8r + 435$

$C(r) = 1.8r + 435$

b. For a rate of 40 yards per minute:

$C(r) = 1.8r + 435$

$C(40) = 1.8(40) + 435$

$= 507$

The number of calories burned is 507.

c. Set the function equal to 600 and solve for r.

$C(r) = 1.8r + 435$

$600 = 1.8r + 435$

$1.8r = 165$

$r \approx 91.7$

The person needs to swim at a speed of about 91.7 yards per minute.

56. $(f+g)(x) = f(x) + g(x)$

$= (x^2 - 3x + 4) + (2x - 5)$

$= x^2 - x - 1$

57. $(f+g)(3) = f(3) + g(3)$

$= (3)^2 - 3 - 1$

$= 9 - 3 - 1$

$= 5$

58. $(g-f)(x) = (2x-5) - (x^2 - 3x + 4)$

$= -x^2 + 5x - 9$

59. $(g-f)(-1) = g(-1) - f(-1)$

$= -(-1)^2 + 5(-1) - 9$

$= -1 - 5 - 9$

$= -15$

60. $(f \cdot g)(-1) = f(-1) \cdot g(-1)$

$= ((-1)^2 - 3(-1) + 4) \cdot (2(-1) - 5)$

$= 8(-7)$

$= -56$

61. $(f \cdot g)(5) = f(5) \cdot g(5)$

$= (5^2 - 3(5) + 4) \cdot (2(5) - 5)$

$= 14(5)$

$= 70$

62. $(f/g)(1) = f(1)/g(1)$

$= (1^2 - 3(1) + 4)/(2(1) - 5)$

$= 2/-3$

$= -\dfrac{2}{3}$

63. $(f/g)(2) = f(2)/g(2)$

$= (2^2 - 3(2) + 4)/(2(2) - 5)$

$= 2(-1)$

$= -2$

64. a. The number of morning newspapers in 1960 was about 300.

b. The number of morning newspapers in 2000 was about 750.

c. The number of evening newspapers in 1960 was about 1450.

d. The number of evening newspapers in 2000 was about 750.

e. The number of daily newspapers in 1960 was about 1750.

f. The number of daily newspapers in 2000 was about 1500.

65. a. $c(2000)$ was about 490 million.

b. $t(2000)$ was about 190 million.

c. $(c+t)(2000)$ was about 680 million.

Practice Test

1. $y = -2x + 1$

x	y
1	$y = -2(-1) + 1 = 3$
0	$y = -2(0) + 1 = 1$
1	$y = -2(1) + 1 = -1$

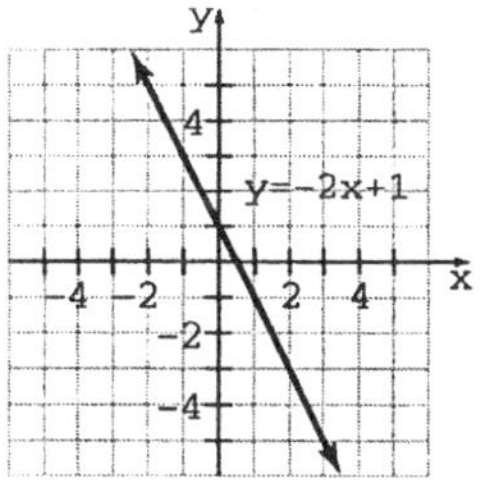

2. $y = \sqrt{x}$

x	y
−1	$y = \sqrt{-1}$ undefined
0	$y = \sqrt{0} = 0$
1	$y = \sqrt{1} = 1$
4	$y = \sqrt{4} = 2$

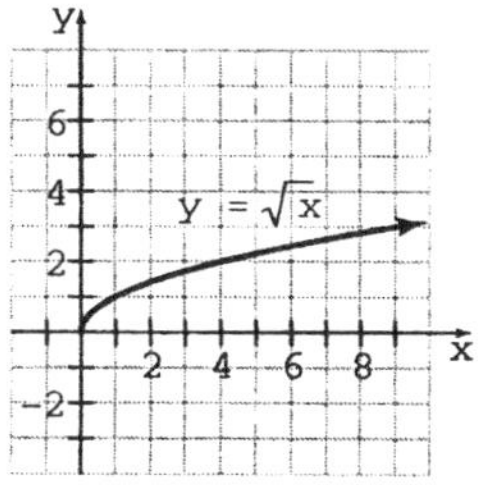

3. $y = x^2 - 4$

x	y
−2	$y = (-2)^2 - 4 = 0$
−1	$y = (-1)^2 - 4 = -3$
0	$y = (0)^2 - 4 = -4$
1	$y = (1)^2 - 4 = -3$
2	$y = (2)^2 - 4 = 0$

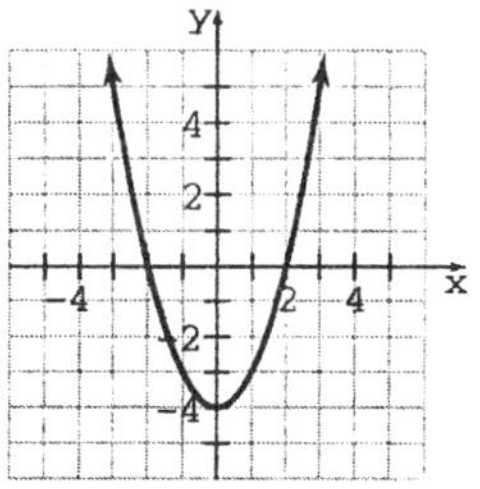

4. $y = |x|$

x	y
−3	$y = \lvert -3 \rvert = 3$
0	$y = \lvert 0 \rvert = 0$
4	$y = \lvert 4 \rvert = 4$

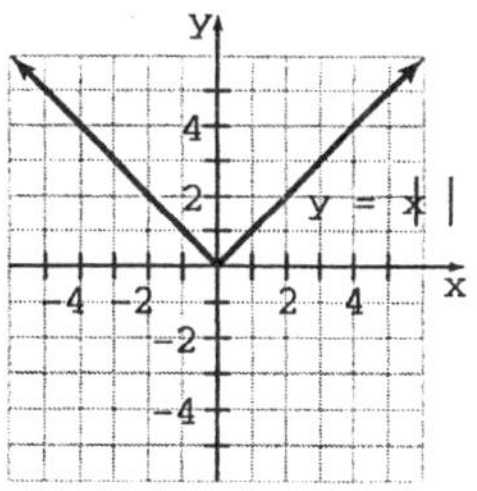

5. A function is a correspondence where each member in the domain corresponds with exactly one member in the range.

6. Yes, because each member in the domain corresponds with exactly one member in the range.

7. Yes, it is a function.
Domain: R
Range: $\{y \mid y \le 4\}$

8. No, it is not a function.
Domain: $\{x \mid -3 \le x \le 3\}$
Range: $\{y \mid -2 \le y \le 2\}$

9. $f(x) = 3x^2 - 6x + 2$
$f(-2) = 3(-2)^2 - 6(-2) + 2 = 12 + 12 + 2 = 26$

10. To write in function notation, solve for y.

$$3x + 7y = 14$$
$$7y = -3x + 14$$
$$y = -\frac{3}{7}x + 2$$
$$f(x) = -\frac{3}{7}x + 2$$

11.

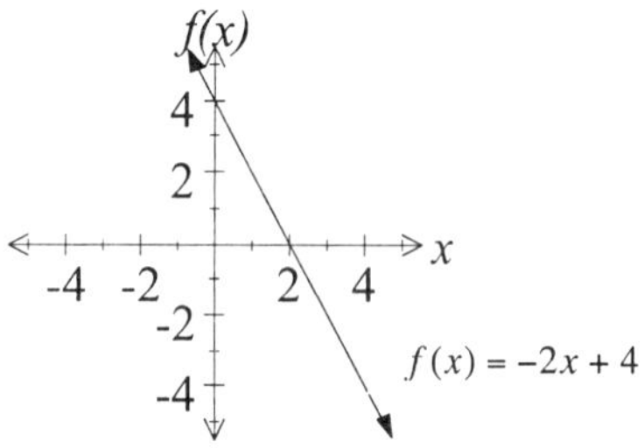

12.

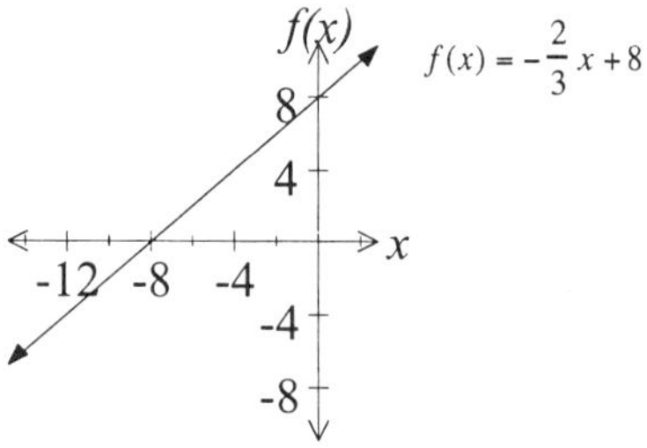

13. $f(x) = -3$ is a horizontal line 3 units below the x-axis.

x	y
−2	−3
0	−3
2	−3

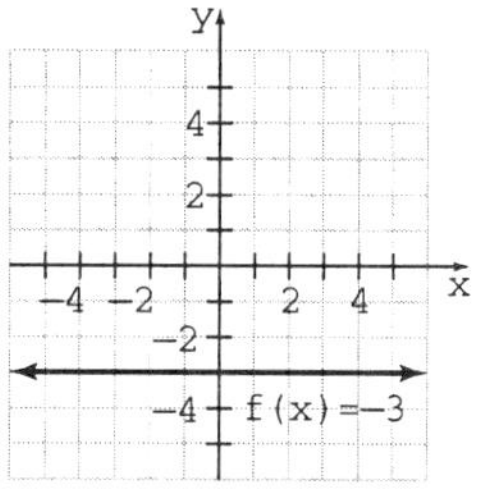

14. a. $p(x) = 10.2x - 50{,}000$

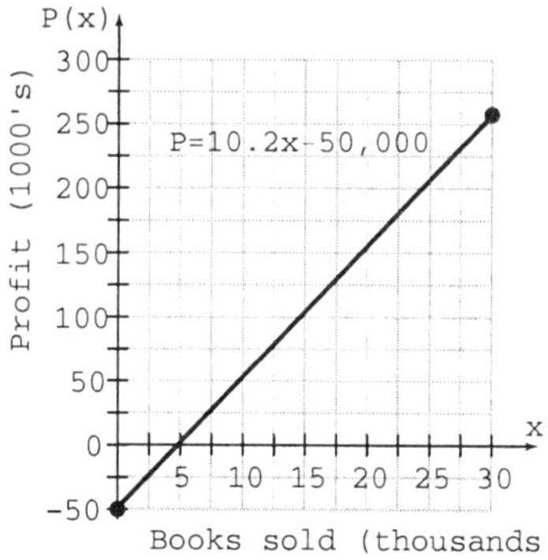

b. The company breaks even when $p(x) = 0$.

$$10.2x - 50{,}000 = 0$$
$$10.2x = 50{,}000$$
$$x = \frac{50{,}000}{10.2} = 4900$$

The company breaks even when it sells 4900 books.

c.

$$10.2x - 50{,}000 = 100{,}000$$
$$10.2x = 150{,}000$$
$$x = \frac{150{,}000}{10.2} \approx 14{,}700$$

The company needs to sell about 14,700 books to break even.

15. The y-intercept is 14 less than the y-value of the y-intercept of (0, 11). $11 - 14 = -3$
The y-intercept of the translated graph is (0, −3).

16. To find the rate of change find the slope between (3, 156) and (11, 132).

$$m = \frac{y_2 - y_1}{x_2 - x_1} = \frac{132-156}{11-3} = \frac{-24}{8} = -3$$

Kim weight was decreasing by 3 pounds per week.

17. For the year 2050, $t = 2050 - 2000 = 50$. The points are (0, 274.634) and (50, 393.931).

$$m = \frac{393.931-274.634}{50-0} = \frac{119.297}{50} = 2.38594 \approx 2.386$$

Use the point-slope form with $m = 2.386$ and $(t_1,\ p_1) = (0,\ 274.634)$.

$$p - p_1 = m(t - t_1)$$
$$p - 274.634 = 2.386(t - 0)$$
$$p - 274.634 = 2.386t$$
$$p = 2.386t + 274.634$$
$$p(t) \approx 2.386t + 274.634$$

18. Write each equation in slope-intercept form by solving for y.

$$2x - 3y = 6$$
$$-3y = -2x + 6$$
$$y = \frac{-2x+6}{-3}$$
$$y = \frac{2}{3}x - 2$$

$$4x + 8 = 6y$$
$$\frac{4x+8}{6} = y$$
$$\frac{4}{6}x + \frac{8}{6} = y$$
$$y = \frac{2}{3}x + \frac{4}{3}$$

Since the slopes are the same, the lines are parallel.

19. First find the slope of $3x - 2y = 6$

$$3x - 2y = 6$$
$$-2y = -3x + 6$$
$$y = \frac{-3x+6}{-2}$$
$$y = \frac{3}{2}x - 3$$
$$m = \frac{3}{2}$$

Therefore the slope of the perpendicular line is $m = -\frac{2}{3}$. Now use point-slope equation.

$$y - (-4) = -\frac{2}{3}(x - 3)$$
$$y + 4 = -\frac{2}{3}x + 2$$
$$y = -\frac{2}{3}x - 2$$

20. a. Let t represent the number of years since 1970. Then $t = 0$ represents 1970 and $t = 30$ represents 2000. Find the slope of the linear function using the points (0, 362) and (30, 266).

$$m = \frac{266-362}{30-0} = \frac{-96}{30} = -3.2$$

Use the point-slope form with $m = -3.2$ and $(t_1,\ r_1) = (0,\ 362)$

$$r - r_1 = m(t - t_1)$$
$$r - 362 = -3.2(t - 0)$$
$$r - 362 = -3.2t$$
$$r = -3.2t + 362$$
$$r(t) = -3.2t + 362$$

b. For 1995, use $t = 25$:

$$r(t) = -3.2t + 362$$
$$r(25) = -3.2(25) + 362 = 282$$

There were 282 deaths due to heart disease per 100,000 people in 1995.

c. For 2010, use $t = 40$:

$$r(t) = -3.2t + 362$$
$$r(40) = -3.2(40) + 362 = 234$$

There will be about 234 deaths due to heart disease per 100,000 people in 2010.

21. $(f + g)(3) = f(3) + g(3)$
$= (2(3)^2 - 3) + (3 - 5)$
$= 15 - 2 = 13$

22. $(f / g)(-1) = f(-1) / g(-1)$
$= (2(-1)^2 - (-1)) / ((-1) - 5)$
$= 3 / (-6) = -\frac{1}{2}$

23. $f(a) = 2a^2 - a$

24. $(f \cdot g)(x) = f(x) \cdot g(x)$
$= (2x^2 - x)(x - 5)$
$= 2x^3 - 10x^2 - x^2 + 5x$
$= 2x^3 - 11x^2 + 5x$

25. a. The total number of tons of paper to be used in 2010 will be about 44 million tons.

b. The number of tons of paper to be used by businesses in 2010 will be about 18 million tons.

c. The number of tons of paper to be used for reference, print media, and household use in 2010 will be about 44 – 18 = 26, or 26

Cumulative Review Test

1. $4x^2 - 7xy^2 + 6$
$= 4(-3)^2 - 7(-3)(-2)^2 + 6$
$= 4(9) - 7(-3)(4) + 6$
$= 36 + (21)(4) + 6$
$= 36 + 84 + 6$
$= 120 + 6$
$= 126$

2. $2 - \{3[6 - 4(6^2 \div 4)]\} = 2 - \{3[6 - 4(36 \div 4)]\}$
$= 2 - \{3[6 - 4(9)]\}$
$= 2 - \{3[6 - 36]\}$
$= 2 - \{3[-30]\}$
$= 2 - \{-90\}$
$= 2 + 90$
$= 92$

3. $2(x + 4) - 5 = -3[x - (2x + 1)]$
$2x + 8 - 5 = -3[x - 2x - 1]$
$2x + 3 = -3[-x - 1]$
$2x + 3 = 3x + 3$
$x = 0$

4. $P = 2E + 3R$
$P - 2E = 3R$
$\frac{P - 2E}{3} = R$

5. Solve for y.
$5x - 2y = 12$
$-2y = -5x + 12$
$y = \frac{-5x + 12}{-2}$
$y = \frac{5}{2}x - 6$

6. $\left(\frac{6x^2y^3}{2x^5y}\right)^3 = \left(\frac{3y^2}{x^3}\right)^3$
$= \frac{3^3 y^{2\cdot 3}}{x^{3\cdot 3}}$
$= \frac{27y^6}{x^9}$

7. $(6x^2 - 3x - 2) - (-2x^2 - 8x - 1)$
$6x^2 - 3x - 2 + 2x^2 + 8x + 1$
$8x^2 + 5x - 1$

8. $(4x^2 - 6x + 3)(3x - 5)$
$12x^3 - 20x^2 - 18x^2 + 30x + 9x - 15$
$12x^3 - 38x^2 + 39x - 15$

9. $6a^2 - 6a - 5a + 5$
$6a(a - 1) - 5(a - 1)$
$(a - 1)(6a - 5)$

10. $2x^2 = x + 15$
$2x^2 - x - 15 = 0$
$(2x + 5)(x - 3) = 0$
$2x + 5 = 0 \quad x - 3 = 0$
$2x = -5 \quad x = 3$
$x = -\frac{5}{2}$
The solutions are $x = -\frac{5}{2}$ and $x = 3$.

11. $\dfrac{x^2-9}{x^2-x-6}\cdot\dfrac{x^2-2x-8}{2x^2-7x-4}$

$=\dfrac{(x+3)(x-3)}{(x-3)(x+2)}\cdot\dfrac{(x-4)(x+2)}{(2x+1)(x-4)}$

$=\dfrac{x+3}{2x+1}$

12. $\dfrac{x}{x+4}-\dfrac{3}{x-5}$

$=\dfrac{x(x-5)}{(x+4)(x-5)}-\dfrac{3(x+4)}{(x+4)(x-5)}$

$=\dfrac{x^2-5x-3x-12}{(x+4)(x-5)}$

$=\dfrac{x^2-8x-12}{(x+4)(x-5)}$

13. $\dfrac{4}{x^2-3x-10}+\dfrac{2}{x^2+5x+6}$

$=\dfrac{4}{(x-5)(x+2)}+\dfrac{2}{(x+2)(x+3)}$

$=\dfrac{4(x+3)}{(x-5)(x+2)(x+3)}+\dfrac{2(x-5)}{(x-5)(x+2)(x+3)}$

$=\dfrac{4x+12}{(x-5)(x+2)(x+3)}+\dfrac{2x-10}{(x-5)(x+2)(x+3)}$

$=\dfrac{4x+12+2x-10}{(x-5)(x+2)(x+3)}$

$=\dfrac{6x+2}{(x-5)(x+2)(x+3)}$

14. $\dfrac{x}{9}-\dfrac{x}{6}=\dfrac{1}{12}$

$36\left(\dfrac{x}{9}-\dfrac{x}{6}\right)=36\left(\dfrac{1}{12}\right)$

$4x-6x=3$

$-2x=3$

$x=-\dfrac{3}{2}$

15. $\dfrac{7}{x+3}+\dfrac{5}{x+2}=\dfrac{5}{x^2+5x+6}$

$\dfrac{7}{x+3}+\dfrac{5}{x+2}=\dfrac{5}{(x+3)(x+2)}$

$(x+3)(x+2)\left(\dfrac{7}{x+3}+\dfrac{5}{x+2}\right)=(x+3)(x+2)\left(\dfrac{5}{(x+3)(x+2)}\right)$

$7(x+2)+5(x+3)=5$

$7x+14+5x+15=5$

$12x+29=5$

$12x=-24$

$x=-2$

Since $x\neq-2$ because it is not in the domain, there is no solution.

16. a. The graph is not a function.

b. Domain: $\{x|x \le 2\}$; Range: R

17. Write each equation in slope-intercept form by solving for y.

$$2x - 5y = 6 \qquad 5x - 2y = 9$$

$$-5y = -2x + 6 \qquad -2y = -5x + 9$$

$$y = \frac{-2x + 6}{-5} \qquad y = \frac{-5x + 9}{-2}$$

$$y = \frac{2}{5}x - \frac{6}{5} \qquad y = \frac{5}{2}x - \frac{9}{2}$$

Since the slopes are $\frac{2}{5}$ and $\frac{5}{2}$ which are neither equal nor negative reciprocals, the lines are neither parallel nor perpendicular.

18.
$$\begin{aligned}(f + g)(x) &= f(x) + g(x)\\ &= (x^2 + 3x - 2) + (4x - 6)\\ &= x^2 + 7x - 8\end{aligned}$$

19.
$$\begin{aligned}(f \cdot g)(4) &= f(4) \cdot g(4)\\ &= \left(4^2 + 3(4) - 2\right)\left(4(4) - 6\right)\\ &= (16 + 12 - 2)(16 - 6)\\ &= 26 \cdot 10\\ &= 260\end{aligned}$$

20. a. Property Taxes:

$= 30.7\%$ of 1.576×10^9

$= 0.307 \times 1.576 \times 10^9$

$= 0.483832 \times 10^9$

$= 4.83832 \times 10^8$ or \$483,832,000

b. Federal Grants:

$= 14.2\%$ of 1.576×10^9

$= 0.142 \times 1.576 \times 10^9$

$= 0.223792 \times 10^9$

$= 2.23792 \times 10^8$ or \$223,792,000

c. State Shared Taxes % – State Grants %:

$= 10.3\% - 9.4\% = 0.9\%$

Difference in Amount of Taxes Obtained:

$= 0.9\% \times 1.576 \times 10^9$

$= 0.009 \times 1.576 \times 10^9$

$= 0.014184 \times 10^9$

$= 1.4184 \times 10^7$ or \$14,184,000

Chapter 9

Exercise Set 9.1

1. The solution to a system of equations represents the ordered pairs that satisfy all the equations in the system.

3. Write the equations in slope-intercept form and compare their slopes and y-intercepts.

5. The point of intersection can only be estimated.

7. a.

$$y = 4x - 6 \qquad y = -2x$$
$$-10 = 4(-1) - 6 \qquad -10 = -2(-1)$$
$$-10 = -4 - 6 \qquad -10 = 2 \text{ False}$$
$$-10 = -10 \text{ True}$$

Since (–1, -10) does not satisfy both equations, it is not a solution to the system of equations.

b.

$$y = 4x - 6$$
$$0 = 4(3) - 6$$
$$0 = 6 \text{ False}$$

Since (3, 0) does not satisfy the first equation, it is not a solution to the system of equations.

c.

$$y = 4x - 6 \qquad y = -2x$$
$$-2 = 4(1) - 6 \qquad -2 = -2(1)$$
$$-2 = -2 \text{ True} \qquad -2 = -2 \text{ True}$$

Since (1, -2) satisfies both equations, it is a solution to the system of linear equations.

9. a.

$$y = 2x - 3 \qquad y = x + 5$$
$$13 = 2(8) - 3 \qquad 13 = 8 + 5$$
$$13 = 13 \text{ True} \qquad 13 = 13 \text{ True}$$

Since (8, 13) satisfies both equations, it is a solution to the system.

b.

$$y = 2x - 3 \qquad y = x + 5$$
$$5 = 2(4) - 3 \qquad 5 = 4 + 5$$
$$5 = 5 \text{ True} \qquad 5 = 9 \text{ False}$$

Since (4, 5) does not satisfy both equations, it is not a solution to the system.

c.

$$y = 2x - 3$$
$$9 = 2(4) - 3$$
$$9 = 5 \text{ False}$$

Since (4, 9) does not satisfy the first equation, it is not a solution to the system.

11. a.

$$2x + y = 9 \qquad 5x + y = 10$$
$$2(3) + 3 = 9 \qquad 5(3) + 3 = 10$$
$$9 = 9 \text{ True} \qquad 18 = 10 \text{ False}$$

Since (3, 3) does not satisfy both equations, it is not a solution to the system.

b.

$$2x + y = 9$$
$$2(2) + (0) = 9$$
$$4 = 9 \text{ False}$$

Since (2, 0) does not satisfy the first equation, it is not a solution to the system.

c.

$$2x + y = 9 \qquad 5x + y = 10$$
$$2(4) + 1 = 6 \qquad 5(4) + 1 = 10$$
$$9 = 9 \text{ True} \qquad 21 = 10 \text{ False}$$

Since (4, 1) does not satisfy both equations, it is not a solution to the system.

13. Solve the first equation for y.

$$2x - 3y = 6$$
$$-3y = -2x + 6$$
$$y = \frac{2}{3}x - 2$$

Notice that it is the same as the second equation. If the ordered pair satisfies the first equation, then it also satisfies the second equation.

a.

$$2x - 3y = 6$$
$$2(3) - 3(0) = 6$$
$$6 = 6 \text{ True}$$

Since (3, 0) satisfies both equations, it is a solution to the system.

b.

$$2x - 3y = 6$$
$$2(3) - 3(-2) = 6$$
$$12 = 6 \text{ False}$$

Since (3, –2) does not satisfy the first equation, it is not a solution to the system.

c.

$$2x - 3y = 6$$
$$2(6) - 3(2) = 6$$
$$6 = 6 \text{ True}$$

Since (6, 2) satisfies both equations, it is a solution to the system.

15. a. $3x - 4y = 8$

$3(0) - 4(-2) = 8$

$8 = 8$ True

$2y = \frac{2}{3}x - 4$

$2(-2) = \frac{2}{3}(0) - 4$

$-4 = -4$ True

Since (0, –2) satisfies both equations, it is a solution to the system.

b. $3x - 4y = 8$

$3(1) - 4(-6) = 8$

$27 = 8$ False

Since (1, –6) does not satisfy the first equation, it is not a solution to the system.

c. $3x - 4y = 8$

$3\left(-\frac{1}{3}\right) - 4\left(-\frac{9}{4}\right) = 8$

$8 = 8$ True

$2y = \frac{2}{3}x - 4$

$2\left(-\frac{9}{4}\right) = \frac{2}{3}\left(-\frac{1}{3}\right) - 4$

$-\frac{9}{2} = -\frac{38}{9}$ False

Since $\left(-\frac{1}{3}, -\frac{9}{4}\right)$ does not satisfy both equations, it is not a solution to the system.

17. consistent—one solution

19. dependent—infinite number of solutions

21. consistent—one solution

23. inconsistent—no solution

25. Write each equation in slope-intercept form.

$3y = 4x - 6$

$y = \frac{4}{3}x - 2$

Since the slopes of the lines are not the same, the lines intersect to produce one solution. This is a consistent system.

27. Write each equation in slope-intercept form.

$2y = 3x + 3$

$y = \frac{3}{2}x + \frac{3}{2}$

$y = \frac{3}{2}x - 2$

Since the lines have the same slope, $\frac{3}{2}$, and different y-intercepts, the lines are parallel. There is no solution. This is an inconsistent system.

29. Write each equation in slope-intercept form.

$2x = y - 6$

$y = 2x + 6$

$4x = 4y + 5$

$4y = 4x - 5$

$y = x - \frac{5}{4}$

Since the slopes of the lines are not the same, the lines intersect to produce one solution. This is a consistent system.

31. Write each equation in slope-intercept form.

$3x + 5y = -7$

$5y = -3x - 7$

$y = -\frac{3}{5}x - \frac{7}{5}$

$-3x - 5y = -7$

$-5y = 3x - 7$

$y = -\frac{3}{5}x + \frac{7}{5}$

Since the lines have the same slope and different y-intercepts, the lines are parallel. There is no solution. This is an inconsistent system.

33. Write each equation in slope-intercept form.

$x = 3y + 4$

$x - 4 = 3y$

$\frac{1}{3}x - \frac{4}{3} = y$

$2x - 6y = 8$

$-6y = -2x + 8$

$y = \frac{1}{3}x - \frac{4}{3}$

Since both equations are identical, the line is the same for both of them. There are an infinite number of solutions. This is a dependent system.

35. Write each equation in slope-intercept form.

$y = \frac{3}{2}x + \frac{1}{2}$

$3x - 2y = -\frac{1}{2}$

$-2y = -3x - \frac{1}{2}$

$y = \frac{3}{2}x + \frac{1}{4}$

Since the lines have the same slope and different y-intercepts, the lines are parallel. There is no solution. This is an inconsistent system.

37. Graph the equations $y = x + 3$ and $y = -x + 3$.

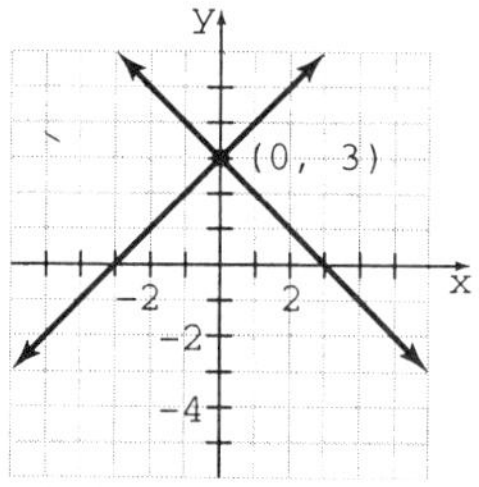

The lines intersect and the point of intersection is (0, 3). This is a consistent system.

39. Graph the equations $y = 3x - 6$ and $y = -x + 6$.

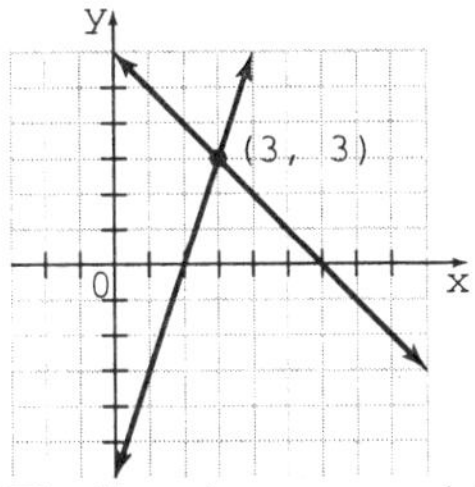

The lines intersect and the point of intersection is (3, 3). This is a consistent system.

41. Graph the equations $2x = 4$ or $x = 2$ and $y = -3$.

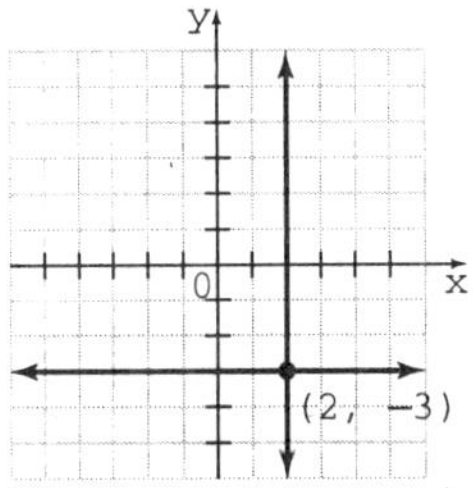

The lines intersect and the point of intersection is (2, –3). This is a consistent system.

43. Graph the equations $y = -x + 5$ and $-x + y = 1$ or $y = x + 1$.

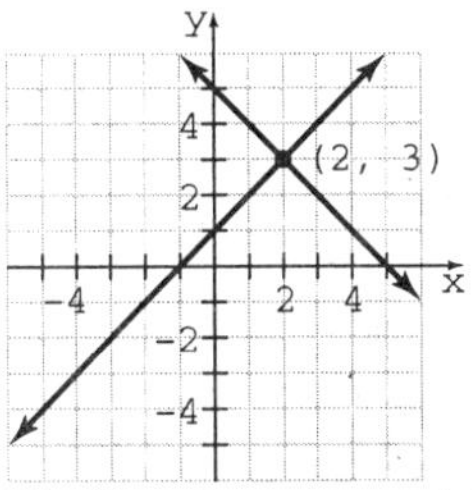

The lines intersect and the point of intersection is (2, 3). This is a consistent system.

45. Graph the equations $y = -\frac{1}{2}x + 4$ and $x + 2y = 6$ or $y = -\frac{1}{2}x + 3$.

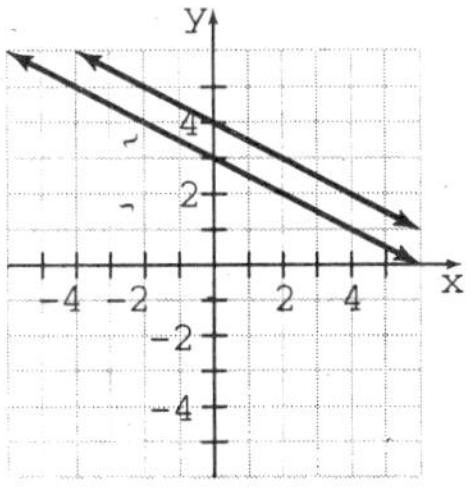

The lines are parallel. The system is inconsistent and there is no solution.

47. Graph the equations $x + 2y = 8$ or $y = -\frac{1}{2}x + 4$ and $5x + 2y = 0$ or $y = -\frac{5}{2}x$.

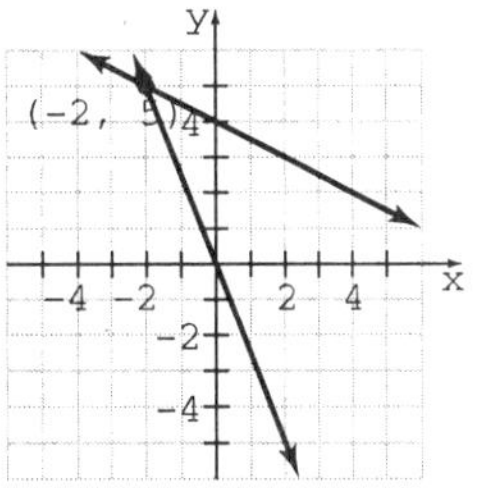

The lines intersect and the point of intersection is (–2, 5). This is a consistent system.

49. Graph the equations $2x + 3y = 6$ or $y = -\frac{2}{3}x + 2$ and $4x = -6y + 12$ or $y = -\frac{2}{3}x + 2$.

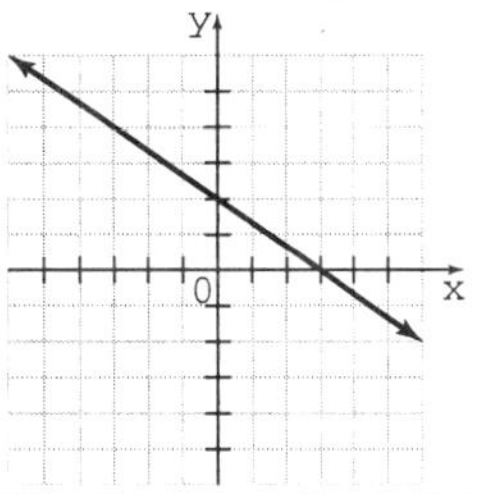

The lines are identical. There are an infinite number of solutions. This is a dependent system.

51. Graph the equations $y = 3$ and $y = 2x - 3$.

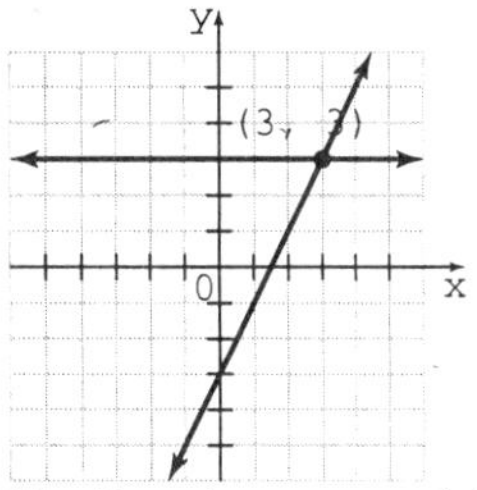

The lines intersect and the point of intersection is (3, 3). This is a consistent system.

53. Graph the equations $x - 2y = 4$ or $y = \frac{1}{2}x - 2$ and $2x - 4y = 8$ or $y = \frac{1}{2}x - 2$.

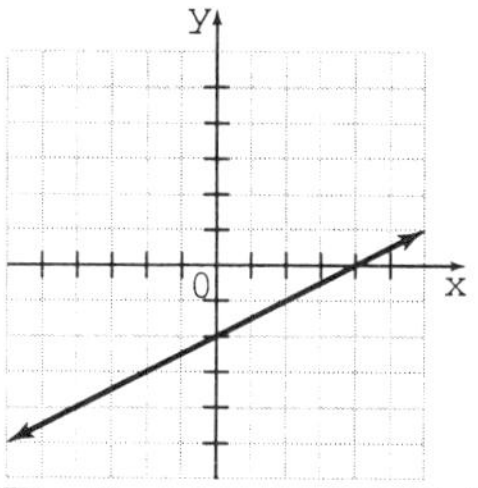

The lines are identical. There are an infinite number of solutions. This is a dependent system.

55. Graph the equations $2x + y = -2$ or $y = -2x - 2$ and $6x + 3y = 6$ or $y = -2x + 2$.

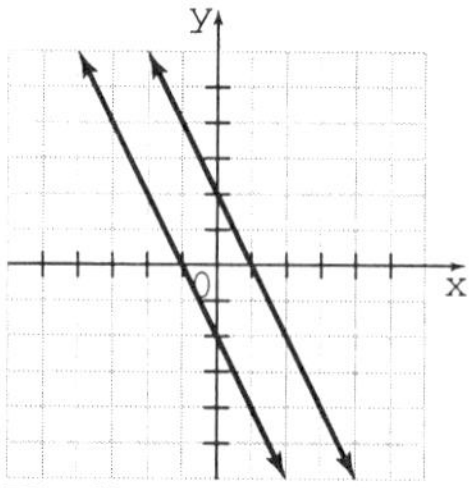

The lines are parallel. The system is inconsistent and there is no solution.

57. Graph the equations $4x - 3y = 6$ or $y = \frac{4}{3}x - 2$ and $2x + 4y = 14$ or $y = -\frac{1}{2}x + \frac{7}{2}$.

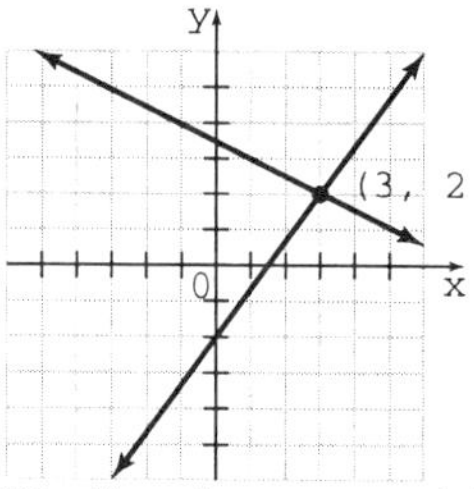

The lines intersect and the point of intersection is (3, 2). This is a consistent system.

59. Graph the equations $2x - 3y = 0$ or $y = \frac{2}{3}x$ and $x + 2y = 0$ or $y = -\frac{1}{2}x$.

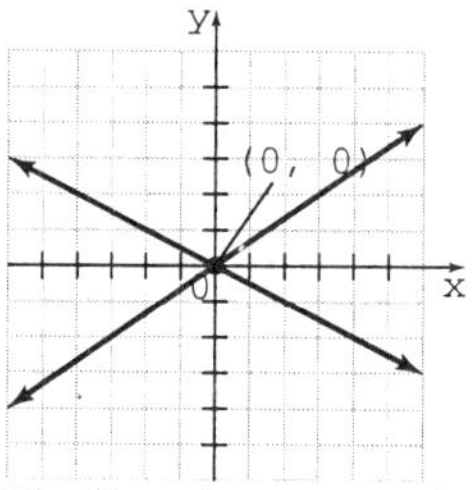

The lines intersect and the point of intersection is (0, 0). This is a consistent system.

61. Write each equation in slope-intercept form.

$$6x - 4y = 12 \qquad 12y = 18x - 24$$
$$-4y = -6x + 12 \qquad y = \frac{3}{2}x - 2$$
$$y = \frac{3}{2}x - 3$$

The lines are parallel because they have the same slope, $\frac{3}{2}$, and different y-intercepts.

63. The system has an infinite number of solutions. If the two lines have two points in common then they must be the same line.

65. The system has no solutions. Distinct parallel lines do not intersect.

67. $x = 5$, $y = 3$ has one solution, (5, 3).

69. (repair) $c = 600 + 650n$

(replacement) $c = 1800 + 450n$

Graph the equations and determine the intersection.

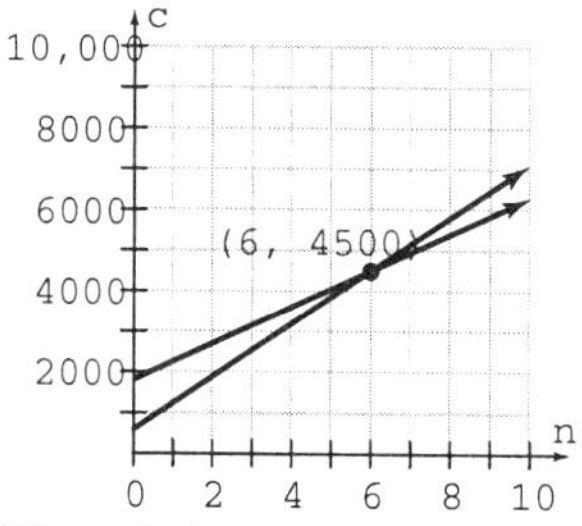

The solution is (6, 4500). Therefore, the total cost of repair equals the total cost of replacement at 6 years.

71. $c = 25h$

$c = 22h + 18$

Graph the equations and determine the intersection.

The solution (6, 150). Therefore, the boats must be rented for 6 hours for the cost to be the same.

79. $3x-(x-6)+4(3-x)=3x-x+6+12-4x$
$$=3x-x-4x+6+12$$
$$=-2x+18$$

80. $2(x+3)-x=5x+2$
$$2x+6-x=5x+2$$
$$x+6=5x+2$$
$$-4x=-4$$
$$x=1$$

81.
$$A=p(1+rt)$$
$$1000=500(1+r\cdot 20)$$
$$1000=500(1+20r)$$
$$1000=500(1)+500(20r)$$
$$1000=500+10{,}000r$$
$$1000-500=500-500+10{,}000r$$
$$500=10000r$$
$$\frac{500}{10{,}000}=\frac{1000r}{10{,}000}$$
$$0.05=r$$

82. a. For the x-intercept, set $y=0$.
$$2x+3y=12$$
$$2x+3(0)=12$$
$$2x=12$$
$$x=6$$
The x-intercept is (6, 0).
For the y-intercept, set $x=0$.
$$2x+3y=12$$
$$2(0)+3y=12$$
$$3y=12$$
$$y=4$$
The y-intercept is (0, 4).

b.

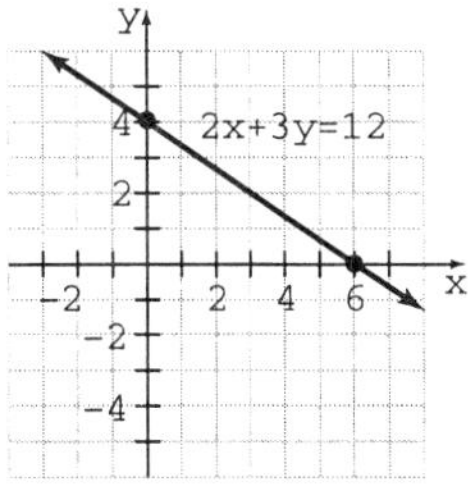

83. $\dfrac{x^2-9x+14}{2-x}=\dfrac{(x-2)(x-7)}{-1(x-2)}=\dfrac{x-7}{-1}=-(x-7)$

84.
$$\frac{4}{b}+2b=\frac{38}{3}$$
$$3b\left(\frac{4}{b}+2b\right)=3b\left(\frac{38}{3}\right)$$
$$12+6b^2=38b$$
$$6b^2-38b+12=0$$
$$2\left(3b^2-19b+6\right)=0$$
$$2(3b-1)(b-6)=0$$
$$3b-1=0 \qquad b-6=0$$
$$3b=1 \qquad b=6$$
$$b=\frac{1}{3}$$
The solutions are $b=\dfrac{1}{3}$ and $b=6$.

Exercise Set 9.2

1. The x in the first equation, since both 6 and 12 are divisible by 3.

3. You will obtain a false statement, such as $3=0$.

5. $x+2y=6$
$2x-3y=5$
Solve the first equation for x, $x=6-2y$.
Substitute $6-2y$ for x in the second equation.
$$2(6-2y)-3y=5$$
$$12-4y-3y=5$$
$$-7y=-7$$
$$y=1$$
Substitute 1 for y in the equation $x=6-2y$.
$$x=6-2(1)$$
$$x=4$$
The solution is (4, 1).

7. $x+y=-2$
$x-y=0$
Solve the first equation for y, $y=-2-x$.
Substitute $-2-x$ for y in the second equation.
$$x-(-2-x)=0$$
$$x+2+x=0$$
$$2x=-2$$
$$x=-1$$
Substitute -1 for x in the equation $y=-2-x$.
$$y=-2-(-1)$$
$$y=-2+1$$
$$y=-1$$
The solution is $(-1, -1)$.

9. $3x + y = 3$

$3x + y + 5 = 0$

Solve the first equation for y, $y = -3x + 3$.

Substitute $-3x + 3$ for y in the second equation.

$3x + y + 5 = 0$

$3x - 3x + 3 + 5 = 0$

$8 = 0$ False

There is no solution.

11. $x = 3$

$x + y + 5 = 0$

Substitute 3 for x in the second equation.

$x + y + 5 = 0$

$3 + y + 5 = 0$

$y = -8$

The solution is (3, –8).

13. $x - \frac{1}{2}y = 6$

$y = 2x - 12$

Substitute $2x$ - 12 for y in the first equation.

$x - \frac{1}{2}y = 6$

$x - \frac{1}{2}(2x - 12) = 6$

$x - x + 6 = 6$

$6 = 6$

Since this is a true statement, there are an infinite number of solutions. This is a dependent system.

15. $2x + y = 9$

$y = 4x - 3$

Substitute $4x - 3$ for y in the first equation.

$2x + y = 9$

$2x + 4x - 3 = 9$

$6x = 12$

$x = 2$

Now substitute 2 for x in the second equation.

$y = 4x - 3$

$y = 4(2) - 3$

$y = 8 - 3$

$y = 5$

The solution is (2, 5).

17. $y = \frac{1}{3}x - 2$

$x - 3y = 6$

Substitute $\frac{1}{3}x - 2$ for y in the second equation.

$x - 3y = 6$

$x - 3\left(\frac{1}{3}x - 2\right) = 6$

$x - x + 6 = 6$

$6 = 6$

Since this is a true statement, there are an infinite number of solutions. This is a dependent system.

19. $2x + 3y = 7$

$6x - 2y = 10$

First solve the second equation for y.

$\frac{1}{2}(6x - 2y) = \frac{1}{2}(10)$

$3x - y = 5$

$y = 3x - 5$

Now substitute $3x - 5$ for y in the first equation.

$2x + 3y = 7$

$2x + 3(3x - 5) = 7$

$2x + 9x - 15 = 7$

$11x = 22$

$x = 2$

Finally substitute 2 for x in the equation $y = 3x - 5$.

$y = 3(2) - 5$

$y = 6 - 5$

$y = 1$

The solution is (2, 1).

21. $3x - y = 14$

$6x - 2y = 10$

First solve the first equation for y.

$3x - y = 14$

$y = 3x - 14$

Now substitute $3x - 14$ for y in the second equation.

$6x - 2y = 10$

$6x - 2(3x - 14) = 10$

$6x - 6x + 28 = 10$

$28 = 10$ False

There is no solution.

23. $4x - 5y = -4$

$3x = 2y - 3$

First solve the second equation for x.

$\frac{1}{3}(3x) = \frac{1}{3}(2y - 3)$

$x = \frac{2}{3}y - 1$

Now substitute $\frac{2}{3}y - 1$ for x in the first equation.

$$4x - 5y = -4$$
$$4\left(\frac{2}{3}y - 1\right) - 5y = -4$$
$$\frac{8}{3}y - 4 - 5y = -4$$
$$\frac{8}{3}y - \frac{15}{3}y = 0$$
$$-\frac{7}{3}y = 0$$
$$y = 0$$

Finally substitute 0 for y in the equation $x = \frac{2}{3}y - 1$.

$$x = \frac{2}{3}(0) - 1$$
$$x = -1$$

The solution is $(-1, 0)$.

25. $4x + 5y = -6$

$$x - \frac{5}{3}y = -2$$

First solve the second equation for x, $x = \frac{5}{3}y - 2$. Now substitute $\frac{5}{3}y - 2$ for x in the first equation.

$$5x + 4y = -7$$
$$4\left(\frac{5}{3}y - 2\right) + 5y = -6$$
$$\frac{20}{3}y - 8 + 5y = -6$$
$$\frac{20}{3}y + \frac{15}{3}y = 8 - 6$$
$$\frac{35}{3}y = 2$$
$$y = \frac{6}{35}$$

Finally, substitute $\frac{9}{37}$ for y in the equation

$$x = \frac{5}{3}y - 2.$$
$$x = \frac{5}{3}\left(\frac{6}{35}\right) - 2$$
$$x = \frac{2}{7} - \frac{14}{7}$$
$$x = -\frac{12}{7}$$

The solution is $\left(-\frac{12}{7}, \frac{6}{35}\right)$.

27. Let x = the smaller number, then y = the larger number.
$x + y = 79$
$x + 7 = y$

Substitute $x + 7$ for y in the first equation.

$$x + y = 79$$
$$x + (x + 7) = 79$$
$$2x + 7 = 79$$
$$2x = 72$$
$$x = 36$$

Now substitute 36 for x in the second equation:

$$x + 7 = y$$
$$36 + 7 = y$$
$$43 = y$$

The two integers are 36 and 43.

29. Let l = the length of the rectangle, then w = the width.
$2l + 2w = 40$
$l = 4 + w$

Substitute $4 + w$ for l in the first equation and solve for w.

$$2l + 2w = 40$$
$$2(4 + w) + 2w = 40$$
$$8 + 2w + 2w = 40$$
$$4w = 32$$
$$w = 8$$

Now substitute 8 for w in the second equation to find l.

$$l = 4 + w$$
$$l = 4 + 8$$
$$l = 12$$

The length of the rectangle is 12 feet and the width is 8 feet.

31. Let b = the amount of money Billy had and j = the amount of money Jean had.

$b + j = 726$
$j = 134 + b$

Substitute $134 + b$ for j in the first equation.

$$b + j = 726$$
$$b + (134 + b) = 726$$
$$2b + 134 = 726$$
$$2b = 592$$
$$b = 296$$

Now substitute 296 for b in the first equation and

solve for j.

$b + j = 726$

$296 + j = 726$

$j = 430$

Billy had $296 and Jean had $430.

33. Let c = the client's portion of the award and a = the attorneys portion.

$c + a = 20{,}000$
$c = 3a$

Now solve the first equation for a and substitute it for a in the second equation.

$c + a = 20000$

$a = 20000 - c$

$c = 3a$

$c = 3(20000 - c)$

$c = 60000 - 3c$

$4c = 60000$

$c = 15000$

The client received $15,000.

35. $c = 1280 + 794n$

$c = 874n$

a. Substitute $874n$ for c in the first equation.

$874n = 1280 + 794n$

$80n = 1280$

$n = 16$

The mortgage plans will have the same total cost at 16 months.

b. $12 \text{ years} \cdot \left(\dfrac{12 \text{ months}}{1 \text{ year}}\right) = 144 \text{ months}$

$c = 1280 + 794(144) = 115{,}616$

$c = 874(144) = 125{,}856$

Yes, since $115,616 is less than $125,856, she should refinance.

37. a. Substitute $65 + 72t$ for m in the first equation and solve for t.

$65 + 72t = 80 + 60t$

$12t = 15$

$t = \dfrac{15}{12} = 1.25$

It will take Roberta 1.25 hours to catch up to Jean.

b. Now substitute 1.25 for t in one of the original equtions.

$m = 80 + 60(1.25)$

$m = 80 + 75$

$m = 155$

They will be at mile marker 155 when they meet.

39. a. $T = 180 - 10t$

b. $T = 20 + 6t$

c. $T = 180 - 10t$

$T = 20 + 6t$

Substitute $20 + 6t$ for T in the first equation.

$20 + 6t = 180 - 10t$

$16t = 160$

$t = 10$

It will take 10 minutes for the ball and oil to reach the same temperature.

d. Substitute 10 for t in one of the original equations.

$T = 180 - 10t$

$T = 180 - 10(10)$

$T = 180 - 100$

$T = 80$

The temperature will be 80°F.

41. $\dfrac{25}{3.5} \approx 7.14$

The willow tree is about 7.14 years old.

42. $4x - 8y = 16$

Let $x = 0$ and solve for y.

$4(0) - 8y = 16$

$y = -2$

Let $y = 0$ and solve for x.

$4x - 8(0) = 16$

$x = 4$

The intercepts are (0, –2) and (4, 0).

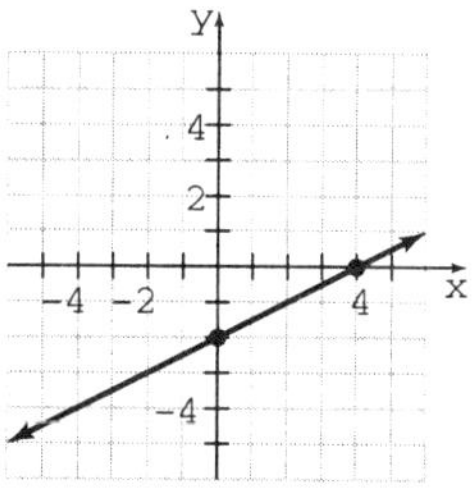

43. $3x - 5y = 8$
Write in slope-intercept form.
$-5y = -3x + 8$
$y = \frac{3}{5}x - \frac{8}{5}$
The slope is $\frac{3}{5}$ and the y intercept is $\left(0, -\frac{8}{5}\right)$.

44. Find the slope of the line using points (–2, 0) and (0, 4).
$m = \frac{4-0}{0-(-2)} = \frac{4}{2} = 2$
The y-intercept is 4, therefore $b = 4$.
$y=mx+b$
$y = 2x + 4$ is the equation of the line

45. $(6x+7)(3x-2) = 18x^2 - 12x + 21x - 14$
$= 18x^2 + 9x - 14$

Exercise Set 9.3

1. Multiply the top equation by 2.
Now the top equation contains a $-2x$ and the bottom equation contains a $2x$. When the equations are added together, the variable x will be eliminated.

3. You will obtain a false statement, such as $0 = 6$

5.
$x + y = 6$
$x - y = 4$
Add: $2x = 10$
$x = 5$
Substitute 5 for x in the first equation.
$x + y = 6$
$5 + y = 6$
$y = 1$
The solution is (5, 1).

7.
$x + y = 5$
$-x + y = 1$
Add: $2y = 6$
$y = 3$
Substitute 3 for y in the first equation.
$x + y = 5$
$x + 3 = 5$
$x = 2$
The solution is (2, 3).

9.
$x + 2y = 15$
$x - 2y = -7$
Add: $2x = 8$
$x = 4$
Substitute 4 for x in the first equation.
$x + 2y = 15$
$4 + 2y = 15$
$2y = 11$
$y = \frac{11}{2}$
The solution is $\left(4, \frac{11}{2}\right)$

11.
$4x + y = 6$
$-8x - 2y = 20$
Multiply the first equation by 2, then add to the second equation.
$2[4x + y = 6]$
gives
$8x + 2y = 12$
$-8x - 2y = 20$
$0 = 32$ False
There is no solution.

13. $-5x + y = 14$
$-3x + y = -2$
To eliminate y, multiply the first equation by –1 and then add.
$-1[-5x + y = 14]$
gives
$5x - y = -14$
$-3x + y = -2$
$2x = -16$
$x = -8$
Substitute –8 for x in the first equation.
$-5x + y = 14$
$-5(-8) + y = 14$
$40 + y = 14$
$y = -26$
The solution is (–8, –26).

15. $2x + y = -6$
$2x - 2y = 3$
To eliminate y, multiply the first equation by 2 and then add.
$2[2x + y = -6]$
gives
$4x + 2y = -12$
$2x - 2y = 3$
$6x = -9$
$x = -\frac{3}{2}$

Substitute $-\frac{3}{2}$ for x in the first equation.

$$2x + y = -6$$
$$2\left(-\frac{3}{2}\right) + y = -6$$
$$-3 + y = -6$$
$$y = -3$$

The solution is $\left(-\frac{3}{2}, -3\right)$

17. $2y = 6x + 16$

$y = -3x - 4$

Rewrite the equations to align the variables on the left hand side of the equal sign.

$-6x + 2y = 16$

$3x + y = -4$

To eliminate x, multiply the second equation by 2, then add.

$2[3x + y = -4]$

gives

$$\begin{array}{r} -6x + 2y = 16 \\ \underline{6x + 2y = -8} \\ 4y = 8 \\ y = 2 \end{array}$$

Substitute 2 for y in the second equation.

$$3x + y = -4$$
$$3x + 2 = -4$$
$$3x = -6$$
$$x = -2$$

The solution is (-2, 2).

19. $5x + 3y = 12$

$2x - 4y = 10$

To eliminate y, multiply the first equation by 4 and the second equation by 3 and then add.

$4[5x + 3y = 12]$

$3[2x - 4y = 10]$

gives

$$\begin{array}{r} 20x + 12y = 48 \\ \underline{6x - 12y = 30} \\ 26x \qquad = 78 \\ x = 3 \end{array}$$

Substitute 3 for x in the first equation.

$$5x + 3y = 12$$
$$5(3) + 3y = 12$$
$$15 + 3y = 12$$
$$3y = -3$$
$$y = -1$$

The solution is (3, –1).

21. $4x - 2y = 6$

$y = 2x - 3$

Align x- and y-terms on the left side.

$4x - 2y = 6$

$-2x + y = -3$

To eliminate x, multiply the second equation by 2 and then add.

$2[-2x + y = -3]$

gives

$$\begin{array}{r} 4x - 2y = 6 \\ \underline{-4x + 2y = -6} \\ 0 = 0 \end{array}$$

Since this is a true statement, there are an infinite number of solutions. This is a dependent system.

23. $5x - 4y = -3$

$5y = 2x + 8$

Align x- and y-terms on the left side.

$5x - 4y = -3$

$-2x + 5y = 8$

To eliminate x, multiply the first equation by 2 and the second equation by 5 and then add.

$2[5x - 4y = -3]$

$5[-2x + 5y = 8]$

gives

$$\begin{array}{r} 10x - 8y = -6 \\ \underline{-10x + 25y = 40} \\ 17y = 34 \\ y = 2 \end{array}$$

Substitute 2 for y in the first equation.

$$5x - 4y = -3$$
$$5x - 4(2) = -3$$
$$5x - 8 = -3$$
$$5x = 5$$
$$x = 1$$

The solution is (1, 2).

25. $5x - 4y = 1$

$-10x + 8y = -4$

Multiply the first equation by 2, then add.

$2[5x - 4y = 1]$

gives

$$\begin{array}{r} 10x - 8y = 1 \\ \underline{-10x + 8y = -4} \\ 0 = -3 \end{array}$$ False

There is no solution.

27. $3x - 5y = 0$
$2x + 3y = 0$
To eliminate *y*, multiply the first equation by 3 and the second equation by 5 and then add.
$3[3x - 5y = 0]$
$5[2x + 3y = 0]$
gives
$$\begin{array}{r} 9x - 15y = 0 \\ \underline{10x + 15y = 0} \\ 19x \qquad = 0 \\ x = 0 \end{array}$$
Substitute 0 for *x* in the second equation.
$$\begin{aligned} 2x + 3y &= 0 \\ 2(0) + 3y &= 0 \\ 3y &= 0 \\ y &= 0 \end{aligned}$$
The solution is (0, 0).

29. $-5x + 4y = -20$
$3x - 2y = 15$
To eliminate *y*, multiply the second equation by 2, then add.
$2[3x - 2y = 15]$
gives
$$\begin{array}{r} -5x + 4y = -20 \\ \underline{6x - 4y = \ \ 30} \\ x \qquad = \ \ 10 \end{array}$$
Substitute 10 for *x* in the first equation.
$$\begin{aligned} -5x + 4y &= -20 \\ -5(10) + 4y &= -20 \\ -50 + 4y &= -20 \\ 4y &= 30 \\ y &= \frac{15}{2} \end{aligned}$$
The solution is $\left(10, \frac{15}{2}\right)$

31. $6x = 4y + 12$
$-3y = -5x + 6$
Align the *x*- and *y*-terms on the left side.
$6x - 4y = 12$
$5x - 3y = 6$
To eliminate *x*, multiply the first equation by 5 and the second equation by -6 and then add.
$5(6x - 4y = 12)$
$-6(5x - 3y = 6)$
gives
$$\begin{array}{r} 30x - 20y = 60 \\ \underline{-30x + 18y = -36} \\ -2y = 24 \\ y = -12 \end{array}$$
Substitute -12 for *y* in the first equation.
$$\begin{aligned} 6x - 4y &= 12 \\ 6x - 4(-12) &= 12 \\ 6x + 48 &= 12 \\ 6x &= -36 \\ x &= -6 \end{aligned}$$
The solution is $(-6, -12)$.

33. $4x + 5y = 0$
$3x = 6y + 4$
Align the *x*- and *y*-terms on the left side.
$4x + 5y = 0$
$3x - 6y = 4$
To eliminate *y*, multiply the first equation by 6 and the second equation by 5 and then add.
$6[4x + 5y = 0]$
$5[3x - 6y - 4]$
gives
$$\begin{array}{r} 24x + 30y = 0 \\ \underline{15x - 30y = 20} \\ 39x \qquad = 20 \\ x = \frac{20}{39} \end{array}$$
Substitute $\frac{20}{39}$ for *x* in the first equation.
$$\begin{aligned} 4x + 5y &= 0 \\ 4\left(\frac{20}{39}\right) + 5y &= 0 \\ 5y &= -\frac{80}{39} \\ y &= -\frac{16}{39} \end{aligned}$$
The solution is $\left(\frac{20}{39}, -\frac{16}{39}\right)$

35. $x - \frac{1}{2}y = 4$
$3x + y = 6$
To eliminate *y*, multiply the first equation by 2 and then add.
$2\left[x - \frac{1}{2}y = 4\right]$
gives
$$\begin{array}{r} 2x - y = 8 \\ \underline{3x + y = 6} \\ 5x \quad = 14 \\ x = \frac{14}{5} \end{array}$$
Substitute $\frac{14}{5}$ for *x* in the second equation.

$$3x + y = 6$$
$$3\left(\frac{14}{5}\right) + y = 6$$
$$\frac{42}{5} + y = 6$$
$$y = 6 - \frac{42}{5}$$
$$y = \frac{30}{5} - \frac{42}{5}$$
$$y = -\frac{12}{5}$$

The solution is $\left(\frac{14}{5}, -\frac{12}{5}\right)$

37. $3x - y = 4$

$2x - \frac{2}{3}y = 6$

To eliminate y, multiply the first equation by $-\frac{2}{3}$ and then add.

$-\frac{2}{3}[3x - y = 4]$

gives

$$-2x + \frac{2}{3}y = \frac{8}{3}$$
$$2x - \frac{2}{3}y = 6$$
$$0 = \frac{26}{3} \quad \text{False}$$

There is no solution.

39. $x + y = 16$

$x - y = 8$

$2x = 24$

$x = 12$

Substitute 12 for x in the first equation to find the second number.

$x + y = 16$

$12 + y = 16$

$y = 4$

The numbers are 12 and 4.

41. $x + 2y = 14$

$x - y = 2$

To eliminate y, multiply the second equation by 2 and then add.

$2[x - y = 2]$

gives

$x + 2y = 14$

$2x - 2y = 4$

$3x = 18$

$x = 6$

Substitute 6 for x in the second equation to find the other number.

$x - y = 2$

$6 - y = 2$

$y = 4$

The numbers are 6 and 4.

43. Let x be the length of the rectangle and y be the width.

$2x + 2y = 18$

$x = 2y$

Align the x and y variables on the left hand side of the equal sign.

$2x + 2y = 18$

$x - 2y = 0$

Now add the equations to eliminate the y.

$3x = 18$

$x = 6$

Substitute 6 for x in the second equation and solve for y.

$x = 2y$

$6 = 2y$

$3 = y$

The width is 3 inches and the length is 6 inches.

45. Let l = the length and w = the width of the photograph.

$2l + 2w = 36$

$l - w = 2$

Multiply the second equation by 2

$2[l - w = 2]$

gives

$2l + 2w = 36$

$2l - 2w = 4$

$4l = 40$

$l = 10$

$10 - w = 2$

$w = 8$

The length is 10 inches and the width is 8 inches.

47. Answers will vary.

49. a. $4x + 2y = 1000$

$2x + 4y = 800$

To eliminate x, multiply the first equation by -2.

$-2[4x+2y=1000]$
gives
$-8x-4y=-2000$
$2x+4y=\ 800$
$-6x\ =-1200$
$x=200$
Substitute 200 for x in the first equation.
$4x+2y=1000$
$4(200)+2y=1000$
$800+2y=1000$
$2y=200$
$y=100$
The solution is (200, 100).

b. They will have the same solution. Dividing an equation by a nonzero number does not change the solutions.
$2x+y=500$
$2x+4y=800$
To eliminate x, multiply the first equation by -1 and then add.
$-1[2x+y=500]$
gives
$-2x-y=-500$
$2x+4y=800$
$3y=300$
$y=100$
Substitute 100 for y in the first equation.
$2x+y=500$
$2x+100=500$
$2x=400$
$x=200$
The solution is (200, 100).

51. $\frac{x+2}{2}-\frac{y+4}{3}=4$
$\frac{x+y}{2}=\frac{1}{2}+\frac{x-y}{3}$
Start by writing each equation in standard form after clearing fractions.
For the first equation:
$6\left[\frac{x+2}{2}-\frac{y+4}{3}=4\right]$
$3(x+2)-2(y+4)=24$
$3x+6-2y-8=24$
$3x-2y-2=24$
$3x-2y=26$
For the second equation:
$6\left[\frac{x+y}{2}=\frac{1}{2}+\frac{x-y}{3}\right]$
$3(x+y)=3+2(x-y)$
$3x+3y=3+2x-2y$
$x+5y=3$
The new system is:
$3x-2y=26$
$x+5y=3$
To eliminate x, multiply the second equation by -3 and then add.
$-3[x+5y=3]$
gives
$3x-2y=26$
$-3x-15y=-9$
$-17y=17$
$y=-1$
Now, substitute -1 for y in the equation $x+5y=3$.
$x+5(-1)=3$
$x-5=3$
$x=8$
The solution is (8, −1).

53. $x+2y-z=2$
$2x-y+z=3$
$2x+y+z=7$
Add the second and third equations to eliminate y.
$2x-y+z=3$
$2x+y+z=7$
$4x+2z=10$
Multiply the second equation by 2 and then add the third equation to eliminate y.
$2[2x-y+z=3]$
gives
$4x-2y+2z=6$
$x+\ 2y-z=2$
$5x+z=8$
Now we have two equations with two unknowns.
$4x+2z=10$
$5x+z=8$
To eliminate z, multiply the second equation by -2 and then add.
$-2[5x+z=8]$
gives
$4x+2z=10$
$-10x-2z=-16$
$-6x=-6$
$x=1$
Substitute 1 for x in the second equation and solve for z
$5x+z=8$
$5(1)+z=8$
$5+z=8$
$z=3$
Now go back to one of the original equations and substitute 1 for x and 3 for z and then find y.

The third equation is used below.

$2x + y + z = 7$

$2(1) + y + 3 = 7$

$2 + y + 3 = 7$

$y + 5 = 7$

$y = 2$

The solution to the system is (1, 2, 3).

55. $5^3 = 5 \cdot 5 \cdot 5 = 125$

56. $2(2x - 3) = 2x + 8$

$4x - 6 = 2x + 8$

$4x - 2x = 8 + 6$

$2x = 14$

$x = 7$

57. $(4x^2y - 3xy + y) - (2x^2y + 6xy - 3y)$

$4x^2y - 3xy + y - 2x^2y - 6xy + 3y$

$2x^2y - 9xy + 4y$

58. $(8a^4b^2c)(4a^2b^7c^4) = 8 \cdot 4a^{4+2}b^{2+7}c^{1+4}$

$= 32a^6b^9c^5$

59. $xy + xc - ay - ac$

$x(y + c) - a(y + c)$

$(y + c)(x - a)$

60. $f(x) = 2x^2 - 4$

$f(-3) = 2(-3)^2 - 4$

$= 2(9) - 4$

$= 18 - 4$

$= 14$

$f(-3) = 14$

Exercise Set 9.4

1. The graph will be a plane.

3. $x = 1$

$2x - y = 4$

$-3x + 2y - 2z = 1$

Substitute 1 for x in the second equation.

$2(1) - y = 4$

$2 - y = 4$

$-y = 2$

$y = -2$

Substitute 1 for x and −2 for y in the third equation.

$-3(1) + 2(-2) - 2z = 1$

$-3 - 4 - 2z = 1$

$-7 - 2z = 1$

$-2z = 8$

$z = -4$

The solution is (1, −2, −4).

5. $5x - 6z = -17$

$3x - 4y + 5z = -1$

$2z = -6$

Solve the third equation for z.

$2z = -6$

$z = -3$

Substitute −3 for z in the first equation.

$5x - 6z = -17$

$5x - 6(-3) = -17$

$5x + 18 = -17$

$5x = -35$

$x = -7$

Substitute −7 for x and −3 for z in the second equation.

$3x - 4y + 5z = -1$

$3(-7) - 4y + 5(-3) = -1$

$-21 - 4y - 15 = -1$

$-4y - 36 = -1$

$-4y = 35$

$y = \frac{35}{-4} = -\frac{35}{4}$

The solution is $\left(-7, -\frac{-35}{4}, -3\right)$.

7. $x + 2y = 6$

$3y = 9$

$x + 2z = 12$

Solve the second equation for y.

$3y = 9$

$y = 3$

Substitute 3 for y in the first equation.

$x + 2y = 6$

$x + 2(3) = 6$

$x + 6 = 6$

$x = 0$

Substitute 0 for x in the third equation.

$x + 2z = 12$

$0 + 2z = 12$

$2z = 12$

$z = 6$

The solution is (0, 3, 6).

9. $x-2y=-3$ (1)

$3x+2y=7$ (2)

$2x-4y+z=-6$ (3)

To eliminate y between equations (1) and (2), add equations (1) and (2).

$$\begin{array}{r} x-2y=-3 \\ \underline{3x+2y=7} \\ \text{Add: } 4x=4 \\ x=1 \end{array}$$

Substitute 1 for x in equation (1).

$$(1)-2y=-3$$
$$-2y=-4$$
$$y=2$$

Substitute 1 for x, and 2 for y in equation (3).

$$2(1)-4(2)+z=-6$$
$$2-8+z=-6$$
$$-6+z=-6$$
$$z=0$$

The solution is (1, 2, 0).

11. $2y+4z=6$ (1)

$x+y+2z=0$ (2)

$2x+y+z=4$ (3)

To eliminate x between equations (2) and (3), multiply equation (2) by –2 and then add.

$-2[x+y+2z=0]$

$2x+y+z=4$

gives

$$\begin{array}{r} -2x-2y-4z=0 \\ \underline{2x+y+z=4} \\ \text{Add: } -y-3z=4 \quad (4) \end{array}$$

To eliminate y between equations (1) and (4), multiply equation (4) by 2 and then add.

$2y+4z=6$

$2[-y-3z=4]$

gives

$$\begin{array}{r} 2y+4z=6 \\ \underline{-2y-6z=8} \\ \text{Add: } -2z=14 \\ z=-7 \end{array}$$

Substitute –7 for z in equation (1).

$$2y+4(-7)=6$$
$$2y-28=6$$
$$2y=34$$
$$y=17$$

Substitute 17 for y, and –7 for z in equation (3).

$$2x+(17)+(-7)=4$$
$$2x+10=4$$
$$2x=-6$$
$$x=-3$$

The solution is (–3, 17, –7).

13. $3p+2q=11$ (1)

$4q-r=6$ (2)

$2p+2r=2$ (3)

To eliminate r between equations (2) and (3), multiply equation (2) by 2 and add to equation (3).

$2[4q-r=6]$

$2p+2r=2$

gives

$$\begin{array}{r} 8q-2r=12 \\ \underline{2p+2r=2} \\ \text{Add: } 2p+8q=14 \quad (4) \end{array}$$

Equations (1) and (4) are two equations in two unknowns. To eliminate q, multiply equation (1) by –4 and add to equation (4).

$4[3p+2q=11]$

$2p+8q=14$

gives

$$\begin{array}{r} -12p-8q=-44 \\ \underline{2p+8q=14} \\ \text{Add: } -10p=-30 \\ p=3 \end{array}$$

Substitute 3 for p in equation (1).

$$3p+2q=11$$
$$3(3)+2q=11$$
$$9+2q=11$$
$$2q=2$$
$$q=1$$

Substitute 3 for p in equation (3).

$$2p+2r=2$$
$$2(3)+2r=2$$
$$6+2r=2$$
$$2r=-4$$
$$r=-2$$

The solution is (3, 1, –2).

15. $p+q+4=4$ (1)

$p-2q-r=1$ (2)

$2p-q-2r=-1$ (3)

To eliminate q between equations (1) and (3), simply add.

$$\begin{array}{r} p+q+r=4 \\ \underline{2p-q-2r=-1} \\ \text{Add: } 3p-r=3 \quad (4) \end{array}$$

To eliminate q between equations (1) and (2), multiply equation (1) by 2 and then add.

$2[p+q+r=4]$
$p-2q-r=1$
gives

$$\begin{array}{lrl} & 2p+2q+2r & =8 \\ & p-2q-r & =1 \\ \hline \text{Add:} & 3p+r & =9 \quad (5) \end{array}$$

Equations (4) and (5) are two equations in two unknowns.
$3p-r=3$
$3p+r=9$
To eliminate r, simply add these two equations.

$$\begin{array}{lrl} & 3p-r & = 3 \\ & 3p+r & = 9 \\ \hline \text{Add:} & 6p & =12 \end{array}$$

$p=2$

Substitute 2 for p in equation (5).
$3p+r=9$
$3(2)+r=9$
$6+r=9$
$r=3$
Substitute 2 for p and 3 for r in equation (1).
$p+q+r=4$
$2+q+3=4$
$q+5=4$
$q=-1$
The solution is (2, –1, 3).

17. $2x-2y+3z=5$ (1)
$2x+y-2z=-1$ (2)
$4x-y-3z=0$ (3)
To eliminate y between equations (2) and (3), simply add.

$$\begin{array}{lrl} & 2x+y-2z & =-1 \\ & 4x-y-3z & = 0 \\ \hline \text{Add:} & 6x-5z & =-1 \quad (4) \end{array}$$

To eliminate y between equations (1) and (2), multiply equation (2) by 2 and then add.
$2x-2y+3z=5$
$2[2x+y-2z=-1]$
gives

$$\begin{array}{lrl} & 2x-2y+3z & = 5 \\ & 4x+2y-4z & =-2 \\ \hline \text{Add:} & 6x-z & = 3 \quad (5) \end{array}$$

Equations (4) and (5) are two equations in two unknowns.
$6x-5z=-1$
$6x-z=3$
To eliminate x, multiply equation (5) by –1 and then add.
$6x-5z=-1$
$-1[6x-z=3]$
gives

$$\begin{array}{lrl} & 6x-5z & =-1 \\ & -6x+z & =-3 \\ \hline \text{Add:} & -4z & =-4 \end{array}$$

$z=1$

Substitute 1 for z in equation (5).
$6x-z=3$
$6x-1=3$
$6x=4$
$x=\frac{4}{6}=\frac{2}{3}$
Substitute $\frac{2}{3}$ for x and 1 for z in equation (2).
$2x+y-2z=-1$
$2\left(\frac{2}{3}\right)+y-2(1)=-1$
$\frac{4}{3}+y-2=-1$
$y-\frac{2}{3}=-1$
$y=-1+\frac{2}{3}=-\frac{1}{3}$
The solution is $\left(\frac{2}{3}, -\frac{1}{3}, 1\right)$.

19. $r-2s+t=2$ (1)
$2r+2s-t=-2$ (2)
$2r-s-2t=1$ (3)
To eliminate s between equations (1) and (2), by adding equations (1) and (2).

$$\begin{array}{lrl} & r-2s+t & =2 \\ & 2r+2s-t & =-2 \\ \hline \text{Add:} & 3r & =0 \end{array}$$

$r=0$

To eliminate s between equations (2) and (3), multiply equation (3) by 2 and then add.
$2r+2s-t=-2$
$2[2r-s-2t=1]$
gives

$$\begin{array}{lrl} & 2r+2s-t & =-2 \\ & 4r-2s-4t & =2 \\ \hline \text{Add:} & 6r-5t & =0 \quad (4) \end{array}$$

Substitute 0 for r into equation (4) and solve for t.
$6(0)-5t=0$
$0-5t=0$
$t=0$
Finally, substitute 0 for r and 0 for t into equation (1)

$(0)-2s+(0)=2$
$-2s=2$
$s=-1$
The solution is (0, −1, 0).

21. $2a+2b-c=2$ (1)
$3a+4b+c=-4$ (2)
$5a-2b-3c=5$ (3)
To eliminate c between equations (1) and (2), simply add.
$2a+2b-c=2$
$\underline{3a+4b+c=-4}$
Add: $5a+6b=-2$ (4)
To eliminate c between equations (2) and (3), multiply equation (2) by 3 and then add.
$3[3a+4b+c=-4]$
$5a-2b-3c=5$
gives
$9a+12b+3c=-12$
$\underline{5a-2b-3c=5}$
Add: $14a+10b=-7$ (5)
Equations (4) and (5) are two equations in two unknowns.
$5a+6b=-2$
$14a+10b=-7$
To eliminate b, multiply equation (4) by −5 and multiply equation (5) by 3 and then add.
$-5[5a+6b=-2]$
$3[14a+10b=-7]$
gives
$-25a-30b=10$
$\underline{42a+30b=-21}$
Add: $17a=-11$
$a=-\frac{11}{17}$
Substitute $-\frac{11}{17}$ for a in equation (4).
$5a+6b=-2$
$5\left(-\frac{11}{17}\right)+6b=-2$
$-\frac{55}{17}+6b=-2$
$6b=-2+\frac{55}{17}$
$b=\frac{1}{6}\cdot\frac{21}{17}=\frac{7}{34}$
Substitute $-\frac{11}{17}$ for a and $\frac{7}{34}$ for b in equation (2).
$3a+4b+c=-4$
$3\left(-\frac{11}{17}\right)+4\left(\frac{7}{34}\right)+c=-4$
$-\frac{33}{17}+\frac{14}{17}+c=-4$
$-\frac{19}{17}+c=-4$
$c=-4+\frac{19}{17}$
$c=-\frac{49}{17}$
The solution is $\left(-\frac{11}{17}, \frac{7}{34}, -\frac{49}{17}\right)$.

23. $-x+3y+z=0$ (1)
$-2x+4y-z=0$ (2)
$3x-y+2z=0$ (3)
To eliminate z between equations (1) and (2), simply add.
$-x+3y+z=0$
$\underline{-2x+4y-z=0}$
Add: $-3x+7y=0$ (4)
To eliminate z between equations (2) and (3), multiply equation (2) by 2 and then add.
$2[-2x+4y-z]=0$
$3x-y+2z=0$
gives
$-4x+8y-2z=0$
$\underline{3x-y+2z=0}$
Add: $-x+7y=0$ (5)
Equations (4) and (5) are two equations in two unknowns.
$-3x+7y=0$
$-x+7y=0$
To eliminate y, multiply equation (4) by −1 and then add.
$-1[-3x+7y=0]$
$-x+7y=0$
gives
$3x-7y=0$
$\underline{-x+7y=0}$
Add: $2x=0$
$x=0$
Substitute 0 for x in equation (5).
$-x+7y=0$
$-0+7y=0$
$7y=0$
$y=0$
Finally, substitute 0 for x and 0 for y into equation (1).

$$
\begin{aligned}
-x+3y+z&=0\\
-0+3(0)+z&=0\\
0+z&=0\\
z&=0
\end{aligned}
$$

The solution is (0, 0, 0).

25. $-\frac{1}{4}x+\frac{1}{2}y-\frac{1}{2}z=-2$ (1)

$\frac{1}{2}x+\frac{1}{3}y-\frac{1}{4}z=2$ (2)

$\frac{1}{2}x-\frac{1}{2}y+\frac{1}{4}z=1$ (3)

To clear fractions, multiply equation (1) by 4, equation (2) by 12, and equation (3) by 4.

$4\left(-\frac{1}{4}x+\frac{1}{2}y-\frac{1}{2}z=-2\right)$

$12\left(\frac{1}{2}x+\frac{1}{3}y-\frac{1}{4}z=2\right)$

$4\left(\frac{1}{2}x-\frac{1}{2}y+\frac{1}{4}z=1\right)$

gives

$-x+2y-2z=-8$ (4)

$6x+4y-3z=24$ (5)

$2x+2y+z=4$ (6)

To eliminate y between equations (4) and (6), simply add.

$$
\begin{aligned}
-x+2y-2z&=-8\\
\underline{2x-2y+\ \ z}&\underline{=\ \ 4}\\
\text{Add: } x\qquad\ -z&=-4 \quad (7)
\end{aligned}
$$

To eliminate y between equations (5) and (6), multiply equation (6) by 2 and then add to equation (5).

$6x+4y-3z=24$

$2[2x-2y+z=4]$

gives

$$
\begin{aligned}
6x+4y-3z&=24\\
\underline{4x-4y+2z}&\underline{=\ \ 8}\\
\text{Add: } 10x\qquad\ -z&=32 \quad (8)
\end{aligned}
$$

Equations (7) and (8) are two equations in two unknowns.

$x-z=-4$

$10x-z=32$

To eliminate z, multiply equation (7) by –1 and then add.

$-1[x-z=-4]$

$10x-z=32$

gives

$$
\begin{aligned}
-x+z&=\ 4\\
\underline{10x-z}&\underline{=32}\\
\text{Add: } 9x\quad&=36\\
x&=\frac{36}{9}=4
\end{aligned}
$$

Substitute 4 for x in equation (7).

$$
\begin{aligned}
x-z&=-4\\
4-z&=-4\\
-z&=-8\\
z&=8
\end{aligned}
$$

Finally, substitute 4 for x and 8 for z in equation (4).

$$
\begin{aligned}
-x+2y-2z&=-8\\
-4+2y-2(8)&=-8\\
-4+2y-16&=-8\\
2y-20&=-8\\
2y&=12\\
y&=\frac{12}{2}=6
\end{aligned}
$$

The solution is (4, 6, 8).

27. $x-\frac{2}{3}y-\frac{2}{3}z=-2$ (1)

$\frac{2}{3}x+y-\frac{2}{3}z=\frac{1}{3}$ (2)

$-\frac{1}{4}x+y-\frac{1}{4}z=\frac{3}{4}$ (3)

To clear fractions, multiply equation (1) by 3, equation (2) by 3, and equation (3) by 4. The resulting system is

$3\left(x-\frac{2}{3}y-\frac{2}{3}z=-2\right)$

$3\left(\frac{2}{3}x+y-\frac{2}{3}z=\frac{1}{3}\right)$

$4\left(-\frac{1}{4}x+y-\frac{1}{4}z=\frac{3}{4}\right)$

gives

$3x-2y-2z=-6$ (4)

$2x+3y-2z=1$ (5)

$-x+4y-z=3$ (6)

To eliminate x between equations (4) and (6), multiply equation (6) by 3 and then add.

$3x-2y-2z=-6$

$3[-x+4y-z=3]$

gives

$$
\begin{aligned}
3x-2y-2z&=-6\\
\underline{-3x+12y-3z}&\underline{=\ \ 9}\\
\text{Add: }\quad 10y-5z&=\ 3 \quad (7)
\end{aligned}
$$

To eliminate x between equations (5) and (6), multiply equation (6) by 2 and then add.

$2x+3y-2z=1$

$2[-x+4y-z=3]$

gives

$2x+3y-2z=1$
$-2x+8y-2z=6$
Add: $11y-4z=7$ (8)

Equations (7) and (8) are two equations in two unknowns.
$10y-5z=3$
$11y-4z=7$

To eliminate z, multiply equation (7) by –4 and equation (8) by 5 and then add.
$-4[10y-5z=3]$
$5[11y-4z=7]$
gives
$-40y+20z=-12$
$55y-20z=35$
Add: $15y=23$

$$y=\frac{23}{15}$$

Substitute $\frac{23}{15}$ for y into equation (7).

$$10y-5z=3$$
$$10\left(\frac{23}{15}\right)-5z=3$$
$$\frac{46}{3}-5z=3$$
$$-5z=3-\frac{46}{3}$$
$$-5z=-\frac{37}{3}$$
$$z=\left(-\frac{1}{5}\right)\left(-\frac{37}{3}\right)=\frac{37}{15}$$

Substitute $\frac{23}{15}$ for y and $\frac{37}{15}$ for z in equation (6).

$$-x+4y-z=3$$
$$-x+4\left(\frac{23}{15}\right)-\frac{37}{15}=3$$
$$-x+\frac{92}{15}-\frac{37}{15}=3$$
$$-x+\frac{55}{15}=3$$
$$-x+\frac{11}{3}=3$$
$$-x=3-\frac{11}{3}$$
$$-x=-\frac{2}{3}$$
$$x=\frac{2}{3}$$

The solution is $\left(\frac{2}{3},\frac{23}{15},\frac{37}{15}\right)$.

29. Multiply each equation by 10.
$10(0.2x+0.3y+0.3z=1.1)$
$10(0.4x-0.2y+0.1z=0.4)$
$10(-0.1x-0.1y+0.3z=0.4)$
gives
$2x+3y+3z=11$ (1)
$4x-2y+z=4$ (2)
$-x-y+3z=4$ (3)

To eliminate multiply equation (2) by –3 and then add.
$2x+3y+3z=11$
$3[4x-2y+z=4]$
gives
$2x+3y+3z=11$
$-12x+6y-3z=-12$
Add: $-10x+9y=-1$ (4)

To eliminate z between equations (1) and (3) multiply equation (1) by –1 and then add.
$-1[2x+3y+3z=11]$
$-x-y+3z=4$
gives
$-2x-3y-3z=-11$
$-x-y+3z=4$
Add: $-3x-4y=-7$ (5)

Equations (4) and (5) are two equations in two unknowns.
$-10x+9y=-1$
$-3x-4y=-7$

To eliminate y, multiply equation (4) by –3 and equation (5) by 10.
$-3[-10+9y=-1]$
$10[-3x-4y=-7]$
gives
$30x-27y=3$
$-30x-40y=-70$
Add: $-67y=-67$
$y=1$

Substitute 1 for y in equation (4).
$-10x+9y=-1$
$-10x+9(1)=-1$
$-10x=-10$
$x=1$

Substitute 1 for x and 1 for y in equation (1).
$2x+3y+3z=11$
$2(1)+3(1)+3z=11$
$5+3z=11$
$3z=6$
$z=2$

The solution is (1, 1, 2).

31. $2x+y+2z=1$ (1)
$x-2y-z=0$ (2)
$3x-y+z=2$ (3)

To eliminate z between equations (2) and (3),

simply add.

$$\begin{aligned} x - 2y - z &= 0 \\ 3x - y + z &= 2 \\ \hline \text{Add: } 4x - 3y \quad &= 2 \quad (4) \end{aligned}$$

To eliminate z between equations (1) and (2), multiply equation (2) by 2 and then add.

$2x + y + 2z = 1$

$2[x - 2y - z = 0]$

gives

$$\begin{aligned} 2x + y + 2z &= 1 \\ 2x - 4y - 2z &= 0 \\ \hline \text{Add: } 4x - 3y \quad &= 1 \quad (5) \end{aligned}$$

Equations (4) and (5) are two equations in two unknowns.

$4x - 3y = 2$

$4x - 3y = 1$

To eliminate x, multiply equation (4) by -1 and then add.

$-1[4x - 3y = 2]$

$4x - 3y = 1$

gives

$$\begin{aligned} -4x + 3y &= -2 \\ 4x - 3y &= 1 \\ \hline \text{Add: } 0 &= -1 \quad \text{False} \end{aligned}$$

Since this is a false statement, there is no solution and the system is inconsistent.

33.

$$\begin{aligned} x - 4y - 3z &= -1 \quad (1) \\ 2x - 10y - 7z &= 5 \quad (2) \\ -3x + 12y + 9z &= 3 \quad (3) \end{aligned}$$

To eliminate x between equations (1) and (2), multiply equation (1) by -2 and then add.

$-2[x - 4y - 3z = -1]$

$2x - 10y - 7z = 5$

gives

$$\begin{aligned} -2x + 8y + 6z &= 2 \\ 2x - 10y - 7z &= 5 \\ \hline \text{Add: } -2y - z &= 7 \quad (4) \end{aligned}$$

To eliminate x between equations (1) and (3), multiply equation (1) by 3 and then add.

$3[x - 4y - 3z = -1]$

$-3x + 12y + 9z = 3$

gives

$$\begin{aligned} 3x - 12y - 9z &= -3 \\ -3x + 12y + 9z &= 3 \\ \hline \text{Add: } 0 &= 0 \end{aligned}$$

Since $0 = 0$ is a true statement, the system is dependent and therefore has infinitely many solutions.

35.

$$\begin{aligned} x + 3y + 2z &= 6 \quad (1) \\ x - 2y - z &= 8 \quad (2) \\ -3x - 9y - 6z &= -4 \quad (3) \end{aligned}$$

To eliminate x between equations (1) and (3), multiply equation (1) by 3 and then add.

$3[x + 3y + 2z = 6]$

$-3x - 9y - 6z = -4$

gives

$$\begin{aligned} 3x + 9y + 6z &= 18 \\ -3x - 9y - 6z &= -4 \\ \hline \text{Add: } 0 &= 14 \end{aligned}$$

Since $0 = 14$ is a false statement, the system is inconsistent.

37. No point is common to all three planes. Therefore, the system is inconsistent.

39. One point is common to all three planes. There is one solution and the system is consistent.

41. **a.** Yes, if two or more of the planes are parallel, there will be no solution.

b. Yes, three planes may intersect at a single point.

c. No, the possibilities are no solution, one solution, or infinitely many solutions.

43. $Ax + By + Cz = -2$

Substitute $(-1, 2, -1)$, $(-1, 1, 2)$, and $(1, -2, 2)$ into the equation forming three equations in the three unknowns A, B, and C.

$A(-1) + B(2) + C(-1) = 1$

$A(-1) + B(1) + C(2) = 1$

$A(1) + B(-2) + C(2) = 1$

gives

$$\begin{aligned} -A + 2B - C &= 1 \quad (1) \\ -A + B + 2C &= 1 \quad (2) \\ A - 2B + 2C &= 1 \quad (3) \end{aligned}$$

To eliminate A between equations (1) and (2), multiply equation (2) by -1 and then add.

$-A + 2B - C = 1$

$-1[-A + B + 2C = 1]$

gives

$$\begin{aligned} -A + 2B - C &= 1 \\ A - B - 2C &= -1 \\ \hline \text{Add: } B - 3C &= 0 \quad (4) \end{aligned}$$

To eliminate A between equations (1) and (3), simply add.

$$\begin{aligned} -A+2B-C&=1 \\ \underline{A-2B+2C}&\underline{=1} \end{aligned}$$

Add: $C=2$

Substitute 2 for C in equation (4).

$B-3(2)=0$

$B-6=0$

$B=6$

Substitute 6 for B and 2 for C in equation (23).

$A-2(6)+2(2)=1$

$A-12+4=1$

$A-8=1$

$A=9$

Therefore, $A=9$, $B=6$, $C=2$. and the equation is $9x+6y+2z=1$.

45. One example is

$x+y+z=10$

$x+2y+z=11$

$x+y+2z=16$

Choose coefficients for x, y, and z, then use the given coordinates to find the constants.

47. a. $y=ax^2+bx+c$

For the point $(1,-1)$, let $y=-1$ and $x=1$.

$-1=a(1)^2+b(1)+c$

$-1=a+b+c$ (1)

For the point $(-1,-5)$, let $y=-5$ and $x=-1$.

$-5=a(-1)^2+b(-1)+c$

$-5=a-b+c$ (2)

For the point (3, 11), let $y=11$ and $x=3$.

$11=a(3)^2+b(3)+c$

$11=9a+3b+c$ (3)

Equations (1), (2), and (3) give us a system of three equations.

$a+b+c=-1$ (1)

$a-b+c=-5$ (2)

$9a+3b+c=11$ (3)

To eliminate a and c between equations (1) and (2) multiply equation (2) by -1 and then add.

$a+b+c=-1$

$-1[a-b+c=-5]$

gives

$$\begin{aligned} a+b+c&=-1 \\ \underline{-a+b-c}&\underline{=\ 5} \end{aligned}$$

Add: $2b=4$

$b=2$

Substitute 2 for b in equations (1) and (3).

Equation (1) becomes

$a+b+c=-1$

$a+2+c=-1$

$a+c=-3$ (4)

Equation (3) becomes

$9a+3b+c=11$

$9a+3(2)+c=11$

$9a+c=5$ (5)

Equations (4) and (5) are two equations in two unknowns. To eliminate c, multiply equation (4) by -1 and then add.

$-1[a+c=-3]$

$9a+c=5$

gives

$$\begin{aligned} -a-c&=3 \\ \underline{9a+c}&\underline{=5} \end{aligned}$$

Add: $8a=8$

$a=1$

Finally, substitute 1 for a in equation (4).

$a+c=-3$

$1+c=-3$

$c=-4$

Thus, $a=1$, $b=2$, and $c=-4$.

b. The quadratic equation is $y=x^2+2x-4$. This is the equation determined by the values found in part a.

49.

$3p+4q=11$ (1)

$2p+r+s=9$ (2)

$q-s=-2$ (3)

$p+2q-r=2$ (4)

To eliminate r between equations (2) and (4), simply add.

$$\begin{aligned} 2p\qquad +r+s&=9 \\ \underline{p+2q-r\qquad}&\underline{=2} \end{aligned}$$

Add: $3p+2q+s=11$ (5)

To eliminate s between equations (3) and (5), simply add.

$$\begin{aligned} q-s&=-2 \\ \underline{3p+2q+s}&\underline{=11} \end{aligned}$$

Add: $3p+3q=9$ (6)

Equations (1) and (6) give us a system of two equations in two unknowns.

To eliminate p, multiply equation (6) by -1 and then add.

$3p+4q=11$

$-1[3p+3q=9]$

gives

$$\begin{aligned} 3p + 4q &= 11 \\ -3p - 3q &= -9 \\ \hline \text{Add:} \quad q &= 2 \end{aligned}$$

Substitute 2 for q in equation (3).

$q - s = -2$
$2 - s = -2$
$-s = -4$
$s = 4$

Substitute 2 for q in equation (1).

$3p + 4q = 11$
$3p + 4(2) = 11$
$3p + 8 = 11$
$3p = 3$
$p = 1$

Finally, substitute 1 for p and 4 for s in equation (2).

$2p + r + s = 9$
$2(1) + r + 4 = 9$
$r + 6 = 9$
$r = 3$

The solution is (1, 2, 3, 4).

51. **a.** commutative property of addition
b. associative property of multiplication
c. distributive property

52. Let t be the time for Margie.
Then, $t - \frac{1}{6}$ is the time for David.

	rate	time	distance
David	5	$t - \frac{1}{6}$	$5\left(t - \frac{1}{6}\right)$
Margie	3	t	$3t$

a. The distances traveled are the same.

$5\left(t - \frac{1}{6}\right) = 3t$

$5t - \frac{5}{6} = 3t$

$5t = 3t + \frac{5}{6}$

$2t = \frac{5}{6}$

$t = \frac{1}{2}\left(\frac{5}{6}\right) = \frac{5}{12}$ hr

or $\frac{5}{12}(60) = 25$ min

b. The distance is

$3t = 3\left(\frac{5}{12}\right) = \frac{15}{12} = 1\frac{1}{4}$

or 1.25 miles.

53. $3x + 4 = -(x - 6)$

$3x + 4 = -x + 6$

$4x = 2$

$x = \frac{2}{4} = \frac{1}{2}$

54. Let w = the width of the rectangle then the length $l = 2w + 2$.

$2l + 2w = \text{perimeter}$
$2(2w + 2) + 2w = 22$
$4w + 4 + 2w = 22$
$6w + 4 = 22$
$6w = 18$
$w = 3$

The width is 3 ft and the length is 2(3) + 2 = 8 ft.

Exercise Set 9.5

1. Let x = number who visited Disneyland
y = number who visited Magic Kingdom

$x + y = 27.1$

$y = x + 2.5$

Substitute $x + 2.5$ for y in the first equation.

$x + (x + 2.5) = 27.1$

$2x + 2.5 = 27.1$

$2x = 24.6$

$x = 12.3$

Substitute 12.3 for x in the second equation.

$y = (12.3) + 2.5$

$y = 14.8$

14.8 million people visited Magic Kingdom and 12.3 million people visited Disneyland.

3. Let F = grams of fat in fries
H = grams of fat in hamburger

$F = 3H + 4$

$F - H = 46$

Substitute $3H + 4$ for F in the second equation.

$F - H = 46$

$3H + 4 - H = 46$

$2H = 42$

$H = 21$

Substitute 21 for H in the first equation.

$F = 3H + 4$
$F = 3(21) + 4$
$F = 63 + 4$
$F = 67$
The hamburger has 21 grams of fat and the fries have 67 grams of fat.

5. Let x be the measure of the larger angle and y be the measure of the smaller angle.
$x + y = 90$
$x = 2y + 15$
Substitute $2y + 15$ for x in the first equation.
$x + y = 90$
$2y + 15 + y = 90$
$3y + 15 = 90$
$3y = 75$
$y = 25$
Now, substitute 25 for y in the second equation.
$x = 2y + 15$
$x = 2(25) + 15$
$x = 50 + 15$
$x = 65$
The two angles measure 25° and 65°.

7. Let A and B be the measures of the two angles.
$A + B = 180$
$A = 3B - 28$
Substitute $3B - 28$ for A in the first equation.
$A + B = 180$
$3B - 28 + B = 180$
$4B - 28 = 180$
$4B = 208$
$B = 52$
Now substitute 52 for B in the second equation.
$A = 3B - 28$
$A = 3(52) - 28$
$A = 128$
The two angles measure 52° and 128°.

9. Let t = team's rowing speed in still water
c = speed of current
$t + c = 15.6$
$t - c = 8.8$
Add the equations to eliminate variable c.
$t + c = 15.6$
$\underline{t - c = 8.8}$
$2t = 24.4$
$t = 12.2$
Substitute 12.2 for t in the first equation.
$(12.2) + c = 15.6$
$c = 3.4$
The team's speed in still air is 12.2 mph and the speed of the current is 3.4 mph.

11. Let x be the weekly salary and y be the commission rate.
$x + 4000y = 660$
$x + 6000y = 740$
Multiply the first equation by -1 and then add.
$-1[x + 4000y = 660]$
$x + 6000y = 740$
gives
$-x - 4000y = -660$
$\underline{x + 6000y = 740}$
Add: $2000y = 80$
$y = \frac{80}{2000} = 0.04$
Substitute 0.04 for y in the first equation.
$x + 4000y = 660$
$x + 4000(0.04) = 660$
$x + 160 = 660$
$x = 500$ dollars
Her weekly salary is \$500 and the commission rate is 4%.

13. Let x be the amount of 5% solution and y be the amount of 30% solution.
$x + y = 3$
$0.05x + 0.30y = 0.20(3)$
Solve the first equation for x.
$x = 3 - y$
Substitute $3 - y$ for x in the second equation.
$0.05x + 0.30y = 0.20(3)$
$0.05(3 - y) + 0.30y = 0.6$
$0.15 - 0.05y + 0.30y = 0.6$
$0.25y = 0.45$
$y = 1.8$
Substitute 1.8 for y in the first equation.
$x + y = 3$
$x + 1.8 = 3$
$x = 1.2$
Pola should mix 1.2 ounces of the 5% solution with 1.8 ounces of the 30% solution.

15. Let x = gallons of concentrate (18% solution) and y = gallons of water (0% solution).
$x + y = 200$
$0.18x + 0y = 0.009(200)$
Solve the second equation for x.
$0.18x + 0y = 0.009(200)$
$0.18x = 1.8$
$x = 10$
Substitute 10 for x in the first equation.
$x + y = 200$
$10 + y = 200$
$y = 190$

The mixture should contain 10 gallons of concentrate and 190 gallons of water.

17. Let x = pounds of birdseed and y = pounds of sunflower seeds

$0.59x + 0.89y = 0.76(40)$

$x + y = 40$

Solve the second equation for y.

$y = 40 - x$

Substitute $40 - x$ for y in the first equation.

$0.59x + 0.89y = 0.76(40)$

$0.59x + 0.89(40 - x) = 30.4$

$0.59x + 35.6 - 0.89x = 30.4$

$-0.3x = -5.2$

$x = 17\frac{1}{3}$

Substitute $17\frac{1}{3}$ for x in the second equation.

$x + y = 40$

$17\frac{1}{3} + y = 40$

$y = 22\frac{2}{3}$

Angela Leinenbach should mix $17\frac{1}{3}$ pounds of birdseed at \$0.59 per pound with $22\frac{2}{3}$ pounds of sunflower seeds at \$0.89 per pound.

19. Let x be the number of the \$4.00 adult tickets sold and y be the number of the \$1.50 children's tickets sold.

$x + y = 225$

$4.00x + 1.50y = 500$

Solve the first equation for x.

$x + y = 225$

$x = 225 - y$

Substitute $225 - y$ for x in the second equation.

$4.00(225 - y) + 1.50y = 500$

$900 - 4.00y + 1.50y = 500$

$-2.50y = -400$

$y = 160$

Substitute 160 for y in the equation $x = 225 - y$.

$x = 225 - (160)$

$x = 65$

65 adult tickets and 160 children's tickets were sold.

20. Let R = orders of regular wings
J = orders of jumbo wings

$5.99R + 8.99J = 1024.66$

$R + J = 134$

Solve the second equation for R.

$R = 134 - J$

Substitute $134 - J$ for R in the first equation.

$5.99R + 8.99J = 1024.66$

$5.99(134 - J) + 8.99J = 1024.66$

$802.66 - 5.99J + 8.99J = 1024.66$

$3J = 222$

$J = 74$

Substitute 74 for J in the second equation

$R + J = 134$

$R + 74 = 134$

$R = 60$

The Wing House sold 60 regular orders and 74 jumbo orders.

21. Let x = amount invested at 5% and y = amount invested at 6%

$x + y = 10,000$

$0.05x + 0.06y = 540$

Solve the first equation for y.

$y = 10,000 - x$

Substitute $10,000 - x$ for y in the second equation.

$0.05x + 0.06y = 540$

$0.05x + 0.06(10,000 - x) = 540$

$0.05x + 600 - 0.06x = 540$

$-0.01x = -60$

$x = 6000$

Substitute 6000 for x in the first equation.

$x + y = 10,000$

$6000 + y = 10,000$

$y = 4000$

Mr. and Mrs. McAdams invested \$6000 at 5% and \$4000 at 6%.

23. Let x be the amount of the whole milk (3.25% fat) and y be the amount of the skim milk (0% fat).

$x + y = 260$

$0.0325x + 0y = 0.02(260)$

Solve the second equation for x.

$0.0325x + 0y = 0.02(260)$

$0.0325x = 5.2$

$x = 160$

Now substitute 160 for x in the first equation.

$(160) + y = 260$

$y = 100$

Becky needs to mix 160 gallons of the whole milk with 100 gallons of skim milk to produce 100 gallons of 2% fat milk.

25. Let x = pounds of *Season's Choice* birdseed at \$1.79/lb and y = pounds of *Garden Mix* birdseed at \$1.19/lb

$1.79x + 1.19y = 28.00$

$x + y = 20$

Solve the second equation for *y*.

$y = 20 - x$

Substitute $20 - x$ for *y* in the first equation.

$1.79x + 1.19(20 - x) = 28.00$

$1.79x + 23.80 - 1.19x = 28.00$

$0.60x = 4.20$

$x = 7$

Substitute 7 for *x* in the second equation.

$(7) + y = 20$

$y = 13$

The class should buy 7 pounds *of Season's Choice* and 13 pounds of *Garden Mix.*

27. Let *x* be the rate of the slower car and *y* the rate of the faster car.

$4x + 4y = 420$

$y = x + 5$

Substitute $x + 5$ for *y* in the first equation.

$4x + 4y = 420$

$4x + 4(x + 5) = 420$

$4x + 4x + 20 = 420$

$8x + 20 = 420$

$8x = 400$

$x = 50$

Now substitute 50 for *x* in the second equation.

$y = x + 5$

$y = 50 + 5$

$y = 55$

The rate of the slower car is 50 mph and the rate of the faster car is 55 mph.

29. Let *x* be the amount of time traveled at 65 mph and *y* be the amount of time traveled at 50 mph.

$x + y = 11.4$

$65x + 50y = 690$

Solve the first equation for *x*

$x = 11.4 - y$

Now substitute $11.4 - y$ into *x* in the second equation.

$65(11.4 - y) + 50y = 690$

$741 - 65y + 50y = 690$

$-15y = -51$

$y = 3.4$

Substitute 3.4 for *y* in the first equation.

$x + (3.4) = 11.4$

$x = 8$

Cabrina traveled for 8 hours at 65 mph and Dabney traveled for 3.4 hours at 50 mph.

31. Let *x* the number of grams of Mix *A* and *y* be the number of grams of Mix *B*.

$0.1x + 0.2y = 20$

$0.06x + 0.02y = 6$

To solve, multiply the second equation by -10 and then add.

$0.1x + 0.2y = 20$

$-10[0.06x + 0.02y = 6]$

gives

$$\begin{aligned} 0.1x + 0.2y &= 20 \\ -0.6x - 0.2y &= -60 \\ \hline \text{Add: } -0.5x \quad &= -40 \end{aligned}$$

$$x = \frac{-40}{-0.5} = 80$$

Now substitute 80 for *x* in the first equation.

$0.1x + 0.2y = 20$

$0.1(80) + 0.2y = 20$

$8 + 0.2y = 20$

$0.2y = 12$

$$y = \frac{12}{0.2} = 60$$

The scientist should feed each animal 80 grams of Mix *A* and 60 grams of Mix *B*.

33. Let *x* be the amount of the first alloy and *y* be the amount of the second alloy.

$0.7x + 0.4y = 0.6(300)$

$0.3x + 0.6y = 0.4(300)$

To solve, multiply the first equation by 3 and the second equation by -2 and then add.

$3[0.7x + 0.4y = 0.6(300)]$

$-2[0.3x + 0.6y = 0.4(300)]$

gives

$$\begin{aligned} 2.1x + 1.2y &= 540 \\ -0.6x - 1.2y &= -240 \\ \hline \text{Add: } 1.5x \quad &= 300 \end{aligned}$$

$$x = \frac{300}{1.5} = 200$$

Now substitute 200 for *x* in the first equation.

$0.7x + 0.4y = 0.6(300)$

$0.7(200) + 0.4y = 0.6(300)$

$140 + 0.4y = 180$

$0.4y = 40$

$$y = \frac{40}{0.4} = 100$$

200 grams of the first alloy should be combined with 100 grams of the second alloy to produce the desired mixture.

35. Let x = speed of Melissa's car and y = speed of Tom's car

$x = y + 15$

$$\frac{150}{x} = \frac{120}{y}$$

Substitute $y+15$ for x in the second equation.

$$\frac{150}{y+15}=\frac{120}{y}$$
$$150y = 120y + 1800$$
$$30y = 1800$$
$$y = 60$$

Substitute 60 for y in the first equation.

$$x = y + 15$$
$$x = 60 + 15$$
$$x = 75$$

Tom traveled at 60 mph and Melissa traveled at 75 mph.

37. $E(t) = 3.62t + 12.6$

$P(t) = -3.62t + 87.4$

$3.62t + 12.6 = -3.62t + 87.4$

$7.24t = 74.8$

$t \approx 10.3$

They will be equal approximately 10 years after 1996 or in 2006.

39. a. Let c = cost
and m = minutes
Plan 1: $c = 0.05m + 8.95$
Plan 2: $c = 0.07m + 5.95$

b.

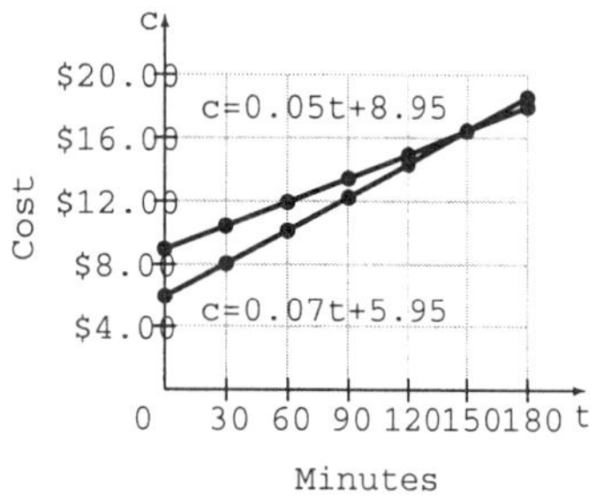

c. The cost is the same at about 150 minutes.

d.
$$0.05m + 8.95 = 0.07m + 5.95$$
$$-0.02m = -3.00$$
$$m = 150$$

41. a. Let x = pieces of personal mail
y = number of bills and statements
z = number of advertisements

$$x + y + z = 24$$
$$y = 2x - 2$$
$$z = 5x + 2$$

b. Substitute $2x - 2$ for y and $5x + 2$ for z in the first equation.

$$x + y + z = 24$$
$$x + 2x - 2 + 5x + 2 = 24$$
$$8x = 24$$
$$x = 3$$

Substitute 3 for x in the second and third equations.

$y = 2x - 2$	$z = 5x + 2$
$y = 2(3) - 2$	$z = 5(3) + 2$
$y = 4$	$z = 17$

An average American household receives 3 pieces of personal mail, 4 bills and statements, and 17 advertisements per week.

43. a. Let x = number of land mines in Iraq
y = number of land mines in Angola
z = number of land mines in Iran

$$x + y + z = 41$$
$$z = 3x - 14$$
$$y = 2x - 5$$

b. Substitute $3x - 14$ for z and $2x - 5$ for y in the first equation.

$$x + y + z = 41$$
$$x + 2x - 5 + 3x - 14 = 41$$
$$6x = 60$$
$$x = 10$$

Substitute 10 for x in the second and third equations.

$z = 3x - 14$	$y = 2x - 5$
$z = 3(10) - 14$	$y = 2(10) - 5$
$z = 16$	$y = 15$

The number of land mines in the countries is
Iraq: 10 million
Angola: 15 million
Iran: 16 million.

45. a. Let x, y, and z be the measures of the three angles.

$$x + y + z = 180$$
$$x = \frac{2}{3}y$$
$$z = 3y - 30$$

b. Substitute $\frac{2}{3}y$ for x and $3y - 30$ for z in the first equation.

$$x+y+z=180$$
$$\frac{2}{3}y+y+3y-30=180$$
$$\frac{14}{3}y-30=180$$
$$\frac{14}{3}y=210$$
$$\frac{3}{14}\left(\frac{14}{3}y\right)=\frac{3}{14}(210)$$
$$y=45$$

Substitute 45 for y in the second equation.

$$x=\frac{2}{3}y$$
$$x=\frac{2}{3}(45)$$
$$x=30$$

Substitute 45 for y in the third equation.

$$z=3y-30$$
$$z=3(45)-30$$
$$z=135-30$$
$$z=105$$

The three angles are 30°, 45°, and 105°.

47. a. Let x be the amount invested at 3%, y be the amount invested at 5%, and z be the amount invested at 6%.

$$y=2x \quad (1)$$
$$x+y+z=10{,}000 \quad (2)$$
$$0.03x+0.05y+0.06z=525 \quad (3)$$

b. Substitute $2x$ for y in equation (2).

$$x+y+z=10{,}000$$
$$x+2x+z=10{,}000$$
$$3x+z=10{,}000 \quad (4)$$

Substitute $2x$ for y in equation (3).

$$0.03x+0.05y+0.06z=525$$
$$0.03x+0.05(2x)+0.06z=525$$
$$0.03x+0.10x+0.06z=525$$
$$0.13x+0.06z=525 \quad (5)$$

Equations (4) and (5) are a system of two equations in two unknowns.

$$3x+z=10{,}000$$
$$0.13x+0.06z=525$$

To eliminate z, multiply equation (4) by –3 and equation (5) by 50 and add.

$$-3[3x+z=10{,}000]$$
$$50[0.13x+0.06x=525]$$

gives

$$-9x-3z=-30{,}000$$
$$6.5x+3z=26{,}250$$

Add: $-2.5x=-3750$

$$x=\frac{-3750}{-2.5}=1500$$

Substitute 1500 for x in equation (4).

$$3x+z=10{,}000$$
$$3(1500)+z=10{,}000$$
$$4500+z=10{,}000$$
$$z=5500$$

Substitute 1500 for x and 5500 for z in equation (2).

$$x+y+z=10{,}000$$
$$1500+y+5500=10{,}000$$
$$y+7000=10{,}000$$
$$y=3000$$

Marion invested \$1500 at 3%, \$3000 at 5%, and \$5500 at 6%.

49. a. Let x be the amount of the 10% solution, y be the amount of the 12% solution, and z be the amount of the 20% solution.

$$x+y+z=8 \quad (1)$$
$$0.10x+0.12y+0.20z=(0.13)8 \quad (2)$$
$$z=x-2 \quad (3)$$

b. Substitute $x-2$ for z in equation (1).

$$x+y+z=8$$
$$x+y+(x-2)=8$$
$$2x+y-2=8$$
$$2x+y=10 \quad (4)$$

Substitute $x-2$ for z in equation (2).

$$0.10x+0.12y+0.20z=(0.13)8$$
$$0.10x+0.12y+0.20(x-2)=(0.13)8$$
$$0.10x+0.12y+0.20x-0.40=1.04$$
$$0.30x+0.12y=1.44 \quad (5)$$

Equations (4) and (5) are a system of two equations in two unknowns.

$$2x+y=10$$
$$0.30x+0.12y=1.44$$

To solve, multiply equation (5) by 100 and equation (4) by –12 and then add.

$$-12[2x+y=10]$$
$$100[0.30x+0.12y=1.44]$$

gives.

$$-24x-12y=-120$$
$$30x+12y=144$$

Add: $6x=24$

$$x=4$$

Substitute 4 for x in equation (4).

$$2x+y=10$$
$$2(4)+y=10$$
$$8+y=10$$
$$y=2$$

Finally, substitute 4 for x in equation (3).

$$z=x-2$$
$$z=4-2$$
$$z=2$$

The mixture consists of 4 liters of the 10% solution, 2 liters of the 12% solution, and 2 liters of the 20% solution.

51. **a.** Let x be the number of children's chairs, y be the number of standard chairs, and z be the number of executive chairs.

$$5x + 4y + 7z = 154 \quad (1)$$
$$3x + 2y + 5 = 94 \quad (2)$$
$$2x + 2y + 4z = 76 \quad (3)$$

b. To eliminate y between equations (1) and (2), multiply equation (2) by −2 and add.

$$5x + 4y + 7z = 154$$
$$-2[3x + 2y + 5z = 94]$$

gives

$$\begin{array}{rl} & 5x + 4y + 7z = 154 \\ & -6x - 4y - 10z = -188 \\ \hline \text{Add:} & -x \quad - 3z = -34 \quad (4) \end{array}$$

To eliminate y between equations (2) and (3), multiply equation (3) by −1 and add.

$$3x + 2y + 5z = 94$$
$$-1[2x + 2y + 4z = 76]$$

gives

$$\begin{array}{rl} & 3x + 2y + 5z = 94 \\ & -2x - 2y - 4z = -76 \\ \hline \text{Add:} & x \quad + z = 18 \quad (5) \end{array}$$

Equations (4) and (5) are a system of two equations in two unknowns. To eliminate x, simply add.

$$\begin{array}{rl} & -x - 3z = -34 \\ & x + z = 18 \\ \hline \text{Add:} & -2z = -16 \end{array}$$

$$z = \frac{-16}{-2} = 8$$

Substitute 8 for z in equation (5).

$$x + z = 18$$
$$x + 8 = 18$$
$$x = 10$$

Substitute 10 for x and 8 for z in equation (3).

$$2x + 2y + 4z = 76$$
$$2(10) + 2y + 4(8) = 76$$
$$20 + 2y + 32 = 76$$
$$2y + 52 = 76$$
$$2y = 24$$
$$y = 12$$

The Donaldson Furniture Company should produce 10 children's chairs, 12 standard chairs, and 8 executive chairs.

53. $I_A + I_B + I_C = 0 \quad (1)$

$-8I_B + 10I_C = 0 \quad (2)$

$4I_A - 8I_B = 6 \quad (3)$

To eliminate I_A between equations (1) and (3), multiply equation (1) by −4 and add.

$$-4[I_A + I_B + C = 0]$$
$$4I_A - 8I_B = 6$$

gives

$$\begin{array}{rl} & -4I_A - 4I_B - 4I_C = 0 \\ & 4I_A - 8I_B \quad = 6 \\ \hline \text{Add:} & -12I_B - 4I_C = 6 \\ \text{or} & -6I_B - 2I_C = 3 \quad (4) \end{array}$$

Equations (4) and (2) are a system of two equations in two unknowns.

$$-8I_B + 10I_C = 0$$
$$-6I_B - 2I_C = 3$$

Multiply equation (4) by 5 and add this result to equation (2).

$$-8I_B + 10I_C = 0$$
$$5[-6I_B - 2I_C = 3]$$

gives

$$\begin{array}{rl} & -8I_B + 10I_C = 0 \\ & -30I_B - 10I_C = 15 \\ \hline \text{Add:} & -38I_B \quad = 15 \end{array}$$

$$I_B = \frac{15}{-38} = -\frac{15}{38}$$

Substitute $-\frac{15}{38}$ for I_B in equation (2).

$$-8I_B + 10I_C = 0$$
$$-8\left(-\frac{15}{38}\right) + 10I_C = 0$$
$$\frac{120}{38} + 10I_C = 0$$
$$10I_C = -\frac{120}{38}$$
$$\frac{1}{10}(10I_C) = \frac{1}{10}\left(-\frac{120}{38}\right)$$
$$I_C = -\frac{12}{38} = -\frac{6}{19}$$

Finally, substitute $-\frac{15}{38}$ for I_B in equation (3).

$$4I_A - 8I_B = 6$$
$$4I_A - 8\left(-\frac{15}{38}\right) = 6$$
$$4I_A + \frac{120}{38} = 6$$
$$4I_A = 6 - \frac{120}{38}$$
$$4I_A = 6 - \frac{60}{19}$$
$$4I_A = \frac{114}{19} - \frac{60}{19}$$
$$4I_A = \frac{54}{19}$$
$$\frac{1}{4}(4I_A) = \frac{1}{4}\left(\frac{54}{19}\right)$$
$$I_A = \frac{27}{38}$$

The current in branch A is $\frac{27}{38}$, the current in branch B is $-\frac{15}{38}$ and the current in branch C is $-\frac{6}{19}$.

56. Substitute –2 for x and 5 for y.

$$\frac{1}{2}x + \frac{2}{5}xy + \frac{1}{8}y = \frac{1}{2}(-2) + \frac{2}{5}(-2)(5) + \frac{1}{8}(5)$$
$$= -1 - 4 + \frac{5}{8}$$
$$= -5 + \frac{5}{8}$$
$$= -\frac{40}{8} + \frac{5}{8}$$
$$= -\frac{35}{8}$$

57. $4 - 2[(x-5) + 2x] = -(x+6)$

$$4 - 2(x - 5 + 2x) = -x - 6$$
$$4 - 2(3x - 5) = -x - 6$$
$$4 - 6x + 10 = -x - 6$$
$$-6x + 14 = -x - 6$$
$$-6x + x = -6 - 14$$
$$-5x = -20$$
$$x = 4$$

58. The slope is

$$m = \frac{-4-(-8)}{6-2} = \frac{-4+8}{6-2} = \frac{4}{4} = 1$$

Use the point-slope form with $m = 1$ and $(x_1, y_1) = (6, -4)$.

$$y - y_1 = m(x - x_1)$$
$$y - (-4) = 1(x - 6)$$
$$y + 4 = x - 6$$
$$y = x - 10$$

59. Use the vertical line test. If a vertical line cannot be drawn to intersect the graph in more than one point, the graph is a function.

Exercise Set 9.6

1. A square matrix has the same number of rows and columns.

3. The next step is to change the –1 in the second row to 1 by multiplying the second row of numbers by –1.

5. Switch row (2) and row (3) in order to continuing placing ones along the diagonal.

7. Dependent

9. $\left[\begin{array}{cc|c} 5 & -10 & -15 \\ 3 & 1 & -4 \end{array}\right] \Rightarrow \left[\begin{array}{cc|c} 1 & -2 & -3 \\ 3 & 1 & -4 \end{array}\right] \frac{1}{5}R_1$

11. $\left[\begin{array}{ccc|c} 4 & 7 & 2 & -1 \\ 3 & 2 & 1 & -5 \\ 1 & 1 & 3 & -8 \end{array}\right] \Rightarrow \left[\begin{array}{ccc|c} 1 & 1 & 3 & -8 \\ 3 & 2 & 1 & -5 \\ 4 & 7 & 2 & -1 \end{array}\right]$ switch R_1 and R_3

13. $\left[\begin{array}{cc|c} 1 & 3 & 12 \\ -3 & 8 & -6 \end{array}\right] \Rightarrow \left[\begin{array}{cc|c} 1 & 3 & 12 \\ 0 & 17 & 30 \end{array}\right] 3R_1 + R_2$

15. $\left[\begin{array}{ccc|c} 1 & 0 & 8 & \frac{1}{4} \\ 5 & 2 & 2 & -2 \\ 6 & -3 & 1 & 0 \end{array}\right] \Rightarrow$

$\left[\begin{array}{ccc|c} 1 & 0 & 8 & \frac{1}{4} \\ 0 & 2 & -38 & -\frac{13}{4} \\ 6 & -3 & 1 & 0 \end{array}\right] -5R_1 + R_2$

17. $x + 3y = 3$

$-x + y = -3$

$\left[\begin{array}{cc|c} 1 & 3 & 3 \\ -1 & 1 & -3 \end{array}\right]$

$\left[\begin{array}{cc|c} 1 & 3 & 3 \\ 0 & 4 & 0 \end{array}\right] R_1 + R_2$

$\left[\begin{array}{cc|c} 1 & 3 & 3 \\ 0 & 1 & 0 \end{array}\right] \frac{1}{4}R_2$

The system is

$x + 3y = 3$

$y = 0$

Substitute 0 for *y* in the first equation.

$x + 3y = 3$

$x + 3(0) = 3$

$x + 0 = 3$

$x = 3$

The solution is (3, 0).

19. $x + 3y = 4$

$-4x - y = 6$

$\left[\begin{array}{cc|c} 1 & 3 & 4 \\ -4 & -1 & 6 \end{array}\right]$

$\left[\begin{array}{cc|c} 1 & 3 & 4 \\ 0 & 11 & 22 \end{array}\right] 4R_1 + R_2$

$\left[\begin{array}{cc|c} 1 & 3 & 4 \\ 0 & 1 & 2 \end{array}\right] \frac{1}{11}R_2$

The system is

$x + 3y = 4$

$y = 2$

Substitute 2 for *y* in the first equation.

$x + 3(2) = 4$

$x + 6 = 4$

$x = -2$

The solution is (–2, 2).

21. $5a - 10b = -10$

$2a + b = 1$

$\left[\begin{array}{cc|c} 5 & -10 & -10 \\ 2 & 1 & 1 \end{array}\right]$

$\left[\begin{array}{cc|c} 1 & -2 & -2 \\ 2 & 1 & 1 \end{array}\right] \frac{1}{5}R_1$

$\left[\begin{array}{cc|c} 1 & -2 & -2 \\ 0 & 5 & 5 \end{array}\right] -2R_1 + R_2$

$\left[\begin{array}{cc|c} 1 & -2 & -2 \\ 0 & 1 & 1 \end{array}\right] \frac{1}{5}R_2$

The system is

$a - 2b = -2$

$b = 1$

Substitute 1 for *b* in the first equation.

$a - 2(1) = -2$

$a - 2 = -2$

$a = 0$

The solution is (0, 1).

23. $2x - 5y = -6$

$-4x + 10y = 12$

$\left[\begin{array}{cc|c} 2 & -5 & -6 \\ -4 & 10 & 12 \end{array}\right]$

$\left[\begin{array}{cc|c} 1 & -\frac{5}{2} & -3 \\ -4 & 10 & 12 \end{array}\right] \frac{1}{2}R_1$

$\left[\begin{array}{cc|c} 1 & -\frac{5}{2} & -3 \\ 0 & 0 & 0 \end{array}\right] 4R_1 + R_2$

Since the last row contains all 0's, this is a dependent system of equations.

25. $12x + 10y = -14$

$4x - 3y = -11$

$\left[\begin{array}{cc|c} 12 & 10 & -14 \\ 4 & -3 & -11 \end{array}\right]$

$\left[\begin{array}{cc|c} 1 & \frac{5}{6} & -\frac{7}{6} \\ 4 & -3 & -11 \end{array}\right] \frac{1}{12}R_1$

$\left[\begin{array}{cc|c} 1 & \frac{5}{6} & -\frac{7}{6} \\ 0 & -\frac{19}{3} & -\frac{19}{3} \end{array}\right] -4R_1 + R_2$

$\left[\begin{array}{cc|c} 1 & \frac{5}{6} & -\frac{7}{6} \\ 0 & 1 & 1 \end{array}\right] -\frac{3}{19}R_2$

The system is

$x + \frac{5}{6}y = -\frac{7}{6}$

$y = 1$

Substitute 1 for *y* in the first equation.

$x + \frac{5}{6}y = -\frac{7}{6}$

$x + \frac{5}{6}(1) = -\frac{7}{6}$

$x + \frac{5}{6} = -\frac{7}{6}$

$x = -\frac{12}{6}$

$x = -2$

The solution is (–2, 1).

27. $-3x + 6y = 5$

$2x - 4y = 8$

$\left[\begin{array}{cc|c} -3 & 6 & 5 \\ 2 & -4 & 8 \end{array}\right]$

$\left[\begin{array}{cc|c} 1 & -2 & -\frac{5}{3} \\ 2 & -4 & 8 \end{array}\right] -\frac{1}{3}R_1$

$\left[\begin{array}{cc|c} 1 & -2 & -\frac{5}{3} \\ 0 & 0 & \frac{34}{3} \end{array}\right] -2R_1 + R_2$

Since the last row contains zeros on the left and a nonzero number on the right, this is an inconsistent system and there is no solution.

29. $9x - 8y = 4$
$-3x + 4y = -1$

$$\left[\begin{array}{cc|c} 9 & -8 & 4 \\ -3 & 4 & -1 \end{array}\right]$$

$$\left[\begin{array}{cc|c} 1 & -\frac{8}{9} & \frac{4}{9} \\ -3 & 4 & -1 \end{array}\right] \frac{1}{9}R_1$$

$$\left[\begin{array}{cc|c} 1 & -\frac{8}{9} & \frac{4}{9} \\ 0 & \frac{4}{3} & \frac{1}{3} \end{array}\right] 3R_1 + R_2$$

$$\left[\begin{array}{cc|c} 1 & -\frac{8}{9} & \frac{4}{9} \\ 0 & 1 & \frac{1}{4} \end{array}\right] \frac{3}{4}R_2$$

The system is

$$x - \frac{8}{9}y = \frac{4}{9}$$
$$y = \frac{1}{4}$$

Substitute $\frac{1}{4}$ for y in the first equation.

$$x - \frac{8}{9}y = \frac{4}{9}$$
$$x - \frac{8}{9}\left(\frac{1}{4}\right) = \frac{4}{9}$$
$$x - \frac{2}{9} = \frac{4}{9}$$
$$x = \frac{4}{9} + \frac{2}{9}$$
$$x = \frac{6}{9}$$
$$x = \frac{2}{3}$$

The solution is $\left(\frac{2}{3}, \frac{1}{4}\right)$.

31. $10m = 8n + 15$
$16n = -15m - 2$
Write the system in standard form.
$10m - 8n = 15$
$15m + 16n = -2$

$$\left[\begin{array}{cc|c} 10 & -8 & 15 \\ 15 & 16 & -2 \end{array}\right]$$

$$\left[\begin{array}{cc|c} 1 & -\frac{4}{5} & \frac{3}{2} \\ 15 & 16 & -2 \end{array}\right] \frac{1}{10}R_1$$

$$\left[\begin{array}{cc|c} 1 & -\frac{4}{5} & \frac{3}{2} \\ 0 & 28 & -\frac{49}{2} \end{array}\right] -15R_1 + R_2$$

$$\left[\begin{array}{cc|c} 1 & -\frac{4}{5} & \frac{3}{2} \\ 0 & 1 & -\frac{7}{8} \end{array}\right] \frac{1}{28}R_2$$

The system is

$$m - \frac{4}{5}n = \frac{3}{2}$$
$$n = -\frac{7}{8}$$

Substitute $-\frac{7}{8}$ for n in the first equation.

$$m - \frac{4}{5}n = \frac{3}{2}$$
$$m - \frac{4}{5}\left(-\frac{7}{8}\right) = \frac{3}{2}$$
$$m + \frac{7}{10} = \frac{3}{2}$$
$$m = \frac{3}{2} - \frac{7}{10}$$
$$m = \frac{15}{10} - \frac{7}{10}$$
$$m = \frac{8}{10}$$
$$m = \frac{4}{5}$$

The solution is $\left(\frac{4}{5}, -\frac{7}{8}\right)$

33. $x - 3y + 2z = 5$
$2x + 5y - 4z = -3$
$-3x + y - 2z = -11$

$$\left[\begin{array}{ccc|c} 1 & -3 & 2 & 5 \\ 2 & 5 & -4 & -3 \\ -3 & 1 & -2 & -11 \end{array}\right]$$

$$\left[\begin{array}{ccc|c} 1 & -3 & 2 & 5 \\ 0 & 11 & -8 & -13 \\ -3 & 1 & -2 & -11 \end{array}\right] -2R_1 + R_2$$

$$\left[\begin{array}{ccc|c} 1 & -3 & 2 & 5 \\ 0 & 11 & -8 & -13 \\ 0 & -8 & 4 & 4 \end{array}\right] 3R_1 + R_3$$

$$\left[\begin{array}{ccc|c} 1 & -3 & 2 & 5 \\ 0 & 1 & -\frac{8}{11} & -\frac{13}{11} \\ 0 & -8 & 4 & 4 \end{array}\right] \frac{1}{11}R_2$$

$$\left[\begin{array}{ccc|c} 1 & -3 & 2 & 5 \\ 0 & 1 & -\frac{8}{11} & -\frac{13}{11} \\ 0 & 0 & -\frac{20}{11} & -\frac{60}{11} \end{array}\right] 8R_2 + R_3$$

$$\left[\begin{array}{ccc|c} 1 & -3 & 2 & 5 \\ 0 & 1 & -\frac{8}{11} & -\frac{13}{11} \\ 0 & 0 & 1 & 3 \end{array}\right] -\frac{11}{20}R_3$$

The system is

$$x - 3y + 2z = 5$$
$$y - \frac{8}{11}z = -\frac{13}{11}$$
$$z = 3$$

Substitute 3 for z in the second equation.

$$y - \frac{8}{11}(3) = -\frac{13}{11}$$
$$y - \frac{24}{11} = -\frac{13}{11}$$
$$y = \frac{11}{11}$$
$$y = 1$$

Substitute 1 for y and 3 for z in the first equation.

$$x - 3(1) + 2(3) = 5$$
$$x - 3 + 6 = 5$$
$$x + 3 = 5$$
$$x = 2$$

The solution is (2, 1, 3).

35. $x + 2y = 5$

$y - z = -1$

$2x - 3z = 0$

Write the system in standard form.

$$x + 2y + 0z = 5$$
$$0x + y - z = -1$$
$$2x + 0y - 3z = 0$$

$$\left[\begin{array}{ccc|c} 1 & 2 & 0 & 5 \\ 0 & 1 & -1 & -1 \\ 2 & 0 & -3 & 0 \end{array}\right]$$

$$\left[\begin{array}{ccc|c} 1 & 2 & 0 & 5 \\ 0 & 1 & -1 & -1 \\ 0 & -4 & -3 & -10 \end{array}\right] -2R_1 + R_3$$

$$\left[\begin{array}{ccc|c} 1 & 2 & 0 & 5 \\ 0 & 1 & -1 & -1 \\ 0 & 0 & -7 & -14 \end{array}\right] 4R_2 + R_3$$

$$\left[\begin{array}{ccc|c} 1 & 2 & 0 & 5 \\ 0 & 1 & -1 & -1 \\ 0 & 0 & 1 & 2 \end{array}\right] -\tfrac{1}{7}R_3$$

The system is

$$x + 2y = 5$$
$$y - z = -1$$
$$z = 2$$

Substitute 2 for z in the second equation.

$$y - z = -1$$
$$y - 2 = -1$$
$$y = 1$$

Substitute 1 for y in the first equation.

$$x + 2y = 5$$
$$x + 2(1) = 5$$
$$x + 2 = 5$$
$$x = 3$$

The solution is (3, 1, 2).

37. $x - 2y + 4z = 5$

$-3x + 4y - 2z = -8$

$4x + 5y - 4z = -3$

$$\left[\begin{array}{ccc|c} 1 & -2 & 4 & 5 \\ -3 & 4 & -2 & -8 \\ 4 & 5 & -4 & -3 \end{array}\right]$$

$$\left[\begin{array}{ccc|c} 1 & -2 & 4 & 5 \\ 0 & -2 & 10 & 7 \\ 4 & 5 & -4 & -3 \end{array}\right] 3R_1 + R_2$$

$$\left[\begin{array}{ccc|c} 1 & -2 & 4 & 5 \\ 0 & -2 & 10 & 7 \\ 0 & 13 & -20 & -23 \end{array}\right] -4R_1 + R_3$$

$$\left[\begin{array}{ccc|c} 1 & -2 & 4 & 5 \\ 0 & 1 & -5 & -\frac{7}{2} \\ 0 & 13 & -20 & -23 \end{array}\right] -\tfrac{1}{2}R_2$$

$$\left[\begin{array}{ccc|c} 1 & -2 & 4 & 5 \\ 0 & 1 & -5 & -\frac{7}{2} \\ 0 & 0 & 45 & \frac{45}{2} \end{array}\right] -13R_2 + R_3$$

$$\left[\begin{array}{ccc|c} 1 & -2 & 4 & 5 \\ 0 & 1 & -5 & -\frac{7}{2} \\ 0 & 0 & 1 & \frac{1}{2} \end{array}\right] \tfrac{1}{45}R_3$$

The system is

$$x - 2y + 4z = 5$$
$$y - 5z = -\frac{7}{2}$$
$$z = \frac{1}{2}$$

Substitute $\frac{1}{2}$ for z in the second equation.

$$y - 5\left(\frac{1}{2}\right) = -\frac{7}{2}$$
$$y - \frac{5}{2} = -\frac{7}{2}$$
$$y = -\frac{2}{2}$$
$$y = -1$$

Substitute -1 for y and $\frac{1}{2}$ for z in the first equation.

$$x-2(-1)+4\left(\frac{1}{2}\right)=5$$
$$x+2+2=5$$
$$x+4=5$$
$$x=1$$

The solution is $\left(1, -1, \frac{1}{2}\right)$.

39.
$$2x-5y+z=1$$
$$3x-5y+z=2$$
$$-4x+10y-2z=-2$$

$$\left[\begin{array}{ccc|c} 2 & -5 & 1 & 1 \\ 3 & -5 & 1 & 2 \\ -4 & 10 & -2 & -2 \end{array}\right]$$

$$\left[\begin{array}{ccc|c} 1 & -\frac{5}{2} & \frac{1}{2} & \frac{1}{2} \\ 3 & -5 & 1 & 2 \\ -4 & 10 & -2 & -2 \end{array}\right] \frac{1}{2}R_1$$

$$\left[\begin{array}{ccc|c} 1 & -\frac{5}{2} & \frac{1}{2} & \frac{1}{2} \\ 0 & \frac{5}{2} & -\frac{1}{2} & \frac{1}{2} \\ -4 & 10 & -2 & -2 \end{array}\right] -3R_1+R_2$$

$$\left[\begin{array}{ccc|c} 1 & -\frac{5}{2} & \frac{1}{2} & \frac{1}{2} \\ 0 & \frac{5}{2} & -\frac{1}{2} & \frac{1}{2} \\ 0 & 0 & 0 & 0 \end{array}\right] 4R_1+R_3$$

Since there is a row of all zeros, the system is dependent.

41.
$$4p-q+r=4$$
$$-6p+3q-2r=-5$$
$$2p+5q-r=7$$

$$\left[\begin{array}{ccc|c} 4 & -1 & 1 & 4 \\ -6 & 3 & -2 & -5 \\ 2 & 5 & -1 & 7 \end{array}\right]$$

$$\left[\begin{array}{ccc|c} 1 & -\frac{1}{4} & \frac{1}{4} & 1 \\ -6 & 3 & -2 & -5 \\ 2 & 5 & -1 & 7 \end{array}\right] \frac{1}{4}R_1$$

$$\left[\begin{array}{ccc|c} 1 & -\frac{1}{4} & \frac{1}{4} & 1 \\ 0 & \frac{3}{2} & -\frac{1}{2} & 1 \\ 2 & 5 & -1 & 7 \end{array}\right] 6R_1+R_2$$

$$\left[\begin{array}{ccc|c} 1 & -\frac{1}{4} & \frac{1}{4} & 1 \\ 0 & \frac{3}{2} & -\frac{1}{2} & 1 \\ 0 & \frac{11}{2} & -\frac{3}{2} & 5 \end{array}\right] -2R_1+R_3$$

$$\left[\begin{array}{ccc|c} 1 & -\frac{1}{4} & \frac{1}{4} & 1 \\ 0 & 1 & -\frac{1}{3} & \frac{2}{3} \\ 0 & \frac{11}{2} & -\frac{3}{2} & 5 \end{array}\right] \frac{2}{3}R_2$$

$$\left[\begin{array}{ccc|c} 1 & -\frac{1}{4} & \frac{1}{4} & 1 \\ 0 & 1 & -\frac{1}{3} & \frac{2}{3} \\ 0 & 0 & \frac{1}{3} & \frac{4}{3} \end{array}\right] -\frac{11}{2}R_2+R_3$$

$$\left[\begin{array}{ccc|c} 1 & -\frac{1}{4} & \frac{1}{4} & 1 \\ 0 & 1 & -\frac{1}{3} & \frac{2}{3} \\ 0 & 0 & 1 & 4 \end{array}\right]$$

The system is

$$x-\frac{1}{4}y+\frac{1}{4}z=1$$
$$y-\frac{1}{3}z=\frac{2}{3}$$
$$z=4$$

Substitute 4 for z in the second equation.

$$y-\frac{1}{3}z=\frac{2}{3}$$
$$y-\frac{1}{3}(4)=\frac{2}{3}$$
$$y-\frac{4}{3}=\frac{2}{3}$$
$$y=\frac{6}{3}$$
$$y=2$$

Substitute 2 for y and 4 for z in the first equation.

$$x-\frac{1}{4}y+\frac{1}{4}z=1$$
$$x-\frac{1}{4}(2)+\frac{1}{4}(4)=1$$
$$x-\frac{1}{2}+1=1$$
$$x+\frac{1}{2}=1$$
$$x=\frac{1}{2}$$

The solution is $\left(\frac{1}{2}, 2, 4\right)$.

43.
$$2x-4y+3z=-12$$
$$3x-y+2z=-3$$
$$-4x+8y-6z=10$$

$$\left[\begin{array}{ccc|c} 2 & -4 & 3 & -12 \\ 3 & -1 & 2 & -3 \\ -4 & 8 & -6 & 10 \end{array}\right]$$

$$\left[\begin{array}{ccc|c} 1 & -2 & \frac{3}{2} & -6 \\ 3 & -1 & 2 & -3 \\ -4 & 8 & -6 & 10 \end{array}\right] \frac{1}{2}R_1$$

$$\left[\begin{array}{ccc|c} 1 & -2 & \frac{3}{2} & -6 \\ 0 & 5 & -\frac{5}{2} & 15 \\ -4 & 8 & -6 & 10 \end{array}\right] -3R_1 + R_2$$

$$\left[\begin{array}{ccc|c} 1 & -2 & \frac{3}{2} & -6 \\ 0 & 5 & -\frac{5}{2} & 15 \\ 0 & 0 & 0 & -14 \end{array}\right] 4R_1 + R_3$$

Since the last row contains zeros on the left and a nonzero number on the right, the system is inconsistent and there is no solution.

45. $5x - 3y + 4z = 22$
$-x - 15y + 10z = -15$
$-3x + 9y - 12z = -6$

$$\left[\begin{array}{ccc|c} 5 & -3 & 4 & 22 \\ -1 & -15 & 10 & -15 \\ -3 & 9 & -12 & -6 \end{array}\right]$$

$$\left[\begin{array}{ccc|c} 1 & -\frac{3}{5} & \frac{4}{5} & \frac{22}{5} \\ -1 & -15 & 10 & -15 \\ -3 & 9 & -12 & -6 \end{array}\right] \frac{1}{5}R_1$$

$$\left[\begin{array}{ccc|c} 1 & -\frac{3}{5} & \frac{4}{5} & \frac{22}{5} \\ 0 & -\frac{78}{5} & \frac{54}{5} & -\frac{53}{5} \\ -3 & 9 & -12 & -6 \end{array}\right] R_1 + R_2$$

$$\left[\begin{array}{ccc|c} 1 & -\frac{3}{5} & \frac{4}{5} & \frac{22}{5} \\ 0 & -\frac{78}{5} & \frac{54}{5} & -\frac{53}{5} \\ 0 & \frac{36}{5} & -\frac{48}{5} & \frac{36}{5} \end{array}\right] 3R_1 + R_3$$

$$\left[\begin{array}{ccc|c} 1 & -\frac{3}{5} & \frac{4}{5} & \frac{22}{5} \\ 0 & 1 & -\frac{9}{13} & \frac{53}{78} \\ 0 & \frac{36}{5} & -\frac{48}{5} & \frac{36}{5} \end{array}\right] -\frac{5}{78}R_2$$

$$\left[\begin{array}{ccc|c} 1 & -\frac{3}{5} & \frac{4}{5} & \frac{22}{5} \\ 0 & 1 & -\frac{9}{13} & \frac{53}{78} \\ 0 & 0 & -\frac{60}{13} & \frac{30}{13} \end{array}\right] -\frac{36}{5}R_2 + R_3$$

$$\left[\begin{array}{ccc|c} 1 & -\frac{3}{5} & \frac{4}{5} & \frac{22}{5} \\ 0 & 1 & -\frac{9}{13} & \frac{53}{78} \\ 0 & 0 & 1 & -\frac{1}{2} \end{array}\right]$$

The system is

$$x - \frac{3}{5}y + \frac{4}{5}z = \frac{22}{5}$$
$$y - \frac{9}{13}z = \frac{53}{78}$$
$$z = -\frac{1}{2}$$

Substitute $-\frac{1}{2}$ for z in the second equation.

$$y - \frac{9}{13}z = \frac{53}{78}$$
$$y - \frac{9}{13}\left(-\frac{1}{2}\right) = \frac{53}{78}$$
$$y + \frac{9}{26} = \frac{53}{78}$$
$$y = \frac{53}{78} - \frac{27}{78}$$
$$y = \frac{26}{78}$$
$$y = \frac{1}{3}$$

Substitute $\frac{1}{3}$ for y and $-\frac{1}{2}$ for z in the first equation.

$$x - \frac{3}{5}y + \frac{4}{5}z = \frac{22}{5}$$
$$x - \frac{3}{5}\left(\frac{1}{3}\right) + \frac{4}{5}\left(-\frac{1}{2}\right) = \frac{22}{5}$$
$$x - \frac{1}{5} - \frac{2}{5} = \frac{22}{5}$$
$$x - \frac{3}{5} = \frac{22}{5}$$
$$x = \frac{25}{5}$$
$$x = 5$$

The solution is $\left(5, \frac{1}{3}, -\frac{1}{2}\right)$.

47. No, this is the same as switching the order of the equations.

49. Let x = smallest angle
y = remaining angle
z = largest angle
$z = x + 55$
$z = y + 20$
$x + y + z = 180$

Write the system in standard form:
$x - z = -55$
$y - z = -20$
$x + y + z = 180$

$$\left[\begin{array}{ccc|c} 1 & 0 & -1 & -55 \\ 0 & 1 & -1 & -20 \\ 1 & 1 & 1 & 180 \end{array}\right]$$

$$\left[\begin{array}{ccc|c} 1 & 0 & -1 & -55 \\ 0 & 1 & -1 & -20 \\ 0 & 1 & 2 & 235 \end{array}\right] -1R_1 + R_3$$

$$\left[\begin{array}{ccc|c} 1 & 0 & -1 & -55 \\ 0 & 1 & -1 & -20 \\ 0 & 0 & 3 & 255 \end{array}\right] -1R_2 + R_3$$

$$\left[\begin{array}{ccc|c} 1 & 0 & -1 & -55 \\ 0 & 1 & -1 & -20 \\ 0 & 0 & 1 & 85 \end{array}\right] \tfrac{1}{3}R_3$$

The system is

$$x - z = -55$$
$$y - z = -20$$
$$z = 85$$

Substitute 85 for z in the second equation.

$$y - z = -20$$
$$y - 85 = -20$$
$$y = 65$$

Substitute 85 for z in the first equation.

$$x - z = -55$$
$$x - 85 = -55$$
$$x = 30$$

The angles are 30°, 65°, and 85°.

51. Let x = amount Chiquita controls,
y = amount Dole controls,
z = amount Del Monte controls

$$x = z + 12$$
$$y = 2z - 3$$
$$x + y + z = 65$$

Write the system in standard form.

$$x - z = 12$$
$$y - 2z = -3$$
$$x + y + z = 65$$

$$\left[\begin{array}{ccc|c} 1 & 0 & -1 & 12 \\ 0 & 1 & -2 & -3 \\ 1 & 1 & 1 & 65 \end{array}\right]$$

$$\left[\begin{array}{ccc|c} 1 & 0 & -1 & 12 \\ 0 & 1 & -2 & -3 \\ 0 & 1 & 2 & 53 \end{array}\right] -1R_1 + R_3$$

$$\left[\begin{array}{ccc|c} 1 & 0 & -1 & 12 \\ 0 & 1 & -2 & -3 \\ 0 & 0 & 4 & 56 \end{array}\right] -1R_2 + R_3$$

$$\left[\begin{array}{ccc|c} 1 & 0 & -1 & 12 \\ 0 & 1 & -2 & -3 \\ 0 & 0 & 1 & 14 \end{array}\right] \tfrac{1}{4}R_3$$

The system is

$$x - z = 12$$
$$y - 2z = -3$$
$$z = 14$$

Substitute 14 for z in the second equation.

$$y - 2z = -3$$
$$y - 2(14) = -3$$
$$y - 28 = -3$$
$$y = 25$$

Substitute 14 for z in the first equation.

$$x - z = 12$$
$$x - 14 = 12$$
$$x = 26$$

Thus, Del Monte controls 14% of the bananas, Dole controls 25% and Chiquita controls 26%, with the remaining 100% − 65% = 35% being controlled by "other."

53.

$$\begin{array}{r} 3x - 8 \\ x + 4 \overline{\smash{\big)}\, 3x^2 + 4x - 23} \\ \underline{3x^2 + 12x} \\ -8x - 23 \\ \underline{-8x - 32} \\ 9 \end{array}$$

$$\frac{3x^2 + 4x - 23}{x + 4} = 3x - 8 + \frac{9}{x + 4}$$

54.

$$2x^2 - x - 36 = 0$$
$$(2x - 9)(x + 4) = 0$$

$$2x - 9 = 0 \qquad x + 4 = 0$$
$$2x = 9 \qquad x = -4$$
$$x = \frac{9}{2}$$

55.

$$\frac{1}{x^2 - 4} - \frac{2}{x - 2} = \frac{1}{(x + 2)(x - 2)} - \frac{2}{x - 2}$$
$$= \frac{1}{(x + 2)(x - 2)} - \frac{2(x + 2)}{(x + 2)(x - 2)}$$
$$= \frac{1}{(x + 2)(x - 2)} - \frac{2x + 4}{(x + 2)(x - 2)}$$
$$= \frac{1 - 2x - 4}{(x + 2)(x - 2)}$$
$$= \frac{-2x - 3}{(x + 2)(x - 2)}$$

56. Let b be the time (in minutes) it takes Terry to stack the wood by herself.

$$\frac{t}{a} + \frac{t}{b} = 1$$
$$\frac{12}{20} + \frac{12}{b} = 1$$
$$20b\left(\frac{12}{20} + \frac{12}{b}\right) = 20b(1)$$
$$12b + 240 = 20b$$
$$240 = 8b$$
$$30 = b$$

It will Terry 30 minutes to stack the wood by herself.

Exercise Set 9.7

1. Answers will vary.

3. If $D = 0$ and either D_x, D_y, or D_z is not equal to 0, the system is inconsistent.

5. $x = \frac{D_x}{D} = \frac{-8}{4} = -2$

$y = \frac{D_y}{D} = \frac{-2}{4} = -\frac{1}{2}$

The solution is $\left(-2, -\frac{1}{2}\right)$.

7. $\begin{vmatrix} 2 & 3 \\ 1 & 5 \end{vmatrix} = (2)(5) - (1)(3) = 10 - 3 = 7$

9. $\begin{vmatrix} \frac{1}{2} & 3 \\ 2 & -4 \end{vmatrix} = \frac{1}{2}(-4) - (2)(3) = -2 - 6 = -8$

11. $\begin{vmatrix} 3 & 2 & 0 \\ 0 & 5 & 3 \\ -1 & 4 & 2 \end{vmatrix} = 3\begin{vmatrix} 5 & 3 \\ 4 & 2 \end{vmatrix} - 0\begin{vmatrix} 2 & 0 \\ 4 & 2 \end{vmatrix} + (-1)\begin{vmatrix} 2 & 0 \\ 5 & 3 \end{vmatrix}$

$= 3(10 - 12) - 0(4 - 0) - 1(6 - 0)$
$= 3(-2) - 0(4) - 1(6)$
$= -6 - 0 - 6$
$= -12$

13. $\begin{vmatrix} 2 & 3 & 1 \\ 1 & -3 & -6 \\ -4 & 5 & 9 \end{vmatrix}$

$= 2\begin{vmatrix} -3 & -6 \\ 5 & 9 \end{vmatrix} - 1\begin{vmatrix} 3 & 1 \\ 5 & 9 \end{vmatrix} + (-4)\begin{vmatrix} 3 & 1 \\ -3 & -6 \end{vmatrix}$
$= 2[-27 - (-30)] - 1(27 - 5) - 4[-18 - (-3)]$
$= 2(3) - 1(22) - 4(-15)$
$= 6 - 22 + 60$
$= 44$

15. $x + 3y = 1$

$-2x - 3y = 4$

To solve, first calculate D, D_x, and D_y.

$D = \begin{vmatrix} 1 & 3 \\ -2 & -3 \end{vmatrix} = (1)(-3) - (-2)(3) = -3 - (-6) = 3$

$D_x = \begin{vmatrix} 1 & 3 \\ 4 & -3 \end{vmatrix}$
$= (1)(-3) - (4)(3)$
$= -3 - 12$
$= -15$

$D_y = \begin{vmatrix} 1 & 1 \\ -2 & 4 \end{vmatrix} = (1)(4) - (-2)(1) = 4 - (-2) = 6$

$x = \frac{D_x}{D} = \frac{-15}{3} = -5$ and $y = \frac{D_y}{D} = \frac{6}{3} = 2$

The solution is $(-5, 2)$.

17. $x - 2y = -1$

$x + 3y = 9$

To solve, first calculate D, D_x, and D_y.

$D = \begin{vmatrix} 1 & -2 \\ 1 & 3 \end{vmatrix} = (1)(3) - (1)(-2) = 3 + 2 = 5$

$D_x = \begin{vmatrix} -1 & -2 \\ 9 & 3 \end{vmatrix}$
$= (-1)(3) - (9)(-2)$
$= -3 + 18$
$= 15$

$D_y = \begin{vmatrix} 1 & -1 \\ 1 & 9 \end{vmatrix} = (1)(9) - (1)(-1) = 9 + 1 = 10$

$x = \frac{D_x}{D} = \frac{15}{5} = 3$ and $y = \frac{D_y}{D} = \frac{10}{5} = 2$

The solution is $(3, 2)$.

19. $5p - 7q = -21$

$-4p + 3q = 22$

To solve, first calculate D, D_p, and D_q.

$D = \begin{vmatrix} 5 & -7 \\ -4 & 3 \end{vmatrix}$
$= (5)(3) - (-4)(-7)$
$= 15 - 28$
$= -13$

$D_p = \begin{vmatrix} -21 & -7 \\ 22 & 3 \end{vmatrix}$
$= (-21)(3) - (22)(-7)$
$= -63 - (-154)$
$= 91$

$D_q = \begin{vmatrix} 5 & -21 \\ -4 & 22 \end{vmatrix}$
$= (5)(22) - (-4)(-21)$
$= 110 - 84$
$= 26$

$p = \frac{D_p}{D} = \frac{91}{-13} = -7$ and $q = \frac{D_q}{D} = \frac{26}{-13} = -2$

The solution is (–7, –2).

21. $4x = -5y - 2$
$-2x = y + 4$

Rewrite the system in standard form:

$4x + 5y = -2$
$2x + y = -4$

Now calculate D, D_x, and D_y.

$$D = \begin{vmatrix} 4 & 5 \\ 2 & 1 \end{vmatrix} = (4)(1) - (2)(5) = 4 - 10 = -6$$

$$D_x = \begin{vmatrix} -2 & 5 \\ -4 & 1 \end{vmatrix} = (-2)(1) - (-4)(5) = -2 - (-20) = 18$$

$$D_y = \begin{vmatrix} 4 & -2 \\ 2 & -4 \end{vmatrix} = (4)(-4) - (2)(-2) = -16 - (-4) = -12$$

$x = \frac{D_x}{D} = \frac{18}{-6} = -3$ and $y = \frac{D_y}{D} = \frac{-12}{-6} = 2$

The solution is (–3, 2).

23. $x + 5y = 3$
$2x + 10y = 6$

To solve, first calculate D, D_x, and D_y.

$$D = \begin{vmatrix} 1 & 5 \\ 2 & 10 \end{vmatrix} = (1)(10) - (2)(5) = 10 - 10 = 0$$

$$D_x = \begin{vmatrix} 3 & 5 \\ 6 & 10 \end{vmatrix} = (3)(10) - (6)(5) = 30 - 30 = 0$$

$$D_y = \begin{vmatrix} 1 & 3 \\ 2 & 6 \end{vmatrix} = (1)(6) - (2)(3) = 6 - 6 = 0$$

Since $D = 0$, $D_x = 0$, and $D_y = 0$, the system is dependent so there are an infinite number of solutions.

25. $3r = -4s - 6$
$3s = -5r + 1$

Rewrite the system in standard form.

$3r + 4s = -6$
$5r + 3s = 1$

Now calculate D, D_r, and D_s.

$$D = \begin{vmatrix} 3 & 4 \\ 5 & 3 \end{vmatrix} = (3)(3) - (5)(4) = 9 - 20 = -11$$

$$D_r = \begin{vmatrix} -6 & 4 \\ 1 & 3 \end{vmatrix} = (-6)(3) - (1)(4) = -18 - 4 = -22$$

$$D_s = \begin{vmatrix} 3 & -6 \\ 5 & 1 \end{vmatrix} = (3)(1) - (5)(-6) = 3 + 30 = 33$$

$r = \frac{D_r}{D} = \frac{-22}{-11} = 2$ and $s = \frac{D_s}{D} = \frac{33}{-11} = -3$

The solution is (2, –3).

27. $5x - 5y = 3$
$x - y = -2$

To solve, first calculate D, D_x, and D_y.

$$D = \begin{vmatrix} 5 & -5 \\ 1 & -1 \end{vmatrix} = (5)(-1) - (1)(-5) = -5 + 5 = 0$$

$$D_x = \begin{vmatrix} 3 & -5 \\ -2 & -1 \end{vmatrix} = (3)(-1) - (-2)(-5) = -3 - 10 = -13$$

Since $D = 0$ and $D_x \neq 0$, the system is inconsistent, so there is no solution.

29. $6.3x - 4.5y = -9.9$
$-9.1x + 3.2y = -2.2$

Here, you can work with decimals in the determinants. If you do not want to use decimals, then you need to multiply each equation by 10 to clear the decimals.

First, calculate D, D_x, and D_y.

$$D = \begin{vmatrix} 6.3 & -4.5 \\ -9.1 & 3.2 \end{vmatrix} = (6.3)(3.2) - (-9.1)(-4.5) = 20.16 - 40.95 = -20.79$$

$$D_x = \begin{vmatrix} -9.9 & -4.5 \\ -2.2 & 3.2 \end{vmatrix} = (-9.9)(3.2) - (-2.2)(-4.5) = -31.68 - 9.90 = -41.58$$

$$D_y = \begin{vmatrix} 6.3 & -9.9 \\ -9.1 & -2.2 \end{vmatrix} = (6.3)(-2.2) - (-9.1)(-9.9) = -13.86 - 90.09 = -103.95$$

$x = \frac{D_x}{D} = \frac{-41.58}{-20.79} = 2$ and

$y = \frac{D_y}{D} = \frac{-103.95}{-20.79} = 5$

The solution is (2, 5).

31. $x + y + z = 2$

$0x - 3y + 4z = 11$

$-3x + 4y - 2z = -11$

To solve, first calculate D, D_x, D_y, and D_z.

$$D = \begin{vmatrix} 1 & 1 & 1 \\ 0 & -3 & 4 \\ -3 & 4 & -2 \end{vmatrix} \text{(using first column)}$$

$$= 1\begin{vmatrix} -3 & 4 \\ 4 & -2 \end{vmatrix} - 0\begin{vmatrix} 1 & 1 \\ 4 & -2 \end{vmatrix} + (-3)\begin{vmatrix} 1 & 1 \\ -3 & 4 \end{vmatrix}$$

$= 1(6 - 16) - 0(-2 - 4) - 3(4 + 3)$

$= 1(-10) - 0(-6) - 3(7)$

$= -10 - 0 - 21$

$= -31$

$$D_x = \begin{vmatrix} 2 & 1 & 1 \\ 11 & -3 & 4 \\ -11 & 4 & -2 \end{vmatrix} \text{(using first row)}$$

$$= 2\begin{vmatrix} -3 & 4 \\ 4 & -2 \end{vmatrix} - 1\begin{vmatrix} 11 & 4 \\ -11 & -2 \end{vmatrix} + 1\begin{vmatrix} 11 & -3 \\ -11 & 4 \end{vmatrix}$$

$= 2(6 - 16) - 1(-22 + 44) + 1(44 - 33)$

$= 2(-10) - 1(22) + 1(11)$

$= -20 - 22 + 11$

$= -31$

$$D_y = \begin{vmatrix} 1 & 2 & 1 \\ 0 & 11 & 4 \\ -3 & -11 & -2 \end{vmatrix} \text{(using first column)}$$

$$= 1\begin{vmatrix} 11 & 4 \\ -11 & -2 \end{vmatrix} - 0\begin{vmatrix} 2 & 1 \\ -11 & -2 \end{vmatrix} + (-3)\begin{vmatrix} 2 & 1 \\ 11 & 4 \end{vmatrix}$$

$= 1(-22 + 44) - 0(-4 + 11) - 3(8 - 11)$

$= 1(22) - 0(7) - 3(-3)$

$= 22 - 0 + 9$

$= 31$

$$D_z = \begin{vmatrix} 1 & 1 & 2 \\ 0 & -3 & 11 \\ -3 & 4 & -11 \end{vmatrix} \text{(using first column)}$$

$$= 1\begin{vmatrix} -3 & 11 \\ 4 & -11 \end{vmatrix} - 0\begin{vmatrix} 1 & 2 \\ 4 & -11 \end{vmatrix} + (-3)\begin{vmatrix} 1 & 2 \\ -3 & 11 \end{vmatrix}$$

$= 1(33 - 44) - 0(-11 - 8) - 3(11 + 6)$

$= 1(-11) - 0(-19) - 3(17)$

$= -11 - 0 - 51$

$= -62$

$x = \frac{D_x}{D} = \frac{-31}{-31} = 1$, $y = \frac{D_y}{D} = \frac{31}{-31} = 1$, and

$z = \frac{D_z}{D} = \frac{-62}{-31} = 2$

The solution is (1, –1, 2).

33. $3x - 5y - 4z = -4$

$4x + 2y + 0z = 1$

$0x + 6y - 4z = -11$

To solve, first calculate D, D_x, D_y, and D_z.

$$D = \begin{vmatrix} 3 & -5 & -4 \\ 4 & 2 & 0 \\ 0 & 6 & -4 \end{vmatrix} \text{(using the first row)}$$

$$= 3\begin{vmatrix} 2 & 0 \\ 6 & -4 \end{vmatrix} - (-5)\begin{vmatrix} 4 & 0 \\ 0 & -4 \end{vmatrix} + (-4)\begin{vmatrix} 4 & 2 \\ 0 & 6 \end{vmatrix}$$

$= 3(-8 - 0) + 5(-16 - 0) - 4(24 - 0)$

$= 3(-8) + 5(-16) - 4(24)$

$= -24 - 80 - 96$

$= -200$

$$D_x = \begin{vmatrix} -4 & -5 & -4 \\ 1 & 2 & 0 \\ -11 & 6 & -4 \end{vmatrix} \text{(using the first row)}$$

$$= -4\begin{vmatrix} 2 & 0 \\ 6 & -4 \end{vmatrix} - (-5)\begin{vmatrix} 1 & 0 \\ -11 & -4 \end{vmatrix} + (-4)\begin{vmatrix} 1 & 2 \\ -11 & 6 \end{vmatrix}$$

$= -4(-8 - 0) + 5(-4 + 0) - 4(6 + 22)$

$= -4(-8) + 5(-4) - 4(28)$

$= 32 - 20 - 112$

$= -100$

$D_y = \begin{vmatrix} 3 & -4 & -4 \\ 4 & 1 & 0 \\ 0 & -11 & -4 \end{vmatrix}$ (using the first row)

$= 3\begin{vmatrix} 1 & 0 \\ -11 & -4 \end{vmatrix} - (-4)\begin{vmatrix} 4 & 0 \\ 0 & -4 \end{vmatrix} + (-4)\begin{vmatrix} 4 & 1 \\ 0 & -11 \end{vmatrix}$

$= 3(-4+0) + 4(-16-0) - 4(-44-0)$

$= 3(-4) + 4(-16) - 4(-44)$

$= -12 - 64 + 176$

$= 100$

$D_z = \begin{vmatrix} 3 & -5 & -4 \\ 4 & 2 & 1 \\ 0 & 6 & -11 \end{vmatrix}$ (using the first row)

$= 3\begin{vmatrix} 2 & 1 \\ 6 & -11 \end{vmatrix} - (-5)\begin{vmatrix} 4 & 1 \\ 0 & -11 \end{vmatrix} + (-4)\begin{vmatrix} 4 & 2 \\ 0 & 6 \end{vmatrix}$

$= 3(-22-6) + 5(-44-0) - 4(24-0)$

$= 3(-28) + 5(-44) - 4(24)$

$= -84 - 220 - 96$

$= -400$

$x = \frac{D_x}{D} = \frac{-100}{-200} = \frac{1}{2}$, $y = \frac{D_y}{D} = \frac{100}{-200} = -\frac{1}{2}$, and

$z = \frac{D_z}{D} = \frac{-400}{-200} = 2$

The solution is $\left(\frac{1}{2}, -\frac{1}{2}, 2\right)$.

35. $x + 4y - 3z = -6$

$2x - 8y + 5z = 12$

$3x + 4y - 2z = -3$

To solve, first calculate D, D_x, D_y, and D_z

$D = \begin{vmatrix} 1 & 4 & -3 \\ 2 & -8 & 5 \\ 3 & 4 & -2 \end{vmatrix}$ (using the first row)

$= 1\begin{vmatrix} -8 & 5 \\ 4 & -2 \end{vmatrix} - 4\begin{vmatrix} 2 & 5 \\ 3 & -2 \end{vmatrix} + (-3)\begin{vmatrix} 2 & -8 \\ 3 & 4 \end{vmatrix}$

$= 1(16-20) - 4(-4-15) - 3(8+24)$

$= 1(-4) - 4(-19) - 3(32)$

$= -4 + 76 - 96$

$= -24$

$D_x = \begin{vmatrix} -6 & 4 & -3 \\ 12 & -8 & 5 \\ -3 & 4 & -2 \end{vmatrix}$ (using the first row)

$= -6\begin{vmatrix} -8 & 5 \\ 4 & -2 \end{vmatrix} - 4\begin{vmatrix} 12 & 5 \\ -3 & -2 \end{vmatrix} + (-3)\begin{vmatrix} 12 & -8 \\ -3 & 4 \end{vmatrix}$

$= -6(16-20) - 4(-24+15) - 3(48-24)$

$= -6(-4) - 4(-9) - 3(24)$

$= 24 + 36 - 72$

$= -12$

$D_y = \begin{vmatrix} 1 & -6 & -3 \\ 2 & 12 & 5 \\ 3 & -3 & -2 \end{vmatrix}$ (using the first row)

$= 1\begin{vmatrix} 12 & 5 \\ -3 & -2 \end{vmatrix} - (-6)\begin{vmatrix} 2 & 5 \\ 3 & -2 \end{vmatrix} + (-3)\begin{vmatrix} 2 & 12 \\ 3 & -3 \end{vmatrix}$

$= 1(-24+15) + 6(-4-15) - 3(-6-36)$

$= 1(-9) + 6(-19) - 3(-42)$

$= -9 - 114 + 126$

$= 3$

$D_z = \begin{vmatrix} 1 & 4 & -6 \\ 2 & -8 & 12 \\ 3 & 4 & -3 \end{vmatrix}$ (using the first row)

$= 1\begin{vmatrix} -8 & 12 \\ 4 & -3 \end{vmatrix} - 4\begin{vmatrix} 2 & 12 \\ 3 & -3 \end{vmatrix} + (-6)\begin{vmatrix} 2 & -8 \\ 3 & 4 \end{vmatrix}$

$= 1(24-48) - 4(-6-36) - 6(8+24)$

$= 1(-24) - 4(-42) - 6(32)$

$= -24 + 168 - 192$

$= -48$

$x = \frac{D_x}{D} = \frac{-12}{-24} = \frac{1}{2}$, $y = \frac{D_y}{D} = \frac{3}{-24} = -\frac{1}{8}$, and

$z = \frac{D_z}{D} = \frac{-48}{-24} = 2.$

The solution is $\left(\frac{1}{2}, -\frac{1}{8}, 2\right)$.

37. $a - b + 2c = 3$

$a - b + c = 1$

$2a + b + 2c = 2$

To solve, first calculate D, D_a, D_b, and D_c.

$$D = \begin{vmatrix} 1 & -1 & 2 \\ 1 & -1 & 1 \\ 2 & 1 & 2 \end{vmatrix}$$
$$= 1\begin{vmatrix} -1 & 1 \\ 1 & 2 \end{vmatrix} - 1\begin{vmatrix} -1 & 2 \\ 1 & 2 \end{vmatrix} + 2\begin{vmatrix} -1 & 2 \\ -1 & 1 \end{vmatrix}$$
$= 1(-2-1) - 1(-2-2) + 2(-1+2)$
$= 1(-3) - 1(-4) + 2(1)$
$= -3 + 4 + 2$
$= 3$

$$D_a = \begin{vmatrix} 3 & -1 & 2 \\ 1 & -1 & 1 \\ 2 & 1 & 2 \end{vmatrix}$$
$$= 3\begin{vmatrix} -1 & 1 \\ 1 & 2 \end{vmatrix} - 1\begin{vmatrix} -1 & 2 \\ 1 & 2 \end{vmatrix} + 2\begin{vmatrix} -1 & 2 \\ -1 & 1 \end{vmatrix}$$
$= 3(-2-1) - 1(-2-2) + 2(-1+2)$
$= 3(-3) - 1(-4) + 2(1)$
$= -9 + 4 + 2$
$= -3$

$$D_b = \begin{vmatrix} 1 & 3 & 2 \\ 1 & 1 & 1 \\ 2 & 2 & 2 \end{vmatrix}$$
$$= 1\begin{vmatrix} 1 & 1 \\ 2 & 2 \end{vmatrix} - 1\begin{vmatrix} 3 & 2 \\ 2 & 2 \end{vmatrix} + 2\begin{vmatrix} 3 & 2 \\ 1 & 1 \end{vmatrix}$$
$= 1(2-2) - 1(6-4) + 2(3-2)$
$= 1(0) - 1(2) + 2(1)$
$= 0 - 2 + 2$
$= 0$

$$D_c = \begin{vmatrix} 1 & -1 & 3 \\ 1 & -1 & 1 \\ 2 & 1 & 2 \end{vmatrix}$$
$$= 1\begin{vmatrix} -1 & 1 \\ 1 & 2 \end{vmatrix} - 1\begin{vmatrix} -1 & 3 \\ 1 & 2 \end{vmatrix} + 2\begin{vmatrix} -1 & 3 \\ -1 & 1 \end{vmatrix}$$
$= 1(-2-1) - 1(-2-3) + 2(-1+3)$
$= 1(-3) - 1(-5) + 2(2)$
$= -3 + 5 + 4$
$= 6$

$a = \dfrac{D_a}{D} = \dfrac{-3}{3} = -1$, $b = \dfrac{D_b}{D} = \dfrac{0}{3} = 0$, and

$c = \dfrac{D_c}{D} = \dfrac{6}{3} = 2$

The solution is (−1, 0, 2).

39. $a + 2b + c = 1$
$a - b + c = 1$
$2a + b + 2c = 2$
To solve, first calculate D, D_a, D_b, and D_c.

$$D = \begin{vmatrix} 1 & 2 & 1 \\ 1 & -1 & 1 \\ 2 & 1 & 2 \end{vmatrix}$$
$$= 1\begin{vmatrix} -1 & 1 \\ 1 & 2 \end{vmatrix} - 1\begin{vmatrix} 2 & 1 \\ 1 & 2 \end{vmatrix} + 2\begin{vmatrix} 2 & 1 \\ -1 & 1 \end{vmatrix}$$
$= 1(-2-1) - 1(4-1) + 2(2+1)$
$= 1(-3) - 1(3) + 2(3)$
$= -3 - 3 + 6$
$= 0$

$$D_a = \begin{vmatrix} 1 & 2 & 1 \\ 1 & -1 & 1 \\ 2 & 1 & 2 \end{vmatrix}$$
$$= 1\begin{vmatrix} -1 & 1 \\ 1 & 2 \end{vmatrix} - 1\begin{vmatrix} 2 & 1 \\ 1 & 2 \end{vmatrix} + 2\begin{vmatrix} 2 & 1 \\ -1 & 1 \end{vmatrix}$$
$= 1(-2-1) - 1(4-1) + 2(2+1)$
$= 1(-3) - 1(3) + 2(3)$
$= -3 - 3 + 6$
$= 0$

$$D_b = \begin{vmatrix} 1 & 1 & 1 \\ 1 & 1 & 1 \\ 2 & 2 & 2 \end{vmatrix}$$
$$= 1\begin{vmatrix} 1 & 1 \\ 2 & 2 \end{vmatrix} - 1\begin{vmatrix} 1 & 1 \\ 2 & 2 \end{vmatrix} + 2\begin{vmatrix} 1 & 1 \\ 1 & 1 \end{vmatrix}$$
$= 1(2-2) - 1(2-2) + 2(1-1)$
$= 1(0) - 1(0) + 2(0)$
$= 0 - 0 + 0$
$= 0$

$$D_c = \begin{vmatrix} 1 & 2 & 1 \\ 1 & -1 & 1 \\ 2 & 1 & 2 \end{vmatrix}$$
$$= 1\begin{vmatrix} -1 & 1 \\ 1 & 2 \end{vmatrix} - 1\begin{vmatrix} 2 & 1 \\ 1 & 2 \end{vmatrix} + 2\begin{vmatrix} 2 & 1 \\ -1 & 1 \end{vmatrix}$$
$= 1(-2-1) - 1(4-1) + 2(2+1)$
$= 1(-3) - 1(3) + 2(3)$
$= -3 - 3 + 6$
$= 0$

Since $D = 0$, $D_a = 0$, $D_b = 0$, and $D_c = 0$, there are an infinite number of solutions to the system and it is a dependent system.

41. $1.1x + 2.3y - 4.0z = -9.2$
$-2.3x + 0y + 4.6z = 6.9$
$0x - 8.2y - 7.5z = -6.8$

Here, you can work with decimals in the determinants. If you do not want to use decimals, then you need to multiply each equation by 10 to clear the decimals. To solve, first calculate D, D_x, D_y, and D_z.

$$D = \begin{vmatrix} 1.1 & 2.3 & -4.0 \\ -2.3 & 0 & 4.6 \\ 0 & -8.2 & -7.5 \end{vmatrix}$$

$$= 1.1\begin{vmatrix} 0 & 4.6 \\ -8.2 & -7.5 \end{vmatrix} - (-2.3)\begin{vmatrix} 2.3 & -4.0 \\ -8.2 & -7.5 \end{vmatrix} + 0\begin{vmatrix} 2.3 & -4.0 \\ 0 & 4.6 \end{vmatrix}$$

$= 1.1(0 + 37.72) + 2.3(-17.25 - 32.8) + 0(10.58 - 0)$
$= 1.1(37.72) + 2.3(-50.05) + 0(10.58)$
$= 41.492 - 115.115 + 0$
$= -73.623$

$$D_x = \begin{vmatrix} -9.2 & 2.3 & -4.0 \\ 6.9 & 0 & 4.6 \\ -6.8 & -8.2 & -7.5 \end{vmatrix}$$

$$= -9.2\begin{vmatrix} 0 & 4.6 \\ -8.2 & -7.5 \end{vmatrix} - 6.9\begin{vmatrix} 2.3 & -4.0 \\ -8.2 & -7.5 \end{vmatrix} + (-6.8)\begin{vmatrix} 2.3 & -4.0 \\ 0 & 4.6 \end{vmatrix}$$

$= -9.2(0 + 37.72) - 6.9(-17.25 - 32.8) - 6.8(10.58 - 0)$
$= -9.2(37.72) - 6.9(-50.05) - 6.8(10.58)$
$= -347.024 + 345.345 - 71.944$
$= -73.623$

$$D_y = \begin{vmatrix} 1.1 & -9.2 & -4.0 \\ -2.3 & 6.9 & 4.6 \\ 0 & -6.8 & -7.5 \end{vmatrix}$$

$$= 1.1\begin{vmatrix} 6.9 & 4.6 \\ -6.8 & -7.5 \end{vmatrix} - (-2.3)\begin{vmatrix} -9.2 & -4.0 \\ -6.8 & -7.5 \end{vmatrix} + 0\begin{vmatrix} -9.2 & -4.0 \\ 6.9 & 4.6 \end{vmatrix}$$

$= 1.1(-51.75 + 31.28) + 2.3(69 - 27.2) + 0(-42.32 + 27.6)$
$= 1.1(-20.47) + 2.3(41.8) + 0(-14.72)$
$= -22.517 + 96.14 + 0$
$= 73.623$

$$D_z = \begin{vmatrix} 1.1 & 2.3 & -9.2 \\ -2.3 & 0 & 6.9 \\ 0 & -8.2 & -6.8 \end{vmatrix}$$

$$= 1.1\begin{vmatrix} 0 & 6.9 \\ -8.2 & -6.8 \end{vmatrix} - (-2.3)\begin{vmatrix} 2.3 & -9.2 \\ -8.2 & -6.8 \end{vmatrix} + 0\begin{vmatrix} 2.3 & -9.2 \\ 0 & 6.9 \end{vmatrix}$$

$= 1.1(0 + 56.58) + 2.3(-15.64 - 75.44) + 0(15.87 - 0)$
$= 1.1(56.58) + 2.3(-91.08) + 0(15.87)$
$= 62.238 - 209.484 + 0$
$= 147.246$

$x = \dfrac{D_x}{D} = \dfrac{-73.623}{-73.623} = 1$, $y = \dfrac{D_y}{D} = \dfrac{73.623}{-73.623} = -1$, and $z = \dfrac{D_z}{D} = \dfrac{-147.246}{-73.623} = 2$

The solution is (1, −1, 2).

43. $-6x+3y-9z=-8$

$5x+2y-3z=1$

$2x-y+3z=-5$

To solve, first calculate D, D_x, D_y, and D_z.

$$D=\begin{vmatrix}-6 & 3 & -9\\ 5 & 2 & -3\\ 2 & -1 & 3\end{vmatrix}\text{ (using the first row)}$$

$$=-6\begin{vmatrix}2 & -3\\ -1 & 3\end{vmatrix}-3\begin{vmatrix}5 & -3\\ 2 & 3\end{vmatrix}+(-9)\begin{vmatrix}5 & 2\\ 2 & -1\end{vmatrix}$$

$$=-6(6-3)-3(15+6)-9(-5-4)$$

$$=-6(3)-3(21)-9(-9)$$

$$=-18-63+81$$

$$=0$$

$$D_x=\begin{vmatrix}-8 & 3 & -9\\ 1 & 2 & -3\\ -5 & -1 & 3\end{vmatrix}\text{ (using the first row)}$$

$$=-8\begin{vmatrix}2 & -3\\ -1 & 3\end{vmatrix}-3\begin{vmatrix}1 & -3\\ -5 & 3\end{vmatrix}+(-9)\begin{vmatrix}1 & 2\\ -5 & -1\end{vmatrix}$$

$$=-8(6-3)-3(3-15)-9(-1+10)$$

$$=-8(3)-3(-12)-9(9)$$

$$=-24+36-81$$

$$=-69$$

Since $D=0$ and $D_x=-69\neq 0$, there is no solution to the system and it is an inconsistent system.

45. $2x+\frac{1}{2}y-3z=5$

$-3x+2y+2z=1$

$4x-\frac{1}{4}y-7z=4$

To clear the system of fractions, multiply the first equation by 2 and the third equation by 4.

$4x+y-6z=10$

$-3x+2y+2z=1$

$16x-y-28z=16$

To solve, first calculate D, D_x, D_y, and D_z.

$$D=\begin{vmatrix}4 & 1 & -6\\ -3 & 2 & 2\\ 16 & -1 & -28\end{vmatrix}\text{ (use the first row)}$$

$$=4\begin{vmatrix}2 & 2\\ -1 & -28\end{vmatrix}-1\begin{vmatrix}-3 & 2\\ 16 & -28\end{vmatrix}+(-6)\begin{vmatrix}-3 & 2\\ 16 & -1\end{vmatrix}$$

$$=4(-56+2)-1(84-32)-6(3-32)$$

$$=4(-54)-1(52)-6(-29)$$

$$=-216-52+174$$

$$=-94$$

$$D_x=\begin{vmatrix}10 & 1 & -6\\ 1 & 2 & 2\\ 16 & -1 & -28\end{vmatrix}\text{ (use the first row)}$$

$$=10\begin{vmatrix}2 & 2\\ -1 & -28\end{vmatrix}-1\begin{vmatrix}1 & 2\\ 16 & -28\end{vmatrix}+(-6)\begin{vmatrix}1 & 2\\ 16 & -1\end{vmatrix}$$

$$=10(-56+2)-1(-28-32)-6(-1-32)$$

$$=10(-54)-1(-60)-6(-33)$$

$$=-540+60+198$$

$$=-282$$

$$D_y=\begin{vmatrix}4 & 10 & -6\\ -3 & 1 & 2\\ 16 & 16 & -28\end{vmatrix}\text{ (use the first row)}$$

$$=4\begin{vmatrix}1 & 2\\ 16 & -28\end{vmatrix}-10\begin{vmatrix}-3 & 2\\ 16 & -28\end{vmatrix}+(-6)\begin{vmatrix}-3 & 1\\ 16 & 16\end{vmatrix}$$

$$=4(-28-32)-10(84-32)-6(-48-16)$$

$$=4(-60)-10(52)-6(-64)$$

$$=-240-520+384$$

$$=-376$$

$$D_z=\begin{vmatrix}4 & 1 & 10\\ -3 & 2 & 1\\ 16 & -1 & 16\end{vmatrix}\text{ (use the first row)}$$

$$=4\begin{vmatrix}2 & 1\\ -1 & 16\end{vmatrix}-1\begin{vmatrix}-3 & 1\\ 16 & 16\end{vmatrix}+10\begin{vmatrix}-3 & 2\\ 16 & -1\end{vmatrix}$$

$$=4(32+1)-1(-48-16)+10(3-32)$$

$$=4(33)-1(-64)+10(-29)$$

$$=132+64-290$$

$$=-94$$

$x = \frac{D_x}{D} = \frac{-282}{-94} = 3$, $y = \frac{D_y}{D} = \frac{-376}{-94} = 4$,

$z = \frac{D_z}{D} = \frac{-94}{-94} = 1$

The solution is (3, 4, 1).

47. $0.2x - 0.1y - 0.3z = -0.1$
$0.2x - 0.1y + 0.1z = -0.9$
$0.1x + 0.2y - 0.4z = 1.7$

To clear decimals multiply each equation by 10.

$2x - y - 3x = -1$
$2x - y + z = -9$
$x + 2y - 4z = 17$

To solve, first calculate D, D_x, D_y, and D_z.

$$D = \begin{vmatrix} 2 & -1 & -3 \\ 2 & -1 & 1 \\ 1 & 2 & -4 \end{vmatrix}$$

$$= 2\begin{vmatrix} -1 & 1 \\ 2 & -4 \end{vmatrix} - 2\begin{vmatrix} -1 & -3 \\ 2 & -4 \end{vmatrix} + 1\begin{vmatrix} -1 & -3 \\ -1 & 1 \end{vmatrix}$$

$= 2(4-2) - 2(4+6) + 1(-1-3)$
$= 2(2) - 2(10) + 1(-4)$
$= 4 - 20 - 4$
$= -20$

$$D_x = \begin{vmatrix} -1 & -1 & -3 \\ -9 & -1 & 1 \\ 17 & 2 & -4 \end{vmatrix}$$

$$= -1\begin{vmatrix} -1 & 1 \\ 2 & -4 \end{vmatrix} - (-9)\begin{vmatrix} -1 & -3 \\ 2 & -4 \end{vmatrix} + 17\begin{vmatrix} -1 & -3 \\ -1 & 1 \end{vmatrix}$$

$= -1(4-2) + 9(4+6) + 17(-1-3)$
$= -1(2) + 9(10) + 17(-4)$
$= -2 + 90 - 68$
$= 20$

$$D_y = \begin{vmatrix} 2 & -1 & -3 \\ 2 & -9 & 1 \\ 1 & 17 & -4 \end{vmatrix}$$

$$= 2\begin{vmatrix} -9 & 1 \\ 17 & -4 \end{vmatrix} - 2\begin{vmatrix} -1 & -3 \\ 17 & -4 \end{vmatrix} + 1\begin{vmatrix} -1 & -3 \\ -9 & 1 \end{vmatrix}$$

$= 2(36-17) - 2(4+51) + 1(-1-27)$
$= 2(19) - 2(55) + 1(-28)$
$= 38 - 110 - 28$
$= -100$

$$D_z = \begin{vmatrix} 2 & -1 & -1 \\ 2 & -1 & -9 \\ 1 & 2 & 17 \end{vmatrix}$$

$$= 2\begin{vmatrix} -1 & -9 \\ 2 & 17 \end{vmatrix} - 2\begin{vmatrix} -1 & -1 \\ 2 & 17 \end{vmatrix} + 1\begin{vmatrix} -1 & -1 \\ -1 & -9 \end{vmatrix}$$

$= 2(-17+18) - 2(-17+2) + 1(9-1)$
$= 2(1) - 2(-15) + 1(8)$
$= 2 + 30 + 8$
$= 40$

$x = \frac{D_x}{D} = \frac{20}{-20} = -1$, $y = \frac{D_y}{D} = \frac{-100}{-20} = 5$, and

$z = \frac{D_z}{D} = \frac{40}{-20} = -2$

The solution is (–1, 5, –2).

49. $\begin{vmatrix} a_1 & b_1 \\ a_2 & b_2 \end{vmatrix} = a_1b_2 - a_2b_1$

$\begin{vmatrix} a_2 & b_2 \\ a_1 & b_1 \end{vmatrix} = a_2b_1 - a_1b_2$

The second result is the negative of the first result. Thus, the second determinant has the opposite sign.

51. 0

53. 0

55. Yes, the determinant will become the opposite of the original value.

57. $\begin{vmatrix} 4 & 6 \\ -2 & y \end{vmatrix} = 32$

$(4)(y) - (-2)(6) = 32$
$4y + 12 = 32$
$4y = 20$
$y = \frac{20}{4} = 5$

59. $\begin{vmatrix} 4 & 7 & y \\ 3 & -1 & 2 \\ 4 & 1 & 5 \end{vmatrix} = -35$

$4\begin{vmatrix} -1 & 2 \\ 1 & 5 \end{vmatrix} - 3\begin{vmatrix} 7 & y \\ 1 & 5 \end{vmatrix} + 4\begin{vmatrix} 7 & y \\ -1 & 2 \end{vmatrix} = -35$

$4(-5-2) - 3(35-y) + 4(14+y) = -35$
$4(-7) - 3(35-y) + 4(14+y) = -35$
$-28 - 105 + 3y + 56 + 4y = -35$
$-77 + 7y = -35$
$7y = 42$
$y = 6$

61. **a.** To eliminate *y*, multiply the first equation by b_2 and the second equation by $-b_1$ and then add.

$b_2[a_1x + b_1y = c_1]$
$-b_1[a_2x + b_2y = c_2]$
gives

$$a_1b_2x + b_1b_2y = c_1b_2$$
$$-a_2b_1x - b_1b_2y = -c_2b_1$$

Add: $(a_1b_2 - a_2b_1)x = c_1b_2 - c_2b_1$

$$x = \frac{c_1b_2 - c_2b_1}{a_1b_2 - a_2b_1}$$

b. To eliminate *x*, multiply the first equation by $-a_2$ and the second equation by a_1 and then add.

$-a_2[a_1x + b_1y = c_1]$
$a_1[a_2x + b_2y = c_2]$
gives

$$-a_1a_2x - a_2b_1y = -a_2c_1$$
$$a_1a_2x + a_1b_2y = a_1c_2$$

Add: $(a_1b_2 - a_2b_1)y = a_1c_2 - a_2c_1$

$$y = \frac{a_1c_2 - a_2c_1}{a_1b_2 - a_2b_1}$$

62. $2x - 5y = 6$

$$-5y = -2x + 6$$
$$\frac{-5y}{-5} = \frac{-2x + 6}{-5}$$
$$y = \frac{-2x}{-5} + \frac{6}{-5}$$
$$y = \frac{2}{5}x - \frac{6}{5}$$

63. $3x + 4y = 8$

Solve for *y*.
$4y = -3x + 8$
$$y = -\frac{3}{4}x + 2$$

x	*y*
–4	$y = -\frac{3}{4}(-4) + 2 = 3 + 2 = 5$
0	$y = -\frac{3}{4}(0) + 2 = 0 + 2 = 2$
4	$y = -\frac{3}{4}(4) + 2 = -3 + 2 = -1$

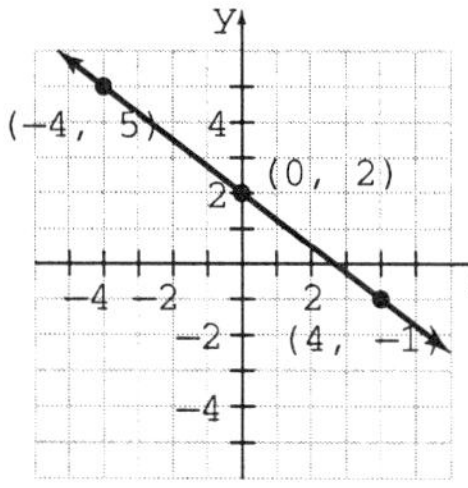

64. $3x + 4y = 8$

For the *x*-intercept, let $y = 0$.

$$3x + 4y = 8$$
$$3x + 4(0) = 8$$
$$3x + 0 = 8$$
$$3x = 8$$
$$x = \frac{8}{3} = 2\frac{2}{3}$$

For the *y*-intercept, let $x = 0$.

$$3x + 4y = 8$$
$$3(0) + 4y = 8$$
$$0 + 4y = 8$$
$$4y = 8$$
$$y = 2$$

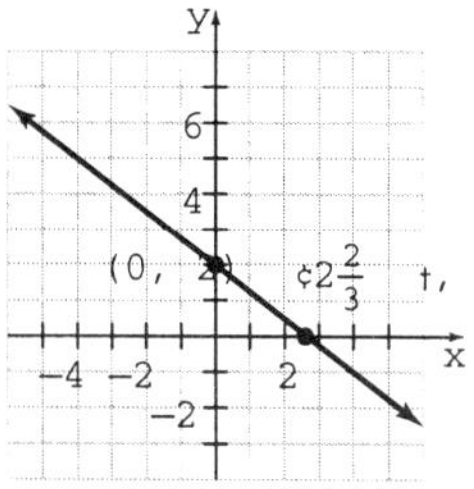

65. $3x + 4y = 8$

Solve for *y*.
$4y = -3x + 8$
$$y = -\frac{3}{4}x + 2$$

The slope is $-\frac{3}{4}$ and the *y*-intercept is 2.

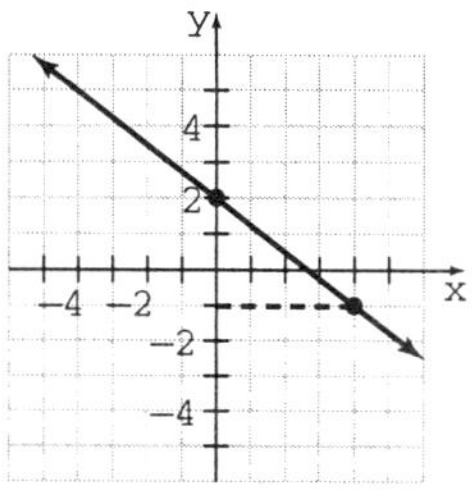

Review Exercises

1. a. $y = 4x - 2$ $\quad 2x + 3y = 8$

$-2 = 4(0) - 2 \quad 2(0) + 3(-2) = 8$

$-2 = -2$ True $\quad -6 = 8$ False

Since (0, –2) does not satisfy both equations, it is not a solution to the system.

b. $y = 4x - 2$

$4 = 4(-2) - 2$

$4 = -10$ False

Since (-2, 4) does not satisfy the first equation, it is not a solution to the system.

c. $y = 4x - 2 \quad 2x + 3y = 8$

$2 = 4(1) - 2 \quad 2(1) + 3(2) = 8$

$2 = 2$ True $\quad 8 = 8$ True

Since (1, 2) satisfies both equations, it is a solution to the system.

2. a. $y = -x + 4$

$\frac{3}{2} = -\frac{5}{2} + 4$

$\frac{3}{2} = \frac{3}{2}$ True

$3x + 5y = 15$

$3\left(\frac{5}{2}\right) + 5\left(\frac{3}{2}\right) = 15$

$15 = 15$ True

Since $\left(\frac{5}{2}, \frac{3}{2}\right)$ satisfies both equations, it is a solution to the system.

b. $y = -x + 4$

$4 = -0 + 4$

$4 = 4$ True

$3x + 5y = 15$

$3(0) + 5(4) = 15$

$20 = 15$ False

Since (0, 4) does not satisfy both equations, it is not a solution to the system.

c. $y = -x + 4$

$\frac{3}{5} = -\frac{1}{2} + 4$

$\frac{3}{5} = \frac{7}{2}$ False

Since $\left(\frac{1}{2}, \frac{3}{5}\right)$ does not satisfy the first equation, it is not a solution to the system.

3. consistent, one solution

4. inconsistent, no solutions

5. dependent, infinite number of solutions

6. consistent, one solution

7. Write each equation in slope-intercept form.

$x + 2y = 10 \qquad 3x = -6y + 12$

$2y = -x + 10 \qquad 6y = -3x + 12$

$y = -\frac{1}{2}x + 5 \qquad y = -\frac{1}{2}x + 2$

Since the slope of each line is $-\frac{1}{2}$ but the y-intercepts are different, the two lines are parallel. There is no solution. This is an inconsistent system.

8. Write each equation in slope-intercept form.

$y = -3x - 6$ is already in this form.

$2x + 5y = 8$

$5y = -2x + 8$

$y = -\frac{2}{5}x + \frac{8}{5}$

Since the slopes of the lines are different, the lines intersect to produce one solution. This is a consistent system.

9. Write each equation in slope-intercept form.

$y = \frac{1}{2}x - 4$ is already in this form.

$x - 2y = 8$

$-2y = -x + 8$

$y = \frac{1}{2}x - 4$

Since both equations are identical, the line is the same for both of them. There are an infinite number of solutions. This is a dependent system.

10. Write each equation in slope-intercept form.

$6x = 4y - 8 \qquad 4x = 6y + 8$

$6x + 8 = 4y \qquad 4x - 8 = 6y$

$\frac{6x + 8}{4} = y \qquad \frac{4x - 8}{6} = y$

$\frac{3}{2}x + 2 = y \qquad \frac{2}{3}x - \frac{4}{3} = y$

Since the slopes of the lines are different, the lines intersect to produce one solution. This is a consistent system.

11. Graph $y = x - 4$ and $y = 2x - 7$.

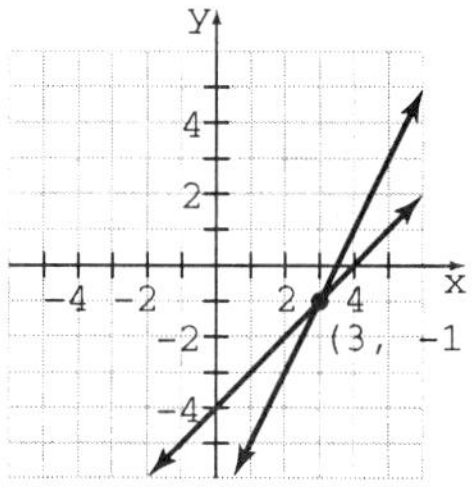

The lines intersect and the point of intersection is $(3, -1)$. This is a consistent system.

12. Graph $x = -2$ and $y = 3$.

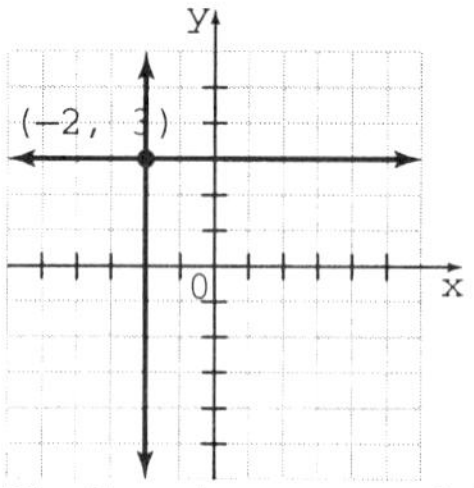

The lines intersect and the point of intersection is $(-2, 3)$. This is a consistent system.

13. Graph $y = 3$ and $y = -2x + 5$.

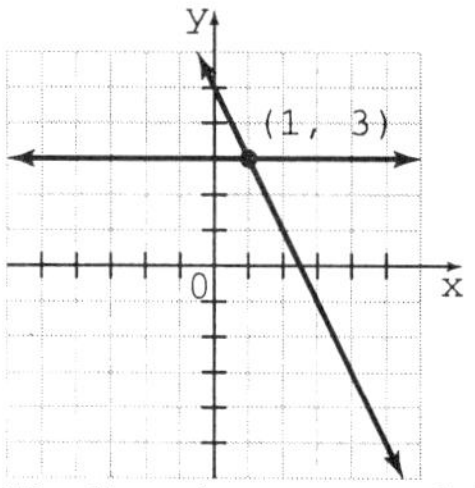

The lines intersect and the point of intersection is $(1, 3)$. This is a consistent system.

14. Graph $x + 3y = 6$ and $y = 2$.

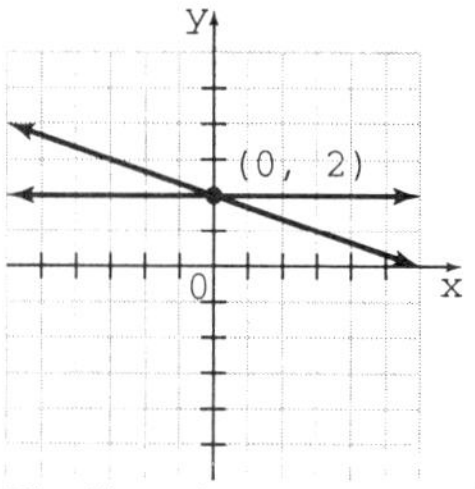

The lines intersect and the point of intersection is $(0, 2)$. This is a consistent system.

15. Graph the equations $x + 2y = 8$ and $2x - y = -4$.

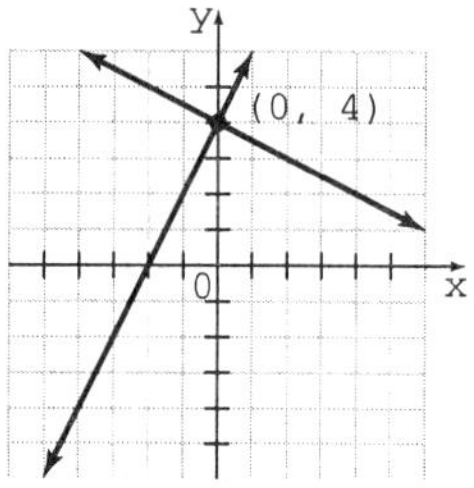

The lines intersect and the point of intersection is $(0, 4)$. This is a consistent system.

16. Graph the equations $y = x - 3$ and $2x - 2y = 6$.

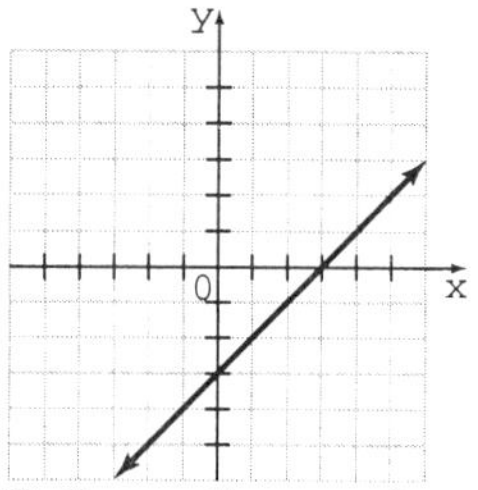

Both equations produce the same line. This is a dependent system. There are an infinite number of solutions.

17. Graph $3x + y = 0$ and $3x - 3y = 12$.

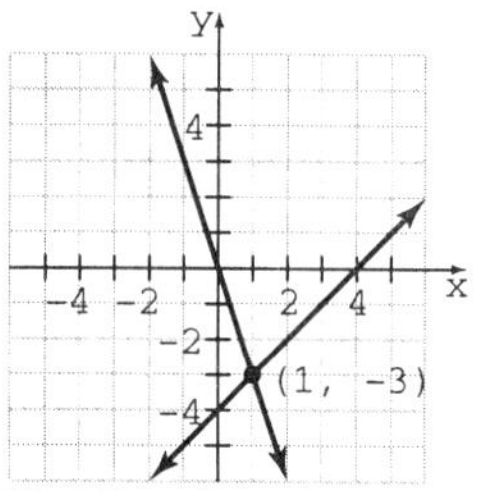

The lines intersect and the point of intersection is $(1, -3)$. This is a consistent system.

18. Graph $x + 2y = 4$ and $\frac{1}{2}x + y = -2$.

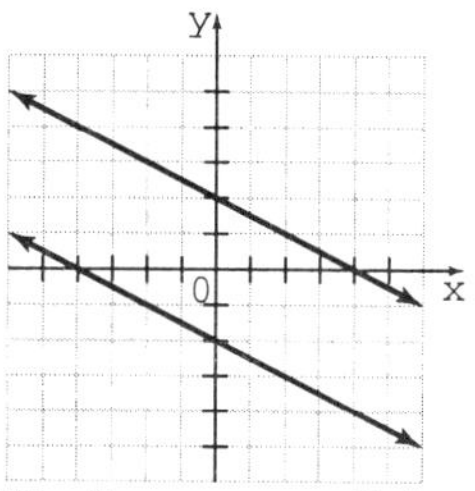

The lines are parallel and do not intersect. The system is inconsistent and there is no solution.

19. $y = 3x - 13$
$2x - 5y = 0$
Substitute $3x - 13$ for y in the second equation.
$$2x - 5y = 0$$
$$2x - 5(3x - 13) = 0$$
$$2x - 15x + 65 = 0$$
$$-13x + 65 = 0$$
$$-13x = -65$$
$$x = 5$$
Now, substitute 5 for x in the first equation.
$$y = 3x - 13$$
$$y = 3(5) - 13$$
$$y = 15 - 3$$
$$y = 2$$
The solution is (5, 2).

20. $x = 3y - 9$
$x + 2y = 1$
Substitute $3y - 9$ for x in the second equation.
$$x + 2y = 1$$
$$3y - 9 + 2y = 1$$
$$5y - 9 = 1$$
$$5y = 10$$
$$y = 2$$
Now substitute 2 for y in the first equation.
$$x = 3y - 9$$
$$x = 3(2) - 9$$
$$x = 6 - 9$$
$$x = -3$$
The solution is (–3, 2).

21. $2x - y = 6$
$x + 2y = 13$
Solve the second equation for x, $x = 13 - 2y$.
Substitute $13 - 2y$ for x in the first equation.
$$2x - y = 6$$
$$2(13 - 2y) - y = 6$$
$$26 - 4y - y = 6$$
$$-5y = -20$$
$$y = 4$$
Substitute 4 for y in the equation $x = 13 - 2y$.
$$x = 13 - 2(4)$$
$$x = 13 - 8$$
$$x = 5$$
The solution is (5, 4).

22. $x = -3y$
$x + 4y = 6$
Substitute $-3y$ for x in the second equation.
$$x + 4y = 6$$
$$-3y + 4y = 6$$
$$y = 6$$
Substitute 6 for y in the first equation.
$$x = -3y$$
$$x = -3(6)$$
$$x = -18$$
The solution is (–18, 6).

23. $4x - 2y = 10$
$y = 2x + 3$
Substitute $2x + 3$ for y in the first equation.
$$4x - 2y = 10$$
$$4x - 2(2x + 3) = 10$$
$$4x - 4x - 6 = 10$$
$$-6 = 10 \quad \text{False}$$
There is no solution.

24. $2x + 4y = 8$
$4x + 8y = 16$
Solve the first equation for x.
$$\frac{1}{2}(2x + 4y) = \frac{1}{2}(8)$$
$$x + 2y = 4$$
$$x = 4 - 2y$$
Substitute $4 - 2y$ for x in the second equation.
$$4x + 8y = 16$$
$$4(4 - 2y) + 8y = 16$$
$$16 - 8y + 8y = 16$$
$$16 = 16 \quad \text{True}$$
There are an infinite number of solutions.

25. $2x - 3y = 8$
$6x + 5y = 10$
Solve the first equation for x.
$$\frac{1}{2}(2x - 3y) = \frac{1}{2}(8)$$
$$x - \frac{3}{2}y = 4$$
$$x = \frac{3}{2}y + 4$$
Substitute $\frac{3}{2}y + 4$ for x in the second equation.
$$6x + 5y = 10$$
$$6\left(\frac{3}{2}y + 4\right) + 5y = 10$$
$$9y + 24 + 5y = 10$$
$$14y = -14$$
$$y = -1$$

Substitute -1 for y in the equation $x = \frac{3}{2}y + 4$.

$$x = \frac{3}{2}(-1) + 4$$
$$x = -\frac{3}{2} + \frac{8}{2}$$
$$x = \frac{5}{2}$$

The solution is $\left(\frac{5}{2}, -1\right)$.

26. $3x - y = -5$
$x + 2y = 8$
Solve the second equation for x, $x = 8 - 2y$.
Substitute $8 - 2y$ for x in the first equation.

$$3x - y = -5$$
$$3(8 - 2y) - y = -5$$
$$24 - 6y - y = -5$$
$$-7y = -29$$
$$y = \frac{29}{7}$$

Substitute $\frac{29}{7}$ for y in the equation $x = 8 - 2y$.

$$x = 8 - 2\left(\frac{29}{7}\right)$$
$$x = \frac{56}{7} - \frac{58}{7}$$
$$x = -\frac{2}{7}$$

The solution is $\left(-\frac{2}{7}, \frac{29}{7}\right)$.

27.
$$\begin{array}{r} -x - y = 6 \\ \underline{-x + y = 10} \\ -2x \quad = 16 \end{array}$$
$$x = -8$$
$$x = \frac{16}{2} = -8$$

Substitute -8 for x in the first equation.

$$-x - y = 6$$
$$-(-8) - y = 6$$
$$8 - y = 6$$
$$-y = -2$$
$$y = 2$$

The solution is (-8, 2).

28.
$$\begin{array}{r} x + 2y = -3 \\ \underline{2x - 2y = \ \ 6} \\ 3x \quad = \ \ 3 \end{array}$$
$$x = 1$$

Substitute 1 for x in the first equation.

$$x + 2y = -3$$
$$1 + 2y = -3$$
$$2y = -4$$
$$y = -2$$

The solution is (1, –2).

29. $x + y = 12$
$2x + y = 5$
To eliminate y, multiply the first equation by –1 and then add.
$-1[x + y = 12]$
gives

$$\begin{array}{r} -x - y = -12 \\ \underline{2x + y = \ \ \ 5} \\ x \quad = -7 \end{array}$$

Substitute –7 for x in the first equation.

$$x + y = 12$$
$$-7 + y = 12$$
$$y = 19$$

The solution is (–7, 19).

30. $4x - 3y = 8$
$2x + 5y = 8$
To eliminate x, multiply the second equation by –2 and then add.
$-2[2x + 5y = 8]$
gives

$$\begin{array}{r} 4x - 3y = \ \ \ 8 \\ \underline{-4x - 10y = -16} \\ -13y = -8 \end{array}$$
$$y = \frac{8}{13}$$

Substitute $\frac{8}{13}$ for y in the first equation.

$$4x - 3y = 8$$
$$4x - 3\left(\frac{8}{13}\right) = 8$$
$$4x - \frac{24}{13} = 8$$
$$4x = \frac{104}{13} + \frac{24}{13}$$
$$4x = \frac{128}{13}$$
$$x = \frac{32}{13}$$

The solution is $\left(\frac{32}{13}, \frac{8}{13}\right)$

31. $-2x + 3y = 15$
$3x + 3y = 10$
To eliminate y, multiply the second equation by -1 and then add.
$-1[3x + 3y = 10]$
gives
$$\begin{aligned} -2x + 3y &= 15 \\ -3x - 3y &= -10 \\ \hline -5x &= 5 \\ x &= -1 \end{aligned}$$
Substitute -1 for x in the second equation.
$$\begin{aligned} 3x + 3y &= 10 \\ 3(-1) + 3y &= 10 \\ -3 + 3y &= 10 \\ 3y &= 13 \\ y &= \frac{13}{3} \end{aligned}$$
The solution is $\left(-1, \frac{13}{3}\right)$.

32. $2x + y = 9$
$-4x - 2y = 4$
Multiply the first equation by 2, and then add.
$2[2x + y = 9]$
gives
$$\begin{aligned} 4x + 2y &= 18 \\ -4x - 2y &= 4 \\ \hline 0 &= 22 \quad \text{False} \end{aligned}$$
There is no solution.

33. $3x = -4y + 10$
$8y = -6x + 20$
Align the x and y terms.

$3x + 4y = 10$
$-6x - 8y = -20$
To eliminate x, multiply the first equation by 2, and then add.
$2[3x + 4y = 10]$
gives
$$\begin{aligned} 6x + 8y &= 20 \\ -6x - 8y &= -20 \\ \hline 0 &= 0 \quad \text{True} \end{aligned}$$
There are an infinite number of solutions.

34. $2x - 5y = 12$
$3x - 4y = -6$
To eliminate x, multiply the first equation by -3 and the second equation by 2 and then add.
$-3[2x - 5y = 12]$
$2[3x - 4y = -6]$
gives
$$\begin{aligned} -6x + 15y &= -36 \\ 6x - 8y &= -12 \\ \hline 7y &= -48 \\ y &= -\frac{48}{7} \end{aligned}$$
Now, substitute $-\frac{48}{7}$ for y in the first equation.
$$\begin{aligned} 2x - 5y &= 12 \\ 2x - 5\left(-\frac{48}{7}\right) &= 12 \\ 2x + \frac{240}{7} &= 12 \\ 2x &= \frac{84}{7} - \frac{240}{7} \\ 2x &= -\frac{156}{7} \\ x &= -\frac{78}{7} \end{aligned}$$
The solution is $\left(-\frac{78}{7}, -\frac{48}{7}\right)$.

35.
$$\begin{aligned} x - 2y - 4z &= 13 \quad (1) \\ 3y + 2z &= -2 \quad (2) \\ 5z &= -20 \quad (3) \end{aligned}$$
Solve equation (3) for z.
$5z = -20$
$z = -4$
Substitute -4 for z in equation (2).
$$\begin{aligned} 3y + 2(-4) &= -2 \\ 3y - 8 &= -2 \\ 3y &= 6 \\ y &= 2 \end{aligned}$$
Substitute -4 for z and 2 for y in equation (1).
$$\begin{aligned} x - 2(2) - 4(-4) &= 13 \\ x - 4 + 16 &= 13 \\ x + 12 &= 13 \\ x &= 1 \end{aligned}$$
The solution is $(1, 2, -4)$.

36.
$$\begin{aligned} 2a + b - 2c &= 5 \\ 3b + 4c &= 1 \\ 3c &= -6 \end{aligned}$$
Solve the third equation for c.
$3c = -6$
$c = -2$
Substitute -2 for c in the second equation.

$3b+4(-2)=1$

$3b-8=1$

$3b=9$

$b=3$

Substitute 3 for b and –2 for c in the first equation.

$2a+(3)-2(-2)=5$

$2a+3+4=5$

$2a+7=5$

$2a=-2$

$a=-1$

The solution is (–1, 3, –2).

37. $x+2y+3z=3$ (1)

$-2x-3y-z=5$ (2)

$4x+2y+5z=-8$ (3)

To eliminate x between equations (1) and (2), multiply equation (1) by 2 and then add.

$2[x+2y+3z=3]$

$-2x-3y-z=5$

gives

$2x+4y+6z=6$

$-2x-3y-z=5$

Add: $y+5z=11$ (4)

To eliminate x between equations (2) and (3), multiply equation (2) by 2 and then add.

$2[-2x-3y-z=5]$

$4x+2y+5z=-8$

gives

$-4x-6y-2z=10$

$4x+2y+5z=-8$

Add: $-4y+3z=2$ (5)

Equations (4) and (5) are two equations in two unknowns.

$y+5z=11$ (4)

$-4y+3z=2$ (5)

To eliminate y, multiply equation (4) by 4 and then add.

$4[y+5z=11]$

$-4y+3z=2$

gives

$4y+20z=44$

$-4y+3z=2$

Add: $23z=46$

$z=2$

Substitute 2 for z in equation (4).

$y+5(2)=11$

$y+10=11$

$y=1$

Finally, substitute 1 for y and 2 for z in equation (1).

$x+2(1)+3(2)=3$

$x+2+6=3$

$x+8=3$

$x=-5$

The solution is (–5, 1, 2).

38. $-x-4y+2z=1$ (1)

$2x+2y+z=0$ (2)

$-3x-2y-5z=5$ (3)

To eliminate y between equations (1) and (2), multiply equation (2) by 2 and add.

$-x-4y+2z=1$

$2[2x+2y+z=0]$

gives

$-x-4y+2z=1$

$4x+4y+2z=0$

Add: $3x+4z=1$ (4)

To eliminate y between equations (2) and (3), simply add.

$2x+2y+z=0$

$-3x-2y-5z=5$

Add: $-x-4z=5$ (5)

Equations (4) and (5) are two equations in two unknowns.

$3x+4z=1$ (4)

$-x-4z=5$ (5)

To eliminate z, simply add equations (4) and (5).

$3x+4z=1$

$-x-4z=5$

Add: $2x=6$

$x=3$

Substitute 3 for x in equation (4).

$3(3)+4z=1$

$9+4z=1$

$4z=-8$

$z=-2$

Finally, substitute 3 for x and –2 for z in equation (1).

$$-(3) - 4y + 2(-2) = 1$$
$$-3 - 4y - 4 = 1$$
$$-7 - 4y = 1$$
$$-4y = 8$$
$$y = -2$$

The solution is (3, –2, –2).

39. $3y - 2z = -4$ (1)
$3x - 5z = -7$ (2)
$2x + y = 6$ (3)

To eliminate y between equations (1) and (3), multiply equation (3) by –3 and then add.

$3y - 2z = -4$
$-3[2x + y = 6]$

gives

$$\begin{array}{rrrl} & 3y & -2z & = -4 \\ -6x & -3y & & = -18 \\ \hline \text{Add: } -6x & & -2z & = -22 \end{array}$$

or

$-3x - z = -11$ (4)

Equations (4) and (2) are two equations into two unknowns. To eliminate x, simply add.

$$\begin{array}{rrl} 3x & -5z & = -7 \\ -3x & - z & = -11 \\ \hline \text{Add: } & -6z & = -18 \end{array}$$
$$z = 3$$

Substitute 3 for z in equation (2).

$$3x - 5z = -7$$
$$3x - 5(3) = -7$$
$$3x - 15 = -7$$
$$3x = 8$$
$$x = \frac{8}{3}$$

Substitute $\frac{8}{3}$ for x in equation (3).

$$2x + y = 6$$
$$2\left(\frac{8}{3}\right) + y = 6$$
$$\frac{16}{3} + y = 6$$
$$y = 6 - \frac{16}{3}$$
$$y = \frac{18}{3} - \frac{16}{3}$$
$$y = \frac{2}{3}$$

The solution is $\left(\frac{8}{3}, \frac{2}{3}, 3\right)$.

40. $3a + 2b - 5c = 19$ (1)
$2a - 3b + 3c = -15$ (2)
$5a - 4b - 2c = -2$ (3)

To eliminate b between equations (1) and (2), multiply equation (1) by 3 and equation (2) by 2 and then add.

$3[3a + 2b - 5c = 19]$
$2[2a - 3b + 3c = -15]$

gives

$$\begin{array}{rrrl} 9a & +6b & -15c & = 57 \\ 4a & -6b & +6c & = -30 \\ \hline \text{Add: } 13a & & -9c & = 27 \quad (4) \end{array}$$

To eliminate b between equations (1) and (3), multiply equation (1) by 2 and then add.

$2[3a + 2b - 5c = 19]$
$5a - 4b - 2c = -2$

gives

$$\begin{array}{rrrl} 6a & +4b & -10c & = 38 \\ 5a & -4b & -2c & = -2 \\ \hline \text{Add: } 11a & & -12c & = 36 \quad (5) \end{array}$$

Equations (4) and (5) are two equations into two unknowns.

$13a - 9c = 27$
$11a - 12c = 36$

To eliminate c, multiply equation (4) by 4 and equation (5) by –3 and then add.

$4[13a - 9c = 27]$
$-3[11a - 12c = 36]$

gives

$$\begin{array}{rrl} 52a & -36c & = 108 \\ -33a & +36c & = -108 \\ \hline \text{Add: } 19a & & = 0 \end{array}$$
$$a = 0$$

Substitute 0 for a in equation (4).

$$13a - 9c = 27$$
$$13(0) - 9c = 27$$
$$-9c = 27$$
$$c = \frac{27}{-9} = -3$$

Finally, substitute 0 for a and –3 for c in equation (1).

$$3a + 2b - 5c = 19$$
$$3(0) + 2b - 5(-3) = 19$$
$$0 + 2a + 15 = 19$$
$$2a + 15 = 19$$
$$2a = 4$$
$$a = \frac{4}{2} = 2$$

The solution is (0, 2, –3).

41. $x - y + 3z = 1$ (1)
$-x + 2y - 2z = 1$ (2)
$x - 3y + z = 2$ (3)

To eliminate x between equations (1) and (2),

simply add.

$$\begin{array}{l} x - y + 3z = 1 \\ -x + 2y - 2z = 1 \\ \hline \text{Add:} \quad y + z = 2 \quad (4) \end{array}$$

To eliminate x between equations (2) and (3), simply add.

$$\begin{array}{l} -x + 2y - 2z = 1 \\ x - 3y + z = 2 \\ \hline \text{Add:} \quad -y - z = 3 \quad (5) \end{array}$$

Equations (4) and (5) are two equations in two unknowns. To eliminate y and z simply add.

$$\begin{array}{l} y + z = 2 \\ -y - z = 3 \\ \hline \text{Add:} \quad 0 = 5 \quad \text{False} \end{array}$$

Since this is a false statement, there is no solution to the system. This is an inconsistent system.

42. $-2x + 2y - 3z = 6$ (1)

$4x - y + 2z = -2$ (2)

$2x + y - z = 4$ (3)

To eliminate x between equations (1) and (2), multiply equation (1) by 2 and then add.

$2[-2x + 2y - 3z = 6]$

$4x - y + 2z = -2$

gives

$$\begin{array}{l} -4x + 4y - 6z = 12 \\ 4x - y + 2z = -2 \\ \hline \text{Add:} \quad 3y - 4z = 10 \quad (4) \end{array}$$

To eliminate x between equations (1) and (3), simply add.

$$\begin{array}{l} -2x + 2y - 3z = 6 \\ 2x + y - z = 4 \\ \hline \text{Add:} \quad 3y - 4z = 10 \quad (5) \end{array}$$

Equations (4) and (5) are two equations in two unknowns.

$3y - 4z = 10$

$3y - 4z = 10$

Since they are identical, there are an infinite number of solutions. This is a dependent system.

43. Let x be Jennifer's age and y be Dennis' age.

$y = x + 10$

$x + y = 66$

Substitute $x + 10$ for y in the second equation.

$x + (x + 10) = 66$

$2x + 10 = 66$

$2x = 56$

$x = 28$

Now substitute 28 for x in the first equation.

$y = (28) + 10$

$y = 38$

Jennifer is 28 years old and Dennis is 38 years old.

44. Let x be the speed of the plane in still air and y be the speed of the wind.

$x + y = 560$

$x - y = 480$

To eliminate y, simply add.

$$\begin{array}{l} x + y = 560 \\ x - y = 480 \\ \hline \text{Add:} \quad 2x = 1040 \\ \quad x = 520 \end{array}$$

Substitute 520 for x in the first equation.

$x + y = 560$

$(520) + y = 560$

$y = 40$

The speed of the plane in still air is 520 mph and the speed of the wind is 40 mph.

45. Let x be the amount of 20% acid solution and y be the amount of 50% acid solution.

$x + y = 6$

$0.2x + 0.5y = 0.4(6)$

To clear decimals, multiply the second equation by 10.

$x + y = 6$

$2x + 5y = 24$

Solve the first equation for y.

$x + y = 6$

$y = -x + 6$

Substitute $-x + 6$ for y in the second equation.

$2x + 5y = 24$

$2x + 5(-x + 6) = 24$

$2x - 5x + 30 = -24$

$-3x + 30 = 24$

$-3x = -6$

$x = \frac{-6}{-3} = 2$

Finally, substitute 3 for x in the equation $y = -x + 6$.

$y = -x + 6$

$y = -3 + 6$

$y = 4$

James should combine 2 liters of the 20% acid solution to 4 liters of the 50% acid solution.

46. Let x be the number of adult tickets and y be the number of children's tickets.

$x + y = 650$

$15x + 11y = 8790$

To solve, multiply the first equation by -11 and then add.

$-11[x + y = 650]$
$15x + 11y = 8790$
gives
$$\begin{array}{rl} -11x - 11y = & -7150 \\ 15x + 11y = & 8790 \\ \hline \text{Add: } 4x \quad = & 1640 \end{array}$$
$$x = \frac{1640}{4} = 410$$
Substitute 410 for x in the first equation.
$x + y = 650$
$410 + y = 650$
$y = 240$
Thus, 410 adult tickets and 240 children's tickets were sold.

47. Let x = age at first time and y = age at second time.
$y = 2x - 5$
$x + y = 118$
Substitute $2x - 5$ for y in the second equation.
$x + y = 118$
$x + 2x - 5 = 118$
$3x - 5 = 118$
$3x = 123$
$x = 41$
Substitute 41 for x in the first equation.
$y = 2x - 5$
$y = 2(41) - 5$
$y = 82 - 5$
$y = 77$
His ages were 41 years and 77 years.

48. Let x be the amount invested at 7%, y the amount invested at 5%, and z the amount invested at 3%.
$x + y + z = 40{,}000$ (1)
$y = x - 5000$ (2)
$0.07x + 0.05y + 0.03z = 2300$ (3)
Substitute $x - 5000$ for y in equations (1) and (3).
Equation (1) becomes
$x + y + z = 40{,}000$
$x + x - 5000 + z = 40{,}000$
$2x + z = 45{,}000$ (4)
Equation (3) becomes
$0.07x + 0.05y + 0.03z = 2300$
$0.07x + 0.05(x - 5000) + 0.03z = 2300$
$0.07x + 0.05x - 250 + 0.03z = 2300$
$0.12x + 0.03z = 2550$ (5)
Equation (4) and (5) are a system of two equations in two unknowns. Solve equation (4) for z.
$2x + z = 45{,}000$
$z = -2x + 45{,}000$
Substitute $-2x + 45{,}000$ for z in equation (5).
$0.12x + 0.03z = 2550$
$0.12x + 0.03(-2x + 45{,}000) = 2550$
$0.12x - 0.06x + 1350 = 2550$
$0.06x = 1200$
$x = 20{,}000$
Now substitute 20,000 for x in equation (2).
$y = x - 5000$
$y = 20{,}000 - 5000 = 15{,}000$
Finally, substitute 20,000 for x and 15,000 for y in equation (1).
$x + y + z = 40{,}000$
$20{,}000 + 15{,}000 + z = 40{,}000$
$35{,}000 + z = 40{,}000$
$z = 5000$
Thus, \$20,000 was invested at 7%, \$15,000 at 5%, and \$5000 at 3%.

49. $x + 5y = 1$
$-2x - 8y = -6$
$$\left[\begin{array}{cc|c} 1 & 5 & 1 \\ -2 & -8 & -6 \end{array}\right]$$
$$\left[\begin{array}{cc|c} 1 & 5 & 1 \\ 0 & 2 & -4 \end{array}\right] 2R_1 + R_2$$
$$\left[\begin{array}{cc|c} 1 & 5 & 1 \\ 0 & 1 & -2 \end{array}\right] \tfrac{1}{2}R_2$$
The system is
$x + 5y = 1$
$y = -2$
Substitute -2 for y in the first equation.
$x + 5(-2) = 1$
$x - 10 = 1$
$x = 11$
The solution is $(11, -2)$.

50. $2x - 3y = 3$
$2x + 4y = 10$
$$\left[\begin{array}{cc|c} 2 & -3 & 3 \\ 2 & 4 & 10 \end{array}\right]$$
$$\left[\begin{array}{cc|c} 1 & \frac{-3}{2} & \frac{3}{2} \\ 2 & 4 & 10 \end{array}\right] \tfrac{1}{2}R_1$$
$$\left[\begin{array}{cc|c} 1 & \frac{-3}{2} & \frac{3}{2} \\ 0 & 7 & 7 \end{array}\right] -2R_1 + R_2$$
$$\left[\begin{array}{cc|c} 1 & \frac{-3}{2} & \frac{3}{2} \\ 0 & 1 & 1 \end{array}\right] \tfrac{1}{7}R_2$$
The system is
$x - \frac{3}{2}y = \frac{3}{2}$
$y = 1$
Substitute 1 for y in the first equation.

$$x-\frac{3}{2}(1)=\frac{3}{2}$$
$$x-\frac{3}{2}=\frac{3}{2}$$
$$x=\frac{6}{2}$$
$$x=3$$

The solution is (3, 1).

51. $y=2x-4$
$4x=2y+8$

Write the system in standard form.

$-2x+y=-4$
$4x-2y=8$

$$\left[\begin{array}{cc|c} -2 & 1 & -4 \\ 4 & -2 & 8 \end{array}\right]$$

$$\left[\begin{array}{cc|c} 1 & -\frac{1}{2} & 2 \\ 4 & -2 & 8 \end{array}\right] -\tfrac{1}{2}R_1$$

$$\left[\begin{array}{cc|c} 1 & -\frac{1}{4} & 2 \\ 0 & 0 & 0 \end{array}\right] -4R_1+R_2$$

Since the last row is all zeros, the system is dependent.

52. $2x-y-z=5$
$x+2y+3z=-2$
$3x-2y+z=2$

$$\left[\begin{array}{ccc|c} 2 & -1 & -1 & 5 \\ 1 & 2 & 3 & -2 \\ 3 & -2 & 1 & 2 \end{array}\right]$$

$$\left[\begin{array}{ccc|c} 1 & -\frac{1}{2} & -\frac{1}{2} & \frac{5}{2} \\ 1 & 2 & 3 & -2 \\ 3 & -2 & 1 & 2 \end{array}\right] \tfrac{1}{2}R_1$$

$$\left[\begin{array}{ccc|c} 1 & -\frac{1}{2} & -\frac{1}{2} & \frac{5}{2} \\ 0 & \frac{5}{2} & \frac{7}{2} & -\frac{9}{2} \\ 3 & -2 & 1 & 2 \end{array}\right] -1R_1+R_2$$

$$\left[\begin{array}{ccc|c} 1 & -\frac{1}{2} & -\frac{1}{2} & \frac{5}{2} \\ 0 & \frac{5}{2} & \frac{7}{2} & -\frac{9}{2} \\ 0 & -\frac{1}{2} & \frac{5}{2} & -\frac{11}{2} \end{array}\right] -3R_1+R_3$$

$$\left[\begin{array}{ccc|c} 1 & -\frac{1}{2} & -\frac{1}{2} & \frac{5}{2} \\ 0 & 1 & \frac{7}{5} & -\frac{9}{5} \\ 0 & -\frac{1}{2} & \frac{5}{2} & -\frac{11}{2} \end{array}\right] \tfrac{2}{5}R_2$$

$$\left[\begin{array}{ccc|c} 1 & -\frac{1}{2} & -\frac{1}{2} & \frac{5}{2} \\ 0 & 1 & \frac{7}{5} & -\frac{9}{5} \\ 0 & 0 & \frac{16}{5} & -\frac{32}{5} \end{array}\right] \tfrac{1}{2}R_2+R_3$$

$$\left[\begin{array}{ccc|c} 1 & -\frac{1}{2} & -\frac{1}{2} & \frac{5}{2} \\ 0 & 1 & \frac{7}{5} & -\frac{9}{5} \\ 0 & 0 & 1 & -2 \end{array}\right] \tfrac{5}{16}R_3$$

The system is

$$x-\frac{1}{2}y-\frac{1}{2}z=\frac{5}{2}$$
$$y+\frac{7}{5}z=-\frac{9}{5}$$
$$z=-2$$

Substitute −2 for z in the second equation.

$$y+\frac{7}{5}z=-\frac{9}{5}$$
$$y+\frac{7}{5}(-2)=-\frac{9}{5}$$
$$y-\frac{14}{5}=-\frac{9}{5}$$
$$y=\frac{5}{5}=1$$

Substitute 1 for y and −2 for z in the first equation.

$$x-\frac{1}{2}y-\frac{1}{2}z=\frac{5}{2}$$
$$x-\frac{1}{2}(1)-\frac{1}{2}(-2)=\frac{5}{2}$$
$$x-\frac{1}{2}+1=\frac{5}{2}$$
$$x+\frac{1}{2}=\frac{5}{2}$$
$$x=\frac{4}{2}=2$$

The solution is (2, 1, −2).

53. $3a-b+c=2$
$2a-3b+4c=4$
$a+2b-3c=-6$

$$\left[\begin{array}{ccc|c} 3 & -1 & 1 & 2 \\ 2 & -3 & 4 & 4 \\ 1 & 2 & -3 & -6 \end{array}\right]$$

$$\left[\begin{array}{ccc|c} 1 & -\frac{1}{3} & \frac{1}{3} & \frac{2}{3} \\ 2 & -3 & 4 & 4 \\ 1 & 2 & -3 & -6 \end{array}\right] \tfrac{1}{3}R_1$$

$$\left[\begin{array}{ccc|c} 1 & -\frac{1}{3} & \frac{1}{3} & \frac{2}{3} \\ 0 & -\frac{7}{3} & \frac{10}{3} & \frac{8}{3} \\ 1 & 2 & -3 & -6 \end{array}\right] -2R_1+R_2$$

$$\left[\begin{array}{ccc|c} 1 & -\frac{1}{3} & \frac{1}{3} & \frac{2}{3} \\ 0 & -\frac{7}{3} & \frac{10}{3} & \frac{8}{3} \\ 0 & \frac{7}{3} & -\frac{10}{3} & -\frac{20}{3} \end{array}\right] -1R_1+R_3$$

$$\left[\begin{array}{ccc|c} 1 & -\frac{1}{3} & \frac{1}{3} & \frac{2}{3} \\ 0 & 1 & -\frac{10}{7} & -\frac{8}{7} \\ 0 & \frac{7}{3} & -\frac{10}{3} & -\frac{20}{3} \end{array}\right] -\frac{3}{7}R_2$$

$$\left[\begin{array}{ccc|c} 1 & -\frac{1}{3} & \frac{1}{3} & \frac{2}{3} \\ 0 & 1 & -\frac{10}{7} & -\frac{8}{7} \\ 0 & 0 & 0 & -4 \end{array}\right]$$

Since the last row has all zeros on the left side and a nonzero number on the right side, the system is inconsistent.

54. $x+y+z=3$
$3x+2y=1$
$y-3z=-10$

$$\left[\begin{array}{ccc|c} 1 & 1 & 1 & 3 \\ 3 & 2 & 0 & 1 \\ 0 & 1 & -3 & -10 \end{array}\right]$$

$$\left[\begin{array}{ccc|c} 1 & 1 & 1 & 3 \\ 0 & -1 & -3 & -8 \\ 0 & 1 & -3 & -10 \end{array}\right] -3R_1+R_2$$

$$\left[\begin{array}{ccc|c} 1 & 1 & 1 & 3 \\ 0 & 1 & 3 & 8 \\ 0 & 1 & -3 & -10 \end{array}\right] -1R_2$$

$$\left[\begin{array}{ccc|c} 1 & 1 & 1 & 3 \\ 0 & 1 & 3 & 8 \\ 0 & 0 & -6 & -18 \end{array}\right] -1R_2+R_3$$

$$\left[\begin{array}{ccc|c} 1 & 1 & 1 & 3 \\ 0 & 1 & 3 & 8 \\ 0 & 0 & 1 & 3 \end{array}\right] -\frac{1}{6}R_3$$

The system is
$x+y+z=3$
$y+3z=8$
$z=3$

Substitute 3 for z in the second equation.
$y+3z=8$
$y+3(3)=8$
$y+9=8$
$y=-1$

Substitute −1 for y and 3 for z in the first equation.
$x+y+z=3$
$x-1+3=3$
$x+2=3$
$x=1$

The solution is (1, −1, 3).

55. $7x-8y=-10$
$-5x+4y=2$

To solve, first calculate D, D_x, and D_y.

$$D=\begin{vmatrix} 7 & -8 \\ -5 & 4 \end{vmatrix} = (7)(4)-(-5)(-8) = 28-40 = -12$$

$$D_x=\begin{vmatrix} -10 & -8 \\ 2 & 4 \end{vmatrix} = (-10)(4)-(2)(-8) = -40+16 = -24$$

$$D_y=\begin{vmatrix} 7 & -10 \\ -5 & 2 \end{vmatrix} = (7)(2)-(-5)(-10) = 14-50 = -36$$

$x=\frac{D_x}{D}=\frac{-24}{-12}=2$ and $y=\frac{D_y}{D}=\frac{-36}{-12}=3$

The solution is (2, 3).

56. $x+4y=5$
$-2x-2y=2$

To solve, first calculate D, D_x, and D_y.

$$D=\begin{vmatrix} 1 & 4 \\ -2 & -2 \end{vmatrix} = (1)(-2)-(-2)(4) = -2+8 = 6$$

$$D_x=\begin{vmatrix} 5 & 4 \\ 2 & -2 \end{vmatrix} = (5)(-2)-(2)(4) = -10-8 = -18$$

$$D_y=\begin{vmatrix} 1 & 5 \\ -2 & 2 \end{vmatrix} = (1)(2)-(-2)(5) = 2+10 = 12$$

$x=\frac{D_x}{D}=\frac{-18}{6}=-3$ and $y=\frac{D_y}{D}=\frac{12}{6}=2$

The solution is (−3, 2).

57. $4m + 3n = 2$
$7m - 2n = -11$
To solve, first calculate D, D_m, and D_n.

$D = \begin{vmatrix} 4 & 3 \\ 7 & -2 \end{vmatrix}$
$= (4)(-2) - (7)(3)$
$= -8 - 21$
$= -29$

$D_m = \begin{vmatrix} 2 & 3 \\ -11 & -2 \end{vmatrix}$
$= (2)(-2) - (-11)(3)$
$= -4 + 33$
$= 29$

$D_n = \begin{vmatrix} 4 & 2 \\ 7 & -11 \end{vmatrix}$
$= (4)(-11) - (7)(2)$
$= -44 - 14$
$= -58$

$m = \dfrac{D_m}{D} = \dfrac{29}{-29} = -1$ and $n = \dfrac{D_n}{D} = \dfrac{-58}{-29} = 2$.
The solution is (–1, 2).

58. $p + q + r = 5$
$2p + q - r = -5$
$-p + 2q - 3r = -4$
To solve, calculate D, D_p, D_q, and D_r.

$D = \begin{vmatrix} 1 & 1 & 1 \\ 2 & 1 & -1 \\ -1 & 2 & -3 \end{vmatrix}$
$= 1\begin{vmatrix} 1 & -1 \\ 2 & -3 \end{vmatrix} - 1\begin{vmatrix} 2 & -1 \\ -1 & -3 \end{vmatrix} + 1\begin{vmatrix} 2 & 1 \\ -1 & 2 \end{vmatrix}$
$= 1(-3 + 2) - 1(-6 - 1) + 1(4 + 1)$
$= 1(-1) - 1(-7) + 1(5)$
$= -1 + 7 + 5$
$= 11$

$D_p = \begin{vmatrix} 5 & 1 & 1 \\ -5 & 1 & -1 \\ -4 & 2 & -3 \end{vmatrix}$
$= 5\begin{vmatrix} 1 & -1 \\ 2 & -3 \end{vmatrix} - 1\begin{vmatrix} -5 & -1 \\ -4 & -3 \end{vmatrix} + 1\begin{vmatrix} -5 & 1 \\ -4 & 2 \end{vmatrix}$
$= 5(-3 + 2) - 1(15 - 4) + 1(-10 + 4)$
$= 5(-1) - 1(11) + 1(-6)$
$= -5 - 11 - 6$
$= -22$

$D_q = \begin{vmatrix} 1 & 5 & 1 \\ 2 & -5 & -1 \\ -1 & -4 & -3 \end{vmatrix}$
$= 1\begin{vmatrix} -5 & -1 \\ -4 & -3 \end{vmatrix} - 5\begin{vmatrix} 2 & -1 \\ -1 & -3 \end{vmatrix} + 1\begin{vmatrix} 2 & -5 \\ -1 & -4 \end{vmatrix}$
$= 1(15 - 4) - 5(-6 - 1) + 1(-8 - 5)$
$= 1(11) - 5(-7) + 1(-13)$
$= 11 + 35 + 13$
$= 33$

$D_r = \begin{vmatrix} 1 & 1 & 5 \\ 2 & 1 & -5 \\ -1 & 2 & -4 \end{vmatrix}$
$= 1\begin{vmatrix} 1 & -5 \\ 2 & -4 \end{vmatrix} - 1\begin{vmatrix} 2 & -5 \\ -1 & -4 \end{vmatrix} + 5\begin{vmatrix} 2 & 1 \\ -1 & 2 \end{vmatrix}$
$= 1(-4 + 10) - 1(-8 - 5) + 5(4 + 1)$
$= 1(6) - 1(-13) + 5(5)$
$= 6 + 13 + 25$
$= 44$

$p = \dfrac{D_p}{D} = \dfrac{-22}{11} = -2$, $q = \dfrac{D_q}{D} = \dfrac{33}{11} = 3$, and
$r = \dfrac{D_r}{D} = \dfrac{44}{11} = 4$
The solution is (–2, 3, 4).

59. $-2a + 3b - 4c = -7$
$a + b + c = 4$
$-2a - 3b + 4c = 3$
To solve, calculate D, D_a, D_b, and D_c.

$D = \begin{vmatrix} -2 & 3 & -4 \\ 1 & 1 & 1 \\ -2 & -3 & 4 \end{vmatrix}$
$= -2\begin{vmatrix} 1 & 1 \\ -3 & 4 \end{vmatrix} - 3\begin{vmatrix} 1 & 1 \\ -2 & 4 \end{vmatrix} + (-4)\begin{vmatrix} 1 & 1 \\ -2 & -3 \end{vmatrix}$
$= -2(4 + 3) - 3(4 + 2) - 4(-3 + 2)$
$= -2(7) - 3(6) - 4(-1)$
$= -14 - 18 + 4$
$= -28$

$$D_a = \begin{vmatrix} -7 & 3 & -4 \\ 4 & 1 & 1 \\ 3 & -3 & 4 \end{vmatrix}$$
$$= -7\begin{vmatrix} 1 & 1 \\ -3 & 4 \end{vmatrix} - 3\begin{vmatrix} 4 & 1 \\ 3 & 4 \end{vmatrix} + (-4)\begin{vmatrix} 4 & 1 \\ 3 & -3 \end{vmatrix}$$
$$= -7(4+3) - 3(16-3) - 4(-12-3)$$
$$= -7(7) - 3(13) - 4(-15)$$
$$= -49 - 39 + 60$$
$$= -28$$

$$D_b = \begin{vmatrix} -2 & -7 & -4 \\ 1 & 4 & 1 \\ -2 & 3 & 4 \end{vmatrix}$$
$$= -2\begin{vmatrix} 4 & 1 \\ 3 & 4 \end{vmatrix} - (-7)\begin{vmatrix} 1 & 1 \\ -2 & 4 \end{vmatrix} + (-4)\begin{vmatrix} 1 & 4 \\ -2 & 3 \end{vmatrix}$$
$$= -2(16-3) + 7(4+2) - 4(3+8)$$
$$= -2(13) + 7(6) - 4(11)$$
$$= -26 + 42 - 44$$
$$= -28$$

$$D_c = \begin{vmatrix} -2 & 3 & -7 \\ 1 & 1 & 4 \\ -2 & -3 & 3 \end{vmatrix}$$
$$= -2\begin{vmatrix} 1 & 4 \\ -3 & 3 \end{vmatrix} - 3\begin{vmatrix} 1 & 4 \\ -2 & 3 \end{vmatrix} + (-7)\begin{vmatrix} 1 & 1 \\ -2 & -3 \end{vmatrix}$$
$$= -2(3+12) - 3(3+8) - 7(-3+2)$$
$$= -2(15) - 3(11) - 7(-1)$$
$$= -30 - 33 + 7$$
$$= -56$$

$a = \frac{D_a}{D} = \frac{-28}{-28} = 1$, $b = \frac{D_b}{D} = \frac{-28}{-28} = 1$, and

$c = \frac{D_c}{D} = \frac{-56}{-28} = 2$

The solution is (1, 1, 2).

60.
$$y + 3z = 4$$
$$-x - y + 2z = 0$$
$$x + 2y + z = 1$$

To solve, first calculate D, D_x, D_y, and D_z.

$$D = \begin{vmatrix} 0 & 1 & 3 \\ -1 & -1 & 2 \\ 1 & 2 & 1 \end{vmatrix}$$
$$= 0\begin{vmatrix} -1 & 2 \\ 2 & 1 \end{vmatrix} - (-1)\begin{vmatrix} 1 & 3 \\ 2 & 1 \end{vmatrix} + 1\begin{vmatrix} 1 & 3 \\ -1 & 2 \end{vmatrix}$$
$$= 0(-1-4) + 1(1-6) + 1(2+3)$$
$$= 0(-5) + 1(-5) + 1(5)$$
$$= 0 - 5 + 5$$
$$= 0$$

$$D_x = \begin{vmatrix} 4 & 1 & 3 \\ 0 & -1 & 2 \\ 1 & 2 & 1 \end{vmatrix}$$
$$= 4\begin{vmatrix} -1 & 2 \\ 2 & 1 \end{vmatrix} - 0\begin{vmatrix} 1 & 3 \\ 2 & 1 \end{vmatrix} + 1\begin{vmatrix} 1 & 3 \\ -1 & 2 \end{vmatrix}$$
$$= -4(-1-4) - 0(1-6) + 1(2+3)$$
$$= 4(-5) - 0(-5) + 1(5)$$
$$= -20 + 0 + 5$$
$$= -15$$

Since $D = 0$ and $D_x = -15$, the system is inconsistent.

Practice Test

1. a.
$$x + 2y = -6$$
$$0 + 2(-6) = -6$$
$$-12 = -6 \quad \text{False}$$

Since (0, 6) does not satisfy the first equation, it is not a solution to the system.

b.
$$x + 2y = -6$$
$$-3 + 2\left(-\frac{3}{2}\right) = -6$$
$$-6 = -6 \quad \text{True}$$

$$3x + 2y = -12$$
$$3(-3) + 2\left(-\frac{3}{2}\right) = -12$$
$$-12 = -12 \quad \text{True}$$

$\left(-3, -\frac{3}{2}\right)$ is a solution to the system.

c.
$$x + 2y = -6$$
$$2 + 2(-4) = -6$$
$$-6 = -6 \quad \text{True}$$

$$3x + 2y = -12$$
$$3(2) + 2(-4) = -12$$
$$-2 = -12 \text{ False}$$

Since (2, –4) does not satisfy both equations, it is not a solution to the system.

2. The system is inconsistent, it has no solution.

3. The system is consistent; it has exactly one solution.

4. The system is dependent; it has an infinite number of solutions.

5. Write both equations in slope-intercept form.

$$5x + 2y = 4 \qquad 5x = 3y - 7$$
$$2y = -5x + 4 \qquad -3y = -5x - 7$$
$$y = \frac{-5x+4}{2} \qquad y = \frac{-5x-7}{-3}$$
$$y = -\frac{5}{2}x + 2 \qquad y = \frac{5}{3}x + \frac{7}{3}$$

Since the slope of the first line is $-\frac{5}{2}$ and the slope of the second line is $\frac{5}{3}$, the slopes are different so that the lines intersect to produce one solution. This is a consistent system.

6. Write both equations in slope-intercept form.

$$5x + 3y = 9 \qquad 2y = -\frac{10}{3}x + 6$$
$$3y = -5x + 9 \qquad \frac{1}{2}(2y) = \frac{1}{2}\left(-\frac{10}{3}x + 6\right)$$
$$y = \frac{-5x+9}{3} \qquad y = -\frac{5}{3}x + 3$$
$$y = -\frac{5}{3}x + 3$$

Since the equations are identical, there is an infinite number of solutions and this is a dependent system.

7. Write both equations in slope-intercept form.

$$5x - 4y = 6 \qquad -10x + 8y = -10$$
$$-4y = -5x + 6 \qquad 8y = 10x - 10$$
$$y = \frac{-5x+6}{-4} \qquad y = \frac{10x-10}{8}$$
$$y = \frac{5}{4}x - \frac{3}{2} \qquad y = \frac{5}{4}x - \frac{5}{4}$$

Since the slope of each line is $\frac{5}{4}$, but the y-intercepts are different $\left(b = -\frac{3}{2} \text{ for the first equation, } b = -\frac{5}{4} \text{ for the second equation}\right)$, the two lines are parallel and produce no solution. This is an inconsistent system.

8. Graph the equations $y = 3x - 2$ and $y = -2x + 8$.

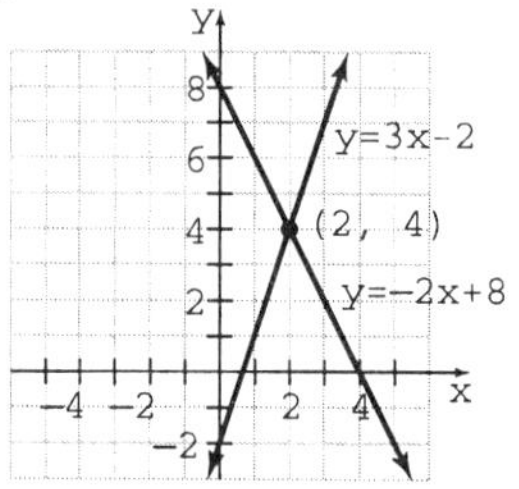

The lines intersect and the point of intersection is (2, 4).

9. Graph the equations $y = -x + 6$ and $y = 2x + 3$.

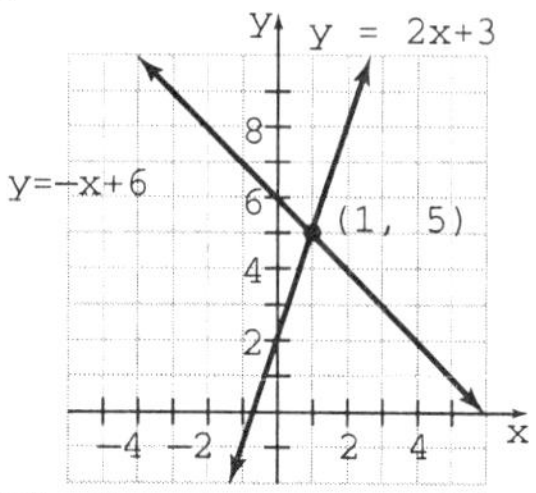

The lines intersect and the point of intersection is (1, 5).

10.
$$y = -3x + 4$$
$$y = 5x - 4$$

Substitute $-3x + 4$ for y in the second equation.

$$-3x + 4 = 5x - 4$$
$$-8x = -8$$
$$x = 1$$

Substitute 1 for x in the first equation.

$$y = -3(1) + 4$$
$$y = -3 + 4$$
$$y = 1$$

The solution is (1, 1).

11.
$$2a + 4b = 2$$
$$5a + b = -13$$

Solve the second equation for b.

$$5a + b = -13$$
$$b = -5a - 13$$

Substitute $-5a - 13$ for b in the first equation.

$$2a + 4(-5a - 13) = 2$$
$$2a - 20a - 52 = 2$$
$$-18a - 52 = 2$$
$$-18a = 54$$
$$a = -3$$

Substitute –3 for a in the equation $b = -5a - 13$.

$b = -5(-3) - 13$

$b = 15 - 13$

$b = 2$

The solution is (–3, 2).

12. $0.3x = 0.2y + 0.4$

$-1.2x + 0.8y = -1.6$

Write the system in standard form.

$0.3x - 0.2y = 0.4$

$-1.2x + 0.8y = -1.6$

To eliminate *x*, multiply the first equation by 4 and then add.

$4[0.3x - 0.2y = 0.4]$

$-1.2x + 0.8y = -1.6$

gives

$$\begin{array}{rl} & 1.2x - 0.8y = 1.6 \\ & \underline{-1.2x + 0.8y = -1.6} \\ \text{Add:} & 0 = 0 \quad \text{True} \end{array}$$

Since this is a true statement, there are an infinite number of solutions and this is a dependent system.

13. $\frac{3}{2}a + b = 6$

$a - \frac{5}{2}b = -4$

To clear fractions, multiply both equations by 2.

$2\left[\frac{3}{2}a + b = 6\right]$

$2\left[a - \frac{5}{2}b = -4\right]$

gives

$3a + 2b = 12$

$2a - 5b = -8$

Now, to eliminate *b*, multiply the first equation by 5 and the second equation by 2 and then add.

$5[3a + 2b = 12]$

$2[2a - 5b = -8]$

gives

$$\begin{array}{rl} & 15a + 10b = 60 \\ & \underline{4a - 10b = -16} \\ \text{Add:} & 19a = 44 \end{array}$$

$a = \frac{44}{19}$

Substitute $\frac{44}{19}$ for *a* in the first equation.

$\frac{3}{2}a + b = 6$

$\frac{3}{2}\left(\frac{44}{19}\right) + b = 6$

$\frac{66}{19} + b = 6$

$b = 6 - \frac{66}{19}$

$b = \frac{114}{19} - \frac{66}{19}$

$b = \frac{48}{19}$

The solution is $\left(\frac{44}{19}, \frac{48}{19}\right)$.

14. $x + y + z = 2$ (1)

$-2x - y + z = 1$ (2)

$x - 2y - z = 1$ (3)

To eliminate *z* between equations (1) and (3) simply add.

$$\begin{array}{rl} & x + y + z = 2 \\ & \underline{x - 2y - z = 1} \\ \text{Add:} & 2x - y = 3 \quad (4) \end{array}$$

To eliminate *z* between equations (2) and (3) simply add.

$$\begin{array}{rl} & -2x - y + z = 1 \\ & \underline{x - 2y - z = 1} \\ \text{Add:} & -x - 3y = 2 \quad (5) \end{array}$$

Equations (4) and (5) are two equations in two unknowns.

$2x - y = 3$

$-x - 3y = 2$

To eliminate *x*, multiply equation (5) by 2 and then add.

$2x - y = 3$

$2[-x - 3y = 2]$

gives

$$\begin{array}{rl} & 2x - y = 3 \\ & \underline{-2x - 6y = 4} \\ \text{Add:} & -7y = 7 \end{array}$$

$y = \frac{7}{-7} = -1$

Substitute –1 for *y* in equation (4).

$2x - y = 3$

$2x - (-1) = 3$

$2x + 1 = 3$

$2x = 2$

$x = \frac{2}{2} = 1$

Finally, substitute 1 for *x* and –1 for *y* in equation (1).

$$\begin{aligned} x+y+z&=2\\ 1-1+z&=2\\ 0+z&=2\\ z&=2 \end{aligned}$$

The solution is (1, –1, 2).

15. $$\begin{aligned} -2x+3y+7z&=5\\ 3x-2y+z&=-2\\ x-6y+5z&=-13 \end{aligned}$$

The augmented matrix is $\left[\begin{array}{ccc|c} -2 & 3 & 7 & 5\\ 3 & -2 & 1 & -2\\ 1 & -6 & 5 & -13 \end{array}\right]$

16. $\left[\begin{array}{ccc|c} 6 & -2 & 4 & 4\\ 4 & 3 & 5 & 6\\ 2 & -1 & 4 & -3 \end{array}\right]$

$\left[\begin{array}{ccc|c} 6 & -2 & 4 & 4\\ 0 & 5 & -3 & 12\\ 2 & -1 & 4 & -3 \end{array}\right] -2R_3+R_2$

17. $$\begin{aligned} x-3y&=7\\ 3x+5y&=7 \end{aligned}$$

$\left[\begin{array}{cc|c} 1 & -3 & 7\\ 3 & 5 & 7 \end{array}\right]$

$\left[\begin{array}{cc|c} 1 & -3 & 7\\ 0 & 14 & -14 \end{array}\right] -3R_1+R_2$

$\left[\begin{array}{cc|c} 1 & -3 & 7\\ 0 & 1 & -1 \end{array}\right] \frac{1}{14}R_2$

The system is

$$\begin{aligned} x-3y&=7\\ y&=-1 \end{aligned}$$

Substitute –1 for *y* in the first equation.

$$\begin{aligned} x-3(-1)&=7\\ x+3&=7\\ x&=4 \end{aligned}$$

The solution is (4, –1).

18. $$\begin{aligned} x-2y+z&=7\\ -2x-y-z&=-7\\ 3x-2y+2z&=15 \end{aligned}$$

$\left[\begin{array}{ccc|c} 1 & -2 & 1 & 7\\ -2 & -1 & -1 & -7\\ 3 & -2 & 2 & 15 \end{array}\right]$

$\left[\begin{array}{ccc|c} 1 & -2 & 1 & 7\\ 0 & -5 & 1 & 7\\ 3 & -2 & 2 & 15 \end{array}\right] 2R_1+R_2$

$\left[\begin{array}{ccc|c} 1 & -2 & 1 & 7\\ 0 & -5 & 1 & 7\\ 0 & 4 & -1 & -6 \end{array}\right] -3R_1+R_3$

$\left[\begin{array}{ccc|c} 1 & -2 & 1 & 7\\ 0 & 1 & -\frac{1}{5} & -\frac{7}{5}\\ 0 & 4 & -1 & -6 \end{array}\right] -\frac{1}{5}R_2$

$\left[\begin{array}{ccc|c} 1 & -2 & 1 & 7\\ 0 & 1 & -\frac{1}{5} & -\frac{7}{5}\\ 0 & 0 & -\frac{1}{5} & -\frac{2}{5} \end{array}\right] -4R_2+R_3$

$\left[\begin{array}{ccc|c} 1 & -2 & 1 & 7\\ 0 & 1 & -\frac{1}{5} & -\frac{7}{5}\\ 0 & 0 & 1 & 2 \end{array}\right] -5R_3$

The system is

$$\begin{aligned} x-2y+z&=7\\ y-\frac{1}{5}z&=-\frac{7}{5}\\ z&=2 \end{aligned}$$

Substitute 2 for *z* in the second equation.

$$\begin{aligned} y-\frac{1}{5}z&=-\frac{7}{5}\\ y-\frac{1}{5}(2)&=-\frac{7}{5}\\ y-\frac{2}{5}&=-\frac{7}{5}\\ y&=-1 \end{aligned}$$

Substitute –1 for *y* and 2 for *z* in the first equation.

$$\begin{aligned} x-2y+z&=7\\ x-2(-1)+2&=7\\ x+2+2&=7\\ x+4&=7\\ x&=3 \end{aligned}$$

The solution is (3, –1, 2).

19. $$\begin{aligned} \begin{vmatrix} 3 & -1\\ 4 & -2 \end{vmatrix} &= (3)(-2)-(4)(-1)\\ &=-6-(-4)\\ &=-6+4\\ &=-2 \end{aligned}$$

20. $$\begin{aligned} \begin{vmatrix} 8 & 2 & -1\\ 3 & 0 & 5\\ 6 & -3 & 4 \end{vmatrix} &= 8\begin{vmatrix} 0 & 5\\ -3 & 4 \end{vmatrix} - 3\begin{vmatrix} 2 & -1\\ -3 & 4 \end{vmatrix} + 6\begin{vmatrix} 2 & -1\\ 0 & 5 \end{vmatrix}\\ &=8(0+15)-3(8-3)+6(10-0)\\ &=8(15)-3(5)+6(10)\\ &=120-15+60\\ &=165 \end{aligned}$$

21. $4x + 3y = -6$

$-2x + 5y = 16$

To solve, first calculate D, D_x, and D_y.

$$D = \begin{vmatrix} 4 & 3 \\ -2 & 5 \end{vmatrix} = (4)(5) - (-2)(3) = 20 + 6 = 26$$

$$D_x = \begin{vmatrix} -6 & 3 \\ 16 & 5 \end{vmatrix} = (-6)(5) - (16)(3) = -30 - 48 = -78$$

$$D_y = \begin{vmatrix} 4 & -6 \\ -2 & 16 \end{vmatrix} = (4)(16) - (-2)(-6) = 64 - 12 = 52$$

$x = \frac{D_x}{D} = \frac{-78}{26} = -3$ and $y = \frac{D_y}{D} = \frac{52}{26} = 2$.

The solution is (–3, 2).

22. $2r - 4s + 3t = -1$

$-3r + 5s - 4t = 0$

$-2r + s - 3t = -2$

To solve, first calculate D, D_r, D_s, and D_t.

$$D = \begin{vmatrix} 2 & -4 & 3 \\ -3 & 5 & -4 \\ -2 & 1 & -3 \end{vmatrix}$$

$$= 2\begin{vmatrix} 5 & -4 \\ 1 & -3 \end{vmatrix} - (-4)\begin{vmatrix} -3 & -4 \\ -2 & -3 \end{vmatrix} + 3\begin{vmatrix} -3 & 5 \\ -2 & 1 \end{vmatrix}$$

$$= 2(-15+4) + 4(9-8) + 3(-3+10)$$

$$= 2(-11) + 4(1) + 3(7)$$

$$= -22 + 4 + 21$$

$$= 3$$

$$D_r = \begin{vmatrix} -1 & -4 & 3 \\ 0 & 5 & -4 \\ -2 & 1 & -3 \end{vmatrix}$$

$$= -1\begin{vmatrix} 5 & -4 \\ 1 & -3 \end{vmatrix} - (-4)\begin{vmatrix} 0 & -4 \\ -2 & -3 \end{vmatrix} + 3\begin{vmatrix} 0 & 5 \\ -2 & 1 \end{vmatrix}$$

$$= -1(-15+4) + 4(0-8) + 3(0+10)$$

$$= -1(-11) + 4(-8) + 3(10)$$

$$= 11 - 32 + 30$$

$$= 9$$

$$D_s = \begin{vmatrix} 2 & -1 & 3 \\ -3 & 0 & -4 \\ -2 & 2 & -3 \end{vmatrix}$$

$$= 2\begin{vmatrix} 0 & -4 \\ -2 & -3 \end{vmatrix} - (-1)\begin{vmatrix} -3 & -4 \\ -2 & -3 \end{vmatrix} + 3\begin{vmatrix} -3 & 0 \\ -2 & -2 \end{vmatrix}$$

$$= 2(0-8) + 1(9-8) + 3(6-0)$$

$$= 2(-8) + 1(1) + 3(6)$$

$$= -16 + 1 + 18$$

$$= 3$$

$$D_t = \begin{vmatrix} 2 & -4 & -1 \\ -3 & 5 & 0 \\ -2 & 1 & -2 \end{vmatrix}$$

$$= 2\begin{vmatrix} 5 & 0 \\ 1 & -2 \end{vmatrix} - (-4)\begin{vmatrix} -3 & 0 \\ -2 & -2 \end{vmatrix} + (-1)\begin{vmatrix} -3 & 5 \\ -2 & 1 \end{vmatrix}$$

$$= 2(-10-0) + 4(6-0) - 1(-3+10)$$

$$= 2(-10) + 4(6) - 1(7)$$

$$= -20 + 24 - 7$$

$$= -3$$

$r = \frac{D_r}{D} = \frac{9}{3} = 3$, $s = \frac{D_s}{D} = \frac{3}{3} = 1$,

$t = \frac{D_t}{D} = \frac{-3}{3} = -1$

The solution is (–1, –1, 2).

23. Let x be the number of pounds of cashews and y be the number of pounds of peanuts.

$x + y = 20$

$7x + 5.5y = 20(6)$

Solve the first equation for y.

$x + y = 20$

$y = -x + 20$

Substitute $-x + 20$ for y in the second equation.

$$7x + 5.5y = 20(6)$$
$$7x + 5.5(-x + 20) = 20(6)$$
$$7x - 5.5x + 110 = 120$$
$$1.5x + 110 = 120$$
$$1.5x = 10$$
$$x = \frac{10}{1.5} \cdot \frac{10}{10} = \frac{100}{15} = \frac{20}{3} \text{ or } 6\frac{2}{3}$$

Thus, Dick should mix $6\frac{2}{3}$ lb of cashews with $20 - 6\frac{2}{3} = 13\frac{1}{3}$ lb of peanuts to obtain the desired mixture.

24. Let x = amount of 6% solution
y = amount of 15% solution
$$x + y = 10$$
$$0.06x + 0.15y = 0.09(10)$$
The system can be written as
$$x + y = 10$$
$$6x + 15y = 90$$
Solve the first equation for y.
$$y = 10 - x$$
Substitute $10 - x$ for y in the second equation.
$$6x + 15y = 90$$
$$6x + 15(10 - x) = 90$$
$$6x + 150 - 15x = 90$$
$$-9x = -60$$
$$x = \frac{-60}{-9} = \frac{20}{3} = 6\frac{2}{3}$$
Substitute $6\frac{2}{3}$ for x into $y = 10 - x$
$$y = 10 - 6\frac{2}{3}$$
$$y = 3\frac{1}{3}$$
She should mix $6\frac{2}{3}$ liters of 6% solution and $3\frac{1}{3}$ liters of 15% solution.

25. Let x = smallest number
y = remaining number
z = largest number
$$x + y + z = 25$$
$$z = 3x$$
$$y = 2x + 1$$
Substitute $2x + 1$ for y and $3x$ for z in the first equation.
$$x + y + z = 25$$
$$x + 2x + 1 + 3x = 25$$
$$6x = 24$$
$$x = 4$$
Substitute 4 for x in the third equation.
$$y = 2x + 1$$
$$y = 2(4) + 1$$
$$y = 9$$
Substitute 4 for x in the second equation.
$$z = 3x$$
$$z = 3(4)$$
$$z = 12$$
The three numbers are 4, 9, and 12.

Cumulative Review Test

1. a. 9 and 1 are natural numbers.

b. $\frac{1}{2}$, –4, 9, 0, –4.63, and 1 are rational numbers.

c. $\frac{1}{2}$, –4, 9, 0, $\sqrt{3}$, –4.63, and 1 are real numbers.

2. $$16 \div \left\{4\left[3 + \left(\frac{5+10}{5}\right)^2\right] - 32\right\}$$
$$16 \div \left\{4\left[3 + \left(\frac{15}{5}\right)^2\right] - 32\right\}$$
$$16 \div \left\{4\left[3 + (3)^2\right] - 32\right\}$$
$$16 \div \{4[3 + 9] - 32\}$$
$$16 \div \{4[12] - 32\}$$
$$16 \div \{48 - 32\}$$
$$16 \div \{16\}$$
$$1$$

3. $$-7(3 - x) = 4(x + 2) - 3x$$
$$-21 + 7x = 4x + 8 - 3x$$
$$-21 + 7x = x + 8$$
$$6x = 29$$
$$x = \frac{29}{6}$$

4. $4x - 6y = 24$
Use intercept method. When $x = 0$, then $y = -4$. When $y = 0$, then $x = 6$.

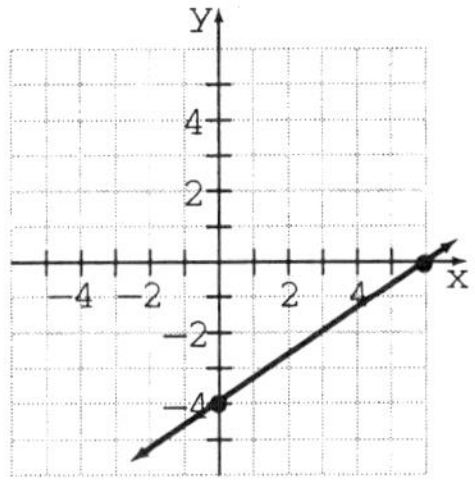

5. $y - y_1 = m(x - x_1)$

$$y - 1 = \frac{2}{5}(x - (-3))$$
$$y - 1 = \frac{2}{5}(x + 3)$$
$$y - 1 = \frac{2}{5}x + \frac{6}{5}$$
$$y = \frac{2}{5}x + \frac{6}{5} + 1$$
$$y = \frac{2}{5}x + \frac{6}{5} + \frac{5}{5}$$
$$y = \frac{2}{5}x + \frac{11}{5}$$

6. The y-intercept is $(0, -2)$. The slope between $(1, 1)$ and $(0, -2)$ must be found.

$$m = \frac{y_2 - y_1}{x_2 - x_1} = \frac{-2-1}{0-1} = \frac{-3}{-1} = 3$$

Substitute $m = 3$ and $b = -2$ into the y-intercept equation. Therefore the equation of the line is $y = 3x - 2$.

7. $$\frac{4a^3b^{-5}}{28a^8b} = \frac{a^{3-8}b^{-5-1}}{7} = \frac{a^{-5}b^{-6}}{7} = \frac{1}{7a^5b^6}$$

8. $3x^3 + 4x^2 + 6x + 8$
$x^2(3x + 4) + 2(3x + 4)$
$(3x + 4)(x^2 + 2)$

9. $x^2 - 16x + 28$
$(x - 14)(x - 2)$

10. $x^2 - 5x = 0$
$x(x - 5) = 0$
$x = 0 \qquad x - 5 = 0$
$x = 5$
The solutions are 0 and 5.

11. $$\frac{y-5}{8} + \frac{2y+7}{8} = \frac{y-5+2y+7}{8} = \frac{3y+2}{8}$$

12. $$\frac{5}{y+2} = \frac{3}{2y-7}$$
$5(2y - 7) = 3(y + 2)$
$10y - 35 = 3y + 6$
$7y = 41$
$y = \frac{41}{7}$

13. a. It is a function since it passes the vertical line test.

b. It is a function since it passes the vertical line test.

c. It is not a function since it fails the vertical line test.

14. a. $$f(x) = \frac{x+3}{x^2-9}$$
$$f(-4) = \frac{(-4)+3}{(-4)^2-9}$$
$$f(-4) = \frac{-1}{16-9}$$
$$f(-4) = -\frac{1}{7}$$

b. $$f(h) = \frac{h+3}{h^2-9}$$

c. $$f(3) = \frac{(3)+3}{(3)^2-9}$$
$$f(3) = \frac{6}{9-9}$$
$f(3) = \frac{6}{0}$ undefined

15. $3x + y = 6$
$y = 2x + 1$
Substitute $2x + 1$ for y in the first equation.
$$3x + y = 6$$
$$3x + 2x + 1 = 6$$
$$5x + 1 = 6$$
$$5x = 5$$
$$x = \frac{5}{5} = 1$$
Substitute 1 for x in the second equation.
$$y = 2x + 1$$
$$y = 2(1) + 1$$
$$y = 2 + 1$$
$$y = 3$$
The solution is (1, 3).

16. $2p + 3q = 11$
$-3p - 5q = -16$
To eliminate p, multiply the first equation by 3 and the second equation by 2 and then add.
$$3[2p + 3q = 11]$$
$$2[-3p - 5q = -16]$$
gives
$$6p + 9q = 33$$
$$-6p - 10q = -32$$
Add: $-q = 1$
$$q = -1$$
Substitute -1 for q in the first equation.
$$2p + 3(-1) = 11$$
$$2p - 3 = 11$$
$$2p = 14$$
$$p = 7$$
The solution is (7, −1).

17. $x - 2y = 0$ (1)
$2x + z = 7$ (2)
$y - 2z = -5$ (3)
To eliminate z between equations (2) and (3), multiply equation (2) by 2 and then add.
$$2[2x + z = 7]$$
$$y - 2z = -5$$
gives
$$4x + 2z = 14$$
$$y - 2z = -5$$
Add: $4x + y = 9$ (4)
Equations (4) and (1) are two equations in two unknowns:
$$x - 2y = 0$$
$$4x + y = 9$$
To eliminate y, multiply equation (4) by 2 and then add.
$$x - 2y = 0$$
$$2[4x + y = 9]$$
gives
$$x - 2y = 0$$
$$8x + 2y = 18$$
Add: $9x = 18$
$$x = \frac{18}{9} = 2$$
Substitute 2 for x in equation (4).
$$4x + y = 9$$
$$4(2) + y = 9$$
$$8 + y = 9$$
$$y = 1$$
Finally, substitute 2 for x in equation (2).
$$2x + z = 7$$
$$2(2) + z = 7$$
$$4 + z = 7$$
$$z = 3$$
The solution is (2, 1, 3).

18. Let x be the measure of the smallest angle. Then $9x$ is the measure of the largest angle and $x + 70$ is the measure of the remaining angle. The sum of the measures of the three angles is 180°.
$$x + (x + 70) + 9x = 180$$
$$11x + 70 = 180$$
$$11x = 110$$
$$x = \frac{110}{11} = 10$$
$$x + 70 = 10 + 70 = 80$$
$$9x = 9(10) = 90$$
The three angles are 10°, 90°, and 80°.

19. Let t be the time for Judy to catch up to Dawn.

	rate	time	distance
Judy	6	t	$6t$
Dawn	4	$t + \frac{1}{2}$	$4\left(t + \frac{1}{2}\right)$

$$6t = 4\left(t + \frac{1}{2}\right)$$
$$6t = 4t + 2$$
$$2t = 2$$
$$t = \frac{2}{2} = 1$$
It takes 1 hour for Judy to catch up to Dawn.

20. Let x be the number of \$20 tickets sold. Then $1000 - x$ is the number of \$16 tickets sold.

$$\begin{aligned} 20x + 16(1000 - x) &= 18,400 \\ 20x + 16,000 - 16x &= 18,400 \\ 4x + 16,000 &= 18,400 \\ 4x &= 2400 \\ x &= \frac{2400}{4} = 600 \end{aligned}$$

Thus, 600 \$20 tickets and $1000 - 600 = 400$ \$16 tickets were sold for the concert.

Chapter 10

Exercise Set 10.1

1. It is necessary to change the direction of the inequality symbol when multiplying or dividing by a negative number.

3. a. Use open circles when the endpoints are not included.

b. Use closed circles when the endpoints are included.

c. Answers may vary. One possible answer is $x < 4$.

d. Answers may vary. One possible answer is $x \geq 4$.

5. $a < x < b$ means $a < x$ and $x < b$.

7. a.

b. $(-2, \infty)$

c. $\{x|x > -2\}$

9. a.

b. $(-\infty, \pi]$

c. $\{w|w \leq \pi\}$

11. a.

b. $\left(-3, \frac{3}{4}\right]$

c. $\left\{q \middle| -3 < q \leq \frac{3}{4}\right\}$

13. a.

b. $(-7, -4]$

c. $\{x|-7 < x \leq -4\}$

15. $x - 7 > -4$

$x > 3$

17. $3 - x < -4$

$-x < -7$

Reverse the inequality

$\frac{-x}{-1} > \frac{-7}{-1}$

$x > 7$

19.

$$4.7x - 5.48 \geq 11.44$$
$$4.7x - 5.48 + 5.48 \geq 11.44 + 5.48$$
$$4.7x \geq 16.92$$
$$\frac{4.7x}{4.7} \geq \frac{16.92}{4.7}$$
$$x \geq 3.6$$

21. $4(x - 2) \leq 4x - 8$

$4x - 8 \leq 4x - 8$

$-8 \leq -8$

Since this is a true statement, the solution is the entire real number line.

23. $5b - 6 \geq 3(b + 3) + 2b$

$5b - 6 \geq 3b + 9 + 2b$

$5b - 6 \geq 5b + 9$

$-6 \geq 9$

Since this is a false statement, there is no solution.

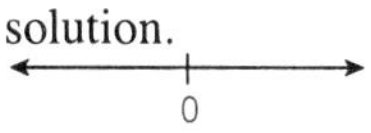

25.

$$\frac{y}{3}+\frac{2}{5}\le 4$$
$$15\left(\frac{y}{3}\right)+15\left(\frac{2}{5}\right)\le 15(4)$$
$$5y+6\le 60$$
$$5y\le 54$$
$$\frac{5y}{5}\le\frac{54}{5}$$
$$y\le\frac{54}{5}$$

27.

$$4+\frac{4x}{3}<6$$
$$\frac{4x}{3}<2$$
$$3\left(\frac{4x}{3}\right)<3(2)$$
$$4x<6$$
$$\frac{4x}{4}<\frac{6}{4}$$
$$x<\frac{3}{2}$$
$$\left(-\infty,\ \frac{3}{2}\right)$$

29.

$$\frac{v-5}{3}-v\ge -3(v-1)$$
$$\frac{v-5}{3}-v\ge -3v+3$$
$$3\left(\frac{v-5}{3}-v\right)\ge 3(-3v+3)$$
$$v-5-3v\ge -9v+9$$
$$-2v-5\ge -9v+9$$
$$7v\ge 14$$
$$v\ge 2 \quad \text{or} \quad [2,\infty)$$

31.

$$\frac{t}{3}-t+2\le -\frac{4t}{3}+3$$
$$3\left(\frac{t}{3}-t+2\right)\le 3\left(-\frac{4t}{3}+3\right)$$
$$t-3t+6\le -4t+9$$
$$-2t+6\le -4t+9$$
$$2t\le 3$$
$$\frac{2t}{2}\le\frac{3}{2}$$
$$t\le\frac{3}{2} \quad \text{or} \quad \left(-\infty,\frac{3}{2}\right]$$

33.

$$-3x+1<3[(x+2)-2x]-1$$
$$-3x+1<3[x+2-2x]-1$$
$$-3x+1<3[2-x]-1$$
$$-3x+1<6-3x-1$$
$$-3x+1<5-3x$$
$$1<5 \quad \Rightarrow \quad \text{a true statement}$$

The solution set is $(-\infty,\ \infty)$.

35. $A\cup B=\{5, 6, 7, 8\}$
$A\cap B=\{6, 7\}$

37. $A\cup B=\{-1, -2, -3, -4, -5, -6\}$
$A\cap B=\{-2, -4\}$

39. $A\cup B=\{0, 1, 2, 3\}$
$A\cap B=\{\ \}$

41. $A\cup B=\{0, 1, 2, 3, 4, 5, 6, 7, 8\}$
$A\cap B=\{\ \}$

43. $A\cup B=\{0.1, 0.2, 0.3, 0.4, ...\}$
$A\cap B=\{0.2, 0.3\}$

45.

$$-2\le q+3<4$$
$$-2-3\le q+3-3<4-3$$
$$-5\le q<1$$
$$[-5,1)$$

47. $-15 \le -3z \le 12$

Divide by -3 and reverse inequalities.

$\frac{-15}{-3} \ge \frac{-3z}{-3} \ge \frac{12}{-3}$

$5 \ge z \ge -4$

$-4 \le z \le 5$

$[-4, 5]$

49. $4 \le 2x - 4 < 7$

$4 + 4 \le 2x - 4 + 4 < 7 + 4$

$8 \le 2x < 11$

$\frac{8}{2} \le \frac{2x}{2} < \frac{11}{2}$

$4 \le x < \frac{11}{2}$

$\left[\frac{4}{2}, \frac{11}{2}\right)$

51. $14 \le 2 - 3g < 20$

$14 - 2 \le 2 - 3g - 2 < 20 - 2$

$12 \le -3g < 18$

Divide by -3 and reverse inequalities.

$\frac{12}{-3} \ge \frac{-3g}{-3} > \frac{18}{-3}$

$-4 \ge g > -6$

$-6 < g \le -4$

$(-6, -4]$

53. $5 \le \frac{3x+1}{2} < 11$

$2(5) \le 2\left(\frac{3x+1}{2}\right) < 2(11)$

$10 \le 3x + 1 < 22$

$10 - 1 \le 3x + 1 - 1 < 22 - 1$

$9 \le 3x < 21$

$\frac{9}{3} \le \frac{3x}{3} < \frac{21}{3}$

$3 \le x < 7$

$\{x | 3 \le x < 7\}$

55. $6 \le -3(2x - 4) < 12$

$6 \le -6x + 12 < 12$

$6 - 12 \le -6x + 12 - 12 < 12 - 12$

$-6 \le -6x < 0$

Divide by -6 and reverse inequalities

$\frac{-6}{-6} \ge \frac{-6x}{-6} > \frac{0}{-6}$

$1 \ge x > 0$

$0 < x \le 1$

$\{x | 0 < x \le 1\}$

57. $0 \le \frac{3(u-4)}{7} \le 1$

$7(0) \le 7\left(\frac{3(u-4)}{7}\right) \le 7(1)$

$0 \le 3(u - 4) \le 7$

$0 \le 3u - 12 \le 7$

$0 + 12 \le 3u - 12 + 12 \le 7 + 12$

$12 \le 3u \le 19$

$\frac{12}{3} \le \frac{3u}{3} \le \frac{19}{3}$

$4 \le u \le \frac{19}{3}$

$\left\{u \middle| 4 \le u \le \frac{19}{3}\right\}$

59. $\{c | -3 < c \le 2\}$

61. $x < 2$

$x > 4$

$x < 2$ and $x > 4$

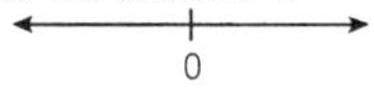

There is no overlap so the solution is the empty set, $\varnothing$.

63. $x + 1 < 3$ and $x + 1 > -4$

$x < 2$ and $x > -5$

$x > -5$

$x < 2$

$x < -2$ and $x > -5$ which is $-5 < x < 2$ or

$\{x | -5 < x < 2\}$

65. $2s + 3 < 7$ or $-3s + 4 \le -17$

$2s < 4$ or $-3s \le -21$

$\frac{2s}{2} < \frac{4}{2}$ or $\frac{-3s}{-3} \ge \frac{-21}{-3}$

$s < 2$ $\quad s \ge 7$

$s < 2$ or $s \ge 7$ which is $(-\infty, 2) \cup [7, \infty)$.

67. $4x + 5 \ge 5$ and $3x - 4 \le 2$

$4x \ge 0$ and $3x \le 6$

$x \ge 0$ and $x \le 2$

$x \ge 0$

$x \le 2$

$x \ge 0$ and $x \le 2$ which is $0 \le x \le 2$

[0, 2]

69. $4 - x < -2$ $\quad 3x - 1 < -1$

$-x < -6$ $\quad 3x < 0$

$x > 6$ $\quad x < 0$

$x > 6$

$x < 0$

$x > 6$ or $x < 0$

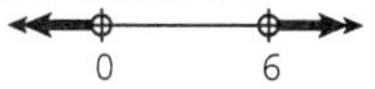

$(-\infty, 0) \cup (6, \infty)$

71. $2k + 5 > -1$ and $7 - 3k \le 7$

$2k > -6$ $\quad -3k \le 0$

$\frac{2k}{2} > \frac{-6}{2}$ $\quad \frac{-3k}{-3} \ge \frac{0}{-3}$

$k > -3$ $\quad k \ge 0$

$k > -3$ and $k \ge 0 \Rightarrow k \ge 0$
In interval notation: $[0, \infty)$

73. **a.** $l + g \le 130$

b. $g = 2w + 2d$

$l + g \le 130$

$l + 2w + 2d \le 130$

c. $l = 40$, $w = 20.5$

$l + 2w + 2d \le 130$

$40 + 2(20.5) + 2d \le 130$

$40 + 41 + 2d \le 130$

$81 + 2d \le 130$

$2d \le 49$

$d \le 24.5$

The maximum depth is 24.5 inches.

75. Let x be the maximum number of boxes.

$70x \le 800$

$x \le \frac{800}{70}$

$x \le 11.43$

The maximum number of boxes is 11.

77. Let x = the number of minutes she talks beyond the first 20 minutes.

$0.99 + 0.07x \le 5.00$

$0.07x \le 4.01$

$\frac{0.07x}{0.07} \le \frac{4.01}{0.07}$

$x \le 57$ (to nearest whole number)

She can talk for 57 minutes beyond the first 20 minutes for a total of 77 minutes.

79. To make a profit, the cost must be less than the revenue: cost < revenue.

$10,025 + 1.09x < 6.42x$

$10,025 < 5.33x$

$\frac{10,025}{5.33} < x$

$1880.86 < x$

She needs to sell a minimum of 1881 books to make a profit.

81. Let x = the number of additional ounces beyond the first ounce.

$0.37 + 0.23x \le 10.00$

$0.23x \le 9.63$

$\frac{0.23x}{0.23} \le \frac{9.63}{0.23}$

$x \le 42$ (rounded up)

The maximum weight is 42 ounces.

83. Let x be the amount of sales in dollars.

$$300 + 0.10x > 400 + 0.08x$$
$$0.10x > 100 + 0.08x$$
$$0.02x > 100$$
$$x > \frac{100}{0.02}$$
$$x > 5000$$

She will earn more by plan 1 if her sales total more than \$5000.

85. Let x be the minimum score for the sixth exam.

$$\frac{65 + 72 + 90 + 47 + 62 + x}{6} \ge 60$$
$$\frac{336 + x}{6} \ge 60$$
$$6\left(\frac{336 + x}{6}\right) \ge 6(60)$$
$$336 + x \ge 360$$
$$x \ge 24$$

She must make a 24 or higher on the sixth exam to pass the course.

87. Let x be the score on the fifth exam.

$$80 \le \frac{87 + 92 + 70 + 75 + x}{5} < 90$$
$$80 \le \frac{324 + x}{5} < 90$$
$$5(80) \le 5\left(\frac{324 + x}{5}\right) < 5(90)$$
$$400 \le 324 + x < 450$$
$$76 \le x < 126$$

To receive a final grade of B, Ms. Mahoney must score 76 or higher on the fifth exam. That is, the score must be $76 \le x \le 100$ (maximum grade is 100).

89. Let x be the value of the third reading.

$$7.2 < \frac{7.48 + 7.15 + x}{3} < 7.8$$
$$7.2 < \frac{14.63 + x}{3} < 7.8$$
$$3(7.2) < 3\left(\frac{14.63 + x}{3}\right) < 3(7.8)$$
$$21.6 < 14.63 + x < 23.4$$
$$6.97 < x < 8.77$$

Any value between 6.97 and 8.77 would result in a normal pH reading.

91. **a.** The taxable income of \$128,479 places a married couple filing jointly in the 30.5% tax bracket. The tax is \$24,393.75 plus 30.5% of the taxable income over \$166,500.

The tax is

$$24{,}393.75 + 0.305(128{,}479 - 109{,}250)$$
$$24{,}393.75 + 0.305(19{,}229)$$
$$24{,}393.75 + 5{,}864.85$$
$$30{,}258.60$$

They will owe \$30,258.60 in taxes.

b. The taxable income of \$175,248 places a married couple filing jointly in the 35.5% tax bracket. The tax is \$41,855.00 plus 35.5% of the taxable income over \$166,500.

The tax is

$$41{,}855.00 + 0.355(175{,}248 - 166{,}500)$$
$$41{,}855.00 + 0.355(8{,}748)$$
$$41{,}855.00 + 3{,}105.54$$
$$44{,}960.54$$

They will owe \$44,960.54 in taxes.

93. **a.** 1995, 1996, 1997, 1998, and 1999

b. 2000, 2001, 2002, 2003, 2004, and 2005

95. Answers may vary.

97. First find the average of 82, 90, 74, 76, and 68.

$$\frac{82 + 90 + 74 + 76 + 68}{5} = \frac{390}{5} = 78$$

This represents $\frac{2}{3}$ of the final grade.

Let x be the score from the final exam. Since this represents $\frac{1}{3}$ of the final grade, the inequality is

$$80 \le \frac{2}{3}(78) + \frac{1}{3}x < 90$$
$$3(80) < 3\left[\frac{2}{3}(78) + \frac{1}{3}x\right] < 3(90)$$
$$240 \le 2(78) + x < 270$$
$$240 \le 156 + x < 270$$
$$84 \le x < 114$$

Russell must score at least 84 points on the final exam to have a final grade of B. The range is $84 \le x \le 100$.

99. **a.** Answers may vary. One possible answer is: Write $x < 2x + 3 < 2x + 5$ as $x < 2x + 3$ and $2x + 3 < 2x + 5$

b. Solve each of the inequalities.

$x < 2x + 3$ and $2x + 3 < 2x + 5$

$-x < 3$ $\quad$ $3 < 5$

$x > -3$ $\quad$ All real numbers

The final answer is $x > -3$ or $(-3, \infty)$.

101. a. 4 is a counting number.

b. 4 and 0 are whole numbers.

c. $-3,\ 4,\ \frac{5}{2},\ 0$ and $-\frac{29}{80}$ are rational numbers.

d. $-3,\ 4,\ \frac{5}{2},\ \sqrt{7},\ 0$ and $-\frac{29}{80}$ are real numbers.

102. Associative property of addition.

103. Commutative property of addition

104.

$$R = L + (V - D)r$$
$$R = L + Vr - Dr$$
$$R - L + Dr = Vr$$
$$\frac{R - L + Dr}{r} = V \text{ or } V = \frac{R - L + Dr}{r}$$

105. a. $A \cup B = \{1, 2, 3, 4, 5, 6, 8, 9\}$

b. $A \cap B = \{1, 8\}$

Exercise Set 10.2

1. $|x| = a,\ a > 0$
Set $x = a$ or $x = -a$.

3. $|x| < a,\ a > 0$
Write $-a < x < a$.

5. $|x| > a,\ a > 0$
Write $x < -a$ or $x > a$.

7. The solution to $|x| > 0$ is all real numbers except 0.
The absolute value of any real number, except 0, is greater than 0, i.e., positive.

9 Set $x = y$ or $x = -y$.

11. If $a \neq 0$, and $k > 0$,

a. $|ax + b| = k$ has 2 solutions.

b. $|ax + b| < k$ has an infinite number of solutions.

c. $|ax + b| > k$ has an infinite number of solutions.

12. a. $|x| = 4$, $x = -4$ or $x = 4$, C

b. $|x| < 4$, $(-4, 4)$, A

c. $|x| > 4$, $(-\infty, 4) \cup (4, \infty)$, D

d. $|x| \geq 4$, $(-\infty, 4] \cup [4, \infty)$, B

e. $|x| \leq 4$, $[-4, 4]$, E

13. a. $|x| = 5, \{-5, 5\}$, D

b. $|x| < 5, \{x | -5 < x < 5\}$, B

c. $|x| > 5,\ \{x | x < -5 \text{ or } x > 5\}$, E

d. $|x| \leq 5,\ \{x | -5 \leq x \leq 5\}$, C

e. $|x| \geq 5, \{x | x \leq -5 \text{ or } x \geq 5\}$, A

15. $|a| = 2$
$a = 2$ or $a = -2$
The solution set is $\{-2, 2\}$.

17. $|c| = \frac{1}{2}$
$c = \frac{1}{2}$ or $c = -\frac{1}{2}$
The solution set is $\left\{-\frac{1}{2}, \frac{1}{2}\right\}$.

19. $|d| = -\frac{3}{4}$
There is no solution since the right side is a negative number and the absolute value can never be equal to a negative number. The solution set $\varnothing$.

21. $|x + 5| = 7$

$x + 5 = 7$ $\quad$ $x + 5 = -7$

$x = 2$ or $x = -12$

The solution set is $\{-12, 2\}$.

23. $|4.5q + 22.5| = 0$
$4.5q + 22.5 = 0$
$4.5q = -22.5$
$q = -5$
The solution set is {–5}.

25. $|5 - 3x| = \frac{1}{2}$

$5 - 3x = \frac{1}{2}$ or $5 - 3x = -\frac{1}{2}$

$-3x = \frac{1}{2} - 5$ $\qquad$ $-3x = -\frac{1}{2} - 5$

$-3x = -\frac{9}{2}$ $\qquad$ $-3x = -\frac{11}{2}$

$-\frac{1}{3}(3x) = -\frac{1}{3}\left(-\frac{9}{2}\right)$ $\qquad$ $-\frac{1}{3}(-3x) = -\frac{1}{3}\left(-\frac{11}{2}\right)$

$x = \frac{3}{2}$ $\qquad$ $x = \frac{11}{6}$

The solution set is $\left\{\frac{3}{2}, \frac{11}{6}\right\}$.

27. $\left|\frac{x-3}{4}\right| = 5$

$\frac{x-3}{4} = 5$ or $\frac{x-3}{4} = -5$

$4\left(\frac{x-3}{4}\right) = 4(5)$ $\qquad$ $4\left(\frac{x-3}{4}\right) = 4(-5)$

$x - 3 = 20$ $\qquad$ $x - 3 = -20$

$x = 23$ $\qquad$ $x = -17$

The solution set is {–17, 23}.

29. $\left|\frac{x-3}{4}\right| + 4 = 4$

$\left|\frac{x-3}{4}\right| = 0$

$\frac{x-3}{4} = 0$

$4\left(\frac{x-3}{4}\right) = 4(0)$

$x - 3 = 0$
$x = 3$
The solution set is {3}.

31. $|w| \le 11$
$-11 \le w \le 11$
The solution set is $\{w \mid -11 \le w \le 11\}$.

33. $|q + 5| \le 8$
$-8 \le q + 5 \le 8$
$-8 - 5 \le q + 5 - 5 \le 8 - 5$
$-13 \le q \le 3$
The solution set is $\{x \mid -13 \le q \le 3\}$.

35. $|5b - 15| < 10$
$-10 < 5b - 15 < 10$
$-10 + 15 < 5b - 15 + 15 < 10 + 15$
$5 < 5b < 25$
$\frac{5}{5} < \frac{5b}{5} < \frac{25}{5}$
$1 < b < 5$
The solution set is $\{b \mid 1 < b < 5\}$.

37. $|2x + 3| - 5 \le 10$
$|2x + 3| \le 15$
$-15 \le 2x + 3 \le 15$
$-15 - 3 \le 2x + 3 - 3 \le 15 - 3$
$-18 \le 2x \le 12$
$\frac{-18}{2} \le \frac{2x}{2} \le \frac{12}{2}$
$-9 \le x \le 6$
The solution set is $\{x \mid -9 \le x \le 6\}$

39. $|3x - 7| + 5 < 11$
$|3x - 7| < 6$
$-6 < 3x - 7 < 6$
$-6 + 7 < 3x - 7 + 7 < 6 + 7$
$1 < 3x < 13$
$\frac{1}{3} < \frac{3x}{3} < \frac{13}{3}$
$\frac{1}{3} < x < \frac{13}{3}$
The solution set is $\left\{x \middle| \frac{1}{3} < x < \frac{13}{3}\right\}$.

41. $|2x - 6| + 5 \le 2$
$|2x - 6| \le -3$
There is no solution since the right side is negative whereas the left side is non-negative; zero or a positive number is never less than a negative number. The solution set is $\varnothing$.

43. $\left|\frac{1}{2}j+3\right|<6$

$-6<\frac{1}{2}j+3<6$

$-6-3<\frac{1}{2}j+3-3<6-3$

$-9<\frac{1}{2}j<-3$

$2(-9)<2\left(\frac{1}{2}j\right)<2(-3)$

$-18<j<-6$

The solution set is $\{j|-18<j<6\}$.

45. $\left|\frac{k}{4}-\frac{3}{8}\right|<\frac{7}{16}$

$-\frac{7}{16}<\frac{k}{4}-\frac{3}{8}<\frac{7}{16}$

$16\left(-\frac{7}{16}\right)<16\left(\frac{k}{4}-\frac{3}{8}\right)<16\left(\frac{7}{16}\right)$

$-7<4k-6<7$

$-7+6<4k-6+6<7+6$

$-1<4k<13$

$-\frac{1}{4}<k<\frac{13}{4}$

The solution is $\left\{k\middle|-\frac{1}{4}<k<\frac{13}{4}\right\}$.

47. $|y|>7$

$y<-7$ or $y>7$

The solution set is $\{y|y<-7 \text{ or } y>7\}$.

49. $|x+4|>5$

$x+4<-5$ or $x+4>5$

$x<-9$ $\quad$ $x>1$

The solution set is $\{x|x<-9 \text{ or } x>1\}$.

51. $|7-3b|>5$

$7-3b<-5$ or $7-3b>5$

$-3b<-12$ $\quad$ $-3b>-2$

$\frac{-3b}{-3}>\frac{-12}{-3}$ $\quad$ $\frac{-3b}{-3}<\frac{-2}{-3}$

$b>4$ $\quad$ $b<\frac{2}{3}$

The solution set is $\left\{b\middle|b<\frac{2}{3} \text{ or } b>4\right\}$.

53. $\left|\frac{2h-5}{3}\right|>1$

$\frac{2h-5}{3}<-1$ or $\frac{2h-5}{3}>1$

$3\left(\frac{2h-5}{3}\right)<3(-1)$ $\quad$ $3\left(\frac{2h-5}{3}\right)>3(1)$

$2h-5<-3$ $\quad$ $2h-5>3$

$2h<2$ $\quad$ $2h>8$

$h<\frac{2}{2}$ $\quad$ $h>\frac{8}{2}$

$h<1$ $\quad$ $h>4$

The solution set is $\{h|h<1 \text{ or } h>4\}$.

55. $|0.1x-0.4|+0.4>0.6$

$|0.1x-0.4|>0.2$

$0.1x-0.4<-0.2$ or $0.1x-0.4>0.2$

$0.1x<0.2$ $\quad$ $0.1x>0.6$

$x<\frac{0.2}{0.1}$ $\quad$ $x>\frac{0.6}{0.1}$

$x<2$ $\quad$ $x>6$

The solution set is $\{x|x<2 \text{ or } x>6\}$.

57. $\left|\frac{x}{2}+4\right|\ge 5$

$\frac{x}{2}+4\le -5$ or $\frac{x}{2}+4\ge 5$

$2\left(\frac{x}{2}+4\right)\le 2(-5)$ $\quad$ $2\left(\frac{x}{2}+4\right)\ge 2(5)$

$x+8\le -10$ $\quad$ $x+8\ge 10$

$x\le -18$ $\quad$ $x\ge 2$

The solution set is $\{x|x\le -18 \text{ or } x\ge 2\}$.

59. $|7w+3|-6\ge -6$

$|7w+3|\ge 0$

Observe that the absolute value of a number is always greater than or equal to 0. Thus, the solution is the set of real numbers or R.

61. $|4-2x|>0$

$4-2x<0$ or $4-2x>0$

$-2x<-4 \qquad -2x>-4$

$x>\frac{-4}{-2} \qquad x<\frac{-4}{-2}$

$x>2 \qquad x<2$

The solution set is $\{x|x<2 \text{ or } x>2\}$.

63. $|3p-5|=|2p+10|$

$3p-5=-(2p+10)$ or $3p-5=2p+10$

$3p-5=-2p-10 \qquad p-5=10$

$5p-5=-10 \qquad p=15$

$5p=-5$

$p=-1$

The solution set is $\{-1, 15\}$.

65. $|6x|=|3x-9|$

$6x=-(3x-9)$ or $6x=3x-9$

$6x=-3x+9 \qquad 3x=-9$

$9x=9 \qquad x=\frac{-9}{3}$

$x=\frac{9}{9} \qquad x=-3$

$x=1$

The solution set is $\{-3, 1\}$.

67. $\left|\frac{2r}{3}+\frac{5}{6}\right|=\left|\frac{r}{2}-3\right|$

$\frac{2r}{3}+\frac{5}{6}=-\left(\frac{r}{2}-3\right)$ or $\frac{2r}{3}+\frac{5}{6}=\frac{r}{2}-3$

$\frac{2r}{3}+\frac{5}{6}=-\frac{r}{2}+3 \qquad 6\left(\frac{2r}{3}+\frac{5}{6}\right)=6\left(\frac{r}{2}-3\right)$

$6\left(\frac{2r}{3}+\frac{5}{6}\right)=6\left(-\frac{r}{2}+3\right) \qquad 4r+5=3r-18$

$4r+5=-3r+18 \qquad r+5=-18$

$7r+5=18 \qquad r=-23$

$7r=13$

$r=\frac{13}{7}$

The solution set is $\left\{-23, \frac{13}{7}\right\}$.

69. $\left|-\frac{3}{4}m+8\right|=\left|7-\frac{3m}{4}\right|$

$-\frac{3}{4}m+8=-\left(7-\frac{3m}{4}\right)$ or $-\frac{3}{4}m+8=7-\frac{3m}{4}$

$-\frac{3}{4}m+8=-7+\frac{3}{4}m \qquad -\frac{3}{4}m+8=7-\frac{3}{4}m$

$-\frac{6}{4}m=-15 \qquad 8=7$ False!

$m=10$

The solution set is $\{10\}$.

71. $|h|=1$

$h=1$ or $h=-1$

The solution set is $\{-1, 1\}$.

73. $|q+6|>2$

$q+6<-2$ or $q+6>2$

$q<-8 \qquad q>-4$

The solution set is $\{q|q<-8 \text{ or } q>-4\}$.

75. $|2w-7|\le 9$

$-9\le 2w-7\le 9$

$-9+7\le 2w-7+7\le 9+7$

$-2\le 2w\le 16$

$\frac{-2}{2}\le\frac{2w}{2}\le\frac{16}{2}$

$-1\le w\le 8$

The solution set is $\{w|-1\le w\le 8\}$.

77. $|5a-1|=9$

$5a-1=-9$ or $5a-1=9$

$5a=-8 \qquad 5a=10$

$a=-\frac{8}{5} \qquad a=2$

The solution set is $\left\{-\frac{8}{5}, 2\right\}$.

79. $|5x+2|>0$

$5+2x<0$ or $5+2x>0$

$2x<-5 \qquad 2x>-5$

$x<-\frac{5}{2} \qquad x>-\frac{5}{2}$

The solution set is $\left\{x\middle|x<-\frac{5}{2} \text{ or } x>-\frac{5}{2}\right\}$.

81. $|4+3x|\le 9$

$-9\le 4+3x\le 9$

$-13\le 3x\le 5$

$$-\frac{13}{3} \le x \le \frac{5}{3}$$

The solution set is $\left\{x \middle| -\frac{13}{3} \le x \le \frac{5}{3}\right\}$.

83. $|3n+8|-4=-10$

$|3n+8|=-6$

Since the right side is negative and the left side is non-negative, there is no solution since the absolute value can never equal a negative number. The solution set is $\varnothing$.

85. $\left|\frac{w+4}{3}\right|-1<3$

$$\left|\frac{w+4}{3}\right|<4$$

$$-4<\frac{w+4}{3}<4$$

$$3(-4)<3\left(\frac{w+4}{3}\right)<3(4)$$

$$-12<w+4<12$$

$$-16<w<8$$

The solution set is $\{w|-16<w<8\}$.

87. $\left|\frac{3x-2}{4}\right|+5 \ge 5$

$$\left|\frac{3x-2}{4}\right| \ge 0$$

Since the absolute value of a number is always greater than or equal to zero, the solution is the set of all real numbers or R.

89. $|2x-8|=\left|\frac{1}{2}x+3\right|$

$2x-8=-\left(\frac{1}{2}x+3\right)$ or $2x-8=\frac{1}{2}x+3$

$2x-8=-\frac{1}{2}x-3$ | $\frac{3}{2}x-8=3$

$\frac{5}{2}x-8=-3$ | $\frac{3}{2}x=11$

$\frac{5}{2}x=5$ | $\frac{2}{3}\left(\frac{3}{2}x\right)=\frac{2}{3}(11)$

$\frac{2}{5}\left(\frac{5}{2}x\right)=\frac{2}{5}(5)$ | $x=\frac{22}{3}$

$x=2$

The solution set is $\left\{2, \frac{22}{3}\right\}$.

91. $|2-3x|=\left|4-\frac{5}{3}x\right|$

$2-3x=-\left(4-\frac{5}{3}x\right)$ or $2-3x=4-\frac{5}{3}x$

$2-3x=-4+\frac{5}{3}x$ | $-3x=2-\frac{5}{3}x$

$-3x=-6+\frac{5}{3}x$ | $-\frac{4}{3}x=2$

$-\frac{14}{3}x=-6$ | $-\frac{3}{4}\left(-\frac{4}{3}x\right)=-\frac{3}{2}(2)$

$\left(-\frac{3}{14}\right)\left(-\frac{14}{3}\right)x=\left(-\frac{3}{14}\right)(-6)$ | $x=-\frac{3}{2}$

$x=\frac{9}{7}$

The solution set is $\left\{-\frac{3}{2}, \frac{9}{7}\right\}$.

93. **a.** $|t-0.089| \le 0.004$

$$-0.004 \le t-0.089 \le 0.004$$

$$-0.004+0.089 \le t-0.089+0.089 \le 0.004+$$

$$0.085 \le t \le 0.093$$

The solution is $[0.085, 0.093]$.

b. 0.085 inches

c. 0.093 inches

95. **a.** $|d-160| \le 28$

$-28 \le d-160 \le 28$

$-28+160 \le d-160+160 \le 28+160$

$132 \le d \le 188$

The solution is [132, 188]

b. The submarine can move between 132 feet and 188 feet below sea level, inclusive.

97. $\{-5, 5\}$ is the solution set of $|x|=5$.

99. $\{x|x \le -5$ or $x \ge 5\}$ is the solution set of $|x| \ge 5$.

101. $|ax+b| \le 0$

$0 \le ax+b \le 0$

which is the same as

$ax+b=0$

$ax=-b$

$x=-\frac{b}{a}$

103. a. Set $ax + b = -c$ or $ax + b = c$ and solve each equation for x.

b. $ax + b = -c$ or $ax + b = c$

$ax = -c - b$ $\quad$ $ax = c - b$

$x = \dfrac{-c-b}{a}$ $\quad$ $x = \dfrac{c-b}{a}$

The solution is $x = \dfrac{-c-b}{a}$ or $x = \dfrac{c-b}{a}$.

105. a. Write $ax + b < -c$ or $ax + b > c$ and solve each inequality for x.

b. $ax + b < -c$ or $ax + b > c$

$ax < -c - b$ $\quad$ $ax > c - b$

$x < \dfrac{-c-b}{a}$ $\quad$ $x > \dfrac{c-b}{a}$

The solution is $x < \dfrac{-c-b}{a}$ or $x > \dfrac{c-b}{a}$.

107. $|x-3| = |3-x|$

$x - 3 = -(3 - x)$ or $x - 3 = 3 - x$

$x - 3 = -3 + x$ $\quad$ $2x - 3 = 3$

$0 = 0$ $\quad$ $2x = 6$

True $\quad$ $x = 3$

Since the first statement is always true all real values work. The solution set is R.

109. $|x| = x$

By definition $|x| = \begin{cases} x, & x \ge 0 \\ -x, & x < 0 \end{cases}$

Thus, $|x| = x$ when $x \ge 0$

The solution set is $\{x | x \ge 0\}$.

111. $|x+1| = 2x - 1$

$x + 1 = -(2x - 1)$ or $x + 1 = 2x - 1$

$x + 1 = -2x + 1$ $\quad$ $1 = x - 1$

$3x + 1 = 1$ $\quad$ $2 = x$

$3x = 0$

$x = 0$

Checking both possible solutions, only $x = 2$ checks. The solution set is $\{2\}$.

113. $|x-2| = -(x-2)$

By the definition,

$|x-2| = \begin{cases} x-2, & x - 2 \ge 0 \\ -(x-2), & x - 2 \le 0 \end{cases}$ or

$\begin{cases} x-2, & x \ge 2 \\ -(x-2), & x \le 2 \end{cases}$

Thus, $|x-2| = -(x-2)$ for $x \le 2$.

The solution set is $\{x | x \le 2\}$.

115. $x + |-x| = 6$

For $x \ge 0$: $x + |-x| = 6$

$x + x = 6$

$2x = 6$

$x = 3$

For $x < 0$: $x + |-x| = 6$

$x - x = 6$

$0 = 6$ False

The solution set is $\{3\}$.

117. $x - |x| = 6$

For $x \ge 0$: $x - |x| = 6$

$x - x = 6$

$0 = 6$ False

For $x < 0$: $x - |x| = 6$

$x - (-x) = 6$

$x + x = 6$

$2x = 6$

$x = 3$ Contradicts $x < 0$

There are no values of x, so the solution set is $\varnothing$.

119. $\dfrac{1}{3} + \dfrac{1}{4} \div \dfrac{2}{5}\left(\dfrac{1}{3}\right)^2 = \dfrac{1}{3} + \dfrac{1}{4} \div \dfrac{2}{5} \cdot \dfrac{1}{9}$

$= \dfrac{1}{3} + \dfrac{1}{4} \cdot \dfrac{5}{2} \cdot \dfrac{1}{9}$

$= \dfrac{1}{3} + \dfrac{5}{72}$

$= \dfrac{1}{3} \cdot \dfrac{24}{24} + \dfrac{5}{72}$

$= \dfrac{24}{72} + \dfrac{5}{72}$

$= \dfrac{29}{72}$

120. Substitute 1 for x and 3 for y.

$4(x + 3y) - 5xy = 4(1 + 3 \cdot 3) - 5(1)(3)$

$= 4(1 + 9) - 5(1)(3)$

$= 4(10) - 5(1)(3)$

$= 40 - 15$

$= 25$

121. Let x be the time needed to swim across the lake. Then $1.5 - x$ is the time needed to make the return trip.

	Rate	**Time**	**Distance**
First Trip	2	x	$2x$
Return Trip	1.6	$1.5 - x$	$1.6(1.5 - x)$

The distances are the same.

$2x = 1.6(1.5 - x)$

$2x = 2.4 - 1.6x$

$3.6x = 2.4$

$x = \frac{2.4}{3.6} = \frac{2}{3}$

The total distance across the lake is

$2x = 2\left(\frac{2}{3}\right) = \frac{4}{3}$ or 1.33 miles.

122. $3(x-2) - 4(x-3) > 2$

$3x - 6 - 4x + 12 > 2$

$-x + 6 > 2$

$-x > -4$

$\frac{-x}{-1} < \frac{-4}{-1}$

$x < 4$

The solution set is $\{x|x < 4\}$.

Exercise Set 10.3

1. The inequalities > and < do not include the corresponding equation; the points on the line satisfy only the equation.

3. (0, 0) cannot be used as a test point if the line passes through the origin.

5. Answers will vary.

7. Yes, the points along the two boundary lines are ordered pairs which are solutions to both inequalities.

9. $x > 1$

Graph the line $x = 1$ (vertical line) using a dashed line. For the check point, select (0, 0):

$x > 1$

$0 > 1 \leftarrow$ Substitute 0 for x

Since this is a false statement, shade the region which does not contain (0, 0).

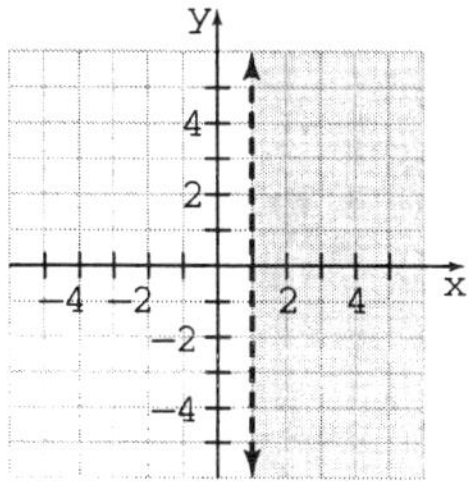

11. $y < -2$

Graph the line $y = -2$ (horizontal line) using a dashed line. For the check point, select (0, 0).

$y < -2$

$0 < -2 \leftarrow$ Substitute 0 for y.

Since this is a false statement, shade the region which does not contain (0, 0).

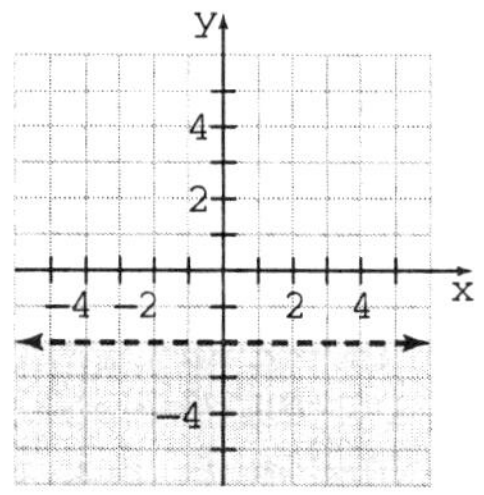

13. $y \geq -\frac{1}{2}x$

Graph the line $y = -\frac{1}{2}x$ using a solid line. For the check point, select (0, 2).

$y \geq -\frac{1}{2}x$

$2 \geq -\frac{1}{2}(0) \leftarrow$ Substitute 0 for x, 2 for y

$2 \geq 0$

Since this is a true statement, shade the region which contains the point (0, 2).

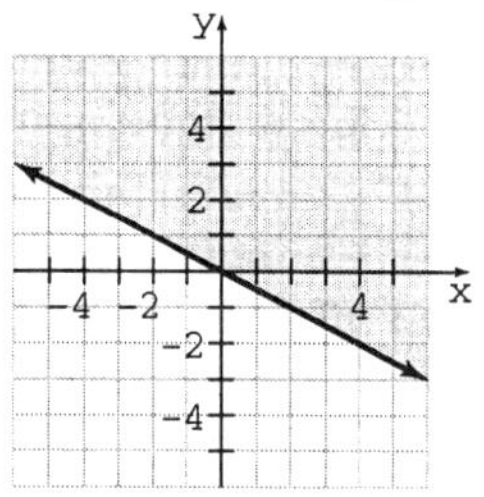

15. $y < 2x + 1$

Graph the line $y = 2x + 1$ using a dashed line. For the check point, select (0, 0).

$y < 2x + 1$

$0 < 2(0) + 1 \leftarrow$ Substitute 0 for x and y

$0 < 1$

Since this is a true statement, shade the region which contains the point (0, 0).

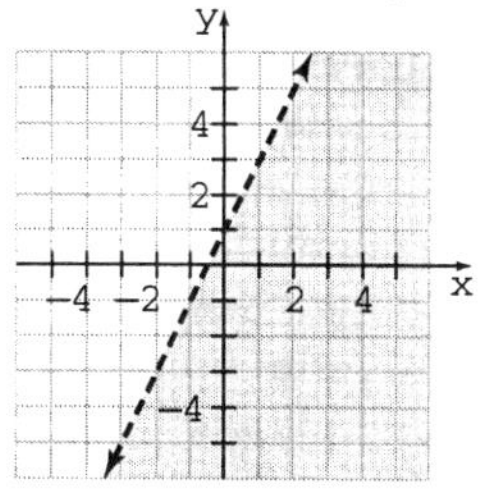

17. $y > 2x - 1$
Graph the line $y = 2x - 1$ using a dashed line.
For the check point, select (0, 0).
$y > 2x - 1$
$0 > 2(0) - 1 \leftarrow$ Substitute 0 for x and y
$0 > -1$
Since this is a true statement, shade the region which contains the point (0, 0).

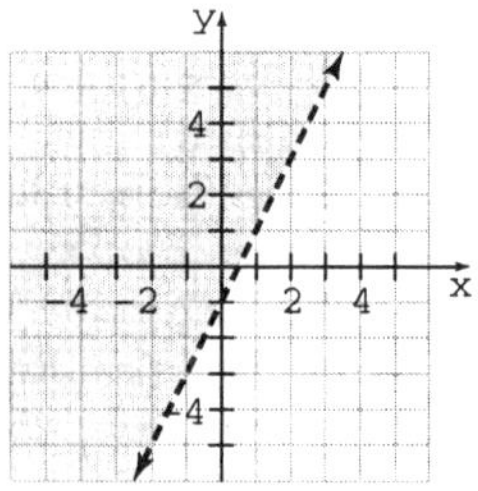

19. $y \geq \frac{1}{2}x - 3$

Graph the line $y = \frac{1}{2}x - 3$ using a solid line. For the check point, select (0, 0).

$y \geq \frac{1}{2}x - 3$

$0 \geq \frac{1}{2}(0) - 3 \leftarrow$ Substitute 0 for x and y

$0 \geq -3$
Since this is a true statement, shade the region which contains the point (0, 0).

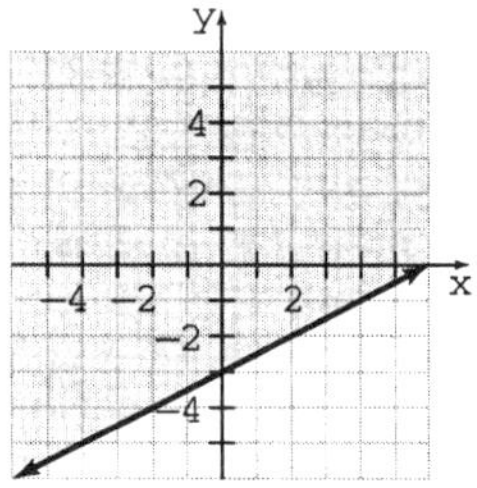

21. $2x - 3y \geq 12$
$-3y \geq -2x + 12$

$\frac{-3y}{-3} \leq \frac{-2x+12}{-3}$

$y \leq \frac{2}{3}x - 4$

Graph the line $y = \frac{2}{3}x - 4$ using a solid line. For the check point, select (0, 0).

$y \leq \frac{2}{3}x - 4$

$(0) \leq \frac{2}{3}(0) - 4 \leftarrow$ Substitute 0 for x and y

$0 \leq -4$

Since this is a false statement, shade the region which does not contain the point (0, 0).

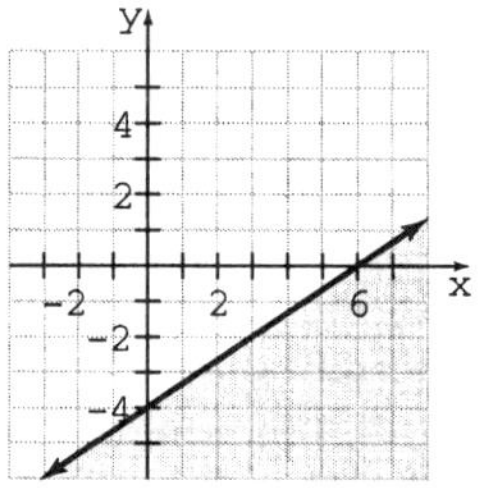

23. $y \leq -3x + 5$
Graph the line $y = -3x + 5$ using a solid line.
For the check point, select (0, 0).
$y \leq -3x + 5$
$0 \leq -3(0) + 5 \leftarrow$ Substitute 0 for x and y
$0 \leq 5$
Since this is a true statement, shade the region which contains the point (0, 0).

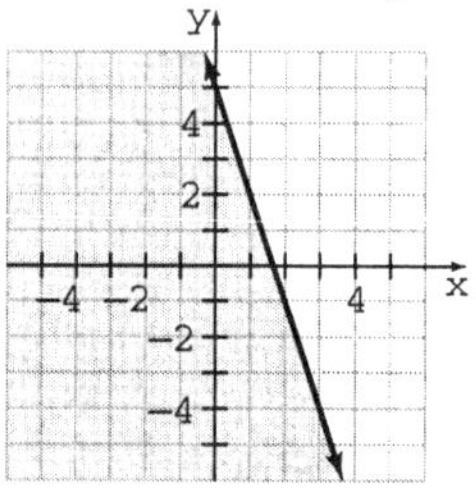

25. $2x + y < 4$
Graph the line $2x + y = 4$ using a dashed line.
For the check point, select (0, 0).
$2x + y < 4$
$2(0) + 0 < 4 \leftarrow$ Substitute 0 for x and y
$0 < 4$
Since this is a true statement, shade the region which contains the point (0, 0).

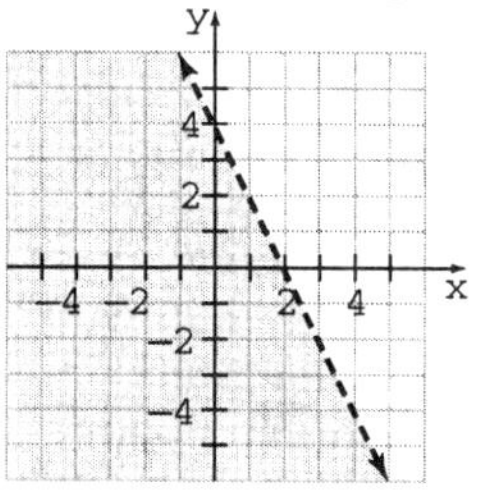

27. $10 \geq 5x - 2y$
Graph the line $10 = 5x - 2y$ using a solid line.
For the check point, select (0, 0).
$10 \geq 5x - 2y$
$10 \geq 5(0) - 2(0) \leftarrow$ Substitute 0 for x and y
$10 \geq 0$
Since this is a true statement, shade the region

which contains the point (0, 0).

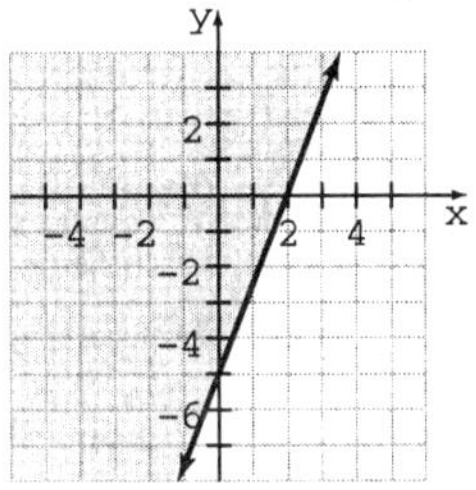

29. $2x - y < 4$
$y \geq -x + 2$
For $2x - y < 4$, graph the line $2x - y = 4$ using a dashed line. For the check point, select (0, 0):
$2x - y < 4$
$2(0) - (0) < 4$
$0 < 4$ True
Since this is a true statement, shade the region which contains the point (0, 0). This is the region "above" the line.
For $y \geq -x + 2$, graph the line $y = -x + 2$ using a solid line. For the check point, select (0, 0):
$y \geq -x + 2$
$(0) \geq -(0) + 2$
$0 \geq 2$ False
Since this is a false statement, shade the region which does not contain the point (0, 0). This is the region "above" the line. To obtain the final region, take the intersection of the above two regions.

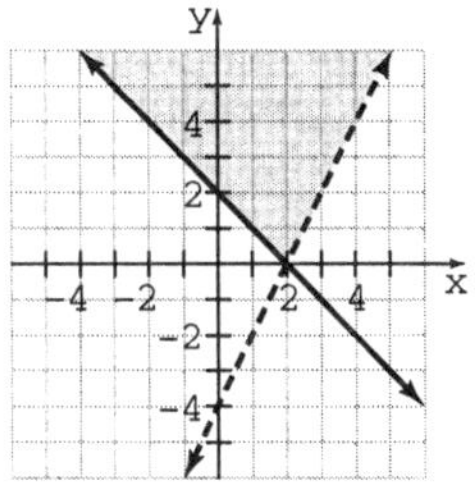

31. $y < 3x - 2$
$y \leq -2x + 3$
For $y < 3x - 2$, graph the line $y = 3x - 2$ using a dashed line. For the check point, select (0, 0):
$y < 3x - 2$
$(0) < 3(0) - 2$
$0 < -2$ False
Since this is a false statement, shade the region which does not contain the point (0, 0). This is the region "below" the line.
For $y \leq -2x + 3$, graph the line $y = -2x + 3$ using a solid line. For the check point, select (0, 0):
$y \leq -2x + 3$
$(0) \leq -2(0) + 3$
$0 \leq 3$ True
Since this is a true statement, shade the region which contains the point (0, 0). This is the region "below" the line. To obtain the final region, take the intersection of the above two regions.

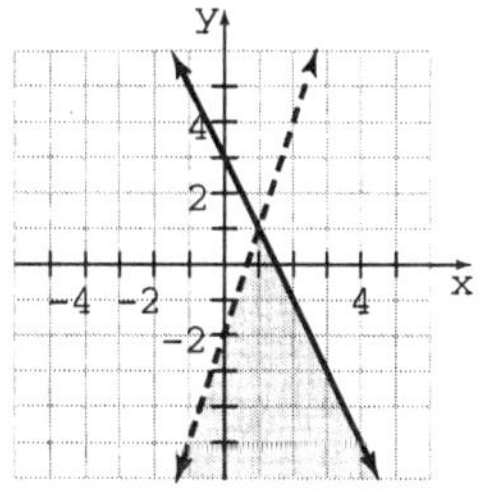

33. $y < x$
$y \geq 3x + 2$
For $y < x$, graph the line $y = x$ using a dashed line. For the check point, select (1, 0):
$y < x$
$0 < 1$ True
Since this is a true statement, shade the region which contains the point (1, 0). This is the region "below" the line.
For $y \geq 3x + 2$, graph the line $y = 3x + 2$ using a solid line. For the check point, select (0, 0):
$y \geq 3x + 2$
$(0) \geq 3(0) + 2$
$0 \geq 2$ False
Since this is a false statement, shade the region which does not contain the point (0, 0). This is the region "above" the line. To obtain the final region, take the intersection of the above two regions.

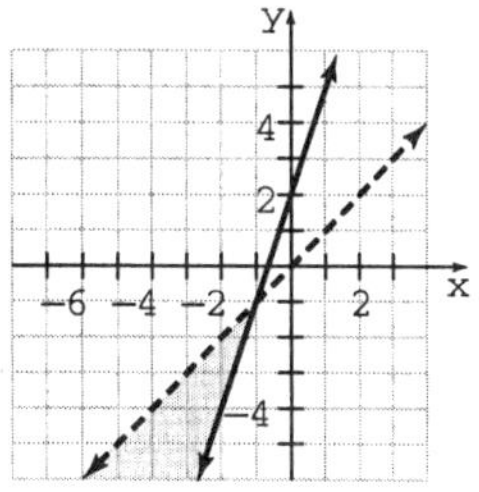

35. $-3x + 2y \geq -5$
$y \leq -4x + 7$
For $-3x + 2y \geq -5$, graph the line $-3x + 2y = -5$ using a solid line. For the check point, select (0, 0):

$-3x + 2y \ge -5$

$-3(0) + 2(0) \ge -5$

$0 \ge -5$ True

Since this is a true statement, shade the region which contains the point (0, 0). This is the region "above" the line.

For $y \le -4x + 7$, graph the line $y = -4x + 7$ using a solid line. For the check point, select (0, 0):

$y \le -4x + 7$

$(0) \le -4(0) + 7$

$0 \le 7$ True

Since this is a true statement, shade the region which contains the point (0, 0). This is the region "below" the line.

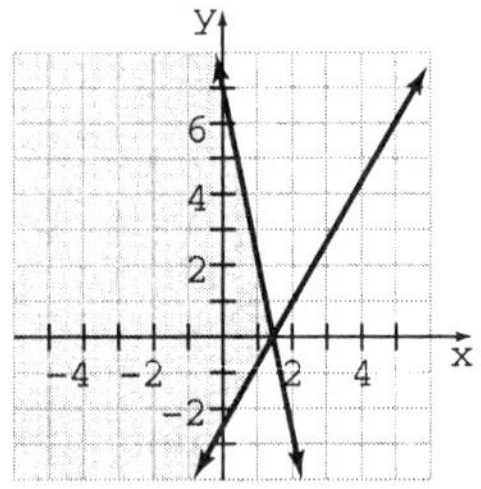

37. $-4x + 5y < 20$

$x \ge -3$

For $-4x + 5y < 20$, graph the line $-4x + 5y = 20$ using a dashed line. For the check point, select (0, 0):

$-4x + 5y < 20$

$-4(0) + 5(0) < 20$

$0 < 20$ True

Since this is a true statement, shade the region which contains the point (0, 0). This is the region "below" the line.

For $x \ge -3$, the graph is the line $x = -3$ along with the region to the right of $x = -3$. To obtain the final region, take the intersection of the above two regions.

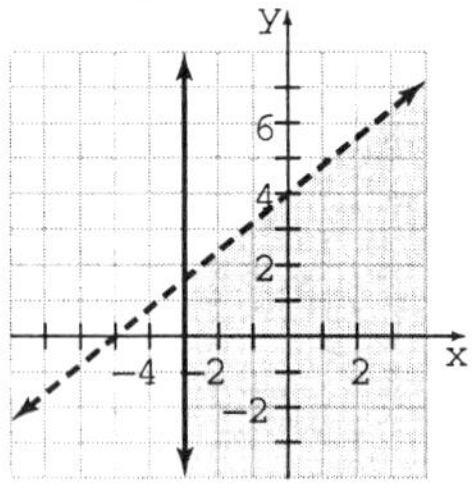

39. $x \le 4$

$y \ge -2$

For $x \le 4$, the graph is the line $x = 4$ along with the region to the left of $x = 4$. For $y \ge -2$, the graph is the line $y = -2$ along with the region above the line $y = -2$. To obtain the final region, take the intersection of the above two regions.

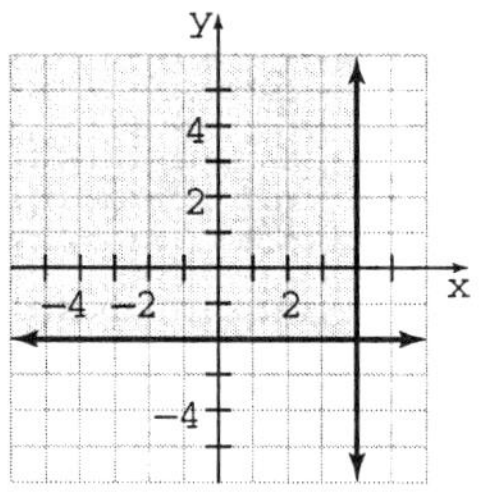

41. $5x + 2y > 10$

$3x - y > 3$

For $5x + 2y > 10$, graph the line $5x + 2y = 10$ using a dashed line. For the check point, select (0, 0):

$5x + 2y > 10$

$5(0) + 2(0) > 10$

$0 > 10$ False

Since this is a false statement, shade the region which does not contain the point (0, 0). This is the region "above" the line.

For $3x - y > 3$, graph the line $3x - y = 3$ using a dashed line. For the check point, select (0, 0):

$3x - y > 3$

$3(0) - 0 > 3$

$0 > 3$ False

Since this is a false statement, shade the region which does not contain the point (0, 0). This is the region "below" the line. To obtain the final region, take the intersection of the above two regions.

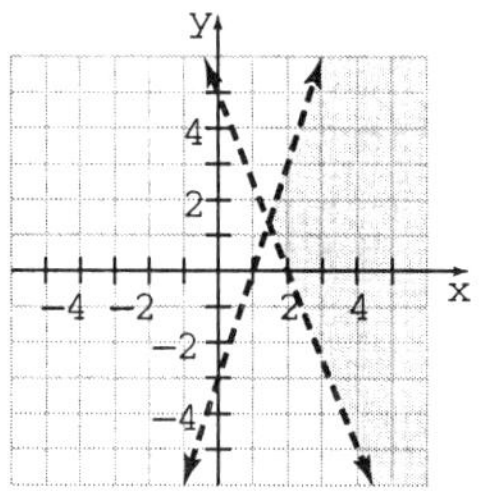

43. $-2x > y + 4$

$-x < \frac{1}{2}y - 1$

For $-2x > y + 4$, graph the line $-2x = y + 4$ using a dashed line. For the check point, select (0, 0):

$-2x > y + 4$

$-2(0) > 0 + 4$

$0 > 4$ False

Since this is a false statement, shade the region which does not contain the point (0, 0). This is the region "below" the line.

For $-x < \frac{1}{2}y - 1$, graph the line $-x = \frac{1}{2}y - 1$ using a dashed line. For the check point, select (0, 0):

$-x < \frac{1}{2}y - 1$

$-0 < \frac{1}{2}(0) - 1$

$0 < -1$ False

Since this is a false statement, shade the region which does not contain the point (0, 0). This is the region "above" the line. To obtain the final region take the intersection of the above two regions. Since the regions do not overlap, the final result is the empty set which means there is no solution.

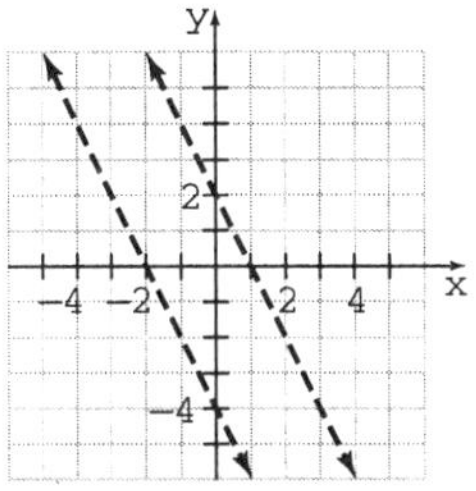

45. $y < 3x - 4$

$6x \geq 2y + 8$

Solve the second inequality for y.

$6x \geq 2y + 8$

$6x - 2y > 8$

$-2y \geq -6x + 8$

$y \leq 3x - 4$

The second inequality is now identical to the first except that the second inequality includes the line.

For $y < 3x - 4$, graph the line $y = 3x - 4$ using a dashed line. For the check point, select (0, 0):

$y < 3x - 4$

$0 < 3(0) - 4$

$0 < -4$ False

Since this is a false statement, shade the region which does not contain the point (0, 0). This is the region "below" the line.

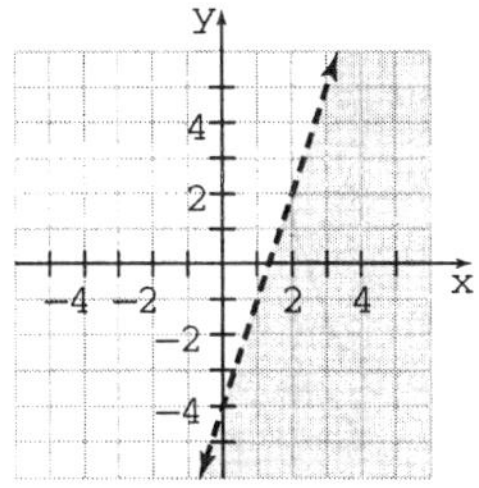

47. $x \geq 0$

$y \geq 0$

$2x + 3y \leq 6$

$4x + y \leq 4$

The first two inequalities indicate that the region must be in the first quadrant. For $2x + 3y \leq 6$, the graph is the line $2x + 3y = 6$ along with the region below this line. For $4x + y \leq 4$, the graph is the line $4x + y = 4$ along with the region below this line. To obtain the final region, take the intersection of these regions.

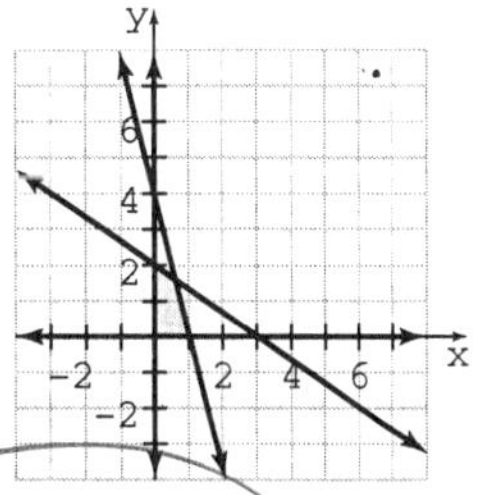

49. $x \geq 0$

$y \geq 0$

$x + y \leq 6$

$7x + 4y \leq 28$

The first two inequalities indicate that the region must be in the first quadrant. For $x + y \leq 6$, the graph is the line $x + y = 6$ along with the region below this line. For $7x + 4y \leq 28$, the graph is the line $7x + 4y = 28$ along with the region below the line. To obtain the final region, take the intersection of these regions.

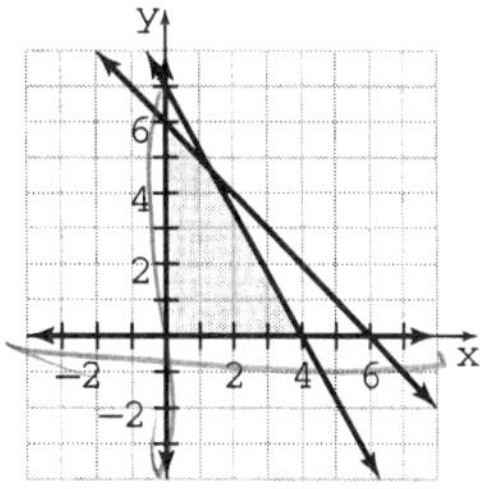

51. $x \geq 0$

$y \geq 0$

$3x + 2y \leq 18$

$2x + 4y \leq 20$

The first two inequalities indicate that the region must be in the first quadrant.

For $3x + 2y \leq 18$, the graph is the line $3x + 2y = 18$ along with the region below this line. For $2x + 4y \leq 20$, the graph is the line $2x + 4y = 20$ along with the region below the line. To obtain the final region, take the intersection of these regions. The final answer is

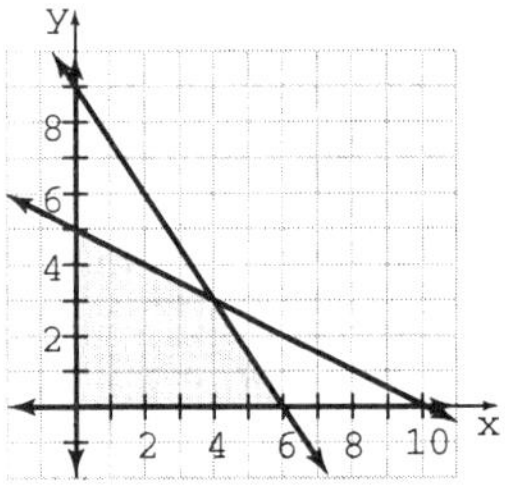

53. $x \geq 0$
$y \geq 0$
$x \leq 4$
$x + y \leq 6$
$x + 2y \leq 8$
The first two inequalities indicate that the region must be in the first quadrant. The third inequality indicates that the region must be on or to the left of the line $x = 4$. For $x + y \leq 6$, the graph is the line $x + y = 6$ along with the region below this line. For $x + 2y \leq 8$, the graph is the line $x + 2y = 8$ along with the region below the line. To obtain the final region, take the intersection of these regions.

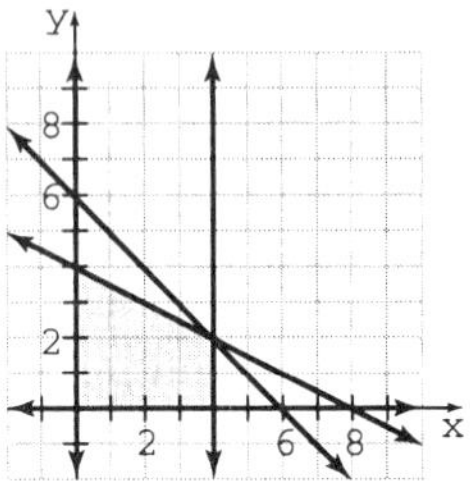

55. $x \geq 0$
$y \geq 0$
$x \leq 15$
$40x + 25y \leq 1000$
$5x + 30y \leq 900$
The first two inequalities indicate that the region must be in the first quadrant. The third inequality indicates that the region must be on or to the left of the line $x = 15$. For $40x + 25y \leq 1000$, the graph is the line $40x + 25y = 1000$ along with the region below this line. For $5x + 30y \leq 900$, the graph is the line $5x + 30y = 900$ along with the region below this line. To obtain the final region, take the intersection of these regions.

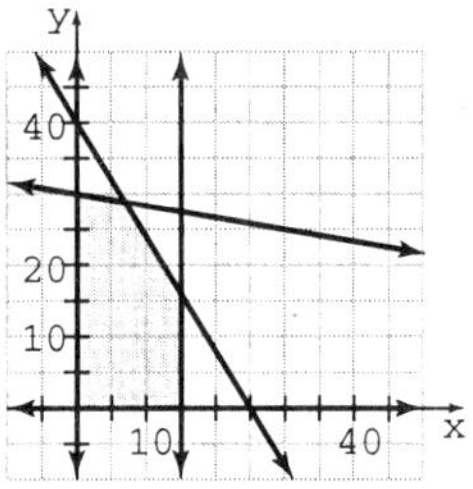

57. $|x| > 1$
$y < x$
For $|x| > 1$, the graph is the region to the right of the dashed line $x = 1$ and to the left of the dashed lines $x = -1$. For $y < x$, the graph is the region below the dashed line $y = x$. To obtain the final region, take the intersection of the above two regions.

59. $|x| \geq 1$
$|y| \geq 2$
For $|x| \geq 1$, the graph is the region to the left of the solid line $x = -1$ along with the region to the right of the solid line $x = 1$. For $|y| \geq 2$, the graph is the region above the solid line $y = 2$ along with the region below the solid line $y = -2$. To obtain the final region, take the intersection of these regions.

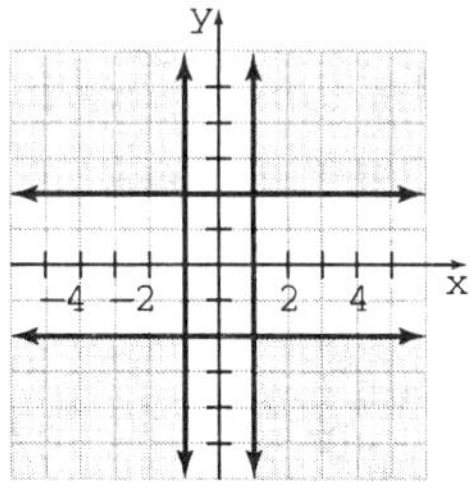

61. $|y| > 2$
$y \leq x + 3$
For $|y| > 2$, the graph is the region above the dashed line $y = 2$ along with the region below the dashed line $y = -2$. For $y \leq x + 3$, the graph is the region below the solid line $y = x + 3$. To obtain the final region, take the intersection of these

regions.

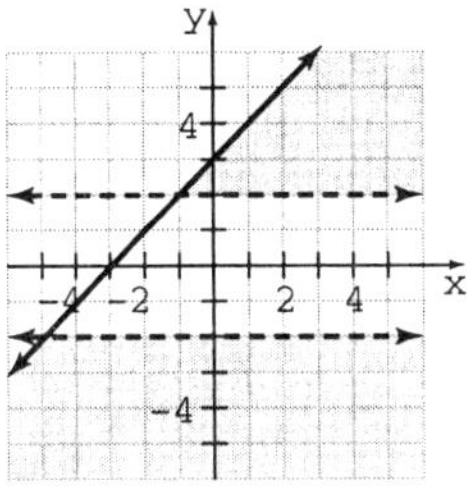

63. $|y| < 4$
$y \geq -2x + 2$
For $|y| < 4$, the graph is the region between the dashed lines $y = -4$ and $y = 4$. For $y \geq -2x + 2$, the graph is the region above the solid line $y = -2x + 2$. To obtain the final region, take the intersection of these regions.

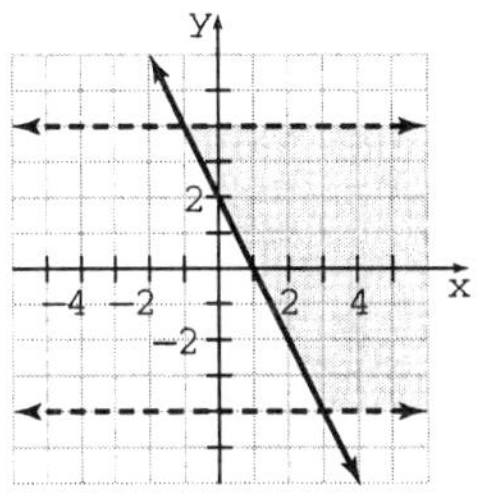

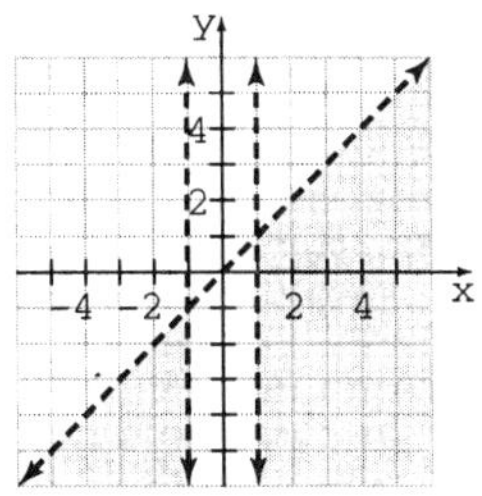

65. $|x + 2| < 3$
$|y| > 4$
$|x + 2| < 3$ can be written as
$-3 < x + 2 < 3$
$-5 < x < 1$
For $|x + 2| < 3$, the graph is the region between the dashed lines $x = -5$ and $x = 1$. For $|y| > 4$, the graph is the region above the dashed line $y = 4$ along with the region below the dashed line $y = -4$. To obtain the final region, take the intersection of these regions.

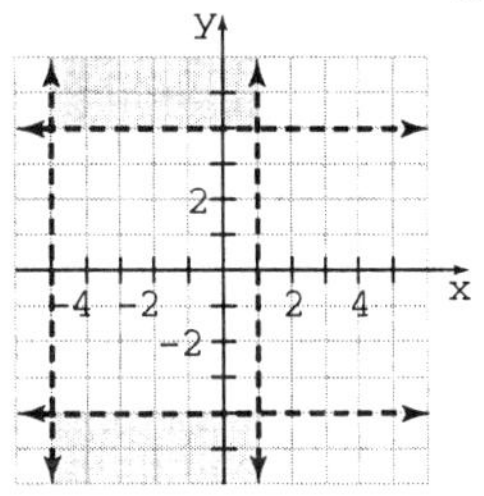

67. $|x - 3| \leq 4$
$|y + 2| \leq 1$
$|x - 3| \leq 4$ can be written as
$-4 \leq x - 3 \leq 4$
$-1 \leq x \leq 7$
For $|x - 3| \leq 4$, the graph is the region between the solid lines $x = -1$ and $x = 7$.
$|y + 2| \leq 1$ can be written as
$-1 \leq y + 2 \leq 1$
$-3 \leq y \leq -1$
For $|y + 2| \leq 1$, the graph is the region between the solid lines $y = -3$ and $y = -1$. To obtain the final region, take the intersection of these regions.

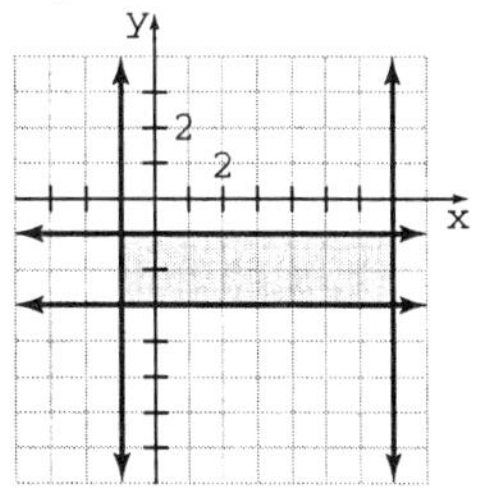

69. a, b.

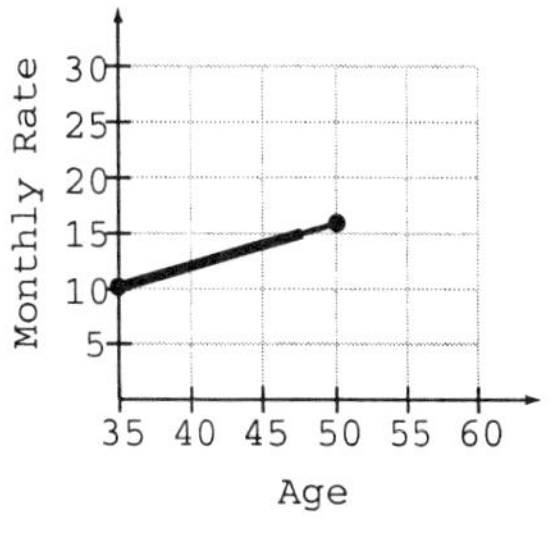

c. The age at which the rate first exceeds \$15 per month is 47.

71. a, b.

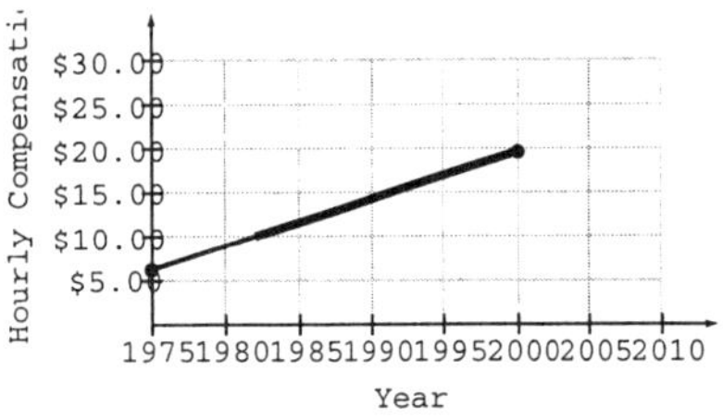

c. 1982 was the year that the average hourly wage first exceeded \$10 per hour.

73. a.

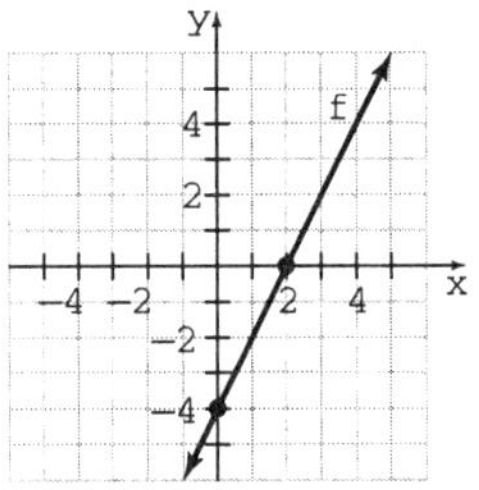

b.

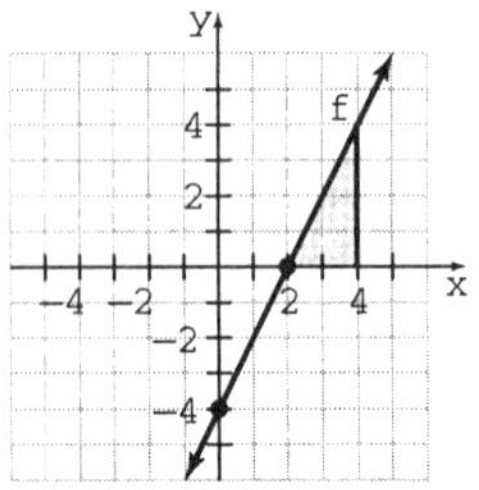

75. If the boundary lines are parallel, there may be no solution. For example, the system $y < x$ and $y > x + 1$ has no solution.

77. There are no solutions. Opposite sides of the same line are being shaded, but not the line itself.

79. There are an infinite number of solutions. Both inequalities include the line $5x - 2y = 3$.

81. There are an infinite number of solutions. The lines are parallel but the same side of each line is being shaded.

83. $y \ge x^2$
$y \le 4$

For $y \ge x^2$, graph the equation $y = x^2$ using a solid line. For the check point, select (0, 1).

$y \ge x^2$

$(1) \ge (0)^2$

$1 \ge 0$ True

Since this is a true statement, shade the region which contains the point (0, 1). This is the region "above" the graph of $y \ge x^2$.

For $y \le 4$, the graph is the region below the solid line $y = 4$. To obtain the final region, take the intersection of these regions.

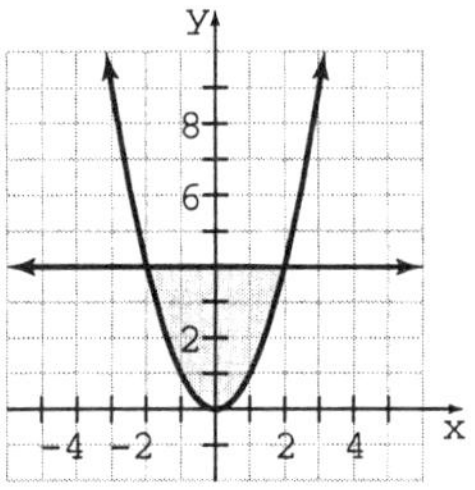

85. $y < |x|$
$y < 4$

For $y < |x|$, graph the equation $y = |x|$ using a dashed line. For the check point, select (0, 3).

$y < |x|$

$3 < |0|$

$3 < 0$ False

Since this is a false statement, shade the region which does not contain the point (0, 3). This is the region below the graph of $y = |x|$.

For $y < 4$, the graph is the region below the dashed line $y = 4$. To obtain the final region, take the intersection of these regions.

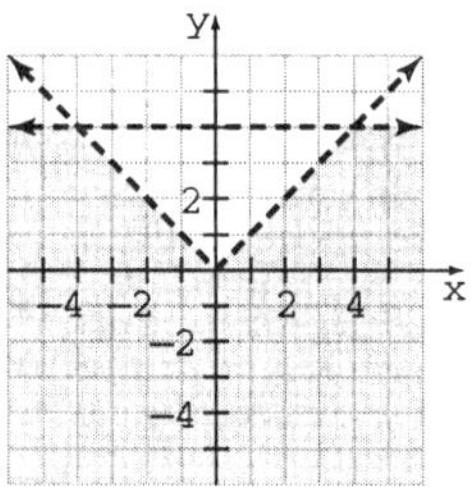

87.
$$f_1d_1 + f_2d_2 = f_3d_3$$
$$f_1d_1 - f_1d_1 + f_2d_2 = f_3d_3 - f_1d_1$$
$$f_2d_2 = f_3d_3 - f_1d_1$$
$$\frac{f_2d_2}{d_2} = \frac{f_3d_3 - f_1d_1}{d_2}$$
$$f_2 = \frac{f_3d_3 - f_1d_1}{d_2}$$

88. Domain: {–1, 0, 4, 5}
Range: (–5, –2, 2, 3}

89. Domain: R
Range: R

90. Domain: R
Range: $\{y|y \ge -1\}$

Review Exercises

1.
$$3z + 7 \le 13$$
$$3z \le 6$$
$$z \le 2$$

2.
$$5 - 2w > -7$$
$$-2w > -12$$
$$\frac{-2w}{-2} < \frac{-12}{-2}$$
$$w < 6$$
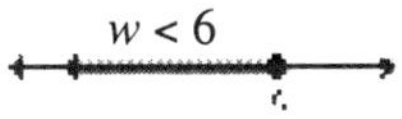

3.
$$2x + 4 > 9$$
$$2x > 5$$
$$x > \frac{5}{2}$$
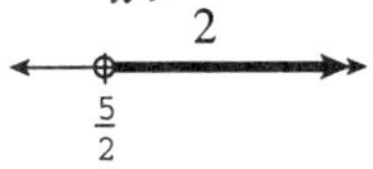

4.
$$16 \le 4x - 5$$
$$21 \le 4x$$
$$\frac{21}{4} \le x$$
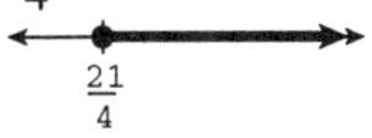

5.
$$\frac{4x + 3}{5} > -3$$
$$5\left(\frac{4x+3}{5}\right) > 5(-3)$$
$$4x + 3 > -15$$
$$4x > -18$$
$$x > \frac{-18}{4}$$
$$x > -\frac{9}{2}$$
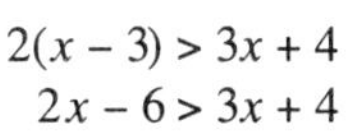

6.
$$2(x - 3) > 3x + 4$$
$$2x - 6 > 3x + 4$$
$$2x - 10 > 3x$$
$$-10 > x$$
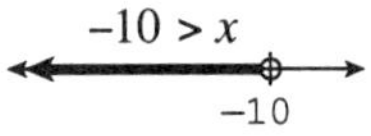

7.
$$-4(x - 2) \ge 6x + 4 - 10x$$
$$-4x + 8 \ge -4x + 4$$
$$8 \ge +4 \quad \text{a true statement}$$
The solution is all real numbers.

8.
$$\frac{x}{2} + \frac{3}{4} > x - \frac{x}{2} + 1$$
$$4\left(\frac{x}{2} + \frac{3}{4}\right) > 4\left(x - \frac{x}{2} + 1\right)$$
$$2x + 3 > 4x - 2x + 4$$
$$2x + 3 > 2x + 4$$
$$3 > 4$$
This is a contradiction, so the solution is { }.
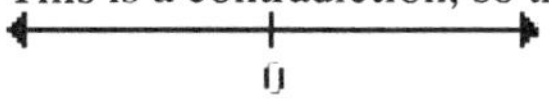

9. Let x be the maximum number of 40-pound boxes. Since the maximum load is 500 pounds, the total weight of the Joseph and boxes must be less than or equal to 500 pounds.
$$180 + 40x \le 500$$
$$40x \le 320$$
$$x \le \frac{320}{40}$$
$$x \le 8$$
The maximum number of boxes that Joseph can carry in the canoe is 8.

10. Let x be the number of additional minutes (beyond 3 minutes) of the phone call.
$$4.50 + 0.95x \le 8.65$$
$$0.95x \le 4.15$$
$$x \le \frac{4.15}{0.95}$$
$$x \le 4.4$$
The customer can talk for 3 minutes plus an additional 4 minutes for a total of 7 minutes.

11. Let x be the number of weeks (after the first week) needed to lose 27 pounds.
$$3 + 1.5x \ge 27$$
$$1.5x \ge 24$$
$$x \ge \frac{24}{1.5}$$
$$x \ge 16$$
The number of weeks is 16 plus the initial week for a total of 17 weeks.

12. Let x be the grade from the 5th exam. The inequality is

$80 \le \dfrac{94+73+72+80+x}{5} < 90$

$80 \le \dfrac{319+x}{5} < 90$

$5(80) \le 5\left(\dfrac{319+x}{5}\right) < 5(90)$

$400 \le 319 + x < 450$

$400 - 319 \le 319 + x < 450 - 319$

$81 \le x < 131$

(must use 100 here since it is not possible to score 131)

Thus, Jekeila needs to score 81 or higher on the 5th exam to receive a B.

$\{x|81 \le x \le 100\}$

13. $A \cup B = \{1,2,3,4,5\}$, $A \cap B = \{2,3,4,5\}$

14. $A \cup B = \{1,2,3,4,5,6,7,8,9\}$, $A \cap B = \{\ \}$

15. $A \cup B = \{1,2,3,4,\ldots\}$, $A \cap B = \{2,4,6,\ldots\}$

16. $A \cup B = \{3,4,5,6,7,8,9,10,11,12\}$,
$A \cap B = \{9,10\}$

17. $1 < x - 4 < 7$

$1 + 4 < x - 4 + 4 < 7 + 4$

$5 < x < 11$

$(5, 11)$

18. $7 < p + 10 \le 15$

$7 - 10 < p + 10 - 10 \le 15 - 10$

$-3 < p \le 5$

$(-3, 5]$

19. $3 < 2x - 4 < 8$

$3 + 4 < 2x - 4 + 4 < 8 + 4$

$7 < 2x < 12$

$\dfrac{7}{2} < \dfrac{2x}{2} < \dfrac{12}{2}$

$\dfrac{7}{2} < x < 6$

$\left(\dfrac{7}{2}, 6\right)$

20. $-12 < 6 - 3x < -2$

$-12 - 6 < 6 - 6 - 3x < -2 - 6$

$18 < -3x < -8$

$\dfrac{-18}{-3} > \dfrac{-3x}{-3} > \dfrac{-8}{-3}$

$6 > x > \dfrac{8}{3}$

$\dfrac{8}{3} < x < 6$

$\left(\dfrac{8}{3}, 6\right)$

21. $-1 < \dfrac{5}{9}x + \dfrac{2}{3} \le \dfrac{11}{9}$

$9(-1) < 9\left(\dfrac{5}{9}x + \dfrac{2}{3}\right) \le 9\left(\dfrac{11}{9}\right)$

$-9 < 5x + 6 \le 11$

$-9 - 6 < 5x + 6 - 6 \le 11 - 6$

$-15 < 5x \le 5$

$\dfrac{-15}{5} < \dfrac{5x}{5} \le \dfrac{5}{5}$

$-3 < x \le 1$

$(-3, 1]$

22. $-8 < \dfrac{4-2x}{3} < 0$

$3(-8) < 3\left(\dfrac{4-2x}{3}\right) < 3(0)$

$-24 < 4 - 2x < 0$

$-24 - 4 < 4 - 4 - 2x < 0 - 4$

$-28 < -2x < -4$

$\dfrac{-28}{-2} > \dfrac{-2x}{-2} > \dfrac{-4}{-2}$

$14 > x > 2$

$2 < x < 14$

$(2, 14)$

23. $h \le 2$ and $7h - 4 > -25$

$h \le 2$ and $7h > -21$

$h \le 2$ and $h > -3$

$h \le 2$

$h > -3$

$x \le 2$ and $x > -3$ which is $-3 < h \le 2$

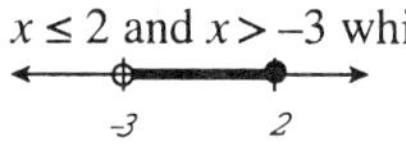

$\{h|-3 < h \le 2\}$

24. $2x-1>5$ or $3x-2\le 7$

$2x>6$ or $3x\le 9$

$x>3$ or $x\le 3$

$x>3$

$x\le 3$

$x>3$ or $x\le 3$

which is the entire real number line or R.

25. $4x-5<11$ and $-3x-4\ge 8$

$4x<16$ and $-3x\ge 12$

$x<4$ and $x\le -4$

$x<4$

$x\le -4$

$x\le -4$ and $x<4$ which is $x\le -4$

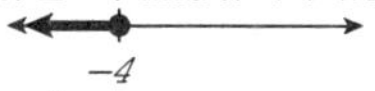

$\{x|x\le -4\}$

26. $\dfrac{7-2g}{3}\le -5$ or $\dfrac{3-g}{8}>1$

$7-2g\le -15$ or $3-g>8$

$-2g\le -22$ or $-g>5$

$g\ge 11$ or $g\le -5$

$g\le -5$

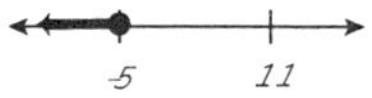

$g\ge 11$

$g\le -5$ or $g\ge 11$

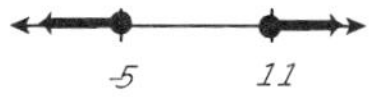

$\{g|g\le -5 \text{ or } g\ge 11\}$

27. $|a|=2$

$a=2$ or $a=-2$

The solution set is $\{-2, 2\}$.

28. $|x|<3$

$-3<x<3$

The solution set is $\{x|-3<x<3\}$.

29. $|x|\ge 4$

$x\le -4$ or $x\ge 4$

The solution set is $\{x|x\le -4 \text{ or } x\ge 4\}$.

30. $|l+5|=11$

$l+5=-11$ or $l+5=11$

$l=-16$ $\quad$ $l=6$

The solution set is $\{-16, 6\}$.

31. $|x-2|\ge 5$

$x-2\le -5$ or $x-2\ge 5$

$x\le -3$ $\quad$ $x\ge 7$

The solution set is $\{x|x\le -3 \text{ or } x\ge 7\}$.

32. $|4-2x|=5$

$4-2x=5$ or $4-2x=-5$

$-2x=1$ $\quad$ $-2x=-9$

$x=\dfrac{1}{-2}$ $\quad$ $x=\dfrac{-9}{-2}$

$x=-\dfrac{1}{2}$ $\quad$ $x=\dfrac{9}{2}$

The solution set is $\left\{-\dfrac{1}{2}, \dfrac{9}{2}\right\}$.

33. $|-2q+9|<7$

$-7<-2q+9<7$

$-7-9<-2q+9-9<7-9$

$-16<-2q<-2$

$\dfrac{-16}{-2}>\dfrac{-2q}{-2}>\dfrac{-2}{-2}$

$8>q>1$

$1<q<8$

The solution set is $\{q|1<q<8\}$.

34. $\left|\dfrac{2x-3}{5}\right|=1$

$\dfrac{2x-3}{5}=1$ or $\dfrac{2x-3}{5}=-1$

$2x-3=5$ $\quad$ $2x-3=-5$

$2x=8$ $\quad$ $2x=-2$

$x=4$ $\quad$ $x=-1$

The solution set is $\{-1, 4\}$.

35. $\left|\frac{x-4}{3}\right| < 6$

$-6 < \frac{x-4}{3} < 6$

$3(-6) < 3\left(\frac{x-4}{3}\right) < 3(6)$

$-18 < x - 4 < 18$

$-14 < x < 22$

The solution set is $\{x \mid -14 < x < 22\}$.

36. $|3x - 4| = |x + 3|$

$3x - 4 = -(x + 3)$ or $3x - 4 = x + 3$

$3x - 4 = -x - 3$ $\quad 2x - 4 = 3$

$4x - 4 = -3$ $\quad 2x = 7$

$4x = 1$ $\quad x = \frac{7}{2}$

$x = \frac{1}{4}$

The solution set is $\left\{\frac{1}{4}, \frac{7}{2}\right\}$.

37. $|2x - 3| + 4 \ge -10$

$|2x - 3| \ge -14$

Since the right side is negative and the left side is non-negative, the solution is the entire real number line since the absolute value of a number is always greater than a negative number. The solution set is all real numbers or R.

38. $|3c + 8| - 5 \le 2$

$|3c + 8| \le 7$

$-7 \le 3c + 8 \le 7$

$-7 - 8 \le 3c + 8 - 8 \le 7 - 8$

$-15 \le 3c \le -1$

$\frac{-15}{3} \le \frac{3c}{3} \le \frac{-1}{3}$

$-5 \le c \le -\frac{1}{3}$

$\left[-5, -\frac{1}{3}\right]$

39. $3 < 2x - 5 \le 9$

$3 + 5 < 2x - 5 + 5 \le 9 + 5$

$8 < 2x \le 14$

$\frac{8}{2} < \frac{2x}{2} \le \frac{14}{2}$

$4 < x \le 7$

The solution is (4, 7].

40. $-6 \le \frac{3-2x}{4} < 5$

$4(-6) \le 4\left(\frac{3-2x}{4}\right) < 4(5)$

$-24 \le 3 - 2x < 20$

$-27 \le -2x < 17$

$\frac{-27}{-2} \ge \frac{-2x}{-2} > \frac{17}{-2}$

$\frac{27}{2} \ge x > -\frac{17}{2}$

$-\frac{17}{2} < x \le \frac{27}{2}$

The solution is $\left(-\frac{17}{2}, \frac{27}{2}\right]$.

41. $2p - 5 < 7$ or $9 - 3p \le 12$

$2p < 12$ or $-3p \le 3$

$p < 6$ or $p \ge -1$

$-1 \le p < 6$

The solution is $[-1, 6)$.

42. $x - 3 \le 4$ or $2x - 5 > 9$

$x - 3 + 3 \le 4 + 3$ $\quad 2x - 5 + 5 > 9 + 5$

$x \le 7$ $\quad 2x > 14$

$x > 7$

The solution is $(-\infty, \infty)$.

43. $-10 < 3(x - 4) \le 12$

$-10 < 3x - 12 \le 12$

$-10 + 12 < 3x - 12 + 12 \le 12 + 12$

$2 < 3x \le 24$

$\frac{2}{3} < x \le 8$

The solution is $\left(\frac{2}{3}, 8\right]$.

44.

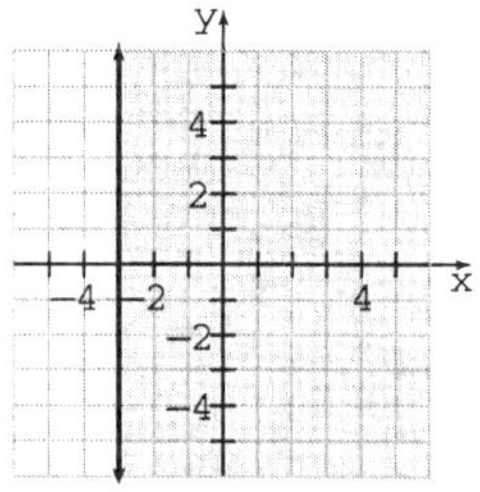

Check (0, 0): $y \ge -3$

$0 \ge -3$ True

45.

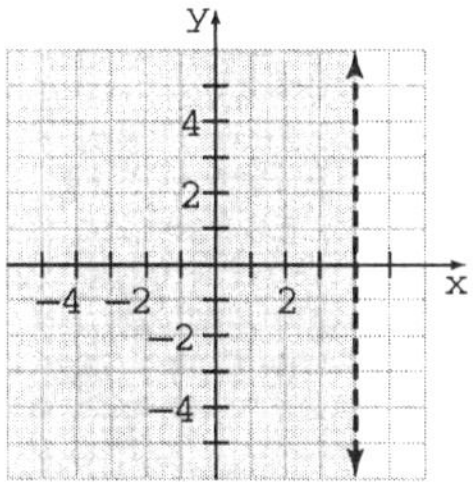

Check (0, 0): $x < 4$

$0 < 4$ True

46.

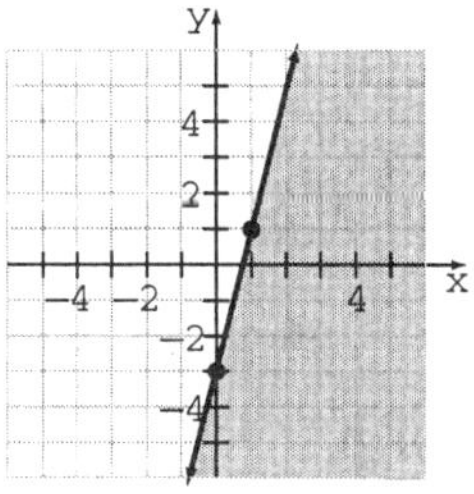

Check (0, 0): $y \le 4x - 3$

$0 \le 4(0) - 3$

$0 \le 0 - 3$

$0 \le -3$ False

47.

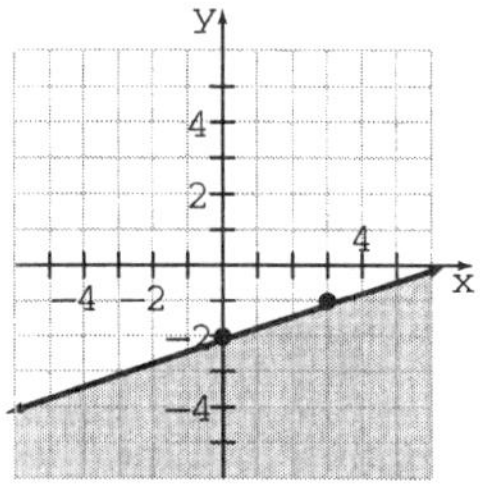

Check (0, 0): $y < \frac{1}{3}x - 2$

$0 < 0 - 2$

$0 < -2$ False

48. $-x + 3y > 6$

$2x - y \le 2$

For $-x + 3y > 6$, graph the line $-x + 3y = 6$ using a dashed line. For the check point, select (0, 0):

$-x + 3y > 6$

$-0 + 3(0) > 6$

$0 > 6$ False

Since this is a false statement, shade the region which does not contain the point (0, 0).

For $2x - y \le 2$, graph the line $2x - y = 2$ using a solid line. For the check point, select (0, 0):

$2x - y \le 2$

$2(0) - 0 \le 2$

$0 \le 2$ True

Since this is a true statement, shade the region which contains the point (0, 0).

To obtain the final region, take the intersection of the above two regions.

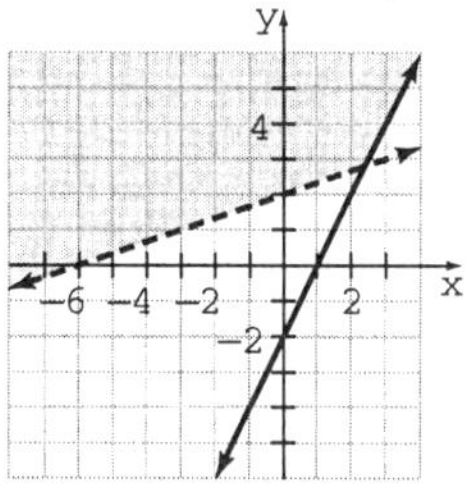

49. $5x - 2y \le 10$

$3x + 2y > 6$

For $5x - 2y \le 10$, graph the line $5x - 2y = 10$ using a solid line. For the check point, select (0, 0):

$5x - 2y \le 10$

$5(0) - 2(0) \le 10$

$0 \le 10$ True

Since this is a true statement, shade the region which contains the point (0, 0).

For $3x + 2y > 6$, graph the line $3x + 2y = 6$ using a dashed line. For the check point, select (0, 0):

$3x + 2y > 6$

$3(0) + 2(0) > 6$

$0 > 6$ False

Since this is a false statement, shade the region which does not contain the point (0, 0). To obtain the final region, take the intersection of the above two regions.

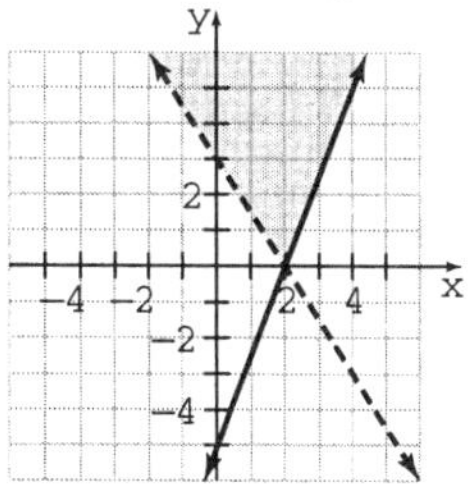

50. $y > 2x + 3$

$y < -x + 4$

For $y > 2x + 3$, graph the line $y = 2x + 3$ using a dashed line. For the check point, select (0, 0):

$y > 2x + 3$

$0 > 2(0) + 3$

$0 > 3$ False

Since this is a false statement, shade the region which does not contain the point (0, 0).

For $y < -x + 4$, graph the line $y = -x + 4$ using a dashed line. For the check point, select (0, 0):

$y < -x + 4$

$0 < -0 + 4$

$0 < 4$ True

Since this is a true statement, shade the region which contains the point (0, 0). To obtain the final region, take the intersection of the above two regions.

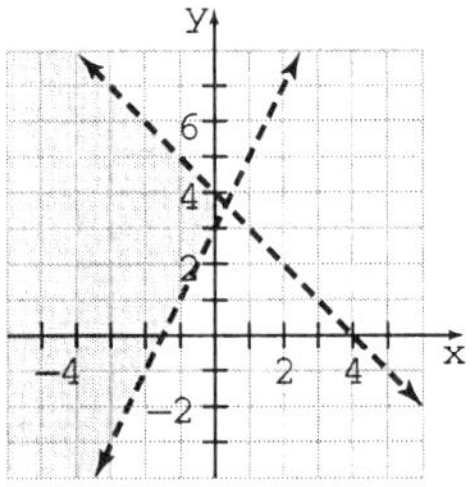

51. $x > -2y + 4$

$y < -\frac{1}{2}x - \frac{3}{2}$

For $x > -2y + 4$, graph the line $x = -2y + 4$ using a dashed line. For the check point, select (0, 0):

$x > -2y + 4$

$0 > -2(0) + 4$

$0 > 4$ False

Since this is a false statement, shade the region which does not contain the point (0, 0).

For $y < -\frac{1}{2}x - \frac{3}{2}$, graph the line $y = -\frac{1}{2}x - \frac{3}{2}$ using a dashed line. For the check point, select (0, 0):

$y < -\frac{1}{2}x - \frac{3}{2}$

$0 < -\frac{1}{2}(0) - \frac{3}{2}$

$0 < -\frac{3}{2}$ False

Since this is a false statement, shade the region which does not contain the point (0, 0). To obtain the final region, take the intersection of the above two regions. The regions do not overlap, so there are no solutions.

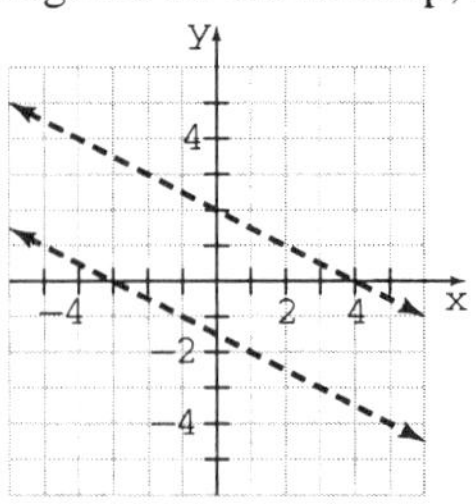

52. $x \geq 0$

$y \geq 0$

$x + y \leq 6$

$4x + y \leq 8$

The first two inequalities indicate that the solution must be in the first quadrant. For $x + y \leq 6$, the graph is the line $x + y = 6$ along with the region below this line. For $4x + y \leq 8$, the graph is the line $4x + y = 8$ along with the region below this line. To obtain the final region, take the intersection of these regions.

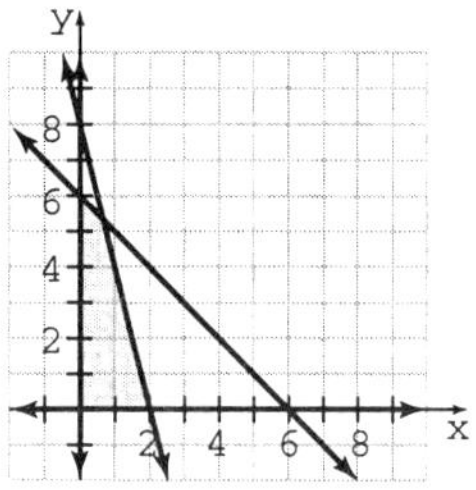

53. $x \geq 0$

$y \geq 0$

$2x + y \leq 6$

$4x + 5y \leq 20$

The first two inequalities indicate that the solution must be in the first quadrant. For $2x + y \leq 6$, graph the line $2x + y = 6$ along with the region below this line. For $4x + 5y \leq 20$, graph the line $4x + 5y = 20$ along with the region below this line. To obtain the final region, take the intersection of these regions.

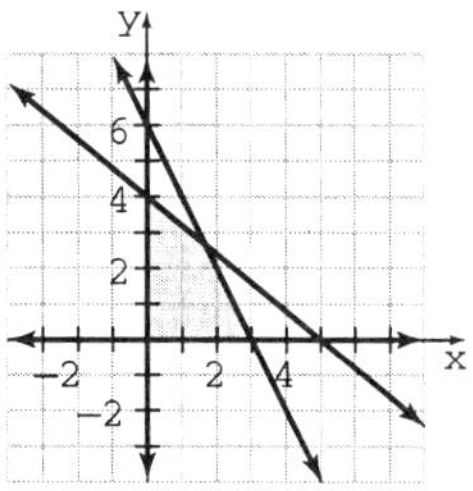

54. $|x| \leq 3$

$|y| > 2$

For $|x| \leq 3$, the graph is the region between the solid lines $x = -3$ and $x = 3$. For $|y| > 2$, the graph is the region above the dashed line $y = 2$ along with the region below the dashed line $y = -2$. To obtain the final region, take the intersection of these regions.

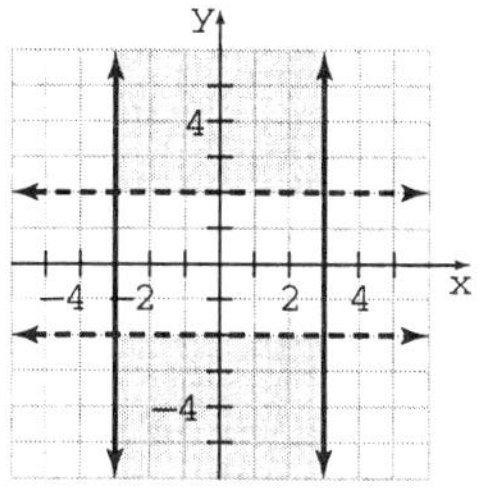

55. $|x| > 4$

$|y-2| \le 3$

For $|x| > 4$, the graph is the region to the left of dashed line $x = -4$ along with the region to the right of the dashed line $x = 4$.

$|y-2| \le 3$ can be written as

$-3 \le y-2 \le 3$

$-1 \le y \le 5$

For $|y-2| \le 3$, the graph is the region between the solid lines $y = -1$ and $y = 5$. To obtain the final region, take the intersection of these regions.

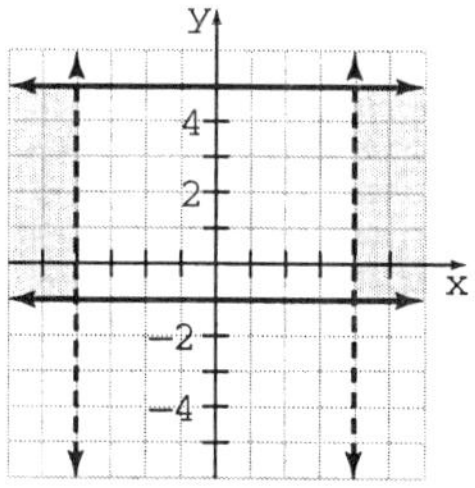

Practice Test

1. $A \cup B = \{5,7,8,9,10,11,13\}$,

$A \cap B = \{8,10\}$

2. $A \cup B = \{1,3,5,7,\ldots\}$,

$A \cap B = \{3,5,7,9,11\}$

3. $x - 2 \ge 7$

$x \ge 9$

4. $4 - x < 7$

$-x < 3$

$x > -3$

5. $4(x-2) < 3(x-2) - 5$

$4x - 8 < 3x - 6 - 5$

$4x - 8 < 3x - 11$

$x - 8 < -11$

$x < -3$

6. $\frac{6-2x}{5} \ge -12$

$5\left(\frac{6-2x}{5}\right) \ge 5(-12)$

$6 - 2x \ge -60$

$-2x \ge -66$

$\frac{-2x}{-2} \le \frac{-66}{-2}$

$x \le 33$

7. $x - 3 \le 4$ and $2x - 4 > 5$

$x - 3 + 3 \le 4 + 3$ $\quad$ $2x - 4 + 4 > 5 + 4$

$x \le 7$ $\quad$ $2x > 9$

$x > \frac{9}{2}$

The solution is $\left(\frac{9}{2}, 7\right]$.

8. $-4 < \frac{x+4}{2} < 7$

$-8 < x + 4 < 14$

$-12 < x < 10$

The solution is $(-12, 10)$.

9. $3x - 1 < 19$ or $3x + 5 > 2x + 3$

$3x < 20$ $\quad$ $x + 5 > 3$

$x < \frac{20}{3}$ $\quad$ $x > -2$

The solution is $(-\infty, \infty)$.

10. $-14 \le -3x + 1 < 10$

$-15 \le -3x < 9$

$5 \ge x > -3$

$-3 < x \le 5$

The solution is $(-3, 5]$.

11. $|x-4|=8$

$x-4=8$ or $x-4=-8$

$x=12 \qquad x=-4$

The solution set is $\{-4, 12\}$.

12. $|2x-3|=\left|\frac{1}{2}x-10\right|$

$2x-3=-\left(\frac{1}{2}x-10\right)$ or $2x-3=\frac{1}{2}x-10$

$2x-3=-\frac{1}{2}x+10 \qquad \frac{3}{2}x-3=-10$

$\frac{5}{2}x-3=10 \qquad \frac{3}{2}x=-7$

$\frac{5}{2}x=13 \qquad \frac{2}{3}\left(\frac{3}{2}x\right)=\frac{2}{3}(-7)$

$\frac{2}{5}\left(\frac{5}{2}x\right)=\frac{2}{5}(13) \qquad x=-\frac{14}{3}$

$x=\frac{26}{5}$

The solution set is $\left\{-\frac{14}{3}, \frac{26}{5}\right\}$.

13. $|3x-1|=0$

$3x-1=0$

$3x=1$

$x=\frac{1}{3}$

The solution set is $\left\{\frac{1}{3}\right\}$.

14. $|2x+3|-7=20$

$|2x+3|=27$

$2x+3=27$ or $2x+3=-27$

$2x=24 \qquad 2x=-30$

$x=12 \qquad x=-15$

The solution set is $\{-15, 12\}$.

15. $|2x-3|+4>9$

$|2x-3|>5$

$2x-3>5$ or $2x-3<-5$

$2x>8 \qquad 2x<-2$

$x>4 \qquad x<-1$

The solution set is $\{x \mid x<-1 \text{ or } x>4\}$.

16. $\left|\frac{2x-3}{4}\right|\le\frac{1}{2}$

$-\frac{1}{2}\le\frac{2x-3}{4}\le\frac{1}{2}$

$4\left(-\frac{1}{2}\right)\le4\left(\frac{2x-3}{4}\right)\le4\left(\frac{1}{2}\right)$

$-2\le2x-3\le2$

$1\le2x\le5$

$\frac{1}{2}\le x\le\frac{5}{2}$

The solution set is $\left\{x\left|\frac{1}{2}\le x\le\frac{5}{2}\right.\right\}$.

17. $|-2x+7|+6\le3$

$|-2x+7|\le-3$

The solution is $\varnothing$. The absolute value of a number is always a non negative number.

18. $|-2x+3|<15$

$-15<-2x+3<15$

$-18<-2x<12$

$-9<-x<6$

$9>x>-6$

$-6<x<9$

$\{x \mid -6<x<9\}$

19.

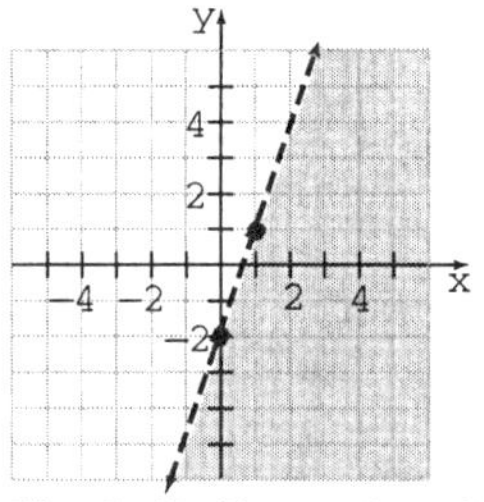

Check (0, 0): $y<3x-2$

$0<3(0)-2$

$0<-2$ False

20.

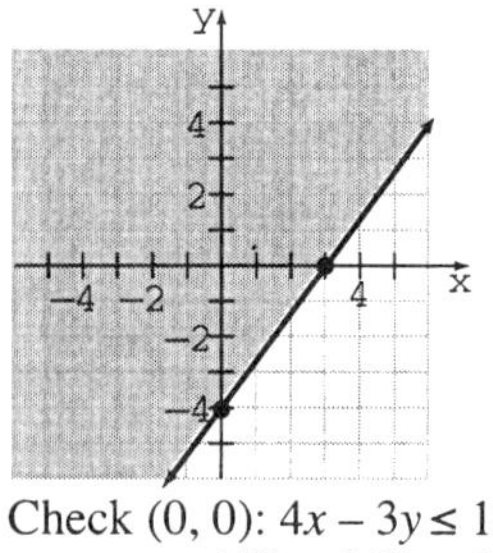

Check (0, 0): $4x-3y\le12$

$4(0)-3(0)\le12$

$0\le12$ True

21. $3x+2y<7$

$-2x+5y\le10$

For $3x+2y<7$, graph the line $3x+2y=7$ using a dashed line. For the check point, select (0, 0).

$$3x + 2y < 7$$
$$3(0) + 2(0) < 7$$
$$0 < 7 \quad \text{True}$$

Since this is a true statement, shade the region which contains the point (0, 0). This is the region "below" the line.

For $-2x + 5y \le 10$, graph the line $-2x + 5y = 10$ using a solid line. For the check point, select (0, 0).

$$-2x + 5y \le 10$$
$$-2(0) + 5(0) \le 10$$
$$0 \le 10 \quad \text{True}$$

Since this is a true statement, shade the region which contains the point (0, 0). This is the region "below" the line. To obtain the final region, take the intersection of the above two regions.

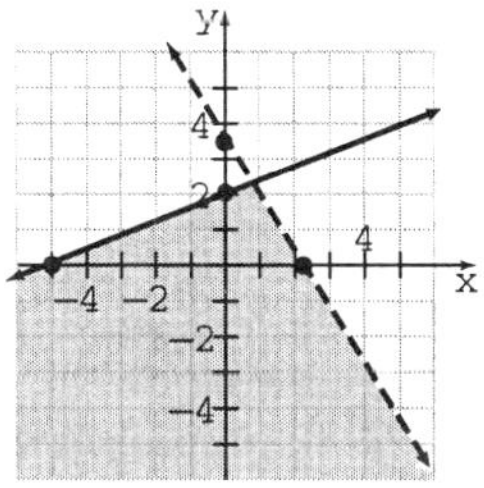

22. $|x| > 3$

$|y| \le 1$

For $|x| > 3$, the graph is the region to the left of the dashed line $x = -3$ along with the region to the right of the dashed line $x = -3$.

For $|y| \le 1$, the graph is the region between the solid lines $y = -1$ and $y = 1$. To obtain the final region, take the intersection of these regions.

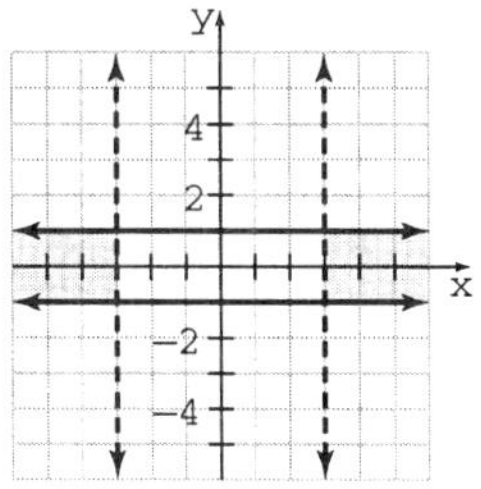

23.
$$x \ge 0$$
$$y \ge 0$$
$$x \le 12$$
$$20x + 18y \le 360$$

For $x \ge 0$, the graph is the region to the right of the line $x = 0$.

For $y \ge 0$, the graph is the region above the line $y = 0$.

For $x \le 12$, the graph is the region to the left of the line $x = 12$.

For $20x + 18y \le 360$, check the point (0, 0).

$$20x + 18y \le 360$$
$$20(0) + 18(0) \le 360$$
$$0 \le 360 \quad \text{True}$$

Since this is a true statement, shade the region which contains the point (0, 0). To obtain the final region, take the intersection of these regions.

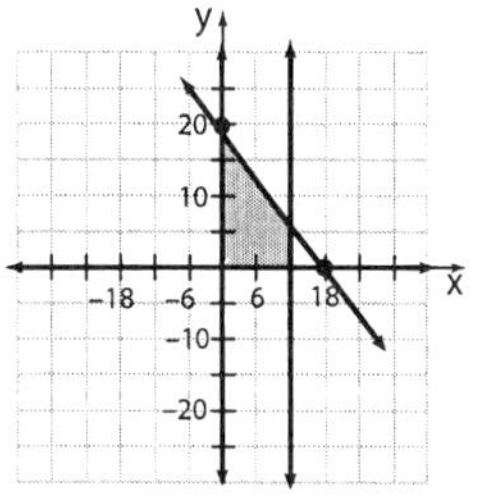

24. Let x = the number of CD's Jason can purchase. The inequality is

$$12.50x \le 65.00$$
$$x \le 5.2$$

Jason can buy a maximum of 5 CD's.

25. Let x = the exam grade(s) needed on the fourth exam to receive a B.

$$80 \le \frac{89 + 83 + 94 + x}{4} < 90$$
$$320 \le 89 + 83 + 94 + x < 360$$
$$320 \le 266 + x < 360$$
$$54 \le x < 94$$

Bianca's test score has to be greater than or equal to 54 but less than 94 to still get a B.

Cumulative Review Test

1. $-40 - 3(4-8)^2 = -40 - 3(-4)^2$
$= -40 - 3(16)$
$= -40 - 48$
$= -88$

2. $2(x + 2y) + 4x - 3y$
$= 2(2 + 2(3)) + 4(2) - 3(3)$
$= 2(2 + 6) + 8 - 9$
$= 2(8) + 8 - 9$
$= 16 + 8 - 9$
$= 24 - 9$
$= 15$

3. $\frac{1}{2}(x+3)+\frac{1}{3}(3x+6)$

$=\frac{1}{2}x+\frac{3}{2}+x+2$

$=\frac{1}{2}x+\frac{3}{2}+\frac{2}{2}x+\frac{4}{2}$

$=\frac{3}{2}x+\frac{7}{2}$

4. $\frac{a-7}{3}=\frac{a+5}{2}-\frac{7a-1}{6}$

$6\left(\frac{a-7}{3}\right)=6\left(\frac{a+5}{2}-\frac{7a-1}{6}\right)$

$2(a-7)=3(a+5)-(7a-1)$

$2a-14=3a+15-7a+1$

$2a-14=-4a+16$

$6a-14=16$

$6a=30$

$a=5$

5. $\frac{40 \text{ pounds of fertilizer}}{5000 \text{ square feet}}=\frac{x \text{ pounds of fertilizer}}{26{,}000 \text{ square feet}}$

$\frac{40}{5000}=\frac{x}{26{,}000}$

$40\cdot 26{,}000=5000x$

$1{,}040{,}000=5000x$

$208=x$

208 pounds of fertilizer will be needed

6. Let x = the customers assets

Plan 1 = Plan 2

$\$1000+0.01x=\$500+0.02x$

$1000+0.01x=500+0.02x$

$500=0.01x$

$50{,}000=x$

If customers assets totaled $50,000, the fees for both plans would be equal.

7. Let c be the cost per pound of the mixture.

$1.80(2.5)+1.40(1)=c(3.5)$

$4.50+1.40=3.5c$

$5.90=3.5c$

$1.685=c$

The cost per pound should be $1.69.

8. $\left(\frac{4a^3b^{-2}}{2a^{-2}b^{-3}}\right)^{-2}=\left(2a^{3-(-2)}b^{-2-(-3)}\right)^{-2}$

$=\left(2a^5b\right)^{-2}$

$=\left(\frac{1}{2a^5b}\right)^2$

$=\frac{1}{4a^{10}b^2}$

9. $(0.003)(0.00015)=\left(3\times 10^{-3}\right)\left(1.5\times 10^{-4}\right)$

$=(3\times 1.5)\left(10^{-3}\times 10^{-4}\right)$

$=4.5\times 10^{-7}$

10. $24p^3q-16p^2q-30pq$

$2pq\left(12p^2-8p-15\right)$

$2pq(2p-3)(6p+5)$

11. $\frac{2b}{b+1}=2-\frac{5}{2b}$

$2b(b+1)\left(\frac{2b}{b+1}\right)=2b(b+1)\left(2-\frac{5}{2b}\right)$

$2b(2b)=2b(b+1)2-(b+1)5$

$4b^2=\left(2b^2+2b\right)2-5b-5$

$4b^2=4b^2+4b-5b-5$

$4b^2=4b^2-b-5$

$0=-b-5$

$b=-5$

12.
$$\frac{4}{r+5}+\frac{1}{r+3}=\frac{2}{r^2+8r+15}$$
$$\frac{4}{r+5}+\frac{1}{r+3}=\frac{2}{(r+5)(r+3)}$$
$$(r+5)(r+3)\left(\frac{4}{r+5}+\frac{1}{r+3}\right)=(r+5)(r+3)\left(\frac{2}{(r+5)(r+3)}\right)$$
$$4(r+3)+1(r+5)=2$$
$$4r+12+r+5=2$$
$$5r+17=2$$
$$5r=-15$$
$$r=-3$$
Since –3 is not in the domain, there is no solution.

13. $y=x^2-2$

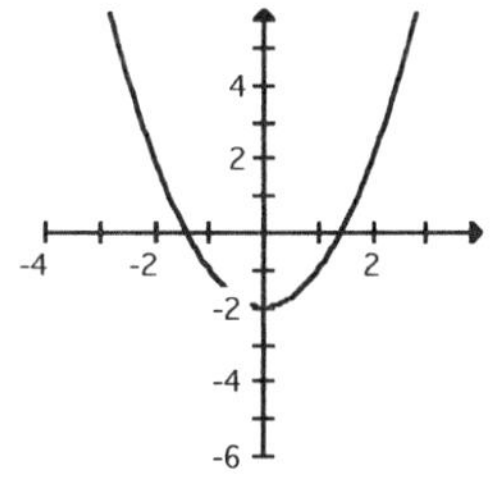

14. a. The relation is a function; each first component corresponds to exactly one second component.

b. The relation is not a function; it does not pass the vertical line test.

c. The relation is a function; it passes the vertical line test.

15. Solve the equation for *y*. $3x-5y=7$
$$-5y=-3x+7$$
$$y=\frac{3}{5}x-\frac{7}{5}$$
The slope of the parallel line is $\frac{3}{5}$ and the *y*-intercept is –2. Therefore the equation of the line is $y=\frac{3}{5}x-2$.

16. Solve the equation for *y*. $4x-3y=12$
$$-3y=-4x+12$$
$$y=\frac{4}{3}x-4$$
The slope of the perpendicular line is $-\frac{3}{4}$.

Use the slope and the point (3, –2) in the point-slope equation to obtain the equation of the line.
$$y-y_1=m(x-x_1)$$
$$y+2=-\frac{3}{4}(x-3)$$
$$y+2=-\frac{3}{4}x+\frac{9}{4}$$
$$y=-\frac{3}{4}x+\frac{9}{4}-\frac{8}{4}$$
$$y=-\frac{3}{4}x+\frac{1}{4}$$

17. a. $(f\cdot g)(x)$
$$=f(x)\cdot g(x)$$
$$=(x^2-11x+30)(x-5)$$
$$=x^3-5x^2-11x^2+55x+30x-150$$
$$=x^3-16x^2+85x-150$$

b. $(f/g)(x)=\dfrac{f(x)}{g(x)}$
$$=\frac{x^2-11x+30}{x-5}$$
$$=\frac{(x-5)(x-6)}{x-5}$$
$$=x-6$$

c. The domain of $(f/g)(x)$ is $\{x \mid x\neq 6\}$.

18. $2a-b-2c=-1$ (1)

$a-2b-c=1$ (2)

$a+b+c=4$ (3)

Add equation (2) and (3) to eliminate *c*.

$a-2b-c=1$ (2)

$a+b+c=4$ (3)

$2a-b\quad=5$

Multiply equation (3) by 2 and add it to equation (1) to eliminate *c*.

$$2a - b - 2c = -1 \quad (1)$$
$$\underline{a + b + \ c = 4 \quad (3) \text{ multiply by 2}}$$

$$2a - \ b - 2c = -1$$
$$\underline{2a + 2b + 2c = 8}$$
$$4a + b \qquad = 7$$

Now we have two equations with two unknowns. Add them together to eliminate *b*.

$$2a - b = 5 \quad (4)$$
$$\underline{4a + b = 7} \quad (5)$$
$$6a \qquad = 12$$
$$a \qquad = 2$$

Substitute $a = 2$ in equation (5) and solve for *b*. $4(2) + b = 7$

$$8 + b = 7$$
$$b = -1$$

Now substitute $a = 2$ *and* b = –1 in equation (3) and solve for *c*.

$$a + b + c = 4$$
$$2 - 1 + c = 4$$
$$1 + c = 4$$
$$c = 3$$

The solution is (2, –1, 3).

19. $$\frac{1}{2} < 3x + 4 < 6$$

$$\frac{1}{2} - 4 < 3x + 4 - 4 < 6 - 4$$
$$-\frac{7}{2} < 3x < 2$$
$$\frac{1}{3}\left(-\frac{7}{2}\right) < \frac{1}{3}(3x) < \frac{1}{3}(2)$$
$$-\frac{7}{6} < x < \frac{2}{3}$$
$$\left(-\frac{7}{6}, \frac{2}{3}\right)$$

20. Graph the line $y = 2x - 3$ using a dashed line. Use the test point (0, 0) to decide where to shade.

$$y < 2x - 3$$
$$0 < 2(0) - 3$$
$$0 < -3 \quad \text{False}$$

Shade the side that (0, 0) does not lie in.

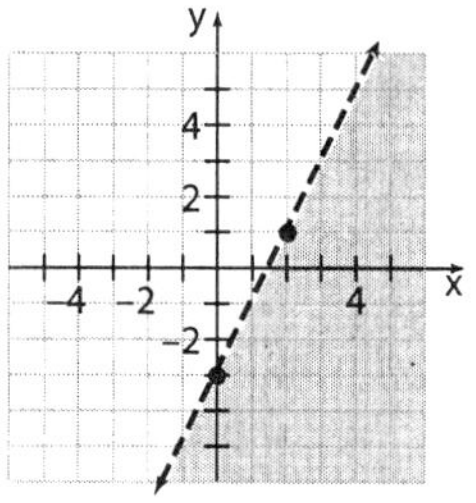

Chapter 11

Exercise Set 11.1

1. a. Every real number has two square roots; a positive or principal square root and a negative square root.

   b. The square roots of 49 are 7 and –7.

   c. When we say square root, we are referring to the principal square root.

   d. $\sqrt{49}=7$

3. There is no real number which, when squared, results in –81.

5. No. If the number under the radical is negative, the answer is not a real number.

7. a. $\sqrt{(1.3)^2}=\sqrt{1.69}=1.3$

   b. $\sqrt{(-1.3)^2}=\sqrt{1.69}=1.3$

9. a. $\sqrt[3]{27}=3$ since $3^3=27$

   b. $-\sqrt[3]{27}=-3$

   c. $\sqrt[3]{-27}=-3$ since $(-3)^3=-27$

11. $\sqrt{64}=8$ since $8^2=64$

13. $\sqrt[3]{-64}=-4$ since $(-4)^3=-64$

15. $\sqrt[3]{-125}=-5$ since $(-5)^3=-125$

17. $\sqrt[5]{1}=1$ since $1^5=1$

19. $-\sqrt[5]{-1}=1$

21. $\sqrt[6]{-64}=2$ is not a real number.

23. $\sqrt[3]{-343}=-7$ since $(-7)^3=-343$

25. $\sqrt{-36}$ is not a real number.

27. $\sqrt{-45.3}$ is not a real number.

29. $\sqrt{\frac{1}{25}}=\frac{1}{5}$ since $\left(\frac{1}{5}\right)^2=\frac{1}{25}$

31. $\sqrt[3]{\frac{1}{8}}=\frac{1}{2}$ since $\left(\frac{1}{2}\right)^3=\frac{1}{8}$

33. $\sqrt{\frac{4}{9}}=\frac{2}{3}$ since $\left(\frac{2}{3}\right)^2=\frac{4}{9}$

35. $\sqrt[3]{-\frac{8}{27}}=-\frac{2}{3}$ since $\left(-\frac{2}{3}\right)^3=-\frac{8}{27}$

37. $-\sqrt[4]{18.2}\approx-2.07$

39. $\sqrt{9^2}=|9|=9$

41. $\sqrt{(-19)^2}=|-19|=19$

43. $\sqrt{152^2}=|152|=152$

45. $\sqrt{(235.23)^2}=|235.23|=235.23$

47. $\sqrt{(0.06)^2}=|0.06|=0.06$

49. $\sqrt{\left(\frac{11}{13}\right)^2}=\left|\frac{11}{13}\right|=\frac{11}{13}$

51. $\sqrt{(x-9)^2}=|x-9|$

53. $\sqrt{(x-3)^2}=|x-3|$

55. $\sqrt{(3x^2-2)^2}=|3x^2-2|$

57. $\sqrt{(6a^3-5b^4)^2}=|6a^3-5b^4|$

59. $\sqrt{a^{10}}=\sqrt{(a^5)^2}=|a^5|$

61. $\sqrt{z^{30}}=\sqrt{(z^{15})^2}=|z^{15}|$

63. $\sqrt{a^2-8a+16}=\sqrt{(a-4)^2}=|a-4|$

65. $\sqrt{9a^2+12ab+4b^2}=\sqrt{(3a+2b)^2}=|3a+2b|$

67. $f(x)=\sqrt{5x-6}$
$f(2)=\sqrt{5\cdot 2-6}$
$=\sqrt{10-6}$
$=\sqrt{4}$
$=2$

69. $g(x)=\sqrt{64-8x}$
$g(-3)=\sqrt{64-8(-3)}$
$=\sqrt{64+24}$
$=\sqrt{88}$
≈ 9.381

71. $h(x)=\sqrt[3]{9x^2+4}$
$h(4)=\sqrt[3]{9(4)^2+4}$
$=\sqrt[3]{144+4}$
$=\sqrt[3]{148}$
≈ 5.290

73. $f(a)=\sqrt[3]{-2a^2+a-6}$
$f(-3)=\sqrt[3]{-2(-3)^2+(-3)-6}$
$=\sqrt[3]{-18-3-6}$
$=\sqrt[3]{-27}$
$=-3$

75. $f(x)=x+\sqrt{x}+5$
$f(81)=81+\sqrt{81}+5$
$=81+9+5$
$=95$

77. $t(x)=\frac{x}{2}+\sqrt{2x}-1$
$t(18)=\frac{18}{2}+\sqrt{2(18)}-1$
$=9+\sqrt{36}-1$
$=9+6-1$
$=14$

79. $k(x)=x^2+\sqrt{\frac{x}{2}}-11$
$k(8)=(8)^2+\sqrt{\frac{8}{2}}-11$
$=64+\sqrt{4}-11$
$=64+2-11$
$=55$

81. Select $x=0$.
$\sqrt{(2(0)-1)^2}\neq 2(0)-1$
$\sqrt{(-1)^2}\neq -1$
$\sqrt{1}\neq -1$
$1\neq -1$
This is true for all $x<\frac{1}{2}$.

83. $\sqrt{(x-1)^2}=x-1$ for all $x\geq 1$. The expression $\sqrt{(x-1)^2}=x-1$, when $(x-1)$ is equal to zero or positive. Therefore, $x-1\geq 0$ and
$x\geq 1$.

85. $\sqrt{(2x-6)^2}=2x-6$ for all $x\geq 3$. The expression $\sqrt{(2x-6)^2}=2x-6$, when $(2x-6)$ is positive or equal to 0. Therefore,
$2x-6\geq 0$
$x\geq 3$

87. **a.** $\sqrt{a^2}=|a|$ for all real values

b. $\sqrt{a^2}=a$ when $a\geq 0$

c. $\sqrt[3]{a^3}=a$ for all real values

89. If n is even, we are finding the even root of a positive number. If n is odd, the expression is also real.

91. $\frac{\sqrt{x+5}}{\sqrt[3]{x+5}}$ The denominator cannot equal zero.
$\sqrt[3]{x+5}\neq 0$
$x\neq -5$
The numerator must be greater than or equal to zero.
$\sqrt{x+5}\geq 0$
$x\geq -5$
Therefore the domain is
$\{x|x$ is a real number $x>-5\}$

93. $f(x)=\sqrt{x}$ matches graph d).
The x-intercept is 0 and the domain is $x\geq 0$.

95. $f(x)=\sqrt{x-5}$ matches graph a). The x-intercept is 5.

97. One answer is $f(x)=\sqrt{x-6}$

99. $f(x) = -\sqrt{x}$

a. No

b. Yes

c. Yes

Explanations will vary.

101. $V = \sqrt{64.4h}$

a. $V = \sqrt{64.4(20)}$
$= \sqrt{1288}$
≈ 35.89
The velocity will be about 35.89 ft/sec.

b. $V = \sqrt{64.4(40)}$
$= \sqrt{2576}$
≈ 50.75
The velocity will be about 50.75 ft/sec.

103. $f(x) = \sqrt{x+1}$

x	$f(x)$
−1	0
0	1
3	2
8	3

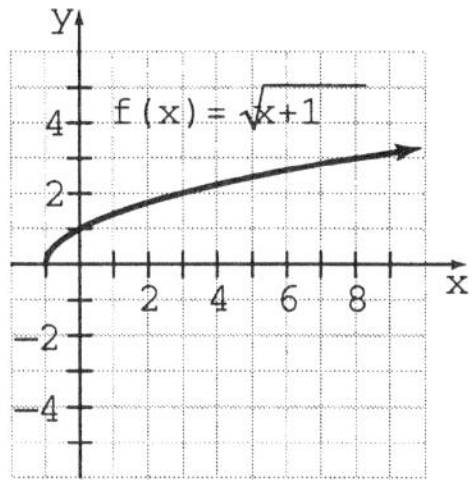

105. $g(x) = \sqrt{x} + 1$

x	$g(x)$
0	1
4	3
9	4

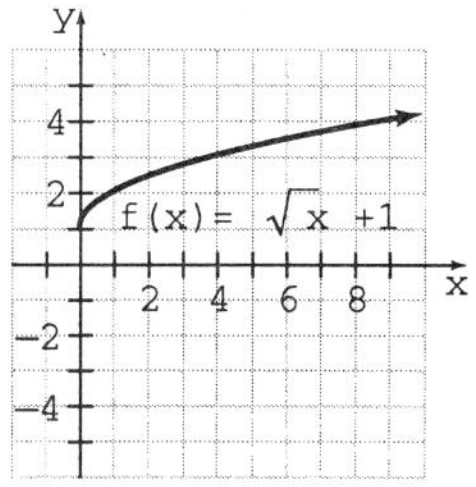

107. $y_1 = \sqrt{x+1}$

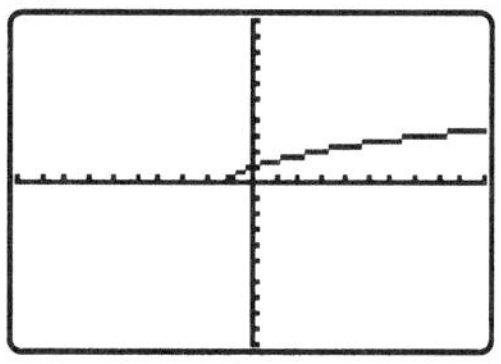

-10, 10, 1, -10, 1C

109. $y_1 = \dfrac{\sqrt{x+5}}{\sqrt[3]{x+5}}$
Yes, $x > -5$.

111. $y = \sqrt[3]{x+4}$

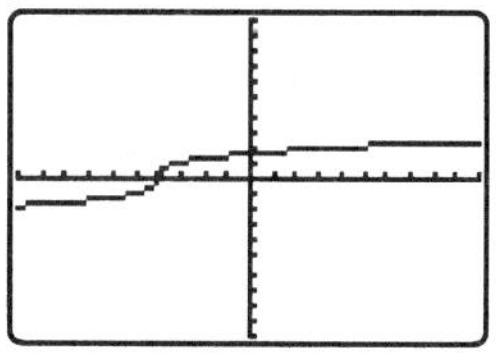

-10, 10, 1, -10, 1C

115. $9ax - 3bx + 12ay - 4by = 3x(3a-b) + 4y(3a-b)$
$= (3a-b)(3x+4y)$

116. $3x^3 - 18x^2 + 24x = 3x(x^2 - 6x + 8)$
$= 3x(x-4)(x-2)$

117. $15x^2 - xy - 6y^2$
$(5x + 3y)(3x - 2y)$

118. $x^3 - \frac{8}{27}y^3 = x^3 - \left(\frac{2}{3}y\right)^3$
$= \left(x - \frac{2}{3}y\right)\left(x^2 + \frac{2}{3}xy + \frac{4}{9}y^2\right)$

Exercise Set 11.2

1. a. $\sqrt[n]{a}$ is a real number when n is even and $a \ge 0$, or n is odd.

b. $\sqrt[n]{a}$ can be expressed with rational exponents as $a^{1/n}$.

3. a. $\sqrt[n]{a^n}$ is always real

b. $\sqrt[n]{a^n} = a$ when $a \ge 0$ and n is even

c. $\sqrt[n]{a^n} = a$ when n is odd

d. $\sqrt[n]{a^n} = |a|$ when n is even and a is any real number

5. **a.** No, $(xy)^{1/2} = x^{1/2}y^{1/2} \neq xy^{1/2}$

b. No, since $(xy)^{-1/2} = \frac{1}{(xy)^{1/2}} = \frac{1}{x^{1/2}y^{1/2}}$

but $\frac{x^{1/2}}{y^{-1/2}} = x^{1/2}y^{1/2}$

7. $\sqrt{a^3} = a^{3/2}$

9. $\sqrt{9^5} = 9^{5/2}$

11. $\sqrt[3]{z^5} = z^{5/3}$

13. $\sqrt[3]{7^{10}} = 7^{10/3}$

15. $\sqrt[4]{9^7} = 7^{7/4}$

17. $\left(\sqrt[3]{y}\right)^{13} = y^{13/3}$

19. $\sqrt[4]{a^3b} = (a^3b)^{1/4}$

21. $\sqrt[4]{x^9y^5} = (x^9y^5)^{1/4}$

23. $\sqrt[3]{5x^6y^7} = (5x^6y^7)^{1/3}$

25. $\sqrt[6]{3a+5b} = (3a+5b)^{1/6}$

27. $a^{1/2} = \sqrt{a}$

29. $c^{5/2} = \sqrt{c^5}$

31. $17^{5/3} = \sqrt[3]{17^5}$

33. $(24x^3)^{1/2} = \sqrt{24x^3}$

35. $(7b^2c)^{3/5} = \left(\sqrt[5]{7b^2c}\right)^3$

37. $(6a+5b)^{1/5} = \sqrt[5]{6a+5b}$

39. $\left(b^3 - c\right)^{-1/3} = \frac{1}{\sqrt[3]{b^3 - c}}$

41. $\sqrt{a^4} = a^{4/2} = a^2$

43. $\sqrt[3]{x^9} = x^{9/3} = x^3$

45. $\sqrt[6]{y^2} = y^{2/6} = y^{1/3} = \sqrt[3]{y}$

47. $\sqrt[8]{b^4} = b^{4/8} = b^{1/2} = \sqrt{b}$

49. $\left(\sqrt{18.1}\right)^2 = (18.1)^{2/2} = (18.1)^1 = 18.1$

51.
$$\begin{aligned}(\sqrt[5]{xy^2})^{15} &= (xy^2)^{15/5}\\ &= (xy^2)^3\\ &= x^3y^6\end{aligned}$$

53.
$$\begin{aligned}(\sqrt[8]{xyz})^4 &= (xyz)^{4/8}\\ &= (xyz)^{1/2}\\ &= \sqrt{xyz}\end{aligned}$$

55.
$$\begin{aligned}\sqrt{\sqrt{x}} &= (\sqrt{x})^{1/2}\\ &= (x^{1/2})^{1/2}\\ &= x^{1/4}\\ &= \sqrt[4]{x}\end{aligned}$$

57.
$$\begin{aligned}\sqrt{\sqrt[4]{y}} &= \left(\sqrt[4]{y}\right)^{1/2}\\ &= \left(y^{1/4}\right)^{1/2}\\ &= y^{1/8}\\ &= \sqrt[8]{y}\end{aligned}$$

59.
$$\begin{aligned}\sqrt[3]{\sqrt[3]{x^2y}} &= (\sqrt[3]{x^2y})^{1/3}\\ &= (x^2y)^{1/3\cdot 1/3}\\ &= (x^2y)^{1/9}\\ &= \sqrt[9]{x^2y}\end{aligned}$$

61. $\sqrt{\sqrt[5]{a^7}} = \left(\sqrt[5]{a^7}\right)^{1/2}$
$= \left(a^{7/5}\right)^{1/2}$
$= a^{7/10}$
$= \sqrt[10]{a^7}$

63. $9^{1/2} = \sqrt{9} = 3$

65. $64^{1/3} = \sqrt[3]{64} = 4$

67. $16^{3/2} = \left(\sqrt{16}\right)^3 = 4^3 = 64$

69. $(-25)^{1/2} = \sqrt{-25}$
Not a real number

71. $\left(\frac{16}{9}\right)^{1/2} = \sqrt{\frac{16}{9}} = \frac{4}{3}$

73. $\left(\frac{1}{8}\right)^{1/3} = \sqrt[3]{\frac{1}{8}} = \frac{1}{2}$

75. $-36^{1/2} = -\sqrt{36} = -6$

77. $-64^{1/3} = -\sqrt[3]{64} = -4$

79. $64^{-1/3} = \frac{1}{64^{1/3}} = \frac{1}{\sqrt[3]{64}} = \frac{1}{4}$

81. $64^{-2/3} = \frac{1}{64^{2/3}} = \frac{1}{\left(\sqrt[3]{64}\right)^2} = \frac{1}{4^2} = \frac{1}{16}$

83. $\left(\frac{64}{27}\right)^{-1/3} = \left(\frac{27}{64}\right)^{1/3} = \sqrt[3]{\frac{27}{64}} = \frac{3}{4}$

85. $(-100)^{3/2} = \left(\sqrt{-100}\right)^3$ Not a real number.

87. $81^{1/2} + 169^{1/2} = \sqrt{81} + \sqrt{169} = 9 + 13 = 22$

89. $343^{-1/3} + 4^{-1/2} = \frac{1}{343^{1/3}} + \frac{1}{4^{1/2}}$
$= \frac{1}{\sqrt[3]{343}} + \frac{1}{\sqrt{4}}$
$= \frac{1}{7} + \frac{1}{2}$
$= \frac{2}{14} + \frac{7}{14}$
$= \frac{9}{14}$

91. $x^4 \cdot x^{1/2} = x^{4+1/2} = x^{9/2}$

93. $\frac{x^{1/2}}{x^{1/3}} = x^{1/2-1/3} = x^{3/6-2/6} = x^{1/6}$

95. $(x^{1/2})^{-2} = x^{1/2(-2)} = x^{-1} = \frac{1}{x}$

97. $(7^{-1/3})^0 = 7^{-1/3(0)} = 7^0 = 1$

99. $\frac{5y^{-1/3}}{60y^{-2}} = \frac{1}{12}y^{-1/3-(-2)} = \frac{1}{12}y^{5/3} = \frac{y^{5/3}}{12}$

101. $4x^{5/3} \cdot 2x^{-7/2} = 4 \cdot 2 \cdot x^{5/3} \cdot x^{-7/2} = 8x^{5/3-7/2}$
$= 8x^{10/6-21/6} = 8x^{-11/6} = \frac{8}{x^{11/6}}$

103. $\left(\frac{8}{64x}\right)^{1/3} = \frac{8^{1/3}}{64^{1/3}x^{1/3}} = \frac{2}{4x^{1/3}} = \frac{1}{2x^{1/3}}$
or $\left(\frac{8}{64x}\right)^{1/3} = \left(\frac{1}{8x}\right)^{1/3} = \frac{1}{8^{1/3}x^{1/3}} = \frac{1}{2x^{1/3}}$

105. $\left(\frac{22x^{3/7}}{2x^{1/2}}\right)^2 = (11x^{3/7-1/2})^2 = (11x^{6/14-7/14})^2$
$= (11x^{-1/14})^2 = (11)^2(x^{-1/14})^2 = 121x^{-1/7}$
$= \frac{121}{x^{1/7}}$

107. $\left(\frac{a^4}{5a^{-2/5}}\right)^{-3} = \frac{a^{-12}}{5^{-3}a^{6/5}}$
$= 5^3 a^{-12-6/5}$
$= 125a^{-66/5}$
$\frac{125}{a^{66/5}}$

109. $\left(\frac{x^{3/4}y^{-2}}{x^{1/2}y^2}\right)^4 = \left(x^{3/4-1/2}y^{-2-2}\right)^4$
$= (x^{1/4}y^{-4})^4$
$= (x^{1/4})^4(y^{-4})^4$
$= xy^{-16}$
$= \frac{x}{y^{16}}$

111. $3z^{-1/2}(2z^4 - z^{1/2}) = 3z^{-1/2} \cdot 2z^4 - 3z^{-1/2}z^{1/2}$
$= 6z^{-1/2+4} - 3z^{-1/2+1/2}$
$= 6z^{7/2} - 3z^0$
$= 6z^{7/2} - 3$

113. $5x^{-1}(x^{-4}+4x^{-1/2})=5x^{-1}\cdot x^{-4}+5x^{-1}\cdot 4x^{-1/2}$
$=5x^{-1-4}+20x^{-1-1/2}$
$=5x^{-5}+20x^{-3/2}$
$=\frac{5}{x^5}+\frac{20}{x^{3/2}}$

115. $-4x^{5/3}(-2x^{1/2}+7x^{1/3})$
$=(-4x^{5/3})(-2x^{1/2})+(-4x^{5/3})(7x^{1/3})$
$=8x^{5/3+1/2}-28x^{5/3+1/3}$
$=8x^{13/6}-28x^{6/3}$
$=8x^{13/6}-28x^2$

117. $\sqrt{140}\approx 11.83$

119. $\sqrt[5]{402.83}\approx 3.32$

121. $45^{2/3}\approx 12.65$

123. $1000^{-1/2}\approx 0.03$

125. $\sqrt[n]{a^n}=\left(\sqrt[n]{a}\right)^n=a$ when n is odd, or n is even with $a\geq 0$.

127. To show $(a^{1/2}+b^{1/2})^2\neq a+b$, use $a=9$ and $b=16$. Then $(a^{1/2}+b^{1/2})^2$ becomes $(9^{1/2}+16^{1/2})^2=(3+4)^2=7^2=49$ whereas $a+b$ becomes $9+16=25$. Since $49\neq 25$, then $(a^{1/2}+b^{1/2})^2\neq a+b$. Answers will vary.

129. To show $(a^{1/3}+b^{1/3})^3\neq a+b$, use $a=1$ and $b=1$. Then $(a^{1/3}+b^{1/3})^3$ becomes $(1^{1/3}+1^{1/3})^3=(\sqrt[3]{1}+\sqrt[3]{1})^3=(1+1)^3=2^3=8$ whereas $a+b$ becomes $1+1=2$. Since $8\neq 2$, then $(a^{1/3}+b^{1/3})^3\neq a+b$. Answers will vary.

131. $x^{5/2}+x^{1/2}=x^{1/2}\cdot x^2+x^{1/2}$
$=x^{1/2}(x^2+1)$

133. $y^{1/3}-y^{7/3}=y^{1/3}-y^{1/3}y^2$
$=y^{1/3}(1-y^2)$
$=y^{1/3}(1-y)(1+y)$

135. $y^{-2/5}+y^{8/5}=y^{-2/5}+y^{-2/5}y^2$
$=y^{-2/5}(1+y^2)$
$=\frac{1+y^2}{y^{2/5}}$

137. a. $E(t)=2^{10}\cdot 2^t$
$E(0)=2^{10}\cdot 2^0$
$=2^{10}\cdot 1$
$=2^{10}$
Initially, there are 2^{10} or 1024 bacteria.

b. $E\left(\frac{1}{2}\right)=2^{10}\cdot 2^{1/2}$
$=2^{10}\sqrt{2}$
≈ 1448.15
After $\frac{1}{2}$ hour there are about 1448 bacteria.

139. $A(t)=2.69t^{3/2}$

a. $t=200-1993=7$
$A(7)=2.69(7)^{3/2}$
≈ 49.82
In 2000, there was about \$49.82 billion in total assets in the U.S. in 401(k) plans.

b. $t=2007-1993=14$
$A(14)=2.69(14)^{3/2}$
≈ 140.91
In 2007, there will be about \$140.91 billion in total assets in the U.S. in 401(k) plans.

141. $(3^{\sqrt{2}})^{\sqrt{2}}=3^{\sqrt{2}\cdot\sqrt{2}}=3^2=9$

142. a. $3^{\pi}=31.5442807$

b. Answers will vary. π is approximately equal to 3, and $3^3=27$, which is fairly close to 31. The value in part a) makes sense.

143. $f(x)=(x-5)^{1/2}(x+3)^{-1/2}$
$=\frac{(x-5)^{1/2}}{(x+3)^{1/2}}$
$=\frac{\sqrt{x-5}}{\sqrt{x+3}}$
The denominator must be greater than zero.
$\sqrt{x+3}>0$
$x>-3$
The numerator must be greater than or equal to zero.

$\sqrt{x-5} \geq 0$

$x \geq 5$

Therefore, the domain is
$\{x|x \text{ is a real number } x \geq 5\}$

145. **a.** If n is even: $\sqrt[n]{(x-4)^{2n}} = (x-4)^2$

b. If n is odd: $\sqrt[n]{(x-4)^{2n}} = (x-4)^2$

147. Let a be the unknown index in the shaded area.

$$\sqrt[4]{\sqrt[5]{\sqrt[a]{\sqrt[3]{z}}}} = z^{1/120}$$
$$\left(\left(\left(z^{1/3}\right)^{1/a}\right)^{1/5}\right)^{1/4} = z^{1/120}$$
$$z^{1/60a} = z^{1/120}$$
$$\frac{1}{60a} = \frac{1}{120} \leftarrow \text{Equate exponents}$$
$$60a = 120$$
$$a = 2$$

149.
$$\frac{a^{-2} + ab^{-1}}{ab^{-2} - a^{-2}b^{-1}} = \frac{\frac{1}{a^2} + \frac{a}{b}}{\frac{a}{b^2} - \frac{1}{a^2b}}$$
$$= \frac{a^2b^2\left(\frac{1}{a^2}\right) + a^2b^2\left(\frac{a}{b}\right)}{a^2b^2\left(\frac{a}{b^2}\right) - a^2b^2\left(\frac{1}{a^2b}\right)}$$
$$= \frac{b^2 + a^3b}{a^3 - b}$$

150.
$$\frac{3x-2}{x+4} = \frac{2x+1}{3x-2}$$
$$(3x-2)(3x-2) = (2x+1)(x+4)$$
$$9x^2 - 12x + 4 = 2x^2 + 9x + 4$$
$$7x^2 - 21x = 0$$
$$7x(x-3) = 0$$
$$7x = 0 \quad \text{or} \quad x - 3 = 0$$
$$x = 0 \qquad x = 3$$

The solution is 0 or 3.

151. Let y be the speed of the plane in still air. The table is

	d	r	$t = \frac{d}{r}$
With wind	560	$y+25$	$\frac{560}{y+25}$
Against wind	500	$y-25$	$\frac{500}{y-25}$

Since the time is the same for both parts of the trip. The equation is
$$\frac{560}{y+25} = \frac{500}{y-25}$$
$$560(y-25) = 500(y+25)$$
$$560y - 14,000 = 500y + 12,500$$
$$560y = 500y + 26,500$$
$$60y = 26,500$$
$$y = \frac{26,500}{60}$$
$$\approx 441.67 \text{ mph}$$

The speed of the plane in still air is about 441.67 mph.

152. **a.** The graph is a relation but not a function.

b. The graph is a relation but not a function.

c. The graph is both a relation and a function.

Exercise Set 11.3

1. **a.** Square the natural numbers.

b. $1^2 = 1,\ 2^2 = 4,\ 3^2 = 9,$
$4^2 = 16,\ 5^2 = 25,\ 6^2 = 36$

3. **a.** Raise natural numbers to the fifth power.

b. $1^5 = 1,\ 2^5 = 32,\ 3^5 = 243,$
$4^5 = 1024,\ 5^5 = 3125$

5. If n is even and a or b is negative, the numbers are not real numbers and the rule does not apply.

7. If n is even and a or b is negative, the numbers are not real numbers and the rule does not apply.

9. $\sqrt{8} = \sqrt{4 \cdot 3} = \sqrt{4}\sqrt{3} = 2\sqrt{3}$

11. $\sqrt{28} = \sqrt{4 \cdot 7} = \sqrt{4}\sqrt{7} = 2\sqrt{7}$

13. $\sqrt{12}=\sqrt{4\cdot 3}=\sqrt{4}\sqrt{3}=2\sqrt{3}$

15. $\sqrt{50}=\sqrt{25\cdot 2}=\sqrt{25}\sqrt{2}=5\sqrt{2}$

17. $\sqrt{75}=\sqrt{25\cdot 3}=\sqrt{25}\sqrt{3}=5\sqrt{3}$

19. $\sqrt{40}=\sqrt{4\cdot 10}=\sqrt{4}\sqrt{10}=2\sqrt{10}$

21. $\sqrt[3]{16}=\sqrt[3]{8\cdot 2}=\sqrt[3]{8}\sqrt[3]{2}=2\sqrt[3]{2}$

23. $\sqrt[3]{54}=\sqrt[3]{27\cdot 2}=\sqrt[3]{27}\sqrt[3]{2}=3\sqrt[3]{2}$

25. $\sqrt[3]{32}=\sqrt[3]{8\cdot 4}=\sqrt[3]{8}\sqrt[3]{4}=2\sqrt[3]{4}$

27. $\sqrt[3]{108}=\sqrt[3]{27\cdot 4}=\sqrt[3]{27}\sqrt[3]{4}=3\sqrt[3]{4}$

29. $\sqrt[4]{162}=\sqrt[4]{81\cdot 2}=\sqrt[4]{81}\sqrt[4]{2}=3\sqrt[4]{2}$

31. $-\sqrt[5]{64}=-\sqrt[5]{32\cdot 2}=-\sqrt[5]{32}\sqrt[5]{2}=-2\sqrt[5]{2}$

33. $\sqrt{x^3}=\sqrt{x^2\cdot x}=\sqrt{x^2}\sqrt{x}=x\sqrt{x}$

35. $\sqrt{a^{11}}=\sqrt{a^{10}\cdot a}=\sqrt{a^{10}}\sqrt{a}=a^5\sqrt{a}$

37. $-\sqrt{z^{21}}=-\sqrt{z^{20}\cdot z}=-\sqrt{z^{20}}\sqrt{z}=-z^{10}\sqrt{z}$

39. $\sqrt[3]{a^7}=\sqrt[3]{a^6\cdot a}=\sqrt[3]{a^6}\sqrt[3]{a}=a^2\sqrt[3]{a}$

41. $\sqrt[3]{b^{13}}=\sqrt[3]{b^{12}\cdot b}=\sqrt[3]{b^{12}}\sqrt[3]{b}=b^4\sqrt[3]{b}$

43. $\sqrt[4]{x^5}=\sqrt[4]{x^4\cdot x}=\sqrt[4]{x^4}\sqrt[4]{x}=x\sqrt[4]{x}$

45. $\sqrt[5]{z^7}=\sqrt[5]{z^5\cdot z^2}=\sqrt[5]{z^5}\sqrt[5]{z^2}=z\sqrt[5]{z^2}$

47. $3\sqrt[5]{y^{23}}=3\sqrt[5]{y^{20}\cdot y^3}=3\sqrt[5]{y^{20}}\sqrt[5]{y^3}=3y^4\sqrt[5]{y^3}$

49.
$$\begin{aligned}\sqrt{24x^3}&=\sqrt{4\cdot 6\cdot x^2\cdot x}\\&=\sqrt{4x^2\cdot 6x}\\&=\sqrt{4x^2}\sqrt{6x}\\&=2x\sqrt{6x}\end{aligned}$$

51.
$$\begin{aligned}\sqrt{8x^4y^7}&=\sqrt{4\cdot 2\cdot x^4\cdot y^6\cdot y}=\sqrt{4x^4y^6\cdot 2y}\\&=\sqrt{4x^4y^6}\sqrt{2y}=2x^2y^3\sqrt{2y}\end{aligned}$$

53.
$$\begin{aligned}-\sqrt{20x^6y^7z^{12}}&=-\sqrt{4\cdot 5\cdot x^6\cdot y^6\cdot y\cdot z^{12}}\\&=-\sqrt{4x^6y^6z^{12}\cdot 5y}\\&=-\sqrt{4x^6y^6z^{12}}\sqrt{5y}\\&=-2x^3y^3z^6\sqrt{5y}\end{aligned}$$

55.
$$\begin{aligned}\sqrt[3]{x^3y^7}&=\sqrt[3]{x^3\cdot y^6\cdot y}\\&=\sqrt[3]{x^3y^6\cdot y}\\&=\sqrt[3]{x^3y^6}\sqrt[3]{y}\\&=xy^2\sqrt[3]{y}\end{aligned}$$

57.
$$\begin{aligned}\sqrt[3]{81a^6b^8}&=\sqrt[3]{27\cdot 3\cdot a^6\cdot b^6\cdot b^2}\\&=\sqrt[3]{27a^6b^6\cdot 3b^2}\\&=\sqrt[3]{27a^6b^6}\sqrt[3]{3b^2}\\&=3a^2b^2\sqrt[3]{3b^2}\end{aligned}$$

59.
$$\begin{aligned}\sqrt[4]{32x^8y^9z^{19}}&=\sqrt[4]{16\cdot 2\cdot x^8\cdot y^8\cdot y\cdot z^{16}\cdot z^3}\\&=\sqrt[4]{16x^8y^8z^{16}\cdot 2yz^3}\\&=\sqrt[4]{16x^8y^8z^{16}}\sqrt[4]{2yz^3}\\&=2x^2y^2z^4\sqrt[4]{2yz^3}\end{aligned}$$

61.
$$\begin{aligned}\sqrt[4]{81a^8b^9}&=\sqrt[4]{81\cdot a^8\cdot b^8\cdot b}\\&=\sqrt[4]{81a^8b^8\cdot b}\\&=\sqrt[4]{81a^8b^8}\sqrt[4]{b}\\&=3a^2b^2\sqrt[4]{b}\end{aligned}$$

63.
$$\begin{aligned}\sqrt[5]{32a^{10}b^{12}}&=\sqrt[5]{32\cdot a^{10}\cdot b^{10}\cdot b^2}\\&=\sqrt[5]{32a^{10}b^{10}\cdot b^2}\\&=\sqrt[5]{32a^{10}b^{10}}\sqrt[5]{b^2}\\&=2a^2b^2\sqrt[5]{b^2}\end{aligned}$$

65. $\sqrt{\frac{36}{4}}=\sqrt{9}=3$

67. $\sqrt{\frac{4}{25}}=\frac{\sqrt{4}}{\sqrt{25}}=\frac{2}{5}$

69. $\frac{\sqrt{27}}{\sqrt{3}}=\sqrt{\frac{27}{3}}=\sqrt{9}=3$

71. $\frac{\sqrt{3}}{\sqrt{48}}=\sqrt{\frac{3}{48}}=\sqrt{\frac{1}{16}}=\frac{1}{4}$

73. $\sqrt[3]{\frac{2}{16}}=\sqrt[3]{\frac{1}{8}}=\frac{\sqrt[3]{1}}{\sqrt[3]{8}}=\frac{1}{2}$

75. $\frac{\sqrt[3]{3}}{\sqrt[3]{81}}=\sqrt[3]{\frac{3}{81}}=\sqrt[3]{\frac{1}{27}}=\frac{1}{3}$

77. $\sqrt[4]{\frac{3}{48}}=\sqrt[4]{\frac{1}{16}}=\frac{1}{2}$

79. $\sqrt[5]{\frac{96}{3}}=\sqrt[5]{32}=2$

81. $\sqrt{\frac{r^4}{25}}=\frac{\sqrt{r^4}}{\sqrt{25}}=\frac{r^2}{5}$

83. $\sqrt{\frac{36x^4}{25y^{10}}}=\frac{\sqrt{36x^4}}{\sqrt{25y^{10}}}=\frac{6x^2}{5y^5}$

85. $\sqrt[3]{\frac{c^6}{27}}=\frac{\sqrt[3]{c^6}}{\sqrt[3]{27}}=\frac{c^2}{3}$

87. $\sqrt[4]{\frac{a^8b^{12}}{b^{-4}}}=\sqrt[4]{a^8b^{12+4}}$

$=\sqrt[4]{a^8b^{16}}$

$=a^2b^4$

89. $\frac{-\sqrt{24}}{\sqrt{3}}=-\sqrt{\frac{24}{3}}=-\sqrt{8}=-\sqrt{4\cdot 2}=-2\sqrt{2}$

91. $\frac{\sqrt{27x^6}}{\sqrt{3x^2}}=\sqrt{\frac{27x^6}{3x^2}}=\sqrt{9x^4}=3x^2$

93. $\frac{\sqrt{40x^6y^9}}{\sqrt{5x^2y^4}}=\sqrt{\frac{40x^6y^9}{5x^2y^4}}$

$=\sqrt{8x^4y^5}$

$=\sqrt{4x^4y^4\cdot 2y}$

$=2x^2y^2\sqrt{2y}$

95. $\sqrt[3]{\frac{7xy}{8x^{13}}}=\sqrt[3]{\frac{7y}{8x^{12}}}=\frac{\sqrt[3]{7y}}{\sqrt[3]{8x^{12}}}=\frac{\sqrt[3]{7y}}{2x^4}$

97. $\sqrt[3]{\frac{25x^2y^9}{5x^8y^2}}=\sqrt[3]{\frac{5y^7}{x^6}}$

$=\frac{\sqrt[3]{5y^7}}{\sqrt[3]{x^6}}$

$=\frac{\sqrt[3]{y^6\cdot 5y}}{x^2}$

$=\frac{y^2\sqrt[3]{5y}}{x^2}$

99. $\sqrt[4]{\frac{20x^4y}{81x^{-8}}}=\sqrt[4]{\frac{20x^{12}y}{81}}$

$=\frac{\sqrt[4]{20x^{12}y}}{\sqrt[4]{81}}$

$=\frac{\sqrt[4]{x^{12}}\sqrt[4]{20y}}{\sqrt[4]{81}}$

$=\frac{x^3\sqrt[4]{20y}}{3}$

101. $\sqrt{a\cdot b}=(a\cdot b)^{1/2}=a^{1/2}\cdot b^{1/2}=\sqrt{a}\cdot\sqrt{b}$

103. No, for example $\frac{\sqrt{18}}{\sqrt{2}}=\sqrt{\frac{18}{2}}=\sqrt{9}=3$.

105. **a.** no

b. $\frac{\sqrt[n]{x}}{\sqrt[n]{x}}$ is equal to 1 when $\sqrt[n]{x}$ is a real number and not equal to 0.

106. $F=\frac{9}{5}C+32$

$F-32=\frac{9}{5}C$

$\frac{5}{9}(F-32)=\frac{5}{9}\left(\frac{9}{5}C\right)$

$\frac{5}{9}(F-32)=C$ or $C=\frac{5}{9}(F-32)$

107. $\frac{15x^{12}-5x^9+30x^6}{5x^6}=\frac{15x^{12}}{5x^6}-\frac{5x^9}{5x^6}+\frac{30x^6}{5x^6}$

$=3x^6-x^3+6$

108. $(x-3)^3+8=(x-3)^3+(2)^3$
$=((x-3)+2)((x-3)^2-(x-3)(2)+(2)^2)$
$=(x-1)(x^2-6x+9-2x+6+4)$
$=(x-1)(x^2-8x+19)$

109. $(2x-3)(x-2)=4x-6$
$2x^2-7x+6=4x-6$
$2x^2-11x+12=0$
$(2x-3)(x-4)=0$
$2x-3=0 \qquad x-4=0$
$2x=3 \qquad x=4$
$x=\frac{3}{2}$
The solution is $\left\{4, \frac{3}{2}\right\}$.

110. $\left|\frac{2x-4}{5}\right|=12$
$\frac{2x-4}{5}=-12$ or $\frac{2x-4}{5}=12$
$2x-4=-60 \qquad 2x-4=60$
$2x=-56 \qquad 2x=64$
$x=-28 \qquad x=32$

The solution is $\{-28, 32\}$.

Exercise Set 11.4

1. Like radicals are radicals with the same radicands and index.

3. $\sqrt{3}+3\sqrt{2}\approx 1.732+3(1.414)$
$\approx 1.732+4.242$
≈ 5.974 or 5.97

5. No. To see this, let $a=16$ and $b=9$. Then, the left side is $\sqrt{a}+\sqrt{b}=\sqrt{16}+\sqrt{9}=4+3=7$ whereas the right side is $\sqrt{a+b}=\sqrt{16+9}=\sqrt{25}=5$.

7. $\sqrt{5}-\sqrt{5}=0$

9. $6\sqrt{3}-2\sqrt{3}=4\sqrt{3}$

11. $2\sqrt{3}-2\sqrt{3}-4\sqrt{3}+5=-4\sqrt{3}+5$

13. $3\sqrt[4]{y}-9\sqrt[4]{y}=-6\sqrt[4]{y}$

15. $3\sqrt{5}-\sqrt[3]{x}+4\sqrt{5}+3\sqrt[3]{x}=7\sqrt{5}+2\sqrt[3]{x}$

17. $5\sqrt{x}-4\sqrt{y}+3\sqrt{x}+2\sqrt{y}-\sqrt{x}=7\sqrt{x}-2\sqrt{y}$

19. $\sqrt{8}-\sqrt{12}=\sqrt{4}\cdot\sqrt{2}-\sqrt{4}\cdot\sqrt{3}$
$=2\sqrt{2}-2\sqrt{3}$
$=2(\sqrt{2}-\sqrt{3})$

21. $-6\sqrt{75}+4\sqrt{125}=-6\sqrt{25}\cdot\sqrt{3}+4\sqrt{25}\cdot\sqrt{5}$
$=-6(5\sqrt{3})+4(5\sqrt{5})$
$=-30\sqrt{3}+20\sqrt{5}$

23. $-4\sqrt{90}+3\sqrt{40}+2\sqrt{10}$
$=-4\sqrt{9}\cdot\sqrt{10}+3\sqrt{4}\cdot\sqrt{10}+2\sqrt{10}$
$=-4(3\sqrt{10})+3(2\sqrt{10})+2(\sqrt{10})$
$=-12\sqrt{10}+6\sqrt{10}+2\sqrt{10}$
$=-4\sqrt{10}$

25. $\sqrt{500xy^2}+y\sqrt{320x}$
$=\sqrt{100y^2}\cdot\sqrt{5x}+y\sqrt{64}\sqrt{5x}$
$=10y\sqrt{5x}+8y\sqrt{5x}$
$=18y\sqrt{5x}$

27. $2\sqrt{5x}-3\sqrt{20x}-4\sqrt{45x}$
$=2\sqrt{5x}-3\sqrt{4}\cdot\sqrt{5x}-4\sqrt{9}\cdot\sqrt{5x}$
$=2\sqrt{5x}-3(2\sqrt{5x})-4(3\sqrt{5x})$
$=2\sqrt{5x}-6\sqrt{5x}-12\sqrt{5x}$
$=-16\sqrt{5x}$

29. $3\sqrt{50a^2}-3\sqrt{72a^2}-8a\sqrt{18}$
$=3\sqrt{25a^2}\cdot\sqrt{2}-3\sqrt{36a^2}\cdot\sqrt{2}-8a\sqrt{9}\cdot\sqrt{2}$
$=3(5a\sqrt{2})-3(6a\sqrt{2})-8a(3\sqrt{2})$
$=15a\sqrt{2}-18a\sqrt{2}-24a\sqrt{2}$
$=-27a\sqrt{2}$

31. $\sqrt[3]{108}+2\sqrt[3]{32}=\sqrt[3]{27}\cdot\sqrt[3]{4}+2\sqrt[3]{8}\cdot\sqrt[3]{4}$
$=3\sqrt[3]{4}+2(2\sqrt[3]{4})$
$=3\sqrt[3]{4}+4\sqrt[3]{4}$
$=7\sqrt[3]{4}$

33. $\sqrt[3]{27}-5\sqrt[3]{8}=3-5\cdot 2=3-10=-7$

35. $2\sqrt[3]{a^4b^2}+3a\sqrt[3]{ab^2} = 2\sqrt[3]{a^3}\cdot\sqrt[3]{ab^2}+3a\sqrt[3]{ab^2}$

$= 2a\sqrt[3]{ab^2}+3a\sqrt[3]{ab^2}$

$= 5a\sqrt[3]{ab^2}$

37. $\sqrt{4r^7s^5}+3r^2\sqrt{r^3s^5}-2rs\sqrt{r^5s^3}$

$=\sqrt{4r^6s^4}\sqrt{rs}+3r^2\sqrt{r^2s^4}\sqrt{rs}-2rs\sqrt{r^4s^2}\sqrt{rs}$

$=2r^3s^2\sqrt{rs}+3r^2(rs^2\sqrt{rs})-2rs(r^2s\sqrt{rs})$

$=2r^3s^2\sqrt{rs}+3r^3s^2\sqrt{rs}-2r^3s^2\sqrt{rs}$

$=3r^3s^2\sqrt{rs}$

39. $\sqrt[3]{128x^9y^{10}}-2x^2y\sqrt[3]{16x^3y^7}$

$=\sqrt[3]{64x^9y^9}\sqrt[3]{2y}-2x^2y\sqrt[3]{8x^3y^6}\sqrt[3]{2y}$

$=4x^3y^3\sqrt[3]{2y}-2x^2y\left(2xy^2\sqrt[3]{2y}\right)$

$=4x^3y^3\sqrt[3]{2y}-4x^3y^3\sqrt[3]{2y}$

$=0$

41. $\sqrt{50}\sqrt{2}=\sqrt{50\cdot 2}=\sqrt{100}=10$

43. $\sqrt[3]{2}\sqrt[3]{28}=\sqrt[3]{56}=\sqrt[3]{8\cdot 7}=2\sqrt[3]{7}$

45. $\sqrt{9m^3n^7}\sqrt{3mn^4}=\sqrt{9m^3n^7\cdot 3mn^4}$

$=\sqrt{27m^4n^{11}}$

$=\sqrt{9\cdot 3\cdot m^4\cdot n^{10}\cdot n}$

$=\sqrt{9m^4n^{10}\cdot 3n}$

$=\sqrt{9m^4n^{10}}\sqrt{3n}$

$=3m^2n^5\sqrt{3n}$

47. $\sqrt[3]{9x^7y^{12}}\sqrt[3]{6x^4y}=\sqrt[3]{9x^7y^{12}\cdot 6x^4y}$

$=\sqrt[3]{54x^{11}y^{13}}$

$=\sqrt[3]{27\cdot 2\cdot x^9\cdot x^2\cdot y^{12}\cdot y}$

$=\sqrt[3]{27x^9y^{12}\cdot 2x^2y}$

$=\sqrt[3]{27x^9y^{12}}\sqrt[3]{2x^2y}$

$=3x^3y^4\sqrt[3]{2x^2y}$

49. $\sqrt[4]{3x^9y^{12}}\sqrt[4]{54x^4y^7}=\sqrt[4]{3x^9y^{12}\cdot 54x^4y^7}$

$=\sqrt[4]{162x^{13}y^{19}}$

$=\sqrt[4]{81\cdot 2\cdot x^{12}\cdot x\cdot y^{16}\cdot y^3}$

$=\sqrt[4]{81x^{12}y^{16}\cdot 2xy^3}$

$=\sqrt[4]{81x^{12}y^{16}}\sqrt[4]{2xy^3}$

$=3x^3y^4\sqrt[4]{2xy^3}$

51. $\sqrt[4]{8x^4yz^3}\sqrt[4]{2x^2y^3z^7}$

$=\sqrt[4]{8x^4yz^3\cdot 2x^2y^3z^7}$

$=\sqrt[4]{16x^6y^4z^{10}}$

$=\sqrt[4]{16\cdot x^4\cdot x^2\cdot y^4\cdot z^8\cdot z^2}$

$=\sqrt[4]{16x^4y^4z^8\cdot x^2z^2}$

$=\sqrt[4]{16x^4y^4z^8}\cdot\sqrt[4]{x^2z^2}$

$=2xyz^2\sqrt[4]{x^2z^2}$

53. $\sqrt{5}\left(\sqrt{5}-\sqrt{3}\right)=\left(\sqrt{5}\right)\left(\sqrt{5}\right)-\left(\sqrt{5}\right)\left(\sqrt{3}\right)$

$=\sqrt{25}-\sqrt{15}$

$=5-\sqrt{15}$

55. $\sqrt[3]{y}\left(2\sqrt[3]{y}-\sqrt[3]{y^8}\right)$

$=\left(\sqrt[3]{y}\right)\left(2\sqrt[3]{y}\right)-\left(\sqrt[3]{y}\right)\left(\sqrt[3]{y^8}\right)$

$=2\sqrt[3]{y^2}-\sqrt[3]{y^9}$

$=2\sqrt[3]{y^2}-y^3$

57. $2\sqrt[3]{x^4y^5}\left(\sqrt[3]{8x^{12}y^4}+\sqrt[3]{16xy^9}\right)$

$=\left(2\sqrt[3]{x^4y^5}\right)\left(\sqrt[3]{8x^{12}y^4}\right)+\left(2\sqrt[3]{x^4y^5}\right)\left(\sqrt[3]{16xy^9}\right)$

$=2\sqrt[3]{8x^{16}y^9}+2\sqrt[3]{16x^5y^{14}}$

$=2\sqrt[3]{8x^{15}y^9}\sqrt[3]{x}+2\sqrt[3]{8x^3y^{12}}\sqrt[3]{2x^2y^2}$

$=2\cdot 2x^5y^3\sqrt[3]{x}+2\cdot 2xy^4\sqrt[3]{2x^2y^2}$

$=4x^5y^3\sqrt[3]{x}+4xy^4\sqrt[3]{2x^2y^2}$

59. $(5+\sqrt{5})(5-\sqrt{5})=5^2-(\sqrt{5})^2$

$=25-5$

$=20$

60. $\left(7-\sqrt{5}\right)\left(7+\sqrt{5}\right)=7^2-(\sqrt{5})^2$

$=49-5$

$=44$

61. $(\sqrt{6}+x)(\sqrt{6}-x)=(\sqrt{6})^2-x^2$
$=6-x^2$

63. $(\sqrt{5}+\sqrt{z})(\sqrt{5}-\sqrt{z})=(\sqrt{5})^2-(\sqrt{z})^2$
$=5-z$

65. $(\sqrt{3}+4)(\sqrt{3}+5)$
$=\sqrt{9}+5\sqrt{3}+4\sqrt{3}+20$
$=3+9\sqrt{3}+20$
$=23+9\sqrt{3}$

67. $(3-\sqrt{2})(4-\sqrt{8})$
$=12-3\sqrt{8}-4\sqrt{2}+\sqrt{16}$
$=12-3\cdot 2\sqrt{2}-4\sqrt{2}+4$
$=12-6\sqrt{2}-4\sqrt{2}+4$
$=16-10\sqrt{2}$

69. $(4\sqrt{3}+\sqrt{2})(\sqrt{3}-\sqrt{2})$
$=4\sqrt{9}-4\sqrt{6}+\sqrt{6}-\sqrt{4}$
$=4\cdot 3-3\sqrt{6}-2$
$=12-3\sqrt{6}-2$
$=10-3\sqrt{6}$

71. $(2\sqrt{5}-3)^2$
$=(2\sqrt{5}-3)(2\sqrt{5}-3)$
$=4\sqrt{25}-6\sqrt{5}-6\sqrt{5}+9$
$=4\cdot 5-12\sqrt{5}+9$
$=20+9-12\sqrt{5}$
$=29-12\sqrt{5}$

73. $(2\sqrt{3x}-\sqrt{y})(3\sqrt{3x}+\sqrt{y})$
$=6(\sqrt{3x})^2+2\sqrt{3x}\sqrt{y}-3\sqrt{3x}\sqrt{y}-(\sqrt{y})^2$
$=6(3x)+2\sqrt{3xy}-3\sqrt{3xy}-y$
$=18x-\sqrt{3xy}-y$

75. $(\sqrt[3]{4}-\sqrt[3]{6})(\sqrt[3]{2}-\sqrt[3]{36})$
$=\sqrt[3]{4}\sqrt[3]{2}-\sqrt[3]{4}\sqrt[3]{36}-\sqrt[3]{6}\sqrt[3]{2}+\sqrt[3]{6}\sqrt[3]{36}$
$=\sqrt[3]{8}-\sqrt[3]{144}-\sqrt[3]{12}+\sqrt[3]{216}$
$=2-2\sqrt[3]{18}-\sqrt[3]{12}+6$
$=8-2\sqrt[3]{18}-\sqrt[3]{12}$

77. $(f\cdot g)(x)=f(x)\cdot g(x)$
$=\sqrt{2x}(\sqrt{8x}-\sqrt{32})$
$=\sqrt{2x}\cdot\sqrt{8x}-\sqrt{2x}\cdot\sqrt{32}$
$=\sqrt{16x^2}-\sqrt{64x}$
$=4x-8\sqrt{x}$

79. $(f\cdot g)(x)=f(x)\cdot g(x)$
$=\sqrt[3]{x}\left(\sqrt[3]{x^5}+\sqrt[3]{x^4}\right)$
$=\sqrt[3]{x}\sqrt[3]{x^5}+\sqrt[3]{x}\sqrt[3]{x^4}$
$=\sqrt[3]{x^6}+\sqrt[3]{x^5}$
$=\sqrt[3]{x^6}+\sqrt[3]{x^3\cdot x^2}$
$=x^2+x\sqrt[3]{x^2}$

81. $(f\cdot g)(x)=f(x)\cdot g(x)$
$=\sqrt[4]{3x^2}\left(\sqrt[4]{9x^4}-\sqrt[4]{x^7}\right)$
$=\sqrt[4]{3x^2}\sqrt[4]{9x^4}-\sqrt[4]{3x^2}\sqrt[4]{x^7}$
$=\sqrt[4]{27x^6}-\sqrt[4]{3x^9}$
$=\sqrt[4]{x^4}\sqrt[4]{27x^2}-\sqrt[4]{x^8}\sqrt[4]{3x}$
$=x\sqrt[4]{27x^2}-x^2\sqrt[4]{3x}$

83. $\sqrt{24}=\sqrt{4\cdot 6}=2\sqrt{6}$

85. $\sqrt{125}+\sqrt{20}=\sqrt{25\cdot 5}+\sqrt{4\cdot 5}$
$=5\sqrt{5}+2\sqrt{5}$
$=7\sqrt{5}$

87. $(3\sqrt{2}-4)(\sqrt{2}+5)$
$=3(\sqrt{2})^2+15\sqrt{2}-4\sqrt{2}-20$
$=6+11\sqrt{2}-20$
$=-14+11\sqrt{2}$

89. $\sqrt{6}(4-\sqrt{2})=\sqrt{6}\cdot 4-\sqrt{6}\cdot\sqrt{2}$
$=4\sqrt{6}-\sqrt{12}$
$=4\sqrt{6}-2\sqrt{3}$

91. $\sqrt{75}\sqrt{6}=\sqrt{450}=\sqrt{225\cdot 2}=15\sqrt{2}$

93. $\sqrt[3]{80x^{11}} = \sqrt[3]{8x^9 \cdot 10x^2} = 2x^3\sqrt[3]{10x^2}$

95. $\sqrt[6]{128ab^{17}c^9} = \sqrt[6]{64b^{12}c^6 \cdot 2ab^5c^3}$
$= 2b^2c\sqrt[6]{2ab^5c^3}$

97. $2b\sqrt[4]{a^4b} + ab\sqrt[4]{16b}$
$= 2b\sqrt[4]{a^4} \cdot \sqrt[4]{b} + ab\sqrt[4]{16} \cdot \sqrt[4]{b}$
$= 2b\left(a\sqrt[4]{b}\right) + ab\left(2\sqrt[4]{b}\right)$
$= 2ab\sqrt[4]{b} + 2ab\sqrt[4]{b}$
$= 4ab\sqrt[4]{b}$

99. $\left(\sqrt[3]{x^2} - \sqrt[3]{y}\right)\left(\sqrt[3]{x} - 2\sqrt[3]{y^2}\right)$
$= \sqrt[3]{x^2}\sqrt[3]{x} - 2\sqrt[3]{x^2}\sqrt[3]{y^2} - \sqrt[3]{y}\sqrt[3]{x} + 2\sqrt[3]{y}\sqrt[3]{y^2}$
$= \sqrt[3]{x^3} - 2\sqrt[3]{x^2y^2} - \sqrt[3]{xy} + 2\sqrt[3]{y^3}$
$= x - 2\sqrt[3]{x^2y^2} - \sqrt[3]{xy} + 2y$

101. $\sqrt[3]{3ab^2}\left(\sqrt[3]{4a^4b^3} - \sqrt[3]{8a^5b^4}\right)$
$= \left(\sqrt[3]{3ab^2}\right)\left(\sqrt[3]{4a^4b^3}\right) - \left(\sqrt[3]{3ab^2}\right)\left(\sqrt[3]{8a^5b^4}\right)$
$= \sqrt[3]{12a^5b^5} - \sqrt[3]{24a^6b^6}$
$= \sqrt[3]{a^3b^3 \cdot 12a^2b^2} - \sqrt[3]{8a^6b^6 \cdot 3}$
$= ab\sqrt[3]{12a^2b^2} - 2a^2b^2\sqrt[3]{3}$

103. $f(x) = \sqrt{2x+5}\sqrt{2x+5} = 2x+5$
No absolute value needed since $x \geq -\frac{5}{2}$.

105. $h(r) = \sqrt{4r^2 - 32r + 64}$
$= \sqrt{4\left(r^2 - 8r + 16\right)}$
$= \sqrt{4(r-4)^2}$
$= 2|r-4|$

107. Perimeter $= \sqrt{45} + \sqrt{45} + \sqrt{80} + \sqrt{80}$
$= 2\sqrt{45} + 2\sqrt{80}$
$= 2\sqrt{9}\sqrt{5} + 2\sqrt{16}\sqrt{5}$
$= 2\left(3\sqrt{5}\right) + 2\left(4\sqrt{5}\right)$
$= 6\sqrt{5} + 8\sqrt{5}$
$= 14\sqrt{5}$

Area $= \sqrt{45}\sqrt{80}$
$= 3\sqrt{5} \cdot 4\sqrt{5}$
$= 12\left(\sqrt{5}\right)^2$
$= 12 \cdot 5$
$= 60$

109. Perimeter $= \sqrt{245} + \sqrt{180} + \sqrt{80}$
$= \sqrt{49}\sqrt{5} + \sqrt{36}\sqrt{5} + \sqrt{16}\sqrt{5}$
$= 7\sqrt{5} + 6\sqrt{5} + 4\sqrt{5}$
$= 17\sqrt{5}$

Area $= \frac{1}{2}\sqrt{245}\sqrt{45}$
$= \frac{1}{2}\sqrt{49}\sqrt{5}\sqrt{9}\sqrt{5}$
$= \frac{1}{2} \cdot 7 \cdot 3\left(\sqrt{5}\right)^2$
$= \frac{21}{2} \cdot 5$
$= 52.5$

111. No, for example $-\sqrt{2} + \sqrt{2} = 0$

113.
a. $s = \sqrt{30FB}$
$s = \sqrt{30(0.85)(80)} \approx 45.17$
The car's speed was about 45.17 mph.

b. $s = \sqrt{30FB}$
$s = \sqrt{30(0.52)(80)} \approx 35.33$
The car's speed was about 35.33 mph.

115. $f(t) = 3\sqrt{t} + 19$

a. $t = 36$
$f(36) = 3\sqrt{36} + 19$
$= 3(6) + 19$
$= 18 + 19$
$= 37$
The length at 36 months is 37 inches.

b.
$$\begin{aligned} t &= 40 \\ f(40) &= 3\sqrt{40} + 19 \\ &= 3\sqrt{4}\sqrt{10} + 19 \\ &= 3\cdot 2\sqrt{10} + 19 \\ &= 6\sqrt{10} + 19 \\ &\approx 37.97 \end{aligned}$$
The length at 40 months is about 37.97 inches.

117. **a.**
$$\begin{aligned} f(x) &= \sqrt{x} \\ g(x) &= 2 \\ (f+g)(x) &= f(x) + g(x) \\ &= \sqrt{x} + 2 \end{aligned}$$

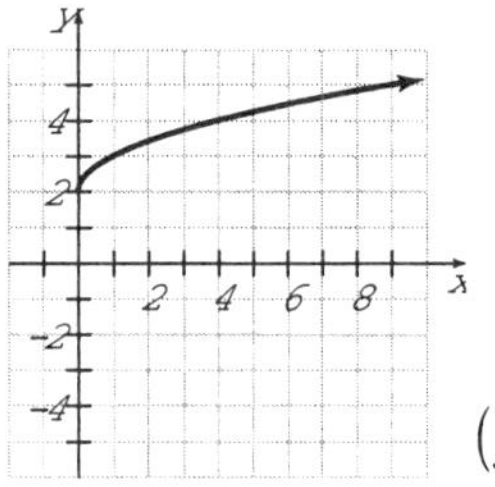

$(f+g)(x) = \sqrt{x} + 2$

b. It raises the graph 2 units.

119. **a.**
$$\begin{aligned} (f-g)(x) &= f(x) - g(x) \\ &= \sqrt{x} - (\sqrt{x} - 2) \\ &= \sqrt{x} - \sqrt{x} + 2 \\ &= 2 \end{aligned}$$

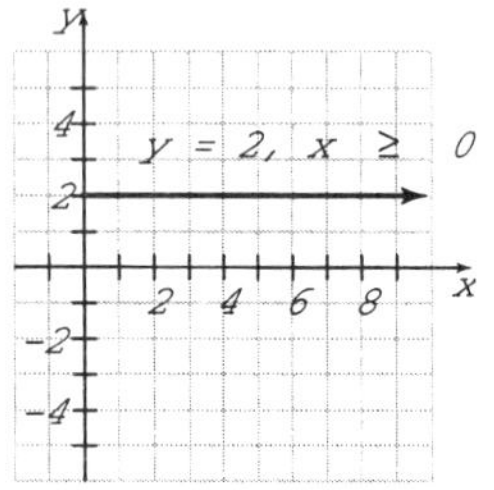

b. $\sqrt{x} \geq 0$, so $x \geq 0$
The domain is $\{x|x$ is a real number, $x \geq 0\}$.

121. $f(x) = \sqrt{x^2}$

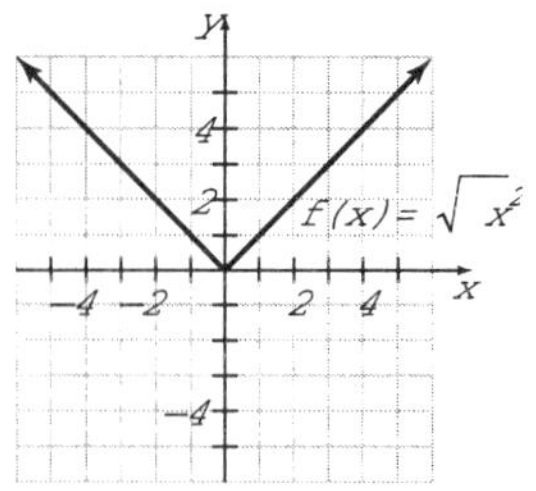

123. A rational number is a number that can be expressed as the quotient of two integers with nonzero denominator.

124. A real number is any number that can represented on the real number line.

125. An irrational number is any real number that cannot be expressed as the quotient of two integers.

126. $|a| = \begin{cases} a, & a \geq 0 \\ -a, & a < 0 \end{cases}$

127.
$$E = \frac{1}{2}mv^2$$
$$2E = 2\left(\frac{1}{2}mv^2\right)$$
$$2E = mv^2$$
$$\frac{2E}{v^2} = \frac{mv^2}{v^2}$$
$$\frac{2E}{v^2} = m \text{ or } m = \frac{2E}{v^2}$$

128. **a.**
$$-4 < 2x - 3 \leq 5$$
$$-4 + 3 < 2x - 3 + 3 \leq 5 + 3$$
$$-1 < 2x \leq 8$$
$$-\frac{1}{2} < x \leq 4$$

b. $\left(-\frac{1}{2}, 4\right]$

c. $\left\{x \middle| -\frac{1}{2} < x \leq 4\right\}$

Exercise Set 11.5

1. **a.** The conjugate of a binomial is a binomial with the same two terms as the original but the sign of the second term is changed. The conjugate of $a + b$ is $a - b$. Also, the conjugate of $a - b$ is $a + b$.

b. The conjugate of $x - \sqrt{3}$ is $x + \sqrt{3}$.

3. **a.** Answers will vary. Possible answer: Multiply the numerator and denominator by a quantity that will result in no radicals in the denominator.

b. $\frac{4}{\sqrt{3y}}=\frac{4}{\sqrt{3y}}\cdot\frac{\sqrt{3y}}{\sqrt{3y}}=\frac{4\sqrt{3y}}{\sqrt{9y^2}}=\frac{4\sqrt{3y}}{3y}$

5. **(1)** No perfect powers are factors of any radicand.

(2) No radicand contains fractions.

(3) No radicals are in any denominator.

7. $\frac{1}{\sqrt{3}}=\frac{1}{\sqrt{3}}\cdot\frac{\sqrt{3}}{\sqrt{3}}=\frac{\sqrt{3}}{3}$

9. $\frac{3}{\sqrt{7}}=\frac{3}{\sqrt{7}}\cdot\frac{\sqrt{7}}{\sqrt{7}}=\frac{3\sqrt{7}}{7}$

11. $\frac{1}{\sqrt{17}}=\frac{1}{\sqrt{17}}\cdot\frac{\sqrt{17}}{\sqrt{17}}=\frac{\sqrt{17}}{17}$

13. $\frac{7}{\sqrt{7}}=\frac{7}{\sqrt{7}}\cdot\frac{\sqrt{7}}{\sqrt{7}}=\frac{7\sqrt{7}}{7}=\sqrt{7}$

15. $\frac{p}{\sqrt{2}}=\frac{p}{\sqrt{2}}\cdot\frac{\sqrt{2}}{\sqrt{2}}=\frac{p\sqrt{2}}{2}$

17. $\frac{\sqrt{y}}{\sqrt{5}}=\frac{\sqrt{y}}{\sqrt{5}}\cdot\frac{\sqrt{5}}{\sqrt{5}}=\frac{\sqrt{5y}}{5}$

19. $\frac{5\sqrt{3}}{\sqrt{5}}=\frac{5\sqrt{3}}{\sqrt{5}}\cdot\frac{\sqrt{5}}{\sqrt{5}}=\frac{5\sqrt{15}}{5}=\sqrt{15}$

21. $\frac{\sqrt{x}}{\sqrt{y}}=\frac{\sqrt{x}}{\sqrt{y}}\cdot\frac{\sqrt{y}}{\sqrt{y}}=\frac{\sqrt{xy}}{y}$

23.
$$\begin{aligned}\sqrt{\frac{5m}{8}}&=\frac{\sqrt{5m}}{\sqrt{8}}\\&=\frac{\sqrt{5m}}{2\sqrt{2}}\\&=\frac{\sqrt{5m}}{2\sqrt{2}}\cdot\frac{\sqrt{2}}{\sqrt{2}}\\&=\frac{\sqrt{10m}}{2\cdot 2}=\frac{\sqrt{10m}}{4}\end{aligned}$$

25.
$$\begin{aligned}\frac{2n}{\sqrt{18n}}&=\frac{2n}{\sqrt{9}\sqrt{2n}}\\&=\frac{2n}{3\sqrt{2n}}\\&=\frac{2n}{3\sqrt{2n}}\cdot\frac{\sqrt{2n}}{\sqrt{2n}}\\&=\frac{2n\sqrt{2n}}{3\cdot 2n}=\frac{\sqrt{2n}}{3}\end{aligned}$$

27.
$$\begin{aligned}\sqrt{\frac{18x^4y^3}{2z^3}}&=\sqrt{\frac{9x^4y^3}{z^3}}=\frac{\sqrt{9x^4y^3}}{\sqrt{z^2\cdot z}}\\&=\frac{\sqrt{9x^4y^2}\sqrt{y}}{\sqrt{z^2}\sqrt{z}}\\&=\frac{3x^2y\sqrt{y}}{z\sqrt{z}}\\&=\frac{3x^2y\sqrt{y}}{z\sqrt{z}}\cdot\frac{\sqrt{z}}{\sqrt{z}}\\&=\frac{3x^2y\sqrt{yz}}{z^2}\end{aligned}$$

29.
$$\begin{aligned}\sqrt{\frac{20y^4z^3}{3xy^{-2}}}&=\sqrt{\frac{20y^6z^3}{3x}}\\&=\frac{\sqrt{20y^6z^3}}{\sqrt{3x}}\\&=\frac{\sqrt{4y^6z^2}\sqrt{5z}}{\sqrt{3x}}\\&=\frac{2y^3z\sqrt{5z}}{\sqrt{3x}}\\&=\frac{2y^3z\sqrt{5z}}{\sqrt{3x}}\cdot\frac{\sqrt{3x}}{\sqrt{3x}}\\&=\frac{2y^3z\sqrt{15xz}}{3x}\end{aligned}$$

31. $\sqrt{\frac{75x^6y^5}{3z^3}} = \sqrt{\frac{25x^6y^5}{z^3}} = \frac{\sqrt{25x^6y^5}}{\sqrt{z^2 \cdot z}}$
$= \frac{\sqrt{25x^6y^4}\sqrt{y}}{\sqrt{z^2}\sqrt{z}}$
$= \frac{5x^3y^2\sqrt{y}}{z\sqrt{z}}$
$= \frac{5x^3y^2\sqrt{y}}{z\sqrt{z}} \cdot \frac{\sqrt{z}}{\sqrt{z}}$
$= \frac{5x^3y^2\sqrt{yz}}{z^2}$

33. $\frac{1}{\sqrt[3]{2}} = \frac{1}{\sqrt[3]{2}} \cdot \frac{\sqrt[3]{4}}{\sqrt[3]{4}} = \frac{\sqrt[3]{4}}{\sqrt[3]{8}} = \frac{\sqrt[3]{4}}{2}$

35. $\frac{3}{\sqrt[3]{x}} = \frac{3}{\sqrt[3]{x}} \cdot \frac{\sqrt[3]{x^2}}{\sqrt[3]{x^2}} = \frac{3\sqrt[3]{x^2}}{x}$

37. $\frac{1}{\sqrt[4]{2}} = \frac{1}{\sqrt[4]{2}} \cdot \frac{\sqrt[4]{8}}{\sqrt[4]{8}} = \frac{\sqrt[4]{8}}{\sqrt[4]{16}} = \frac{\sqrt[4]{8}}{2}$

39. $\frac{z}{\sqrt[4]{8}} = \frac{z}{\sqrt[4]{8}} \cdot \frac{\sqrt[4]{2}}{\sqrt[4]{2}} = \frac{z\sqrt[4]{2}}{\sqrt[4]{16}} = \frac{z\sqrt[4]{2}}{2}$

41. $\frac{x}{\sqrt[4]{z^2}} = \frac{x}{\sqrt[4]{z^2}} \cdot \frac{\sqrt[4]{z^2}}{\sqrt[4]{z^2}} = \frac{x\sqrt[4]{z^2}}{\sqrt[4]{z^4}} = \frac{x\sqrt[4]{z^2}}{z}$

43. $\frac{13}{\sqrt[5]{y^2}} = \frac{13}{\sqrt[5]{y^2}} \cdot \frac{\sqrt[5]{y^3}}{\sqrt[5]{y^3}} = \frac{13\sqrt[5]{y^3}}{\sqrt[5]{y^5}} = \frac{13\sqrt[5]{y^3}}{y}$

45. $\frac{3}{\sqrt[7]{a^3}} = \frac{3}{\sqrt[7]{a^3}} \cdot \frac{\sqrt[7]{a^4}}{\sqrt[7]{a^4}} = \frac{3\sqrt[7]{a^4}}{\sqrt[7]{a^7}} = \frac{3\sqrt[7]{a^4}}{a}$

47. $\sqrt[3]{\frac{1}{4x}} = \frac{\sqrt[3]{1}}{\sqrt[3]{4x}} = \frac{1}{\sqrt[3]{4x}} \cdot \frac{\sqrt[3]{2x^2}}{\sqrt[3]{2x^2}} = \frac{\sqrt[3]{2x^2}}{2x}$

49. $\frac{5m}{\sqrt[4]{2}} = \frac{5m}{\sqrt[4]{2}} \cdot \frac{\sqrt[4]{2^3}}{\sqrt[4]{2^3}} = \frac{5m\sqrt[4]{8}}{2}$

51. $\sqrt[4]{\frac{5}{3x^3}} = \frac{\sqrt[4]{5}}{\sqrt[4]{3x^3}} \cdot \frac{\sqrt[4]{3^3x}}{\sqrt[4]{3^3x}} = \frac{\sqrt[4]{135x}}{3x}$

53. $\sqrt[3]{\frac{3x^2}{2y^2}} = \frac{\sqrt[3]{3x^2}}{\sqrt[3]{2y^2}} \cdot \frac{\sqrt[3]{4y}}{\sqrt[3]{4y}}$
$= \frac{\sqrt[3]{12x^2y}}{2y}$

55. $\sqrt[3]{\frac{8xy^2}{2z^2}} = \sqrt[3]{\frac{4xy^2}{z^2}}$
$= \frac{\sqrt[3]{4xy^2}}{\sqrt[3]{z^2}}$
$= \frac{\sqrt[3]{4xy^2}}{\sqrt[3]{z^2}} \cdot \frac{\sqrt[3]{z}}{\sqrt[3]{z}}$
$= \frac{\sqrt[3]{4xy^2z}}{z}$

57. $\frac{1}{\sqrt{3}+1} = \frac{1}{\sqrt{3}+1} \cdot \frac{(\sqrt{3}-1)}{(\sqrt{3}-1)}$
$= \frac{\sqrt{3}-1}{(\sqrt{3})^2 - 1^2}$
$= \frac{\sqrt{3}-1}{3-1}$
$= \frac{\sqrt{3}-1}{2}$

59. $\frac{1}{2+\sqrt{3}} = \frac{1}{2+\sqrt{3}} \cdot \frac{(2-\sqrt{3})}{(2-\sqrt{3})}$
$= \frac{2-\sqrt{3}}{2^2 - (\sqrt{3})^2}$
$= \frac{2-\sqrt{3}}{4-3}$
$= \frac{2-\sqrt{3}}{1}$
$= 2-\sqrt{3}$

61. $\frac{4}{\sqrt{2}-7} = \frac{4}{\sqrt{2}-7} \cdot \frac{(\sqrt{2}+7)}{(\sqrt{2}+7)}$
$= \frac{4\sqrt{2}+28}{2-49}$
$= \frac{4\sqrt{2}+28}{-47}$
$= \frac{-4\sqrt{2}-28}{47}$

63. $\dfrac{\sqrt{5}}{2\sqrt{5}-\sqrt{6}} = \dfrac{\sqrt{5}}{2\sqrt{5}-\sqrt{6}} \cdot \dfrac{\left(2\sqrt{5}+\sqrt{6}\right)}{\left(2\sqrt{5}+\sqrt{6}\right)}$

$= \dfrac{10+\sqrt{30}}{20-6} = \dfrac{10+\sqrt{30}}{14}$

65. $\dfrac{2}{6+\sqrt{x}} = \dfrac{2}{6+\sqrt{x}} \cdot \dfrac{\left(6-\sqrt{x}\right)}{\left(6-\sqrt{x}\right)}$

$= \dfrac{12-2\sqrt{x}}{36-x}$

67. $\dfrac{4\sqrt{x}}{\sqrt{x}-y} = \dfrac{4\sqrt{x}}{\sqrt{x}-y} \cdot \dfrac{\left(\sqrt{x}+y\right)}{\left(\sqrt{x}+y\right)}$

$= \dfrac{4x+4y\sqrt{x}}{x-y^2}$

69. $\dfrac{\sqrt{2}-2\sqrt{3}}{\sqrt{2}+4\sqrt{3}} = \dfrac{\sqrt{2}-2\sqrt{3}}{\sqrt{2}+4\sqrt{3}} \cdot \dfrac{\left(\sqrt{2}-4\sqrt{3}\right)}{\left(\sqrt{2}-4\sqrt{3}\right)}$

$= \dfrac{2-4\sqrt{6}-2\sqrt{6}+8\cdot 3}{2-16\cdot 3}$

$= \dfrac{26-6\sqrt{6}}{-46} = \dfrac{-13+3\sqrt{6}}{23}$

71. $\dfrac{\sqrt{a^3}+\sqrt{a^7}}{\sqrt{a}} = \dfrac{\sqrt{a^3}+\sqrt{a^7}}{\sqrt{a}} \cdot \dfrac{\left(\sqrt{a}\right)}{\left(\sqrt{a}\right)}$

$= \dfrac{\sqrt{a^4}+\sqrt{a^8}}{a}$

$= \dfrac{a^2+a^4}{a} = a+a^3$

73. $\dfrac{2}{\sqrt{x+2}-3} = \dfrac{2}{\sqrt{x+2}-3} \cdot \dfrac{\left(\sqrt{x+2}+3\right)}{\left(\sqrt{x+2}+3\right)}$

$= \dfrac{2\sqrt{x+2}+6}{\left(\sqrt{x+2}\right)^2-3^2}$

$= \dfrac{2\sqrt{x+2}+6}{x+2-9} = \dfrac{2\sqrt{x+2}+6}{x-7}$

75. $\sqrt{\dfrac{x}{9}} = \dfrac{\sqrt{x}}{\sqrt{9}} = \dfrac{\sqrt{x}}{3}$

77. $\sqrt{\dfrac{2}{5}} = \dfrac{\sqrt{2}}{\sqrt{5}} = \dfrac{\sqrt{2}}{\sqrt{5}} \cdot \dfrac{\sqrt{5}}{\sqrt{5}} = \dfrac{\sqrt{10}}{5}$

79. $\left(\sqrt{5}+\sqrt{6}\right)\left(\sqrt{5}-\sqrt{6}\right) = \left(\sqrt{5}\right)^2 - \left(\sqrt{6}\right)^2$

$= 5-6 = -1$

81. $\sqrt{\dfrac{24x^3y^6}{5z}} = \dfrac{\sqrt{24x^3y^6}}{\sqrt{5z}}$

$= \dfrac{\sqrt{4x^2y^6}\sqrt{6x}}{\sqrt{5z}}$

$= \dfrac{2xy^3\sqrt{6x}}{\sqrt{5z}} \cdot \dfrac{\sqrt{5z}}{\sqrt{5z}} = \dfrac{2xy^3\sqrt{30xz}}{5z}$

83. $\sqrt{\dfrac{12xy^4}{2x^3y^4}} = \sqrt{\dfrac{6}{x^2}} = \dfrac{\sqrt{6}}{\sqrt{x^2}} = \dfrac{\sqrt{6}}{x}$

85. $\dfrac{1}{\sqrt{a}+3} = \dfrac{1}{\sqrt{a}+3} \cdot \dfrac{\sqrt{a}-3}{\sqrt{a}-3}$

$= \dfrac{\sqrt{a}-3}{\left(\sqrt{a}+3\right)\left(\sqrt{a}-3\right)}$

$= \dfrac{\sqrt{a}-3}{\left(\sqrt{a}\right)^2-(3)^2}$

$= \dfrac{\sqrt{a}-3}{a-9}$

87. $-\dfrac{7\sqrt{x}}{\sqrt{98}} = -\dfrac{7\sqrt{x}}{7\sqrt{2}}$

$= -\dfrac{\sqrt{x}}{\sqrt{2}} \cdot \dfrac{\sqrt{2}}{\sqrt{2}}$

$= -\dfrac{\sqrt{2x}}{\sqrt{4}} = -\dfrac{\sqrt{2x}}{2}$

89. $\sqrt[4]{\dfrac{3y^2}{2x}} = \dfrac{\sqrt[4]{3y^2}}{\sqrt[4]{2x}} \cdot \dfrac{\sqrt[4]{8x^3}}{\sqrt[4]{8x^3}}$

$= \dfrac{\sqrt[4]{24x^3y^2}}{\sqrt[4]{16x^4}} = \dfrac{\sqrt[4]{24x^3y^2}}{2x}$

91. $\sqrt[3]{\dfrac{32y^{12}z^{10}}{2x}} = \sqrt[3]{\dfrac{16y^{12}z^{10}}{x}}$

$$= \frac{\sqrt[3]{16y^{12}z^{10}}}{\sqrt[3]{x}}$$
$$= \frac{\sqrt[3]{8y^{12}z^{9}}\sqrt[3]{2z}}{\sqrt[3]{x}}$$
$$= \frac{2y^4z^3\sqrt[3]{2z}}{\sqrt[3]{x}} \cdot \frac{\sqrt[3]{x^2}}{\sqrt[3]{x^2}}$$
$$= \frac{2y^4z^3\sqrt[3]{2x^2z}}{\sqrt[3]{x^3}}$$
$$= \frac{2y^4z^3\sqrt[3]{2x^2z}}{x}$$

93. $$\frac{\sqrt{ar}}{\sqrt{a}-2\sqrt{r}} \cdot \frac{(\sqrt{a}+2\sqrt{r})}{(\sqrt{a}+2\sqrt{r})} = \frac{\sqrt{ar}(\sqrt{a}+2\sqrt{r})}{(\sqrt{a})^2-(2\sqrt{r})^2}$$
$$= \frac{a\sqrt{r}+2r\sqrt{a}}{a-4r}$$

95. $$\frac{\sqrt[3]{6x}}{\sqrt[3]{5xy}} = \sqrt[3]{\frac{6x}{5xy}}$$
$$= \sqrt[3]{\frac{6}{5y}}$$
$$= \frac{\sqrt[3]{6}}{\sqrt[3]{5y}} \cdot \frac{\sqrt[3]{25y^2}}{\sqrt[3]{25y^2}} = \frac{\sqrt[3]{150y^2}}{5y}$$

97. $$\sqrt[4]{\frac{2x^7y^{12}z^4}{3x^9}} = \sqrt[4]{\frac{2y^{12}z^4}{3x^2}}$$
$$= \frac{\sqrt[4]{2y^{12}z^4}}{\sqrt[4]{3x^2}}$$
$$= \frac{\sqrt[4]{y^{12}z^4}\sqrt[4]{2}}{\sqrt[4]{3x^2}}$$
$$= \frac{y^3z\sqrt[4]{2}}{\sqrt[4]{3x^2}} \cdot \frac{\sqrt[4]{27x^2}}{\sqrt[4]{27x^2}}$$
$$= \frac{y^3z\sqrt[4]{54x^2}}{\sqrt[4]{81x^4}}$$
$$= \frac{y^3z\sqrt[4]{54x^2}}{3x}$$

99. $$\frac{1}{\sqrt{2}} + \frac{\sqrt{2}}{2} = \frac{1}{\sqrt{2}} \cdot \frac{\sqrt{2}}{\sqrt{2}} + \frac{\sqrt{2}}{2}$$
$$= \frac{\sqrt{2}}{\sqrt{4}} + \frac{\sqrt{2}}{2}$$
$$= \frac{\sqrt{2}}{2} + \frac{\sqrt{2}}{2}$$

101. $$\sqrt{5} - \frac{1}{\sqrt{5}} = \sqrt{5} - \frac{1}{\sqrt{5}} \cdot \frac{\sqrt{5}}{\sqrt{5}}$$
$$= \sqrt{5} - \frac{\sqrt{5}}{5}$$
$$= \frac{5\sqrt{5}-\sqrt{5}}{5}$$
$$= \frac{4\sqrt{5}}{5}$$

103. $$\sqrt{\frac{1}{6}} + \sqrt{24} = \frac{1}{\sqrt{6}} + 2\sqrt{6}$$
$$= \frac{1}{\sqrt{6}} \cdot \frac{\sqrt{6}}{\sqrt{6}} + 2\sqrt{6}$$
$$= \frac{\sqrt{6}}{6} + 2\sqrt{6}$$
$$= \frac{\sqrt{6}}{6} + \frac{(2\sqrt{6})6}{6}$$
$$= \frac{\sqrt{6}}{6} + \frac{12\sqrt{6}}{6}$$
$$= \frac{13\sqrt{6}}{6}$$

105. $$3\sqrt{2} - \frac{2}{\sqrt{8}} + \sqrt{50}$$
$$= 3\sqrt{2} - \frac{2}{\sqrt{8}} \cdot \frac{\sqrt{8}}{\sqrt{8}} + \sqrt{25}\sqrt{2}$$
$$= 3\sqrt{2} - \frac{2\sqrt{4}\sqrt{2}}{\sqrt{64}} + 5\sqrt{2}$$
$$= \frac{3\sqrt{2}}{1} - \frac{4\sqrt{2}}{8} + \frac{5\sqrt{2}}{1}$$
$$= \frac{6\sqrt{2}}{2} - \frac{\sqrt{2}}{2} + \frac{10\sqrt{2}}{2}$$
$$= \frac{(6-1+10)\sqrt{2}}{2}$$
$$= \frac{15\sqrt{2}}{2}$$

107. $$\sqrt{\frac{1}{2}} + 3\sqrt{2} + \sqrt{18}$$
$$= \frac{\sqrt{1}}{\sqrt{2}} \cdot \frac{\sqrt{2}}{\sqrt{2}} + 3\sqrt{2} + \sqrt{9} \cdot \sqrt{2}$$
$$= \frac{\sqrt{2}}{\sqrt{4}} + 3\sqrt{2} + 3\sqrt{2}$$
$$= \frac{\sqrt{2}}{2} + 6\sqrt{2}$$
$$= \frac{\sqrt{2}}{2} + \frac{12\sqrt{2}}{2}$$

$= \frac{13\sqrt{2}}{2}$

109. $\frac{2}{\sqrt{50}} - 3\sqrt{50} - \frac{1}{\sqrt{8}}$

$= \frac{2}{5\sqrt{2}} - 3(5\sqrt{2}) - \frac{1}{2\sqrt{2}}$

$= \frac{2}{5\sqrt{2}} \cdot \frac{\sqrt{2}}{\sqrt{2}} - 15\sqrt{2} - \frac{1}{2\sqrt{2}} \cdot \frac{\sqrt{2}}{\sqrt{2}}$

$= \frac{2\sqrt{2}}{10} - 15\sqrt{2} - \frac{\sqrt{2}}{4}$

$= \frac{\sqrt{2}}{5} - 15\sqrt{2} - \frac{\sqrt{2}}{4}$

$= \frac{4\sqrt{2}}{20} - \frac{300\sqrt{2}}{20} - \frac{5\sqrt{2}}{20}$

$= \frac{-301\sqrt{2}}{20}$

111. $\sqrt{\frac{3}{8}} + \sqrt{\frac{3}{2}} = \frac{\sqrt{3}}{\sqrt{8}} + \frac{\sqrt{3}}{\sqrt{2}}$

$= \frac{\sqrt{3}}{2\sqrt{2}} + \frac{\sqrt{3}}{\sqrt{2}}$

$= \frac{\sqrt{3}}{2\sqrt{2}} \cdot \frac{\sqrt{2}}{\sqrt{2}} + \frac{\sqrt{3}}{\sqrt{2}} \cdot \frac{2\sqrt{2}}{2\sqrt{2}}$

$= \frac{\sqrt{6}}{4} + \frac{2\sqrt{6}}{4}$

$= \frac{3\sqrt{6}}{4}$

113. $-2\sqrt{\frac{x}{y}} + 3\sqrt{\frac{y}{x}} = -2\frac{\sqrt{x}}{\sqrt{y}} + 3\frac{\sqrt{y}}{\sqrt{x}}$

$= -2\frac{\sqrt{x}}{\sqrt{y}} \cdot \frac{\sqrt{y}}{\sqrt{y}} + 3\frac{\sqrt{y}}{\sqrt{x}} \cdot \frac{\sqrt{x}}{\sqrt{x}}$

$= -2\frac{\sqrt{xy}}{y} + 3\frac{\sqrt{xy}}{x}$

$= \left(-\frac{2}{y} + \frac{3}{x}\right)\sqrt{xy}$

115. $\frac{3}{\sqrt{a}} - \sqrt{\frac{9}{a}} + \sqrt{a} = \frac{3}{\sqrt{a}} - \frac{\sqrt{9}}{\sqrt{a}} + \sqrt{a}$

$= \frac{3}{\sqrt{a}}\left(\frac{\sqrt{a}}{\sqrt{a}}\right) - \frac{\sqrt{9}}{\sqrt{a}}\left(\frac{\sqrt{a}}{\sqrt{a}}\right) + \sqrt{a}$

$= \frac{3\sqrt{a}}{a} - \frac{3\sqrt{a}}{a} + \sqrt{a}$

$= \sqrt{a}$

117. $\frac{\sqrt{(a+b)^4}}{\sqrt[3]{a+b}} = \frac{(a+b)^{4/2}}{(a+b)^{1/3}}$

$= (a+b)^{6/3-1/3}$

$= (a+b)^{5/3} = \sqrt[3]{(a+b)^5}$

119. $\frac{\sqrt[5]{(a+2b)^4}}{\sqrt[3]{(a+2b)^2}} = \frac{(a+2b)^{4/5}}{(a+2b)^{2/3}}$

$= (a+2b)^{4/5-2/3}$

$= (a+2b)^{2/15} = \sqrt[15]{(a+2b)^2}$

121. $\frac{\sqrt[3]{r^2s^4}}{\sqrt{rs}} = \frac{(r^2s^4)^{1/3}}{(rs)^{1/2}}$

$= \frac{r^{2/3}s^{4/3}}{r^{1/2}s^{1/2}}$

$= r^{2/3-1/2}s^{4/3-1/2}$

$= r^{1/6}s^{5/6}$

$= (rs^5)^{1/6}$

$= \sqrt[6]{rs^5}$

123. $\frac{\sqrt[5]{x^4y^6}}{\sqrt[3]{(xy)^2}} = \frac{(x^4y^6)^{1/5}}{(xy)^{2/3}}$

$= \frac{x^{4/5}y^{6/5}}{x^{2/3}y^{2/3}}$

$= x^{4/5-2/3}y^{6/5-2/3}$

$= x^{2/15}y^{8/15}$

$= (x^2y^8)^{1/15}$

$= \sqrt[15]{x^2y^8}$

125. $d = \sqrt{\frac{72}{I}}$

$d = \sqrt{\frac{72}{5.3}} \approx 3.69$

The person is about 3.69 m from the light source.

127. $r = \sqrt[3]{\frac{3V}{4\pi}}$

$r = \sqrt[3]{\frac{3(7238.23)}{4\pi}} = 12$

The radius of the tank is 12 inches.

129. $N(t) = \frac{6.21}{\sqrt[4]{t}}$

a. $t = 1960 - 1959 = 1$

$N(1) = \frac{6.21}{\sqrt[4]{1}} = 6.21$

The number of farms in 1960 was 6.21 million.

b. $t = 2007 - 1959 = 48$

$N(48) = \frac{6.21}{\sqrt[4]{48}} \approx 2.36$

The number of farms in 2007 will be about 2.36 million.

131. $\frac{2}{\sqrt{2}} = \frac{2}{\sqrt{2}} \cdot \frac{\sqrt{2}}{\sqrt{2}} = \frac{2\sqrt{2}}{\sqrt{4}} = \frac{2\sqrt{2}}{2} = \sqrt{2}$

$\frac{3}{\sqrt{3}} = \frac{3}{\sqrt{3}} \cdot \frac{\sqrt{3}}{\sqrt{3}} = \frac{3\sqrt{3}}{\sqrt{9}} = \frac{3\sqrt{3}}{3} = \sqrt{3}$

Since $3 > 2$, then $\sqrt{3} > \sqrt{2}$ and we conclude that $\frac{3}{\sqrt{3}} > \frac{2}{\sqrt{2}}$.

133.
$$\begin{aligned}\frac{1}{\sqrt{3}+2} &= \frac{1}{\sqrt{3}+2} \cdot \frac{\sqrt{3}-2}{\sqrt{3}-2}\\ &= \frac{\sqrt{3}-2}{\left(\sqrt{3}\right)^2 - 2^2}\\ &= \frac{\sqrt{3}-2}{3-4}\\ &= \frac{\sqrt{3}-2}{-1}\\ &= -\sqrt{3}+2\\ &= 2-\sqrt{3}\end{aligned}$$

$2 + \sqrt{3} > 2 - \sqrt{3}$

Therefore, $2 + \sqrt{3} > \frac{1}{\sqrt{3}+2}$.

135. $f(x) = x^{a/2}$, $g(x) = x^{b/3}$

a. $x^{4/2} = x^2$

$x^{12/2} = x^6$

$x^{8/2} = x^4$

Therefore $x^{a/2}$ is a perfect square when $a = 4, 8, 12$.

b. $x^{9/3} = x^3$

$x^{18/3} = x^6$

$x^{27/3} = x^9$

Therefore, $x^{b/3}$ is a perfect cube when $b = 9, 18, 27$.

137.
$$\begin{aligned}\frac{3}{\sqrt{2a-3b}} &= \frac{3}{\sqrt{2a-3b}} \cdot \frac{\sqrt{2a-3b}}{\sqrt{2a-3b}}\\ &= \frac{3\sqrt{2a-3b}}{2a-3b}\end{aligned}$$

139.
$$\begin{aligned}\frac{5-\sqrt{5}}{6} &= \frac{5-\sqrt{5}}{6} \cdot \frac{5+\sqrt{5}}{5+\sqrt{5}}\\ &= \frac{25-5}{30+6\sqrt{5}}\\ &= \frac{20}{2(15+3\sqrt{5})}\\ &= \frac{10}{15+3\sqrt{5}}\end{aligned}$$

141.
$$\begin{aligned}\frac{\sqrt{x+h}-\sqrt{x}}{h} &= \frac{\sqrt{x+h}-\sqrt{x}}{h} \cdot \frac{\sqrt{x+h}+\sqrt{x}}{\sqrt{x+h}+\sqrt{x}}\\ &= \frac{x+h-x}{h\left(\sqrt{x+h}+\sqrt{x}\right)}\\ &= \frac{h}{h\left(\sqrt{x+h}+\sqrt{x}\right)}\\ &= \frac{1}{\sqrt{x+h}+\sqrt{x}}\end{aligned}$$

144.
$$\begin{aligned}A &= \frac{1}{2}h(b_1 + b_2)\\ 2A &= h(b_1 + b_2)\\ \frac{2A}{h} &= b_1 + b_2\\ \frac{2A}{h} - b_1 &= b_2\\ b_2 &= \frac{2A}{h} - b_1\end{aligned}$$

145. Let r be the rate of the slower car and $r + 10$ be the rate of the faster.
Distance the first traveled plus distance the second traveled is 270 miles.

$$3r + 3(r + 10) = 270$$
$$3r + 3r + 30 = 270$$
$$6r + 30 = 270$$
$$6r = 240$$
$$r = 40$$

The rate of the slower car is 40 mph and the rate of the faster car is $r + 10 = 50$ mph.

146.
$$\begin{array}{r} 4x^2+9x-2 \\ x-2 \\ \hline -8x^2-18x+4 \\ 4x^3+9x^2-2x \quad \\ \hline 4x^3+x^2-20x+4 \end{array}$$

147.
$$\frac{x}{2}-\frac{4}{x}=-\frac{7}{2}$$
$$2x\left(\frac{x}{2}-\frac{4}{x}\right)=2x\left(-\frac{7}{2}\right)$$
$$x^2-8=-7x$$
$$x^2+7x-8=0$$
$$(x+8)(x-1)=0$$
$$x=-8 \text{ or } 1$$

Exercise Set 11.6

1. a. Answers will vary.

b.
$$\sqrt{2x+26}-2=4$$
$$\sqrt{2x+26}-2+2=4+2$$
$$\sqrt{2x+26}=6$$
$$\left(\sqrt{2x+26}\right)^2=6^2$$
$$2x+26=36$$
$$2x=10$$
$$x=\frac{10}{2}$$
$$x=5$$

3. 0 is the only solution to the equation. For all other values, the left side of the equation is negative and the right side is positive.

5. Answers will vary. Possible answer:
The equation has no solution since the left side is a positive number whereas the right side is 0. A positive number is never equal to 0.
Also, the equation can be written as $\sqrt{x-3}=-3$ for which the left side is positive and the right side is negative. It is impossible for $\sqrt{x-3}$ to equal a negative number.

7. One solution, $x = 9$

9.
$$\sqrt{x}=4$$
$$\left(\sqrt{x}\right)^2=4^2$$
$$x=16$$

11. $\sqrt{x}=-9$ No solution.

13.
$$\sqrt[3]{x}=-4$$
$$\left(\sqrt[3]{x}\right)^3=(-4)^3$$
$$x=-64$$

15.
$$-\sqrt{2x+4}=-6$$
$$\left(-\sqrt{2x+4}\right)^2=(-6)^2$$
$$2x+4=36$$
$$2x=32$$
$$x=16$$

Check:
$$-\sqrt{2x+4}=-6$$
$$-\sqrt{2(16)+4}=-6$$
$$-\sqrt{32+4}=-6$$
$$-\sqrt{36}=-6$$
$$-6=-6 \quad \text{True}$$

17.
$$\sqrt[3]{3x}+4=7$$
$$\sqrt[3]{3x}=3$$
$$\left(\sqrt[3]{3x}\right)^3=(3)^3$$
$$3x=27$$
$$x=9$$

19.
$$\sqrt[3]{2x+29}=3$$
$$\left(\sqrt[3]{2x+29}\right)^3=3^3$$
$$2x+29=27$$
$$2x=-2$$
$$x=-1$$

21. $\sqrt[4]{x} = 2$

$\left(\sqrt[4]{x}\right)^4 = 2^4$

$x = 16$

23. $\sqrt[4]{x+5} = 3$

$\left(\sqrt[4]{x+5}\right)^4 = 3^4$

$x + 5 = 81$

$x = 76$

25. $\sqrt[4]{2x+1} + 5 = 1$

$\sqrt[4]{2x+1} = -4 \Rightarrow$ no solution

27. $\sqrt{x+8} = \sqrt{x-8}$

$\left(\sqrt{x+8}\right)^2 = \left(\sqrt{x-8}\right)^2$

$x + 8 = x - 8$

$8 = -8 \Rightarrow$ no solution

29. $\sqrt[3]{6t+1} = \sqrt[3]{2t+5}$

$\left(\sqrt[3]{6t+1}\right)^3 = \left(\sqrt[3]{2t+5}\right)^3$

$6t + 1 = 2t + 5$

$4t = 4$

$t = 1$

31. $\sqrt[4]{x+8} = \sqrt[4]{2x}$

$\left(\sqrt[4]{x+8}\right)^4 = \left(\sqrt[4]{2x}\right)^4$

$x + 8 = 2x$

$x = 8$

33. $\sqrt[4]{3x+1} - 4 = 0$

$\sqrt[4]{3x+1} = 4$

$\left(\sqrt[4]{3x+1}\right)^4 = (4)^4$

$3x + 1 = 16$

$3x = 15$

$x = 5$

35. $\sqrt{m^2 + 6m - 4} = m$

$\left(\sqrt{m^2 + 6m - 4}\right)^2 = (m)^2$

$m^2 + 6m - 4 = m^2$

$6m - 4 = 0$

$6m = 4$

$m = \frac{2}{3}$

37. $\sqrt{6c+1} - 11 = 0$

$\sqrt{6c+1} = 11$

$\left(\sqrt{6c+1}\right)^2 = 11^2$

$6c + 1 = 121$

$6c = 120$

$c = 20$

39. $\sqrt{z^2 + 3} = z + 1$

$\left(\sqrt{z^2+3}\right)^2 = (z+1)^2$

$z^2 + 3 = z^2 + 2z + 1$

$3 = 2z + 1$

$2 = 2z$

$1 = z$

40. $\sqrt{x} + 2x = 1$

$\sqrt{x} = 1 - 2x$

$\left(\sqrt{x}\right)^2 = (1 - 2x)^2$

$x = 4x^2 - 4x + 1$

$0 = 4x^2 - 5x + 1$

$0 = (x - 1)(4x - 1)$

$x - 1 = 0$ or $4x - 1 = 0$

$x = 1$ $\quad 4x = 1$

$x = \frac{1}{4}$

(see next page for check)

Check:	Check:
$-\sqrt{x} = 2x - 1$	$-\sqrt{x} = 2x - 1$
$-\sqrt{1} = 2(1) - 1$	$-\sqrt{\frac{1}{4}} = 2\left(\frac{1}{4}\right) - 1$
$-1 = 1$	$-\frac{1}{2} = \frac{1}{2} - 1$
False	$-\frac{1}{2} = \frac{1}{2} - \frac{2}{2}$
	$-\frac{1}{2} = -\frac{1}{2}$ True

Only $\frac{1}{4}$ is a solution. 1 is an extraneous root.

41. $\sqrt{2y+5} = y-5$

$\left(\sqrt{2y+5}\right)^2 = (y-5)^2$

$2y+5 = y^2-10y+25$

$0 = y^2-12y+20$

$0 = (y-10)(y-2)$

$y = 10$ or $y = 2$

Check:

$\sqrt{2y+5} = y-5$

$\sqrt{2(10)+5} = (10)-5$

$\sqrt{25} = 5$

$5 = 5$ True

Check:

$\sqrt{2y+5} = y-5$

$\sqrt{2(2)+5} = (2)-5$

$\sqrt{9} = -3$

$3 = -3$ False

This check shows that 10 is the only solution to this equation.

43. $\sqrt{5x+6} = 2x-6$

$\left(\sqrt{5x+6}\right)^2 = (2x-6)^2$

$5x+6 = 4x^2-24x+36$

$0 = 4x^2-29x+30$

$0 = (4x-5)(x-6)$

$x = \frac{5}{4}$ or $y = 6$

Check:

$\sqrt{5x+6} = 2x-6$

$\sqrt{5\left(\frac{5}{4}\right)+6} = 2\left(\frac{5}{4}\right)-6$

$\sqrt{\frac{49}{4}} = -\frac{14}{4}$

$\frac{7}{2} = -\frac{7}{2}$ False

Check:

$\sqrt{5x+6} = 2x-6$

$\sqrt{5(6)+6} = 2(6)-6$

$\sqrt{36} = 12-6$

$6 = 6$ True

This check shows that 6 is the only solution to this equation.

45. $(2a+9)^{1/2} - a + 3 = 0$

$(2a+9)^{1/2} = a-3$

$[(2a+9)^{1/2}]^2 = (a-3)^2$

$2a+9 = a^2-6a+9$

$0 = a^2-8a$

$0 = a(a-8)$

$a = 0$ or $a = 8$

Check:

$(2a+9)^{1/2} - a + 3 = 0$

$(2\cdot 0+9)^{1/2} - 0 + 3 = 0$

$3+3 = 0$

$6 = 0$ False

Check: $(2a+9)^{1/2} - a + 3 = 0$

$(2\cdot 8+9)^{1/2} - 8 + 3 = 0$

$5-8+3 = 0$

$0 = 0$ True

Only 8 is a solution. 0 is an extraneous solution.

47. $(2x^2+4x+6)^{1/2} = \sqrt{2x^2+6}$

$[(2x^2+4x+6)^{1/2}]^2 = \left(\sqrt{2x^2+6}\right)^2$

$2x^2+4x+6 = 2x^2+6$

$4x = 0$

$x = 0$

49. $(r+2)^{1/3} = (3r+8)^{1/3}$

$[(r+2)^{1/3}]^3 = [(3r+8)^{1/3}]^3$

$r+2 = 3r+8$

$2 = 2r+8$

$-6 = 2r$

$-3 = r$

(see next page for check)

Check: $(r+2)^{1/3} = (3r+8)^{1/3}$

$(-3+2)^{1/3} = [3(-3)+8]^{1/3}$

$(-1)^{1/3} = (-1)^{1/3}$

$-1 = -1$ True

51. $(5x+8)^{1/4} = (9x+2)^{1/4}$

$[(5x+8)^{1/4}]^4 = [(9x+2)^{1/4}]^4$

$5x+8 = 9x+2$

$8 = 4x+2$

$6 = 4x$

$\frac{3}{2} = x$

Check:

$$(5x+8)^{1/4} = (9x+2)^{1/4}$$
$$\left(5\cdot\frac{3}{2}+8\right)^{1/4} = \left(9\cdot\frac{3}{2}+2\right)^{1/4}$$
$$\left(\frac{15}{2}+8\right)^{1/4} = \left(\frac{27}{2}+2\right)^{1/4}$$
$$\left(\frac{31}{2}\right)^{1/4} = \left(\frac{31}{2}\right)^{1/4} \quad \text{True}$$

53.
$$\sqrt[4]{x+5} = -3$$
$$\left(\sqrt[4]{x+5}\right)^4 = (-3)^4$$
$$x+5 = 81$$
$$x = 76$$

Check:
$$\sqrt[4]{x+5} = -3$$
$$\sqrt[4]{76+5} = -3$$
$$\sqrt[4]{81} = -3$$
$$3 = -3 \quad \text{False}$$

Thus, 76 is not a solution to this equation and we conclude that there is no solution.

55.
$$\sqrt{3x-5} = \sqrt{2x}-1$$
$$\left(\sqrt{3x-5}\right)^2 = \left(\sqrt{2x}-1\right)^2$$
$$3x-5 = 2x-2\sqrt{2x}+1$$
$$x-6 = -2\sqrt{2x}$$
$$(x-6)^2 = \left(-2\sqrt{2x}\right)^2$$
$$x^2-12x+36 = 4(2x)$$
$$x^2-20x+36 = 0$$
$$(x-18)(x-2) = 0$$
$$x = 18 \text{ or } x = 2$$

Upon checking these values, only $x = 2$ satisfies the equation. The solution is $x = 2$.

57.
$$2\sqrt{b}-1 = \sqrt{b+16}$$
$$\left(2\sqrt{b}-1\right)^2 = \left(\sqrt{b+16}\right)^2$$
$$\left(2\sqrt{b}\right)^2 - 2\cdot1\cdot2\sqrt{b}+1^2 = b+16$$
$$4b-4\sqrt{b}+1 = b+16$$
$$3b-15 = 4\sqrt{b}$$
$$(3b-15)^2 = \left(4\sqrt{b}\right)^2$$
$$9b^2-90b+225 = 16b$$
$$9b^2-106b+225 = 0$$
$$(9b-25)(b-9) = 0$$
$$b = \frac{25}{9} \text{ or } b = 9$$

Upon checking these values, only $x = 9$

59.
$$\sqrt{x+3} = \sqrt{x}-3$$
$$\left(\sqrt{x+3}\right)^2 = \left(\sqrt{x}-3\right)^2$$
$$x+3 = x-6\sqrt{x}+9$$
$$-6 = -6\sqrt{x}$$
$$1 = \sqrt{x}$$
$$(1)^2 = \left(\sqrt{x}\right)^2$$
$$1 = x$$

Check:
$$\sqrt{x+3} = \sqrt{x}-3$$
$$\sqrt{1+3} = \sqrt{1}-3$$
$$\sqrt{4} = 1-3$$
$$2 = -2 \quad \text{False}$$

Thus, 1 is not a solution to this equation and we conclude that there is no solution.

61.
$$\sqrt{x+7} = 6-\sqrt{x-5}$$
$$\left(\sqrt{x+7}\right)^2 = \left(6-\sqrt{x-5}\right)^2$$
$$x+7 = 36-12\sqrt{x-5}+x-5$$
$$12\sqrt{x-5} = 24$$
$$\sqrt{x-5} = 2$$
$$x-5 = 4 \Rightarrow x = 9$$

Check:
$$\sqrt{x+7} = 6-\sqrt{x-5}$$
$$\sqrt{9+7} = 6-\sqrt{9-5}$$
$$\sqrt{16} = 6-\sqrt{4}$$
$$4 = 6-2$$
$$4 = 4 \quad \text{True}$$

63.
$$\sqrt{4x-3} = 2+\sqrt{2x-5}$$
$$\left(\sqrt{4x-3}\right)^2 = \left(2+\sqrt{2x-5}\right)^2$$
$$4x-3 = 4+4\sqrt{2x-5}+2x-5$$
$$2x-2 = 4\sqrt{2x-5}$$
$$x-1 = 2\sqrt{2x-5}$$
$$(x-1)^2 = \left(2\sqrt{2x-5}\right)^2$$
$$x^2-2x+1 = 4(2x-5)$$
$$x^2-2x+1 = 8x-20$$
$$x^2-10x+21 = 0$$
$$(x-7)(x-3) = 0$$
$$x-7 = 0 \text{ or } x-3 = 0$$
$$x = 7 \qquad x = 3$$

Check:
$$\sqrt{4x-3}=2+\sqrt{2x-5}$$
$$\sqrt{4(7)-3}=2+\sqrt{2(7)-5}$$
$$\sqrt{25}=2+\sqrt{9}$$
$$5=2+3$$
$$5=5 \text{ True}$$

Check:
$$\sqrt{4x-3}=2+\sqrt{2x-5}$$
$$\sqrt{4(3)-3}=2+\sqrt{2(3)-5}$$
$$\sqrt{9}=2+\sqrt{1}$$
$$3=3 \text{ True}$$

Both 7 and 3 are solutions.

65.
$$\sqrt{y+1}=\sqrt{y+5}-2$$
$$\left(\sqrt{y+1}\right)^2=\left(\sqrt{y+5}-2\right)^2$$
$$y+1=\left(\sqrt{y+5}\right)^2-4\sqrt{y+5}+4$$
$$y+1=y+5-4\sqrt{y+5}+4$$
$$4\sqrt{y+5}=8$$
$$\sqrt{y+5}=2$$
$$\left(\sqrt{y+5}\right)^2=2^2$$
$$y+5=4 \Rightarrow y=-1$$

67.
$$f(x)=g(x)$$
$$\sqrt{x+5}=\sqrt{2x-2}$$
$$\left(\sqrt{x+5}\right)^2=\left(\sqrt{2x-2}\right)^2$$
$$x+5=2x-2$$
$$7=x$$

69.
$$f(x)=g(x)$$
$$\sqrt[3]{5x-17}=\sqrt[3]{6x-21}$$
$$\left(\sqrt[3]{5x-17}\right)^3=\left(\sqrt[3]{6x-21}\right)^3$$
$$5x-17=6x-21$$
$$x=4$$

71.
$$f(x)=g(x)$$
$$2(8x+24)^{1/3}=4(2x-2)^{1/3}$$
$$[2(8x+24)^{1/3}]^3=[4(2x-2)^{1/3}]^3$$
$$8(8x+24)=64(2x-2)$$
$$64x+192=128x-128$$
$$64x=320$$
$$x=5$$

73.
$$p=\sqrt{2v}$$
$$p^2=\left(\sqrt{2v}\right)^2$$
$$p^2=2v$$
$$\frac{p^2}{2}=v$$

75.
$$v=\sqrt{2gh}$$
$$v^2=\left(\sqrt{2gh}\right)^2$$
$$v^2=2gh$$
$$g=\frac{v^2}{2h}$$

77.
$$v=\sqrt{\frac{FR}{M}}$$
$$v^2=\left(\sqrt{\frac{FR}{M}}\right)^2$$
$$v^2=\frac{FR}{M}$$
$$Mv^2=FR$$
$$F=\frac{Mv^2}{R}$$

79.
$$x=\sqrt{\frac{m}{k}}V_0$$
$$x^2=\left(\sqrt{\frac{m}{k}}V_0\right)^2$$
$$x^2=\frac{mV_0^2}{k}$$
$$x^2k=mV_0^2$$
$$m=\frac{x^2k}{V_0^2}$$

81.
$$r=\sqrt{\frac{A}{\pi}}$$
$$r^2=\left(\sqrt{\frac{A}{\pi}}\right)^2$$
$$r^2=\frac{A}{\pi}$$
$$\pi r^2=A \text{ or } A=\pi r^2$$

83.
$$a^2+b^2=c^2$$
$$\left(\sqrt{6}\right)^2+9^2=x^2$$
$$6+81=x^2$$
$$87=x^2 \Rightarrow x=\sqrt{87}$$

85.
$$a^2+b^2=c^2$$
$$x^2+5^2=\left(\sqrt{57}\right)^2$$
$$x^2+25=57$$
$$x^2=32$$
$$x=\sqrt{32} \Rightarrow x=4\sqrt{2}$$

87.
$$\sqrt{x+5}-\sqrt{x}=\sqrt{x-3}$$
$$\left(\sqrt{x+5}-\sqrt{x}\right)^2=\left(\sqrt{x-3}\right)^2$$
$$x+5-2\sqrt{x(x+5)}+x=x-3$$
$$x+8=2\sqrt{x^2+5x}$$
$$(x+8)^2=\left(2\sqrt{x^2+5x}\right)^2$$
$$x^2+16x+64=4\left(x^2+5x\right)$$
$$x^2+16x+64=4x^2+20x$$
$$3x^2+4x-64=0$$
$$(3x+16)(x-4)=0 \Rightarrow x=-\frac{16}{3} \text{ or } x=4$$

Upon checking, only $x=4$ satisfies the equation.

89.
$$\sqrt{4y+6}+\sqrt{y+5}=\sqrt{y+1}$$
$$\left(\sqrt{4y+6}+\sqrt{y+5}\right)^2=\left(\sqrt{y+1}\right)^2$$
$$4y+6+2\sqrt{(4y+6)(y+5)}+y+5=y+1$$
$$2\sqrt{(4y+6)(y+5)}=-4y-10$$
$$\left(2\sqrt{4y^2+26y+30}\right)^2=(-4y-10)^2$$
$$4\left(4y^2+26y+30\right)=16y^2+80y+100$$
$$16y^2+104y+120=16y^2+80y+100$$
$$24y=-20 \Rightarrow y=-\frac{5}{6}$$

Upon checking, this value does not satisfy the equation. There is no solution.

91.
$$\sqrt{a+2}-\sqrt{a-3}=\sqrt{a-6}$$
$$\left(\sqrt{a+2}-\sqrt{a-3}\right)^2=\left(\sqrt{a-6}\right)^2$$
$$a+2-2\sqrt{(a+2)(a-3)}+a-3=a-6$$
$$a+5=2\sqrt{(a+2)(a-3)}$$
$$(a+5)^2=\left(2\sqrt{a^2-a-6}\right)^2$$
$$a^2+10a+25=4\left(a^2-a-6\right)$$
$$a^2+10a+25=4a^2-4a-24$$
$$3a^2-14a-49=0$$
$$(3a+7)(a-7)=0 \Rightarrow a=-\frac{7}{3} \text{ or } a=7$$

Upon checking, only $a=7$ satisfies the equation.

93.
$$\sqrt{2t-1}+\sqrt{t-4}=\sqrt{3t+1}$$
$$\left(\sqrt{2t-1}+\sqrt{t-4}\right)^2=\left(\sqrt{3t+1}\right)^2$$
$$2t-1+2\sqrt{(2t-1)(t-4)}+t-4=3t+1$$
$$2\sqrt{(2t-1)(t-4)}=6$$
$$\left(2\sqrt{2t^2-9t+4}\right)^2=(6)^2$$
$$4\left(2t^2-9t+4\right)=36$$
$$8t^2-36t+16=36$$
$$8t^2-36t-20=0$$
$$4(2t+1)(t-5)=0 \Rightarrow t=-\frac{1}{2} \text{ or } t=5$$

Upon checking, only $t=5$ satisfies the equation.

95. $\sqrt{2-\sqrt{x}}=\sqrt{x}$

$\left(\sqrt{2-\sqrt{x}}\right)^2=\left(\sqrt{x}\right)^2$

$2-\sqrt{x}=x$

$2-x=\sqrt{x}$

$(2-x)^2=\left(\sqrt{x}\right)^2$

$4-4x+x^2=x$

$x^2-5x+4=0$

$(x-4)(x-1)=0 \Rightarrow x=4 \text{ or } x=1$

Upon checking, only $x=1$ satisfies the equation.

97. $\sqrt{2+\sqrt{x+1}}=\sqrt{7-x}$

$\left(\sqrt{2+\sqrt{x+1}}\right)^2=\left(\sqrt{7-x}\right)^2$

$2+\sqrt{x+1}=7-x$

$\sqrt{x+1}=5-x$

$\left(\sqrt{x+1}\right)^2=(5-x)^2$

$x+1=25-10x+x^2$

$x^2-11x+24=0$

$(x-8)(x-3)=0 \Rightarrow x=8 \text{ or } x=3$

Upon checking, only $x=3$ satisfies the equation.

99. $l=\sqrt{20^2+40^2}$

$l=\sqrt{400+1600}$

$=\sqrt{2000}$

≈ 44.7 ft

101. $s=\sqrt{A}$

$s=\sqrt{144}$

$=12$ feet

103. $T=2\pi\sqrt{\frac{l}{32}}$

a. Let $l = 10$

$T=2\pi\sqrt{\frac{l}{32}}$

$=2\pi\sqrt{\frac{10}{32}}$

$=2\pi\sqrt{0.3125}$

$\approx 2\pi(0.55901)$

≈ 3.51 seconds

b. Replace l with $2l$:

$T_D=2\pi\sqrt{\frac{2l}{32}}$

$=2\pi\sqrt{2}\sqrt{\frac{l}{32}}$

$=\sqrt{2}\left(2\pi\sqrt{\frac{l}{32}}\right)$

$=\sqrt{2}T$

c. This part must be solved in two phases. First, we need to find the length of the pendulum:

$T=2\pi\sqrt{\frac{l}{g}}$

$2=2\pi\sqrt{\frac{l}{32}}$

$\frac{1}{\pi}=\sqrt{\frac{l}{32}}$

$\left(\frac{1}{\pi}\right)^2=\frac{l}{32}$

$l=\frac{32}{\pi^2}$

Now, find T using $g=\frac{32}{6}$ and $l=\frac{32}{\pi^2}$

$T=2\pi\sqrt{\frac{l}{g}}$

$=2\pi\sqrt{\frac{\frac{32}{\pi^2}}{\frac{32}{6}}}$

$=2\pi\sqrt{\frac{6}{\pi^2}}$

$=2\pi\frac{\sqrt{6}}{\sqrt{\pi^2}}$

$=2\pi\frac{\sqrt{6}}{\pi}$

$=2\sqrt{6}$

≈ 4.90 seconds

105. $r=\sqrt[4]{\frac{8\mu l}{\pi R}}$

$r^4=\left(\sqrt[4]{\frac{8\mu l}{\pi R}}\right)^4$

$$r^4 = \frac{8\mu l}{\pi R^4}$$
$$\pi R r^4 = 8\mu l$$
$$R = \frac{8\mu l}{\pi r^4}$$

107. $N = 0.2\left(\sqrt{R}\right)^3$

$N = 0.2\left(\sqrt{149.4}\right)^3$

$= 0.2(12.223)^3$

$= 0.2(1826.106)$

≈ 365.2 days

109. $R = \sqrt{F_1^2 + F_2^2}$

$R = \sqrt{60^2 + 80^2}$

$= \sqrt{10{,}000}$

$= 100$ lb

111. $c = \sqrt{gH} = \sqrt{32 \cdot 10} = \sqrt{320} \approx 17.89$ ft / sec

113. The diagonal and the two given sides form a right triangle. Use the Pythagorean formula to solve for the diagonal.

$a^2 + b^2 = c^2$

$25^2 + 32^2 = c^2$

$625 + 1024 = c^2$

$1649 = c^2 \Rightarrow c = \sqrt{1649} \approx 40.6$

The diagonal is about 37.7 inches in length.

115. $x = \dfrac{-b \pm \sqrt{b^2 - 4ac}}{2a}$

$x = \dfrac{-0 \pm \sqrt{0^2 - 4(1)(-4)}}{2(1)} = \dfrac{\pm\sqrt{16}}{2}$

Now, $x = \dfrac{\sqrt{16}}{2} = \dfrac{4}{2} = 2$ or

$x = -\dfrac{\sqrt{16}}{2} = -\dfrac{4}{2} = -2$

117. $x = \dfrac{-b \pm \sqrt{b^2 - 4ac}}{2a}$

$x = \dfrac{-5 \pm \sqrt{5^2 - 4(2)(-12)}}{2(2)}$

$= \dfrac{-5 \pm \sqrt{121}}{4}$

$= \dfrac{-5 \pm 11}{4}$

Now, $x = \dfrac{-5+11}{4} = \dfrac{6}{4} = \dfrac{3}{2}$ or

119. $f(x) = \sqrt{x-5}$

$4 = \sqrt{x-5}$

$4^2 = \left(\sqrt{x-5}\right)^2$

$16 = x - 5$

$21 = x$

121. $f(x) = \sqrt{3x^2 - 11} + 4$

$12 = \sqrt{3x^2 - 11} + 4$

$8 = \sqrt{3x^2 - 11}$

$8^2 = \left(\sqrt{3x^2 - 11}\right)^2$

$64 = 3x^2 - 11$

$75 = 3x^2$

$25 = x^2$

$\pm 5 = x$

123. **a.** $y = \sqrt{4x-12}$, $y = x - 3$

The points of intersection are (3, 0) and (7, 4). The x-values are 3 and 7.

b. $\sqrt{4x-12} = x - 3$

For $x = 3$:

$\sqrt{4 \cdot 3 - 12} = 3 - 3$

$\sqrt{12 - 12} = 0$

$\sqrt{0} = 0$

$0 = 0$ True

For $x = 7$:

$\sqrt{4x - 12} = x - 3$

$\sqrt{4 \cdot 7 - 12} = 7 - 3$

$\sqrt{16} = 4$

$4 = 4$ True

c. $\sqrt{4x-12} = x - 3$

$\left(\sqrt{4x-12}\right)^2 = (x-3)^2$

$4x - 12 = x^2 - 6x + 9$

$0 = x^2 - 10x + 21$

$0 = (x-3)(x-7)$

$x - 3 = 0$ or $x - 7 = 0$

$x = 3$ $\quad$ $x = 7$

125. At $x = 4$, $g(x) = 0$ or $y = 0$.

Therefore, the graph must have an x-intercept at 4.

127. $L_1 = p - 1.96\sqrt{\dfrac{p(1-p)}{n}}$

$L_1 = 0.60 - 1.96\sqrt{\dfrac{0.60(1-0.60)}{36}}$

$\approx 0.60 - 0.16$

≈ 0.44

$$L_2 = p + 1.96\sqrt{\frac{p(1-p)}{n}}$$
$$L_2 = 0.60 + 1.96\sqrt{\frac{0.60(1-0.60)}{36}}$$
$$= 0.60 + 0.16$$
$$\approx 0.76$$

129. $$\sqrt{x^2+9} = (x^2+9)^{1/2}$$
$$\left(\sqrt{x^2+9}\right)^2 = [(x^2+9)^{1/2}]^2$$
$$x^2+9 = x^2+9$$
$$9 = 9$$
All real numbers, x, satisfy this equation.

131. Graph:
$y_1 = \sqrt{x+8}$
$y_2 = \sqrt{3x+5}$

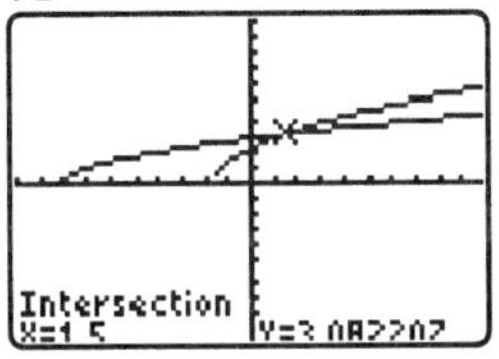

-10, 10, 1, -10, 10

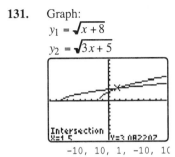

The graphs of the equations intersect when $x = 1.5$.

133. Graph:
$y = \sqrt[3]{5x^2-6} - 4$

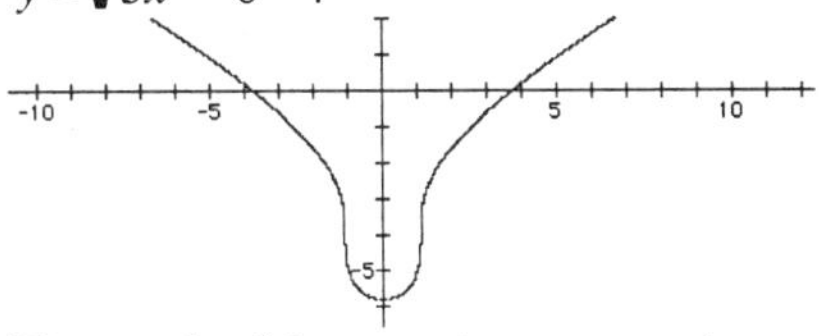

The graph of the equation crosses the x-axis at
$x \approx -3.74$ and $x \approx 3.74$.

135. $$\sqrt{\sqrt{x+25} - \sqrt{x}} = 5$$
$$\left(\sqrt{\sqrt{x+25} - \sqrt{x}}\right)^2 = 5^2$$
$$\sqrt{x+25} - \sqrt{x} = 25$$
$$\sqrt{x+25} = 25 + \sqrt{x}$$
$$\left(\sqrt{x+25}\right)^2 = \left(25+\sqrt{x}\right)^2$$
$$x + 25 = 625 + 50\sqrt{x} + x$$
$$-600 = 50\sqrt{x}$$
$$-12 = \sqrt{x}$$
$$(-12)^2 = \left(\sqrt{x}\right)^2$$
$$144 = x$$

Check: $$\sqrt{\sqrt{x+25} - \sqrt{x}} = 5$$
$$\sqrt{\sqrt{144+25} - \sqrt{144}} = 5$$
$$\sqrt{\sqrt{169} - \sqrt{144}} = 5$$
$$\sqrt{13-12} = 5$$
$$\sqrt{1} = 5$$
$$1 = 5 \text{ False}$$
Thus, 144 is not a solution and we conclude that there is no solution.

137. $$z = \frac{\bar{x} - \mu}{\frac{\sigma}{\sqrt{n}}}$$
$$z\left(\frac{\sigma}{\sqrt{n}}\right) = \bar{x} - \mu$$
$$\left(z\frac{\sigma}{\sqrt{n}}\right)^2 = (\bar{x} - \mu)^2$$
$$\frac{z^2\sigma^2}{n} = (\bar{x} - \mu)^2$$
$$z^2\sigma^2 = n(\bar{x} - \mu)^2$$
$$\frac{z^2\sigma^2}{(\bar{x} - \mu)^2} = n \text{ or } n = \frac{z^2\sigma^2}{(\bar{x} - \mu)^2}$$

140. $$P_1P_2 - P_1P_3 = P_2P_3 \text{ Solve for } P_2.$$
$$P_1P_2 - P_1P_2 - P_1P_3 = P_2P_3 - P_1P_2$$
$$-P_1P_3 = P_2P_3 - P_1P_2$$
$$-P_1P_3 = P_2(P_3 - P_1)$$
$$\frac{-P_1P_3}{P_3 - P_1} = \frac{P_2(P_3 - P_1)}{P_3 - P_1}$$
or $$\frac{P_1P_3}{P_1 - P_3} = P_2 \text{ or } P_2 = \frac{P_1P_3}{P_1 - P_3}$$

141. $\frac{x(x-3)+x(x-4)}{2x-7}=\frac{x^2-3x+x^2-4x}{2x-7}$
$=\frac{2x^2-7x}{2x-7}$
$=\frac{x(2x-7)}{2x-7}$
$=x$

142. $\frac{4a^2-9b^2}{4a^2+12ab+9b^2}\cdot\frac{6a^2b}{8a^2b^2-12ab^3}$
$=\frac{(2a-3b)(2a+3b)}{(2a+3b)(2a+3b)}\cdot\frac{6a^2b}{4ab^2(2a-3b)}$
$=\frac{3a}{2b(2a+3b)}$

143. $(t^2-t-12)\div\frac{t^2-9}{t^2-3t}$
$=(t^2-t-12)\cdot\frac{t^2-3t}{t^2-9}$
$=(t-4)(t+3)\cdot\frac{t(t-3)}{(t+3)(t-3)}$
$=t(t-4)$

144. $\frac{2}{x+3}-\frac{1}{x-3}+\frac{2x}{x^2-9}$
$=\frac{2}{x+3}-\frac{1}{x-3}+\frac{2x}{(x+3)(x-3)}$
$=\frac{2}{x+3}\cdot\frac{x-3}{x-3}-\frac{1}{x-3}\cdot\frac{x+3}{x+3}+\frac{2x}{(x+3)(x-3)}$
$=\frac{2(x-3)}{(x+3)(x-3)}-\frac{x+3}{(x+3)(x-3)}+\frac{2x}{(x+3)(x-3)}$
$=\frac{2x-6}{(x+3)(x-3)}-\frac{x+3}{(x+3)(x-3)}+\frac{2x}{(x+3)(x-3)}$
$=\frac{2x-6-(x+3)+2x}{(x+3)(x-3)}$
$=\frac{2x-6-x-3+2x}{(x+3)(x-3)}$
$=\frac{3x-9}{(x+3)(x-3)}$
$=\frac{3(x-3)}{(x+3)(x-3)}$
$=\frac{3}{x+3}$

145. $2+\frac{3x}{x-1}=\frac{8}{x-1}$
$(x-1)(2)+(x-1)\left(\frac{3x}{x-1}\right)=(x-1)\left(\frac{8}{x-1}\right)$
$2(x-1)+3x=8$
$2x-2+3x=8$
$5x-2=8$
$5x=10$
$x=2$

Exercise Set 11.7

1. $i=\sqrt{-1}$

3. Yes

5. Yes

7. The conjugate of $a+bi$ is $a-bi$.

9. Answers will vary. Possible answers:
a. $\sqrt{2}$
b. 2
c. $\sqrt{-3}$
d. 6
e. Every number we have studied is a complex

11. $5=5+0i$

13. $\sqrt{49}=7=7+0i$

15. $21-\sqrt{-36}=21-\sqrt{36}\sqrt{-1}$
$=21-6i$

17. $\sqrt{-24}=\sqrt{4}\sqrt{-1}\sqrt{6}=2i\sqrt{6}=0+2i\sqrt{6}$

19. $8-\sqrt{-12}=8-\sqrt{12}\sqrt{-1}$
$=8-\sqrt{4}\sqrt{3}\sqrt{-1}$
$=8-2i\sqrt{3}$

21. $1+\sqrt{-98}=1+\sqrt{98}\sqrt{-1}$
$=1+\sqrt{49}\sqrt{2}\sqrt{-1}$
$=1+7i\sqrt{2}$

23. $9-\sqrt{-25}=9-\sqrt{25}\sqrt{-1}$
$=9-5i$

25. $2i-\sqrt{-45}=0+2i-\sqrt{9}\sqrt{5}\sqrt{-1}$
$=0+2i-3i\sqrt{5}$
$=0+\left(2-3\sqrt{5}\right)i$

27. $(19-i)+(2+6i)=19-i+2+6i$
$=21+5i$

29. $(8-3i)+(-8+3i)=8-3i-8+3i$
$=0$

31. $\left(1+\sqrt{-1}\right)+\left(-13-\sqrt{-169}\right)=1+\sqrt{-1}-13-\sqrt{-169}$
$=1+i-13-13i$
$=-12-12i$

33. $\left(\sqrt{3}+\sqrt{2}\right)+\left(3\sqrt{2}-\sqrt{-8}\right)$
$=\sqrt{3}+\sqrt{2}+3\sqrt{2}-\sqrt{-8}$
$=\sqrt{3}+\sqrt{2}+3\sqrt{2}-\sqrt{-4\cdot 2}$
$=\sqrt{3}+\sqrt{2}+3\sqrt{2}-2i\sqrt{2}$
$=\sqrt{3}+4\sqrt{2}-2i\sqrt{2}$
$=\left(\sqrt{3}+4\sqrt{2}\right)-2i\sqrt{2}$

35. $\left(3-\sqrt{-72}\right)+\left(6+\sqrt{-8}\right)$
$=3-\sqrt{-72}+6+\sqrt{-8}$
$=3-\sqrt{36}\sqrt{2}\sqrt{-1}+6+\sqrt{4}\sqrt{2}\sqrt{-1}$
$=3-6i\sqrt{2}+6+2i\sqrt{2}$
$=9-4i\sqrt{2}$

37. $\left(\sqrt{4}-\sqrt{-45}\right)+\left(-\sqrt{81}+\sqrt{-5}\right)$
$=\sqrt{4}-\sqrt{-45}-\sqrt{81}+\sqrt{-5}$
$=\sqrt{4}-\sqrt{9}\sqrt{5}\sqrt{-1}-\sqrt{81}+\sqrt{-1}\sqrt{5}$
$=2-3i\sqrt{5}-9+i\sqrt{5}$
$=-7-2i\sqrt{5}$

39. $2(5-i)=10-2i$

41. $i(2+9i)=2i+9i^2=2i+9(-1)=-9+2i$

43. $\sqrt{-9}(7+11i)=3i(7+11i)$
$=21i+33i^2$
$=21i+33(-1)$
$=-33+21i$

45. $\sqrt{-16}\left(\sqrt{3}-5i\right)=4i\left(\sqrt{3}-5i\right)$
$=4i\sqrt{3}-20i^2$
$=4i\sqrt{3}-20(-1)$
$=4i\sqrt{3}+20$
$=20+4i\sqrt{3}$

47. $\sqrt{-32}\left(\sqrt{2}+\sqrt{-8}\right)$
$=\sqrt{-16}\sqrt{2}\left(\sqrt{2}+\sqrt{-4}\sqrt{2}\right)$
$=4i\sqrt{2}\left(\sqrt{2}+2i\sqrt{2}\right)$
$=4i\cdot 2+8i^2\cdot 2$
$=8i+16i^2$
$=8i+16(-1)$
$=8i-16$ or $-16+8i$

49. $(3+2i)(1+i)$
$=3(1)+3(i)+2i(1)+2i(i)$
$=3+3i+2i+2i^2$
$=3+3i+2i+2(-1)$
$=3+3i+2i-2$
$=1+5i$

51. $(20-3i)(20+3i)=400+60i-60i-9i^2$
$=400+60i-60i-9(-1)$
$=400+9$
$=409$

53. $\left(7+\sqrt{-2}\right)\left(5-\sqrt{-8}\right)$
$=\left(7+i\sqrt{2}\right)\left(5-2i\sqrt{2}\right)$
$=35-14i\sqrt{2}+5i\sqrt{2}-2i^2\cdot 2$
$=35-14i\sqrt{2}+5i\sqrt{2}-2(-1)\cdot 2$
$=35-14i\sqrt{2}+5i\sqrt{2}+4$
$=39-9i\sqrt{2}$

55. $\left(\frac{1}{2}-\frac{1}{3}i\right)\left(\frac{1}{4}+\frac{2}{3}i\right)$

$=\frac{1}{8}+\frac{1}{3}i-\frac{1}{12}i-\frac{2}{9}i^2$

$=\frac{1}{8}+\frac{1}{3}i-\frac{1}{12}i-\frac{2}{9}(-1)$

$=\frac{1}{8}+\frac{1}{3}i-\frac{1}{12}i+\frac{2}{9}$

$=\frac{1}{8}+\frac{2}{9}+\left(\frac{1}{3}-\frac{1}{12}\right)i$

$=\frac{25}{72}+\frac{1}{4}i$

57. $\frac{2}{3i}=\frac{2}{3i}\cdot\frac{-i}{-i}=\frac{-2i}{-3i^2}=\frac{-2i}{-3(-1)}=-\frac{2i}{3}$

59. $\frac{2+3i}{2i}=\frac{2+3i}{2i}\cdot\frac{-i}{-i}$

$=\frac{(2+3i)(-i)}{-2i^2}$

$=\frac{-2i-3i^2}{-2i^2}$

$=\frac{-2i-3(-1)}{-2(-1)}$

$=\frac{-2i+3}{2}$ or $\frac{3-2i}{2}$

61. $\frac{4}{2-i}=\frac{4}{2-i}\cdot\frac{2+i}{2+i}$

$=\frac{4(2+i)}{(2-i)(2+i)}$

$=\frac{8+4i}{4+2i-2i-i^2}$

$=\frac{8+4i}{4+2i-2i-(-1)}$

$=\frac{8+4i}{5}$

63. $\frac{9}{1-2i}=\frac{9}{1-2i}\cdot\frac{1+2i}{1+2i}$

$=\frac{9(1+2i)}{(1-2i)(1+2i)}$

$=\frac{9+18i}{1+2i-2i-4i^2}$

$=\frac{9+18i}{1+2i-2i-4(-1)}$

$=\frac{9+18i}{5}$

65. $\frac{6-3i}{4+2i}=\frac{6-3i}{4+2i}\cdot\frac{4-2i}{4-2i}$

$=\frac{(6-3i)(4-2i)}{16-4i^2}$

$=\frac{24-12i-12i+6i^2}{16-4i^2}$

$=\frac{24-12i-12i-6}{16+4}$

$=\frac{18-24i}{20}$

$=\frac{2(9-12i)}{20}$

$=\frac{9-12i}{10}$

67. $\frac{4}{6-\sqrt{-4}}=\frac{4}{6-\sqrt{4}\sqrt{-1}}$

$=\frac{4}{6-2i}\cdot\frac{6+2i}{6+2i}$

$=\frac{4(6+2i)}{36-4i^2}$

$=\frac{24+8i}{36-4(-1)}$

$=\frac{24+8i}{36+4}$

$=\frac{8(3+i)}{40}$

$=\frac{3+i}{5}$

69. $\frac{\sqrt{2}}{5+\sqrt{-12}}=\frac{\sqrt{2}}{5+\sqrt{4}\sqrt{3}\sqrt{-1}}$

$=\frac{\sqrt{2}}{5+2i\sqrt{3}}\cdot\frac{5-2i\sqrt{3}}{5-2i\sqrt{3}}$

$=\frac{\sqrt{2}\left(5-2i\sqrt{3}\right)}{25-4i^2\sqrt{3}^2}$

$=\frac{5\sqrt{2}-2i\sqrt{6}}{25-4(-1)(3)}$

$=\frac{5\sqrt{2}-2i\sqrt{6}}{25+12}$

$=\frac{5\sqrt{2}-2i\sqrt{6}}{37}$

71. $\frac{\sqrt{10}+\sqrt{-3}}{5-\sqrt{-20}}$

$=\frac{\sqrt{10}+\sqrt{3}\sqrt{-1}}{5-\sqrt{4}\sqrt{5}\sqrt{-1}}$

$=\frac{\sqrt{10}+i\sqrt{3}}{5-2i\sqrt{5}}\cdot\frac{5+2i\sqrt{5}}{5+2i\sqrt{5}}$

$=\frac{(\sqrt{10}+i\sqrt{3})(5+2i\sqrt{5})}{5^2-4i^2\sqrt{5}^2}$

$=\frac{5\sqrt{10}+2i\sqrt{50}+5i\sqrt{3}+2i^2\sqrt{15}}{5^2-4(-1)(5)}$

$$= \frac{5\sqrt{10} + 2i\sqrt{25}\sqrt{2} + 5i\sqrt{3} + 2(-1)\sqrt{15}}{25 + 20}$$
$$= \frac{(5\sqrt{10} - 2\sqrt{15}) + (10\sqrt{2} + 5\sqrt{3})i}{45}$$

73. $\frac{\sqrt{-75}}{\sqrt{-3}} = \frac{\sqrt{25}\sqrt{3}\sqrt{-1}}{\sqrt{3}\sqrt{-1}} = \frac{5i\sqrt{3}}{i\sqrt{3}} = 5$

75. $\frac{\sqrt{-32}}{\sqrt{-18}\sqrt{2}} = \frac{\sqrt{16}\sqrt{2}\sqrt{-1}}{\sqrt{9}\sqrt{2}\sqrt{-1}\sqrt{2}}$
$= \frac{4i\sqrt{2}}{3 \cdot 2i} = \frac{2\sqrt{2}}{3}$

77. $(4 - 2i) + (3 - 5i) = 4 + 3 - 2i - 5i = 7 - 7i$

79. $\left(\sqrt{8} - \sqrt{2}\right) - \left(\sqrt{-12} - \sqrt{-48}\right)$
$= \left(2\sqrt{2} - \sqrt{2}\right) - \left(2i\sqrt{3} - 4i\sqrt{3}\right)$
$= \sqrt{2} - \left(-2i\sqrt{3}\right)$
$= \sqrt{2} + 2i\sqrt{3}$

81. $5.2(4 - 3.2i)$
$= 5.2(4) - 5.2(3.2i)$
$= 20.8 - 16.64i$

83. $(9 + 2i)(3 - 5i)$
$= 27 - 45i + 6i - 10i^2$
$= 27 - 39i - 10(-1)$
$= 27 + 10 - 39i$
$= 37 - 39i$

85. $\frac{5+4i}{2i} = \frac{5+4i}{2i} \cdot \frac{-2i}{-2i}$
$= \frac{(5+4i)(-2i)}{-4i^2}$
$= \frac{-10i - 8i^2}{-4i^2}$
$= \frac{-10i - 8(-1)}{-4(-1)}$
$= \frac{8 - 10i}{4}$ or $\frac{4 - 5i}{2}$

87. $\frac{5}{\sqrt{3} - \sqrt{-4}} = \frac{5}{\sqrt{3} - 2i}$
$= \frac{5}{\sqrt{3} - 2i} \cdot \frac{\sqrt{3} + 2i}{\sqrt{3} + 2i}$
$= \frac{5(\sqrt{3} + 2i)}{(\sqrt{3})^2 - (2i)^2}$
$= \frac{5\sqrt{3} + 10i}{3 - 4i^2}$

89. $\left(5 - \frac{5}{9}i\right) - \left(4 - \frac{3}{5}i\right) = 5 - \frac{5}{9}i - 4 + \frac{3}{5}i$
$= 1 - \frac{5}{9}i + \frac{3}{5}i$
$= 1 - \frac{25}{45}i + \frac{27}{45}i$
$= 1 + \frac{2}{45}i$

91. $\left(\frac{2}{3} - \frac{1}{5}i\right)\left(\frac{3}{5} - \frac{3}{4}i\right)$
$= \left(\frac{2}{3}\right)\left(\frac{3}{5}\right) - \frac{2}{3}\left(\frac{3}{4}i\right) - \left(\frac{1}{5}i\right)\left(\frac{3}{5}\right) + \left(\frac{1}{5}i\right)\left(\frac{3}{4}i\right)$
$= \frac{2}{5} - \frac{1}{2}i - \frac{3}{25}i + \frac{3}{20}i^2$
$= \frac{2}{5} - \frac{1}{2}i - \frac{3}{25}i + \frac{3}{20}(-1)$
$= \left(\frac{2}{5} - \frac{3}{20}\right) + \left(-\frac{1}{2} - \frac{3}{25}\right)i$
$= \left(\frac{8}{20} - \frac{3}{20}\right) + \left(-\frac{25}{50} - \frac{6}{50}\right)i$
$= \frac{5}{20} - \frac{31}{50}i$
$= \frac{1}{4} - \frac{31}{50}i$

93. $\frac{\sqrt{-96}}{\sqrt{-24}} = \frac{\sqrt{96}\sqrt{-1}}{\sqrt{24}\sqrt{-1}}$
$= \frac{i\sqrt{96}}{i\sqrt{24}}$
$= \frac{\sqrt{96}}{\sqrt{24}}$
$= \sqrt{\frac{96}{24}}$
$= \sqrt{4}$
$= 2$

95. $(5.23 - 6.41i) - (9.56 + 4.5i)$
$= 5.23 - 6.41i - 9.56 - 4.5i$
$= -4.33 - 10.91i$

97. $i^{10} = i^8 i^2$
$= (i^4)^2 i^2 = 1^2 \cdot i^2 = 1(i^2) = i^2 = -1$

99. $i^{200} = (i^4)^{50} = 1^{50} = 1$

101. $i^{93} = i^{92} \cdot i^1$
$= (i^4)^{23} i = 1^{23} \cdot i = 1(i) = i$

103. $i^{807} = i^{804} \cdot i^3$
$= (i^4)^{201} \cdot i^3$
$= 1^{201} \cdot i^3$
$= 1 \cdot (i^3)$
$= i^3$
$= -i$

105. **a.** The additive inverse of $2+3i$ is its opposite, $-2-3i$. Note that $(2+3i)+(-2-3i)=0$.

b. The multiplicative inverse of $2+3i$ is its reciprocal, $\frac{1}{2+3i}$. To simplify this, multiply the numerator and denominator by the conjugate of the denominator.

$$\frac{1}{2+3i} = \frac{1}{2+3i} \cdot \frac{2-3i}{2-3i}$$
$$= \frac{2-3i}{(2+3i)(2-3i)}$$
$$= \frac{2-3i}{4-6i+6i-9i^2}$$
$$= \frac{2-3i}{4+9} \text{ or } \frac{2-3i}{13}$$

107. True. The product of two pure imaginary numbers is always a real number. Consider two pure imaginary numbers *bi* and *di* where *b*, *d* are non-zero real numbers whose product
$(bi)(di) = bdi^2$
$= bd(-1)$
$= -bd$
which is a real number.

109. False. The product of two complex numbers is not always a real number. For example,
$(1+i)(1+i) = 1+i+i+i^2$
$= 1+2i+(-1)$
$= 0+2i$
which is not a real number.

111. Even values of n will result in i^n being a real number. $i^2 = -1$, $i^{2n} = (i^2)^n = (-1)^n$

113. $f(x) = x^2$
$f(i) = i^2 = -1$

115. $f(x) = x^4 - 2x$
$f(2i) = (2i)^4 - 2(2i)$
$= 2^4 i^4 - 4i$
$= 16(1) - 4i$
$= 16 - 4i$

117. $f(x) = x^2 + x$
$f(3+i) = (3+i)^2 + (3+i)$
$= 9+6i+i^2+3+i$
$= 9+6i-1+3+i$
$= 11+7i$

119. $x^2 - 2x + 5$
$= (1-2i)^2 - 2(1-2i) + 5$
$= 1^2 - 2(1)(2i) + (2i)^2 - 2 + 4i + 5$
$= 1 - 4i - 4 - 2 + 4i + 5$
$= 0 + 0i$
$= 0$

121. $x^2 + 2x + 7$
$= (-1+i\sqrt{5})^2 + 2(-1+i\sqrt{5}) + 7$
$= (-1)^2 - 2(1)(i\sqrt{5}) + (i\sqrt{5})^2 - 2 + 2i\sqrt{5} + 7$
$= 1 - 2i\sqrt{5} - 5 - 2 + 2i\sqrt{5} + 7$
$= 1 + 0i$
$= 1$

123. $x^2 - 4x + 5 = 0$
$(2-i)^2 - 4(2-i) + 5 = 0$
$2^2 - 2(2)(i) + (i)^2 - 8 + 4i + 5 = 0$
$4 - 4i - 1 - 8 + 4i + 5 = 0$
$0 + 0i = 0$
$0 = 0$ True

$2 - i$ is a solution.

125.
$$x^2-6x+11=0$$
$$(-3+i\sqrt{3})^2-6(-3+i\sqrt{3})+11=0$$
$$(-3)^2-2(3)(i\sqrt{3})+(i\sqrt{3})^2+18-6i\sqrt{3}+11=0$$
$$9-6i\sqrt{3}-3+18-6i\sqrt{3}+11=0$$
$$35-12i\sqrt{3}=0 \quad \text{False}$$
$-3+i\sqrt{3}$ is not a solution.

127.
$$Z=\frac{V}{I}$$
$$Z=\frac{1.8+0.5i}{0.6i}$$
$$=\frac{1.8+0.5i}{0.6i}\cdot\frac{-0.6i}{-0.6i}$$
$$=\frac{-1.08i-0.3i^2}{-0.36i^2}$$
$$=\frac{-1.08i-0.3(-1)}{-0.36(-1)}$$
$$=\frac{0.3-1.08i}{0.36}$$
$$\approx 0.83-3i$$

129.
$$Z_T=\frac{Z_1Z_2}{Z_1+Z_2}$$
$$=\frac{(2-i)(4+i)}{(2-i)+(4+i)}$$
$$=\frac{8+2i-4i-i^2}{6}$$
$$=\frac{8-2i-(-1)}{6}$$
$$=\frac{9-2i}{6}$$
$$\approx 1.5-0.33i$$

131. $i^{-1}=\frac{1}{i}=\frac{1}{i}\cdot\frac{i}{i}=\frac{i}{i^2}=\frac{i}{-1}=-i$

133.
$$x^2-4x+6=0$$
$$a=1,\ b=-4,\ c=6$$
$$x=\frac{-(-4)\pm\sqrt{(-4)^2-4(1)(6)}}{2(1)}$$
$$=\frac{4\pm\sqrt{16-24}}{2}$$
$$=\frac{4\pm\sqrt{-8}}{2}$$
$$=\frac{4\pm 2i\sqrt{2}}{2}$$
$$=\frac{2\left(2\pm i\sqrt{2}\right)}{2}$$
$$=2\pm i\sqrt{2}$$

135.
$$a+b=5+2i\sqrt{3}+1+i\sqrt{3}$$
$$=5+1+2i\sqrt{3}+i\sqrt{3}$$
$$=6+3i\sqrt{3}$$

137.
$$ab=(5+2i\sqrt{3})(1+i\sqrt{3})$$
$$=5(1)+(5)(i\sqrt{3})+(2i\sqrt{3})(1)+(2i\sqrt{3})(i\sqrt{3})$$
$$=5+5i\sqrt{3}+2i\sqrt{3}+2i^2(\sqrt{3})^2$$
$$=5+5i\sqrt{3}+2i\sqrt{3}-6$$
$$=-1+7i\sqrt{3}$$

139.
$$4c+9\overline{\smash{)}8c^2+6c-25}\quad\text{quotient: } 2c-3$$
$$\underline{8c^2+18c}\quad\leftarrow 2c(4c+9)$$
$$-12c-25$$
$$\underline{-12c-27}\quad\leftarrow -3(4c+9)$$
$$2\quad\leftarrow \text{Remainder}$$
Thus, $\frac{8c^2+6c-25}{4c+9}=2c-3+\frac{2}{4c+9}$

141.
$$\frac{x}{4}+\frac{1}{2}=\frac{x-1}{2}$$
$$4\left(\frac{x}{4}+\frac{1}{2}\right)=4\left(\frac{x-1}{2}\right)$$
$$x+2=2(x-1)$$
$$x+2=2x-2$$
$$x=4$$

142. This problem can be solved using a single variable. To do this, let x be the amount that is \$5.50 per pound. Then $40-x$ is the amount that is \$6.30 per pound and the equation is
$$5.50(x)+6.30(40-x)=6(40)$$
$$5.5x+252-6.3x=240$$
$$252-0.8x=240$$
$$-0.8x=-12$$
$$x=15$$
Thus, combine 15 lb of the \$5.50 per pound coffee with 40 – 15 = 25 lb of the \$6.30 per pound coffee to obtain 40 lb of \$6.00 per pound coffee.

Review Exercises

1. $\sqrt{49} = 7$

2. $\sqrt[3]{-27} = -3$

3. $\sqrt[4]{256} = 4$

4. $\sqrt[3]{-125} = -5$

5. $\sqrt{(-7)^2} = |-7| = 7$

6. $\sqrt{(-93.4)^2} = |-93.4| = 93.4$

7. $\sqrt{x^2} = |x|$

8. $\sqrt{(x-2)^2} = |x-2|$

9. $\sqrt{(x-y)^2} = |x-y|$

10. $\sqrt{(x^2-4x+12)^2} = |x^2-4x+12|$

11. $f(x) = \sqrt{10x+9}$
$$\begin{aligned} f(4) &= \sqrt{10(4)+9} \\ &= \sqrt{40+9} \\ &= \sqrt{49} \\ &= 7 \end{aligned}$$

12. $k(x) = 2x + \sqrt{\frac{x}{3}}$
$$\begin{aligned} k(27) &= 2(27) + \sqrt{\frac{27}{3}} \\ &= 54 + \sqrt{9} \\ &= 54 + 3 \\ &= 57 \end{aligned}$$

13. $g(x) = \sqrt[3]{2x+3}$
$$\begin{aligned} g(4) &= \sqrt[3]{2(4)+3} \\ &= \sqrt[3]{11} \\ &\approx 2.2 \end{aligned}$$

14. $\text{Area} = (\text{side})^2$
$121 = s^2$
$\pm 11 = s$
Disregard the negative value since lengths must be positive. The length of each side is 11 m.

15. $\sqrt{x^5} = x^{5/2}$

16. $\sqrt[3]{x^5} = x^{5/3}$

17. $\left(\sqrt[4]{y}\right)^{15} = y^{15/4}$

18. $\sqrt[7]{5^2} = 5^{2/7}$

19. $x^{1/2} = \sqrt{x}$

20. $a^{4/5} = \sqrt[5]{a^4}$

21. $(3m^2n)^{7/4} = \left(\sqrt[4]{3m^2n}\right)^7$

22. $(x+3y)^{-5/3} = \dfrac{1}{\left(\sqrt[3]{x+3y}\right)^5}$

23. $\sqrt[3]{3^6} = 3^{6/3} = 3^2 = 9$

24. $\sqrt{x^{10}} = x^{10/2} = x^5$

25. $\left(\sqrt[4]{7}\right)^8 = 7^{8/4} = 7^2 = 49$

26. $\sqrt[20]{a^5} = a^{5/20} = a^{1/4} = \sqrt[4]{a}$

27. $-25^{1/2} = -\sqrt{25} = -5$

28. $(-25)^{1/2}$ is not a real number.

29. $\left(\frac{64}{27}\right)^{-1/3} = \left(\frac{27}{64}\right)^{1/3} = \sqrt[3]{\frac{27}{64}} = \frac{\sqrt[3]{27}}{\sqrt[3]{64}} = \frac{3}{4}$

30. $64^{-1/2}+8^{-2/3}=\frac{1}{64^{1/2}}+\frac{1}{8^{2/3}}$
$=\frac{1}{\sqrt{64}}+\frac{1}{\left(\sqrt[3]{8}\right)^2}$
$=\frac{1}{8}+\frac{1}{2^2}$
$=\frac{1}{8}+\frac{1}{4}$
$=\frac{1}{8}+\frac{2}{8}$
$=\frac{3}{8}$

31. $x^{3/5}x^{-1/3}=x^{3/5-1/3}=x^{9/15-5/15}=x^{4/15}$

32. $\left(\frac{64}{y^6}\right)^{1/3}=\sqrt[3]{\frac{64}{y^6}}=\frac{\sqrt[3]{64}}{\sqrt[3]{y^6}}=\frac{4}{y^2}$

33. $\left(\frac{a^{-6/5}}{a^{1/5}}\right)^{2/3}=\frac{a^{-6/5\cdot 2/3}}{a^{1/5\cdot 2/3}}$
$=\frac{a^{-12/15}}{a^{2/15}}$
$=a^{-12/15-(2/15)}$
$=a^{-14/15}$
$=\frac{1}{a^{14/15}}$

34. $\left(\frac{28x^5y^{-3}}{4y^{1/2}}\right)^2=\left(\frac{7x^5}{y^{7/2}}\right)^2$
$=\frac{7^2x^{5\cdot 2}}{y^{(7/2)\cdot 2}}$
$=\frac{49x^{10}}{y^7}$

35. $a^{1/2}\left(5a^{3/2}-2a^2\right)=a^{1/2}\left(5a^{3/2}\right)-a^{1/2}\left(2a^2\right)$
$=5a^{1/2+3/2}-2a^{1/2+2}$
$=5a^{4/2}-2a^{1/2+4/2}$
$=5a^2-2a^{5/2}$

36. $4x^{-2/3}\left(x^{-1/2}+\frac{9}{4}x^{2/3}\right)$
$=4x^{-2/3}\left(x^{-1/2}\right)+4x^{-2/3}\left(\frac{9}{4}x^{2/3}\right)$
$=4x^{-2/3+(-1/2)}+9x^{-2/3+(2/3)}$
$=4x^{-4/6+(-3/6)}+9x^{0/3}$
$=4x^{-7/6}+9x^0$
$=\frac{4}{x^{7/6}}+9$

37. $x^{2/5}+x^{7/5}=x^{2/5}+x^{2/5}\cdot x^1$
$=x^{2/5}(1+x)$

38. $a^{-1/2}+a^{5/2}=a^{-1/2}+a^{-1/2}\cdot a^{6/2}$
$=a^{-1/2}(1+a^3)$
$=\frac{1+a^3}{a^{1/2}}$

39. $f(x)=\sqrt{7x-17}$
$f(6)=\sqrt{7(6)-17}$
$=\sqrt{42-17}$
$=\sqrt{25}$
$=5$

40. $g(x)=\sqrt[3]{9r-17}$
$g(4)=\sqrt[3]{9(4)-17}$
$=\sqrt[3]{36-17}$
$=\sqrt[3]{19}$
≈ 2.668

41. $f(x)=\sqrt{x}$

x	$f(x)$
0	0
1	1
4	2
9	3

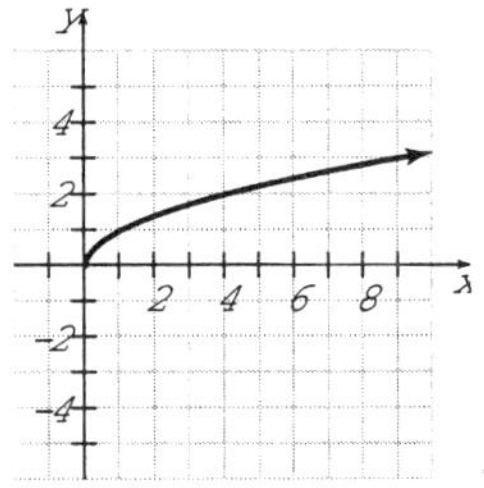

$f(x) = \sqrt{x}$

42. $f(x) = \sqrt{x} - 4$

x	$f(x)$
0	−4
1	−3
4	−2
9	−1
16	0

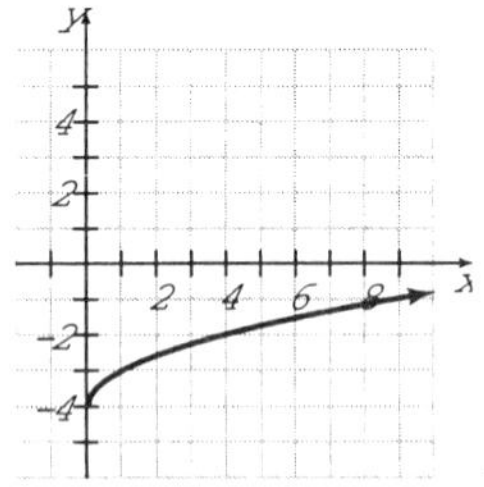

$f(x) = \sqrt{x} - 4$

43. $\sqrt{75} = \sqrt{25}\sqrt{3} = 5\sqrt{3}$

44. $\sqrt[3]{128} = \sqrt[3]{64}\sqrt[3]{2} = 4\sqrt[3]{2}$

45. $\sqrt{\frac{49}{4}} = \frac{\sqrt{49}}{\sqrt{4}} = \frac{7}{2}$

46. $\sqrt[3]{\frac{27}{125}} = \frac{\sqrt[3]{27}}{\sqrt[3]{125}} = \frac{3}{5}$

47. $-\sqrt{\frac{81}{64}} = -\frac{\sqrt{81}}{\sqrt{64}} = -\frac{9}{8}$

48. $\sqrt[3]{-\frac{8}{125}} = \frac{\sqrt[3]{-8}}{\sqrt[3]{125}} = \frac{-2}{5} = -\frac{2}{5}$

49. $\sqrt{20}\sqrt{5} = \sqrt{20 \cdot 5} = \sqrt{100} = 10$

50. $\sqrt[3]{16} \cdot \sqrt[3]{4} = \sqrt[3]{16 \cdot 4} = \sqrt[3]{64} = 4$

51. $\sqrt{8x^2y^3z^4} = \sqrt{4x^2y^2z^4}\sqrt{2y}$
$= 2xyz^2\sqrt{2y}$

52. $\sqrt{50x^3y^7} = \sqrt{25x^2y^6}\sqrt{2xy}$
$= 5xy^3\sqrt{2xy}$

53. $\sqrt[3]{54a^7b^{10}} = \sqrt[3]{27a^6b^9}\sqrt[3]{2ab}$
$= 3a^2b^3\sqrt[3]{2ab}$

54. $\sqrt[3]{125x^8y^9z^{16}} = \sqrt[3]{125x^6y^9z^{15}}\sqrt[3]{x^2z}$
$= 5x^2y^3z^5\sqrt[3]{x^2z}$

55. $\left(\sqrt[6]{x^2y^3z^5}\right)^{42} = (x^2y^3z^5)^{42/6}$
$= (x^2y^3z^5)^7$
$= x^{14}y^{21}z^{35}$

56. $\left(\sqrt[5]{2ab^4c^6}\right)^{15} = \left(2ab^4c^6\right)^{15/5}$
$= \left(2ab^4c^6\right)^3$
$= 8a^3b^{12}c^{18}$

57. $\sqrt{5x}\sqrt{8x^5} = \sqrt{40x^6}$
$= \sqrt{4x^6}\sqrt{10}$
$= 2x^3\sqrt{10}$

58. $\sqrt[3]{2x^2y}\,\sqrt[3]{4x^9y^4} = \sqrt[3]{8x^{11}y^5}$
$= 2x^3y\,\sqrt[3]{x^2y^2}$

59. $\sqrt[3]{2x^4y^5}\sqrt[3]{16x^4y^4} = \sqrt[3]{32x^8y^9}$
$= \sqrt[3]{8x^6y^9}\sqrt[3]{4x^2}$
$= 2x^2y^3\sqrt[3]{4x^2}$

60. $\sqrt[4]{8x^4y^7}\sqrt[4]{2x^5y^9} = \sqrt[4]{16x^9y^{16}}$
$= \sqrt[4]{16x^8y^{16}}\sqrt[4]{x}$
$= 2x^2y^4\sqrt[4]{x}$

61. $\sqrt{3x}(\sqrt{12x} - \sqrt{20}) = \sqrt{36x^2} - \sqrt{60x}$
$= \sqrt{36x^2} - \sqrt{4}\sqrt{15x}$
$= 6x - 2\sqrt{15x}$

62. $\sqrt[3]{2x^2y}\left(\sqrt[3]{4x^4y^7}+\sqrt[3]{9x}\right)$
$=\sqrt[3]{8x^6y^8}+\sqrt[3]{18x^3y}$
$=\sqrt[3]{8x^6y^6}\sqrt[3]{y^2}+\sqrt[3]{x^3}\sqrt[3]{18y}$
$=2x^2y^2\sqrt[3]{y^2}+x\sqrt[3]{18y}$

63. $\sqrt{\sqrt{a^3b^2}}=\left(\sqrt{a^3b^2}\right)^{1/2}$
$=\left[\left(a^3b^2\right)^{1/2}\right]^{1/2}$
$=\left(a^3b^2\right)^{1/4}$
$=\sqrt[4]{a^3b^2}$

64. $\sqrt{\sqrt[3]{x^5y^2}}=\left(\sqrt[3]{x^5y^2}\right)^{1/2}$
$=\left[\left(x^5y^2\right)^{1/3}\right]^{1/2}$
$=\left(x^5y^2\right)^{1/6}$
$=\sqrt[6]{\left(x^5y^2\right)}$

65. $\left(\dfrac{3r^2p^{1/3}}{r^{1/2}p^{4/3}}\right)^3=(3r^{2-1/2}p^{1/3-4/3})^3$
$=(3r^{3/2}p^{-1})^3$
$=3^3(r^{3/2})^3(p^{-1})^3$
$=27r^{9/2}p^{-3}$
$=\dfrac{27r^{9/2}}{p^3}$

66. $\left(\dfrac{4y^{2/5}z^{1/3}}{x^{-1}y^{3/5}}\right)^{-1}=\left(\dfrac{4y^{2/5-3/5}z^{1/3}}{x^{-1}}\right)^{-1}$
$=\left(\dfrac{4y^{-1/5}z^{1/3}}{x^{-1}}\right)^{-1}$
$=\left(\dfrac{4xz^{1/3}}{y^{1/5}}\right)^{-1}$
$=\dfrac{y^{1/5}}{4xz^{1/3}}$

67. $\sqrt{\dfrac{3}{5}}=\dfrac{\sqrt{3}}{\sqrt{5}}\cdot\dfrac{\sqrt{5}}{\sqrt{5}}=\dfrac{\sqrt{15}}{5}$

68. $\sqrt[3]{\dfrac{7}{4}}=\dfrac{\sqrt[3]{7}}{\sqrt[3]{4}}=\dfrac{\sqrt[3]{7}}{\sqrt[3]{4}}\cdot\dfrac{\sqrt[3]{2}}{\sqrt[3]{2}}=\dfrac{\sqrt[3]{14}}{\sqrt[3]{8}}=\dfrac{\sqrt[3]{14}}{2}$

69. $\sqrt[4]{\dfrac{2}{3}}=\dfrac{\sqrt[4]{2}}{\sqrt[4]{3}}=\dfrac{\sqrt[4]{2}}{\sqrt[4]{3}}\cdot\dfrac{\sqrt[4]{3^3}}{\sqrt[4]{3^3}}=\dfrac{\sqrt[4]{54}}{\sqrt[4]{3^4}}=\dfrac{\sqrt[4]{54}}{3}$

70. $\dfrac{x}{\sqrt{11}}=\dfrac{x}{\sqrt{11}}\cdot\dfrac{\sqrt{11}}{\sqrt{11}}=\dfrac{x\sqrt{11}}{11}$

71. $\dfrac{6}{\sqrt{x}}=\dfrac{6}{\sqrt{x}}\cdot\dfrac{\sqrt{x}}{\sqrt{x}}=\dfrac{6\sqrt{x}}{x}$

72. $\dfrac{a}{\sqrt[3]{25}}=\dfrac{a}{\sqrt[3]{5^2}}\cdot\dfrac{\sqrt[3]{5}}{\sqrt[3]{5}}=\dfrac{a\sqrt[3]{5}}{\sqrt[3]{5^3}}=\dfrac{a\sqrt[3]{5}}{5}$

73. $\dfrac{9}{\sqrt[3]{y^2}}=\dfrac{9}{\sqrt[3]{y^2}}\cdot\dfrac{\sqrt[3]{y}}{\sqrt[3]{y}}=\dfrac{9\sqrt[3]{y}}{\sqrt[3]{y^3}}=\dfrac{9\sqrt[3]{y}}{y}$

74. $\dfrac{7}{\sqrt[4]{z}}=\dfrac{7}{\sqrt[4]{z}}\cdot\dfrac{\sqrt[4]{z^3}}{\sqrt[4]{z^3}}=\dfrac{7\sqrt[4]{z^3}}{\sqrt[4]{z^4}}=\dfrac{7\sqrt[4]{z^3}}{z}$

75. $\sqrt[3]{\dfrac{x^3}{8}}=\dfrac{\sqrt[3]{x^3}}{\sqrt[3]{8}}=\dfrac{x}{2}$

76. $\dfrac{\sqrt[3]{2x^9}}{\sqrt[3]{16x^6}}=\sqrt[3]{\dfrac{2x^9}{16x^6}}=\sqrt[3]{\dfrac{x^3}{8}}=\dfrac{\sqrt[3]{x^3}}{\sqrt[3]{8}}=\dfrac{x}{2}$

77. $\sqrt{\dfrac{32x^2y^5}{2x^4y}}=\sqrt{\dfrac{16y^4}{x^2}}=\dfrac{\sqrt{16y^4}}{\sqrt{x^2}}=\dfrac{4y^2}{x}$

78. $\sqrt[4]{\dfrac{32x^9y^{15}}{2xy^3}}=\sqrt[4]{16x^8y^{12}}=2x^2y^3$

79. $\sqrt{\dfrac{5x^4}{y}}=\dfrac{\sqrt{5x^4}}{\sqrt{y}}$
$=\dfrac{\sqrt{x^4}\sqrt{5}}{\sqrt{y}}$
$=\dfrac{x^2\sqrt{5}}{\sqrt{y}}\cdot\dfrac{\sqrt{y}}{\sqrt{y}}$
$=\dfrac{x^2\sqrt{5y}}{y}$

80. $\sqrt{\dfrac{12a}{7b}} = \dfrac{\sqrt{12a}}{\sqrt{7b}}$
$= \dfrac{2\sqrt{3a}}{\sqrt{7b}}$
$= \dfrac{2\sqrt{3a}}{\sqrt{7b}} \cdot \dfrac{\sqrt{7b}}{\sqrt{7b}}$
$= \dfrac{2\sqrt{21ab}}{7b}$

81. $\sqrt{\dfrac{18x^4y^5}{3z}} = \dfrac{\sqrt{18x^4y^5}}{\sqrt{3z}}$
$= \dfrac{\sqrt{9x^4y^4}\sqrt{2y}}{\sqrt{3z}}$
$= \dfrac{3x^2y^2\sqrt{2y}}{\sqrt{3z}}$
$= \dfrac{3x^2y^2\sqrt{2y}}{\sqrt{3z}} \cdot \dfrac{\sqrt{3z}}{\sqrt{3z}}$
$= \dfrac{3x^2y^2\sqrt{6yz}}{3z}$
$= \dfrac{x^2y^2\sqrt{6yz}}{z}$

82. $\sqrt{\dfrac{125x^2y^5}{3z}} = \dfrac{\sqrt{125x^2y^5}}{\sqrt{3z}}$
$= \dfrac{\sqrt{25x^2y^4}\sqrt{5y}}{\sqrt{3z}}$
$= \dfrac{5xy^2\sqrt{5y}}{\sqrt{3z}} \cdot \dfrac{\sqrt{3z}}{\sqrt{3z}}$
$= \dfrac{5xy^2\sqrt{15yz}}{3z}$

83. $\sqrt[3]{\dfrac{108x^3y^7}{2y^3}} = \sqrt[3]{54x^3y^4} = 3xy\sqrt[3]{2y}$

84. $\sqrt[3]{\dfrac{3x}{5y}} = \dfrac{\sqrt[3]{3x}}{\sqrt[3]{5y}} \cdot \dfrac{\sqrt[3]{25y^2}}{\sqrt[3]{25y^2}} = \dfrac{\sqrt[3]{75xy^2}}{5y}$

85. $\sqrt[3]{\dfrac{4x^5y^3}{x^6}} = \sqrt[3]{\dfrac{4y^3}{x}}$
$= \dfrac{\sqrt[3]{4y^3}}{\sqrt[3]{x}}$
$= \dfrac{\sqrt[3]{y^3}\ \sqrt[3]{4}}{\sqrt[3]{x}}$
$= \dfrac{y\sqrt[3]{4}}{\sqrt[3]{x}} \cdot \dfrac{\sqrt[3]{x^2}}{\sqrt[3]{x^2}}$
$= \dfrac{y\sqrt[3]{4x^2}}{x}$

86. $\sqrt[3]{\dfrac{y^6}{2x^2}} = \dfrac{\sqrt[3]{y^6}}{\sqrt[3]{2x^2}}$
$= \dfrac{y^2}{\sqrt[3]{2x^2}} \cdot \dfrac{\sqrt[3]{4x}}{\sqrt[3]{4x}}$
$= \dfrac{y^2\sqrt[3]{4x}}{2x}$

87. $\sqrt[4]{\dfrac{3a^2b^{11}}{a^5b}} = \sqrt[4]{\dfrac{3b^{10}}{a^3}}$
$= \dfrac{\sqrt[4]{3b^{10}}}{\sqrt[4]{a^3}} \cdot \dfrac{\sqrt[4]{a}}{\sqrt[4]{a}}$
$= \dfrac{\sqrt[4]{3ab^{10}}}{\sqrt[4]{a^4}}$
$= \dfrac{b^2\sqrt[4]{3ab^2}}{a}$

88. $\sqrt[4]{\dfrac{2x^2y^6}{8x^3}} = \sqrt[4]{\dfrac{y^6}{4x}}$
$= \dfrac{\sqrt[4]{y^6}}{\sqrt[4]{4x}}$
$= \dfrac{y\sqrt[4]{y^2}}{\sqrt[4]{4x}}$
$= \dfrac{y\sqrt[4]{y^2}}{\sqrt[4]{4x}} \cdot \dfrac{\sqrt[4]{4x^3}}{\sqrt[4]{4x^3}}$
$= \dfrac{y\sqrt[4]{4x^3y^2}}{2x}$

89. $\left(3-\sqrt{2}\right)\left(3+\sqrt{2}\right)=3^2-\left(\sqrt{2}\right)^2=9-2=7$

90. $\left(\sqrt{x}+y\right)\left(\sqrt{x}-y\right)=\left(\sqrt{x}\right)^2-y^2=x-y^2$

91. $\left(x-\sqrt{y}\right)\left(x+\sqrt{y}\right)=x^2-\left(\sqrt{y}\right)^2=x^2-y$

92.
$$\begin{aligned}\left(\sqrt{3}+5\right)^2&=\left(\sqrt{3}\right)^2+2(5)\left(\sqrt{3}\right)+5^2\\&=3+10\sqrt{3}+25\\&=28+10\sqrt{3}\end{aligned}$$

93.
$$\begin{aligned}&\left(\sqrt{x}-\sqrt{3y}\right)\left(\sqrt{x}+\sqrt{5y}\right)\\&=\left(\sqrt{x}\right)^2+\sqrt{x}\sqrt{5y}-\sqrt{x}\sqrt{3y}-\sqrt{3y}\sqrt{5y}\\&=x+\sqrt{5xy}-\sqrt{3xy}-\sqrt{15y^2}\\&=x+\sqrt{5xy}-\sqrt{3xy}-y\sqrt{15}\end{aligned}$$

94.
$$\begin{aligned}&\left(\sqrt[3]{2x}-\sqrt[3]{3y}\right)\left(\sqrt[3]{3x}-\sqrt[3]{2y}\right)\\&=\sqrt[3]{2x}\left(\sqrt[3]{3x}\right)-\left(\sqrt[3]{2x}\right)\left(\sqrt[3]{2y}\right)\\&\quad-\sqrt[3]{3y}\left(\sqrt[3]{3x}\right)+\sqrt[3]{3y}\sqrt[3]{2y}\\&=\sqrt[3]{6x^2}-\sqrt[3]{4xy}-\sqrt[3]{9xy}+\sqrt[3]{6y^2}\end{aligned}$$

95.
$$\begin{aligned}\frac{5}{2+\sqrt{5}}&=\frac{5}{2+\sqrt{5}}\cdot\frac{2-\sqrt{5}}{2-\sqrt{5}}\\&=\frac{5\left(2-\sqrt{5}\right)}{2^2-\left(\sqrt{5}\right)^2}\\&=\frac{10-5\sqrt{5}}{4-5}\\&=\frac{10-5\sqrt{5}}{-1}\\&=-10+5\sqrt{5}\end{aligned}$$

96.
$$\begin{aligned}\frac{x}{3+\sqrt{x}}&=\frac{x}{3+\sqrt{x}}\cdot\frac{3-\sqrt{x}}{3-\sqrt{x}}\\&=\frac{x\left(3-\sqrt{x}\right)}{3^2-\left(\sqrt{x}\right)^2}\\&=\frac{3x-x\sqrt{x}}{9-x}\end{aligned}$$

97.
$$\begin{aligned}\frac{a}{7-\sqrt{b}}&=\frac{a}{7-\sqrt{b}}\cdot\frac{7+\sqrt{b}}{7+\sqrt{b}}\\&=\frac{a\left(7+\sqrt{b}\right)}{\left(7-\sqrt{b}\right)\left(7+\sqrt{b}\right)}\\&=\frac{7a+a\sqrt{b}}{49+7\sqrt{b}-7\sqrt{b}-b}\\&=\frac{7a+a\sqrt{b}}{49-b}\end{aligned}$$

98.
$$\begin{aligned}\frac{x}{\sqrt{y}-8}&=\frac{x}{\sqrt{y}-8}\cdot\frac{\sqrt{y}+8}{\sqrt{y}+8}\\&=\frac{x\left(\sqrt{y}+8\right)}{\left(\sqrt{y}-8\right)\left(\sqrt{y}+8\right)}\\&=\frac{x\sqrt{y}+8x}{y+8\sqrt{y}-8\sqrt{y}-64}\\&=\frac{x\sqrt{y}+8x}{y-64}\end{aligned}$$

99.
$$\begin{aligned}\frac{\sqrt{x}}{\sqrt{x}+\sqrt{y}}&=\frac{\sqrt{x}}{\sqrt{x}+\sqrt{y}}\cdot\frac{\sqrt{x}-\sqrt{y}}{\sqrt{x}-\sqrt{y}}\\&=\frac{\sqrt{x}\left(\sqrt{x}-\sqrt{y}\right)}{\left(\sqrt{x}\right)^2-\left(\sqrt{y}\right)^2}\\&=\frac{\sqrt{x^2}-\sqrt{xy}}{x-y}\\&=\frac{x-\sqrt{xy}}{x-y}\end{aligned}$$

100.
$$\begin{aligned}\frac{\sqrt{x}-2\sqrt{y}}{\sqrt{x}-\sqrt{y}}&=\frac{\sqrt{x}-2\sqrt{y}}{\sqrt{x}-\sqrt{y}}\cdot\frac{\sqrt{x}+\sqrt{y}}{\sqrt{x}+\sqrt{y}}\\&=\frac{x+\sqrt{xy}-2\sqrt{xy}-2y}{\left(\sqrt{x}\right)^2-\left(\sqrt{y}\right)^2}\\&=\frac{x-\sqrt{xy}-2y}{x-y}\end{aligned}$$

101. $\dfrac{3}{\sqrt{a-1}-2}=\dfrac{3}{\sqrt{a-1}-2}\cdot\dfrac{\sqrt{a-1}+2}{\sqrt{a-1}+2}$

$=\dfrac{3\left(\sqrt{a-1}+2\right)}{\left(\sqrt{a-1}-2\right)\left(\sqrt{a-1}+2\right)}$

$=\dfrac{3\sqrt{a-1}+6}{a-1+2\sqrt{a-1}-2\sqrt{a-1}-4}$

$=\dfrac{3\sqrt{a-1}+6}{a-5}$

102. $\dfrac{4}{\sqrt{y+2}-3}=\dfrac{4}{\sqrt{y+2}-3}\cdot\dfrac{\sqrt{y+2}+3}{\sqrt{y+2}+3}$

$=\dfrac{4\sqrt{y+2}+12}{\left(\sqrt{y+2}\right)^2-3^2}$

$=\dfrac{4\sqrt{y+2}+12}{y+2-9}$

$=\dfrac{4\sqrt{y+2}+12}{y-7}$

103. $\sqrt[3]{x}+3\sqrt[3]{x}-2\sqrt[3]{x}=2\sqrt[3]{x}$

104. $\sqrt{3}+\sqrt{27}-\sqrt{192}=\sqrt{3}+3\sqrt{3}-\sqrt{64}\sqrt{3}$

$=\sqrt{3}+3\sqrt{3}-8\sqrt{3}$

$=-4\sqrt{3}$

105. $\sqrt[3]{16}-5\sqrt[3]{54}+2\sqrt[3]{64}$

$=\sqrt[3]{8}\sqrt[3]{2}-5\sqrt[3]{27}\sqrt[3]{2}+2\sqrt[3]{64}$

$=2\sqrt[3]{2}-5\left(3\sqrt[3]{2}\right)+2(4)$

$=2\sqrt[3]{2}-15\sqrt[3]{2}+8$

$=8-13\sqrt[3]{2}$

106. $4\sqrt{2}-\dfrac{3}{\sqrt{32}}+\sqrt{50}$

$=4\sqrt{2}-\dfrac{3}{4\sqrt{2}}+5\sqrt{2}$

$=4\sqrt{2}-\dfrac{3}{4\sqrt{2}}\cdot\dfrac{\sqrt{2}}{\sqrt{2}}+5\sqrt{2}$

$=4\sqrt{2}-\dfrac{3\sqrt{2}}{8}+5\sqrt{2}$

$=\dfrac{8}{8}\left(4\sqrt{2}\right)-\dfrac{3\sqrt{2}}{8}+\left(\dfrac{8}{8}\right)5\sqrt{2}$

$=\dfrac{32\sqrt{2}}{8}-\dfrac{3\sqrt{2}}{8}+\dfrac{40\sqrt{2}}{8}$

$=\dfrac{69\sqrt{2}}{8}$

107. $3\sqrt{x^5y^6}-\sqrt{16x^7y^8}$

$=3\sqrt{x^4y^6}\sqrt{x}-\sqrt{16x^6y^8}\sqrt{x}$

$=3x^2y^3\sqrt{x}-4x^3y^4\sqrt{x}$

$=\left(3x^2y^3-4x^3y^4\right)\sqrt{x}$

108. $2\sqrt[3]{x^7y^8}-\sqrt[3]{x^4y^2}+3\sqrt[3]{x^{10}y^2}$

$=2\sqrt[3]{x^6y^6}\sqrt[3]{xy^2}-\sqrt[3]{x^3}\sqrt[3]{xy^2}+3\sqrt[3]{x^9}\sqrt[3]{xy^2}$

$=2x^2y^2\sqrt[3]{xy^2}-x\sqrt[3]{xy^2}+3x^3\sqrt[3]{xy^2}$

$=(2x^2y^2-x+3x^3)\sqrt[3]{xy^2}$

109. $(f\cdot g)(x)=f(x)\cdot g(x)$

$=\sqrt{3x}\cdot\left(\sqrt{6x}-\sqrt{10}\right)$

$=\sqrt{3x}\sqrt{6x}-\sqrt{3x}\sqrt{10}$

$=\sqrt{18x^2}-\sqrt{30x}$

$=\sqrt{9x^2}\sqrt{2}-\sqrt{30x}$

$=3x\sqrt{2}-\sqrt{30x}$

110. $(f\cdot g)(x)=f(x)\cdot g(x)$

$=\sqrt[3]{2x^2}\left(\sqrt[3]{4x^4}+\sqrt[3]{8x^5}\right)$

$=\sqrt[3]{2x^2}\sqrt[3]{4x^4}+\sqrt[3]{2x^2}\sqrt[3]{8x^5}$

$=\sqrt[3]{8x^6}+\sqrt[3]{16x^7}$

$=\sqrt[3]{8x^6}+\sqrt[3]{8x^6}\sqrt[3]{2x}$

$=2x^2+2x^2\sqrt[3]{2x}$

111. $f(x)=\sqrt{2x+4}\sqrt{2x+4},\quad x\ge -2$

$=\sqrt{(2x+4)^2}$

$=|2x+4|$

$=2x+4\quad \text{since } x\ge -2$

112. $g(a) = \sqrt{20a^2 + 50a + 125}$
$= \sqrt{5(4a^2 + 10a + 25)}$
$= \sqrt{5(2a+5)^2}$
$= \sqrt{5}|2a+5|$

113. $\dfrac{\sqrt[3]{(x+5)^5}}{\sqrt{(x+5)^3}} = \dfrac{(x+5)^{5/3}}{(x+5)^{3/2}}$
$= (x+5)^{5/3-3/2}$
$= (x+5)^{1/6}$
$= \sqrt[6]{x+5}$

114. $\dfrac{\sqrt[3]{a^3b^2}}{\sqrt[4]{a^4b}} = \dfrac{a\sqrt[3]{b^2}}{a\sqrt[4]{b}}$
$= \dfrac{\sqrt[3]{b^2}}{\sqrt[4]{b}}$
$= \dfrac{b^{2/3}}{b^{1/4}}$
$= b^{2/3-1/4}$
$= b^{5/12}$
$= \sqrt[12]{b^5}$

115. **a.** $P = 2l + 2w$
$P = 2\sqrt{48} + 2\sqrt{12}$
$= 2\sqrt{16\cdot 3} + 2\sqrt{4\cdot 3}$
$= 8\sqrt{3} + 4\sqrt{3}$
$= 12\sqrt{3}$

b. $A = lw$
$A = (\sqrt{48})(\sqrt{12})$
$= \sqrt{576}$
$= 24$

116. **a.** $P = s_1 + s_2 + s_3$
$P = \sqrt{125} + \sqrt{45} + \sqrt{130}$
$= 5\sqrt{5} + 3\sqrt{5} + \sqrt{130}$
$= 8\sqrt{5} + \sqrt{130}$

b. $A = \frac{1}{2}bh$
$A = \frac{1}{2}(\sqrt{130})(\sqrt{20})$
$= \frac{1}{2}\sqrt{2600}$
$= \frac{1}{2}\sqrt{100\cdot 26}$
$= \frac{10}{2}\sqrt{26}$
$= 5\sqrt{26}$

117. **a.** $f(x) = \sqrt{x} + 2$
$g(x) = -3$
$(f+g)(x) = f(x) + g(x)$
$= \sqrt{x} + 2 - 3$
$= \sqrt{x} - 1$

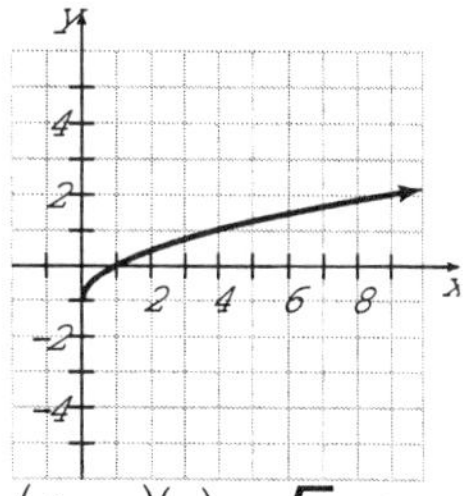

$(f+g)(x) = \sqrt{x} - 1$

b. $\sqrt{x} \ge 0,\ x \ge 0$
The domain is $\{x \mid x \ge 0\}$.

118. **a.** $f(x) = -\sqrt{x}$
$g(x) = \sqrt{x} + 2$
$(f+g)(x) = f(x) + g(x)$
$= -\sqrt{x} + \sqrt{x} + 2$
$= 2$

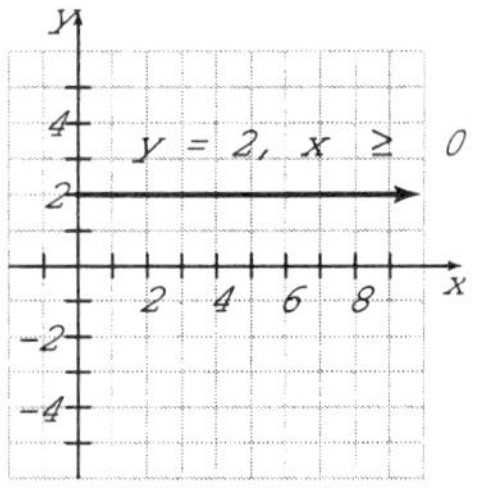

b. $\sqrt{x} \ge 0,\ x \ge 0$

The domain is $\{x \mid x \ge 0\}$.

119. $\sqrt{x} = 8$

$\left(\sqrt{x}\right)^2 = 8^2$

$x = 64$

Check: $\sqrt{64} = 8$ True

64 is the solution.

120. $\sqrt{x} = -8$

$\left(\sqrt{x}\right)^2 = (-8)^2$

$x = 64$

Check: $\sqrt{64} \ne -8$ False

no solution

121. $\sqrt[3]{x} = 5$

$\left(\sqrt[3]{x}\right)^3 = 5^3$

$x = 125$

Check: $\sqrt[3]{x} = 5$

$\sqrt[3]{125} = 5$

$5 = 5$ True

125 is a solution.

122. $\sqrt[3]{x} = -5$

$\left(\sqrt[3]{x}\right)^3 = (-5)^3$

$x = -125$

Check: $\sqrt[3]{x} = -5$

$\sqrt[3]{-125} = -5$

$-5 = -5$ True

−125 is a solution.

123. $3 + \sqrt{x} = 10$

$\sqrt{x} = 7$

$\left(\sqrt{x}\right)^2 = 7^2$

$x = 49$

Check: $3 + \sqrt{x} = 10$

$3 + \sqrt{49} = 10$

$3 + 7 = 10$

$10 = 10$ True

49 is a solution.

124. $8 + \sqrt[3]{x} = 12$

$\sqrt[3]{x} = 4$

$\left(\sqrt[3]{x}\right)^3 = 4^3$

$x = 64$

Check: $8 + \sqrt[3]{x} = 12$

$8 + \sqrt[3]{64} = 12$

$8 + 4 = 12$

$12 = 12$ True

64 is a solution.

125 $\sqrt{3x+4} = \sqrt{5x+12}$

$\left(\sqrt{3x+4}\right)^2 = \left(\sqrt{5x+12}\right)^2$

$3x + 4 = 5x + 12$

$-8 = 2x$

$-4 = x$

Check: $\sqrt{3x+4}$ becomes

$\sqrt{3(-4)+4} = \sqrt{-12+4} = \sqrt{-8}$

which is not a real number. −4 is not a solution and there is no real solution.

126. $\sqrt{x^2+2x-4}=x$

$\left(\sqrt{x^2+2x-4}\right)^2=(x)^2$

$x^2+2x-4=x^2$

$2x-4=0$

$x=2$

Check: $\sqrt{2^2+2\cdot 2-4}=2$

$\sqrt{4+4-4}=2$

$\sqrt{4}=2$

$2=2$ True

2 is a solution.

127. $\sqrt[3]{x-9}=\sqrt[3]{5x+3}$

$\left(\sqrt[3]{x-9}\right)^3=\left(\sqrt[3]{5x+3}\right)^3$

$x-9=5x+3$

$-4x=12$

$x=-3$

Check: $\sqrt[3]{-3-9}=\sqrt[3]{5(-3)+3}$

$\sqrt[3]{-12}=\sqrt[3]{-15+3}$

$\sqrt[3]{-12}=\sqrt[3]{-12}$ True

–3 is a solution.

128. $(x^2+5)^{1/2}=x+1$

$\left[(x^2+5)^{1/2}\right]^2=(x+1)^2$

$x^2+5=x^2+2x+1$

$5=2x+1$

$4=2x$

$x=2$

Check: $(2^2+5)^{1/2}=2+1$

$(4+5)^{1/2}=3$

$9^{1/2}=3$

$3=3$ True

2 is a solution.

129. $\sqrt{x}+3=\sqrt{3x+9}$

$\left(\sqrt{x}+3\right)^2=\left(\sqrt{3x+9}\right)^2$

$\left(\sqrt{x}+3\right)\left(\sqrt{x}+3\right)=3x+9$

$x+6\sqrt{x}+9=3x+9$

$6\sqrt{x}=2x$

$\left(6\sqrt{x}\right)^2=(2x)^2$

$36x=4x^2$

$4x^2-36x=0$

$4x(x-9)=0$

$4x=0$ or $x-9=0$

$x=0$ $\quad$ $x=9$

A check shows that 0 and 9 are solutions.

130. $\sqrt{6x-5}-\sqrt{2x+6}-1=0$

$\sqrt{6x-5}=\sqrt{2x+6}+1$

$\left(\sqrt{6x-5}\right)^2=\left(\sqrt{2x+6}+1\right)^2$

$6x-5=2x+6+2\sqrt{2x+6}+1$

$6x-5=2x+7+2\sqrt{2x+6}$

$4x-12=2\sqrt{2x+6}$

$\frac{4x}{2}-\frac{12}{2}=\frac{2}{2}\sqrt{2x+6}$

$2x-6=\sqrt{2x+6}$

$(2x-6)^2=\left(\sqrt{2x+6}\right)^2$

$4x^2-24x+36=2x+6$

$4x^2-26x+30=0$

$2x^2-13x+15=0$

Express the middle term, $-13x$, as $-10x-3x$.

$2x^2-10x-3x+15=0$

$2x(x-5)-3(x-5)=0$

$(2x-3)(x-5)=0$

$2x-3=0$ or $x-5=0$

$x=\frac{3}{2}$ $\quad$ $x=5$

The solution $x=5$ checks in the original equation but $x=\frac{3}{2}$ does not check.

Therefore the only solution is $x=5$.

131.
$$f(x) = g(x)$$
$$\sqrt{3x+4} = 2\sqrt{2x-4}$$
$$\left(\sqrt{3x+4}\right)^2 = \left(2\sqrt{2x-4}\right)^2$$
$$3x+4 = 4(2x-4)$$
$$3x+4 = 8x-16$$
$$20 = 5x$$
$$4 = x$$

132.
$$f(x) = g(x)$$
$$(4x+3)^{1/3} = (6x-9)^{1/3}$$
$$[(4x+3)^{1/3}]^3 = [(6x-9)^{1/3}]^3$$
$$4x+3 = 6x-9$$
$$12 = 2x$$
$$6 = x$$

133. $V = \sqrt{\frac{2L}{w}}$ Solve for L.
$$V^2 = \frac{2L}{w}$$
$$V^2w = 2L$$
$$\frac{V^2w}{2} = L \text{ or } L = \frac{V^2w}{2}$$

134. $r = \sqrt{\frac{A}{\pi}}$ Solve for A.
$$r^2 = \left(\sqrt{\frac{A}{\pi}}\right)^2$$
$$r^2 = \frac{A}{\pi}$$
$$\pi r^2 = A \text{ or } A = \pi r^2$$

135. Pythagorean Theorem: $a^2 + b^2 = c^2$
$$\left(\sqrt{20}\right)^2 + 6^2 = x^2$$
$$20 + 36 = x^2$$
$$56 = x^2$$
$$\sqrt{56} = x \text{ or } x = 2\sqrt{14}$$

136. Pythagorean Theorem: $a^2 + b^2 = c^2$
$$\left(\sqrt{26}\right)^2 + x^2 = \left(\sqrt{101}\right)^2$$
$$26 + x^2 = 101$$
$$x^2 = 75$$
$$x = \sqrt{75} \text{ or } x = 5\sqrt{3}$$

137.
$$l = \sqrt{a^2 + b^2}$$
$$= \sqrt{5^2 + 2^2}$$
$$= \sqrt{29}$$
$$\approx 5.39 \text{ m}$$

138.
$$v = \sqrt{2gh}$$
$$= \sqrt{2(32)(20)}$$
$$= \sqrt{1280} \approx 35.78 \text{ ft/sec}$$

139.
$$T = 2\pi\sqrt{\frac{L}{32}}$$
$$= 2\pi\sqrt{\frac{64}{32}}$$
$$= 2\pi\sqrt{2}$$
$$\approx 8.89 \text{ sec}$$

140.
$$V = \sqrt{\frac{2K}{m}}$$
$$= \sqrt{\frac{2(45)}{0.145}}$$
$$\approx 25 \text{ meters per second}$$

141.
$$m = \frac{m_0}{\sqrt{1 - \frac{v^2}{c^2}}}$$
$$= \frac{m_0}{\sqrt{1 - \frac{(0.98c)^2}{c^2}}}$$
$$= \frac{m_0}{\sqrt{1 - \frac{0.9604c^2}{c^2}}}$$
$$= \frac{m_0}{\sqrt{1 - 0.9604}}$$
$$= \frac{m_0}{\sqrt{0.0396}}$$
$$\approx 5m_0$$
It is about 5 times its original mass.

142. $5 = 5 + 0i$

143. $-6 = -6 + 0i$

144. $2 - \sqrt{-256} = 2 - \sqrt{-1}\sqrt{256}$
$$= 2 - 16i$$

145. $3 + \sqrt{-16} = 3 + \sqrt{16}\sqrt{-1}$
$$= 3 + 4i$$

146. $(3 + 2i) + (4 - i) = 7 + i$

147. $(4 - 6i) - (3 - 4i) = 4 - 6i - 3 + 4i$
$$= 1 - 2i$$

148. $\left(\sqrt{3}+\sqrt{-5}\right)+\left(2\sqrt{3}-\sqrt{-7}\right)$
$=\sqrt{3}+\sqrt{5}\sqrt{-1}+2\sqrt{3}-\sqrt{7}\sqrt{-1}$
$=\sqrt{3}+i\sqrt{5}+2\sqrt{3}-i\sqrt{7}$
$=3\sqrt{3}+\left(\sqrt{5}-\sqrt{7}\right)i$

149. $\sqrt{-6}\left(\sqrt{6}+\sqrt{-6}\right)=\sqrt{6}\sqrt{-1}\left(\sqrt{6}+\sqrt{6}\sqrt{-1}\right)$
$=i\sqrt{6}\left(\sqrt{6}+i\sqrt{6}\right)$
$=i\sqrt{36}+i^2\sqrt{36}$
$=6i+6(-1)=-6+6i$

150. $(4+3i)(2-3i)=8-12i+6i-9i^2$
$=8-6i-9(-1)$
$=8-6i+9$
$=17-6i$

151. $\left(6+\sqrt{-3}\right)\left(4-\sqrt{-15}\right)$
$=\left(6+\sqrt{3}\sqrt{-1}\right)\left(4-\sqrt{15}\sqrt{-1}\right)$
$=\left(6+i\sqrt{3}\right)\left(4-i\sqrt{15}\right)$
$=24-6i\sqrt{15}+4i\sqrt{3}-i^2\sqrt{45}$
$=24-6i\sqrt{15}+4i\sqrt{3}-(-1)\sqrt{9}\sqrt{5}$
$=\left(24+3\sqrt{5}\right)+\left(4\sqrt{3}-6\sqrt{15}\right)i$

152. $\dfrac{2}{3i}=\dfrac{2}{3i}\cdot\dfrac{-3i}{-3i}$
$=\dfrac{-6i}{-9i^2}$
$=\dfrac{-6i}{-9(-1)}$
$=\dfrac{-6i}{9}$
$=\dfrac{-2i}{3}$

153. $\dfrac{2+\sqrt{3}}{2i}=\dfrac{2+\sqrt{3}}{2i}\cdot\dfrac{-2i}{-2i}$
$=\dfrac{-4i-2i\sqrt{3}}{-4i^2}$
$=\dfrac{-4i-2i\sqrt{3}}{-4(-1)}$
$=\dfrac{2\left(-2i-i\sqrt{3}\right)}{4}$
$=\dfrac{-2i-i\sqrt{3}}{2}$
$=\dfrac{\left(-2-\sqrt{3}\right)i}{2}$

154. $\dfrac{5}{3+2i}=\dfrac{5}{3+2i}\cdot\dfrac{3-2i}{3-2i}$
$=\dfrac{5(3-2i)}{9-4i^2}$
$=\dfrac{15-10i}{9-4(-1)}$
$=\dfrac{15-10i}{9+4}$
$=\dfrac{15-10i}{13}$

155. $\dfrac{\sqrt{3}}{5-\sqrt{-6}}=\dfrac{\sqrt{3}}{5-i\sqrt{6}}$
$=\dfrac{\sqrt{3}}{\left(5-i\sqrt{6}\right)}\cdot\dfrac{5+i\sqrt{6}}{5+i\sqrt{6}}$
$=\dfrac{5\sqrt{3}+i\sqrt{18}}{(5)^2-\left(i\sqrt{6}\right)^2}$
$=\dfrac{5\sqrt{3}+3i\sqrt{2}}{(5)^2+\left(\sqrt{6}\right)^2}$
$=\dfrac{5\sqrt{3}+3i\sqrt{2}}{25+6}$
$=\dfrac{5\sqrt{3}+3i\sqrt{2}}{31}$

156. x^2-2x+9
$=\left(1+2i\sqrt{2}\right)^2-2\left(1+2i\sqrt{2}\right)+9$
$=1^2+2(1)\left(2i\sqrt{2}\right)+\left(2i\sqrt{2}\right)^2-2-4i\sqrt{2}+9$
$=1+4i\sqrt{2}-8-2-4i\sqrt{2}+9$
$=0+0i$
$=0$

157. $x^2-2x+12$
$=(1-2i)^2-2(1-2i)+12$
$=1^2-2(1)(2i)+(2i)^2-2+4i+12$
$=1-4i-4-2+4i+12$
$=7+0i$
$=7$

158. $i^{53}=i^{52}i=(i^4)^{13}=1^{13}\cdot i=i$

159. $i^{19}=i^{16}i^3=(i^4)^4i^3=1^4\,?i^3=1(i^3)=i^3=-i$

160. $i^{404}=(i^4)^{101}=1^{101}=1$

161. $i^{5326} = i^{5324}i^2$
$= (i^4)^{1331}i^2$
$= 1^{1331} \cdot i^2$
$= 1(i^2)$
$= i^2$
$= -1$

Practice Test

1. $\sqrt{(5x-1)^2} = |5x-1|$

2. $\left(\frac{x^{2/5} \cdot x^{-1}}{x^{3/5}}\right)^2 = \left(x^{2/5-3/5-1}\right)^2$
$= \left(x^{2/5-3/5-5/5}\right)^2$
$= \left(x^{-6/5}\right)^2$
$= x^{-12/5}$
$= \frac{1}{x^{12/5}}$

3. $x^{-2/3} + x^{4/3} = x^{-2/3}(1) + x^{-2/3}\left(x^{6/3}\right)$
$= x^{-2/3}\left(1 + x^{6/3}\right)$
$= x^{-2/3}\left(1 + x^2\right)$
$= \frac{1+x^2}{x^{2/3}}$

4. $g(x) = \sqrt{x}$

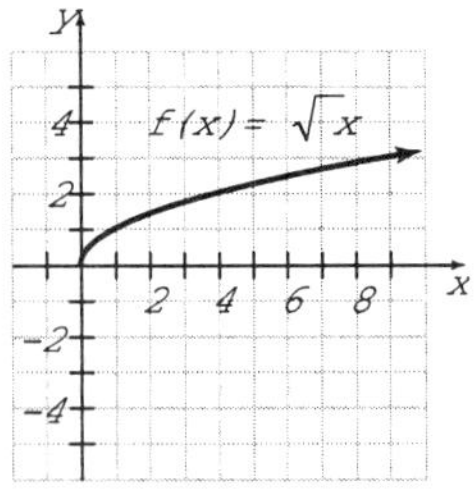

5. $\sqrt{48x^7y^{10}} = \sqrt{16x^6y^{10}}\sqrt{3x}$
$= 4x^3y^5\sqrt{3x}$

6. $\sqrt[3]{4x^5y^2}\sqrt[3]{10x^6y^8} = \sqrt[3]{40x^{11}y^{10}}$
$= \sqrt[3]{8x^9y^9} \cdot \sqrt[3]{5x^2y}$
$= 2x^3y^3\sqrt[3]{5x^2y}$

7. $\sqrt{\frac{3x^6y^3}{8z}} = \frac{\sqrt{3x^6y^3}}{\sqrt{8z}}$
$= \frac{\sqrt{x^6y^2}\sqrt{3y}}{\sqrt{4}\sqrt{2z}}$
$= \frac{x^3y\sqrt{3y}}{2\sqrt{2z}} \cdot \frac{\sqrt{2z}}{\sqrt{2z}}$
$= \frac{x^3y\sqrt{6yz}}{2(2z)}$
$= \frac{x^3y\sqrt{6yz}}{4z}$

8. $\frac{4}{\sqrt[3]{x}} = \frac{4}{\sqrt[3]{x}} \cdot \frac{\sqrt[3]{x^2}}{\sqrt[3]{x^2}} = \frac{4\sqrt[3]{x^2}}{x}$

9. $\frac{\sqrt{3}}{3+\sqrt{27}} = \frac{\sqrt{3}}{3+\sqrt{27}} \cdot \frac{3-\sqrt{27}}{3-\sqrt{27}}$
$= \frac{\sqrt{3}\left(3-\sqrt{27}\right)}{\left(3+\sqrt{27}\right)\left(3-\sqrt{27}\right)}$
$= \frac{3\sqrt{3}-\sqrt{81}}{9-3\sqrt{27}+3\sqrt{27}-27}$
$= \frac{3\sqrt{3}-9}{-18}$
$= \frac{\sqrt{3}-3}{-6}$ or $\frac{3-\sqrt{3}}{6}$

10. $2\sqrt{24} - 5\sqrt{6} + 3\sqrt{54}$
$= 2\sqrt{4}\sqrt{6} - 5\sqrt{6} + 3\sqrt{9}\sqrt{6}$
$= 4\sqrt{6} - 5\sqrt{6} + 9\sqrt{6}$
$= 8\sqrt{6}$

11. $\sqrt[3]{8x^3y^5} + 2\sqrt[3]{x^6y^8}$
$= \sqrt[3]{8x^3y^3}\sqrt[3]{y^2} + 2\sqrt[3]{x^6y^6}\sqrt[3]{y^2}$
$= 2xy\sqrt[3]{y^2} + 2x^2y^2\sqrt[3]{y^2}$
$= (2xy + 2x^2y^2)\sqrt[3]{y^2}$

12. $(\sqrt{3}-5)(6-\sqrt{8}) = \sqrt{3}(6)-\sqrt{3}\sqrt{8}-5(6)+5\sqrt{8}$
$= 6\sqrt{3}-\sqrt{24}-30+5\sqrt{8}$
$= 6\sqrt{3}-\sqrt{4\cdot 6}-30+5\sqrt{4\cdot 2}$
$= 6\sqrt{3}-2\sqrt{6}-30+10\sqrt{2}$

13. $\sqrt[4]{\sqrt{x^5y^3}} = \sqrt[4]{(x^5y^3)^{1/2}}$
$= \left[(x^5y^3)^{1/2}\right]^{1/4}$
$= (x^5y^3)^{1/8}$
$= \sqrt[8]{x^5y^3}$

14. $\dfrac{\sqrt[4]{(7x+2)^5}}{\sqrt[3]{(7x+2)^2}} = \dfrac{(7x+2)^{5/4}}{(7x+2)^{2/3}}$
$= (7x+2)^{5/4-2/3}$
$= (7x+2)^{7/12}$
$= \sqrt[12]{(7x+2)^7}$

15. $\sqrt{3x+4} = 5$
$(\sqrt{3x+4})^2 = 5^2$
$3x+4 = 25$
$3x = 21$
$x = 7$

Check: $\sqrt{3x+4} = 5$
$\sqrt{3\cdot 7+4} = 5$
$\sqrt{25} = 5$
$5 = 5$ True

7 is the solution.

16. $\sqrt{x^2-x-12} = x+3$
$(\sqrt{x^2-x-12})^2 = (x+3)^2$
$x^2-x-12 = x^2+6x+9$
$-x-12 = 6x+9$
$-12 = 7x+9$
$-21 = 7x$
$x = -3$

Check:
$\sqrt{(-3)^2-(-3)-12}\ 0 -3+3$
$\sqrt{9+3-12} = -3+3$
$\sqrt{0} = 0$
$0 = 0$
-3 is the solution.

17. $\sqrt{a-8} = \sqrt{a}-2$
$(\sqrt{a-8})^2 = (\sqrt{a}-2)^2$
$a-8 = a-4\sqrt{a}+4$
$-12 = -4\sqrt{a}$
$\sqrt{a} = 3$
$a = 3^2 = 9$

$\sqrt{a-8} = \sqrt{a}-2$
Check: $\sqrt{9-8} = \sqrt{9}-2$
$\sqrt{1} = 3-2$
$1 = 1$ True

9 is the solution.

18. $f(x) = g(x)$
$(9x+37)^{1/3} = 2(2x+2)^{1/3}$
$[(9x+37)^{1/3}]^3 = [2(2x+2)^{1/3}]^3$
$9x+37 = 8(2x+2)$
$9x+37 = 16x+16$
$21 = 7x$
$3 = x$

19. $w = \dfrac{\sqrt{2gh}}{4}$ Solve for g.
$4w = \sqrt{2gh}$
$(4w)^2 = 2gh$
$\dfrac{16w^2}{2h} = \dfrac{2gh}{2h}$
$\dfrac{8w^2}{h} = g$

20. $V = \sqrt{64.4h}$
$V = \sqrt{64.4(200)}$
$= \sqrt{12{,}880}$
≈ 113.49 ft/sec

21. Let x be the length of the ladder.

$$x = \sqrt{12^2 + 5^2}$$
$$= 169$$
$$= 13 \text{ feet}$$

22. $$T = 2\pi\sqrt{\frac{m}{k}}$$
$$T = 2\pi\sqrt{\frac{1400}{65{,}000}}$$
$$\approx 0.92 \text{ sec}$$

23. $$\left(6-\sqrt{-4}\right)\left(2+\sqrt{-9}\right) = (6-2i)(2+3i)$$
$$= 12 + 18i - 4i - 6i^2$$
$$= 12 + 14i - 6(-1)$$
$$= 12 + 14i + 6$$
$$= 18 + 14i$$

24. $$\frac{5-i}{7+2i} = \frac{5-i}{7+2i}\cdot\frac{7-2i}{7-2i}$$
$$= \frac{(5-i)(7-2i)}{(7+2i)(7-2i)}$$
$$= \frac{35-10i-7i+2i^2}{49-14i+14i-4i^2}$$
$$= \frac{35-17i+2(-1)}{49-4(-1)}$$
$$= \frac{35-17i-2}{49+4}$$
$$= \frac{33-17i}{53}$$

25. $$x^2 + 6x + 12$$
$$= (-3+i)^2 + 6(-3+i) + 12$$
$$= (-3)^2 - 2(3)(i) + (i)^2 - 18 + 6i + 12$$
$$= 9 - 6i - 1 - 18 + 6i + 12$$
$$= 2 + 0i$$
$$= 2$$

Cumulative Review Test

1. $$\frac{1}{5}(x-3) = \frac{3}{4}(x+3) - x$$
$$20\left[\frac{1}{5}(x-3)\right] = 20\left[\frac{3}{4}(x+3)\right] - 20x$$
$$4(x-3) = 5(3)(x+3) - 20x$$
$$4x - 12 = 15x + 45 - 20x$$
$$4x - 12 = -5x + 45$$
$$9x - 12 = 45$$
$$9x = 57$$
$$x = \frac{57}{9}$$

2. $$3(x-4) = 6x - (4-5x)$$
$$3x - 12 = 6x - 4 + 5x$$
$$3x - 12 = 11x - 4$$
$$-8x = 8$$
$$x = -1$$

3. Let x be the original price of the sweater.

$$x - 60\%x = 20$$
$$x - .60x = 20$$
$$.40x = 20$$
$$x = \frac{20}{.40}$$
$$x = 50$$

The original price of the sweater is $50.

4. $$y = \frac{3}{2}x - 3$$

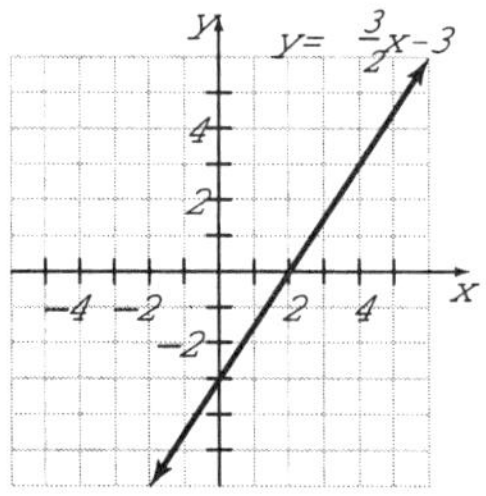

5. $$(5xy-6)(5xy+6)$$
$$= 25x^2y^2 + 30xy - 30xy - 36$$
$$= 25x^2y^2 - 36$$

6. $V = lwh$

$6r^3 + 5r^2 + r = (3r+1)(w)(r)$

$\dfrac{6r^3 + 5r^2 + r}{(3r+1)(r)} = \dfrac{(3r+1)(w)(r)}{(3r+1)(r)}$

$\dfrac{r\left(6r^2 + 5r + 1\right)}{r(3r+1)} = w$

$\dfrac{r(3r+1)(2r+1)}{r(3r+1)} = w \text{ or } w = 2r+1$

7. $4x^3 - 9x^2 + 5x = x\left(4x^2 - 9x + 5\right)$

$= x(4x-5)(x-1)$

8. $(x+1)^3 - 8$

$= (x+1)^3 - 2^3$

$= (x+1-2)\left((x+1)^2 + 2(x+1) + 2^2\right)$

$= (x-1)\left(x^2 + 2x + 1 + 2x + 2 + 4\right)$

$= (x-1)\left(x^2 + 4x + 7\right)$

9. $8x^2 - 3 = -10x$

$8x^2 + 10x - 3 = 0$

$(4x-1)(2x+3) = 0$

$4x - 1 = 0$ or $2x + 3 = 0$

$4x = 1$ $\quad$ $2x = -3$

$x = \dfrac{1}{4}$ $\quad$ $x = -\dfrac{3}{2}$

10. $\dfrac{4x+4y}{x^2y} \cdot \dfrac{y^3}{8x} = \dfrac{4(x+y)}{x^2y} \cdot \dfrac{y^3}{8x}$

$= \dfrac{x+y}{x^2} \cdot \dfrac{y^2}{2x}$

$= \dfrac{(x+y)y^2}{2x^3}$

11. $\dfrac{x-4}{x-5} - \dfrac{3}{x+5} - \dfrac{10}{x^2-25}$

$= \dfrac{x-4}{x-5} - \dfrac{3}{x+5} - \dfrac{10}{(x+5)(x-5)}$

$= \dfrac{x-4}{x-5} \cdot \dfrac{x+5}{x+5} - \dfrac{3}{x+5} \cdot \dfrac{x-5}{x-5} - \dfrac{10}{(x+5)(x-5)}$

$= \dfrac{(x-4)(x+5)}{(x+5)(x-5)} - \dfrac{3(x-5)}{(x+5)(x-5)} - \dfrac{10}{(x+5)(x-5)}$

$= \dfrac{(x-4)(x+5) - 3(x-5) - 10}{(x+5)(x-5)}$

$= \dfrac{x^2 + x - 20 - 3x + 15 - 10}{(x+5)(x-5)}$

$= \dfrac{x^2 \;\; 2x - 15}{(x+5)(x-5)}$

$= \dfrac{(x-5)(x+3)}{(x+5)(x-5)}$ or $\dfrac{x+3}{x+5}$

12. $\dfrac{4}{x} - \dfrac{1}{6} = \dfrac{1}{x}$

$6x\left(\dfrac{4}{x} - \dfrac{1}{6} = \dfrac{1}{x}\right)$

$24 - x = 6$

$-x = -18$

$x = 18$

Upon checking, this value satisfies the equation. The solution is 18.

13. $y = 3x - 6 \;\Rightarrow\; m_1 = 3$

$6y = 18x + 6 \;\Rightarrow\; y = 3x + 1 \;\Rightarrow\; m_2 = 3$

The slopes of both lines is 3. Since the slopes of the two lines are the same, the lines are parallel.

14. First find the slope of the given line.

$3x - 2y = 6$

$-2y = -3x + 6$

$y = \dfrac{3}{2}x - 3 \;\Rightarrow$ The slope is $\dfrac{3}{2}$.

The slope of any line parallel to this line

must have an opposite reciprocal slope. Therefore, the slope of the line perpendicular to the given line is $-\frac{2}{3}$. Finally, use the point-slope formula to find the equation of the line.

$y - y_1 = m(x - x_1)$ with $m = -\frac{2}{3}$ an $(3,-4)$

$$y-(-4) = -\frac{2}{3}(x-3)$$
$$y+4 = -\frac{2}{3}x+2$$
$$y = -\frac{2}{3}x-2$$

15. $f(x) = x^2 - 3x + 4$ and $g(x) = 2x - 5$

$$(g-f)(x) = (2x-5)-(x^2-3x+4)$$
$$= 2x-5-x^2+3x-4$$
$$= -x^2+5x-9$$

16.
$$x + 2y = 12 \quad (1)$$
$$4x = 8 \quad (2)$$
$$3x - 4y + 5z = 20 \quad (3)$$

Using equation (2), solve for x.

$4x = 8 \Rightarrow x = \frac{8}{4} \Rightarrow x = 2$

Substitute 2 for x in equation (1) in order to solve for y.

$2 + 2y = 12 \Rightarrow 2y = 10 \Rightarrow y = 5$

Substitute 2 for x and 5 for y in equation (3) in order to solve for z.

$$3(2)-4(5)+5z = 20$$
$$6-20+5z = 20$$
$$-14+5z = 20$$
$$5z = -34$$
$$z = -\frac{34}{5}$$

The solution is $\left(2, 5, -\frac{34}{5}\right)$.

17. $\begin{vmatrix} 3 & -6 & -1 \\ 2 & 1 & -2 \\ 1 & 3 & 1 \end{vmatrix}$

$$= 3\begin{vmatrix} 1 & -2 \\ 3 & 1 \end{vmatrix} - (-6)\begin{vmatrix} 2 & -2 \\ 1 & 1 \end{vmatrix} + (-1)\begin{vmatrix} 2 & 1 \\ 1 & 3 \end{vmatrix}$$
$$= 3(1-(-6)) + 6(2-(-2)) - (6-1)$$
$$= 3(1+6) + 6(2+2) - (6-1)$$
$$= 3(7) + 6(4) - (5)$$
$$= 21 + 24 - 5$$
$$= 40$$

18.
$$|3-2x| < 7$$
$$-7 < 3-2x < 7$$
$$-7-3 < 3-2x-3 < 7-3$$
$$-10 < -2x < 4$$
$$\frac{-10}{-2} > \frac{-2x}{-2} > \frac{4}{-2}$$
$$5 > x > -2 \text{ or } -2 < x < 5$$

The solution set is $\{x|-2 < x < 5\}$

19.
$$\sqrt{2x^2+7}+3 = 8$$
$$\sqrt{2x^2+7} = 5$$
$$\left(\sqrt{2x^2+7}\right)^2 = 5^2$$
$$2x^2+7 = 25$$
$$2x^2 = 18$$
$$x^2 = 9$$
$$x = 3 \text{ or } -3$$

$\sqrt{2(3)^2+7}+3 = 8$ or $\sqrt{2(-3)^2+7}+3 = 8$

$\sqrt{25} = 5$ $\quad\quad$ $\sqrt{25} = 5$

$5 = 5$ $\quad\quad$ $5 = 5$

Both values check. The solution is 3 and –3.

20. $d = kt^2$

$16 = k(1)^2 \quad \Rightarrow \quad k = 16$

$$\begin{aligned} d &= 16(5)^2 \\ &= 16 \cdot 25 \\ &= 400 \end{aligned}$$

The object will fall 400 feet in 5 seconds.

Chapter 12

Exercise Set 12.1

1. The two square roots of 36 are $\pm\sqrt{36} = \pm 6$.

3. The square root property is: If $x^2 = a$, where a is a real number, then $x = \pm\sqrt{a}$.

5. A trinomial, $x^2 + bx + c$, is a perfect square trinomial if $\left(\frac{b}{2}\right)^2 = c$.

7. **a.** Yes, $x = 9$ is the solution to the equation. It is the only real number that satisfies the equation.

   b. No, $x = 3$ is not the solution. Both –3 and 3 satisfy the equation.

9. Multiply the equation by 1/2.

11. You should add the square of half the coefficient of the first degree term: $\left(\frac{-6}{2}\right)^2 = (-3)^2 = 9$.

13.
$$x^2 = 49$$
$$x = \pm\sqrt{49} = \pm 7$$

15.
$$x^2 + 49 = 0$$
$$x^2 = -49$$
$$x = \pm\sqrt{-49} = \pm 7i$$

17.
$$y^2 + 48 = 0$$
$$y^2 = -48$$
$$y = \pm\sqrt{-48} = \pm 4i\sqrt{3}$$

19.
$$y^2 + 11 = -50$$
$$y^2 = -61$$
$$y = \pm\sqrt{-61} = \pm i\sqrt{61}$$

21.
$$(p-4)^2 = 16$$
$$p - 4 = \pm\sqrt{16}$$
$$p - 4 = \pm 4$$
$$p = 4 \pm 4$$
$$p = 4 + 4 \quad \text{or} \quad P = 4 - 4$$
$$P = 8 \qquad\qquad P = 0$$

23.
$$(x+3)^2 + 49 = 0$$
$$(x+3)^2 = -49$$
$$x + 3 = \pm\sqrt{-49}$$
$$x + 3 = \pm\sqrt{-49}$$
$$x = -3 \pm 7i$$

25.
$$(a-4)^2 + 45 = 0$$
$$(a-4)^2 = -45$$
$$a - 4 = \pm\sqrt{-45}$$
$$a - 4 = \pm 3i\sqrt{5}$$
$$a = 4 \pm 3i\sqrt{5}$$

27.
$$\left(b + \frac{1}{3}\right)^2 = \frac{4}{9}$$
$$b + \frac{1}{3} = \pm\sqrt{\frac{4}{9}}$$
$$b + \frac{1}{3} = \pm\frac{2}{3}$$
$$b = -\frac{1}{3} \pm \frac{2}{3}$$
$$b = -\frac{1}{3} + \frac{2}{3} \quad \text{or} \quad b = -\frac{1}{3} - \frac{2}{3}$$
$$b = \frac{1}{3} \qquad\qquad b = -\frac{3}{3}$$
$$b = -1$$

29.
$$\left(b - \frac{1}{3}\right)^2 + \frac{4}{9} = 0$$
$$\left(b - \frac{1}{3}\right)^2 = -\frac{4}{9}$$
$$b - \frac{1}{3} = \pm\sqrt{-\frac{4}{9}}$$
$$b - \frac{1}{3} = \pm\frac{2}{3}i$$
$$b = \frac{1}{3} \pm \frac{2}{3}i \quad \text{or} \quad b = \frac{1 \pm 2i}{3}$$

31.
$$(x + 1.8)^2 = 0.81$$
$$x + 1.8 = \pm\sqrt{0.81}$$
$$x + 1.8 = \pm 0.9$$
$$x = -1.8 \pm 0.9$$
$$x = -1.8 + 0.9 \quad \text{or} \quad x = -1.8 - 0.9$$
$$x = -0.9 \qquad\qquad x = -2.7$$

33. $(2a-5)^2 = 12$

$2a-5 = \pm\sqrt{12}$

$2a-5 = \pm 2\sqrt{3}$

$2a = 5 \pm 2\sqrt{3}$

$a = \dfrac{5 \pm 2\sqrt{3}}{2}$

35. $\left(2y+\dfrac{1}{2}\right)^2 = \dfrac{4}{25}$

$2y+\dfrac{1}{2} = \pm\sqrt{\dfrac{4}{25}}$

$2y+\dfrac{1}{2} = \pm\dfrac{2}{5}$

$2y+\dfrac{1}{2} = \dfrac{2}{5}$ or $2y+\dfrac{1}{2} = -\dfrac{2}{5}$

$2y = -\dfrac{1}{2}+\dfrac{2}{5}$ $\quad$ $2y = -\dfrac{1}{2}-\dfrac{2}{5}$

$2y = -\dfrac{1}{10}$ $\quad$ $2y = -\dfrac{9}{10}$

$y = -\dfrac{1}{20}$ $\quad$ $y = -\dfrac{9}{20}$

37. $x^2+3x-4=0$

$x^2+3x=4$

$x^2+3x+\dfrac{9}{4} = 4+\dfrac{9}{4}$

$\left(x+\dfrac{3}{2}\right)^2 = \dfrac{25}{4}$

$x+\dfrac{3}{2} = \pm\sqrt{\dfrac{25}{4}}$

$x+\dfrac{3}{2} = \pm\dfrac{5}{2}$

$x = -\dfrac{3}{2}\pm\dfrac{5}{2}$

$x = -\dfrac{3}{2}+\dfrac{5}{2}$ or $x = -\dfrac{3}{2}-\dfrac{5}{2}$

$x = \dfrac{2}{2}$ $\quad$ $x = \dfrac{-8}{2}$

$x = 1$ $\quad$ $x = -4$

39. $x^2+2x-15=0$

$x^2+2x=15$

$x^2+2x+1 = 15+1$

$(x+1)^2 = 16$

$x+1 = \pm 4$

$x = \pm 4 - 1$

$x = 4-1$ or $x = -4-1$

$x = 3$ $\quad$ $x = -5$

41. $x^2-6x+8=0$

$x^2-6x=-8$

$x^2-6x+9 = -8+9$

$x^2-6x+9 = 1$

$(x-3)^2 = 1$

$x-3 = \pm 1$

$x = \pm 1 + 3$

$x = 1+3$ or $x = -1+3$

$x = 4$ $\quad$ $x = 2$

43. $x^2-6x+5=0$

$x^2-6x=-5$

$x^2-6x+9=-5+9$

$x^2-6x+9=4$

$(x-3)^2 = 4$

$x-3=\pm 2$

$x=\pm 2+3$

$x=2+3$ or $x=-2+3$

$x=5$ $\quad$ $x=1$

45. $2x^2+x-1=0$

$\dfrac{1}{2}\left(2x^2+x-1\right) = \dfrac{1}{2}(0)$

$x^2+\dfrac{1}{2}x-\dfrac{1}{2}=0$

$x^2+\dfrac{1}{2}x=\dfrac{1}{2}$

$x^2+\dfrac{1}{2}x+\dfrac{1}{16}=\dfrac{1}{2}+\dfrac{1}{16}$

$\left(x+\dfrac{1}{4}\right)^2 = \dfrac{9}{16}$

$x+\dfrac{1}{4} = \pm\sqrt{\dfrac{9}{16}}$

$x+\dfrac{1}{4} = \pm\dfrac{3}{4}$

$x = \pm\dfrac{3}{4}-\dfrac{1}{4}$

$x = \dfrac{3}{4}-\dfrac{1}{4}$ or $x = -\dfrac{3}{4}-\dfrac{1}{4}$

$x = \dfrac{2}{4}$ $\quad$ $x = -\dfrac{4}{4}$

$x = \dfrac{1}{2}$ $\quad$ $x = -1$

47.
$$2z^2-7z-4=0$$
$$\frac{1}{2}\left(2z^2-7z-4\right)=\frac{1}{2}(0)$$
$$z^2-\frac{7}{2}z-2=0$$
$$z^2-\frac{7}{2}z=2$$
$$z^2-\frac{7}{2}z+\frac{49}{16}=2+\frac{49}{16}$$
$$\left(z-\frac{7}{4}\right)^2=\frac{81}{16}$$
$$z-\frac{7}{4}=\pm\sqrt{\frac{81}{16}}$$
$$z-\frac{7}{4}=\pm\frac{9}{4}$$
$$z=\pm\frac{9}{4}+\frac{7}{4}$$
$z=\frac{9}{4}+\frac{7}{4}$ or $z=-\frac{9}{4}+\frac{7}{4}$

$z=\frac{16}{4}$ $\quad z=-\frac{2}{4}$

$z=4$ $\quad z=-\frac{1}{2}$

49.
$$x^2-11x+28=0$$
$$x^2-11x=-28$$
$$x^2-11x+\frac{121}{4}=-28+\frac{121}{4}$$
$$\left(x-\frac{11}{2}\right)^2=\frac{9}{4}$$
$$x-\frac{11}{2}=\pm\sqrt{\frac{9}{4}}$$
$$x-\frac{11}{2}=\pm\frac{3}{2}$$
$$x=\pm\frac{3}{2}+\frac{11}{2}$$
$x=\frac{3}{2}+\frac{11}{2}$ or $x=-\frac{3}{2}+\frac{11}{2}$

$x=\frac{14}{2}$ $\quad x=\frac{8}{2}$

$x=7$ $\quad x=4$

51.
$$-x^2+3x+4=0 \Leftarrow \text{multiply by } -1$$
$$x^2-3x-4=0$$
$$x^2-3x=4$$
$$x^2-3x+\frac{9}{4}=4+\frac{9}{4}$$
$$\left(x-\frac{3}{2}\right)^2=\frac{25}{4}$$
$$x-\frac{3}{2}=\pm\sqrt{\frac{25}{4}}$$
$$x-\frac{3}{2}=\pm\frac{5}{2}$$
$$x=\frac{3}{2}\pm\frac{5}{2}$$
$x=\frac{3}{2}+\frac{5}{2}$ or $x=\frac{3}{2}-\frac{5}{2}$

$x=\frac{8}{2}$ $\quad x=\frac{-2}{2}$

$x=4$ $\quad x=-1$

53.
$$-z^2+9z-20=0 \Leftarrow \text{multiply by } -1$$
$$z^2-9z+20=0$$
$$z^2-9z=-20$$
$$z^2-9z+\frac{81}{4}=-20+\frac{81}{4}$$
$$\left(z-\frac{9}{2}\right)^2=\frac{1}{4}$$
$$z-\frac{9}{2}=\pm\frac{1}{2}$$
$$z=\pm\frac{1}{2}+\frac{9}{2}$$
$z=\frac{1}{2}+\frac{9}{2}$ or $z=-\frac{1}{2}+\frac{9}{2}$

$z=\frac{10}{2}$ $\quad z=\frac{8}{2}$

$z=5$ $\quad z=4$

55.
$$b^2=3b+28$$
$$b^2-3b=28$$
$$b^2-3b+\frac{9}{4}=\frac{112}{4}+\frac{9}{4}$$
$$\left(b-\frac{3}{2}\right)^2=\frac{121}{4}$$
$$b-\frac{3}{2}=\pm\frac{11}{2}$$
$$b=\pm\frac{11}{2}+\frac{3}{2}$$

$b=-\frac{11}{2}+\frac{3}{2}$ or $b=\frac{11}{2}+\frac{3}{2}$

$b=-\frac{8}{2}$ $\quad$ $b=\frac{14}{2}$

$b=-4$ $\quad$ $b=7$

57. $x^2+9x=10$

$x^2+9x+\frac{81}{4}=10+\frac{81}{4}$

$\left(x+\frac{9}{2}\right)^2=\frac{40}{4}+\frac{81}{4}$

$\left(x+\frac{9}{2}\right)^2=\frac{121}{4}$

$x+\frac{9}{2}=\pm\frac{11}{2}$

$x=\pm\frac{11}{2}-\frac{9}{2}$

$x=\frac{11}{2}-\frac{9}{2}$ or $x=-\frac{11}{2}-\frac{9}{2}$

$x=\frac{2}{2}$ $\quad$ $x=-\frac{20}{2}$

$x=1$ $\quad$ $x=-10$

59. $x^2-4x-10=0$

$x^2-4x=10$

$x^2-4x+4=10+4$

$(x-2)^2=14$

$x-2=\pm\sqrt{14}$

$x=2\pm\sqrt{14}$

61. $r^2+8r+5=0$

$r^2+8r=-5$

$r^2+8r+16=-5+16$

$(r+4)^2=11$

$r+4=\pm\sqrt{11}$

$r=-4\pm\sqrt{11}$

63. $c^2-c-3=0$

$c^2-c=3$

$c^2-c+\frac{1}{4}=3+\frac{1}{4}$

$\left(c-\frac{1}{2}\right)^2=\frac{13}{4}$

$c-\frac{1}{2}=\pm\sqrt{\frac{13}{4}}$

$c=\frac{1}{2}\pm\frac{\sqrt{13}}{2}$

$c=\frac{1\pm\sqrt{13}}{2}$

65. $x^2+3x+6=0$

$x^2+3x=-6$

$x^2+3x+\frac{9}{4}=-\frac{24}{4}+\frac{9}{4}$

$\left(x+\frac{3}{2}\right)^2=\frac{-15}{4}$

$x+\frac{3}{2}=\pm\sqrt{\frac{-15}{4}}$

$x+\frac{3}{2}=\pm\frac{i\sqrt{15}}{2}$

$x=-\frac{3}{2}\pm\frac{i\sqrt{15}}{2}$

$x=\frac{-3\pm i\sqrt{15}}{2}$

67. $2x^2-2x=0$

$x^2-x=0$

$x^2-x+\frac{1}{4}=\frac{1}{4}$

$\left(x-\frac{1}{2}\right)^2=\frac{1}{4}$

$x-\frac{1}{2}=\pm\frac{1}{2}$

$x=\pm\frac{1}{2}+\frac{1}{2}$

$x=\frac{1}{2}+\frac{1}{2}$ or $x=-\frac{1}{2}+\frac{1}{2}$

$x=\frac{2}{2}$ $\quad$ $x=0$

$x=1$ $\quad$ $x=0$

69. $-\frac{1}{4}b^2-\frac{1}{2}b=0$

$-4\left(-\frac{1}{4}b^2-\frac{1}{2}b=0\right)$

$b^2+2b=0$

$b^2+2b+1=0+1$

$(b+1)^2=1$

$b+1=\pm 1$

$b+1=1$ or $b+1=-1$

$b=0$ $\quad$ $b=-2$

71. $18z^2-6z=0$

$z^2-\frac{1}{3}z=0$

$z^2-\frac{1}{3}z+\frac{1}{36}=0+\frac{1}{36}$

$\left(z-\frac{1}{6}\right)^2=\frac{1}{36}$

$z-\frac{1}{6}=\pm\frac{1}{6}$

$z=\pm\frac{1}{6}+\frac{1}{6}$

$z=\frac{1}{6}+\frac{1}{6}$ or $z=-\frac{1}{6}+\frac{1}{6}$

$z=\frac{2}{6}$ $\quad$ $z=0$

$z=\frac{1}{3}$ $\quad$ $z=0$

73. $-\frac{1}{2}p^2-p+\frac{3}{2}=0$

$p^2+2p-3=0$

$p^2+2p=3$

$p^2+2p+1=3+1$

$(p+1)^2=4$

$p+1=\pm 2$

$p=\pm 2-1$

$p=2-1$ or $p=-2-1$

$p=1$ $\quad$ $p=-3$

75. $2x^2=8x+90$

$x^2=4x+45$

$x^2-4x=45$

$x^2-4x+4=45+4$

$(x-2)^2=49$

$x-2=\pm 7$

$x=\pm 7+2$

$x=7+2$ or $x=-7+2$

$x=9$ or $x=-5$

77. $3x^2+33x+72=0$

$x^2+11x+24=0$

$x^2+11x=-24$

$x^2+11x+\frac{121}{4}=-24+\frac{121}{4}$

$\left(x+\frac{11}{2}\right)^2=-\frac{96}{4}+\frac{121}{4}$

$\left(x-\frac{11}{2}\right)^2=\frac{25}{4}$

$x+\frac{11}{2}=\pm\frac{5}{2}$

$x=\pm\frac{5}{2}-\frac{11}{2}$

$x=-\frac{5}{2}-\frac{11}{2}$ or $x=\frac{5}{2}-\frac{11}{2}$

$x=-\frac{16}{2}$ or $x=-\frac{6}{2}$

$x=-8$ or $x=-3$

79. $3w^2+2w-1=0$

$3w^2+2w=1$

$w^2+\frac{2}{3}w=\frac{1}{3}$

$w^2+\frac{2}{3}w+\frac{1}{9}=\frac{1}{3}+\frac{1}{9}$

$\left(w+\frac{1}{3}\right)^2=\frac{4}{9}$

$w+\frac{1}{3}=\pm\frac{2}{3}$

$w=\pm\frac{2}{3}-\frac{1}{3}$

$w=-\frac{2}{3}-\frac{1}{3}$ or $w=\frac{2}{3}-\frac{1}{3}$

$w=-\frac{3}{3}$ $\quad$ $w=\frac{1}{3}$

$w=-1$

81.
$$2x^2 - x = -5$$
$$x^2 - \frac{1}{2}x = -\frac{5}{2}$$
$$x^2 - \frac{1}{2}x + \frac{1}{16} = -\frac{40}{16} + \frac{1}{16}$$
$$\left(x - \frac{1}{4}\right)^2 = -\frac{39}{16}$$
$$x - \frac{1}{4} = \pm\frac{i\sqrt{39}}{4}$$
$$x = \frac{1}{4} \pm \frac{i\sqrt{39}}{4}$$
$$x = \frac{1 \pm i\sqrt{39}}{4}$$

83.
$$\frac{5}{2}x^2 + \frac{3}{2}x - \frac{5}{4} = 0$$
$$\frac{2}{5}\left[\frac{5}{2}x^2 + \frac{3}{2}x - \frac{5}{4} = 0\right]$$
$$x^2 + \frac{3}{5}x - \frac{1}{2} = 0$$
$$x^2 + \frac{3}{5}x = \frac{1}{2}$$
$$x^2 + \frac{3}{5}x + \frac{9}{100} = \frac{1}{2} + \frac{9}{100}$$
$$\left(x + \frac{3}{10}\right)^2 = \frac{59}{100}$$
$$x + \frac{3}{10} = \pm\frac{\sqrt{59}}{10}$$
$$x = -\frac{3}{10} \pm \frac{\sqrt{59}}{10}$$
$$x = \frac{-3 \pm \sqrt{59}}{10}$$

85. **a.** $21 = (x+2)(x-2)$

b.
$$21 = (x+2)(x-2)$$
$$21 = x^2 - 2x + 2x - 4$$
$$0 = x^2 - 25$$
$$0 = (x+5)(x-5)$$
$$x + 5 = 0 \quad \text{or} \quad x - 5 = 0$$
$$x = -5 \qquad\qquad x = 5$$

Disregard the negative answer since x represents a distance. $x = 5$.

87. **a.** $18 = (x+4)(x+2)$

b.
$$18 = (x+4)(x+2)$$
$$18 = x^2 + 2x + 4x + 8$$
$$0 = x^2 + 6x - 10$$

Using the quadratic formula:
$$x = \frac{-(6) \pm \sqrt{6^2 - 4(1)(-10)}}{2(1)}$$
$$x = \frac{-6 \pm \sqrt{76}}{2}$$
$$x = \frac{-6 \pm 2\sqrt{19}}{2}$$
$$x = -3 + \sqrt{19}$$

Disregard the negative answer since x represents a distance. $x = -3 + \sqrt{19}$.

89.
$$d = \frac{1}{6}x^2$$
$$24 = \frac{1}{6}x^2$$
$$6 \cdot 24 = x^2$$
$$144 = x^2$$
$$x = 12$$

The car's speed was about 12 mph.

91. Let x be the first integer. Then $x + 2$ is the next consecutive odd integer.
$$x(x+2) = 63$$
$$x^2 + 2x = 63$$
$$x^2 + 2x + 1 = 63 + 1$$
$$(x+1)^2 = 64$$
$$x + 1 = \pm 8$$
$$x = -1 \pm 8$$
$$x = -1 + 8 \quad \text{or} \quad x = -1 - 8$$
$$x = 7 \qquad\qquad x = -9$$

Since it was given that the integers are positive, one integer is 7 and the other is $7 + 2 = 9$.

93. Let x be the width of the rectangle. Then $2x + 2$ is the length. Use length · width = area.

$$(2x+2)x = 60$$
$$2x^2 + 2x = 60$$
$$x^2 + x = 30$$
$$x^2 + x + \frac{1}{4} = 30 + \frac{1}{4}$$
$$\left(x+\frac{1}{2}\right)^2 = \frac{120}{4} + \frac{1}{4}$$
$$\left(x+\frac{1}{2}\right)^2 = \frac{121}{4}$$
$$x + \frac{1}{2} = \pm\frac{11}{2}$$
$$x = -\frac{1}{2} \pm \frac{11}{2}$$
$$x = -\frac{1}{2} + \frac{11}{2} \quad \text{or} \quad x = -\frac{1}{2} - \frac{11}{2}$$
$$x = \frac{10}{2} = 5 \qquad x = -\frac{12}{2} = 6$$

Since the width cannot be negative, the width is 5 ft.

Length $= 2(5) + 2 = 10 + 2 = 12$ ft.

The rectangle is 5 ft by 12 ft.

95. Let s be the length of the side. Then $s + 6$ is the length of the diagonal (d). Use $s^2 + s^2 = d^2$.

$$2s^2 = (s+6)^2$$
$$2s^2 = s^2 + 12s + 36$$
$$s^2 = 12s + 36$$
$$s^2 - 12s = 36$$
$$s^2 - 12s + 36 = 36 + 36$$
$$(s-6)^2 = 72$$
$$s - 6 = \pm 6\sqrt{2}$$
$$s = 6 \pm 6\sqrt{2}$$

Length is never negative. Thus, $s = 6 + 6\sqrt{2} \approx 14.49$.

The room is about 14.49 ft by 14.49 ft.

97. Since the radius is 10 inches, the diameter (d) is 20 inches. Use the formula $s^2 + s^2 = d^2$ to find the length (s) of the other two sides.

$$s^2 + s^2 = d^2$$
$$2s^2 = 20^2$$
$$2s^2 = 400$$
$$s^2 = 200$$
$$s = \pm\sqrt{200} = \pm 10\sqrt{2}$$

Length is never negative.

Thus $s = 10\sqrt{2} \approx 14.14$ inches.

99.
$$A = \pi r^2$$
$$24\pi = \pi r^2$$
$$24 = r^2$$
$$\pm\sqrt{24} = r$$
$$\pm 2\sqrt{6} = r$$

Length is never negative.

Thus $r = 2\sqrt{6} \approx 4.90$ feet.

101.
$$A = P\left(1 + \frac{r}{n}\right)^{nt}$$
$$540.80 = 500\left(1+\frac{r}{1}\right)^{1(2)}$$
$$540.80 = 500(1+r)^2$$
$$1.0816 = (1+r)^2$$
$$\pm 1.04 = 1 + r$$
$$-1 \pm 1.04 = r$$

An interest rate is never negative. Thus $r = -1 + 1.04 = 0.04 = 4\%$.

103.
$$A = P\left(1 + \frac{r}{n}\right)^{nt}$$
$$1432.86 = 1200\left(1+\frac{r}{2}\right)^{2(3)}$$
$$1432.86 = 1200\left(1+\frac{r}{2}\right)^{6}$$
$$1.19405 = \left(1+\frac{r}{2}\right)^{6}$$
$$\pm 1.03 \approx 1 + \frac{r}{2}$$
$$-1 \pm 1.03 \approx \frac{r}{2}$$
$$-2 \pm 2.06 \approx r$$

An interest rate is never negative.

Thus Steve Rodi's annual interest rate is about $-2 + 2.06 = 0.06 = 6\%$.

105. **a.** To find the surface area, we must first determine the radius. Use $V = \pi r^2 h$ with $V = 160$ and $h = 10$ to get

$$160 = \pi r^2(10)$$
$$16 = \pi r^2$$
$$\frac{16}{\pi} = r^2$$
$$\frac{4}{\sqrt{\pi}} = r$$

Since the radius equals $\frac{4}{\sqrt{\pi}}$, use the formula $S = 2\pi r^2 + 2\pi rh$ to calculate the surface area.

$$S = 2\pi\left(\frac{4}{\sqrt{\pi}}\right)^2 + 2\pi\left(\frac{4}{\sqrt{\pi}}\right)(10)$$
$$= 2\pi\left(\frac{16}{\pi}\right) + \frac{80\pi}{\sqrt{\pi}}$$
$$= 32 + 80\sqrt{\pi}$$
$$\approx 173.80$$

The surface area is about 173.80 square inches.

b. Use $V = \pi r^2 h$ with $V = 160$ and $h = 10$ to obtain $160 = \pi r^2(10)$. In part (a) this was solved for r to get

$$r = \frac{4}{\sqrt{\pi}} = \frac{4}{\sqrt{\pi}} \cdot \frac{\sqrt{\pi}}{\sqrt{\pi}} = \frac{4\sqrt{\pi}}{\pi} \approx 2.26$$

The radius is about 2.26 inches.

c. Use $S = 2\pi r^2 + 2\pi rh$ with $S = 160$ and $h = 10$.

$$160 = 2\pi r^2 + 2\pi r(10)$$
$$160 = 2\pi r^2 + 20\pi r$$
$$\frac{160}{2\pi} = \frac{2\pi r^2}{2\pi} + \frac{20\pi r}{2\pi}$$
$$\frac{80}{\pi} = r^2 + 10r$$
$$\frac{80}{\pi} + 25 = r^2 + 10r + 25$$
$$\frac{80 + 25\pi}{\pi} = (r+5)^2$$
$$\pm\sqrt{\frac{80 + 25\pi}{\pi}} = r + 5$$
$$\pm\sqrt{\frac{80 + 25\pi}{\pi}} - 5 = r$$

The radius is never negative. Thus $r \approx 2.1$ inches.

107.
$$-4(2z-6) = -3(z-4) + z$$
$$-8z + 24 = -3z + 12 + z$$
$$-8z + 24 = -2z + 12$$
$$-6z = -12$$
$$z = 2$$

108. Let x = the amount invested at 7%. Then the amount invested at $6\frac{1}{4}\%$ will be $10{,}000 - x$. The interest earned at 7% will be $0.07x$. and the amount of interest earned at 6.25% will be $.0625(10{,}000 - x)$. The total interest earned is \$656.50.

$$0.07x + 0.0625(10{,}000 - x) = 656.50$$
$$0.07x + 625 - 0.0625x = 656.50$$
$$0.0075x = 31.5$$
$$x = 4200$$

\$4200 was invested at 7% and \$10,000 – \$4200 = \$5800 was invested at $6\frac{1}{4}\%$

109.
$$m = \frac{y_2 - y_1}{x_2 - x_1}$$
$$m = \frac{4-4}{-1-(-3)} = \frac{0}{-1+3} = \frac{0}{2} = 0$$

110.
$$\begin{array}{r} 4x^2 + 9x - 2 \\ \underline{x - 2} \\ -8x^2 - 18x + 4 \\ \underline{4x^3 + 9x^2 - 2x \quad} \\ 4x^3 + x^2 - 20x + 4 \end{array}$$

111.
$$|x+3| = |2x-7|$$

$x + 3 = 2x - 7$ or $x + 3 = -(2x - 7)$

$-x = -10$ $x + 3 = -2x + 7$

$x = 10$ $3x = 4$

$x = \frac{4}{3}$

Exercise Set 12.2

1. The quadratic formula is $x = \dfrac{-b \pm \sqrt{b^2 - 4ac}}{2a}$ which gives the values of x where $ax^2 + bx + c = 0$.

3. $a = -3$, $b = 6$, and $c = 5$

5. Yes, multiply both sides of the equation $-6x^2 + \frac{1}{2}x - 5 = 0$ by -1 to obtain $6x^2 - \frac{1}{2}x + 5 = 0$. The equations are equivalent so they will have the same solutions.

7. **a.** For a quadratic equation in standard form, $ax^2 + bx + c = 0$, the discriminant is the expression under the square root symbol in the quadratic formula, $b^2 - 4ac$.

   b. $3x^2 - 6x + 20 = 0$, $a = 3$, $b = -6$, and $c = 20$.
   $$\begin{aligned} b^2 - 4ac &= (-6)^2 - 4(3)(20) \\ &= 36 - 240 \\ &= -204 \end{aligned}$$

   c. If $b^2 - 4ac > 0$, then the quadratic equation will have two distinct real solutions. Since there is a positive number under the radical sign in the quadratic formula, the value of the radical will be real and there will be two real solutions. If $b^2 - 4ac = 0$, then the equation has the single real solution $\dfrac{-b}{2a}$. If $b^2 - 4ac < 0$, the expression under the radical sign in the quadratic formula is negative. Thus the equation has no real solution.

9. $$x^2 + 3x + 2 = 0$$
   $$\begin{aligned} b^2 - 4ac &= (3)^2 - 4(1)(2) \\ &= 9 - 8 \\ &= 1 \end{aligned}$$
   Since $1 > 0$, there are two real solutions.

11. $$3z^2 + 4z + 5 = 0$$
    $$\begin{aligned} b^2 - 4ac &= 4^2 - 4(3)(5) \\ &= 16 - 60 \\ &= -44 \end{aligned}$$
    Since $-44 < 0$, there is no real solution.

13. $$5p^2 + 3p - 7 = 0$$
    $$\begin{aligned} b^2 - 4ac &= 3^2 - 4(5)(-7) \\ &= 9 + 140 \\ &= 149 \end{aligned}$$
    Since $149 > 0$ there are two real solutions.

15. $$-5x^2 + 5x - 6 = 0$$
    $$\begin{aligned} b^2 - 4ac &= 5^2 - 4(-5)(-6) \\ &= 25 - 120 \\ &= -95 \end{aligned}$$
    Since $-95 < 0$, there is no real solution.

17. $$x^2 + 10.2x + 26.01 = 0$$
    $$\begin{aligned} b^2 - 4ac &= (10.2)^2 - 4(1)(26.01) \\ &= 104.04 - 104.04 \\ &= 0 \end{aligned}$$
    There is one real solution.

19. $$b^2 = -3b - \frac{9}{4}$$
    $$b^2 + 3b + \frac{9}{4} = 0$$
    $$\begin{aligned} b^2 - 4ac &= 3^2 - 4(1)\left(\frac{9}{4}\right) \\ &= 9 - 9 \\ &= 0 \end{aligned}$$
    There is one real solution.

21. $$x^2 - 9x + 18 = 0$$
    $$\begin{aligned} x &= \frac{9 \pm \sqrt{9^2 - 4(1)(18)}}{2(1)} \\ &= \frac{9 \pm \sqrt{81 - 72}}{2} \\ &= \frac{9 \pm \sqrt{9}}{2} \\ &= \frac{9 \pm 3}{2} \end{aligned}$$
    $$x = \frac{9+3}{2} = \frac{12}{2} = 6 \quad \text{or} \quad x = \frac{9-3}{2} = \frac{6}{2} = 3$$
    The solutions are 6 and 3.

23. $a^2-6a+8=0$

$$a=\frac{-b\pm\sqrt{b^2-4ac}}{2a}$$
$$=\frac{6\pm\sqrt{(-6)^2-4(1)(8)}}{2(1)}$$
$$=\frac{6\pm\sqrt{36-32}}{2}$$
$$=\frac{6\pm\sqrt{4}}{2}$$
$$=\frac{6\pm2}{2}$$
$$a=\frac{6-2}{2}=\frac{4}{2}=2 \quad\text{or}\quad a=\frac{6+2}{2}=\frac{8}{2}=4$$

The solutions are 2 and 4.

25. $x^2=-2x+3$

$x^2+2x-3=0$

$$x=\frac{-2\pm\sqrt{2^2-4(1)(-3)}}{2(1)}$$
$$=\frac{-2\pm\sqrt{4+12}}{2}$$
$$=\frac{-2\pm\sqrt{16}}{2}$$
$$=\frac{-2\pm4}{2}$$
$$x=\frac{-2+4}{2}=\frac{2}{2}=1 \quad\text{or}\quad x=\frac{-2-4}{2}=\frac{-6}{2}=-3$$

The solutions are 1 and –3.

27. $-b^2=4b-20$

$b^2+4b-20=0$

$$b=\frac{-4\pm\sqrt{(4)^2-4(1)(-20)}}{2(1)}$$
$$=\frac{-4\pm\sqrt{16+80}}{2}$$
$$=\frac{-4\pm\sqrt{96}}{2}$$
$$=\frac{-4\pm4\sqrt{6}}{2}$$
$$=-2\pm\sqrt{6}$$

The solutions are $-2+2\sqrt{6}$ and $-2-2\sqrt{6}$.

29. $b^2-49=0$

$$b=\frac{0\pm\sqrt{0^2-4(1)(-49)}}{2(1)}$$
$$=\frac{\pm\sqrt{196}}{2}$$
$$=\frac{\pm14}{2}$$
$$b=\frac{14}{2}=7 \quad\text{or}\quad b=\frac{-14}{2}=-7$$

The solutions are 7 and –7.

31. $3w^2-4w+5=0$

$$w=\frac{-(-4)\pm\sqrt{(\ 4)^2-4(3)(5)}}{2(3)}$$
$$=\frac{4\pm\sqrt{16-60}}{6}$$
$$=\frac{4\pm\sqrt{-44}}{6}$$
$$=\frac{4\pm2i\sqrt{11}}{6}$$
$$=\frac{2\left(2\pm i\sqrt{11}\right)}{6}$$
$$=\frac{2\pm i\sqrt{11}}{3}$$

The solutions are $\frac{2-i\sqrt{11}}{3}$ and $\frac{2+i\sqrt{11}}{3}$.

33. $c^2-3c=0$

$$c=\frac{3\pm\sqrt{(-3)^2-4(1)(0)}}{2(1)}$$
$$=\frac{3\pm\sqrt{9}}{2}$$
$$=\frac{3\pm3}{2}$$
$$c=\frac{3+3}{2}=\frac{6}{2}=3 \quad\text{or}\quad c=\frac{3-3}{2}=\frac{0}{2}=0$$

The solutions are 3 and 0.

35. $4s^2 - 8s + 6 = 0$

$\frac{1}{2}(4s^2 - 8s + 6 = 0)$

$2s^2 - 4s + 3 = 0$

$$s = \frac{-(-4) \pm \sqrt{(-4)^2 - 4(2)(3)}}{2(2)} = \frac{4 \pm \sqrt{16 - 24}}{4} = \frac{4 \pm \sqrt{-8}}{4} = \frac{4 \pm 2i\sqrt{2}}{4} = \frac{2 \pm i\sqrt{2}}{2}$$

The solutions are $\frac{2 - i\sqrt{2}}{2}$ and $\frac{2 + i\sqrt{2}}{2}$.

37. $a^2 + 2a + 1 = 0$

$$a = \frac{-(2) \pm \sqrt{(2)^2 - 4(1)(1)}}{2(1)} = \frac{-2 \pm \sqrt{4 - 4}}{2} = \frac{-2 \pm \sqrt{0}}{2} = \frac{-2 \pm 0}{2} = \frac{-2}{2} = -1$$

The solution is −1.

39. $x^2 - 10x + 25 = 0$

$$x = \frac{-(-10) \pm \sqrt{(-10)^2 - 4(1)(25)}}{2(1)} = \frac{10 \pm \sqrt{100 - 100}}{2} = \frac{10 \pm \sqrt{0}}{2} = \frac{10 \pm 0}{2} = \frac{10}{2} = 5$$

The solution is 5.

41. $x^2 - 2x - 1 = 0$

$$x = \frac{-(-2) \pm \sqrt{(-2)^2 - 4(1)(-1)}}{2(1)} = \frac{2 \pm \sqrt{4 + 4}}{2} = \frac{2 \pm \sqrt{8}}{2} = \frac{2 \pm 2\sqrt{2}}{2} = 1 \pm \sqrt{2}$$

The solutions are $1 - \sqrt{2}$ and $1 + \sqrt{2}$.

43. $2 - 3r^2 = -4r$

$3r^2 - 4r - 2 = 0$

$$r = \frac{-(-4) \pm \sqrt{(-4)^2 - 4(3)(-2)}}{2(3)} = \frac{4 \pm \sqrt{16 + 24}}{6} = \frac{4 \pm \sqrt{40}}{6} = \frac{4 \pm 2\sqrt{10}}{6} = \frac{2 \pm \sqrt{10}}{3}$$

The solutions are $\frac{2 + \sqrt{10}}{3}$ and $\frac{2 - \sqrt{10}}{3}$.

45. $2x^2 + 5x - 3 = 0$

$$x = \frac{-(5) \pm \sqrt{(5)^2 - 4(2)(-3)}}{2(2)} = \frac{-5 \pm \sqrt{25 + 24}}{4} = \frac{-5 \pm \sqrt{49}}{4} = \frac{-5 \pm 7}{4}$$

$x = \frac{-5 + 7}{4} = \frac{2}{4} = \frac{1}{2}$ or $x = \frac{-5 - 7}{4} = \frac{-12}{4} = -3$

The solutions are $\frac{1}{2}$ and −3.

47. $(2a+3)(3a-1)=2$

$6a^2+7a-3=2$

$6a^2+7a-5=0$

$a=\frac{-(7)\pm\sqrt{(7)^2-4(6)(-5)}}{2(6)}$

$=\frac{-7\pm\sqrt{49+120}}{12}$

$=\frac{-7\pm\sqrt{169}}{12}$

$=\frac{-7\pm13}{12}$

$a=\frac{-7-13}{12}$ or $a=\frac{-7+13}{12}$

$=\frac{-20}{12}$ $\quad=\frac{6}{12}$

$=-\frac{5}{3}$ $\quad=\frac{1}{2}$

The solutions are $\frac{1}{2}$ and $-\frac{5}{3}$.

49. $\frac{1}{2}t^2+t-12=0$

$2\left(\frac{1}{2}t^2+t-12=0\right)$

$t^2+2t-24=0$

$t=\frac{-(2)\pm\sqrt{(2)^2-4(1)(-24)}}{2(1)}$

$=\frac{-2\pm\sqrt{4+96}}{2}$

$=\frac{-2\pm\sqrt{100}}{2}$

$=\frac{-2\pm10}{2}$

$t=\frac{-2+10}{2}$ or $t=\frac{-2-10}{2}$

$=\frac{8}{2}$ $\quad=\frac{-12}{2}$

$=4$ $\quad=-6$

The solutions are 4 and –6.

51. $9r^2-9r+2=0$

$r=\frac{-(-9)\pm\sqrt{(-9)^2-4(9)(2)}}{2(9)}$

$=\frac{9\pm\sqrt{81-72}}{18}$

$=\frac{9\pm\sqrt{9}}{18}$

$=\frac{9\pm3}{18}$

$r=\frac{9+3}{18}$ or $r=\frac{9-3}{18}$

$=\frac{12}{18}$ $\quad=\frac{6}{18}$

$=\frac{2}{3}$ $\quad=\frac{1}{3}$

The solutions are $\frac{2}{3}$ and $\frac{1}{3}$..

53. $\frac{1}{2}x^2+2x+\frac{2}{3}=0$

$6\left(\frac{1}{2}x^2+2x+\frac{2}{3}=0\right)$

$3x^2+12x+4=0$

$x=\frac{-12\pm\sqrt{(12)^2-4(3)(4)}}{2(3)}$

$=\frac{-12\pm\sqrt{144-48}}{6}$

$=\frac{-12\pm\sqrt{96}}{6}$

$=\frac{-12\pm4\sqrt{6}}{6}$

$=\frac{2(-6\pm2\sqrt{6})}{2(3)}$

$=\frac{-6\pm2\sqrt{6}}{3}$

The solutions are $\frac{-6+2\sqrt{6}}{3}$ and $\frac{-6-2\sqrt{6}}{3}$.

55. $a^2-\frac{a}{5}-\frac{1}{3}=0$

$15\left(a^2-\frac{a}{5}-\frac{1}{3}=0\right)$

$15a^2-3a-5=0$

$a=\frac{-(-3)\pm\sqrt{(-3)^2-4(15)(-5)}}{2(15)}$

$=\frac{3\pm\sqrt{9+300}}{30}$

$=\frac{3\pm\sqrt{309}}{30}$

The solutions are $\frac{3-\sqrt{309}}{30}$ and $\frac{3+\sqrt{309}}{30}$.

57.
$$c = \frac{6-c}{c-4}$$
$$c(c-4) = -c+6$$
$$c^2 - 4c = -c+6$$
$$c^2 - 3c - 6 = 0$$
$$c = \frac{-(-3) \pm \sqrt{(-3)^2 - 4(1)(-6)}}{2(1)}$$
$$= \frac{3 \pm \sqrt{9+24}}{2}$$
$$= \frac{3 \pm \sqrt{33}}{2}$$
The solutions are $\frac{3+\sqrt{33}}{2}$ and $\frac{3-\sqrt{33}}{2}$.

59.
$$2x^2 - 4x + 3 = 0$$
$$x = \frac{-(-4) \pm \sqrt{(-4)^2 - 4(2)(3)}}{2(2)}$$
$$= \frac{4 \pm \sqrt{16-24}}{4}$$
$$= \frac{4 \pm \sqrt{-8}}{4}$$
$$= \frac{4 \pm 2i\sqrt{2}}{4}$$
$$= \frac{2 \pm i\sqrt{2}}{2}$$
The solutions are $\frac{2+i\sqrt{2}}{2}$ and $\frac{2-i\sqrt{2}}{2}$.

61.
$$2y^2 + y = -3$$
$$2y^2 + y + 3 = 0$$
$$y = \frac{-1 \pm \sqrt{(1)^2 - 4(2)(3)}}{2(2)}$$
$$= \frac{-1 \pm \sqrt{1-24}}{4}$$
$$= \frac{-1 \pm \sqrt{-23}}{4}$$
$$= \frac{-1 \pm i\sqrt{23}}{4}$$
The solutions are $\frac{-1+i\sqrt{23}}{4}$ and $\frac{-1-i\sqrt{23}}{4}$.

63.
$$0.1x^2 + 0.6x - 1.2 = 0$$
$$10(0.1x^2 + 0.6x - 1.2 = 0)$$
$$x^2 + 6x - 12 = 0$$
$$x = \frac{-6 \pm \sqrt{6^2 - 4(1)(-12)}}{2(1)}$$
$$= \frac{-6 \pm \sqrt{36+48}}{2}$$
$$= \frac{-6 \pm \sqrt{84}}{2}$$
$$= \frac{-6 \pm 2\sqrt{21}}{2}$$
$$= -3 \pm \sqrt{21}$$
The solutions are $-3+\sqrt{21}$ or $-3-\sqrt{21}$.

65.
$$f(x) = x^2 - 2x + 4,\ f(x) = 4$$
$$x^2 - 2x + 4 = 4$$
$$x^2 - 2x = 0$$
$$x = \frac{2 \pm \sqrt{(-2)^2 - 4(1)(0)}}{2(1)}$$
$$= \frac{2 \pm \sqrt{4}}{2}$$
$$= \frac{2 \pm 2}{2}$$
$$x = \frac{2+2}{2} = \frac{4}{2} = 2 \quad \text{or} \quad x = \frac{2-2}{2} = \frac{0}{2} = 0$$
The values of x are 2 and 0.

67.
$$k(x) = x^2 - x - 10,\ k(x) = 20$$
$$x^2 - x - 10 = 20$$
$$x^2 - x - 30 = 0$$
$$x = \frac{1 \pm \sqrt{(-1)^2 - 4(1)(-30)}}{2(1)}$$
$$= \frac{1 \pm \sqrt{1+120}}{2}$$
$$= \frac{1 \pm \sqrt{121}}{2}$$
$$= \frac{1 \pm 11}{2}$$
$$x = \frac{1+11}{2} = \frac{12}{2} = 6 \quad \text{or} \quad x = \frac{1-11}{2} = \frac{-10}{2} = -5$$

The values of x are 6 and -5.

69. $h(t) = 2t^2 - 7t + 1$, $h(t) = -3$

$$2t^2 - 7t + 1 = -3$$
$$2t^2 - 7t + 4 = 0$$
$$t = \frac{7 \pm \sqrt{(-7)^2 - 4(2)(4)}}{2(2)}$$
$$= \frac{7 \pm \sqrt{49 - 32}}{4}$$
$$= \frac{7 \pm \sqrt{17}}{4}$$

The values of t are $\frac{7 + \sqrt{17}}{4}$ and $\frac{7 - \sqrt{17}}{4}$.

71. $g(a) = 2a^2 - 3a + 16$, $g(a) = 14$

$$2a^2 - 3a + 16 = 14$$
$$2a^2 - 3a + 2 = 0$$
$$a = \frac{3 \pm \sqrt{(-3)^2 - 4(2)(2)}}{2(2)}$$
$$= \frac{3 \pm \sqrt{9 - 16}}{4}$$
$$= \frac{3 \pm \sqrt{-7}}{4}$$

There are no real values of a for which $g(a) = 14$.

73. If 2 and 5 are solutions, the factors must be $(x-2)$ and $(x-5)$.

$$f(x) = (x-2)(x-5)$$
$$f(x) = x^2 - 5x - 2x + 10$$
$$f(x) = x^2 - 7x + 10$$

75. If 4 and -6 are solutions, the factors must be $(x-4)$ and $(x+6)$.

$$f(x) = (x-4)(x+6)$$
$$f(x) = x^2 + 6x - 4x - 24$$
$$f(x) = x^2 + 2x - 24$$

77. If $-\frac{3}{5}$ and $\frac{2}{3}$ are solutions, the factors must be $(5x+3)$ and $(3x-2)$.

$$f(x) = (5x+3)(3x-2)$$
$$f(x) = 15x^2 - 10x + 9x - 6$$
$$f(x) = 15x^2 - x - 6$$

79. If $\sqrt{3}$ and $-\sqrt{3}$ are solutions, the factors must be $(x - \sqrt{3})$ and $(x + \sqrt{3})$.

$$f(x) = (x - \sqrt{3})(x + \sqrt{3})$$
$$f(x) = x^2 - 3$$

81. $2i$ and $-2i$ are solutions, the factors must be $(x - 2i)$ and $(x + 2i)$.

$$f(x) = (x - 2i)(x + 2i)$$
$$f(x) = x^2 - 4i^2$$
$$f(x) = x^2 + 4$$

83. If $3 + \sqrt{2}$ and $3 - \sqrt{2}$ are solutions, the factors must be $\left(x - (3 + \sqrt{2})\right)$ and $\left(x - (3 - \sqrt{2})\right)$.

$$f(x) = \left(x - (3 + \sqrt{2})\right)\left(x - (3 - \sqrt{2})\right)$$
$$f(x) = (x - 3 - \sqrt{2})(x - 3 + \sqrt{2})$$
$$f(x) = (x-3)^2 - (\sqrt{2})^2$$
$$f(x) = x^2 - 6x + 9 - 2$$
$$f(x) = x^2 - 6x + 7$$

85. If $2 + 3i$ and $2 - 3i$ are solutions, the factors must be $(x - (2 + 3i))$ and $(x - (2 - 3i))$.

$$f(x) = (x - (2 + 3i))(x - (2 - 3i))$$
$$f(x) = (x - 2 - 3i)(x - 2 + 3i)$$
$$f(x) = (x-2)^2 - 9i^2$$
$$f(x) = x^2 - 4x + 4 + 9$$
$$f(x) = x^2 - 4x + 13$$

87. **a.** $n(10-0.02n)=450$

b. $n(10-0.02n)=450$

$10n-0.02n^2=450$

$0.02n^2-10n+450=0$

$n=\dfrac{10\pm\sqrt{(-10)^2-4(0.02)(450)}}{2(0.02)}$

$n=\dfrac{10\pm\sqrt{100-36}}{0.04}$

$n=\dfrac{10\pm\sqrt{64}}{0.04}$

$n=\dfrac{10\pm 8}{0.04} \Rightarrow n=450 \text{ or } n=50$

Since $n \le 65$, the number of lamps that must be sold is 50.

89. **a.** $n(50-0.2n)=1680$

b. $n(50-0.2n)=1680$

$50n-0.2n^2=1680$

$0.2n^2-50n+1680=0$

$n=\dfrac{50\pm\sqrt{(-50)^2-4(0.2)(1680)}}{2(0.2)}$

$n=\dfrac{50\pm\sqrt{2500-1344}}{0.4}$

$n=\dfrac{50\pm\sqrt{1156}}{0.4}$

$n=\dfrac{50\pm 34}{0.4} \Rightarrow n=210 \text{ or } n=40$

Since $n \le 50$, the number of chairs that must be sold is 40.

91. Any quadratic equation for which the discriminant is a non-negative perfect square can be solved by factoring. Any quadratic equation for which the discriminant is a positive number but not a perfect square can be solved by the quadratic formula but not by factoring over the set of integers.

93. Yes. If the discriminant is a perfect square, the simplified expression will not contain a radical and the quadratic equation can be solved by factoring.

95. Let x be the number.

$2x^2+3x=14$

$2x^2+3x-14=0$

$x=\dfrac{-3\pm\sqrt{3^2-4(2)(-14)}}{2(2)}$

$=\dfrac{-3\pm\sqrt{9+112}}{4}$

$=\dfrac{-3\pm\sqrt{121}}{4}$

$=\dfrac{-3\pm 11}{4}$

$x=\dfrac{-3+11}{4}$ since x is positive

$x=\dfrac{8}{4}$

$x=2$

97. Let x be the width. Then $3x-2$ is the length. Use A = (length)(width).

$21=(3x-2)(x)$

$21=3x^2-2x$

$3x^2-2x-21=0$

$x=\dfrac{-(-2)\pm\sqrt{(-2)^2-4(3)(-21)}}{2(3)}$

$=\dfrac{2\pm\sqrt{4+252}}{6}$

$=\dfrac{2\pm\sqrt{256}}{6}$

$=\dfrac{2\pm 16}{6}$

Since width is positive, use

$x=\dfrac{2+16}{6}=\dfrac{18}{6}=3$

$3x-2=3(3)-2=9-2=7$

The width is 3 feet and the length is 7 feet.

99. Let x be the amount by which each side is to be reduced.

Then $6-x$ is the new width

and $8-x$ is the new length

new area $=\dfrac{1}{2}$ (old area)

$= \frac{1}{2}(6 \cdot 8)$
$= \frac{1}{2}(48)$
$= 24$

new area = (new width)(new length)

$24 = (6 - x)(8 - x)$
$0 = 48 - 14x + x^2 - 24$
$0 = x^2 - 14x + 24$

$$x = \frac{-(-14) \pm \sqrt{(-14)^2 - 4(1)(24)}}{2(1)} = \frac{14 \pm \sqrt{196 - 96}}{2} = \frac{14 \pm \sqrt{100}}{2} = \frac{14 \pm 10}{2}$$

$x = \frac{14 + 10}{2} = \frac{24}{2} = 12$ or $x = \frac{14 - 10}{2} = \frac{4}{2} = 2$

We reject $x = 12$, since this would give negative values for width and length. The only meaningful value is $x = 2$ inches since this gives positive values for the new width and length.

101. **a.** $h = \frac{1}{2}at^2 + v_0t + h_0$

$20 = \frac{1}{2}(-32)t^2 + 60t + 80$
$20 = -16t^2 + 60t + 80$
$0 = -16t^2 + 60t + 60$
$0 = 16t^2 - 60t - 60$

$$t = \frac{-(-60) \pm \sqrt{(-60)^2 - 4(16)(-60)}}{2(16)}$$

$$t = \frac{60 \pm \sqrt{7440}}{32}$$

Since time must be positive, use

$$t = \frac{60 + \sqrt{7440}}{32} \approx 4.57$$

The horseshoe is 20 feet from the ground after about 4.57 seconds.

b. $0 = \frac{1}{2}(-32)t^2 + 60t + 80$

$0 = -16t^2 + 60t + 80$

$$t = \frac{-60 \pm \sqrt{60^2 - 4(-16)(80)}}{2(-16)}$$

$$t = \frac{-60 \pm \sqrt{8720}}{-32}$$

$$t = \frac{60 \pm \sqrt{8720}}{32}$$

Since time must be positive, use

$$t = \frac{60 + \sqrt{8720}}{32} \approx 4.79$$

The horseshoes strike the ground after about 4.79 seconds.

103. $x^2 - \sqrt{5}x - 10 = 0,\quad a = 1,\ b = -\sqrt{5},\ c = -10$

$$x = \frac{-\left(-\sqrt{5}\right) \pm \sqrt{\left(-\sqrt{5}\right)^2 - 4(1)(-10)}}{2(1)} = \frac{\sqrt{5} \pm \sqrt{5 + 40}}{2} = \frac{\sqrt{5} \pm \sqrt{45}}{2} = \frac{\sqrt{5} \pm 3\sqrt{5}}{2}$$

$x = \frac{\sqrt{5} + 3\sqrt{5}}{2} = \frac{4\sqrt{5}}{2} = 2\sqrt{5}$ or $x = \frac{\sqrt{5} - 3\sqrt{5}}{2} = \frac{-2\sqrt{5}}{2} = -\sqrt{5}$

The solutions are $2\sqrt{5}$ and $-\sqrt{5}$.

105. Let s be the length of the side of the original cube. Then $s + 0.2$ is the length of the side of the expanded cube.

$(s + 0.2)^3 = s^3 + 6$
$s^3 + 0.6s^2 + 0.12s + 0.008 = s^3 + 6$
$0.6s^2 + 0.12s + 0.008 = 6$
$0.6s^2 + 0.12s - 5.992 = 0$

$$s = \frac{-0.12 \pm \sqrt{(0.12)^2 - 4(0.6)(-5.992)}}{2(0.6)} = \frac{-0.12 \pm \sqrt{0.0144 + 14.3803}}{1.2} = \frac{-0.12 \pm \sqrt{14.3952}}{1.2}$$

Use the positive value since a length cannot be negative.

$$s = \frac{-0.12 + \sqrt{14.3952}}{1.2} \approx \frac{-0.12 + 3.7941}{1.2} \approx 3.0618$$

The original side was about 3.0618 millimeters long.

107. **a.** $h = \frac{1}{2}at^2 + v_0t + h_0$

$0 = \frac{1}{2}(-32)t^2 + 0t + 60$

$0 = -16t^2 + 0t + 60$

$t = \frac{-(0) \pm \sqrt{0^2 - 4(-16)(60)}}{2(-16)}$

$t = \frac{0 \pm \sqrt{3840}}{-32}$

$t \approx -1.94$ or $t \approx 1.94$

Since time must be positive, use 1.94 sec.

b. $h = \frac{1}{2}at^2 + v_0t + h_0$

$0 = \frac{1}{2}(-32)t^2 + 0t + 120$

$0 = -16t^2 + 0t + 120$

$t = \frac{-(0) \pm \sqrt{0^2 - 4(-16)(120)}}{2(-16)}$

$t = \frac{0 \pm \sqrt{7680}}{-32}$

$t \approx -2.74$ or $t \approx 2.74$

Since time must be positive, use 2.74 sec.

c. Courtney's rock will strike the ground first.

d. The height of Travis' rock is given by $h(t) = -16t^2 + 100t + 60$. The height of Courtney's rock is given by $h(t) = -16t^2 + 60t + 120$. We want to know when these will be equal.

$-16t^2 + 100t + 60 = -16t^2 + 60t + 120$

$100t + 60 = 60t + 120$

$40t = 60$

$t = \frac{60}{40} \Rightarrow t = 1.5$

The rocks will be the same distance above the ground after 1.5 seconds.

108. $\frac{3.33 \times 10^3}{1.11 \times 10^1} = \frac{3.33}{1.11} \times \frac{10^3}{10^1} = 3 \times 10^2$ or 300

109. $f(x) = x^2 + 2x - 5$

$f(3) = (3)^2 + 2(3) - 5$

$= 9 + 6 - 5$

$= 10$

110. $3x + 4y = 2$

$2x = -5y - 1$

Rewrite the system in standard form.

$3x + 4y = 2$

$2x + 5y = -1$

To eliminate the x variable, multiply the first equation by 2 and the second equation by -3, and then add.

$6x + 8y = 4$

$\underline{-6x - 15y = 3}$

$-7y = 7 \Rightarrow y = -1$

Substitute -1 for y in the first equation to find x.

$3x + 4(-1) = 2$

$3x - 4 = 2$

$3x = 6 \Rightarrow x = 2$

The solution is (2, −1).

111. $\frac{x + \sqrt{y}}{x - \sqrt{y}} \cdot \frac{x + \sqrt{y}}{x + \sqrt{y}}$

$= \frac{x^2 + x\sqrt{y} + x\sqrt{y} + y}{x^2 + x\sqrt{y} - x\sqrt{y} - y}$

$= \frac{x^2 + 2x\sqrt{y} + y}{x^2 - y}$

112. $\sqrt{x^2 + 6x - 4} = x$

$x^2 + 6x - 4 = x^2$

$6x - 4 = 0$

$6x = 4$

$x = \frac{4}{6}$ or $\frac{2}{3}$

Exercise Set 12.3

1. Answers will vary.

3. $A = s^2$, for s.

$\sqrt{A} = s$

5. $A = S^2 - s^2$, for S

$A + s^2 = S^2$

$\sqrt{A + s^2} = S \Rightarrow S = \sqrt{A + s^2}$

7. $E = i^2 r$, for i

$$\frac{E}{r} = i^2$$

$$\sqrt{\frac{E}{r}} = i$$

9. $d = 16t^2$, for t

$$\frac{d}{16} = t^2$$

$$\sqrt{\frac{d}{16}} = t \Rightarrow t = \frac{\sqrt{d}}{4}$$

11. $E = mc^2$, for c

$$\frac{E}{m} = c^2$$

$$\sqrt{\frac{E}{m}} = c$$

13. $V = \frac{1}{3}\pi r^2 h$, for r

$$3V = \pi r^2 h$$

$$\frac{3V}{\pi h} = r^2$$

$$\sqrt{\frac{3V}{\pi h}} = r$$

15. $d = \sqrt{L^2 + W^2}$, for W

$$d^2 = L^2 + W^2$$

$$d^2 - L^2 = W^2$$

$$\sqrt{d^2 - L^2} = W$$

17. $a^2 + b^2 = c^2$, for b

$$b^2 = c^2 - a^2$$

$$b = \sqrt{c^2 - a^2}$$

19. $d = \sqrt{L^2 + W^2 + H^2}$, for H

$$d^2 = L^2 + W^2 + H^2$$

$$d^2 - L^2 - W^2 = H^2$$

$$\sqrt{d^2 - L^2 - W^2} = H$$

21. $h = -16t^2 + s_0$, for t

$$16t^2 = s_0 - h$$

$$t^2 = \frac{s_0 - h}{16}$$

$$t = \sqrt{\frac{s_0 - h}{16}}$$

$$t = \frac{\sqrt{s_0 - h}}{4}$$

23. $E = \frac{1}{2}mv^2$, for v

$$2E = mv^2$$

$$\frac{2E}{m} = v^2$$

$$\sqrt{\frac{2E}{m}} = v$$

25. $a = \frac{v_2^2 - v_1^2}{2d}$, for v_1

$$2ad = v_2^2 - v_1^2$$

$$2ad + v_1^2 = v_2^2$$

$$v_1^2 = v_2^2 - 2ad$$

$$v_1 = \sqrt{v_2^2 - 2ad}$$

27. $v' = \sqrt{c^2 - v^2}$, for c

$$(v')^2 = c^2 - v^2$$

$$(v')^2 + v^2 = c^2$$

$$c = \sqrt{(v')^2 + v^2}$$

29. a. $P(n) = 2.4n^2 + 9n - 3$

$$P(6) = 2.4(6)^2 + 9(6) - 3$$

$$= 137.4 \Rightarrow \$13{,}740$$

The profit would be \$13,740.

b. $P(n) = 2.4n^2 + 9n - 3$

$$200 = 2.4n^2 + 9n - 3$$

$$2.4n^2 + 9n - 203 = 0$$

$$x = \frac{-9 \pm \sqrt{9^2 - 4(2.4)(-203)}}{2(2.4)}$$

$$x = \frac{-9 \pm \sqrt{2029.8}}{4.8}$$

$$x \approx 8 \text{ or } x \approx -11$$

Eight tractors must be sold.

31. $T = 6.2t^2 + 12t + 32$

a. When the car is turned on, $t = 0$.
$T = 6.2(0)^2 + 12(0) + 32 = 32$
The temperature is 32°F.

b. $T = 6.2(1)^2 + 12(1) + 32 = 50.2$
The temperature after 1 minute is 50.2°F.

c. $120 = 6.2t^2 + 12t + 32$
$0 = 6.2t^2 + 12t - 88$
$$t = \frac{-12 \pm \sqrt{12^2 - 4(6.2)(-88)}}{2(6.2)}$$
$$t \approx \frac{-12 \pm 48.23}{12.4}$$
$t \approx 2.92$ or $t \approx -4.86$
The radiator temperature will reach 120°F about 2.92 min. after the engine is started.

33. a. $C = -0.01(1500)^2 + 80(1500) + 20{,}000$
$= 117{,}500$
The cost is about $117,500.

b. $150{,}000 = -0.01s^2 + 80s + 20{,}000$
$0 = -0.01s^2 + 80s - 130{,}000$
$$s = \frac{-80 \pm \sqrt{80^2 - 4(-0.01)(-130{,}000)}}{2(-0.01)}$$
$$s \approx \frac{-80 \pm 34.64}{-0.02}$$
$s \approx 2268$ or $s \approx 5732$
Notice that s must fall between 1200 and 4000. Therefore, Mr. Boyle can purchase a 2268 sq ft house for $150,000.

35. a. $0 = -3.3t^2 - 2.3t + 62$
$$t = \frac{-(-2.3) \pm \sqrt{(-2.3)^2 - 4(-3.3)(62)}}{2(-3.3)}$$
$$t = \frac{2.3 \pm 28.7}{-6.6}$$
$t \approx -4.7$ or $t = 4$
Since time must be positive, the car takes 4 seconds for the drop.

b. $s = 6.74(4) + 2.3 = 29.26$
The speed is 29.26 feet per second.

37. a. $m(t) = 0.05t^2 - 0.32t + 3.15$
In 2003, $t = 21$.
$m(21) = 0.05(21^2) - 0.32(21) + 3.15$
$= 22.05 - 6.72 + 3.15$
≈ 18.5
Veterinary bills for dogs in 2003 amounted to about $18.5 billion.

b. $m(t) = 0.05t^2 - 0.32t + 3.15$, $m(t) = 12$
$25 = 0.05t^2 - 0.32t + 3.15$
$0 = 0.05t^2 - 0.32t - 21.85$
$$t = \frac{0.32 \pm \sqrt{(0.32)^2 - 4(0.05)(-21.85)}}{2(0.05)}$$
$$= \frac{0.32 \pm \sqrt{0.1024 + 4.37}}{0.1}$$
$$= \frac{0.32 \pm \sqrt{4.4724}}{0.1}$$
$$\approx \frac{0.32 \pm 2.1148050}{0.1}$$
$$t \approx \frac{0.32 + 2.1148050}{0.1} = \frac{2.4348050}{0.1} \approx 24.3$$
or
$$t \approx \frac{0.32 - 2.1148050}{0.1} = \frac{-1.794805}{0.1} \approx -17.9$$
Since $1 \le t \le 28$, the only solution is $t \approx 24.3$. Thus $25 billion will be spent on veterinary bills for dogs approximately 24.3 years after 1982 or in 2006.

39. Let x be the width of the playground. Then the length is given by $x + 5$.
Area = length × width
$500 = x(x + 5)$
$x^2 + 5x - 500 = 0$
$(x - 20)(x + 25) = 0$
$x = 20$ or $x = -25$
Disregard the negative value. The width of the playground is 20 meters and the length is 20 + 5 = 25 meters.

41. Let r be the rate at which the present equipment drills.

	d	r	$t = \frac{d}{r}$
present equipment	64	r	$\frac{64}{r}$
new equipment	64	$r + 1$	$\frac{64}{r+1}$

They would have hit water in 3.2 hours less time with the new equipment.

$$\frac{64}{r+1} = \frac{64}{r} - 3.2$$

$$r(r+1)\left(\frac{64}{r+1}\right) = r(r+1)\left(\frac{64}{r}\right) - r(r+1)(3.2)$$

$$64r = 64(r+1) - 3.2r(r+1)$$

$$64r = 64r + 64 - 3.2r^2 - 3.2r$$

$$0 = 64 - 3.2r^2 - 3.2r$$

$$3.2r^2 + 3.2r - 64 = 0$$

$$r^2 + r - 20 = 0$$

$$(r+5)(r-4) = 0$$

$r + 5 = 0$ or $r - 4 = 0$

$r = -5$ $\quad$ $r = 4$

Use the positive value. The present equipment drills at a rate of 4 ft/hr.

43. Let x be Latoya's rate going uphill so $x + 2$ is her rating going downhill. Using $\frac{d}{r} = t$ gives

$$t_{\text{uphill}} + t_{\text{downhill}} = 1.75$$

$$\frac{6}{x} + \frac{6}{x+2} = 1.75$$

$$x(x+2)\left(\frac{6}{x}\right) + x(x+2)\left(\frac{6}{x+2}\right) = x(x+2)(1.75)$$

$$6(x+2) + 6x = 1.75x(x+2)$$

$$6x + 12 + 6x = 1.75x^2 + 3.5x$$

$$0 = 1.75x^2 - 8.5x - 12$$

$$x = \frac{-(-8.5) \pm \sqrt{(-8.5)^2 - 4(1.75)(-12)}}{2(1.75)}$$

$$= \frac{8 \pm \sqrt{156.25}}{3.5}$$

$$= \frac{8.5 \pm 12.5}{3.5}$$

$x = 6$ or $x \approx -1.14$

Since the time must be positive, Latoya's uphill rate is 6 mph and her downhill rate is $x + 2 = 8$ mph.

45. Let x be the time of the experienced mechanic then $x + 1$ is the time of the inexperienced mechanic.

$$\frac{6}{x} + \frac{6}{x+1} = 1$$

$$x(x+1)\left(\frac{6}{x}\right) + x(x+1)\left(\frac{6}{x+1}\right) = x(x+1)(1)$$

$$6(x+1) + 6x = x^2 + x$$

$$6x + 6 + 6x = x^2 + x$$

$$0 = x^2 - 11x - 6$$

$$x = \frac{-(-11) \pm \sqrt{(-11)^2 - 4(1)(-6)}}{2(1)}$$

$$= \frac{11 \pm \sqrt{121 + 24}}{2}$$

$$= \frac{11 \pm \sqrt{145}}{2}$$

$x = \frac{11 + \sqrt{145}}{2} \approx 11.52$ or $x = \frac{11 - \sqrt{145}}{2} \approx -0.52$

Since the time must be positive, it takes Bonita about 11.52 hours and Pamela about 12.52 hours to rebuild the engine.

47. Let r be the speed of the plane in still air.

	d	r	$t = \frac{d}{r}$
With wind	80	$r + 30$	$\frac{80}{r+30}$
Against wind	80	$r - 30$	$\frac{80}{r-30}$

The total time is 1.3 hours

$$\frac{80}{r+30} + \frac{80}{r-30} = 1.3$$

$$(r+30)(r-30)\left(\frac{80}{r+30} + \frac{80}{r-30} = 1.3\right)$$

$$80(r-3) + 80(r+30) = 1.3(r^2 - 900)$$

$$80r - 240 + 80r + 240 = 1.3r^2 - 1170$$

$$160r = 1.3r^2 - 1170$$

$$0 = 1.3r^2 - 160r - 1170$$

$$r = \frac{-(-160) \pm \sqrt{(-160)^2 - 4(1.3)(-1170)}}{2(1.3)}$$

$$= \frac{160 \pm \sqrt{25{,}600 + 6084}}{2.6}$$

$$= \frac{160 \pm \sqrt{31{,}684}}{2.6}$$

$$= \frac{160 \pm 178}{2.6}$$

$r = \frac{160 + 178}{2.6}$ or $r = \frac{160 - 178}{2.6}$

$= \frac{338}{2.6} \qquad = \frac{-18}{2.6}$

$= 130 \qquad \approx -6.92$

Since speed must be positive, the speed of the plane in still air is 130 mph.

49. Let t be the number of hours for Chris to clean alone. Then $t + 0.5$ is the number of hours for John to clean alone.

	Rate of work	Time worked	Part of Task completed
Chris	$\frac{1}{t}$	6	$\frac{6}{t}$
John	$\frac{1}{t+0.5}$	6	$\frac{6}{t+0.5}$

$$\frac{6}{t} + \frac{6}{t+0.5} = 1$$

$$t(t+0.5)\left(\frac{6}{t}\right) + t(t+0.5)\left(\frac{6}{t+0.5}\right) = t(t+0.5)(1)$$

$$6(t+0.5) + 6t = t(t+0.5)$$

$$6t + 3 + 6t = t^2 + 0.5t$$

$$12t + 3 = t^2 + 0.5t$$

$$0 = t^2 - 11.5t - 3$$

$$t = \frac{11.5 \pm \sqrt{(-11.5)^2 - 4(1)(-3)}}{2(1)}$$

$$= \frac{11.5 \pm \sqrt{132.25 + 12}}{2}$$

$$= \frac{11.5 \pm \sqrt{144.25}}{2}$$

$t = \frac{11.5 + \sqrt{144.25}}{2}$ or $t = \frac{11.5 - \sqrt{144.25}}{2}$

$\approx 11.76 \qquad \approx -0.26$

Since the time must be positive, it takes Chris about 11.76 hours and John about 11.76 + 0.5 = 12.26 hours to clean alone.

51. Let x be the speed of the trip from Lubbock to Plainview. Then $x - 10$ is the speed from Plainview to Amarillo.

	d	r	t
first part	60	x	$\frac{60}{x}$
second part	100	$x - 10$	$\frac{100}{x-10}$

Including the 2.5 hours she spent in Plainview, the entire trip took Lisa 5.5 hours.

$$\frac{60}{x} + 2.5 + \frac{100}{x-10} = 5.5$$

$$\frac{60}{x} + \frac{100}{x-10} = 3$$

$$x(x-10)\left(\frac{60}{x} + \frac{100}{x-10} = 3\right)$$

$$60(x-10) + 100x = 3(x^2 - 10x)$$

$$60x - 600 + 100x = 3x^2 - 30x$$

$$-600 + 160x = 3x^2 - 30x$$

$$0 = 3x^2 - 190x + 600$$

$$0 = (x-60)(3x-10)$$

$x - 60 = 0$ or $3x - 10 = 0$

$x = 60 \qquad x = \frac{10}{3}$

$\frac{10}{3}$ miles per hour is too slow for a car, so the speed of the trip from Lubbock to Plainview was 60 mph.

53. Answers will vary.

55. Let l = original length and w = original width. A system of equations that describes this situation is

$l \cdot w = 18$

$(l+2)(w+3) = 48$

If you solve for l in the first equation you get $l = \frac{18}{w}$. Substitute $\frac{18}{w}$ into the l in the second equation. The result is an equation in only one variable which can be solved.

$$(l+2)(w+3) = 48$$

$$\left(\frac{18}{w} + 2\right)(w+3) = 48$$

$$18 + \frac{54}{w} + 2w + 6 = 48$$

$$2w - 24 + \frac{54}{w} = 0$$

$$w\left(2w - 24 + \frac{54}{w} = 0\right)$$

$$2w^2 - 24w + 54 = 0$$

$$2(w-3)(w-9) = 0$$

$$w = 3 \text{ or } w = 9$$

If $w = 3$, then $l = \frac{18}{3} = 6$. One possible set of dimensions for the original rectangle is 6 m by 3 m.

If $w = 9$, then $l = \frac{18}{9} = 2$. Another possible set of dimensions for the original rectangle is 2 m by 9 m.

57. $-\left[4(5-3)^3\right]+2^3$

$-\left[4(2)^3\right]+2^3$

$-\left[4(8)\right]+8$

$-32+8$

-24

58. $IR + Ir = E$, for R

$IR = E - Ir$

$R = \frac{E-Ir}{I}$

59. $\frac{2r}{r-4} - \frac{2r}{r+4} + \frac{64}{r^2-16}$

$= \frac{2r}{r-4} \cdot \frac{r+4}{r+4} - \frac{2r}{r+4} \cdot \frac{r-4}{r-4} + \frac{64}{(r+4)(r-4)}$

$= \frac{2r^2+8r}{(r+4)(r-4)} - \frac{2r^2-8r}{(r+4)(r-4)} + \frac{64}{(r+4)(r-4)}$

$= \frac{2r^2+8r-(2r^2-8r)+64}{(r+4)(r-4)}$

$= \frac{16r+64}{(r+4)(r-4)}$

$= \frac{16(r+4)}{(r+4)(r-4)}$ or $\frac{16}{r-4}$

60. $\left(\frac{x^{3/4}y^{-2}}{x^{1/2}y^2}\right)^4 = \left(x^{(3/4)-(1/2)}y^{-2-2}\right)^4$

$= \left(x^{1/4}y^{-4}\right)^4$

$= x^1y^{-16}$

$= \frac{x}{y^{16}}$

61. $\sqrt{x^2+3x+9} = x$

$x^2+3x+9 = x^2$

$3x+9 = 0$

$3x = -9 \Rightarrow x = -3$

Upon checking, this value does not satisfy the equation. There is no real solution.

Exercise Set 12.4

1. A given equation can be expressed as an equation in quadratic form if the equation can be written in the form $au^2 + bu + c = 0$.

3. Let $u = x^2$. Then $3x^4 - 5x^2 + 1 = 0 \Rightarrow 3(x^2)^2 - 5x^2 + 1 = 0 \Rightarrow 3u^2 - 5u + 1 = 0$

5. Let $u = z^{-1}$. Then $z^{-2} - z^{-1} = 56 \Rightarrow (z^{-1})^2 - z^{-1} = 56 \Rightarrow u^2 - u = 56$

7. $x^4 + x^2 - 6$

Let $u = x^2$

$(x^2)^2 + x^2 - 6$

$u^2 + u - 6$

$(u + 3)(u - 2)$

Back substitute x^2 for u

$(x^2 + 3)(x^2 - 2)$

9. $x^4 + 5x^2 + 6$

Let $u = x^2$

$(x^2)^2 + 5x^2 + 6$

$u^2 + 5u + 6$

$(u + 2)(u + 3)$

Back substitute x^2 for u

$(x^2 + 2)(x^2 + 3)$

11. $6a^4 + 5a^2 - 25$

Let $u = a^2$

$6(a^2)^2 + 5a^2 - 25$

$6u^2 + 5u - 25$

$(2u + 5)(3u - 5)$

Back substitute a^2 for u

$(2a^2 + 5)(3a^2 - 5)$

13. $4(x + 1)^2 + 8(x + 1) + 3$

Let $u = x + 1$

$4u^2 + 8u + 3$

$(2u + 3)(2u + 1)$

Back substitute $x + 1$ for u

$(2(x + 1) + 3)(2(x + 1) + 1)$

$(2x + 2 + 3)(2x + 2 + 1)$

$(2x + 5)(2x + 3)$

15. $6(a + 2)^2 - 7(a + 2) - 5$

Let $u = a + 2$

$6u^2 - 7u - 5$

$(2u + 1)(3u - 5)$

Back substitute $a + 2$ for u

$(2(a + 2) + 1)(3(a + 2) - 5)$

$(2a + 4 + 1)(3a + 6 - 5)$

$(2a + 5)(3a + 1)$

17. $a^2b^2 + 8ab + 15$

Let $u = ab$

$(ab)^2 + 8ab + 15$

$u^2 + 8u + 15$

$(u + 3)(u + 5)$

Back substitute ab for u

$(ab + 3)(ab + 5)$

19. $3x^2y^2 - 2xy - 5$

Let $u = xy$

$3(xy)^2 - 2xy - 5$

$3u^2 - 2u - 5$

$(3u - 5)(u + 1)$

Back substitute xy for u

$(3xy - 5)(xy + 1)$

21. $2a^2(5 - a) - 7a(5 - a) + 5(5 - a)$

Factor out $(5 - a)$

$(5 - a)(2a^2 - 7a + 5)$

$(5 - a)(2a - 5)(a - 1)$

23. $2x^2(x - 3) + 7x(x - 3) + 6(x - 3)$

Factor out $x - 3$

$(x - 3)(2x^2 + 7x + 6)$

$(x - 3)(2x + 3)(x + 2)$

25. $y^4 + 13y^2 + 30$

Let $u = y^2$

$(y^2)^2 + 13y^2 + 30$

$u^2 + 13u + 30$

$(u+3)(u+10)$

Back substitute y^2 for u

$(y^2+3)(y^2+10)$

27. $x^2(x+3) + 3x(x+3) + 2(x+3)$

Factor out $x+3$

$(x+3)(x^2+3x+2)$

$(x+3)(x+1)(x+2)$

31. $x^4 - 10x^2 + 9 = 0$

$(x^2)^2 - 10x^2 + 9 = 0$

$u^2 - 10u + 9 = 0$ ← Replace x^2 with u

$(u-9)(u-1) = 0$

$u - 9 = 0$ or $u - 1 = 0$

$u = 9$ $\quad u = 1$

$x^2 = 9$ $\quad x^2 = 1$ ← Replace u with x^2

$x = \pm\sqrt{9} = \pm 3$ $\quad x = \pm\sqrt{1} = \pm 1$

The solutions are 3, –3, 1, and –1.

33. $x^4 - 26x^2 + 25 = 0$

$(x^2)^2 - 26x^2 + 25 = 0$

$u^2 - 26u + 25 = 0$ ← Replace x^2 with u

$(u-25)(u-1) = 0$

$u - 25 = 0$ $\quad u - 1 = 0$

$u = 25$ $\quad u = 1$

$x^2 = 25$ $\quad x^2 = 1$ ← Replace u with x^2

$x = \pm\sqrt{25} = \pm 5$ $\quad x = \pm\sqrt{1} = \pm 1$

The solutions are 5, –5, 1, and –1.

35. $x^4 - 13x^2 + 36 = 0$

$(x^2)^2 - 13x^2 + 36 = 0$

$u^2 - 13u + 36 = 0$ ← Replace x^2 with u

$(u-9)(u-4) = 0$

$u - 9 = 0$ $\quad u - 4 = 0$

$u = 9$ $\quad u = 4$

$x^2 = 9$ $\quad x^2 = 4$ ← Replace u with x^2

$x = \pm\sqrt{9} = \pm 3$ $\quad x = \pm\sqrt{4} = \pm 2$

The solutions are 3, –3, 2, and –2.

29. $5a^5b^2 - 8a^4b^3 + 3a^3b^4$

Factor out a^3b^2

$a^3b^2(5a^2 - 8ab + 3b^2)$

$a^3b^2(5a - 3b)(a - b)$

37. $a^4 - 7a^2 + 12 = 0$

$(a^2)^2 - 7a^2 + 12 = 0$

$u^2 - 7u + 12 = 0 \leftarrow$ Replace a^2 with u

$(u-4)(u-3) = 0$

$u - 4 = 0$ or $u - 3 = 0$

$u = 4$ $\quad u = 3$

$a^2 = 4$ $\quad a^2 = 3 \leftarrow$ Replace u with a^2

$a = \pm\sqrt{4} = \pm 2$ $\quad a = \pm\sqrt{3}$

The solutions are 2, –2, $\sqrt{3}$, and $-\sqrt{3}$.

39. $4x^4 - 5x^2 + 1 = 0$

$4(x^2)^2 - 5x^2 + 1 = 0$

$4u^2 - 5u + 1 = 0 \leftarrow$ Replace x^2 with u

$(4u-1)(u-1) = 0$

$4u - 1 = 0$ or $u - 1 = 0$

$u = \frac{1}{4}$ $\quad u = 1$

$x^2 = \frac{1}{4}$ $\quad x^2 = 1 \leftarrow$ Replace u with x^2

$x = \pm\sqrt{\frac{1}{4}} = \pm\frac{1}{2}$ $\quad x = \pm\sqrt{1} = \pm 1$

The solutions are $\frac{1}{2}$, $-\frac{1}{2}$, 1, and –1.

41. $r^4 - 8r^2 = -15$

$r^4 - 8r^2 + 15 = 0$

$(r^2)^2 - 8r^2 + 15 = 0$

$u^2 - 8u + 15 = 0 \leftarrow$ Replace r^2 with u

$(u-3)(u-5) = 0$

$u - 3 = 0$ $\quad u - 5 = 0$

$u = 3$ $\quad u = 5$

$r^2 = 3$ $\quad r^2 = 5 \leftarrow$ Replace u with r^2

$r = \pm\sqrt{3}$ $\quad r = \pm\sqrt{5}$

The solutions are $\sqrt{3}$, $-\sqrt{3}$, $\sqrt{5}$, and $-\sqrt{5}$.

43. $z^4 - 7z^2 = 18$

$z^4 - 7z^2 - 18 = 0$

$(z^2)^2 - 7z^2 - 18 = 0$

$u^2 - 7u - 18 = 0 \leftarrow$ Replace z^2 with u

$(u-9)(u+2) = 0$

$u - 9 = 0$ $\quad u + 2 = 0$

$u = 9$ $\quad u = -2$

$z^2 = 9$ $\quad z^2 = -2 \leftarrow$ Replace u with z^2

$z = \pm 3$ $\quad z = \pm\sqrt{-2} = \pm i\sqrt{2}$

The solutions are $3, -3, i\sqrt{2}$, and $-i\sqrt{2}$.

45. $-c^4 = 4c^2 - 5$

$c^4 + 4c^2 - 5 = 0$

$(c^2)^2 + 4c^2 - 5 = 0$

$u^2 + 4u - 5 = 0 \leftarrow$ Replace c^2 with u

$(u-1)(u+5) = 0$

$u - 1 = 0$ $\quad u + 5 = 0$

$u = 1$ $\quad u = -5$

$c^2 = 1$ $\quad c^2 = -5 \leftarrow$ Replace u with c^2

$c = \pm 1$ $\quad c = \pm\sqrt{-5} = \pm i\sqrt{5}$

The solutions are $1, -1, i\sqrt{5}$, and $-i\sqrt{5}$.

47. $\sqrt{x} = 2x - 6$

$2x - \sqrt{x} - 6 = 0$

$2(x^{1/2})^2 - x^{1/2} - 6 = 0$

$2u^2 - u - 6 = 0 \leftarrow$ Replace $x^{1/2}$ with u

$(2u+3)(u-2) = 0$

$2u + 3 = 0$ or $u - 2 = 0$

$u = -\frac{3}{2}$ or $u = 2$

$x^{1/2} = -\frac{3}{2}$ or $x^{1/2} = 2 \leftarrow$ Replace u with $x^{1/2}$

$x = 2^2 = 4$

$x^{1/2} = -\frac{3}{2}$ has no solution since there is no value of x for which $x^{1/2} = -\frac{3}{2}$.

The solution is 4.

49. $x + \sqrt{x} = 6$

$x + \sqrt{x} - 6 = 0$

$(x^{1/2})^2 + x^{1/2} - 6 = 0$

$u^2 + u - 6 = 0 \leftarrow$ Replace $x^{1/2}$ with u

$(u+3)(u-2) = 0$

$u + 3 = 0$ or $u - 2 = 0$

$u = -3$ or $u = 2$

$x^{1/2} = -3$ or $x^{1/2} = 2 \leftarrow$ Replace u with $x^{1/2}$

$x = 2^2 = 4$
$x^{1/2} = -3$ has no solution since there is no value of x for which $x^{1/2} = -3$.
The solution is 4.

51.
$$9x + 3\sqrt{x} = 2$$
$$9x + 3\sqrt{x} - 2 = 0$$
$$9\left(x^{1/2}\right)^2 + 3x^{1/2} - 2 = 0$$
$$9u^2 + 3u - 2 = 0 \leftarrow \text{Replace } x^{1/2} \text{ with } u$$
$$(3u - 1)(3u + 2) = 0$$
$3u - 1 = 0$ or $3u + 2 = 0$
$u = \frac{1}{3}$ $\quad u = -\frac{2}{3}$
$x^{1/2} = \frac{1}{3}$ $\quad x^{1/2} = -\frac{2}{3}$ $\leftarrow$ Replace u with $x^{1/2}$
$x = \frac{1}{9}$

$x^{1/2} = -\frac{2}{3}$ has no solution since there is no value of x for which $x^{1/2} = -\frac{2}{3}$.
The solution is $\frac{1}{9}$.

53.
$$(x+3)^2 + 2(x+3) = 24$$
$$(x+3)^2 + 2(x+3) - 24 = 0$$
$$u^2 + 2u - 24 = 0 \leftarrow \text{Replace } x + 3 \text{ with } u$$
$$(u-4)(u+6) = 0$$
$u - 4 = 0$ or $u + 6 = 0$
$u = 4$ $\quad u = -6$
$x + 3 = 4$ $\quad x + 3 = -6$ $\leftarrow$ Replace u with $x + 3$
$x = 1$ $\quad x = -9$

The solutions are 1 and –9.

55.
$$6(a-2)^2 = -19(a-2) - 10$$
$$6(a-2)^2 + 19(a-2) + 10 = 0$$
$$6u^2 + 19u + 10 = 0 \leftarrow \text{Replace } a - 2 \text{ with } u$$
$$(3u+2)(2u+5) = 0$$
$3u + 2 = 0$ or $2u + 5 = 0$
$u = -\frac{2}{3}$ $\quad u = -\frac{5}{2}$
$a - 2 = -\frac{2}{3}$ $\quad a - 2 = -\frac{5}{2}$ $\leftarrow$ Replace u with $a - 2$
$a = \frac{4}{3}$ $\quad a = -\frac{1}{2}$

The solutions are $\frac{4}{3}$ and $-\frac{1}{2}$.

57.
$$(x^2 - 1)^2 - (x^2 - 1) - 6 = 0$$
$$u^2 - u - 6 = 0 \leftarrow \text{Replace } x^2 - 1 \text{ with } u$$
$$(u+2)(u-3) = 0$$

$$u+2=0 \quad \text{or} \quad u-3=0$$
$$u=-2 \qquad u=3$$
$$x^2-1=-2 \qquad x^2-1=3 \leftarrow \text{Replace } u \text{ with } x^2-1$$
$$x^2=-1 \qquad x^2=4$$
$$x=\sqrt{-1}=\pm i \qquad x=\pm\sqrt{4}=\pm 2$$

The solutions are i, $-i$, 2, and -2.

59. $2(b+2)^2+5(b+2)-3=0$

$$2u^2+5u-3=0 \leftarrow \text{Replace } b+2 \text{ with } u$$
$$(u+3)(2u-1)=0$$
$$u+3=0 \quad \text{or} \quad 2u-1=0$$
$$u=-3 \qquad u=\frac{1}{2}$$
$$b+2=-3 \qquad b+2=\frac{1}{2} \leftarrow \text{Replace } u \text{ with } b+2$$
$$b=-5 \qquad b=-\frac{3}{2}$$

The solutions are -5 and $-\frac{3}{2}$.

61. $18(x^2-5)^2+27(x^2-5)+10=0$

$$18u^2+27u+10=0 \leftarrow \text{Replace } x^2-5 \text{ with } u$$
$$(3u+2)(6u+5)=0$$
$$3u+2=0 \quad \text{or} \quad 6u+5=0$$
$$u=-\frac{2}{3} \qquad u=-\frac{5}{6}$$
$$x^2-5=-\frac{2}{3} \qquad x^2-5=-\frac{5}{6} \leftarrow \text{Replace } u \text{ with } x^2-5$$
$$x^2=\frac{13}{3} \qquad x^2=\frac{25}{6}$$
$$x=\pm\sqrt{\frac{13}{3}} \qquad x=\pm\sqrt{\frac{25}{6}}$$
$$=\pm\frac{\sqrt{13}}{\sqrt{3}}\cdot\frac{\sqrt{3}}{\sqrt{3}} \qquad =\pm\frac{5}{\sqrt{6}}\cdot\frac{\sqrt{6}}{\sqrt{6}}$$
$$=\pm\frac{\sqrt{39}}{3} \qquad =\pm\frac{5\sqrt{6}}{6}$$

The solutions are $\frac{\sqrt{39}}{3}$, $-\frac{\sqrt{39}}{3}$, $\frac{5\sqrt{6}}{6}$, and $-\frac{5\sqrt{6}}{6}$.

63. $x^{-2}+10x^{-1}+25=0$

$$\left(x^{-1}\right)^2+10\left(x^{-1}\right)+25=0$$
$$u^2+10u+25=0 \leftarrow \text{Replace } x^{-1} \text{ with } u$$
$$(u+5)(u+5)=0$$
$$u+5=0$$
$$u=-5$$
$$x^{-1}=-5 \leftarrow \text{Replace } u \text{ with } x^{-1}$$
$$x=-\frac{1}{5}$$

The solution is $-\frac{1}{5}$.

65.

$$6b^{-2}-5b^{-1}+1=0$$
$$6(b^{-1})^2-5(b^{-1})+1=0$$
$$6u^2-5u+1=0 \leftarrow \text{Replace } b^{-1} \text{ with } u$$
$$(2u-1)(3u-1)=0$$
$$2u-1=0 \quad \text{or} \quad 3u-1=0$$
$$u=\frac{1}{2} \qquad u=\frac{1}{3}$$
$$b^{-1}=\frac{1}{2} \qquad b^{-1}=\frac{1}{3} \leftarrow \text{Replace } u \text{ with } b^{-1}$$
$$b=2 \qquad b=3$$

The solutions are 2 and 3.

67.

$$2b^{-2}=7b^{-1}-3$$
$$2b^{-2}-7b^{-1}+3=0$$
$$2(b^{-1})^2-7(b^{-1})+3=0$$
$$2u^2-7u+3=0 \leftarrow \text{Replace } b^{-1} \text{ with } u$$
$$(2u-1)(u-3)=0$$
$$2u-1=0 \quad \text{or} \quad u-3=0$$
$$u=\frac{1}{2} \qquad u=3$$
$$b^{-1}=\frac{1}{2} \qquad b^{-1}=3 \leftarrow \text{Replace } u \text{ with } b^{-1}$$
$$b=2 \qquad b=\frac{1}{3}$$

The solutions are 2 and $\frac{1}{3}$.

69.

$$x^{-2}+9x^{-1}=10$$
$$x^{-2}+9x^{-1}-10=0$$
$$(x^{-1})^2+9(x^{-1})-10=0$$
$$u^2+9u-10=0 \leftarrow \text{Replace } x^{-1} \text{ with } u$$
$$(u+10)(u-1)=0$$
$$u+10=0 \quad \text{or} \quad u-1=0$$
$$u=-10 \qquad u=1$$
$$x^{-1}=-10 \qquad x^{-1}=1 \leftarrow \text{Replace } u \text{ with } x^{-1}$$
$$x=-\frac{1}{10} \qquad x=1$$

The solutions are $-\frac{1}{10}$ and 1.

71.

$$x^{-2}=4x^{-1}+12$$
$$x^{-2}-4x^{-1}-12=0$$
$$(x^{-1})^2-4(x^{-1})-12=0$$
$$u^2-4u-12=0 \leftarrow \text{Replace } x^{-1} \text{ with } u$$
$$(u+2)(u-6)=0$$

$u+2=0$ or $u-6=0$

$u=-2$ $\quad u=6$

$x^{-1}=-2$ $\quad x^{-1}=6 \leftarrow$ Replace u with x^{-1}

$x=-\frac{1}{2}$ $\quad x=\frac{1}{6}$

The solutions are $-\frac{1}{2}$ and $\frac{1}{6}$.

73.

$$x^{2/3}-3x^{1/3}=-2$$
$$\left(x^{1/3}\right)^2-3x^{1/3}+2=0$$
$$u^2-3u+2=0 \leftarrow \text{Replace } x^{1/3} \text{ with } u$$
$$(u-1)(u-2)=0$$

$u-1=0$ or $u-2=0$

$u=1$ $\quad u=2$

$x^{1/3}=1$ $\quad x^{1/3}=2 \leftarrow$ Replace u with $x^{1/3}$

$x=1^3$ $\quad x=2^3$

$=1$ $\quad =8$

The solutions are 1 and 8.

75.

$$b^{2/3}+11b^{1/3}+28=0$$
$$\left(b^{1/3}\right)^2+11b^{1/3}+28=0$$
$$u^2+11u+28=0 \leftarrow \text{Replace } b^{1/3} \text{ with } u$$
$$(u+7)(u+4)=0$$

$u+7=0$ or $u+4=0$

$u=-7$ or $u=-4$

$b^{1/3}=-7$ or $b^{1/3}=-4 \leftarrow$ Replace u with $b^{1/3}$

$b=(-7)^3$ or $b=(-4)^3$

$=-343$ $\quad =-64$

The solutions are –343 and –64.

77.

$$-2a-5a^{1/2}+3=0$$
$$-2\left(a^{1/2}\right)^2-5a^{1/2}+3=0$$
$$-2u^2-5u+3=0 \leftarrow \text{Replace } a^{1/2} \text{ with } u$$
$$2u^2+5u-3=0$$
$$(2u-1)(u+3)=0$$

$2u-1=0$ or $u+3=0$

$u=\frac{1}{2}$ $\quad u=-3$

$a^{1/2}=2$ $\quad a^{1/2}=-3 \leftarrow$ Replace u with $a^{1/2}$

$a=\left(\frac{1}{2}\right)^2=\frac{1}{4}$

$a^{1/2}=-3$ has no solution since there is no value of a for which $a^{\frac{1}{2}}=-3$. The solution is $\frac{1}{4}$.

79.

$$c^{2/5} - 3c^{1/5} + 2 = 0$$
$$\left(c^{1/5}\right)^2 - 3c^{1/5} + 2 = 0$$
$$u^2 - 3u + 2 = 0 \leftarrow \text{ Replace } c^{1/5} \text{ with } u$$
$$(u-2)(u-1) = 0$$
$$u - 2 = 0 \qquad u - 1 = 0$$
$$u = 2 \qquad u = 1$$
$$c^{1/5} = 2 \qquad c^{1/5} = 1 \leftarrow \text{ Replace } u \text{ with } c^{1/5}$$
$$c = 2^5 \qquad c = 1^5$$
$$= 32 \qquad = 1$$

The solutions are 32 and 1.

81.

$$f(x) = x - 5\sqrt{x} + 4,\ f(x) = 0$$
$$0 = \left(x^{1/2}\right)^2 - 5x^{1/2} + 4$$
$$0 = u^2 - 5u + 4 \leftarrow \text{ Replace } x^{1/2} \text{ with } u$$
$$0 = (u-1)(u-4)$$
$$u - 1 = 0 \quad \text{or} \quad u - 4 = 0$$
$$u = 1 \qquad u = 4$$
$$x^{1/2} = 1 \qquad x^{1/2} = 4 \leftarrow \text{ Replace } u \text{ with } x^{1/2}$$
$$x = 1 \qquad x = 16$$

The x-intercepts are (1, 0) and (16, 0).

83.

$$h(x) = x + 13\sqrt{x} + 36,\ h(x) = 0$$
$$0 = \left(x^{1/2}\right)^2 + 13x^{1/2} + 36$$
$$0 = u^2 + 13u + 36 \leftarrow \text{ Replace } x^{1/2} \text{ with } u$$
$$0 = (u+9)(u+4)$$
$$u + 9 = 0 \quad \text{or} \quad u + 4 = 0$$
$$u = -9 \qquad u = -4$$
$$x^{1/2} = -9 \qquad x^{1/2} = -4 \leftarrow \text{ Replace } u \text{ with } x^{1/2}$$

There are no values of x for which $x^{1/2} = -9$ or $x^{1/2} = -4$. There are no x-intercepts.

85.

$$p(x) = 4x^{-2} - 19x^{-1} - 5,\ p(x) = 0$$
$$0 = 4\left(x^{-1}\right)^2 - 19x^{-1} - 5$$
$$0 = 4u^2 - 19u - 5 \leftarrow \text{ Replace } x^{-1} \text{ with } u$$
$$0 = (4u+1)(u-5)$$

$4u+1=0$ or $u-5=0$

$u=-\frac{1}{4}$ $u=5$

$x^{-1}=-\frac{1}{4}$ $x^{-1}=5$ ← Replace u with x^{-1}

$x=-4$ $x=\frac{1}{5}$

The x-intercepts are $(-4, 0)$ and $\left(\frac{1}{5}, 0\right)$.

87. $f(x)=x^{2/3}+x^{1/3}-6$, $f(x)=0$

$0=\left(x^{1/3}\right)^2+x^{1/3}-6$

$0=u^2+u-6$ ← Replace $x^{1/3}$ with u

$0=(u+3)(u-2)$

$u+3=0$ or $u-2=0$

$u=-3$ $u=2$

$x^{1/3}=-3$ $x^{1/3}=2$ ← Replace u with $x^{1/3}$

$x=-27$ $x=8$

The x-intercepts are $(-27, 0)$ and $(8, 0)$.

89. $g(x)=\left(x^2-3x\right)^2+2\left(x^2-3x\right)-24$, $g(x)=0$

$0=u^2+2u-24$ ← Replace x^2-3x with u

$0=(u+6)(u-4)$

$u+6=0$ or $u-4=0$

$x^2-3x+6=0$ $x^2-3x-4=0$ ← Replace u with x^2-3x

$(x-4)(x+1)=0$

$x-4=0$ or $x+1=0$

$x=4$ $x=-1$

There are no x-intercepts for $x^2-3x+6=0$ since $b^2-4ac=(-3)^2-4(1)(6)=9-24=-15$.
The x-intercepts are $(4, 0)$ or $(-1, 0)$.

91.

$f(x)=x^4-20x+64$, $f(x)=0$

$0=\left(x^2\right)^2-20x^2+64$

$0=u^2-20u+64$ ← Replace x^2 with u

$0=(u-16)(u-4)$

$u-16=0$ or $u-4=0$

$u=16$ $u=4$

$x^2=16$ $x^2=4$ ← Replace u with x^2

$x=\pm\sqrt{16}=\pm4$ $x=\pm\sqrt{4}=\pm2$

The x-intercepts are $(4, 0)$, $(-4, 0)$, $(2, 0)$, and $(-2, 0)$.

93. When solving an equation of the form $ax^4 + bx^2 + c = 0$, let $u = x^2$.

95. When solving an equation of the form $ax^{-2} + bx^{-1} + c = 0$, let $u = x^{-1}$.

97. If the solutions are ±2 and ±4, the factors must be $(x-2)$, $(x+2)$, $(x-4)$ and $(x+4)$.

$$0 = (x-2)(x+2)(x-4)(x+4)$$
$$0 = (x^2-4)(x^2-16)$$
$$0 = x^4 - 20x^2 + 64$$

99. If the solutions are $\pm\sqrt{2}$ and $\pm\sqrt{3}$, the factors must be $(x+\sqrt{2})$, $(x-\sqrt{2})$, $(x+\sqrt{3})$, $(x-\sqrt{3})$.

$$0=(x+\sqrt{2})(x-\sqrt{2})(x+\sqrt{3})(x-\sqrt{3})$$
$$0=(x^2-2)(x^2-3)$$
$$0=x^4-5x^2+6$$

101. No. An equation of the form $ax^4 + bx^2 + c = 0$ can have no imaginary solutions, two imaginary solutions, or four imaginary solutions.

103. **a.** $\frac{3}{x^2} - \frac{3}{x} = 60$ The LCD is x^2

$$x^2\left(\frac{3}{x^2}\right) - x^2\left(\frac{3}{x}\right) = x^2(60)$$
$$3 - 3x = 60x^2$$
$$0 = 60x^2 + 3x - 3$$
$$0 = 3(20x^2 + x - 1)$$
$$0 = 3(5x-1)(4x+1)$$
$$5x - 1 = 0 \quad \text{or} \quad 4x + 1 = 0$$
$$x = \frac{1}{5} \qquad x = -\frac{1}{4}$$

The solutions are $\frac{1}{5}$ and $-\frac{1}{4}$

b. $\frac{3}{x^2} - \frac{3}{x} = 60$

$$3x^{-2} - 3x^{-1} = 60$$
$$3(x^{-1})^2 - 3x^{-1} - 60 = 0$$
$$3u^2 - 3u - 60 = 0 \leftarrow \text{Replace } x^{-1} \text{ with } u$$
$$3(u^2 - u - 20) = 0$$
$$3(u-5)(u+4) = 0$$

$$u - 5 = 0 \quad \text{or} \quad u + 4 = 0$$
$$u = 5 \qquad u = -4$$
$$x^{-1} = 5 \qquad x^{-1} = -4 \leftarrow \text{Replace } u \text{ with } x^{-1}$$
$$x = \frac{1}{5} \qquad x = -\frac{1}{4}$$

The solutions are $\frac{1}{5}$ and $-\frac{1}{4}$.

112. $15(r+2) + 22 = -\frac{8}{r+2}$

$$15(r+2)(r+2) + 22(r+2) = -\frac{8}{r+2}(r+2)$$
$$15(r+2)^2 + 22(r+2) = -8$$
$$15(r+2)^2 + 22(r+2) + 8 = 0$$
$$15u^2 + 22u + 8 = 0 \leftarrow \text{Replace } r+2 \text{ with } u$$
$$(5u+4)(3u+2) = 0$$
$$5u + 4 = 0 \quad \text{or} \quad 3u + 2 = 0$$
$$u = -\frac{4}{5} \qquad u = -\frac{2}{3}$$
$$r + 2 = -\frac{4}{5} \qquad r + 2 = -\frac{2}{3} \leftarrow \text{Replace } u \text{ with } r+2$$
$$r = -\frac{14}{5} \qquad r = -\frac{8}{3}$$

The solutions are $-\frac{14}{5}$ and $-\frac{8}{3}$.

107. $4 - (x-2)^{-1} = 3(x-2)^{-2}$

$$4 - u^{-1} = 3u^{-2} \leftarrow \text{Replace } x-2 \text{ by } u$$
$$4 - \frac{1}{u} = \frac{3}{u^2}$$
$$u^2\left(4 - \frac{1}{u}\right) = u^2\left(\frac{3}{u^2}\right)$$
$$4u^2 - u = 3$$
$$4u^2 - u - 3 = 0$$
$$(4u+3)(u-1) = 0$$
$$4u + 3 = 0 \quad \text{or} \quad u - 1 = 0$$
$$u = -\frac{3}{4} \qquad u = 1$$
$$x - 2 = -\frac{3}{4} \qquad x - 2 = 1 \leftarrow \text{Replace } u \text{ by } x-2$$
$$x = \frac{5}{4} \qquad x = 3$$

The solutions are $\frac{5}{4}$ and 3.

109. $x^6 - 9x^3 + 8 = 0$

$(x^3)^2 - 9x^3 + 8 = 0$

$u^2 - 9u + 8 = 0$ ← Replace x^3 with u

$(u-8)(u-1) = 0$

$u - 8 = 0$ or $u - 1 = 0$

$u = 8$ $\quad$ $u = 1$

$x^3 = 8$ $\quad$ $x^3 = 1$ ← Replace u with x^3

$x = 2$ $\quad$ $x = 1$

The solutions are 2 and 1.

111. $(x^2 + 2x - 2)^2 - 7(x^2 + 2x - 2) + 6 = 0$

$u^2 - 7u + 6 = 0$ ← Replace $x^2 + 2x - 2$ with u

$(u-6)(u-1) = 0$

$u - 6 = 0$ or $u - 1 = 0$

$u = 6$ $\quad$ $u = 1$

$x^2 + 2x - 2 = 6$ $\quad$ $x^2 + 2x - 2 = 1$ ← Replace u with $x^2 + 2x - 2$

$x^2 + 2x - 8 = 0$ $\quad$ $x^2 + 2x - 3 = 0$

$(x+4)(x-2) = 0$ $\quad$ $(x+3)(x-1) = 0$

$x + 4 = 0$ or $x - 2 = 0$ $\quad$ $x + 3 = 0$ or $x - 1 = 0$

$x = -4$ $\quad$ $x = 2$ $\quad$ $x = -3$ $\quad$ $x = 1$

The solutions are –4, 2, –3, and 1.

113. $2n^4 - 6n^2 - 3 = 0$

$2(n^2)^2 - 6n^2 - 3 = 0$

$2u^2 - 6u - 3 = 0$ ← Replace n^2 with u

$$u = \frac{6 \pm \sqrt{(-6)^2 - 4(2)(-3)}}{2(2)}$$

$$= \frac{6 \pm \sqrt{60}}{4}$$

$$= \frac{6 \pm 2\sqrt{15}}{4}$$

$$= \frac{3 \pm \sqrt{15}}{2}$$

$n^2 = \dfrac{3 \pm \sqrt{15}}{2}$ ← Replace u with n^2

$$n = \pm\sqrt{\frac{3 \pm \sqrt{15}}{2}}$$

115.
$$\frac{4}{5} - \left(\frac{3}{4} - \frac{2}{3}\right) = \frac{4}{5} - \left(\frac{9}{12} - \frac{8}{12}\right)$$
$$= \frac{4}{5} - \left(\frac{1}{12}\right)$$
$$= \frac{48}{60} - \frac{5}{60}$$
$$= \frac{43}{60}$$

116.
$$6(x+4) - 4(3x+3) = 6$$
$$6x + 24 - 12x - 12 = 6$$
$$-6x + 12 = 6$$
$$-6x = -6$$
$$x = 1$$

117. D: all real numbers

R: $\{y \mid y \geq 0\}$

118. $\sqrt[3]{16x^3y^6} = \sqrt[3]{8 \cdot 2 \cdot x^3 \left(y^2\right)^3}$
$= 2xy^2\sqrt[3]{2}$

119. $\sqrt{75} + \sqrt{108} = \sqrt{25 \cdot 3} + \sqrt{36 \cdot 3}$
$= 5\sqrt{3} + 6\sqrt{3}$
$= 11\sqrt{3}$

Exercise Set 12.5

1. The graph of a quadratic equation is called a parabola.

3. The axis of symmetry of a parabola is the line where, if the graph is folded, the two sides overlap.

5. For $f(x) = ax^2 + bx + c$, the vertex of the graph is $\left(-\dfrac{b}{2a}, \dfrac{4ac - b^2}{4}\right)$.

7. a. For $f(x) = ax^2 + bx + c$, $f(x)$ will have a minimum if $a > 0$ since the graph opens upward.

b. For $f(x) = ax^2 + bx + c$, $f(x)$ will have a maximum if $a < 0$ since the graph opens downward.

9. To find the y-intercepts of the graph of a quadratic function, set $x = 0$ and solve for y.

11. a. For $f(x) = ax^2$, the general shape of $f(x)$ if $a > 0$ is

b. For $f(x) = ax^2$, the general shape of $f(x)$ if $a < 0$ is

13. Since $a = 3$ is greater than 0, the graph opens upward, and therefore has a minimum value.

15. $P(x) = x^2 - 6x + 4$
$P(2) = 2^2 - 6(2) + 4$
$= 4 - 12 + 4$
$= -8 + 4$
$= -4$

17. $P(x) = 2x^2 - 3x - 6$
$P\left(\frac{1}{2}\right) = 2\left(\frac{1}{2}\right)^2 - 3\left(\frac{1}{2}\right) - 6$
$= 2\left(\frac{1}{4}\right) - \frac{3}{2} - 6$
$= \frac{1}{2} - \frac{3}{2} - \frac{12}{2}$
$= \frac{-14}{2} = -7$

19. $P(x) = 0.2x^3 + 1.6x^2 - 2.3$
$P(0.4) = 0.2(0.4)^3 + 1.6(0.4)^2 - 2.3$
$= 0.2(0.064) + 1.6(0.16) - 2.3$
$= 0.0128 + 0.256 - 2.3$
$= -2.0312$

21. $f(x) = x^2 + 8x + 15$

a. Since $a = 1$, the parabola opens upward.

b. $y = 0^2 + 8(0) + 15 = 15$
The y-intercept is (0, 15).

c. $x = -\dfrac{b}{2a} = -\dfrac{8}{2(1)} = -\dfrac{8}{2} = -4$
$y = \dfrac{4ac - b^2}{4a}$
$= \dfrac{4(1)(15) - 8^2}{4(1)}$
$= \dfrac{60 - 64}{4}$
$= \dfrac{-4}{4}$
$= -1$
The vertex is (–4, –1).

d. $0 = x^2 + 8x + 15 = (x + 5)(x + 3)$
$x + 5 = 0$ or $x + 3 = 0$
$x = -5$ $\quad$ $x = -3$
The x-intercepts are (–5, 0) and (–3, 0).

e.

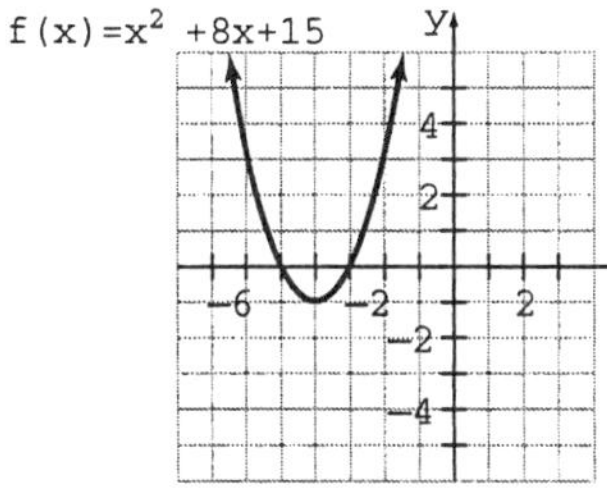

23. $g(x)=x^2+2x-3$

a. Since $a=1$ the parabola opens upward.

b. $y=0^2+2(0)-3=-3$
The y-intercept is (0, –3).

c. $x=-\frac{b}{2a}=-\frac{2}{2(1)}=-\frac{2}{2}=-1$

$y=\frac{4ac-b^2}{4a}=\frac{4(1)(-3)-(2)^2}{4(1)}=-4$

The vertex is (–1, –4).

d. $0=x^2+2x-3$
$(x-1)(x+3)=0$
$x=1$ or $x=-3$
The x-intercepts are (1, 0) and (–3, 0).

e.

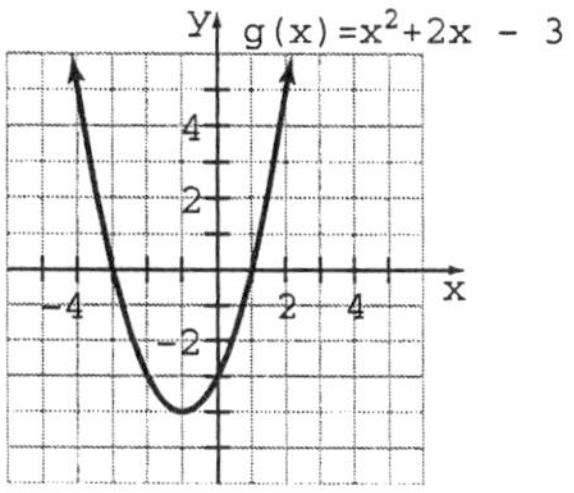

25. $f(x)=-x^2-2x+8$

a. Since $a=-1$, the parabola opens downward.

b. $y=-(0)^2-2(0)+8=8$
The y-intercept is (0, 8).

c. $x=-\frac{b}{2a}=-\frac{-2}{2(-1)}=-\frac{2}{2}=-1$

$y=\frac{4ac-b^2}{4a}$

$=\frac{4(-1)(8)-(-2)^2}{4(-1)}$

$=\frac{-32-4}{-4}$

$=\frac{-36}{-4}$

$=9$

The vertex is (1, 9).

d. $0=-x^2-2x+8=-(x^2+2x-8)$

$=-(x+4)(x-2)$

$x+4=0$ or $x-2=0$

$x=-4$ $\quad x=2$

The x-intercepts are (–4, 0) and (2, 0).

e.

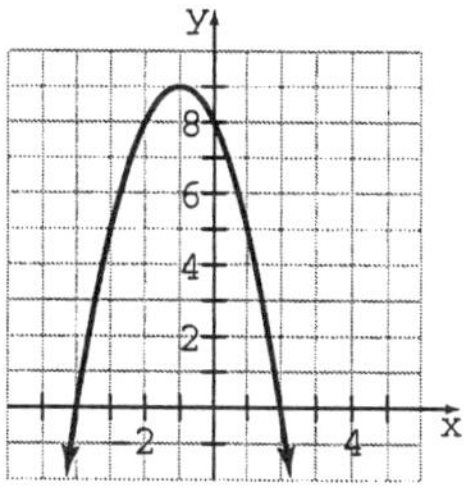

27. $p(x)=-x^2+8x-15$

a. Since $a=-1$, the parabola opens downward.

b. $y=-(0)^2+8(0)-15=-15$
The y-intercept is (0, –15).

c. $x=-\frac{b}{2a}=-\frac{8}{2(-1)}=\frac{8}{2}=4$

$y=\frac{4ac-b^2}{4a}$

$=\frac{4(-1)(-15)-(8)^2}{4(-1)}$

$=\frac{60-64}{-4}$

$=\frac{-4}{-4}$

$=1$

The vertex is (4, 1).

d. $0=-x^2+8x-15=-(x^2-8x+15)$

$=-(x-5)(x-3)$

$x-5=0$ or $x-3=0$

$x=5$ $\quad x=3$

The x-intercepts are (5, 0) and (3, 0).

e.

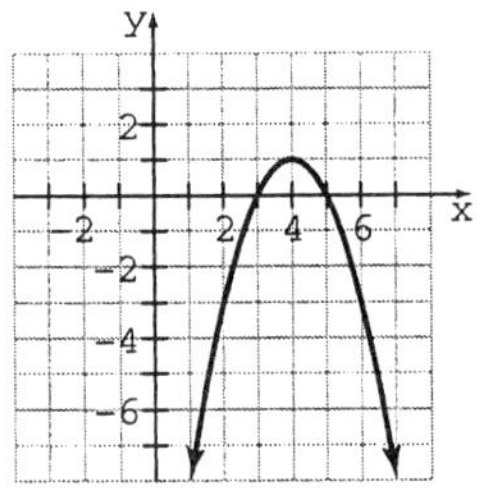

29. $t(x)=-x^2+4x-5$

a. Since $a=-1$, the parabola opens downward.

b. $y=-0^2+4(0)-5=-5$; The y-intercept is (0, –5).

c. $x=-\frac{b}{2a}=-\frac{4}{2(-1)}=-\frac{4}{-2}=2$

$y=\frac{4ac-b^2}{4a}$

$=\frac{4(-1)(-5)-(4)^2}{4(-1)}$

$=\frac{20-16}{-4}$

$=-1$

The vertex is (2, –1)

d. $0=-x^2+4x-5$

Since this is not factorable, check the discriminant.

$b^2-4ac=4^2-4(-1)(-5)$

$=16-20$

$=-4$

Since $-4<0$ there are no real roots. Thus, there are no x-intercepts.

e.

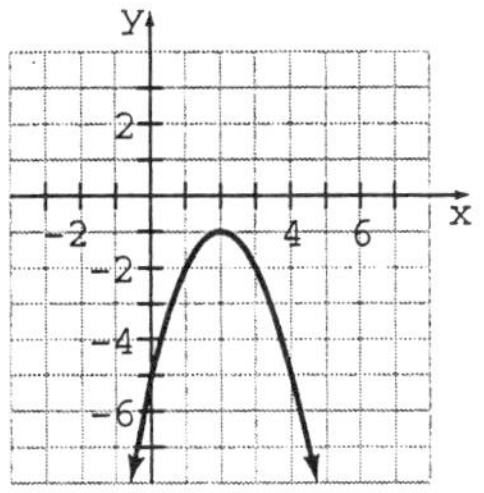

$t(x)=-x^2+4x-5$

31. $f(x)=x^2-4x+4$

a. Since $a=1$, the parabola opens upward.

b. $y=0^2-4(0)+4=4$

The y-intercept is (0, 4).

c. $x=-\frac{b}{2a}=-\frac{-4}{2(1)}=\frac{4}{2}=2$

$y=\frac{4ac-b^2}{4a}$

$=\frac{4(1)(4)-(-4)^2}{4(1)}$

$=\frac{16-16}{4}$

$=\frac{0}{4}$

$=0$

The vertex is (2, 0).

d. $0=x^2-4x+4=(x-2)(x-2)$

$x-2=0$

$x=2$

The x-intercept (2, 0).

e.

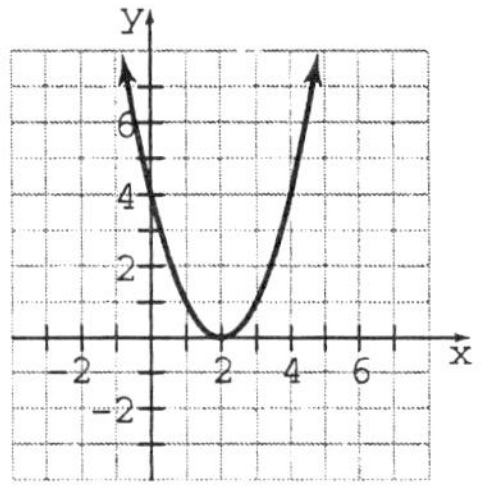

33. $r(x)=x^2+2$

a. Since $a=1$ the parabola opens upward.

b. $y=0^2+2=2$

The y-intercept is (0, 2).

c. $x = -\frac{b}{2a} = -\frac{0}{2(1)} = \frac{0}{2} = 0$

$y = \frac{4ac - b^2}{4a}$

$= \frac{4(1)(2) - 0^2}{4(1)}$

$= \frac{8}{4}$

$= 2$

The vertex is (0, 2).

d. $0 = x^2 + 2$

Since this is not factorable, check the discriminant.

$b^2 - 4ac = 0 - 4(1)2 = -8$

There are no real roots. Thus, there are no x-intercepts.

e.

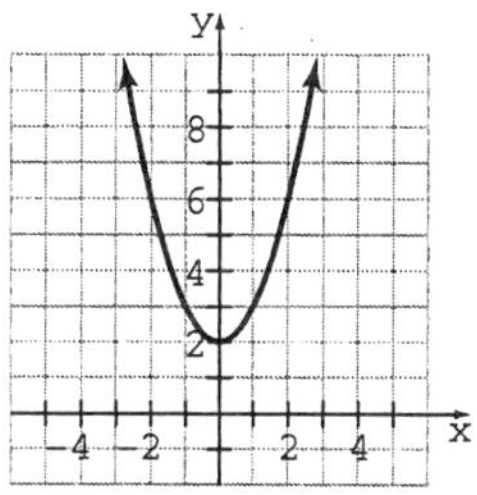

35. $g(x) = -x^2 + 6x$

a. Since $a = -1$ the parabola opens downward.

b. $y = -0^2 + 6(0) = 0$

The y-intercept is (0, 0).

c. $x = -\frac{b}{2a} = -\frac{6}{2(-1)} = -\frac{6}{-2} = 3$

$y = \frac{4ac - b^2}{4a}$

$= \frac{4(-1)(0) - 6^2}{4(-1)}$

$= \frac{0 - 36}{-4}$

$= \frac{-36}{-4}$

$= 9$

The vertex is (3, 9).

d. $0 = -x^2 + 6x = -x(x - 6)$

$-x = 0$ or $x - 6 = 0$

$x = 0$ $\quad$ $x = 6$

The x-intercepts are (0, 0) and (6, 0).

e.

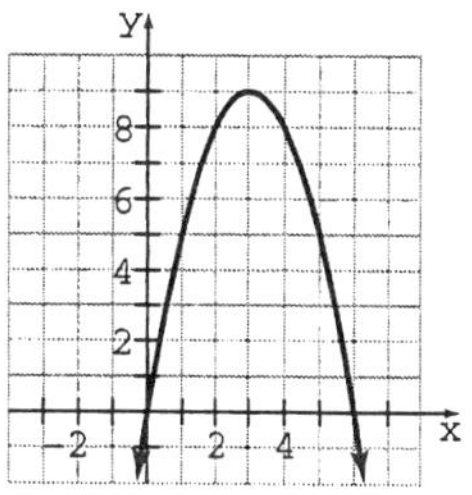

37. $y = -2x^2 + 4x - 8$

a. Since $a = -2$ the parabola opens downward.

b. $y = -2(0)^2 + 4(0) - 8 = -8$

The y-intercept is (0, –8).

c. $x = -\frac{b}{2a} = -\frac{4}{2(-2)} = -\frac{4}{-4} = 1$

$y = \frac{4ac - b^2}{4a}$

$= \frac{4(-2)(-8) - (4)^2}{4(-2)}$

$= \frac{64 - 16}{-8}$

$= -6$

The vertex is (1, –6).

d. $0 = -2x^2 + 4x - 8 = -2(x^2 - 2x + 4)$

Since this is not factorable, check the discriminant.

$b^2 - 4ac = 4^2 - 4(-2)(-8)$

$= 16 - 64$

$= -48$

Since $-48 < 0$, there are no real roots. Thus, there are no x-intercepts.

e.

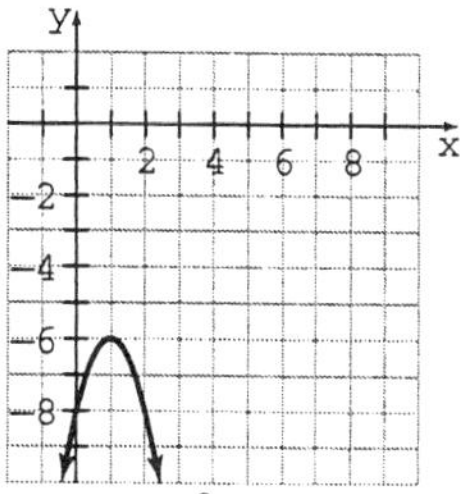

$f(x) = -2x^2 + 4x - 8$

39. $g(x) = -2x^2 - 6x + 4$

a. Since $a = -2$ the parabola opens downward.

b. $y = -2(0)^2 - 6(0) + 4 = 4$
The y-intercept is (0, 4).

c. $x = -\frac{b}{2a} = -\frac{(-6)}{2(-2)} = \frac{6}{-4} = -\frac{3}{2}$

$$y = \frac{4ac - b^2}{4a} = \frac{4(-2)(4) - (-6)^2}{4(-2)} = \frac{-32 - 36}{-8} = \frac{-68}{-8} = \frac{17}{2}$$

The vertex is $\left(-\frac{3}{2}, \frac{17}{2}\right)$.

d. $0 = -2x^2 - 6x + 4 = -2(x^2 + 3x - 2)$
Since this is not factorable, check the discriminant.

$$b^2 - 4ac = (-6)^2 - 4(-2)(4) = 36 + 32 = 68$$

There are two real roots.

$$x = \frac{-b \pm \sqrt{b^2 - 4ac}}{2a} = \frac{-(-6) \pm \sqrt{68}}{2(-2)} = \frac{6 \pm 2\sqrt{17}}{-4} = \frac{3 \pm \sqrt{17}}{-2} = \frac{-3 \pm \sqrt{17}}{2}$$

The x-intercepts are $\left(\frac{-3+\sqrt{17}}{2}, 0\right)$ and $\left(\frac{-3-\sqrt{17}}{2}, 0\right)$.

e.

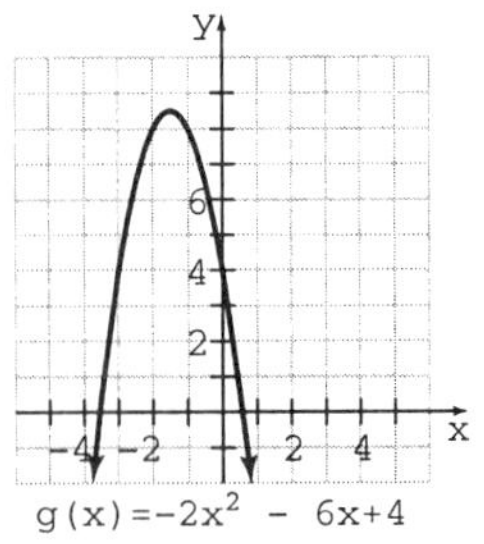

41. $y = 3x^2 + 4x - 6$

a. Since $a = 3$ the parabola opens upward.

b. $y = 3(0)^2 + 4(0) - 6 = -6$
The y-intercept is (0, –6).

c. $x = -\frac{b}{2a} = -\frac{4}{2(3)} = -\frac{4}{6} = -\frac{2}{3}$

$$y = \frac{4ac - b^2}{4a} = \frac{4(3)(-6) - 4^2}{4(3)} = \frac{-72 - 16}{12} = \frac{-88}{12} = -\frac{22}{3}$$

The vertex is $\left(-\frac{2}{3}, -\frac{22}{3}\right)$.

d. $0 = 3x^2 + 4x - 6$
Since this is not factorable, check the discriminant.

$$b^2 - 4ac = 4^2 - 4(3)(-6) = 16 + 72 = 88$$

Since $88 > 0$ there are two real roots.

$$x = \frac{-b \pm \sqrt{b^2 - 4ac}}{2a} = \frac{-4 \pm \sqrt{88}}{2(3)} = \frac{-4 \pm 2\sqrt{22}}{6} = \frac{-2 \pm \sqrt{22}}{3}$$

The x-intercepts are $\left(\frac{-2+\sqrt{22}}{3}, 0\right)$ and $\left(\frac{-2-\sqrt{22}}{3}, 0\right)$.

e.

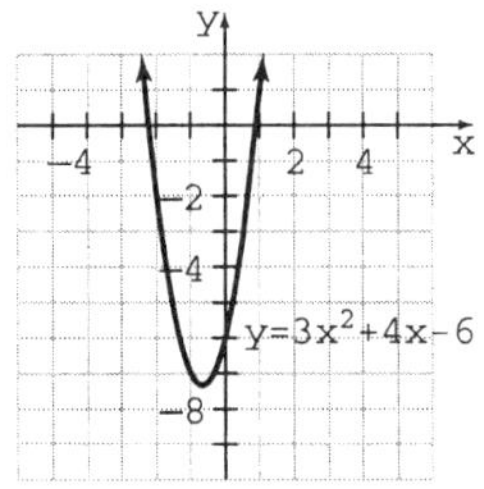

43. $y = 2x^2 - x - 6$

a. Since $a = 2$ the parabola opens upward.

b. $y = 2(0)^2 - 0 - 6 = -6$
The y-intercept is (0, –6).

c. $x = -\frac{b}{2a} = -\frac{-1}{2(2)} = \frac{1}{4}$

$$y = \frac{4ac - b^2}{4a} = \frac{4(2)(-6) - (-1)^2}{4(2)} = \frac{-48 - 1}{8} = -\frac{49}{8}$$

The vertex is $\left(\frac{1}{4}, -\frac{49}{8}\right)$.

d. $0 = 2x^2 - x - 6 = (2x + 3)(x - 2)$

$2x + 3 = 0$ or $x - 2 = 0$

$x = -\frac{3}{2}$ $\quad x = 2$

The x-intercepts are $\left(-\frac{3}{2}, 0\right)$ and (2, 0).

e.

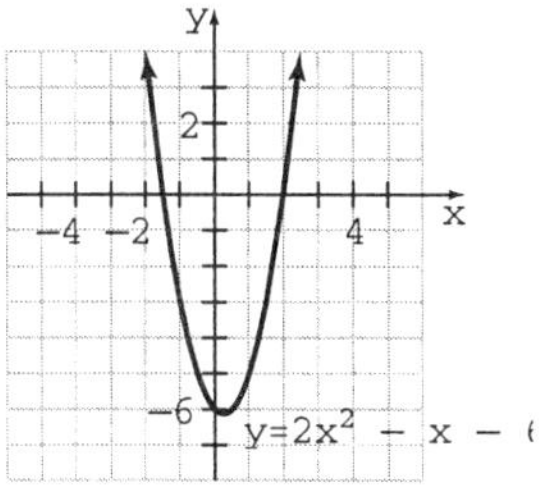

45. $f(x) = -x^2 + 3x - 5$

a. Since $a = -1$ the parabola opens downward.

b. $y = -0^2 + 3(0) - 5 = -5$
The y-intercept is (0, –5).

c. $x = -\frac{b}{2a} = -\frac{3}{2(-1)} = -\frac{3}{-2} = \frac{3}{2}$

$$y = \frac{4ac - b^2}{4a} = \frac{4(-1)(-5) - 3^2}{4(-1)} = -\frac{11}{4}$$

The vertex is $\left(\frac{3}{2}, -\frac{11}{4}\right)$.

d. $0 = -x^2 + 3x - 5$
Since this is not factorable, check the discriminant.
$b^2 - 4ac = 3^2 - 4(-1)(-5) = 9 - 20 = -11$
Since $-11 < 0$ there are no real roots. Thus, there are no x-intercepts.

e.

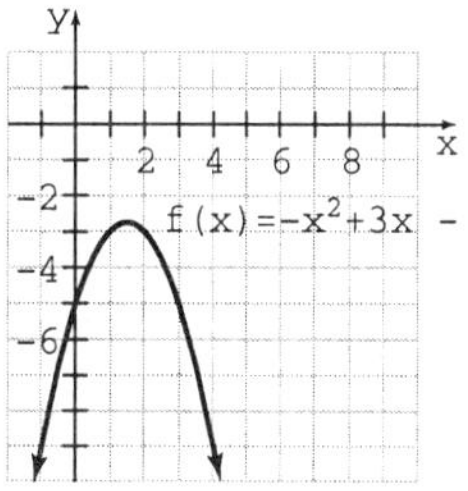

47. In the function $f(x) = (x - 3)^2$, h has a value of 3. The graph of $f(x)$ is the graph of $g(x) = x^2$ shifted 3 units to the right.

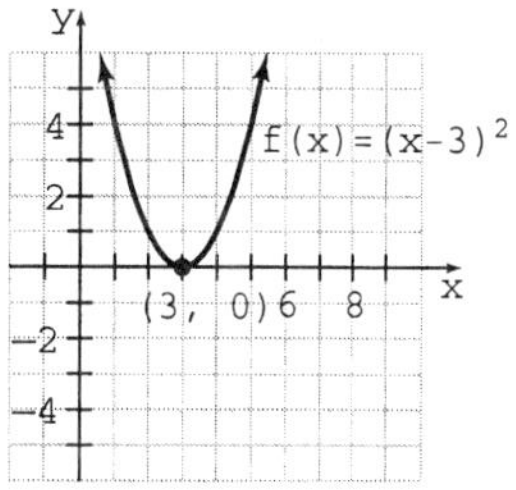

49. In the function $f(x) = (x+1)^2$, h has a value of –1. The graph of $f(x)$ is the graph of $g(x) = x^2$ shifted 1 unit to the left.

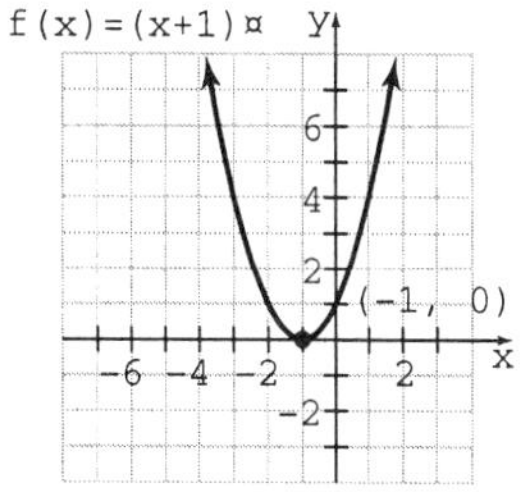

51. In the function $f(x) = x^2 + 3$, k has a value of 3. The graph $f(x)$ will be the graph of $g(x) = x^2$

shifted 3 units up.

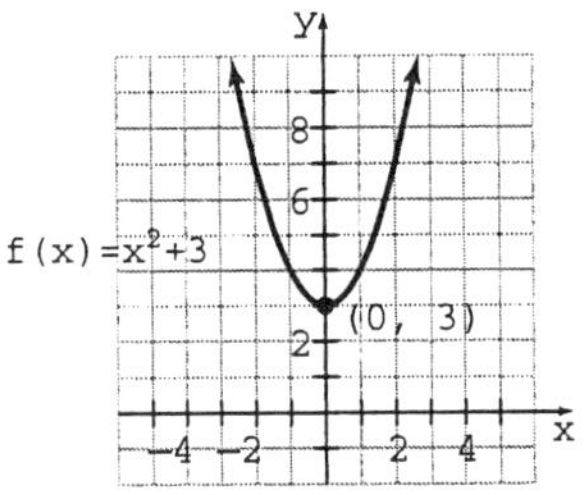

53. In the function $f(x) = x^2 - 1$, k has a value of -1. The graph $f(x)$ will be the graph of $g(x) = x^2$ shifted 1 units down.

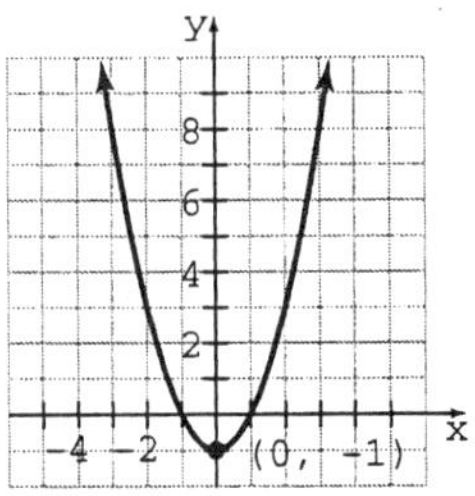

55. 55In the function $f(x) = (x-2)^2 + 3$, h has a value of 2 and k has a value of 3. The graph of $f(x)$ will be the graph of $g(x) = x^2$ shifted 2 units to the right and 3 units up.

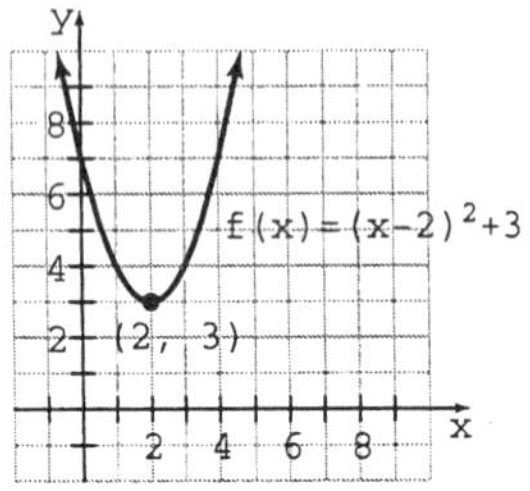

57. In the function $f(x) = (x+4)^2 + 4$, h has a value of −4 and k has a value of 4. The graph of $f(x)$ will be the graph of $g(x) = x^2$ shifted 4 units to the left and 4 units up.

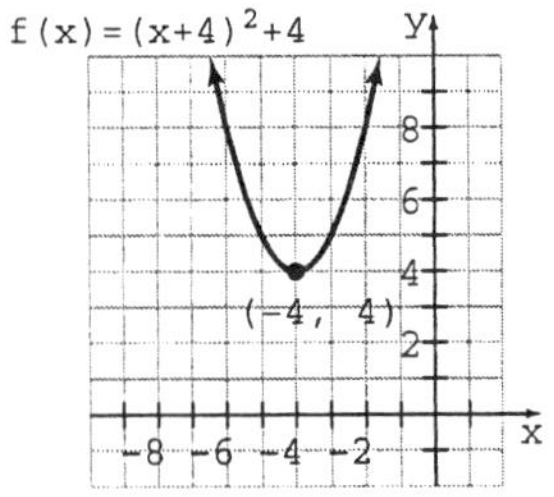

59. In the function $g(x) = -(x+3)^2 - 2$, a has the value -1, h has the value -3, and k has the value -2. Since $a < 0$, the parabola opens downward. The graph of $g(x)$ will be the graph of $f(x) = -x^2$ shifted 3 units to the left and 2 units down.

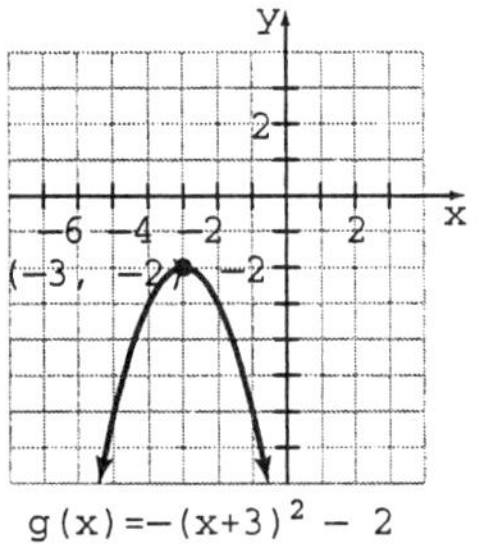

61. In the function $y = -2(x-2)^2 + 2$, a has a value of -2, h has a value of 2, and k has a value of 2. The graph of y will be the graph of $g(x) = -2x^2$ shifted 2 units to the right and 2 units up.

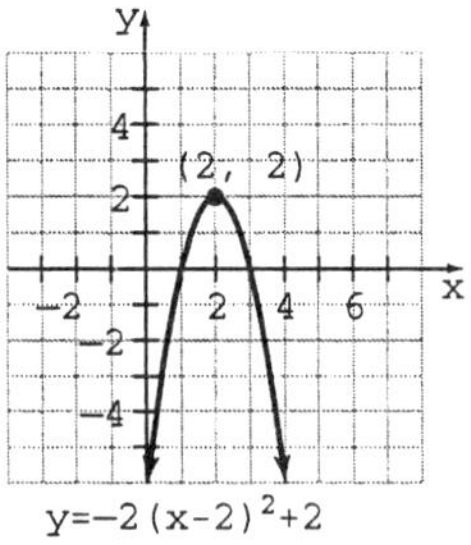

63. In the function $h(x) = -2(x+1)^2 - 3$, a has a value of -2, h has a value of -1, and k has a value of -3. Since $a < 0$, the parabola opens downward. The graph of $h(x)$ will be the graph of $f(x) = -2x^2$ shifted 1 unit left and 3 units down.

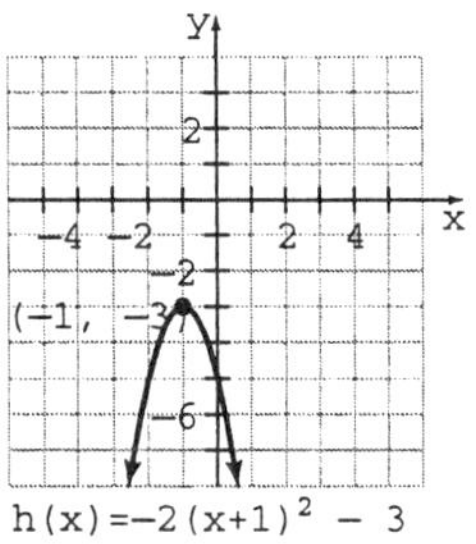

65. **a.** $f(x) = x^2 - 6x + 8$
$= (x^2 - 6x + 9) + 8 - 9$
$f(x) = (x-3)^2 - 1$

b. Since $h = 3$ and $k = -1$, the vertex is $(3, -1)$

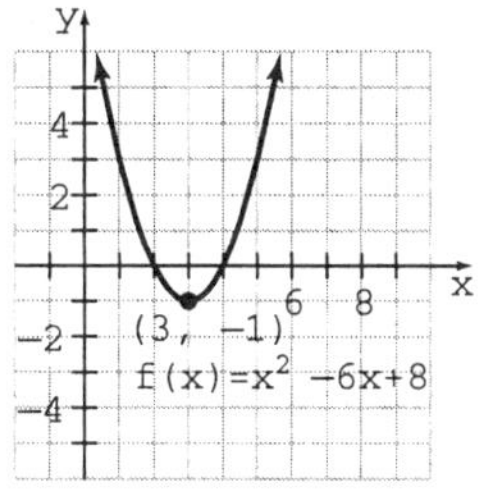

67. **a.** $f(x) = x^2 - x + 1$
$= \left(x^2 - x + \frac{1}{4}\right) + 1 - \frac{1}{4}$
$f(x) = \left(x - \frac{1}{2}\right)^2 + \frac{3}{4}$

b. Since $h = \frac{1}{2}$ and $k = \frac{3}{4}$, the vertex is $\left(\frac{1}{2}, \frac{3}{4}\right)$.

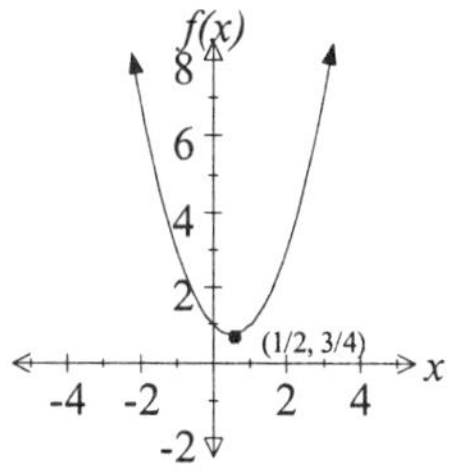

69. **a.** $f(x) = -x^2 - 4x - 6$
$= -(x^2 + 4x) - 6$
$= -(x^2 + 4x + 4) - 6 - (-4)$
$= -(x^2 + 4x + 4) - 6 + 4$
$f(x) = -(x+2)^2 - 2$

b. Since $h = -2$ and $k = -2$, the vertex is $(-2, -2)$. Since $a < 0$, the parabola opens downward.

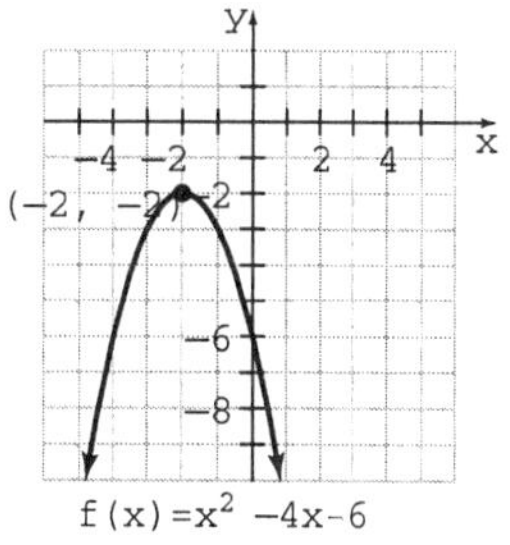

71. **a.** $g(x) = x^2 - 4x - 5$
$= (x^2 - 4x + 4) - 5 - 4$
$g(x) = (x-2)^2 - 9$

b. Since $h = 2$ and $k = -9$, the vertex is $(2, -9)$.

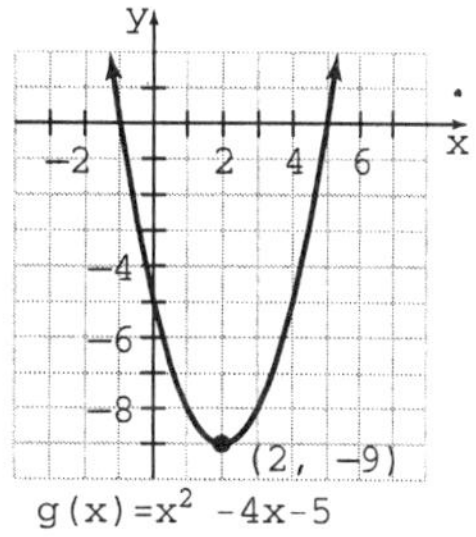

73. **a.** $f(x) = 2x^2 + 5x - 3$
$= 2\left(x^2 + \frac{5}{2}x\right) - 3$
$= 2\left(x^2 + \frac{5}{2}x + \frac{25}{16}\right) - 3 - 2\left(\frac{25}{16}\right)$
$= 2\left(x^2 + \frac{5}{2}x + \frac{25}{16}\right) - 3 - \frac{25}{8}$
$f(x) = 2\left(x + \frac{5}{4}\right)^2 - \frac{49}{8}$

b. Since $h = -\frac{5}{4}$ and $k = -\frac{49}{8}$, the vertex is $\left(-\frac{5}{4}, -\frac{49}{8}\right)$. The graph of $f(x)$ will be the graph of $g(x) = 2x^2$ shifted $\frac{5}{4}$ units left and $\frac{49}{8}$ units down.

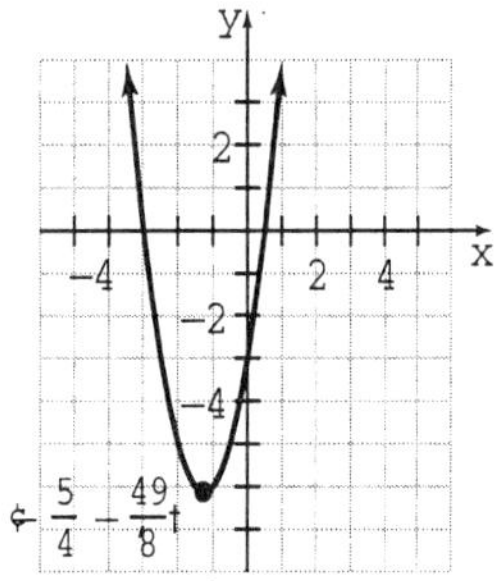

$f(x) = 2x^2 + 5x - 3$

75. $y = x^2 + 3x - 4$

$x = -\frac{b}{2a} = -\frac{3}{2}$

The parabola opens up and the x-coordinate of the vertex is $-\frac{3}{2}$. The graph is c).

77. The leading coefficient is negative. The function will decrease as x increases. The y-intercept is (0, –6). The graph is c).

79. **a.**

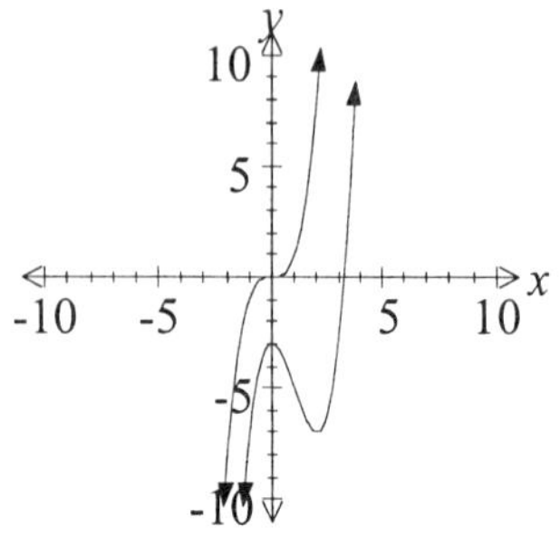

b. As x increases the functions increase.

c. Answers will vary.

d. As x decreases, the functions decrease.

e. Answers will vary.

81. $f(x) = 2(x+3)^2 - 1$. The vertex is $(h, k) = (-3, -1)$. Since $a > 0$, the parabola opens up. The graph is d).

83. $f(x) = 2(x-1)^2 + 3$ The vertex is $(h, k) = (1, 3)$. Since $a > 0$, the parabola opens up. The graph is b).

85. **a.** $f(x) = (x+4)(18-x)$

$= 18x - x^2 + 72 - 4x$

$= -x^2 + 14x + 72$

Since $a = -1$, the graph of this function is a parabola that opens downward and thus has a maximum value at its vertex.

$x = -\frac{b}{2a} = -\frac{14}{2(-1)} = 7$

b. $y = \frac{4ac - b^2}{4a}$

$= \frac{4(-1)(72) - (14)^2}{4(-1)}$

$= \frac{-288 - 196}{-4}$

$= 121$

The maximum area is 121 square units.

87. **a.** $f(x) = (x+5)(26-x)$

$= 26x - x^2 + 130 - 5x$

$= -x^2 + 21x + 130$

Since $a = -1$, the graph of this function is a parabola that opens downward and thus has a maximum value at its vertex.

$x = -\frac{b}{2a} = -\frac{21}{2(-1)} = 10.5$

b. $y = \frac{4ac - b^2}{4a}$

$= \frac{4(-1)(130) - (21)^2}{4(-1)}$

$= \frac{-520 - 441}{-4}$

$= 240.25$

The maximum area is 240.25 square units.

89. **a.** $R(n) = -0.02n^2 + 10n$

Since $a = -0.02$, the graph of this function is a parabola that opens downward and thus has a maximum value at its vertex.

$$x = -\frac{b}{2a} = -\frac{10}{2(-0.02)} = 250$$

The maximum revenue will be achieved when 250 batteries are sold.

b. $$y = \frac{4ac - b^2}{4a}$$
$$= \frac{4(-0.02)(0) - (10)^2}{4(-0.02)}$$
$$= \frac{0 - 100}{-0.08}$$
$$= 1250$$

The maximum revenue is \$1250.

91. $N(t) = -0.043t^2 + 1.22t + 46.0$

Since $a = -0.043$, the graph of this function is a parabola that opens downward and thus has a maximum value at its vertex.

$$x = -\frac{b}{2a} = -\frac{1.22}{2(-0.043)} \approx 14.2$$

The maximum enrollment will be obtained about 14 years after 1989 which is the year 2003.

93. For $f(x) = (x-2)^2 + \frac{5}{2}$, the vertex is $\left(2, \frac{5}{2}\right)$. For $g(x) = (x-2)^2 - \frac{3}{2}$, the vertex is $\left(2, -\frac{3}{2}\right)$. These points are on the vertical line $x = 2$. The distance between the two points is

$$\frac{5}{2} - \left(-\frac{3}{2}\right) = \frac{5}{2} + \frac{3}{2} = \frac{8}{2} = 4 \text{ units.}$$

95. For $f(x) = 2(x+4)^2 - 3$, the vertex is $(-4, -3)$. For $g(x) = -(x+1)^2 - 3$, the vertex is $(-1, -3)$. These points are on the horizontal line $y = -3$. The distance between the two points is $-1 - (-4) = -1 + 4 = 3$ units.

97. A function that has the shape of $f(x) = 2x^2$ will have the form $f(x) = 2(x-h)^2 + k$. If $(h, k) = (3, -2)$, the function is $f(x) = 2(x-3)^2 - 2$.

99. A function that has the shape of $f(x) = -4x^2$ will have the form $f(x) = -4(x-h)^2 + k$. If $(h, k) = \left(-\frac{3}{5}, -\sqrt{2}\right)$, the function is

$$f(x) = -4\left(x + \frac{3}{5}\right)^2 - \sqrt{2}.$$

101. **a.** The graphs will have the same x-intercepts but $f(x) = x^2 - 8x + 12$ will open up and $g(x) = -x^2 + 8x - 12$ will open down.

b. Yes, because the x-intercepts are located by setting $x^2 - 8x + 12$ and $-x^2 + 8x - 12$ equal to zero. They have the same solution set, therefore the x-intercepts are equal. The x-intercepts for both are (6, 0), and (2, 0).

c. No. The vertex for $f(x) = x^2 - 8x + 12$ is $(4, -4)$ and the vertex for $g(x) = -x^2 + 8x - 12$ is $(4, 4)$.

d.

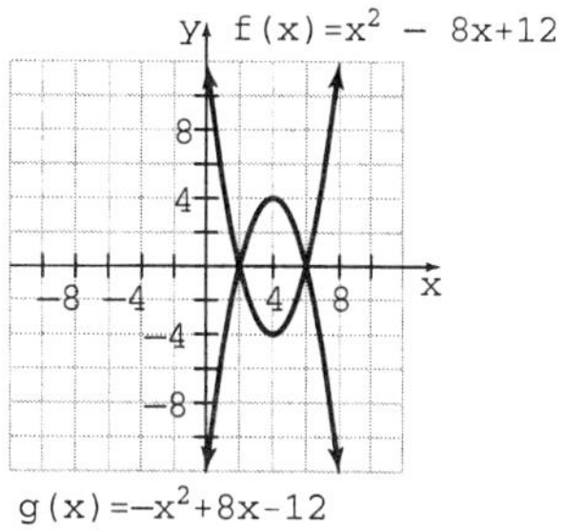

103. **a.** The vertex $x = -\frac{b}{2a} = -\frac{24}{2(-1)} = 12$

$$I = -(12)^2 + 24(12) - 44$$
$$= -144 + 288 - 44$$
$$= 100$$

The vertex is at (12, 100). To find the roots set $I = 0$.

$$0 = -x^2 + 24x - 44$$
$$= -1(x^2 - 24x + 44)$$
$$= -1(x-2)(x-22)$$
$$x - 2 = 0 \quad \text{or} \quad x - 22 = 0$$
$$x = 2 \qquad\qquad x = 22$$

The roots are 2 and 22.

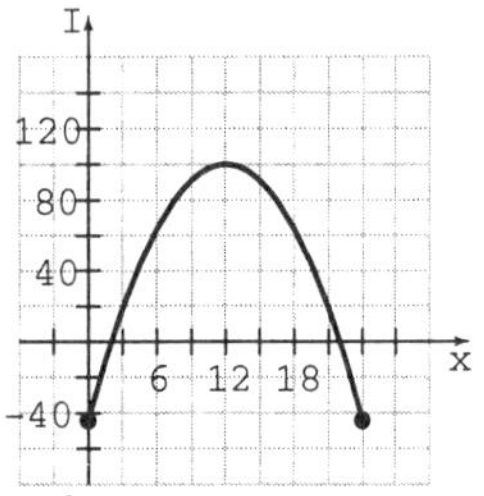

$I=-x^2+24x-44,\ 0 \le x \le 24$

b. The minimum cost will be \$2 since the smaller root is 2.

c. The maximum cost is \$22 since the larger root is 22.

d. The maximum value will occur at the vertex of the parabola, (12, 100). Therefore, they should charge \$12.

e. The maximum value will occur at the vertex of the parabola, (12, 100). Since I is in hundreds of dollars the maximum income is 100(\$100) = \$10,000.

105. **a.** The number of bird feeders sold for the maximum profit will be the x-coordinate of the vertex.

$$f(x) = -0.4x^2 + 80x - 200$$
$$x = -\frac{b}{2a} = -\frac{80}{2(-0.4)} = -\frac{80}{-0.8} = 100$$

The company must sell 100 bird feeders for maximum profit.

b. The maximum profit will be the y-coordinate of the vertex, $y = f(100)$.

$$f(100) = -0.4(100^2) + 80(100) - 200$$
$$= -0.4(10{,}000) + 8000 - 200$$
$$= -4000 + 8000 - 200$$
$$= 3800$$

The maximum profit will be \$3800.

107. **a.** The maximum height is h, the y-coordinate of the vertex.

$$h(t) = -4.9t^2 + 24.5t + 9.8$$
$$h = \frac{4ac - b^2}{4a}$$
$$= \frac{4(-4.9)(9.8) - (24.5)^2}{4(-4.9)}$$
$$= \frac{-192.08 - 600.25}{-19.6}$$
$$= \frac{-792.33}{-19.6}$$
$$= 40.425$$

The maximum height obtained by the cannonball is 40.425 meters.

b. The time the cannonball reaches the maximum height is t, the x-coordinate of the vertex,. $t = -\frac{b}{2a} = -\frac{24.5}{2(-4.9)} = 2.5$

The cannonball will reach the maximum height after 2.5 seconds.

c. $f(x) = 0$ when the cannonball hits the ground.

$$h(t) = -4.9t^2 + 24.5t + 9.8$$
$$0 = -4.9t^2 + 24.5t + 9.8$$
$$t = \frac{-24.5 \pm \sqrt{(24.5)^2 - 4(-4.9)(9.8)}}{2(-4.9)}$$
$$t = \frac{-24.5 \pm \sqrt{792.33}}{-9.8}$$
$$t \approx \frac{-24.5 \pm 28.14836}{-9.8}$$

$t \approx -0.37$ and $t \approx 5.37$

Disregard the negative value. The cannonball will hit the ground after about 5.37 seconds.

109. **a.** $2004 \Rightarrow x = 13$

$$C(x) = 0.19x^2 - 0.657x + 16.6$$
$$C(13) = 0.19(13)^2 - 0.657(13) + 16.6$$
$$\approx 40.17$$

The average annual rent per square foot in 2004 should be about \$40.17.

b. The function will reach a minimum at it's vertex:

$$x=-\frac{b}{2a}=-\frac{-0.657}{2(0.19)}\approx 1.7$$

The annual rent per square foot reached a minimum about 1.7 years after 1991 which is in 1992.

c. $$h=\frac{4ac-b^2}{4a}$$

$$=\frac{4(0.19)(16.6)-(-0.657)^2}{4(0.19)}$$

$$=\frac{12.616-0.431649}{.76}$$

$$\approx 16.03$$

The minimum annual rent per square foot was about \$16.

111. If the perimeter of the room is 60 ft., then $60=2l+2w$, where l is the length and w is the width. Then $60=2(l+w)$ and $30=l+w$. Therefore $l=30-w$. The area of the room is $A=lw=(30-w)w=30w-w^2=-w^2+30w$. The maximum area is the *y*-coordinate of the vertex.

$$A=\frac{4ac-b^2}{4a}$$

$$=\frac{4(-1)(0)-30^2}{4(-1)}$$

$$=\frac{0-900}{-4}$$

$$=225$$

The maximum area is 225 ft^2.

113. If two numbers differ by 8 and x is one of the numbers, then $x+8$ is the other number. The product is $f(x)=x(x+8)=x^2+8x$. The maximum product is the *y*-coordinate of the vertex.

$$x=-\frac{b}{2a}=-\frac{8}{2(1)}=-4$$

$$y=f(-4)$$

$$=(-4)^2+8(-4)$$

$$=16-32$$

$$=-16$$

The maximum product is –16. The numbers are –4 and –4 + 8 = 4.

115. If two numbers add to 40 and x is one of the numbers, then $40-x$ is the other number. The product is

$f(x)=x(40-x)=40x-x^2=-x^2+40x$.

The maximum product is the *y*-coordinate of the vertex.

$$x=-\frac{b}{2a}=-\frac{40}{2(-1)}=-\frac{40}{-2}=20$$

$$y=f(20)=-20^2+40(20)=-400+800=400$$

The maximum product is 400. The numbers are 20 and 40 – 20 = 20.

117. $C(x)=2000+40x,\ R(x)=800x-x^2$

$$P(x)=R(x)-C(x)$$

$$P(x)=\left(800x-x^2\right)-(2000+40x)$$

$$=800x-x^2-2000-40x$$

$$=-x^2+760x-2000$$

The maximum profit is $P(x)$, the *y*-coordinate of the vertex. The number of items that must be produced and sold to obtain maximum profit is the *x* coordinate of the vertex.

$$x=-\frac{b}{2a}=-\frac{760}{2(-1)}=380$$

a. $$P(380)=-380^2+760(380)-2000$$

$$=-144{,}400+288800-2000$$

$$=142{,}400$$

The maximum profit is \$142,400.

b. The number of items that must be produced and sold to obtain maximum profit is 380.

119. The *y*-intercept is (0, –5) and as *x* increases the functions decreases because of the leading coefficient. The graph is b).

121. **a.** $$f(t)=-16t^2+52t+3$$

$$=-16\left(t^2-3.25t\right)+3$$

$$=-16\left(t^2-3.25t+\left(\frac{3.25}{2}\right)^2\right)+3+(16)\left(\left(\frac{3.25}{2}\right)^2\right)$$

$$=-16(t-1.625)^2+3+42.25$$

$$=-16(t-1.625)^2+45.25$$

b. The maximum height was 45.25 feet obtained at 1.625 seconds.

c. It is the same answer.

125. The radius of the outer circle is $r = 15$ ft. The area $A = \pi r^2 = \pi(15^2) = 225\pi$ ft^2. The radius of the inner circle is $r = 5$ ft. The area is $A = \pi r^2 = \pi(5^2) = 25\pi \text{ ft}^2$. The blue shaded area is $225\pi \text{ ft}^2 - 25\pi \text{ ft}^2 = 200\pi \text{ ft}^2$.

126.
$$(x-3) \div \frac{x^2+3x-18}{x} = \frac{x-3}{1} \cdot \frac{x}{x^2+3x-18}$$
$$= \frac{x-3}{1} \cdot \frac{x}{(x+6)(x-3)}$$
$$= \frac{x}{x+6}$$

127.
$$x - y = -5 \quad (1)$$
$$2x + 2y - z = 0 \quad (2)$$
$$x + y + z = 3 \quad (3)$$

First eliminate the variable z from equations (2) and (3) by adding these equations.
$$2x + 2y - z = 0$$
$$x + y + z = 3$$
$$3x + 3y = 3 \quad (4)$$

Equations (1) and (4) form a system of equations in two variables. Multiply equation (1) by 3 and add the result to equation (4) to eliminate the variable y.
$$3(x - y = -5) \Rightarrow 3x - 3y = -15$$
$$3x + 3y = 3 \Rightarrow 3x + 3y = 3$$
$$6x = -12$$
$$x = -2$$

Substitute -2 for x in equation (1) to find y.
$$(-2) + y = -5$$
$$y = 3$$

Substitute -2 for x and 3 for y in equation (3) to find z.
$$(-2) + (3) + z = 3$$
$$1 + z = 3$$
$$z = 2$$

The solution is (–2, 3, 2).

128.
$$\begin{vmatrix} \frac{1}{2} & 3 \\ 2 & -4 \end{vmatrix} = \left(\frac{1}{2}\right)(-4) - (2)(3)$$
$$= -2 - 6$$
$$= -8$$

129. $y \le \frac{2}{3}x + 3$

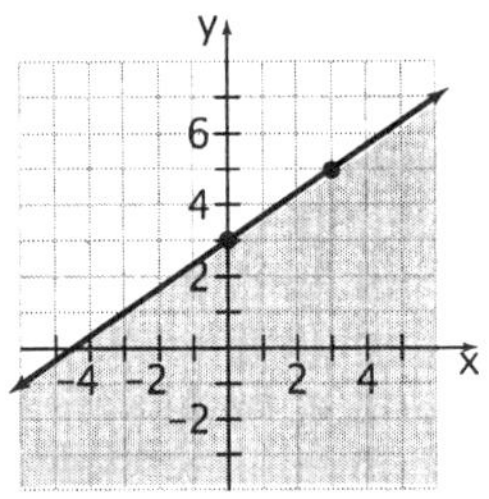

Exercise Set 12.6

1. a. For $f(x) = x^2 - 7x + 10$, $f(x) > 0$ when the graph is above the x-axis. The solution is $x < 2$ or $x > 5$.

b. For $f(x) = x^2 - 7x + 10$ $f(x) < 0$ when the graph is below the x-axis. The solution is $2 < x < 5$.

3. Yes. The boundary values 5 and –3 are included in the solution set since this is a greater than *or equal to* inequality. These values make the expression equal to zero.

5. The boundary values –2 and 1 are included in the solution set since this is a less than *or equal to* inequality. These values make the expression equal to zero. However, the boundary value –1 is not included in the solution set since it would result in a zero in the denominator, which is undefined.

7. $x^2 - 2x - 8 \ge 0$
$(x+2)(x-4) \ge 0$

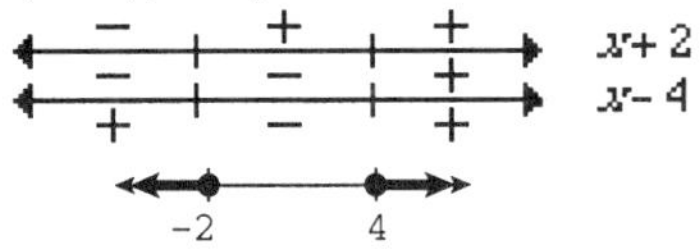

9. $x^2 + 8x + 7 > 0$
$(x+7)(x+1) > 0$

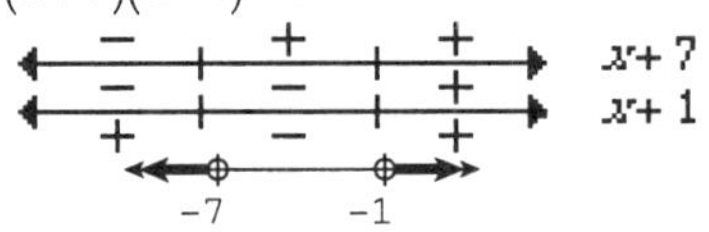

11. $p^2 + 4p > 0$
$p(p+4) > 0$

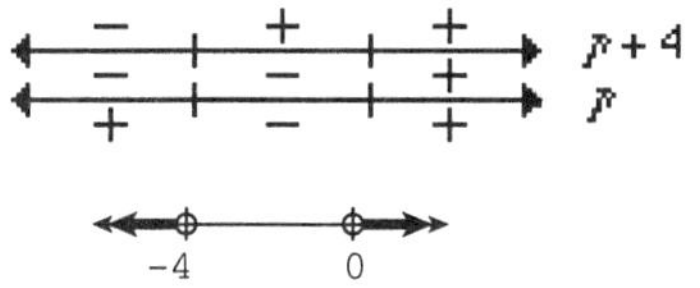

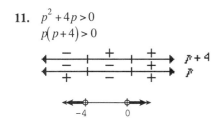

13. $x^2 - 16 < 0$
$(x+4)(x-4) < 0$

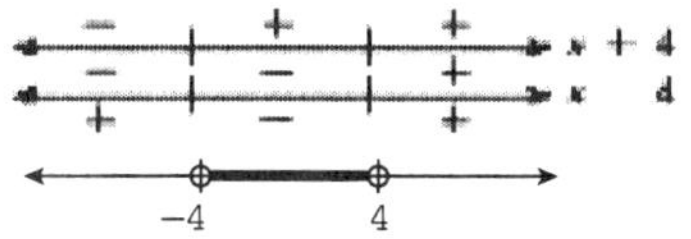

15. $2x^2 + 5x \quad 3 \ge 0$
$(2x-1)(x+3) \ge 0$

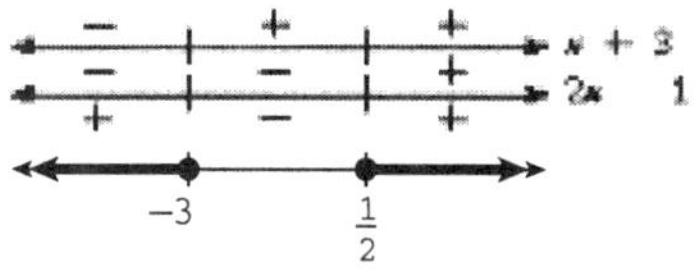

17. $5x^2 + 19x \le 4$
$5x^2 + 19x - 4 \le 0$
$(x+4)(5x-1) \le 0$

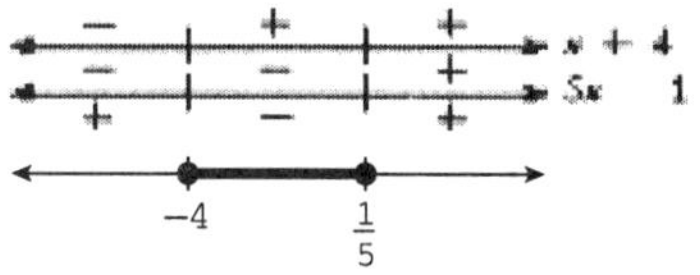

19. $2x^2 - 12x + 9 \le 0$
$2x^2 - 12x + 9 = 0$

$$x = \frac{-(-12) \pm \sqrt{(-12)^2 - 4(2)(9)}}{2(2)}$$
$$= \frac{12 \pm \sqrt{144 - 72}}{4}$$
$$= \frac{12 \pm \sqrt{72}}{4}$$
$$= \frac{12 \pm 6\sqrt{2}}{4}$$
$$= \frac{6 \pm 3\sqrt{2}}{2}$$

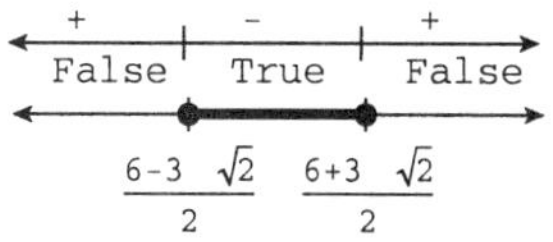

21. $(x-2)(x+1)(x+3) \ge 0$
$x - 2 = 0 \quad x + 1 = 0 \quad x + 3 = 0$
$x = 2 \quad x = -1 \quad x = -3$

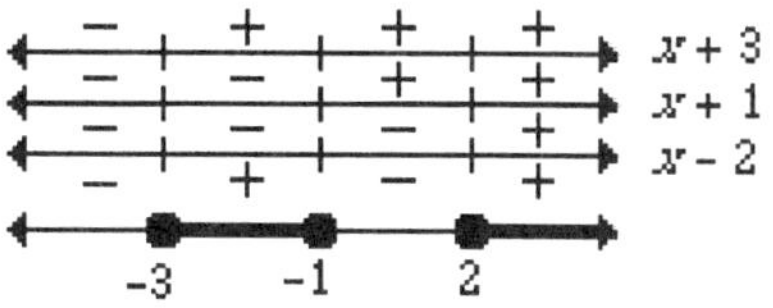

$[-3, -1] \cup [2, \infty)$

23. $(a-3)(a+2)(a+4) < 0$
$a - 3 = 0 \quad a + 2 = 0 \quad a + 4 = 0$
$a = 3 \quad a = -2 \quad a = -4$

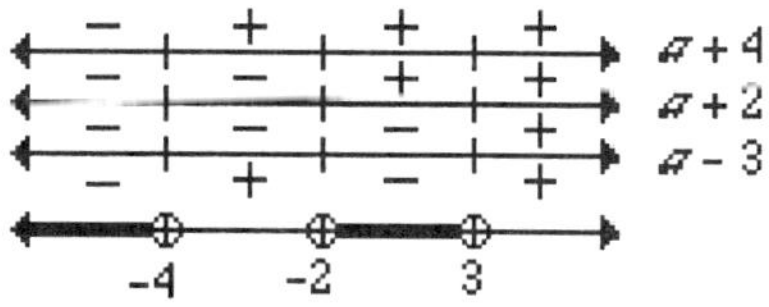

$(-\infty, -4) \cup (-2, 3)$

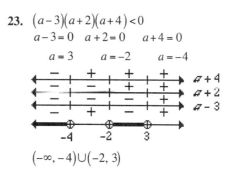

25. $(2c+5)(3c-6)(c+6) > 0$
$2c + 5 = 0 \quad 3c - 6 = 0 \quad c + 6 = 0$
$2c = -5 \quad 3c = 6 \quad c = -6$
$c = -\frac{5}{2} \quad c = 2$

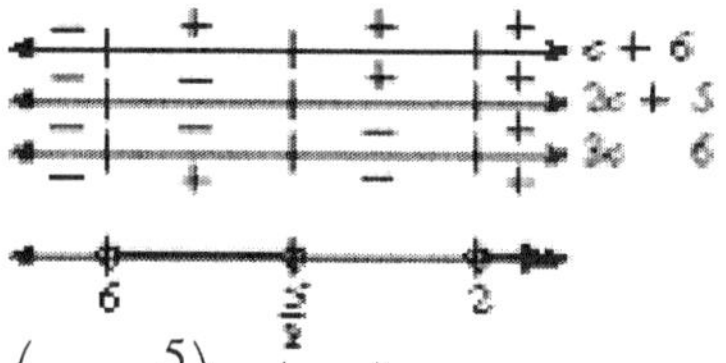

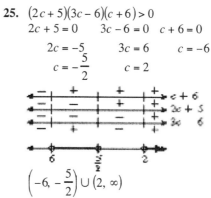

$\left(-6, -\frac{5}{2}\right) \cup (2, \infty)$

27. $(3x+5)(x-3)(x+4) > 0$
$3x + 5 = 0 \quad x - 3 = 0 \quad x + 4 = 0$
$3x = -5 \quad x = 3 \quad x = -4$
$x = -\frac{5}{3}$

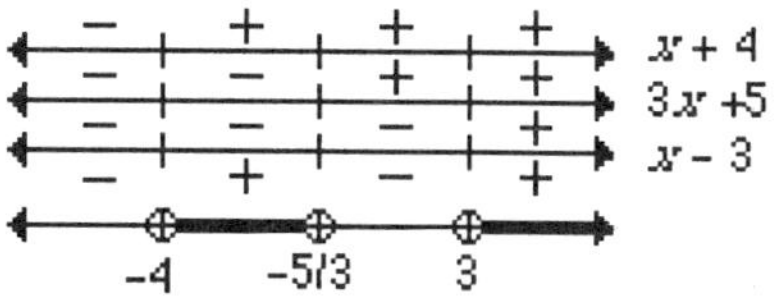

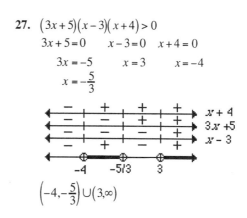

$\left(-4, -\frac{5}{3}\right) \cup (3, \infty)$

29. $(x+2)(x+2)(3x-8) \ge 0$

$x+2=0 \quad 3x-8=0$

$x=-2 \quad x=\frac{8}{3}$

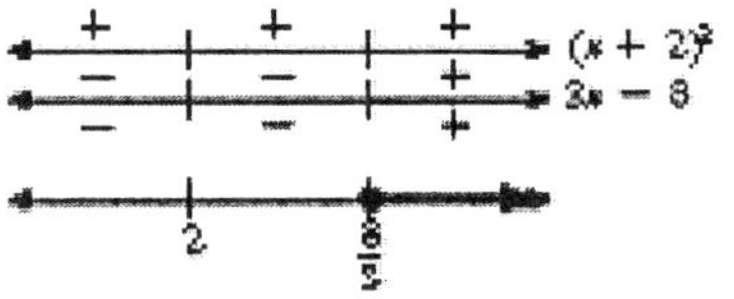

$\left[\frac{8}{3}, \infty\right)$

31. $x^3-4x^2+4x<0$

$x\left(x^2-4x-4\right)<0$

$x(x-2)^2<0$

$x=0 \quad x-2=0$

$x=2$

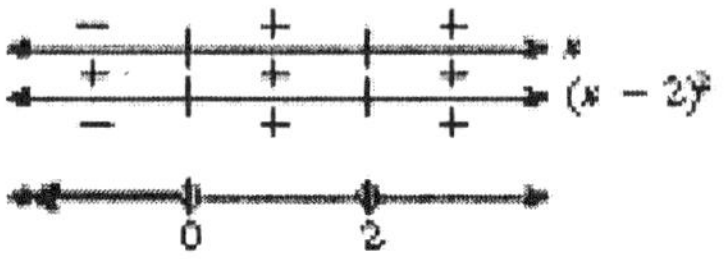

$(-\infty, 0)$

33. $f(x)=x^2-3x, f(x)\ge 0$

$x^2-3x\ge 0$

$x(x-3)\ge 0$

$x=0 \quad x-3=0$

$x=3$

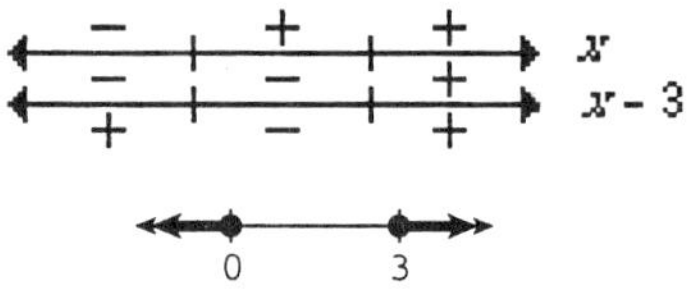

35. $f(x)=x^2+4x, f(x)>0$

$x^2+4x>0$

$x(x+4)>0$

$x=0 \quad x+4=0$

$x=-4$

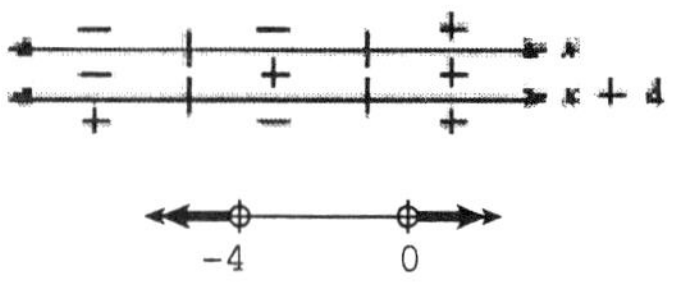

37. $f(x)=x^2-14x+48, f(x)<0$

$x^2-14x+48<0$

$(x-6)(x-8)<0$

$x-6=0 \quad x-8=0$

$x=6 \quad x=8$

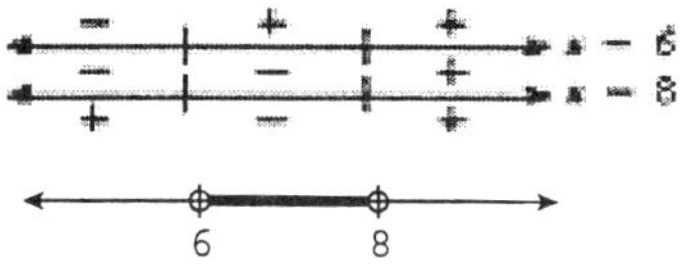

39. $f(x)=2x^2+9x-4, f(x)\le 2$

$2x^2+9x-4\le 2$

$2x^2+9x-6\le 0$

$x=\frac{-9\pm\sqrt{9^2-4(2)(-6)}}{2(2)}$

$=\frac{-9\pm\sqrt{129}}{4}$

$x=\frac{-9-\sqrt{129}}{4} \quad x=\frac{-98+\sqrt{129}}{4}$

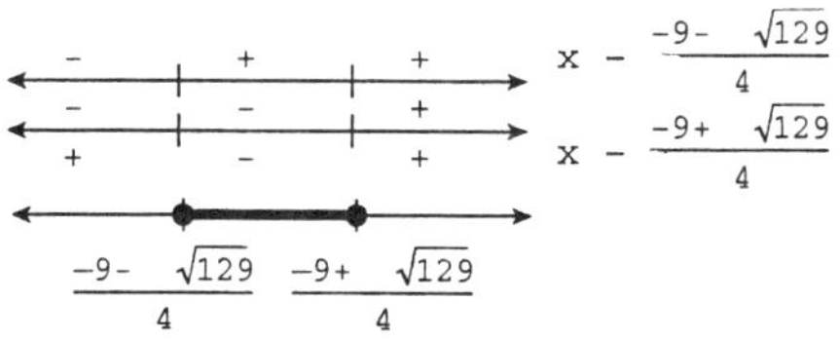

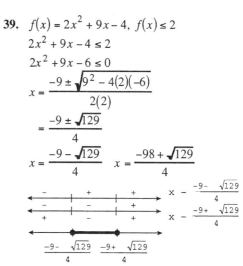

41. $f(x)=2x^3+9x^2-35x, f(x)\ge 0$

$2x^3+9x^2-35x\ge 0$

$x\left(2x^2+9x-35\right)\ge 0$

$x(2x-5)(x+7)\ge 0$

$x=0 \quad 2x-5=0 \quad x+7=0$

$x=\frac{5}{2} \quad x=-7$

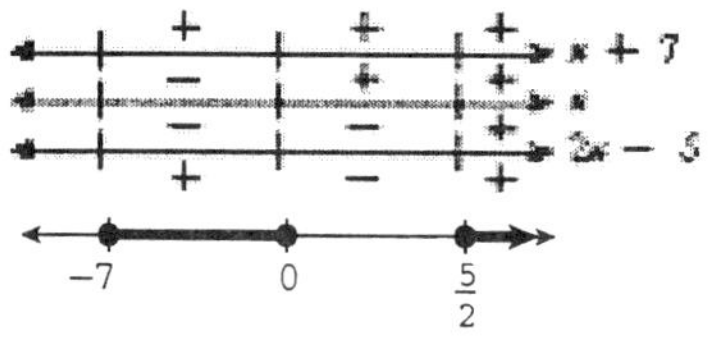

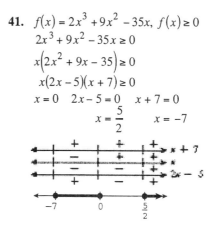

43. $\frac{x+2}{x-4}>0$

$x\ne 4$

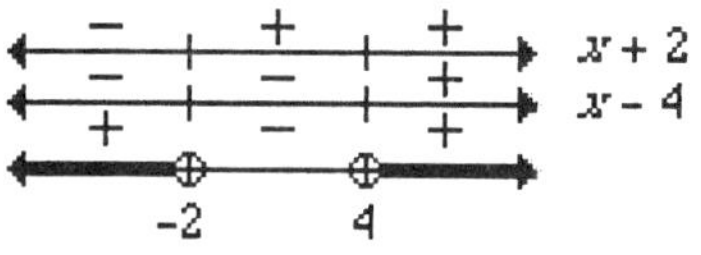

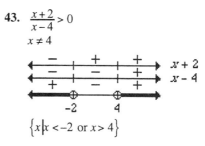

$\{x|x<-2 \text{ or } x>4\}$

45. $\frac{x-3}{x+5} < 0$

$x \neq -5$

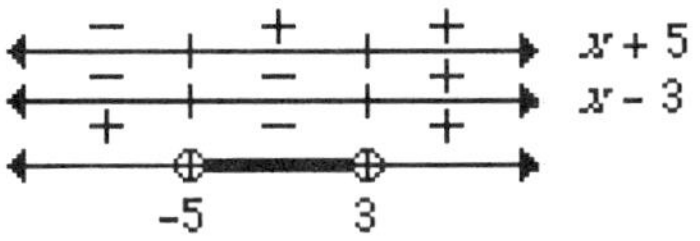

$\{x \mid -5 < x < 3\}$

47. $\frac{x+1}{x-1} \geq 0$

$x \neq 1$

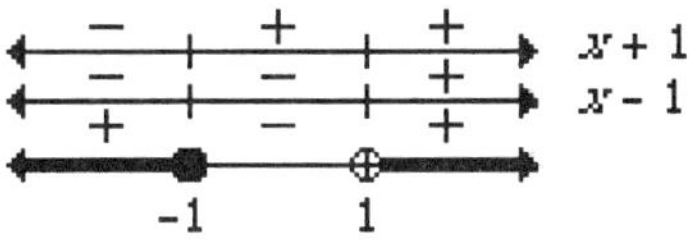

$\{x \mid x \leq -1 \text{ or } x > 1\}$

49. $\frac{a-4}{a+5} < 0$

$a \neq -5$

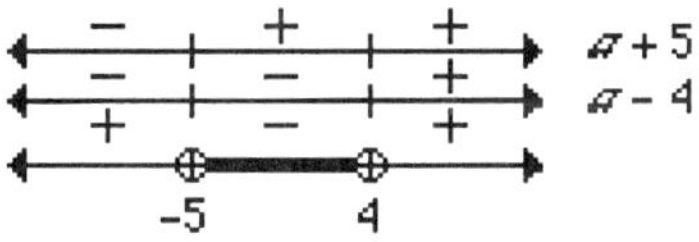

$\{a \mid -5 < a < 4\}$

51. $\frac{c-8}{c-4} > 0$

$c \neq 4$

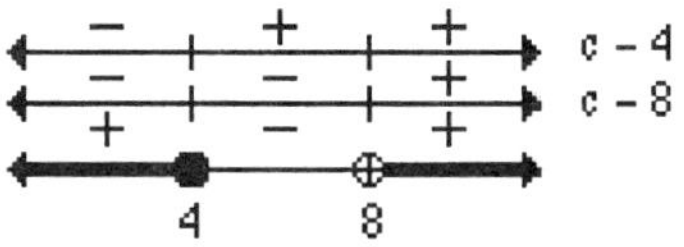

$\{c \mid c < 4 \text{ or } c > 8\}$

53. $\frac{3y+6}{y+6} \leq 0$

$y \neq -6,$

$3y + 6 = 0 \Rightarrow y = -2$

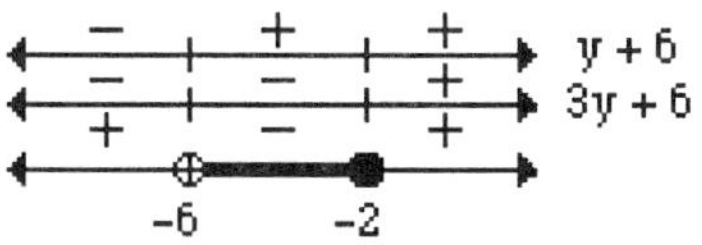

$\{y \mid -6 < y \leq -2\}$

55. $\frac{3a+6}{2a-1} \geq 0$

$a \neq \frac{1}{2}$

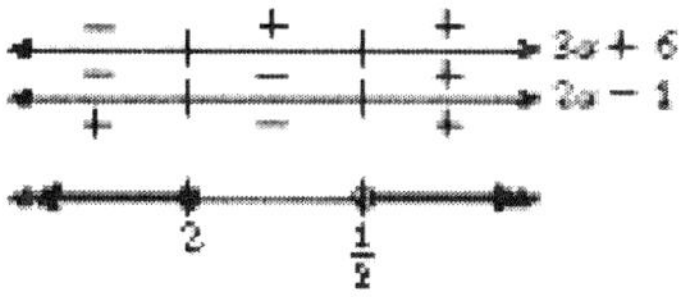

$\left\{a \middle| a \leq -2 \text{ or } a > \frac{1}{2}\right\}$

57. $\frac{3x+4}{2x-1} < 0$

$x \neq \frac{1}{2}$

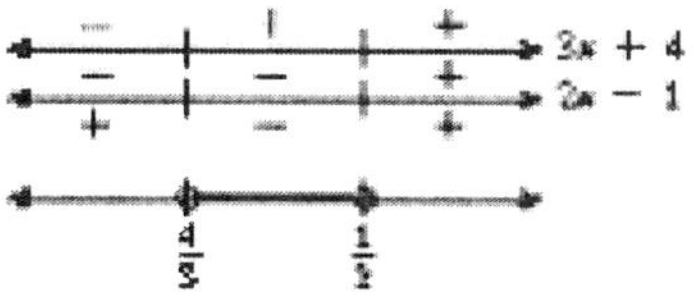

$\left\{x \middle| -\frac{4}{3} < x < \frac{1}{2}\right\}$

59. $\frac{3x+5}{x-2} \leq 0$

$x \neq 2$

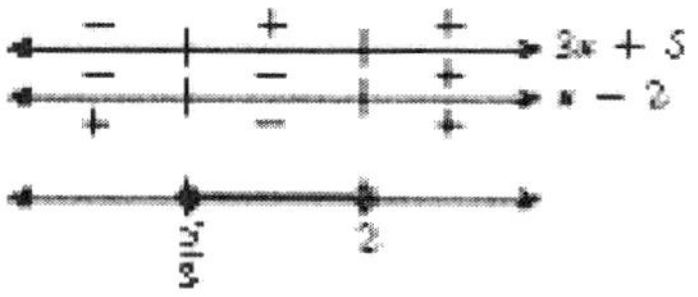

$\left\{x \middle| \frac{-5}{3} \leq x < 2\right\}$

61. $\frac{(x+1)(x-5)}{x+3} < 0$

$x \neq -3$

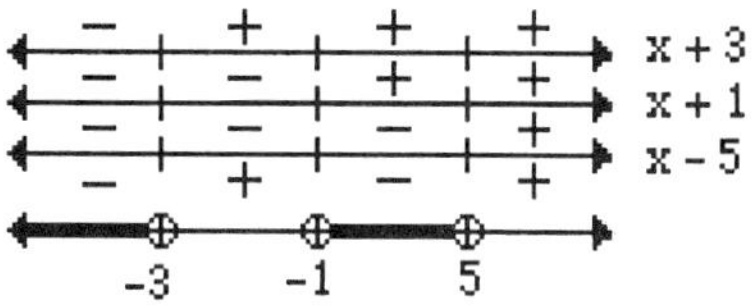

$(-\infty, -3) \cup (-1, 5)$

63. $\dfrac{(x-2)(x+3)}{x-4}>0$

$x \neq 4$

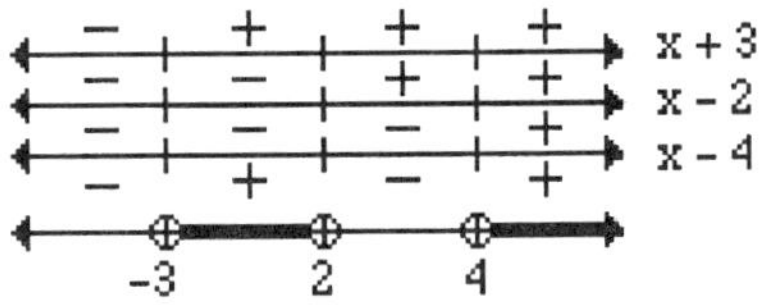

$(-3, 2)\cup(4, \infty)$

65. $\dfrac{(a-1)(a-6)}{a+2}\geq 0$

$a \neq -2$

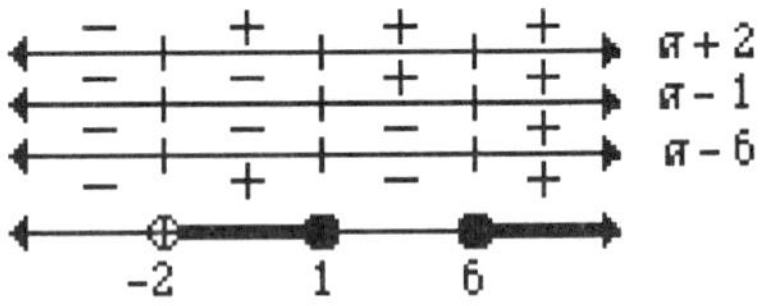

$(-2,1]\cup[6,\infty)$

67. $\dfrac{c}{(c-3)(c+7)}\leq 0$

$c \neq 3,\ c \neq -7$

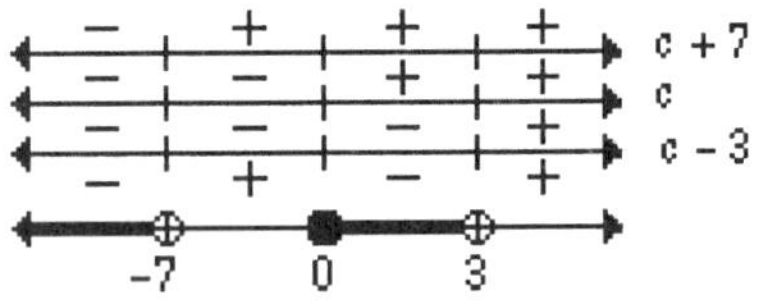

$(-\infty,-7)\cup[0, 3)$

69. $\dfrac{x-6}{(x+4)(x-1)}\leq 0$

$x \neq -4,\ x \neq 1$

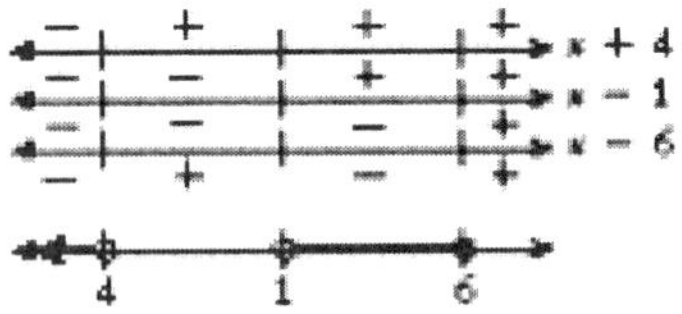

$(-\infty, -4)\cup(1, 6]$

71. $\dfrac{(x-3)(2x+5)}{x-6}>0$

$x \neq 6$

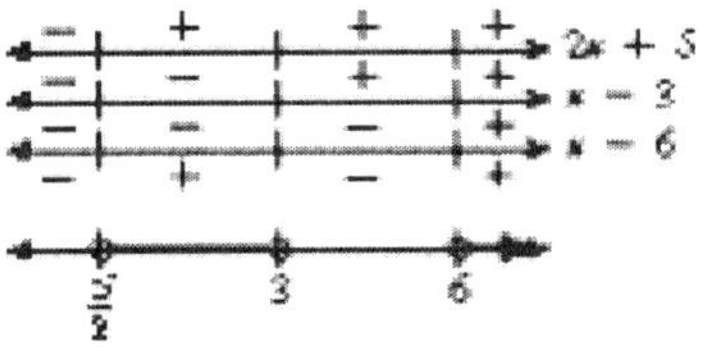

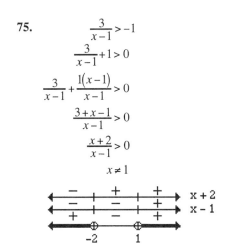

$\left(-\dfrac{5}{2}, 3\right)\cup(6, \infty)$

73.

$$\frac{2}{x-4}\geq 1$$
$$\frac{2}{x-4}-1\geq 0$$
$$\frac{2}{x-4}+\frac{-1(x-4)}{x-4}\geq 0$$
$$\frac{2-x+4}{x-4}\geq 0$$
$$\frac{6-x}{x-4}\geq 0$$

$x \neq 4$

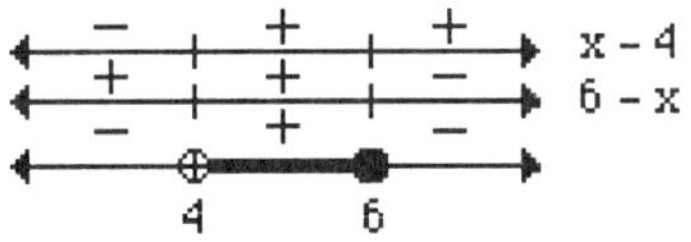

75.

$$\frac{3}{x-1}>-1$$
$$\frac{3}{x-1}+1>0$$
$$\frac{3}{x-1}+\frac{1(x-1)}{x-1}>0$$
$$\frac{3+x-1}{x-1}>0$$
$$\frac{x+2}{x-1}>0$$

$x \neq 1$

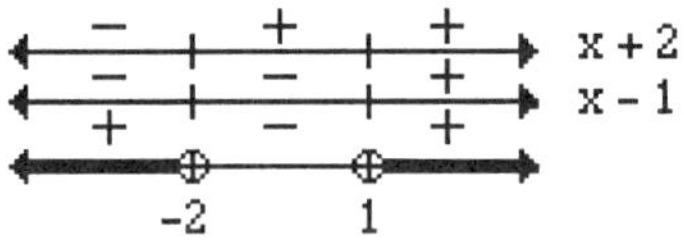

77.
$$\frac{4}{x+2} \le 1$$
$$\frac{4}{x+2} - 1 \le 0$$
$$\frac{4}{x+2} + \frac{-1(x+2)}{x+2} \le 0$$
$$\frac{4-x-2}{x+2} \le 0$$
$$\frac{2-x}{x+2} \le 0$$
$$x \ne -2$$

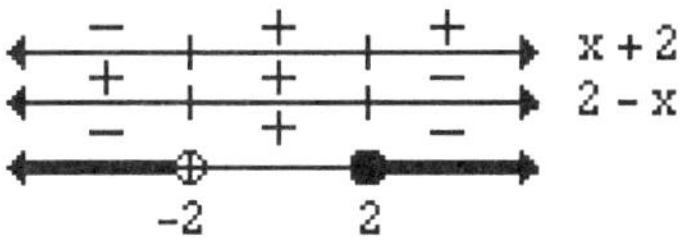

79.
$$\frac{2p-5}{p-4} \le 1$$
$$\frac{2p-5}{p-4} - 1 \le 0$$
$$\frac{2p-5}{p-4} - \frac{1(p-4)}{p-4} \le 0$$
$$\frac{2p-5-p+4}{p-4} \le 0$$
$$\frac{p-1}{p-4} \le 0$$
$$p \ne 4$$

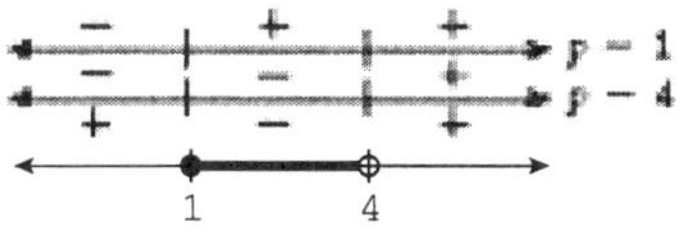

81.
$$\frac{4}{x-2} \ge 2$$
$$\frac{4}{x-2} - 2 \ge 0$$
$$\frac{4}{x-2} - \frac{2(x-2)}{x-2} \ge 0$$
$$\frac{4-2x+4}{x-2} \ge 0$$
$$\frac{8-2x}{x-2} \ge 0$$
$$x \ne 2$$

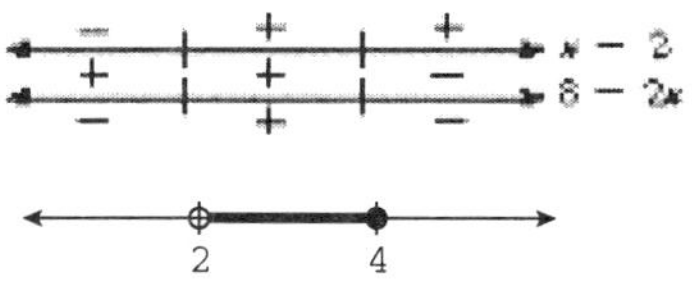

83.
$$\frac{w}{3w-2} > -2$$
$$\frac{w}{3w-2} + 2 > 0$$
$$\frac{w}{3w-2} + \frac{2(3w-2)}{3w-2} > 0$$
$$\frac{w+6w-4}{3w-2} > 0$$
$$\frac{7w-4}{3w-2} > 0$$
$$w \ne \frac{2}{3}$$

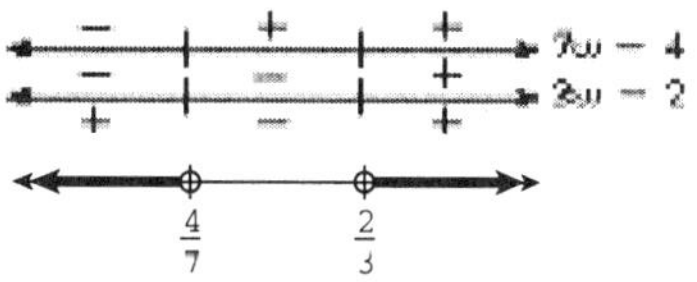

85. a. $y = \frac{x^2-4x+4}{x-4} > 0$ where the graph of y is above the x-axis, on the interval $(4, \infty)$.

b. $y = \frac{x^2-4x+4}{x-4} < 0$ where the graph of y is below the x-axis, on the interval $(-\infty, 2) \cup (2, 4)$.

87. A quadratic inequality with the union of the two outer regions of the number line as its solution, not including the boundary values, will be of the form $ax^2 + bx + c > 0$ with $a > 0$. Since the boundary values are $x = -4$ and $x = 2$, the factors are $x + 4$ and $x - 2$. Therefore one quadratic inequality is $(x+4)(x-2) > 0$ or $x^2 + 2x - 8 > 0$.

89. Since the solution set is $x \le -3$ and $x > 4$, the factors are $x + 3$ and $x - 4$. Because -3 is included in the solution set, $x + 3$ is the numerator. Since 4 is not included in the solution set, $x - 4$ is the denominator. The inequality symbol will be $\ge$ because the union of the outer regions of number line is the solution set. Therefore, the rational inequality is $\frac{x+3}{x-4} \ge 0$.

91. $(x+3)^2(x-1)^2 \ge 0$
The solution is all real numbers since any nonzero number squared is positive and zero squared is zero.

93. $\dfrac{x^2}{(x+1)^2} \geq 0$

This statement is true for all real numbers, except −1, since any nonzero number squared is negative. It is undefined when $x = -1$. Therefore, −1 is not a solution.

95. If $f(x) = ax^2 + bx + c$ and $a > 0$, the graph of $f(x)$ opens upward. If the discriminant is negative, the graph of $f(x)$ has no x-intercepts. Therefore, the graph lies above the x-axis and $f(x) < 0$ has no solution.

97. $(x+1)(x-3)(x+5)(x+9) \geq 0$

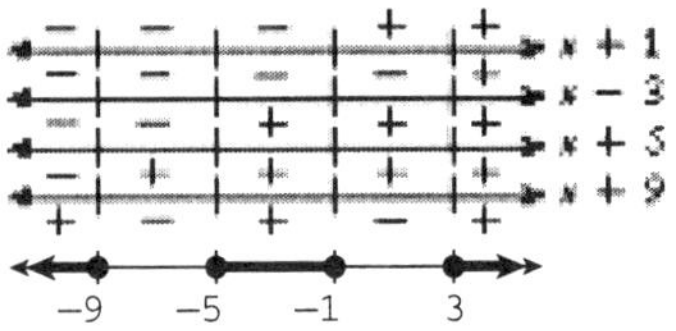

99. One possible answer is: Use a parabola that opens upward and has x-intercepts of (0, 0) and (3, 0). The x-values for which the parabola lies above the x-axis are $(-\infty, 0) \cup (3, \infty)$.

$x^2 - 3x > 0$

101. One possible answer is: Use a parabola that opens upward and has its vertex on or above the x-axis. Then there are no x-values for which the parabola lies below the x-axis.

$x^2 < 0$

103.

$$x^4 - 10x^2 + 9 > 0$$
$$\left(x^2\right)^2 - 10x^2 + 9 > 0$$
$$u^2 - 10u + 9 > 0$$
$$(u-9)(u-1) > 0$$
$$\left(x^2-9\right)\left(x^2-1\right) > 0$$
$$(x+3)(x-3)(x+1)(x-1) > 0$$

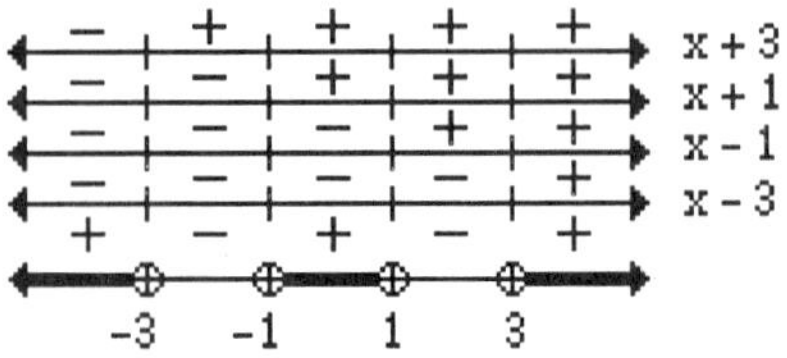

$(-\infty, -3) \cup (-1, 1) \cup (3, \infty)$

105.

$$x^3 + x^2 - 4x - 4 \geq 0$$
$$\left(x^3 + x^2\right) - (4x+4) \geq 0$$
$$x^2(x+1) - 4(x+1) \geq 0$$
$$\left(x^2 - 4\right)(x+1) \geq 0$$
$$(x+2)(x-2)(x+1) \geq 0$$

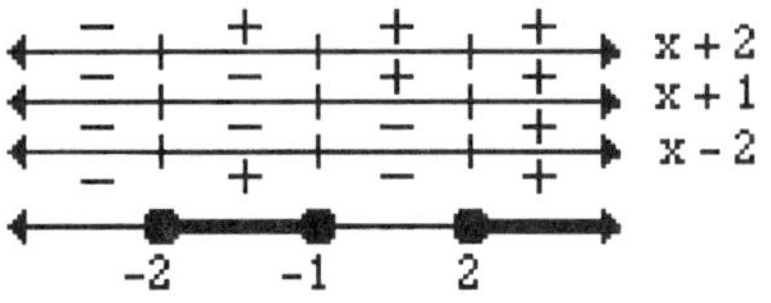

$[-2, -1] \cup [2, \infty)$

109. Let x = the number of quarts of 100% antifreeze added. The equation below describes the situation.

$$100\%(x) + 20\%(10) = 50\%(x+10)$$
$$x + 0.2(10) = 0.5(x+10)$$
$$x + 2 = 0.5x + 5$$
$$0.5x = 3$$
$$x = 6$$

Paul should add 6 quarts of 100% antifreeze.

110.

$$(6r + 5s - t) + (-3r - 2s - 5t)$$
$$= 6r + 5s - t - 3r - 2s - 5t$$
$$= 3r + 3s - 6t$$

111.

$$\frac{\dfrac{1 + \dfrac{x}{x+1}}{2x+1}}{x-1} = \frac{\dfrac{\dfrac{1}{1}\cdot\dfrac{x+1}{x+1} + \dfrac{x}{x+1}}{2x+1}}{x-1}$$
$$= \frac{\dfrac{\dfrac{2x+1}{x+1}}{2x+1}}{x-1}$$
$$= \frac{2x+1}{x+1} \cdot \frac{x-1}{2x+1}$$
$$= \frac{x-1}{x+1}$$

112. $h(x) = \frac{x^2 + 4x}{x + 6}$

$$\begin{aligned} h(-3) &= \frac{(-3)^2 + 4(-3)}{(-3) + 6} \\ &= \frac{9 - 12}{3} \\ &= \frac{-3}{3} \\ &= -1 \end{aligned}$$

113.

$$\begin{aligned} (3-4i)(6+5i) &= 18 + 15i - 24i - 20i^2 \\ &= 18 - 9i - 20(-1) \\ &= 18 - 9i + 20 \\ &= 38 - 9i \end{aligned}$$

Review Exercises

1.

$$\begin{aligned} (x-3)^2 &= 24 \\ x - 3 &= \pm\sqrt{24} \\ x - 3 &= \pm 2\sqrt{6} \\ x &= 3 \pm 2\sqrt{6} \end{aligned}$$

$x = 3 + 2\sqrt{6}$ or $x = 3 - 2\sqrt{6}$

2.

$$\begin{aligned} (2x+5)^2 &= 60 \\ 2x + 5 &= \pm\sqrt{60} \\ 2x + 5 &= \pm 2\sqrt{15} \\ 2x &= -5 \pm 2\sqrt{15} \\ x &= \frac{-5 \pm 2\sqrt{15}}{2} \end{aligned}$$

$x = \frac{-5 + 2\sqrt{15}}{2}$ or $x = \frac{-5 - 2\sqrt{15}}{2}$

3.

$$\begin{aligned} \left(x - \frac{1}{3}\right)^2 &= \frac{4}{9} \\ x - \frac{1}{3} &= \pm\sqrt{\frac{4}{9}} \\ x - \frac{1}{3} &= \pm\frac{2}{3} \\ x &= \frac{1}{3} \pm \frac{2}{3} \end{aligned}$$

$x = \frac{1}{3} + \frac{2}{3} = \frac{3}{3} = 1$ or $x = \frac{1}{3} - \frac{2}{3} = -\frac{1}{3}$

$x = 1$ or $x = -\frac{1}{3}$

4.

$$\begin{aligned} \left(2x - \frac{1}{2}\right)^2 &= 4 \\ 2x - \frac{1}{2} &= \pm\sqrt{4} \\ 2x - \frac{1}{2} &= \pm 2 \\ 2x &= \frac{1}{2} \pm 2 \\ 2x &= \frac{1 \pm 4}{2} \\ x &= \frac{1 \pm 4}{4} \end{aligned}$$

$x = \frac{1+4}{4} = \frac{5}{4}$ or $x = \frac{1-4}{4} = -\frac{3}{4}$

$x = \frac{5}{4}$ or $x = -\frac{3}{4}$

5.

$$\begin{aligned} x^2 - 7x + 12 &= 0 \\ x^2 - 7x &= -12 \\ x^2 - 7x + \frac{49}{4} &= -\frac{48}{4} + \frac{49}{4} \\ x^2 - 7x + \frac{49}{4} &= \frac{1}{4} \\ \left(x - \frac{7}{2}\right)^2 &= \frac{1}{4} \\ x - \frac{7}{2} &= \pm\frac{1}{2} \\ x &= \frac{7}{2} \pm \frac{1}{2} \end{aligned}$$

$x = \frac{7}{2} + \frac{1}{2} = 4$ or $x = \frac{7}{2} - \frac{1}{2} = 3$

$x = 4$ or $x = 3$

6.
$$x^2-4x-21=0$$
$$x^2-4x=21$$
$$x^2-4x+4=21+4$$
$$(x-2)^2=25$$
$$x-2=\pm 5$$
$$x=2\pm 5$$
$$x=7 \text{ or } x=-3$$

7.
$$a^2+2a-5=0$$
$$a^2+2a=5$$
$$a^2+2a+1=5+1$$
$$(a+1)^2=6$$
$$a+1=\pm\sqrt{6}$$
$$a=-1\pm\sqrt{6}$$
$$a=-1+\sqrt{6} \text{ or } a=-1-\sqrt{6}$$

8.
$$z^2+6z=10$$
$$z^2+6z+9=10+9$$
$$(z+3)^2=19$$
$$z+3=\pm\sqrt{19}$$
$$z=-3\pm\sqrt{19}$$
$$z=-3+\sqrt{19} \text{ or } z=-3-\sqrt{19}$$

9.
$$x^2-2x+10=0$$
$$x^2-2x=-10$$
$$x^2-2x+1=-10+1$$
$$(x-1)^2=-9$$
$$(x-1)^2=\sqrt{-9}$$
$$x-1=\pm 3i$$
$$x=1\pm 3i$$
$$x=1+3i \text{ or } x=1-3i$$

10.
$$2r^2-8r=-64$$
$$r^2-4r=-32$$
$$r^2-4r+4=-32+4$$
$$(r-2)^2=-28$$
$$r-2=\pm\sqrt{-28}$$
$$r=2\pm\sqrt{4}\sqrt{7}\sqrt{-1}$$
$$r=2\pm 2i\sqrt{7}$$
$$r=2+2i\sqrt{7} \text{ or } r=2-2i\sqrt{7}$$

11. Area = length × width
$$32=(x+5)(x+1)$$
$$32=x^2+x+5x+5$$
$$0=x^2+6x-27$$
$$0=(x-3)(x+9)$$
$$x-3=0 \quad \text{or} \quad x+9=0$$
$$x=3 \qquad x=-9$$
Disregard the negative value. $x=3$.

12. Area = length × width
$$63=(x+2)(x+4)$$
$$63=x^2+4x+2x+8$$
$$0=x^2+6x-55$$
$$0=(x-5)(x+11)$$
$$x-5=0 \quad \text{or} \quad x+11=0$$
$$x=5 \qquad x=-11$$
Disregard the negative value. $x=5$.

13. Let x = the smaller integer. The larger will then be $x+1$. Their product is 72.
$$x(x+1)=72$$
$$x^2+x=72$$
$$x^2+x-72=0$$
$$(x+9)(x-8)=0$$
$$x+9=0 \qquad x-8=0$$
$$x=-9 \qquad x=8$$
Since the integers must be positive, disregard the negative value. The smaller integer is 8 and the larger is 9.

14. Let x = the length of side of the square room. The diagonal can then be described by $x+7$. Two of the adjacent sides and the diagonal make up a right triangle. Use the Pythagorean theorem to solve the problem.
$$a^2+b^2=c^2$$
$$x^2+x^2=(x+7)^2$$
$$2x^2=x^2+14x+49$$
$$x^2-14x-49=0$$
$$x=\frac{-(-14)\pm\sqrt{(-14)^2-4(1)(-49)}}{2(1)}$$
$$x=\frac{14\pm\sqrt{392}}{2}$$
$$x\approx 16.90 \text{ or } x\approx -2.90$$

Disregard the negative value. The room is about 16.90 feet by 16.90 feet.

15. $2x^2 - 5x - 7 = 0$

$a = 2,\ b = -5,\ c = -7$

$$\begin{aligned} b^2 - 4ac &= (-5)^2 - 4(2)(-7) \\ &= 25 + 56 \\ &= 81 \end{aligned}$$

Since the discriminant is positive, this equation has two distinct real solutions.

16. $3x^2 + 2x = -4$

$3x^2 + 2x + 4 = 0$

$a = 3,\ b = 2,\ c = 4$

$$\begin{aligned} b^2 - 4ac &= (2)^2 - 4(3)(4) \\ &= 4 - 48 \\ &= -44 \end{aligned}$$

Since the discriminant is negative, this equation has no real solutions.

17. $r^2 + 12n = -36$

$r^2 + 12n + 36 = 0$

$a = 1,\ b = 12,\ c = 36$

$$\begin{aligned} b^2 - 4ac &= (12)^2 - 4(1)(36) \\ &= 144 - 144 \\ &= 0 \end{aligned}$$

Since the discriminant is 0, the equation has one real solution.

18. $5x^2 - x + 1 = 0$

$a = 5,\ b = -1,\ c = 1$

$$\begin{aligned} b^2 - 4ac &= (-1)^2 - 4(5)(1) \\ &= 1 - 20 \\ &= -19 \end{aligned}$$

Since the discriminant is negative, this equation has no real solutions.

19. $a^2 - 14n = -49$

$a^2 - 14n + 49 = 0$

$a = 1,\ b = -14,\ c = 49$

$$\begin{aligned} b^2 - 4ac &= (-14)^2 - 4(1)(49) \\ &= 196 - 196 \\ &= 0 \end{aligned}$$

Since the discriminant is 0, the equation has one real solution.

20. $\frac{1}{2}x^2 - 3x = 10$

$\frac{1}{2}x^2 - 3x - 10 = 0$

$a = \frac{1}{2},\ b = -3,\ c = -10$

$$\begin{aligned} b^2 - 4ac &= (-3)^2 - 4\left(\frac{1}{2}\right)(-10) \\ &= 9 + 20 \\ &= 29 \end{aligned}$$

Since the discriminant is positive, this equation has two real solutions.

21. $3x^2 + 7x = 0 \quad a = 3,\ b = 7,\ c = 0$

$$\begin{aligned} x &= \frac{-b \pm \sqrt{b^2 - 4ac}}{2a} \\ x &= \frac{-(7) \pm \sqrt{(7)^2 - 4(3)(0)}}{2(3)} \\ &= \frac{-7 \pm \sqrt{49 - 0}}{6} \\ &= \frac{-7 \pm \sqrt{49}}{6} \\ &= \frac{-7 \pm 7}{6} \end{aligned}$$

$x = \frac{-7+7}{6}$ or $x = \frac{-7-7}{6}$

$x = \frac{0}{6}$ $\qquad x = \frac{-14}{6}$

$x = 0$ $\qquad x = -\frac{7}{3}$

22. $x^2 - 11x = -30$

$x^2 - 11x + 30 = 0$

$a = 1,\ b = -11,\ c = 30$

$$\begin{aligned} x &= \frac{-b \pm \sqrt{b^2 - 4ac}}{2a} \\ x &= \frac{-(-11) \pm \sqrt{(-11)^2 - 4(1)(30)}}{2(1)} \\ &= \frac{11 \pm \sqrt{121 - 120}}{2} \\ &= \frac{11 \pm \sqrt{1}}{2} \\ &= \frac{11 \pm 1}{2} \end{aligned}$$

$x = \frac{11+1}{2}$ or $x = \frac{11-1}{2}$

$= \frac{12}{2}$ $\qquad = \frac{10}{2}$

$= 6$ $\qquad = 5$

23. $r^2 = 7r + 8$

$r^2 - 7r - 8 = 0$

$a = 1, b = -7, c = -8$

$$r = \frac{-b \pm \sqrt{b^2 - 4ac}}{2a}$$

$$r = \frac{-(-7) \pm \sqrt{(-7)^2 - 4(1)(-8)}}{2(1)}$$

$$= \frac{7 \pm \sqrt{49 + 32}}{2}$$

$$= \frac{7 \pm \sqrt{81}}{2}$$

$$= \frac{7 \pm 9}{2}$$

$r = \frac{7+9}{2}$ or $r = \frac{7-9}{2}$

$= \frac{16}{2}$ $\quad = \frac{-2}{2}$

$= 8$ $\quad = -1$

24. $4x^2 = 9x$

$4x^2 - 9x = 0$

$a = 4, b = -9, c = 0$

$$a = \frac{-b \pm \sqrt{b^2 - 4ac}}{2a}$$

$$x = \frac{-(-9) \pm \sqrt{(-9)^2 - 4(4)(0)}}{2(4)}$$

$$= \frac{9 \pm \sqrt{81 - 0}}{8}$$

$$= \frac{9 \pm \sqrt{81}}{8}$$

$$= \frac{9 \pm 9}{8}$$

$x = \frac{9-9}{8}$ or $x = \frac{9+9}{8}$

$= \frac{0}{8}$ $\quad = \frac{18}{8}$

$= 0$ $\quad = \frac{9}{4}$

25. $6a^2 + a - 15 = 0$

$a = 6, b = 1, c = -15$

$$a = \frac{-b \pm \sqrt{b^2 - 4ac}}{2a}$$

$$a = \frac{-1 \pm \sqrt{1^2 - 4(6)(-15)}}{2(6)}$$

$$= \frac{-1 \pm \sqrt{1 + 360}}{12}$$

$$= \frac{-1 \pm \sqrt{361}}{12}$$

$$= \frac{-1 \pm 19}{12}$$

$a = \frac{-1+19}{12}$ or $a = \frac{-1-19}{12}$

$= \frac{18}{12}$ $\quad = \frac{-20}{12}$

$= \frac{3}{2}$ $\quad = -\frac{5}{3}$

26. $4x^2 + 11x = 3$

$4x^2 + 11x - 3 = 0$

$a = 4, b = 11, c = -3$

$$x = \frac{-b \pm \sqrt{b^2 - 4ac}}{2a}$$

$$x = \frac{-(11) \pm \sqrt{(11)^2 - 4(4)(-3)}}{2(4)}$$

$$= \frac{-11 \pm \sqrt{121 + 48}}{8}$$

$$= \frac{-11 \pm \sqrt{169}}{8}$$

$$= \frac{-11 \pm 13}{8}$$

$x = \frac{-11+13}{8}$ or $x = \frac{-11-13}{8}$

$= \frac{2}{8}$ $\quad = \frac{-24}{8}$

$= \frac{1}{4}$ $\quad = -3$

27. $x^2 - 6x + 7 = 0$

$a = 1, b = -6, c = 7$

$$x = \frac{-b \pm \sqrt{b^2 - 4ac}}{2a}$$

$$x = \frac{-(-6) \pm \sqrt{(-6)^2 - 4(1)(7)}}{2(1)}$$

$$= \frac{6 \pm \sqrt{36 - 28}}{2}$$

$$= \frac{6 \pm \sqrt{8}}{2}$$

$$= \frac{6 \pm \sqrt{4}\sqrt{2}}{2}$$

$$= \frac{6 \pm 2\sqrt{2}}{2}$$

$$= \frac{2(3 \pm \sqrt{2})}{2}$$

$x = 3 + \sqrt{2}$ or $x = 3 - \sqrt{2}$

28. $b^2 + 4b = 8$

$b^2 + 4b - 8 = 0$

$a = 1, b = 4, c = -8$

$$b = \frac{-b \pm \sqrt{b^2 - 4ac}}{2a}$$

$$b = \frac{-(4) \pm \sqrt{(4)^2 - 4(1)(-8)}}{2(1)}$$

$$= \frac{-4 \pm \sqrt{16 + 32}}{2}$$

$$= \frac{-4 \pm \sqrt{48}}{2}$$

$$= \frac{-4 \pm \sqrt{16}\sqrt{3}}{2}$$

$$= \frac{-4 \pm 4\sqrt{3}}{2}$$

$$= \frac{2(-2 \pm 2\sqrt{3})}{2}$$

$= -2 \pm 2\sqrt{3}$

$x = -2 + 2\sqrt{3}$ or $x = -2 - 2\sqrt{3}$

29. $2x^2 + 4x - 3 = 0$

$a = 2, b = 4, c = -3$

$$x = \frac{-b \pm \sqrt{b^2 - 4ac}}{2a}$$

$$x = \frac{-4 \pm \sqrt{4^2 - 4(2)(-3)}}{2(2)}$$

$$= \frac{-4 \pm \sqrt{16 + 24}}{4}$$

$$= \frac{-4 \pm \sqrt{40}}{4}$$

$$= \frac{-4 \pm \sqrt{4}\sqrt{10}}{4}$$

$$= \frac{-4 \pm 2\sqrt{10}}{4}$$

$$= \frac{2(-2 \pm \sqrt{10})}{2(2)}$$

$x = \frac{-2 + \sqrt{10}}{2}$ or $x = \frac{-2 - \sqrt{10}}{2}$

30. $3x^2 - 6x - 8 = 0$

$a = 3, b = -6, c = -8$

$$x = \frac{-b \pm \sqrt{b^2 - 4ac}}{2a}$$

$$x = \frac{-(-6) \pm \sqrt{(-6)^2 - 4(3)(-8)}}{2(3)}$$

$$= \frac{6 \pm \sqrt{36 + 96}}{6}$$

$$= \frac{6 \pm \sqrt{132}}{6}$$

$$= \frac{6 \pm \sqrt{4}\sqrt{33}}{6}$$

$$= \frac{6 \pm 2\sqrt{33}}{6}$$

$$= \frac{2(3 \pm \sqrt{33})}{6}$$

$x = \frac{3 + \sqrt{33}}{3}$ or $x = \frac{3 - \sqrt{33}}{3}$

31. $x^2 - x + 30 = 0$

$a = 1,\ b = -1,\ c = 30$

$$x = \frac{-b \pm \sqrt{b^2 - 4ac}}{2a}$$

$$x = \frac{-(-1) \pm \sqrt{(-1)^2 - 4(1)(30)}}{2(1)}$$

$$= \frac{1 \pm \sqrt{1 - 120}}{2}$$

$$= \frac{1 \pm \sqrt{-119}}{2}$$

$$= \frac{1 \pm \sqrt{119}\sqrt{-1}}{2}$$

$$= \frac{1 \pm i\sqrt{119}}{2}$$

$$x = \frac{1 + i\sqrt{119}}{2} \quad \text{or} \quad x = \frac{1 - i\sqrt{119}}{2}$$

32. $1.2r^2 + 5.7r = 2.3$

$1.2r^2 + 5.7r - 2.3 = 0$

$a = 1.2,\ b = 5.7,\ c = -2.3$

$$r = \frac{-b \pm \sqrt{b^2 - 4ac}}{2a}$$

$$r = \frac{-5.7 \pm \sqrt{(5.7)^2 - 4(1.2)(-2.3)}}{2(1.2)}$$

$$= \frac{-5.7 \pm \sqrt{32.49 + 11.04}}{2.4}$$

$$= \frac{-5.7 \pm \sqrt{43.53}}{2.4}$$

$$x = \frac{-5.7 + \sqrt{43.53}}{2.4} \quad \text{or} \quad x = \frac{-5.7 - \sqrt{43.53}}{2.4}$$

33. $x^2 - \frac{5}{6}x = \frac{25}{6}$

$6x^2 - 5x = 25$

$6x^2 - 5x - 25 = 0$

$a = 6,\ b = -5,\ c = -25$

$$x = \frac{-b \pm \sqrt{b^2 - 4ac}}{2a}$$

$$x = \frac{-(-5) \pm \sqrt{(-5)^2 - 4(6)(-25)}}{2(6)}$$

$$x = \frac{5 \pm \sqrt{25 + 600}}{12}$$

$$= \frac{5 \pm \sqrt{625}}{12}$$

$$= \frac{5 \pm 25}{12}$$

$$x = \frac{5 + 25}{12} = \frac{30}{12} = \frac{5}{2} \quad \text{or} \quad x = \frac{5 - 25}{12} = \frac{-20}{12} = -\frac{5}{3}$$

34. $2x^2 + \frac{5x}{2} - \frac{3}{4} = 0$

$8x^2 + 10x - 3 = 0$

$a = 8,\ b = 10,\ c = -3$

$$x = \frac{-b \pm \sqrt{b^2 - 4ac}}{2a}$$

$$= \frac{-10 \pm \sqrt{10^2 - 4(8)(-3)}}{2(8)}$$

$$= \frac{-10 \pm \sqrt{100 + 96}}{16}$$

$$= \frac{-10 \pm \sqrt{196}}{16}$$

$$= \frac{-10 \pm 14}{16}$$

$$x = \frac{-10 + 14}{16} = \frac{4}{16} = \frac{1}{4} \quad \text{or} \quad x = \frac{-10 - 14}{16} = \frac{-24}{16} = -\frac{3}{2}$$

35. $f(x) = x^2 - 4x - 45,\ f(x) = 15$

$x^2 - 4x - 45 = 15$

$x^2 - 4x - 60 = 0$

$(x - 10)(x + 6) = 0$

$x - 10 = 0$ or $x + 6 = 0$

$x = 10$　　　$x = -6$

The solutions are 10 and –6.

36. $g(x) = 6x^2 + 5x$, $g(x) = 6$

$$6x^2 + 5x = 6$$
$$6x^2 + 5x - 6 = 0$$
$$(2x+3)(3x-2) = 0$$
$$2x + 3 = 0 \quad \text{or} \quad 3x - 2 = 0$$
$$x = -\frac{3}{2} \qquad x = \frac{2}{3}$$

The solutions are $-\frac{3}{2}$ and $\frac{2}{3}$.

37. $h(r) = 5r^2 - 7r - 6$, $h(r) = -4$

$$5r^2 - 7r - 6 = -4$$
$$5r^2 - 7r - 2 = 0$$
$$r = \frac{7 \pm \sqrt{(-7)^2 - 4(5)(-2)}}{2(5)} = \frac{7 \pm \sqrt{49+40}}{10} = \frac{7 \pm \sqrt{89}}{10}$$

The solutions are $\frac{7+\sqrt{89}}{10}$ and $\frac{7-\sqrt{89}}{10}$.

38. $f(x) = -2x^2 + 6x + 5$, $f(x) = -4$

$$-2x^2 + 6x + 5 = -4$$
$$-2x^2 + 6x + 9 = 0$$
$$x = \frac{-6 \pm \sqrt{6^2 - 4(-2)(9)}}{2(-2)} = \frac{-6 \pm \sqrt{36+72}}{-4} = \frac{-6 \pm \sqrt{108}}{-4} = \frac{-6 \pm 6\sqrt{3}}{-4} = \frac{3 \pm 3\sqrt{3}}{2}$$

The solutions are $\frac{3+3\sqrt{3}}{2}$ and $\frac{3-3\sqrt{3}}{2}$.

39. Solutions are 4 and -1.
Factors are $(x-4)$ and $(x+1)$.

$$f(x) = (x-4)(x+1)$$
$$f(x) = x^2 - 3x - 4$$

40. Solutions are $\frac{2}{3}$ and -2.
Factors are $(3x-2)$ and $(x+2)$.

$$f(x) = (3x-2)(x+2)$$
$$f(x) = 3x^2 + 4x - 4$$

41. Solutions are $x = -\sqrt{7}$ and $x = \sqrt{7}$. Factors are $(x+\sqrt{7})$ and $(x-\sqrt{7})$.

$$f(x) = (x+\sqrt{7})(x-\sqrt{7})$$
$$f(x) = x^2 - 7$$

42. Solutions are $x = 3 - 2i$ and $x = 3 + 2i$.
Factors are $x - (3 - 2i) = x - 3 + 2i$ and $x - (3 + 2i) = x - 3 - 2i$.

$$f(x) = (x - 3 + 2i)(x - 3 - 2i)$$
$$= (x-3)^2 - (2i)^2$$
$$= x^2 - 6x + 9 - 4i^2$$
$$= x^2 - 6x + 9 + 4$$
$$f(x) = x^2 - 6x + 13$$

43. Let x = the width of the garden. Then the length is $x+3$.
Area = length × width

$$88 = (x+3)x$$
$$88 = x^2 + 3x$$
$$x^2 + 3x - 88 = 0$$
$$(x+11)(x-8) = 0$$
$$x + 11 = 0 \quad \text{or} \quad x - 8 = 0$$
$$x = -11 \qquad x = 8$$

Disregard the negative value. The width is 8 feet and the length is 8 + 3 = 11 feet.

44. Using the Pythagorean Theorem,

$$a^2 + b^2 = c^2$$
$$8^2 + 8^2 = x^2$$
$$64 + 64 = x^2$$
$$128 = x^2$$
$$\sqrt{128} = x$$
$$x = 8\sqrt{2} \approx 11.31$$

45. $A = P\left(1+\frac{r}{n}\right)^{nt}$

$882 = 800\left(1+\frac{r}{1}\right)^{1(2)}$

$1323 = 1200(1+r)^2$

$1.1025 = (1+r)^2$

$\pm 1.05 = 1+r$

$r = -1 \pm 1.05$

Since the rate must be positive,
$r = -1 + 1.05 = 0.05$.
The annual interest is 5%.

46. Let x be the smaller positive number. Then $x + 2$ is the larger positive number.

$x(x+2) = 63$

$x^2 + 2x - 63 = 0$

$(x+9)(x-7) = 0$

$x = -9$ or $x = 7$

Since x must be positive, $x = 7$ and $7+2 = 9$.

47. Let x be the width. Then $2x - 4$ is the length and the equation is $A = lw$.

$96 = (2x-4)(x)$

$96 = 2x^2 - 4x$

$0 = 2x^2 - 4x - 96$

$0 = (2x+12)(x-8)$

$0 = 2x+12$ or $0 = x-8$

$-12 = 2x$ $\quad 8 = x$

$-6 = x$

Since the width must be positive, $x = 8$.
The width is 8 inches and the length $2x - 4$ is $2(8) - 4 = 16 - 4 = 12$ inches.

48. $V = 12d - 0.05d^2$, $d = 50$

$V = 12(50) - 0.05(50)^2$

$V = 12(50) - 0.05(2500)$

$V = 600 - 125$

$V = 475$

The value is $475.

49. $d = -16t^2 + 784$

a. $d = -16(3)^2 + 784$

$d = -16(9) + 784$

$d = -144 + 784$

$d = 640$

The object is 640 feet from the ground 3 seconds after being dropped.

b. $0 = -16t^2 + 784$

$16t^2 = 784$

$t^2 = 49$

$t = \pm\sqrt{49}$

$t \approx \pm 7$

Since the time must be positive, $t = 7$ seconds.

50. $h = -16t^2 + 16t + 100$

a. $h = -16(2)^2 + 16(2) + 100$

$= -16(4) + 32 + 100$

$= -64 + 32 + 100$

$= 68$

The height is 68 feet.

b. $0 = -16t^2 + 16t + 100$

$0 = 4t^2 - 4t - 25$

$t = \dfrac{-(-4) \pm \sqrt{(-4)^2 - 4(4)(-25)}}{2(4)}$

$= \dfrac{4 \pm \sqrt{16 + 400}}{8}$

$= \dfrac{4 \pm \sqrt{416}}{8}$

Since the time must be positive,

$t = \dfrac{4 + \sqrt{416}}{8}$

≈ 3.05 seconds

The object hits the ground in about 3.05 seconds.

51. a. $L = 0.0004t^2 + 0.16t + 20$

$L = 0.0004(100)^2 + 0.16(100) + 20$

$= 40$

40 milliliters will leak out at 100°C.

b. $53 = 0.0004t^2 + 0.16t + 20$

$0 = 0.0004t^2 + 0.16t - 33$

$$t = \frac{-0.16 \pm \sqrt{(0.16)^2 - 4(0.0004)(-33)}}{2(0.0004)}$$

$$= \frac{-0.16 \pm \sqrt{0.0784}}{0.0008}$$

$$t = \frac{-0.16 + \sqrt{0.0784}}{0.0008} = 150$$

$$\text{or } t = \frac{-0.16 - \sqrt{0.0784}}{0.0008} = -550$$

Since the temperature must be positive, $t = 150$. The operating temperature is 150°C.

52. Let x be the time in which the smaller machine can do the job then $x - 1$ is the time for the larger machine.

$$\frac{12}{x} + \frac{12}{x-1} = 1$$

$$x(x-1)\left(\frac{12}{x}\right) + x(x-1)\left(\frac{12}{x-1}\right) = x(x-1)$$

$$12(x-1) + 12x = x^2 - x$$

$$12x - 12 + 12x = x^2 - x$$

$$-12 + 24x = x^2 - x$$

$$0 = x^2 - 25x + 12$$

$$x = \frac{-(-25) \pm \sqrt{(-25)^2 - 4(1)(12)}}{2(1)}$$

$$= \frac{25 \pm \sqrt{577}}{2}$$

$$x = \frac{25 + \sqrt{577}}{2} \approx 24.5 \text{ or}$$

$$x = \frac{25 - \sqrt{577}}{2} \approx 0.49$$

x cannot equal 0.49 since this would mean the smaller machine could do the work in 0.49 hours and the larger can do the work in $x - 1$ or -0.51 hours. Therefore the smaller machine does the work in 24.51 hours and the larger machine does the work in 23.51 hours.

53. Let x be the speed (in miles per hour) for the first 25 miles. Then, the speed for the next 65 miles is $x + 15$. The time for the first 20 miles is $\frac{d}{r} = \frac{25}{x}$ and the time for the next 30 miles is $\frac{d}{r} = \frac{65}{x+15}$. The total time is 1.5 hours.

$$\frac{25}{x} + \frac{65}{x+15} = 1.5$$

$$x(x+15)\left(\frac{25}{x} + \frac{65}{x+15}\right) = x(x+15)(1.5)$$

$$25(x+15) + x(65) = (x^2 + 15x)(1.5)$$

$$90x + 375 = 1.5x^2 + 22.5x$$

$$0 = 1.5x^2 - 67.5x - 375$$

$$0 = 3x^2 - 135x - 750$$

$$0 = 3(x-50)(x+5)$$

$$x - 50 = 0 \quad \text{or} \quad x + 5 = 0$$

$$x = 50 \quad \text{or} \quad x = -5$$

Since the speed must be positive, $x = 50$. Thus, the speed was 50 mph.

54. Let r be the speed (in miles per hour) of the canoe in still water. For the trip downstream, the rate is $r + 0.4$ and the distance 3 miles so that the time is $t = \frac{3}{r+0.4}$. For the trip upstream the rate is $r - 0.4$ and the distance is 3 miles so that the time is $t = \frac{3}{r-0.4}$. The total time is 4 hours.

$$\frac{3}{r+0.4} + \frac{3}{r-0.4} = 4$$

$$(r+0.4)(r-0.4)\left[\frac{3}{r+0.4} + \frac{3}{r-0.4} = 4\right]$$

$$3(r-0.4) + 3(r+0.4) = 4(r+0.4)(r-0.4)$$

$$3r - 1.2 + 3r + 1.2 = 4(r^2 - 0.16)$$

$$6r = 4r^2 - 0.64$$

$$0 = 4r^2 - 6r - 0.64$$

$$0 = 2r^2 - 3r - 0.32$$

$$r = \frac{-(-3) \pm \sqrt{(-3)^2 - 4(2)(-0.32)}}{2(2)}$$

$$= \frac{3 \pm \sqrt{9 + 2.56}}{4}$$

$$= \frac{3 \pm \sqrt{11.56}}{4}$$

$$= \frac{3 \pm 3.4}{4}$$

$$r = \frac{3+3.4}{4} = \frac{6.4}{4} = 1.6 \quad \text{or} \quad r = \frac{3-3.4}{4} = \frac{-0.4}{4} = -0.1$$

Since the rate must be positive, $r = 1.6$. Rachel canoes 1.6 miles per hour in still water.

55. Let x be the length. The width is $x - 2$.
Area = length × width

$$73 = x(x-2)$$

$$73 = x^2 - 2x$$

$$0 = x^2 - 2x - 73$$

$$x = \frac{-(-2) \pm \sqrt{(-2)^2 - 4(1)(-73)}}{2(1)}$$

$$x = \frac{2 \pm \sqrt{296}}{2}$$

$$x \approx 9.6 \quad \text{or} \quad x \approx -7.6$$

Since the width must be positive, $x = 9.6$. The length is 9.6 inches and the width is $9.6 - 2 = 7.6$ inches.

56. If the business sells n tables at a price of $(60 - 0.3n)$ dollars per table, then the revenue is given by $R(n) = n(60 - 0.3n)$ with $n \le 40$. Set this equal to 1530 and solve for n.

$$R(n) = n(60 - 0.3n)$$

$$1530 = n(60 - 0.3n)$$

$$1530 = 60n - 0.3n^2$$

$$0.3n^2 - 60n + 1530 = 0 \quad \Leftarrow \text{divide by } 0.3$$

$$n^2 - 200n + 5100 = 0$$

$$(n-30)(n-170) = 0$$

$$n = 30 \quad \text{or} \quad n = 170$$

Disregard $n = 170$ since $n \le 40$. The business must sell 30 tables.

57.

$$a^2 + b^2 = c^2$$

$$a^2 = c^2 - b^2$$

$$\sqrt{a^2} = \pm\sqrt{c^2 - b^2}$$

$$a = \pm\sqrt{c^2 - b^2}$$

Since a refers to a length, it cannot be negative. Therefore disregard the negative sign.

$$a = \sqrt{c^2 - b^2}$$

58. $h = -4.9t^2 + c$

$h - c = -4.9t^2$

$\frac{h-c}{-4.9} = \frac{-4.9t^2}{-4.9}$

$\frac{h-c}{-4.9} = t^2$

$\pm\sqrt{\frac{h-c}{-4.9}} = t$ or $t = \pm\sqrt{\frac{c-h}{4.9}}$

Disregard the negative value for time.

$t = \sqrt{\frac{c-h}{4.9}}$

59. $V_x^2 + V_y^2 = V^2$ for V_y

$V_y^2 = V^2 - V_x^2$

$V_y - \sqrt{V^2 \quad V_x^2}$

60. $a = \frac{v_2^2 - v_1^2}{2d}$ for v_2

$2ad = v_2^2 - v_1^2$

$2ad + v_1^2 = v_2^2$

$v_2 = \sqrt{v_1^2 + 2ad}$

61. $x^4 - x^2 - 20$

$\left(x^2\right)^2 - x^2 - 20$

Let $u = x^2$

$u^2 - u - 20$

$(u-5)(u+4)$

Backsubstitute $x^2 = u$

$\left(x^2 - 5\right)\left(x^2 + 4\right)$

62. $4x^4 + 4x^2 - 3$

$4\left(x^2\right)^2 + 4x^2 - 3$

Let $u = x^2$

$4u^2 + 4u - 3$

$(2u-1)(2u+3)$

Backsubstitute $x^2 = u$

$\left(2x^2 - 1\right)\left(2x^2 + 3\right)$

63. $(x+5)^2 + 10(x+5) + 24$

Let $u = x + 5$

$u^2 + 10u + 24$

$(u+6)(u+4)$

Backsubstitute ˆ$x + 5 = u$

$(x+5+6)(x+5+4)$

$(x+11)(x+9)$

64. $4(2x-3)^2 - 12(2x+3) + 5$

Let $u = 2x + 3$

$4u^2 - 12u + 5$

$(2u-1)(2u-5)$

Backsubstitute ˆ$2x + 3 = u$

$(2(2x+3)-1)(2(2x+3)-5)$

$(4x+6-1)(4x+6-5)$

$(4x+5)(4x+1)$

65. $x^4 - 13x^2 + 36 = 0$

$(x^2)^2 - 13x^2 + 36 = 0$

$u^2 - 13u + 36 = 0$

$(u-9)(u-4) = 0$

$u - 9 = 0$ or $u - 4 = 0$

$u = 9$ $\quad$ $u = 4$

$x^2 = 9$ $\quad$ $x^2 = 4$

$x = \pm 3$ $\quad$ $x = \pm 2$

The solutions are±3 and ±2.

66. $x^4 - 19x^2 + 48 = 0$

$(x^2)^2 - 19x^2 + 48 = 0$

$u^2 - 19u + 48 = 0$

$(u-16)(u-3) = 0$

$u - 16 = 0$ or $u - 3 = 0$

$u = 16$ $\quad$ $u = 3$

$x^2 = 16$ $\quad$ $x^2 = 3$

$x = \pm 4$ $\quad$ $x = \pm\sqrt{3}$

The solutions are±4 and $\pm\sqrt{3}$.

67.
$$a^4 = 5a^2 + 24$$
$$a^4 - 5a^2 - 24 = 0$$
$$(a^2)^2 - 5a^2 - 24 = 0$$
$$u^2 - 5u - 24 = 0$$
$$(u-8)(u+3) = 0$$
$$u - 8 = 0 \quad \text{or} \quad u + 3 = 0$$
$$u = 8 \qquad u = -3$$
$$a^2 = 8 \qquad a^2 = -3$$
$$a = \pm\sqrt{8} \qquad a = \pm\sqrt{-3}$$
$$= \pm 2\sqrt{2} \qquad = \pm i\sqrt{3}$$
The solutions are $\pm 2\sqrt{2}$ and $\pm i\sqrt{3}$.

68.
$$6y^{-2} + 11y^{-1} - 10 = 0$$
$$\frac{6}{y^2} + \frac{11}{y} - 10 = 0$$
$$y^2\left(\frac{6}{y^2} + \frac{11}{y} - 10\right) = y^2(0)$$
$$6 + 11y - 10y^2 = 0$$
$$10y^2 - 11y - 6 = 0$$
$$(2y-3)(5y+2) = 0$$
$$2y - 3 = 0 \quad \text{or} \quad 5y + 2 = 0$$
$$2y = 3 \qquad 5y = -2$$
$$y = \frac{3}{2} \qquad y = -\frac{2}{5}$$
The solutions are $\frac{3}{2}$ and $-\frac{2}{5}$.

69.
$$4r + 23\sqrt{r} - 6 = 0$$
$$4\left(r^{1/2}\right)^2 + 23r^{1/2} - 6 = 0$$
$$4u^2 + 23u - 6 = 0$$
$$(4u-1)(u+6) = 0$$
$$4u - 1 = 0 \quad \text{or} \quad u + 6 = 0$$
$$4u = 1 \qquad u = -6$$
$$u = \frac{1}{4}$$
$$r^{1/2} = \frac{1}{4} \qquad r^{1/2} = -6$$
$$r = \left(\frac{1}{4}\right)^2$$
$$= \frac{1}{16}$$
There are no solutions for $r^{1/2} = -6$ since there is no real number x for which $r^{1/2} = -6$.

The solution is $\frac{1}{16}$.

70.
$$2p^{2/3} - 7p^{1/3} + 6 = 0$$
$$2\left(p^{1/3}\right)^2 - 7p^{1/3} + 6 = 0$$
$$2u^2 - 7u + 6 = 0$$
$$(2u-3)(u-2) = 0$$
$$2u - 3 = 0 \quad \text{or} \quad u - 2 = 0$$
$$u = \frac{3}{2} \qquad u = 2$$
$$p^{1/3} = \frac{3}{2} \qquad p^{1/3} = 2$$
$$p = \left(\frac{3}{2}\right)^3 \qquad p = 2^3$$
$$= \frac{27}{8} \qquad = 8$$
The solutions are $\frac{27}{8}$ and 8.

71.

$$6(x-2)^{-2} = -13(x-2)^{-1} + 8$$
$$6\left[(x-2)^{-1}\right]^2 = -13(x-2)^{-1} + 8$$
$$6u^2 = -13u + 8$$
$$6u^2 + 13u - 8 = 0$$
$$(2u-1)(3u+8) = 0$$
$$2u - 1 = 0 \quad \text{or} \quad 3u + 8 = 0$$
$$u = \frac{1}{2} \qquad u = -\frac{8}{3}$$
$$(x-2)^{-1} = \frac{1}{2} \qquad (x-2)^{-1} = -\frac{8}{3}$$
$$x - 2 = 2 \qquad x - 2 = -\frac{3}{8}$$
$$x = 4 \qquad x = \frac{13}{8}$$

The solutions are 4 and $\frac{13}{8}$.

72.

$$10(r+2) = \frac{12}{r+2} - 7$$
$$10(r+2)^2 + 7(r+2) = 12$$
$$10(r+2)^2 + 7(r+2) - 12 = 0$$
$$10u^2 + 7u - 12 = 0$$
$$(5u-4)(2u+3) = 0$$
$$5u - 4 = 0 \quad \text{or} \quad 2u + 3 = 0$$
$$u = \frac{4}{5} \qquad u = -\frac{3}{2}$$
$$r + 2 = \frac{4}{5} \qquad r + 2 = -\frac{3}{2}$$
$$r = -\frac{6}{5} \qquad r = -\frac{7}{2}$$

The solutions are $-\frac{6}{5}$ and $-\frac{7}{2}$.

73. $f(x) = x^4 - 29x^2 + 100$

To find the x-intercepts, set $f(x) = 0$.

$$0 = x^4 - 29x^2 + 100$$
$$0 = \left(x^2\right)^2 - 29x^2 + 100$$
$$0 = u^2 - 29u + 100$$
$$0 = (u-25)(u-4)$$
$$u - 25 = 0 \qquad u - 4 = 0$$
$$u = 25 \qquad u = 4$$
$$x^2 = 25 \qquad x^2 = 4$$
$$x = \pm 5 \qquad x = \pm 2$$

The x-intercepts are (5, 0), (–5, 0), (2, 0), and (–2, 0).

74. $f(x) = 30x + 13\sqrt{x} - 10$

To find the x-intercepts, set $f(x) = 0$.

$$0 = 30x + 13\sqrt{x} - 10$$
$$0 = 30\left(\sqrt{x}\right)^2 + 13\sqrt{x} - 10$$
$$0 = 30u^2 + 13u - 10$$
$$0 = (6u+5)(5u-2)$$
$$6u + 5 = 0 \qquad 5u - 2 = 0$$
$$u = -\frac{5}{6} \qquad u = \frac{2}{5}$$
$$\sqrt{x} = -\frac{5}{6} \qquad \sqrt{x} = \frac{2}{5}$$
$$x = \frac{4}{25}$$

Since $\sqrt{x}$ cannot be negative, the solution is $\frac{4}{25}$. The only x-intercept is $\left(\frac{4}{25}, 0\right)$.

75. $f(x) = x - 6\sqrt{x} + 10$

To find the x-intercepts, set $f(x) = 0$.

$$0 = x - 6\sqrt{x} + 10$$
$$0 = \left(\sqrt{x}\right)^2 - 6\sqrt{x} + 10$$
$$0 = u^2 - 6u + 10$$
$$u = \frac{-(-6) \pm \sqrt{(-6)^2 - 4(1)(10)}}{2(1)}$$
$$u = \frac{6 \pm \sqrt{-4}}{2}$$
$$u = \frac{6 \pm 2i}{2} \quad \text{or} \quad u = 3 \pm i$$
$$\sqrt{x} = 3 \pm i \Rightarrow x = (3 \pm i)^2 = 8 \pm 6i$$

Since x-intercepts must be real numbers, this function has no x-intercepts.

76. $g(x) = (x^2 - 6x)^2 - 5(x^2 - 6x) - 24$

To find the x-intercepts, set $g(x) = 0$.

$0 = (x^2 - 6x)^2 - 5(x^2 - 6x) - 24$

$0 = u^2 - 5u - 24$

$0 = (u + 3)(u - 8)$

$u + 3 = 0$ or $u - 8 = 0$

$u = -3$ $\quad$ $u = 8$

$x^2 - 6x = -3$ $\quad$ $x^2 - 6x = 8$

$x^2 - 6x + 3 = 0$ $\quad$ $x^2 - 6x - 8 = 0$

$\dfrac{6 \pm \sqrt{(-6)^2 - 4(1)(3)}}{2(1)}$ $\quad$ $\dfrac{6 \pm \sqrt{(-6)^2 - 4(1)(-8)}}{2(1)}$

$\dfrac{6 \pm \sqrt{24}}{2}$ $\quad$ $\dfrac{6 \pm \sqrt{68}}{2}$

$\dfrac{6 \pm 2\sqrt{6}}{2}$ $\quad$ $\dfrac{6 \pm 2\sqrt{17}}{2}$

$3 \pm \sqrt{6}$ $\quad$ $3 \pm \sqrt{17}$

The x-intercepts are

$(3 + \sqrt{6}, 0)$, $(3 - \sqrt{6}, 0)$,

$(3 + \sqrt{17}, 0)$ and $(3 - \sqrt{17}, 0)$.

77. $y = x^2 + 5x$

a. Since $a = 1$ the parabola opens upward.

b. $y = 0^2 + 5(0) = 0$

The y-intercept is (0, 0).

c. $x = -\dfrac{b}{2a} = -\dfrac{5}{2(1)} = -\dfrac{5}{2}$

$y = \dfrac{4ac - b^2}{4a}$

$= \dfrac{4(1)(0) - 5^2}{4(1)}$

$= -\dfrac{25}{4}$

The vertex is $\left(-\frac{5}{2}, -\frac{25}{4}\right)$.

d. $0 = x^2 + 5x$

$0 = x(x + 5)$

$0 = x$ or $0 = x + 5$

$x = 0$ $\quad$ $x = -5$

The x-intercepts are (0, 0) and (–5, 0).

e.

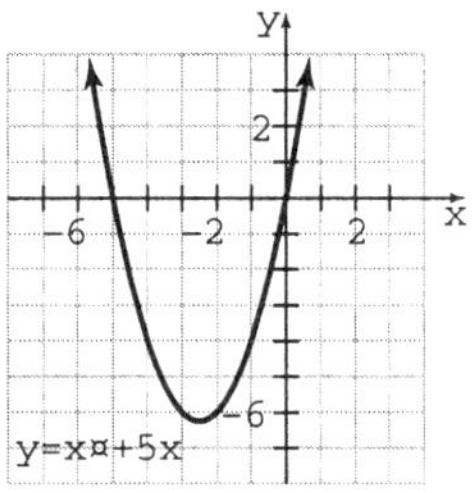

78. $y = x^2 - 2x - 8$

a. Since $a = 1$ the parabola opens upward.

b. $y = (0)^2 - 2(0) - 8 = -8$

The y-intercept is (0, –8).

c. $x = -\dfrac{b}{2a} = -\dfrac{-2}{2(1)} = \dfrac{2}{2} = 1$

$y = \dfrac{4ac - b^2}{4a}$

$= \dfrac{4(1)(-8) - (-2)^2}{4(1)}$

$= \dfrac{-32 - 4}{4}$

$= \dfrac{-36}{4}$

$= -9$

The vertex is (1, –9).

d. $0 = x^2 - 2x - 8$

$0 = (x - 4)(x + 2)$

$0 = x - 4$ or $0 = x + 2$

$4 = x$ or $-2 = x$

The x-intercepts are (4, 0) and (–2, 0).

e.

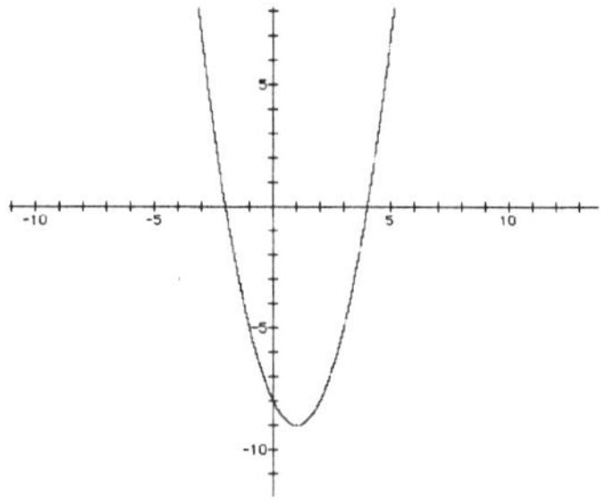

79. $y = -x^2 - 9$

a. Since $a = -1$ the parabola opens downward.

b. $y = -(0)^2 - 9 = -9$
The y-intercept is (0, –9).

c. $x = -\frac{b}{2a} = -\frac{0}{2(-1)} = -\frac{0}{-2} = 0$

$y = \frac{4ac - b^2}{4a}$

$= \frac{4(-1)(-9) - 0^2}{4(-1)} = \frac{36}{-4} = -9$

The vertex is (0, –9).

d. $0 = -x^2 - 9$
$x^2 = -9$
$x = \pm\sqrt{-9}$ or $\pm 3i$
There are no real roots. Thus, there are no x-intercepts.

e.

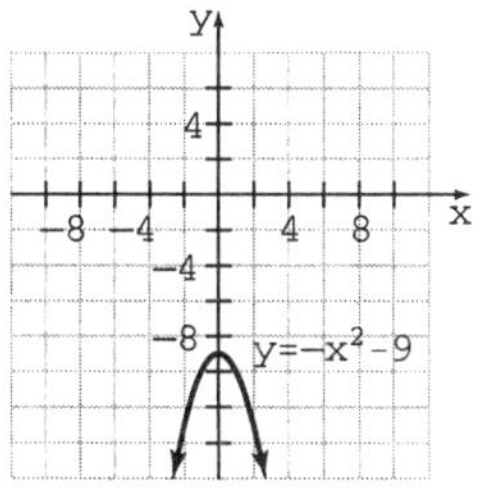

80. $y = -2x^2 - x + 15$

a. Since $a = -2$ the parabola opens downward.

b. $y = -2(0)^2 - 0 + 15 = 15$
The y-intercept is (0, 15).

c. $x = -\frac{b}{2a} = -\frac{-1}{2(-2)} = \frac{1}{-4} = -\frac{1}{4}$

$y = \frac{4ac - b^2}{4a}$

$= \frac{4(-2)(15) - (-1)^2}{4(-2)} = \frac{121}{8}$

The vertex is $\left(-\frac{1}{4}, \frac{121}{8}\right)$.

d. $0 = -1(2x^2 + x - 15)$
$= -1(2x - 5)(x + 3)$
$0 = (2x - 5)$ or $0 = x + 3$
$5 = 2x$
$\frac{5}{2} = x$ $\quad -3 = x$
The x-intercepts are (–3, 0) and $\left(\frac{5}{2}, 0\right)$.

e.

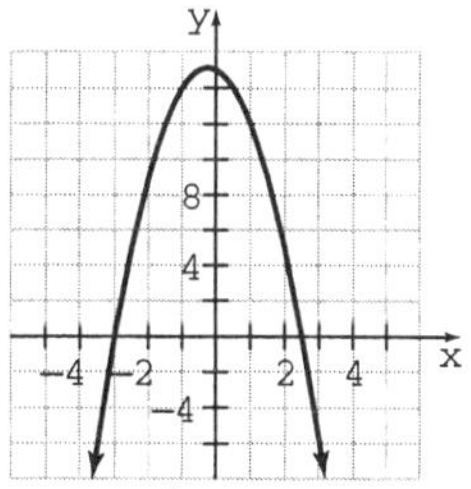

y=–2x²–x+15

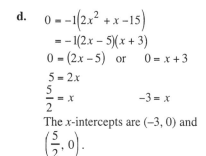

81. a. $I = -x^2 + 22x - 30$, $2 \le x \le 20$
The x-coordinate of the vertex will be the cost per ticket to maximize profit.
$x = -\frac{b}{2a} = -\frac{22}{2(-1)} = 11$
They should charge $11 per ticket.

b. The maximum profit in hundreds is the
y-coordinate of the vertex.
$I(11) = -11^2 + 22(11) - 30$
$= -121 + 242 - 30$
$= 91$
The maximum profit is $91 hundred or $9100.

82. a. $s(t) = -16t^2 + 80t + 60$
The ball will attain maximum height at the
x-coordinate of the vertex.
$t = -\frac{b}{2a} = -\frac{80}{2(-16)} = -\frac{80}{-32} = 2.5$
The ball will attain maximum height 2.5 seconds after being thrown.

b. The maximum height is the y-coordinate of the vertex.
$s(2.5) = -16(2.5)^2 + 80(2.5) + 60$
$= -100 + 200 + 60$
$= 160$
The maximum height is 160 feet.

83. The graph of $f(x) = (x-3)^2$ has vertex (3, 0). The graph will be $g(x) = x^2$ shifted right 3 units.

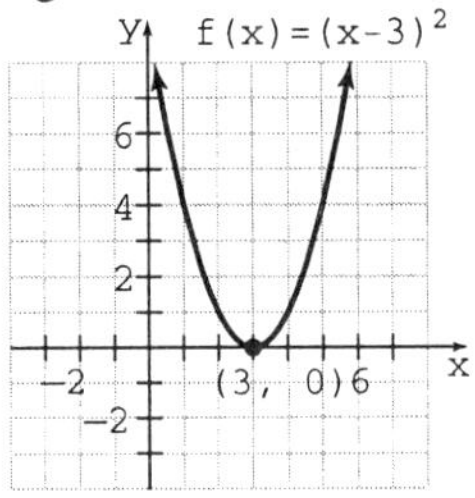

84. The graph of $f(x) = -(x+2)^2 - 3$ has vertex (−2, −3). Since $a < 0$, the parabola opens downward. The graph will be $g(x) = -x^2$ shifted left 2 units and down 3 units.

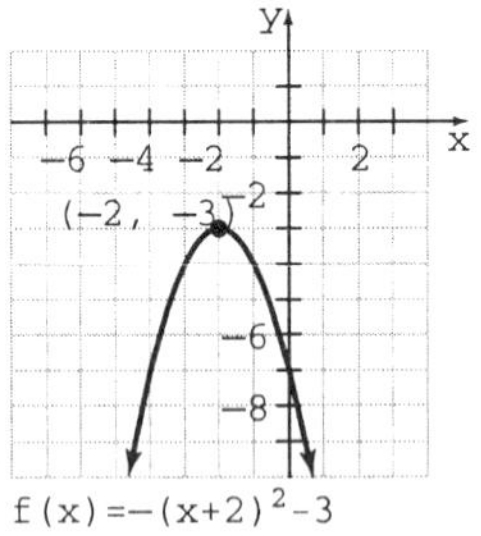

85. The graph of $g(x) = -2(x+4)^2 - 1$ has vertex $(-4, -1)$. Since $a < 0$, the parabola opens downward. The graph will be $f(x) = -2x^2$ shifted left 4 units and down 1 unit.

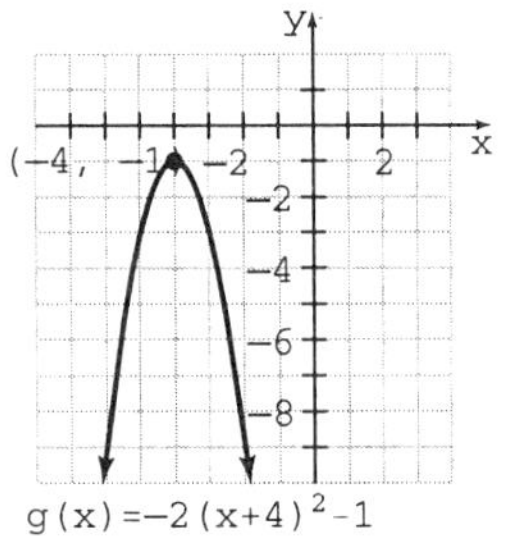

86. The graph of $h(x) = \frac{1}{2}(x-1)^2 + 3$ has vertex (1, 3). The graph will be $f(x) = \frac{1}{2}x^2$ shifted right 1 unit and up 3 units.

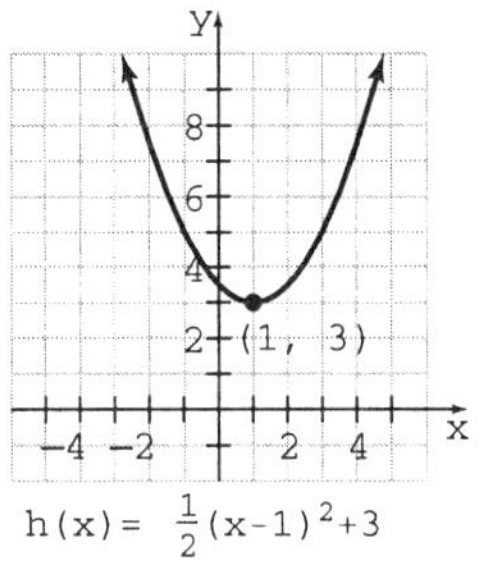

87. $x^2 + 7x + 6 \geq 0$
$(x+1)(x+6) \geq 0$

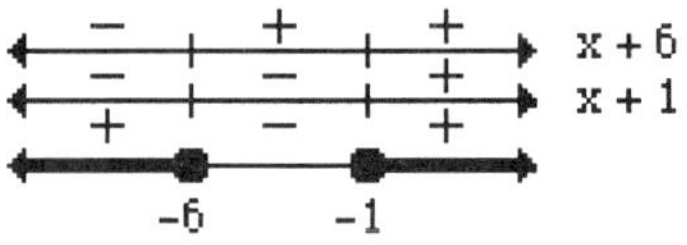

88. $x^2 + 3x - 10 \leq 0$
$(x+5)(x-2) \leq 0$

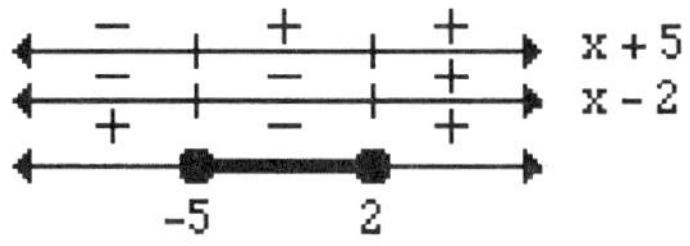

89. $x^2 \leq 11x - 20$
$x^2 - 11x + 20 \leq 0$
$x^2 - 11x + 20 = 0$

$$x = \frac{-(-11) \pm \sqrt{(-11)^2 - 4(1)(20)}}{2(1)} = \frac{11 \pm \sqrt{121 - 80}}{2} = \frac{11 \pm \sqrt{41}}{2}$$

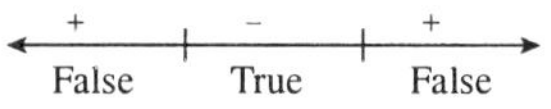

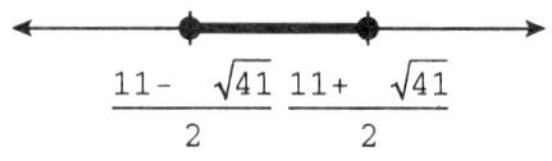

90. $3x^2 + 8x > 16$
$3x^2 + 8x - 16 > 0$
$(3x-4)(x+4) > 0$
$3x - 4 = 0$ or $x + 4 = 0$
$x = \frac{4}{3}$ $\quad x = -4$

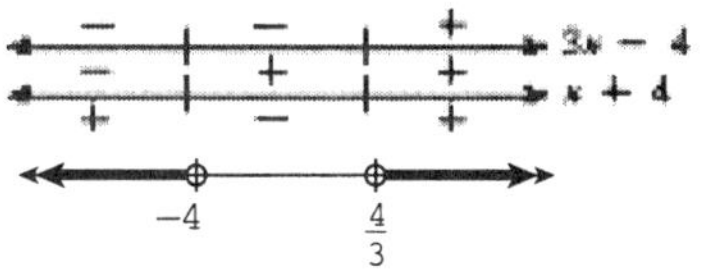

91. $4x^2 - 9 \le 0$
$(2x-3)(2x+3) \le 0$
$2x - 3 = 0$ or $2x + 3 = 0$
$x = \frac{3}{2}$ $\quad x = -\frac{3}{2}$

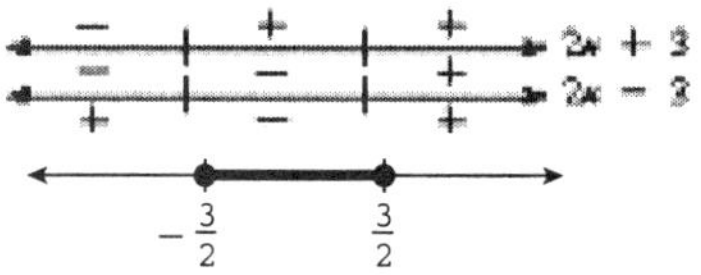

92. $5x^2 - 25 > 0$
$5(x^2 - 5) > 0$
$5(x+\sqrt{5})(x-\sqrt{5}) > 0$
$x + \sqrt{5} = 0$ or $x - \sqrt{5} = 0$
$x = -\sqrt{5}$ $\quad x = \sqrt{5}$

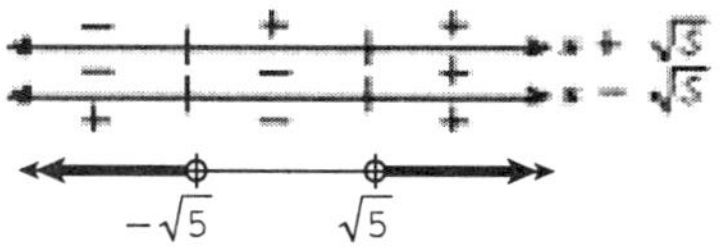

93. $\frac{x+1}{x-3} > 0$
$x \ne 3$

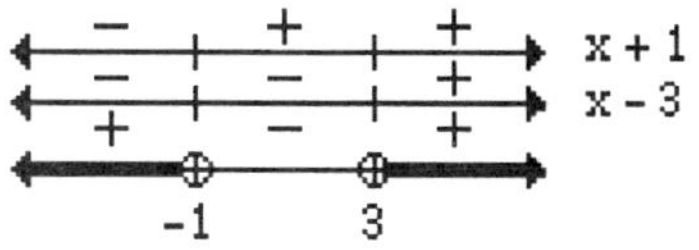

$\{x | x < -1 \text{ or } x > 3\}$

94. $\frac{x-5}{x+2} \le 0$
$x = -2$

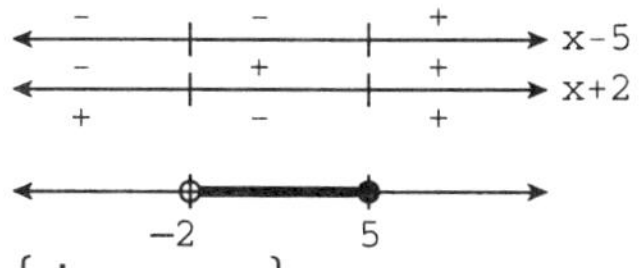

$\{x | -2 < x \le 5\}$

95. $\frac{2x-4}{x+3} \ge 0$
$\frac{2(x-2)}{x+3} \ge 0$
$x \ne -3$

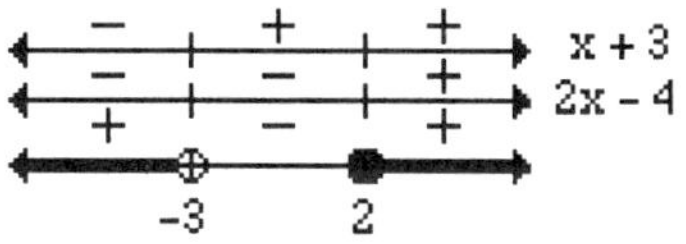

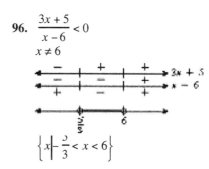

$\{x | x < -3 \text{ or } x \ge 2\}$

96. $\frac{3x+5}{x-6} < 0$
$x \ne 6$

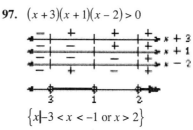

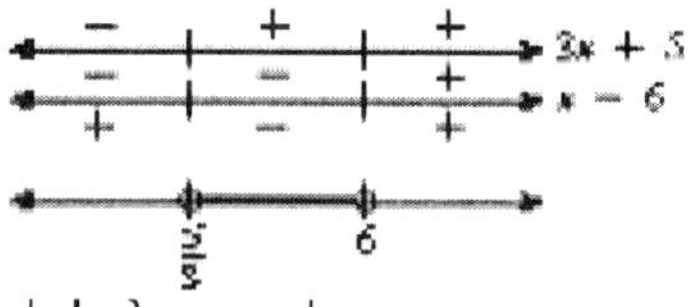

$\left\{x \middle| -\frac{5}{3} < x < 6\right\}$

97. $(x+3)(x+1)(x-2) > 0$

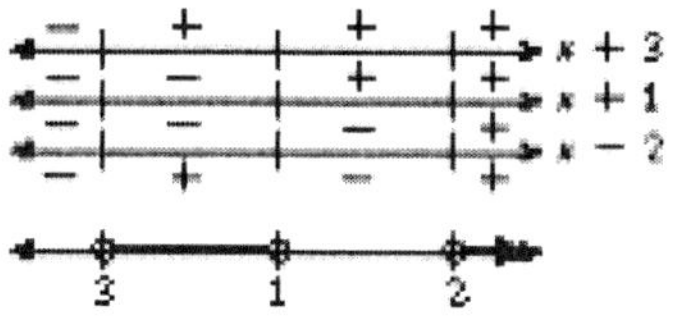

$\{x | -3 < x < -1 \text{ or } x > 2\}$

98. $x(x-3)(x-5) \le 0$

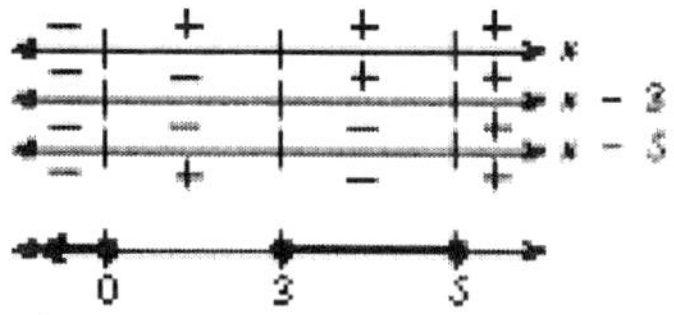

$\{x | x \le 0 \text{ or } 3 \le x \le 5\}$

99. $(3x+4)(x-1)(x-3) \ge 0$

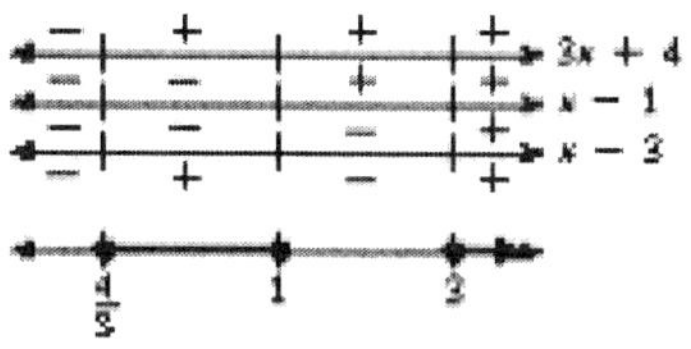

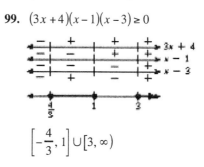

$\left[-\frac{4}{3}, 1\right] \cup [3, \infty)$

100. $2x(x+2)(x+5)<0$

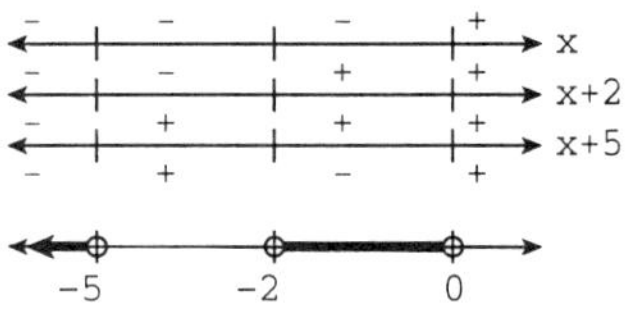

$(-\infty, -5)\cup(-2, 0)$

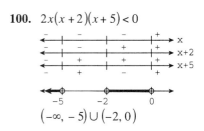

101. $\dfrac{x(x-4)}{x+2}>0$

$x\neq -2$

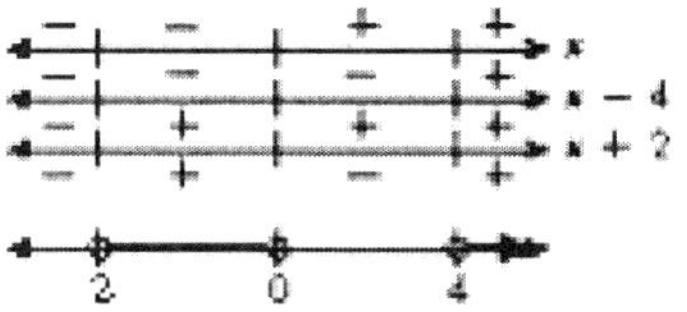

$(-2, 0)\cup(4, \infty)$

102. $\dfrac{(x-2)(x-5)}{x+3}<0$

$x\neq -3$

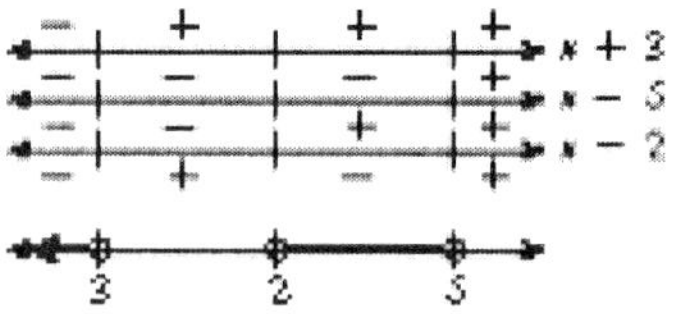

$(-\infty, -3)\cup(2, 5)$

103. $\dfrac{x-3}{(x+2)(x-5)}\geq 0$

$x\neq -2, x\neq 5$

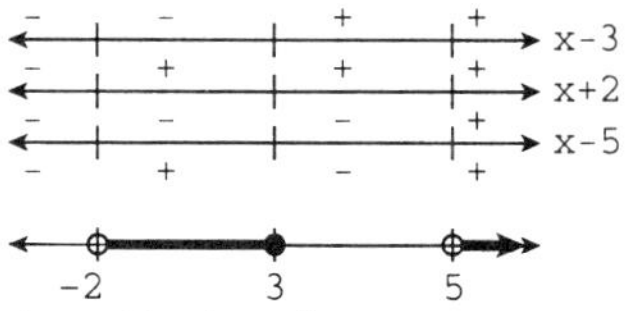

$(-2, 3]\cup(5, \infty)$

104. $\dfrac{x(x-5)}{x+3}\leq 0$

$x\neq -3$

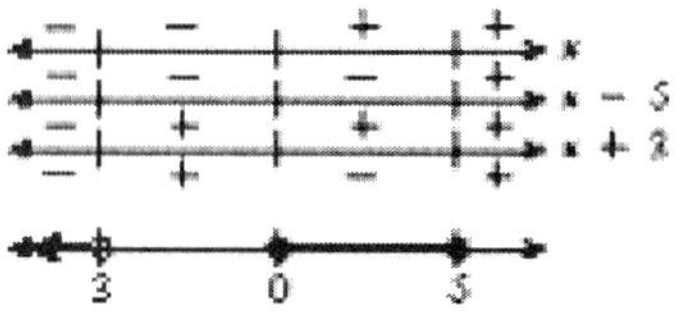

$(-\infty, -3)\cup[0, 5]$

105. $\dfrac{3}{x+4}\geq -1$

$\dfrac{3}{x+4}+1\geq 0$

$\dfrac{3+1(x+4)}{x+4}\geq 0$

$\dfrac{3+x+4}{x+4}\geq 0$

$\dfrac{x+7}{x+4}\geq 0$

$x\neq -4$

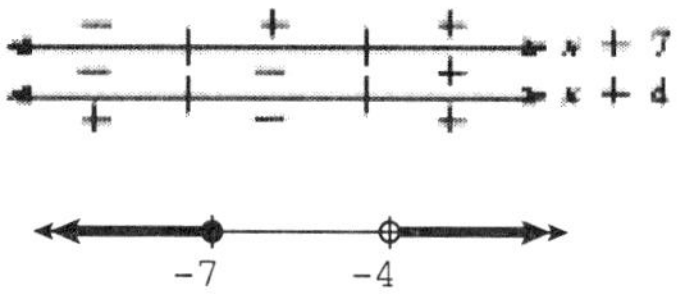

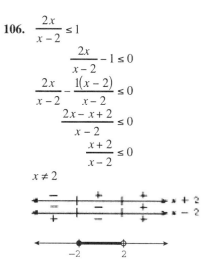

106. $\dfrac{2x}{x-2}\leq 1$

$\dfrac{2x}{x-2}-1\leq 0$

$\dfrac{2x}{x-2}-\dfrac{1(x-2)}{x-2}\leq 0$

$\dfrac{2x-x+2}{x-2}\leq 0$

$\dfrac{x+2}{x-2}\leq 0$

$x\neq 2$

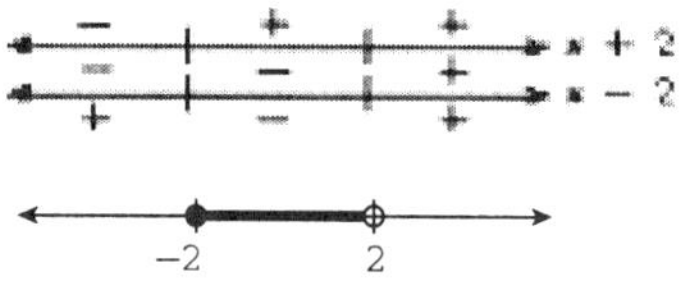

107. $\dfrac{2x+3}{3x-5}<4$

$\dfrac{2x+3}{3x-5}-4<0$

$\dfrac{2x+3-4(3x-5)}{3x-5}<0$

$\dfrac{2x+3-12x+20}{3x-5}<0$

$\dfrac{-10x+23}{3x-5}<0$

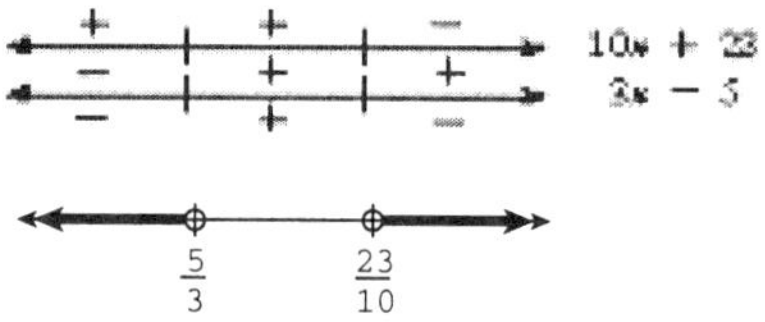

Practice Test

1. $x^2+2x-15=0$

$x^2+2x=15$

$x^2+2x+1=15+1$

$(x+1)^2=16$

$x+1=\pm 4$

$x=-1\pm 4$

$x=3$ or $x=-5$

2. $x^2-4x=-17$

$x^2-4x+4=-17+4$

$(x-2)^2=-13$

$x-2=\pm\sqrt{-13}$

$x-2=\pm i\sqrt{13}$

$x=2\pm i\sqrt{13}$

$x=2+i\sqrt{13}$ or $x=2-i\sqrt{13}$

3. $x^2-5x-14=0$ $\quad a=1, b=-5, c=-14$

$x=\dfrac{-b\pm\sqrt{b^2-4ac}}{2a}$

$x=\dfrac{-(-5)\pm\sqrt{(-5)^2-4(1)(-14)}}{2(1)}$

$=\dfrac{5\pm\sqrt{25+56}}{2}$

$=\dfrac{5\pm\sqrt{81}}{2}$

$=\dfrac{5\pm 9}{2}$

$x=\dfrac{5+9}{2}=\dfrac{14}{2}=7$ or $x=\dfrac{5-9}{2}=\dfrac{-4}{2}=-2$

4. $a^2+8a=-5$

$a^2+8a+5=0$

$a=1, b=8, c=5$

$a=\dfrac{-b\pm\sqrt{b^2-4ac}}{2a}$

$a=\dfrac{-8\pm\sqrt{8^2-4(1)(5)}}{2(1)}$

$=\dfrac{-8\pm\sqrt{64-20}}{2}$

$=\dfrac{-8\pm\sqrt{44}}{2}$

$=\dfrac{-8\pm 2\sqrt{11}}{2}$

$=\dfrac{2\left(-4\pm\sqrt{11}\right)}{2}$

$=-4\pm\sqrt{11}$

$a=-4+\sqrt{11}$ or $a=-4-\sqrt{11}$

5. $3r^2+r=2$

$3r^2+r-2=0$

$(3r-2)(r+1)=0$

$3r-2=0$ or $r+1=0$

$3r=2$ $\qquad r=-1$

$r=\dfrac{2}{3}$

6. $p^2+5=-7p$

$p^2+7p+5=0$

$p=\dfrac{-7\pm\sqrt{(7)^2-4(1)(5)}}{2(1)}$

$p=\dfrac{-7\pm\sqrt{29}}{2}$

$p=\dfrac{-7+\sqrt{29}}{2}$ or $p=\dfrac{-7-\sqrt{29}}{2}$

7. x-intercepts are 4 and $-\dfrac{2}{5}$

Factors are $(x-4)$ and $(5x+2)$

$f(x)=(x-4)(5x+2)$

$f(x)=5x^2-18x-8$

8. $K=\dfrac{1}{2}mv^2$ for v

$2K=mv^2$

$\dfrac{2K}{m}=v^2$

$v=\sqrt{\dfrac{2K}{m}}$

9. a. $c(s) = -0.01s^2 + 78s + 22,000$

$c(2000) = -0.01(2000)^2 + 78(2000) + 22,000$
$= -40,000 + 156,000 + 22,000$
$= 138,000$

The cost is \$138,000.

b. $160,000 = -0.01s^2 + 78s + 22,000$

$0 = -0.01s^2 + 78s - 138,000$

$s = \dfrac{-78 \pm \sqrt{78^2 - 4(-0.01)(-138,000)}}{2(-0.01)}$

$= \dfrac{-78 \pm \sqrt{564}}{-0.02}$

$s = \dfrac{-78 + \sqrt{564}}{-0.02} \approx 2712.57$

$s = \dfrac{-78 - \sqrt{564}}{-0.02} \approx 5087.43$

Since $1300 \le s \le 3900$, the house should have about 2712.57 square feet.

10. The formula $d = rt$ can be written $t = \dfrac{d}{r}$.

Let r = his actual rate.

	distance	rate	time $= \dfrac{d}{r}$
actual trip	520	r	$\dfrac{520}{r}$
faster trip	520	$r + 15$	$\dfrac{520}{r+15}$

The faster trip would have taken 2.4 hours less time than the actual trip.

$\dfrac{520}{r+15} = \dfrac{520}{r} - 2.4$

$r(r+15)\left(\dfrac{520}{r+15}\right) = r(r+15)\left(\dfrac{520}{r} - 2.4\right)$

$520r = 520(r+15) - 2.4r(r+15)$

$520r = 520r + 7800 - 2.4r^2 - 36r$

$0 = -2.4r^2 - 36r + 7800$

$0 = r^2 + 15r - 3250$

$0 = (r+65)(r-50)$

$r + 65 = 0 \qquad r - 50 = 0$

$r = -65 \qquad r = 50$

Since speed is never negative, Tom drove an average speed of 50 mph.

11. $2x^4 + 15x^2 - 50 = 0$

$2(x^2)^2 + 15x^2 - 50 = 0$

$2u^2 + 15u - 50 = 0$

$(u+10)(2u-5) = 0$

$u + 10 = 0$ or $2u - 5 = 0$

$u = -10 \qquad u = \dfrac{5}{2}$

$x^2 = -10 \qquad x^2 = \dfrac{5}{2}$

$x = \pm\sqrt{-10} \qquad x = \pm\sqrt{\dfrac{5}{2}}$

$= \pm i\sqrt{10} \qquad = \pm\dfrac{\sqrt{5}}{\sqrt{2}} \cdot \dfrac{\sqrt{2}}{\sqrt{2}}$

$= \pm\dfrac{\sqrt{10}}{2}$

12. $3r^{2/3} + 11r^{1/3} - 42 = 0$

$3(r^{1/3})^2 + 11r^{1/3} - 42 = 0$

$3u^2 + 11u - 42 = 0$

$(3u - 7)(u + 6) = 0$

$3u - 7 = 0$ or $u + 6 = 0$

$u = \dfrac{7}{3} \qquad u = -6$

$r^{1/3} = \dfrac{7}{3} \qquad r^{1/3} = -6$

$r = \left(\dfrac{7}{3}\right)^3 \qquad r = (-6)^3$

$= \dfrac{343}{27} \qquad r = -216$

13. $f(x) = 16x - 40\sqrt{x} + 25$

$0 = 16\left(\sqrt{x}\right)^2 - 40\sqrt{x} + 25$

$0 = 16u^2 - 40u + 25$

$0 = (4u - 5)^2$

$4u - 5 = 0$

$u = \dfrac{5}{4}$

$\sqrt{x} = \dfrac{5}{4}$

$x = \dfrac{25}{16}$

The x-intercept is $\left(\dfrac{25}{16}, 0\right)$.

14. $f(x) = (x-3)^2 + 2$
The vertex is (3, –2). The graph will be the graph of $g(x) = x^2$ shifted 3 units right and 2 units down.

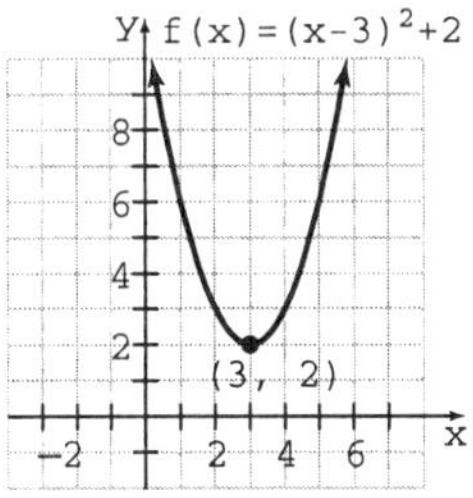

15. $h(x) = -\frac{1}{2}(x-2)^2 - 2$
The vertex is (2, 2). The graph will be the graph of $g(x) = -\frac{1}{2}x^2$ shifted 2 units right and 2 units down.

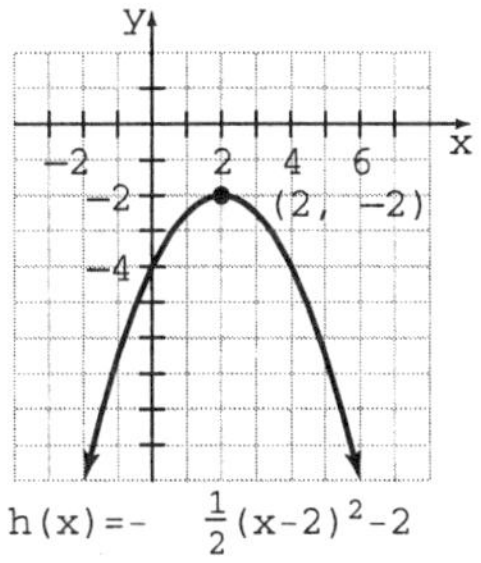

16. Begin by rewriting the equation in standard form.
$5x^2 - 4x - 3 = 0$ $a = 5, b = -4, c = -3$
The discriminant is
$b^2 - 4ac = (-4)^2 - 4(5)(-3)$
$= 76$
Since the discriminant is greater than 0, the quadratic equation has two distinct real solutions.

17. $y = x^2 + 2x - 8$

a. Since $a = 1$ the parabola opens upward.

b. $y = 0^2 + 2(0) - 8$
$y = -8$
The y-intercept is (0, –8).

c. $x = -\frac{b}{2a} = -\frac{2}{2(1)} = -1$
$y = (-1)^2 + 2(-1) - 8 = -9$
The vertex is (–1, –9).

d. The x-intercepts occur when $y = 0$.
$0 = x^2 + 2x - 8$
$0 = (x+4)(x-2)$
$x + 4 = 0$ or $x - 2 = 0$
$x = -4$ $x = 2$
The x-intercepts are (2, 0) and (–4, 0).

e.

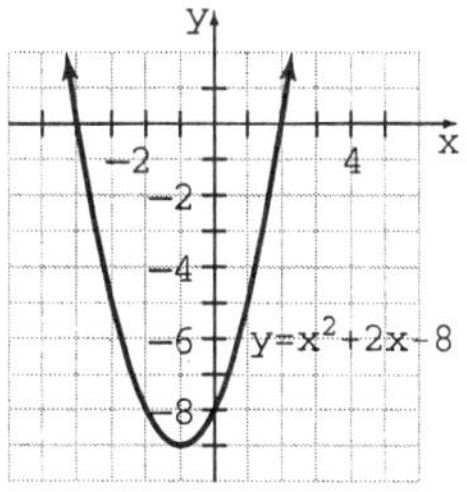

18. Since –6 and $\frac{1}{2}$ are the x-intercepts, the factors are $(x + 6)$ and $(2x - 1)$.
$f(x) = (x + 6)(2x - 1)$
$f(x) = 2x^2 + 11x - 6$

19.
$x^2 - x \geq 42$
$x^2 - x - 42 \geq 0$
$(x-7)(x+6) \geq 0$
$x - 7 = 0$ or $x + 6 = 0$
$x = 7$ $x = -6$

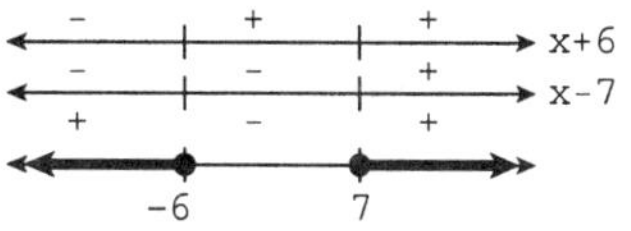

20. $\frac{(x+3)(x-4)}{x+1} \geq 0$
$x \neq -1$
$x + 3 = 0$ or $x - 4 = 0$ or $x + 1 = 0$
$x = -3$ $x = 4$ $x = -1$

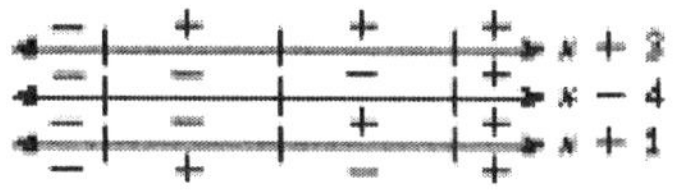

21. $\frac{x+3}{x+2} \le -1$

$\frac{x+3}{x+2} + 1 \le 0$

$\frac{x+3}{x+2} + \frac{x+2}{x+2} \le 0$

$\frac{2x+5}{x+2} \le 0$

$x \ne -2$

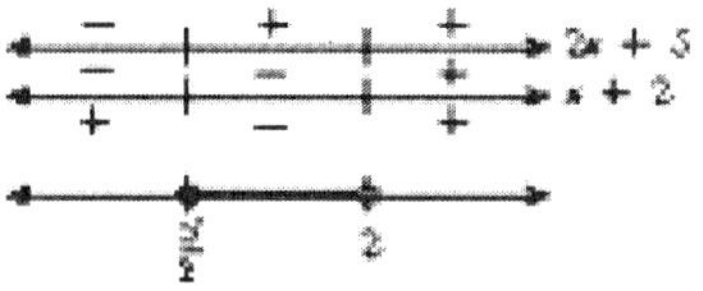

a. $\left[-\frac{5}{2}, -2\right)$

b. $\left\{x \middle| -\frac{5}{2} \le x < -2\right\}$

22. Let x be the width of the carpet. Then $2x + 4$ is the length $A = lw$.

$48 = x(2x+4)$

$48 = 2x^2 + 4x$

$0 = 2x^2 + 4x - 48$

$0 = x^2 + 2x - 24$

$0 = (x-4)(x+6)$

$x - 4 = 0$ or $x + 6 = 0$

$x = 4$ $\quad$ $x = -6$

The width is 4 feet and the length is $2 \cdot 4 + 4 = 8 + 4 = 12$ feet.

23. $d = -16t^2 + 64t + 80$

$d = 0$ when the ball strikes the ground.

$0 = -16t^2 + 64t + 80$

$0 = -16(t^2 - 4t - 5)$

$0 = -16(t-5)(t+1)$

$t - 5 = 0$ or $t + 1 = 0$

$t = 5$ $\quad$ $t = -1$

The time must be positive, so $t = 5$. Thus, the ball strikes the ground in 5 seconds.

24. a. $f(x) = -1.4x^2 + 56x - 60$

$$x = -\frac{b}{2a} = -\frac{56}{2(-1.4)} = -\frac{56}{-2.8} = 20$$

The company must sell 20 carvings.

b. $f(20) = -1.4(20)^2 + 56(20) - 60$

$= -560 + 1120 - 60$

$= 500$

The maximum weekly profit is \$500.

25. If the business sells n brooms at a price of $(10 - 0.1n)$ dollars per broom, then the revenue is given by $R(n) = n(10 - 0.1n)$ with $n \le 32$. Set this equal to 160 and solve for n.

$R(n) = n(10 - 0.1n)$

$160 = n(10 - 0.1n)$

$160 = 10n - 0.1n^2$

$0.1n^2 - 10n + 160 = 0 \Leftarrow$ divide by 0.1

$n^2 - 100n + 1600 = 0$

$(n-20)(n-80) = 0$

$n = 20$ or $n = 80$

Disregard $n = 80$ since $n \le 32$. The business must sell 20 brooms.

Cumulative Review Test

1. $-4 \div (-2) + 16 - \sqrt{49}$

$-4 \div (-2) + 16 - 7$

$2 + 16 - 7$

$18 - 7$

11

2. Evaluate $2x^2 + 3x + 1$ when $x = 2$.

$2(2)^2 + 3(2) + 1 = 8 + 6 + 1 = 15$

3. $4x - \{3 - [2(x-2) - 5x]\}$

$4x - \{3 - [2x - 4 - 5x]\}$

$4x - \{3 - [-4 - 3x]\}$

$4x - \{3 + 4 + 3x\}$

$4x - \{7 + 3x\}$

$4x - 7 - 3x$

$x - 7$

4. $-\frac{1}{2}(4x-6)=\frac{1}{3}(3-6x)+2$

$-2x+3=1-2x+2$

$-2x+3=-2x+3$

This is an identity. The solution is all real numbers.

5. a. $x=-4$ is a vertical line.

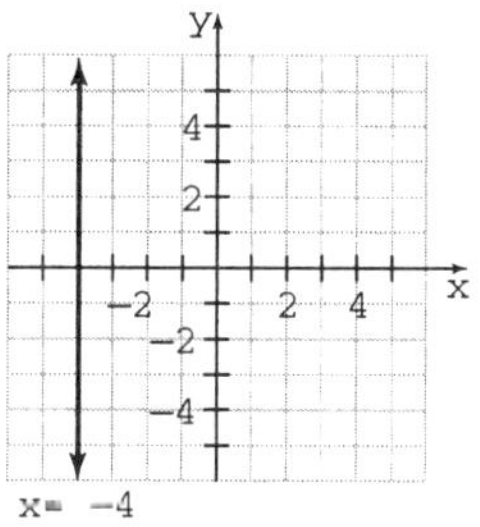

b. $y=2$ is a horizontal line.

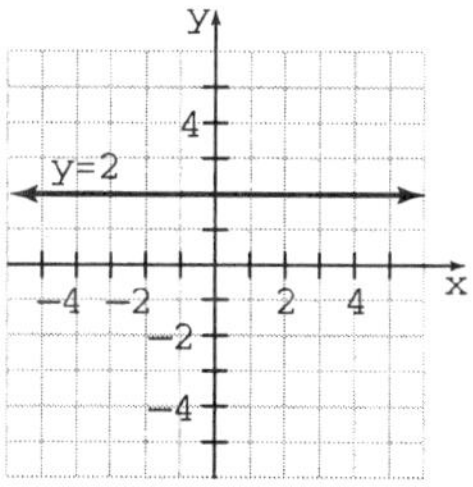

6. $m=\frac{y_2-y_1}{x_2-x_1}=\frac{1-3}{2-4}=\frac{-2}{-2}=1$

$y-y_1=m(x-x_1)$

$y-3=1(x-4)$

$y-3=x-4 \Rightarrow y=x-1$

7. $183{,}000=1.83\times10^5$

8. $\frac{x+2}{x^2-x-6}+\frac{x-3}{x^2-8x+15}$

$=\frac{x+2}{(x-3)(x+2)}+\frac{x-3}{(x-5)(x-3)}$

$=\frac{1}{(x-3)}+\frac{1}{(x-5)}$

$=\frac{1}{x-3}\cdot\frac{x-5}{x-5}+\frac{1}{x-5}\cdot\frac{x-3}{x-3}$

$=\frac{x-5+x-3}{(x-3)(x-5)}$

$=\frac{2x-8}{(x-3)(x-5)}$ or $\frac{2(x-4)}{(x-3)(x-5)}$

9. $\frac{1}{a-2}=\frac{4a-1}{a^2+5a-14}+\frac{2}{a+7}$

$\frac{1}{a-2}=\frac{4a-1}{(a+7)(a-2)}+\frac{2}{a+7}$

$(a+7)(a-2)\left[\frac{1}{a-2}=\frac{4a-1}{(a+7)(a-2)}+\frac{2}{a+7}\right]$

$a+7=4a-1+2(a-2)$

$a+7=4a-1+2a-4$

$a+7=6a-5$

$-5a+7=-5$

$-5a=-12$

$a=\frac{12}{5}$

10. $w=kI^2R$, $w=12$, $I=2$, $R=100$

$12=k(2^2)(100)$

$12=400k$

$\frac{12}{400}=k$

$k=\frac{3}{100}$

$w=\frac{3}{100}I^2R$, $I=0.8$, $R=600$

$w=\frac{3}{100}(0.8)^2(600)$

$w=11.52$

The wattage is 11.52 watts.

11. $N(x)=-0.2x^2+40x$

$N(50)=-0.2(50)^2+40(50)$

$=1500$

50 trees would produce about 1500 baskets of apples.

12. a. No, the graph is not a function since each x-value does not have a unique y-value.

b. The domain is the set of x-values, Domain: $\{x|x\ge-2\}$. The range is the set of y-values, Range: R

13. $4x-3y=10 \quad (1)$
$2x+y=5 \quad (2)$

In order to eliminate the y variable ,multiply equation (2) by 3 and add the result to equation (1).

$$4x-3y=10 \qquad 4x-3y=10$$
$$3(2x+y=5) \Rightarrow \underline{6x+3y=15}$$
$$10x = 25$$
$$x=\frac{5}{2}$$

Substitute $\frac{5}{2}$ for x in equation (2) and solve for y.

$$2\left(\frac{5}{2}\right)+y=5$$
$$5+y=5 \Rightarrow y=0$$

The solution is $\left(\frac{5}{2}, 0\right)$.

14. $\begin{vmatrix} 4 & 0 & -2 \\ 3 & 5 & 1 \\ 1 & -1 & 7 \end{vmatrix}$ (use row one)

$$= 4\begin{vmatrix} 5 & 1 \\ -1 & 7 \end{vmatrix} - 0\begin{vmatrix} 3 & 1 \\ 1 & 7 \end{vmatrix} + (-2)\begin{vmatrix} 3 & 5 \\ 1 & -1 \end{vmatrix}$$
$$= 4(35+1)-0(21-1)-2(-3-5)$$
$$= 4(36)-0(20)-2(-8)$$
$$= 144-0+16$$
$$= 160$$

15. $-4<\frac{x+4}{2}<8$

$$-8<x+4<16$$
$$-12<x<12$$

In interval notation the solution is (–12, 12).

16. $|4-2x|=5$

$$4-2x=5 \qquad 4-2x=-5$$
$$-2x=1 \qquad -2x=-9$$
$$x=-\frac{1}{2} \qquad x=\frac{9}{2}$$

The solution set is $\left\{-\frac{1}{2}, \frac{9}{2}\right\}$

17. $\frac{3-4i}{2+3i}=\left(\frac{3-4i}{2+3i}\right)\left(\frac{2-3i}{2-3i}\right)$

$$=\frac{6-17i+12i^2}{4-9i^2}$$
$$=\frac{6-17i-12}{4+9}$$
$$=\frac{-6-17i}{13}$$

18. $4x^2=-3x-12$

$$4x^2+3x+12=0$$

Use quadratic formula.

$$x=\frac{-b\pm\sqrt{b^2-4ac}}{2a}$$
$$=\frac{-3\pm\sqrt{9-192}}{8}$$
$$=\frac{-3\pm\sqrt{-183}}{8}$$
$$=\frac{-3\pm i\sqrt{183}}{8}$$

19. $V=\frac{1}{3}\pi r^2 h$

$$3V=3\left(\frac{1}{3}\pi r^2 h\right)$$
$$3V=\pi r^2 h$$
$$\frac{3V}{\pi h}=\frac{\pi r^2 h}{\pi h}$$
$$\frac{3V}{\pi h}=r^2$$
$$\sqrt{\frac{3V}{\pi h}}=\sqrt{r^2}$$
$$\sqrt{\frac{3V}{\pi h}}=r$$

20. $(x+5)^2+10(x+5)+24$

$$=((x+5)+4)((x+5)+6)$$
$$=(x+9)(x+11)$$

Chapter 13

Exercise Set 13.1

1. To find $(f \circ g)(x)$, substitute $g(x)$ for x in $f(x)$.

3. **a.** Each y has a unique x in a one-to-one function.

   b. Use the horizontal line test to determine whether a graph is one-to-one.

5. **a.** Yes; each first coordinate is paired with only one second coordinate.

   b. Yes; each second coordinate is paired with only one first coordinate.

   c. $\{(5, 3), (2, 4), (3, -1), (-2, 0)\}$
   Reverse each ordered pair.

7. The domain of f is the range of f^{-1} and the range of f is the domain of f^{-1}

9. $f(x) = x^2 + 1$, $g(x) = x + 5$

   a. $(f \circ g)(x) = (x+5)^2 + 1$
   $= x^2 + 10x + 25 + 1$
   $= x^2 + 10x + 26$

   b. $(f \circ g)(4) = 4^2 + 10(4) + 26 = 82w$

   c. $(g \circ f)(x) = (x^2 + 1) + 5 = x^2 + 6$

   d. $(g \circ f)(4) = 4^2 + 6 = 22$

11. $f(x) = x + 3,\ g(x) = x^2 + x - 4$

   a. $(f \circ g)(x) = (x^2 + x - 4) + 3 = x^2 + x - 1$

   b. $(f \circ g)(4) = 4^2 + 4 - 1 = 19$

   c. $(g \circ f)(x) = (x+3)^2 + (x+3) - 4$
   $= x^2 + 6x + 9 + x + 3 - 4$
   $= x^2 + 7x + 8$

   d. $(g \circ f)(4) = 4^2 + 7(4) + 8 = 52$

13. $f(x) = \frac{1}{x},\ g(x) = 2x + 3$

   a. $(f \circ g)(x) = \frac{1}{2x+3}$

   b. $(f \circ g)(4) = \frac{1}{2(4)+3} = \frac{1}{11}$

   c. $(g \circ f)(x) = 2\left(\frac{1}{x}\right) + 3 = \frac{2}{x} + 3$

   d. $(g \circ f)(4) = \frac{2}{4} + 3 = 3\frac{1}{2}$

15. $f(x) = \frac{2}{x},\ g(x) = x^2 + 1$

   a. $(f \circ g)(x) = \frac{2}{x^2+1}$

   b. $(f \circ g)(4) = \frac{2}{4^2+1} = \frac{2}{17}$

   c. $(g \circ f)(x) = \left(\frac{2}{x}\right)^2 + 1 = \frac{4}{x^2} + 1$

   d. $(g \circ f)(4) = \frac{4}{4^2} + 1 = 1\frac{1}{4}$

17. $f(x) = x^2 + 1,\ g(x) = x^2 + 5$

   a. $(f \circ g)(x) = (x^2 + 5)^2 + 1$
   $= x^4 + 10x^2 + 25 + 1$
   $= x^4 + 10x^2 + 26$

   b. $(f \circ g)(4) = 4^4 + 10(4)^2 + 26 = 442$

   c. $(g \circ f)(x) = (x^2 + 1)^2 + 5$
   $= x^4 + 2x^2 + 1 + 5$
   $= x^4 + 2x^2 + 6$

   d. $(g \circ f)(4) = 4^4 + 2(4)^2 + 6 = 294$

19. $f(x) = x - 4,\ g(x) = \sqrt{x+5},\ x \ge -5$

   a. $(f \circ g)(x) = \sqrt{x+5} - 4$

b. $(f \circ g)(4) = \sqrt{4+5} - 4$
$= \sqrt{9} - 4$
$= 3 - 4$
$= -1$

c. $(g \circ f)(x) = \sqrt{(x-4)+5} = \sqrt{x+1}$

d. $(g \circ f)(4) = \sqrt{4+1} = \sqrt{5}$

21. This function is not a one-to-one function since it does not pass the horizontal line test.

23. This function is a one-to-one function since it passes both the vertical line test and the horizontal line test.

25. Yes, the ordered pairs represent a one-to-one function. For each value of x there is a unique value for y and each y-value has a unique x-value.

27. No, the ordered pairs do not represent a one-to-one function. For each value of x there is a unique y, but for each y-value there is not a unique x since (–4, 2) and (0, 2) are ordered pairs in the given set.

29. $y = 2x + 4$ is a line with a slope of 2 and having a y-intercept of 4. It is a one-to-one function since it passes both the vertical line test and the horizontal line test.

31. $y = x^2 - 4$ is a parabola with vertex at (0, –4). It is not a one-to-one function since it does not pass the horizontal line test. Horizontal lines above $y = -4$ intersect the graph in 2 different points.

33. $y = x^2 - 2x + 6$ is a parabola with vertex at (1, 5). It is not a one-to-one function since it does not pass the horizontal line test. Horizontal lines above $y = 5$ intersect the graph in two different points.

35. $y = x^2 - 4, x \ge 0$ is the right side of the parabola from Exercise 31. It is a one-to-one function since it passes both the vertical line test and the horizontal line test.

37. $y = |x|$. It is not a one-to-one function since it does not pass the vertical line test and the horizontal line test.

39. $y = -\sqrt{x}$ is a one-to-one function since it passes both the vertical line test and the horizontal line test.

41. $y = x^3$ is a one-to-one function since it passes both the vertical line test and the horizontal line test.

43. For $f(x)$: Domain: {–2, –1, 2, 4, 9}
Range: {0, 3, 4, 6, 7}
For $f^{-1}(x)$: Domain:{0, 3, 4, 6, 7}
Range: {–2, –1, 2, 4, 9}

45. For $f(x)$: Domain:{ –1, 1, 2, 4}
Range: {–3, –1, 0, 2}
For $f^{-1}(x)$: Domain: {–3, –1, 0, 2}
Range: {–1, 1, 2, 4}

47. For $f(x)$: Domain: $\{x | x \ge 2\}$
Range: $\{y | y \ge 0\}$
For $f^{-1}(x)$: Domain: $\{x | x \ge 0\}$
Range: $\{y | y \ge 2\}$

49. **a.** Yes, $f(x) = x - 3$ is a one-to-one function.

b.
$$y = x - 3$$
$$x = y - 3$$
$$x + 3 = y$$
$$y = x + 3$$
$$f^{-1}(x) = x + 3$$

51. **a.** Yes, $h(x) = 5x$ is a one-to-one function.

b.
$$y = 5x$$
$$x = 5y$$
$$y = \frac{x}{5}$$
$$h^{-1}(x) = \frac{x}{5}$$

53. **a.** No, $p(x) = x^2$ is not a one-to-one function.

b. Does not exist

55. **a.** No; $t(x) = x^2 + 3$ is not a one-to-one function.

b. Does not exist

57. **a.** Yes; $g(x) = \frac{1}{x}$ is a one-to-one function.

b.
$$y = \frac{1}{x}$$
$$x = \frac{1}{y}$$
$$y = \frac{1}{x}$$
$$g^{-1}(x) = \frac{1}{x}$$

59`. **a.** No; $f(x) = x^2 + 4$ is not a one-to-one function.

b. Does not exist

61. **a.** Yes, $g(x) = x^3 - 5$ is a one-to-one function.

b.
$$y = x^3 - 5$$
$$x = y^3 - 5$$
$$x + 5 = y^3$$
$$\sqrt[3]{x+5} = y$$
$$g^{-1}(x) = \sqrt[3]{x+5}$$

63. **a.** Yes, $g(x) = \sqrt{x+2}$, $x \ge -2$ is a one-to-one function.

b.
$$y = \sqrt{x+2}$$
$$x = \sqrt{y+2}$$
$$x^2 = y + 2$$
$$x^2 - 2 = y$$
$$g^{-1}(x) = x^2 - 2,\ x \ge 0$$

65. **a.** Yes, $h(x) = x^2 - 4$, $x \ge 0$ is a one-to-one function.

b.
$$y = x^2 - 4$$
$$x = y^2 - 4$$
$$x + 4 = y^2$$
$$y = \sqrt{x+4}$$
$$h^{-1}(x) = \sqrt{x+4},\ x \ge -4$$

67. $f(x) = 2x + 8$

a.
$$y = 2x + 8$$
$$x = 2y + 8$$
$$x - 8 = 2y$$
$$\frac{x-8}{2} = y$$
$$f^{-1}(x) = \frac{x-8}{2}$$

b.

x	$f(x)$
0	8
–4	0

x	$f^{-1}(x)$
0	–4
8	0

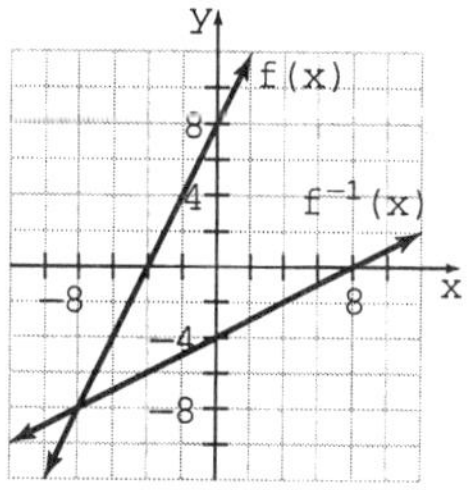

69. $f(x) = \sqrt{x}$, $x \ge 0$

a.
$$y = \sqrt{x}$$
$$x = \sqrt{y}$$
$$x^2 = \left(\sqrt{y}\right)^2$$
$$x^2 = y$$
$$f^{-1}(x) = x^2 \text{ for } x \ge 0$$

b.

x	$f(x)$
0	0
1	1
4	2

x	$f^{-1}(x)$
0	0
1	1
2	4

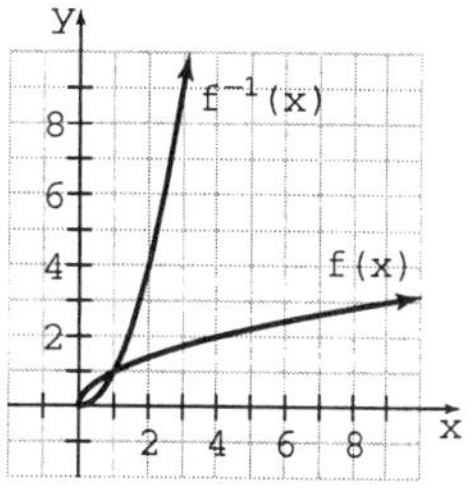

71. $f(x) = \sqrt{x+4},\ x \geq -4$

a.
$$y = \sqrt{x+4}$$
$$x = \sqrt{y+4}$$
$$x^2 = \left(\sqrt{y+4}\right)^2$$
$$x^2 = y+4$$
$$x^2 - 4 = y$$
$$f^{-1}(x) = x^2 - 4 \text{ for } x \geq 0$$

b.

x	$f(x)$
–4	0
–3	1
0	2

x	$f^{-1}(x)$
0	–4
1	–3
2	0

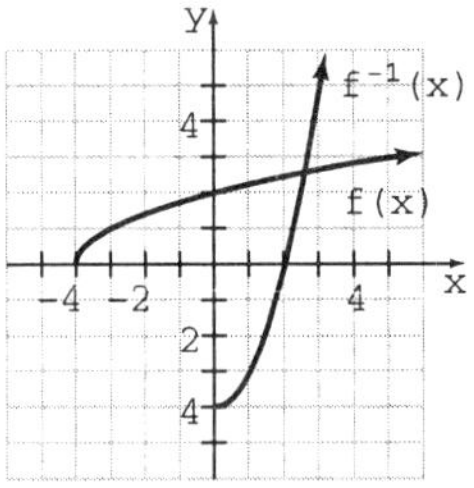

73. $f(x) = \sqrt[3]{x}$

a.
$$y = \sqrt[3]{x}$$
$$x = \sqrt[3]{y}$$
$$x^3 = \left(\sqrt[3]{y}\right)^3$$
$$x^3 = y$$
$$f^{-1}(x) = x^3$$

b.

x	$f(x)$
–8	–2
–1	–1
0	0
1	1
8	2

x	$f^{-1}(x)$
–2	–8
–1	–1
0	0
1	1
2	8

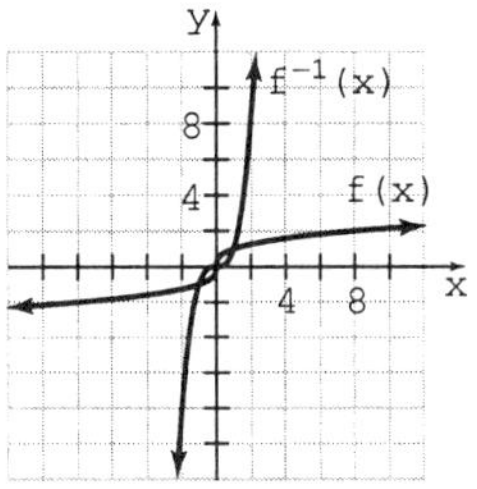

75. $f(x) = \dfrac{1}{x},\ x > 0$

a.
$$y = \frac{1}{x}$$
$$x = \frac{1}{y}$$
$$xy = 1$$
$$y = \frac{1}{x}$$
$$f^{-1}(x) = \frac{1}{x},\ x > 0$$

b.

x	$f(x)$
$\frac{1}{2}$	2
1	1
3	$\frac{1}{3}$

x	$f^{-1}(x)$
2	$\frac{1}{2}$
1	1
$\frac{1}{3}$	3

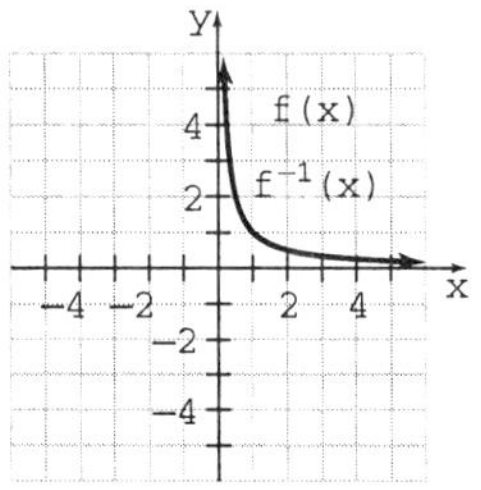

77. $(f \circ f^{-1})(x) = (x+4) - 4 = x$
$(f^{-1} \circ f)(x) = (x-4) + 4 = x$

79. $(f \circ f^{-1})(x) = \frac{1}{2}(2x - 10) + 5$
$= x - 5 + 5$
$= x$
$(f^{-1} \circ f)(x) = 2\left(\frac{1}{2}x + 5\right) - 10$
$= x + 10 - 10$
$= x$

81. $(f \circ f^{-1})(x) = \sqrt[3]{(x^3 + 2) - 2}$
$= \sqrt[3]{x^3}$
$= x$
$(f^{-1} \circ f)(x) = \left(\sqrt[3]{x - 2}\right)^3 + 2$
$= x - 2 + 2$
$= x$

83. $(f \circ f^{-1})(x) = \frac{2}{\frac{2}{x}} = 2 \cdot \frac{x}{2} = x$
$(f^{-1} \circ f)(x) = \frac{2}{\frac{2}{x}} = 2 \cdot \frac{x}{2} = x$

85. No, composition of functions is not commutative.
Let $f(x) = x^2$ and $g(x) = x + 1$.
Then $(f \circ g)(x) = (x + 1)^2 = x^2 + 2x + 1$ while $(g \circ f)(x) = x^2 + 1$.

87. a. $(f \circ g)(x) = f[g(x)]$
$= \left(\sqrt[3]{x - 2}\right)^3 + 2$
$= x - 2 + 2$
$= x$
$(g \circ f)(x) = g[f(x)]$
$= \sqrt[3]{\left(x^3 + 2\right) - 2}$
$= \sqrt[3]{x^3}$
$= x$

b. The domain of *f* is all real numbers and the domain of *g* is all real numbers. The domains of $(f \circ g)(x)$ and $(g \circ f)(x)$ are also all real numbers.

89. The range of $f^{-1}(x)$ is the domain of $f(x)$.

91. $f(x) = 3x$ converts yards, x, into feet, y.
$y = 3x$
$x = 3y$
$\frac{x}{3} = y$
$f^{-1}(x) = \frac{x}{3}$
Here, x is feet and $f^{-1}(x)$ is yards. The inverse function converts feet to yards.

93. $f(x) = \frac{5}{9}(x - 32)$
$y = \frac{5}{9}(x - 32)$
$x = \frac{5}{9}(y - 32)$
$\frac{9}{5}x = \frac{9}{5}\left[\frac{5}{9}(y - 32)\right]$
$\frac{9}{5}x = y - 32$
$\frac{9}{5}x + 32 = y$
$f^{-1}(x) = \frac{9}{5}x + 32$
Here, x is degrees Celsius and $f^{-1}(x)$ is degrees Fahrenheit. The inverse function converts Celsius to Fahrenheit.

95.
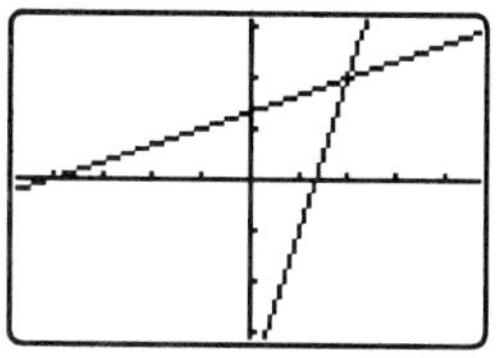
-4.7, 4.7, 1, -3.1,
Yes, the functions are inverses.

97.
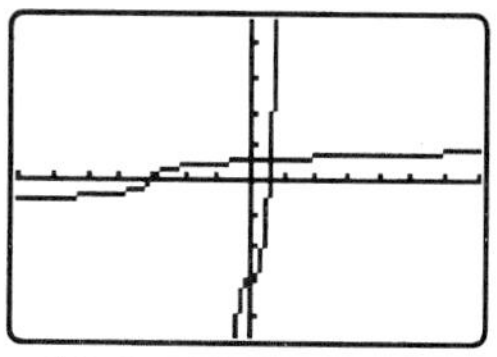
-28.2, 28.2, 4, -18.6
Yes, the functions are inverses.

99. a. $r(3) = 2(3) = 6$
The radius is 6 feet.

b. $A = \pi r^2$
$A = \pi(6)^2$
$A = 36\pi \approx 113.10$
The surface area is $36\pi \approx 113.10$ square feet.

c. $(A \circ r)(t) = \pi(2t)^2 = \pi(4t^2) = 4\pi t^2$

d. $4\pi(3)^2 = 4\pi(9) = 36\pi$

e. The answers agree.

102. First find the slope of the given line.
$2x + 3y - 9 = 0$
$3y = -2x + 9$
$y = -\frac{2}{3}x + 3 \Rightarrow m = -\frac{2}{3}$
Now use this slope together with the given point $\left(\frac{1}{2}, 3\right)$ to find the equation.
point - slope form:
$y - y_1 = m(x - x_1)$
$y - 3 = -\frac{2}{3}\left(x - \frac{1}{2}\right)$
$y - 3 = -\frac{2}{3}x + \frac{1}{3}$
$3\left(y - 3 = -\frac{2}{3}x + \frac{1}{3}\right)$
$3y - 9 = -2x + 1$
$2x + 3y = 10$

103.

-2	1	6	6	-8
		-2	-8	4
	1	4	-2	-4

$x^2 + 4x - 2 - \frac{4}{x + 2}$

104. $\dfrac{\frac{3}{x^2} - \frac{2}{x}}{\frac{x}{4}} = \dfrac{4x^2\left(\frac{3}{x^2} - \frac{2}{x}\right)}{4x^2\left(\frac{x}{4}\right)} = \dfrac{12 - 8x}{x^3}$

105.
$$\sqrt{\frac{24x^3y^2}{3xy^3}} = \sqrt{\frac{8x^2}{y}}$$
$$= \frac{\sqrt{4x^2}\sqrt{2}}{\sqrt{y}}$$
$$= \frac{2x\sqrt{2}}{\sqrt{y}} \cdot \frac{\sqrt{y}}{\sqrt{y}}$$
$$= \frac{2x\sqrt{2y}}{\sqrt{y^2}}$$
$$= \frac{2x\sqrt{2y}}{y}$$

106. $x^2 + 2x - 6 = 0$
$x^2 + 2x = 6$
$x^2 + 2x + 1 = 6 + 1$
$(x + 1)^2 = 7$
$x + 1 = \pm\sqrt{7}$
$x = -1 \pm \sqrt{7}$

Exercise Set 13.2

1. Exponential functions are functions of the form $f(x) = a^x, a > 0, a \neq 1$.

3. a. $y = \left(\frac{1}{2}\right)^x$; as x increases, y decreases.

b. No, y can never be zero because $\left(\frac{1}{2}\right)^x$ can never be 0.

c. No, y can never be negative because $\left(\frac{1}{2}\right)^x$ is never negative.

5. $y = 2^x$ and $y = 3^x$

a. Let $x = 0$
$y = 2^0 \quad y = 3^0$
$y = 1 \quad y = 1$
They have the same y-intercepts at (0, 1).

b. $y = 3^x$ will be steeper than $y = 2^x$ for $x > 0$.

7. $y = 2^x$

x	−2	−1	0	1	2
y	$\frac{1}{4}$	$\frac{1}{2}$	1	2	4

Domain: R
R: $\{y|y > 0\}$

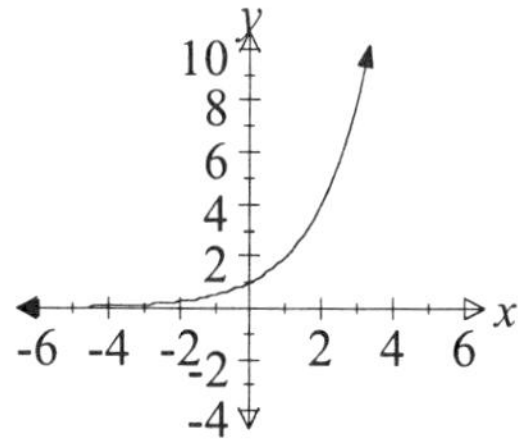

9. $y = \left(\frac{1}{2}\right)^x$

x	−2	−1	0	1	2
y	4	2	1	$\frac{1}{2}$	$\frac{1}{4}$

Domain: R
R: $\{y|y > 0\}$

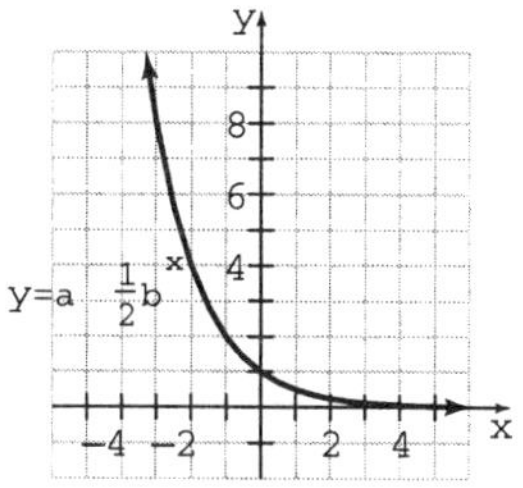

11. $y = 4^x$

x	−2	−1	0	1	2
y	$\frac{1}{16}$	$\frac{1}{4}$	1	4	16

Domain: R
R: $\{y|y > 0\}$

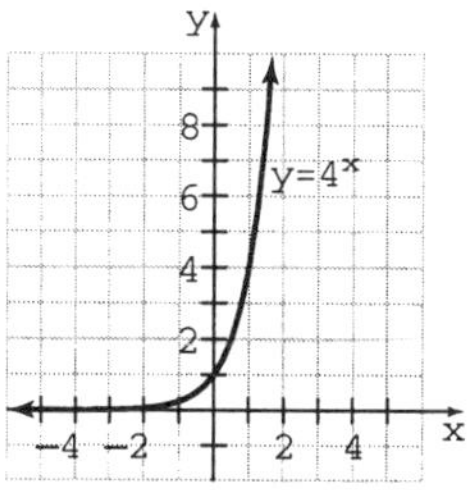

13. $y = \left(\frac{1}{4}\right)^x$

x	−2	−1	0	1	2
y	16	4	1	$\frac{1}{4}$	$\frac{1}{16}$

Domain: R
R: $\{y|y > 0\}$

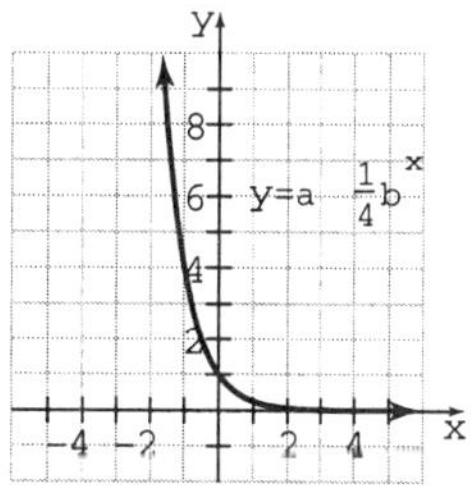

15. $y = 3^{-x} = \frac{1}{3^x} = \left(\frac{1}{3}\right)^x$

x	−2	−1	0	1	2
y	9	3	1	$\frac{1}{3}$	$\frac{1}{9}$

Domain: R
R: $\{y|y > 0\}$

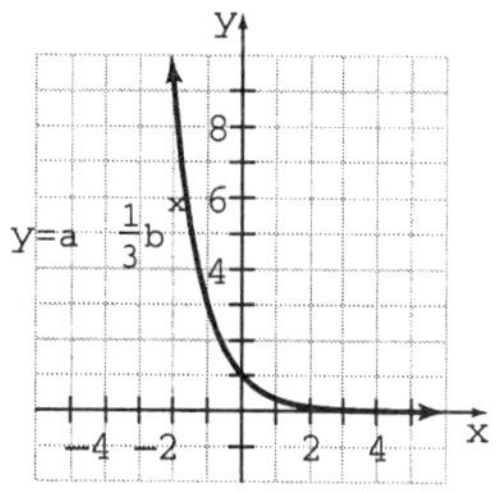

17. $y = \left(\frac{1}{3}\right)^{-x} = 3^x$

x	−2	−1	0	1	2
y	$\frac{1}{9}$	$\frac{1}{3}$	1	3	9

Domain: R
R: $\{y|y > 0\}$

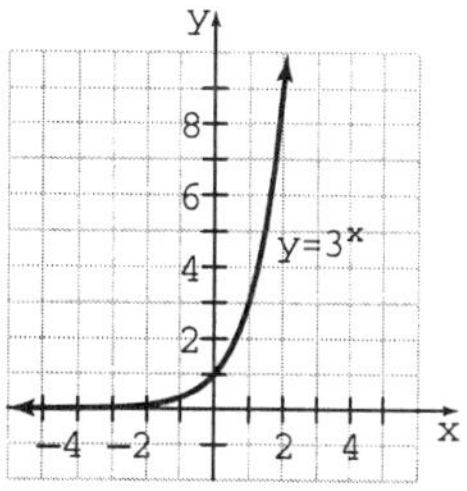

19. $y = 2^{x-1}$

x	-2	0	2	4	6
y	$\frac{1}{8}$	$\frac{1}{2}$	2	8	32

Domain: R
Range: $\{y \mid y > 0\}$

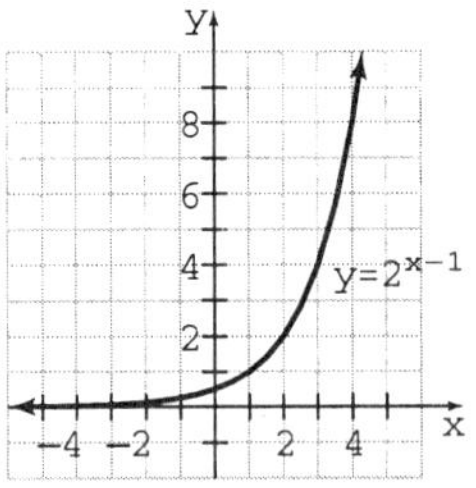

21. $y = \left(\frac{1}{3}\right)^{x+1}$

x	-3	-2	-1	0	1
y	9	3	1	$\frac{1}{3}$	$\frac{1}{9}$

Domain: R
R: $\{y \mid y > 0\}$

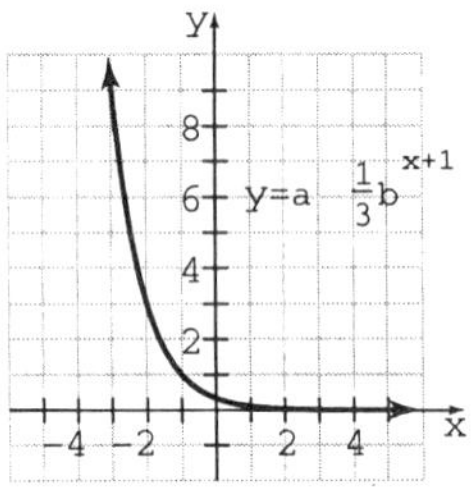

23. $y = 2^x - 1$

x	-2	-1	0	1	2	3
y	$-\frac{3}{4}$	$-\frac{1}{2}$	0	1	3	7

Domain: R
Range: $\{y \mid y > -1\}$

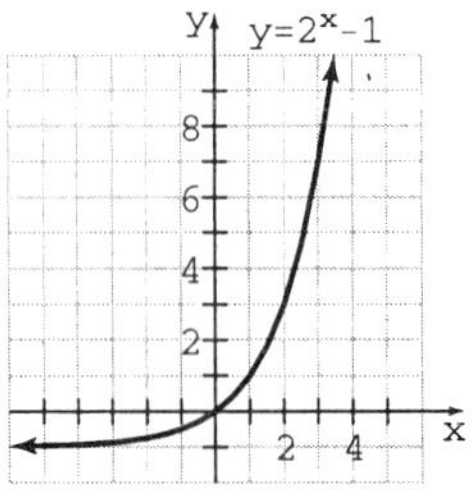

25. $y = 3^x - 1$

x	-2	-1	0	1	2
y	$-\frac{8}{9}$	$-\frac{2}{3}$	0	2	8

Domain: R
Range: $\{y \mid y > -1\}$

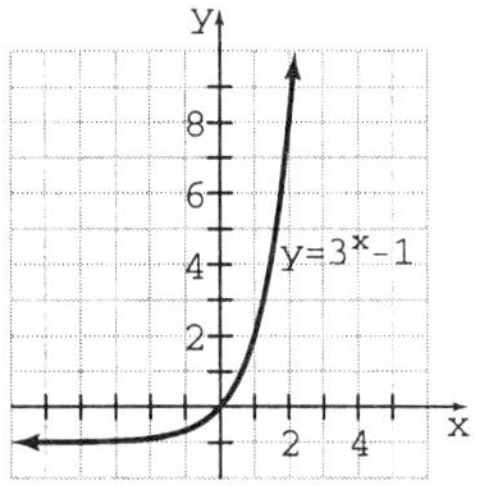

27. **a.** The graph is the horizontal line through $y = 1$.

b. Yes. A horizontal line will pass the vertical line test.

c. No. $f(x)$ is not one-to-one and therefore does not have an inverse function.

29. $y = a^x - k$ will have the same basic shape as the graph $y = a^x$. However, $y = a^x - k$ will be k units lower than $y = a^x$.

31. The graph of $y = a^{x+2}$ is the graph of $y = a^x$ shifted 2 units to the left.

33. **a.**

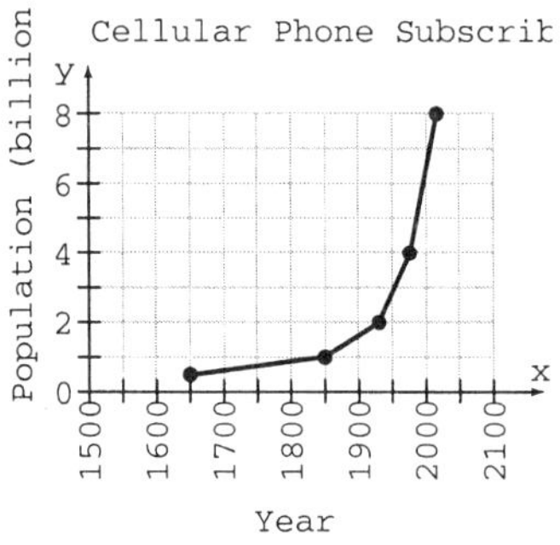

b. 200 years

c. 40 years

35. $g = 2^n,\ n = 8$
$g = 2^8 = 256$
The plant has 256 gametes.

37. $N(t) = 4(3)^t,\ t = 2$

$N(2) = 4(3)^2 = 4 \cdot 9 = 36$

There will be 36 bacteria in the petri dish after two days.

39. $A = p\left(1 + \frac{r}{n}\right)^{nt}$.

Use $p = 5000$,

$r = 6\% = 0.06$ and $n = 4$ and $t = 4$.

$A = 5000\left(1 + \frac{0.06}{4}\right)^{4 \cdot 4}$

$A = 5000(1 + 0.015)^{16}$

$A = 5000(1.015)^{16}$

$A \approx 5000(1.2689855)$

$A \approx 6344.93$

He has \$6344.93 after 4 years.

41. $A = A_0 2^{-t/5600}$

Use $A_0 = 12$ and $t = 1000$.

$A = 12(2^{-1000/5600})$

$A \approx 12(2^{-0.18})$

$A \approx 12(0.88)$

$A \approx 10.6$ grams

There are about 10.6 grams left.

43. $y = 80(2)^{-0.4t}$

a. $t = 10$

$y = 80(2)^{-0.4(10)}$

$y = 80(2)^{-4} = 80\left(\frac{1}{16}\right)$

$y = 5$

After 10 years, 5 grams remain.

b. $t = 100$

$y = 80(2)^{-0.4(100)}$

$y = 80(2)^{-40}$

$y \approx 80(9.094947 \times 10^{-13})$

$y \approx 7.28 \times 10^{-11}$

After 100 years, about 7.28×10^{-11} grams are left.

45. $y = 2000(1.2)^{0.1x}$

a. $x = 10$

$y = 2000(1.2)^{0.1(10)}$

$y = 2000(1.2)^1$

$y = 2400$

In 10 years, the population is expected to be 2400.

b. $x = 100$

$y = 2000(1.2)^{0.1(100)}$

$y = 2000(1.2)^{10}$

$y \approx 2000(6.1917364)$

$y \approx 12{,}383$

In 100 years, the population is expected to be about 12,383.

47. $V(t) = 24{,}000(0.82)^t,\ t = 4$

$V(t) = 24{,}000(0.82)^4 \approx 10{,}850.92$

The SUV will be worth about \$10,850.92 in 4 years..

49. a. Answers will vary. One possible answer is: Since the amount is reduced by 5%, the consumption is 95% of the previous year, or 0.95. Thus, $A(t) = 116{,}000(0.95)^t$.

b. $t = 2003 - 1998 = 5$

$A(5) = 116{,}000(0.95)^5$

$A \approx 116{,}000(0.7737809)$

$\approx 89{,}758.6$

The expected average use in 2003 is about 89,758.6 gallons.

51. $A = 41.97(0.996)^x$

$A(389) = 41.97(0.996)^{389}$

$A \approx 8.83$

The altitude at the top of Mt. Everest is about 8.83 kilometers.

53. a. $A = p\left(1 + \frac{r}{n}\right)^{nt}$

$A = 100\left(1 + \frac{0.07}{365}\right)^{365 \cdot 10}$

$A \approx 100(1.0001918)^{3650}$

$A = 201.36$

The amount is \$201.36.

b. For simple interest,

$A = 100 + 100(0.07)t$

$A = 100 + 100(0.07)(10)$

$A = 100 + 70$

$A = 170$

$\$201.36 - \$170 = \$31.36$

55. a. $y_1 = 3^{x-5}$

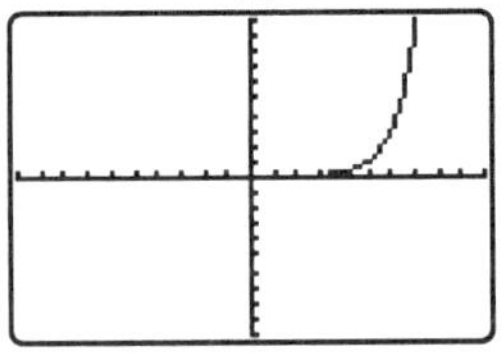

-10, 10, 1, -10, 10

b. $4 = 3^{x-5}$ when $x \approx 6.26$.

57. a. Day 15: $2^{15-1} = 2^{14} = \$16{,}384$

b. Day 20: $2^{20-1} = 2^{19} = \$524{,}288$

c. nth Day: 2^{n-1}

d. Day 30: $2^{30-1} = \$2^{29} = \$536{,}870{,}912$

e. $2^0 + 2^1 + 2^2 + \cdots + 2^{29}$

59. a. $2.3x^4y - 6.2x^6y^2 + 9.2x^5y^2$
$= -6.2x^6y^2 + 9.2x^5y^2 + 2.3x^4y$

b. $-6.2x^6y^2$ is the leading term.
$6 + 2 = 8$ is the degree of the polynomial.

c. $-6.2x^6y^2$ is the leading term, so -6.2 is the leading coefficient.

60.
$$\begin{aligned}(f \cdot g)(x) &= f(x) \cdot g(x)\\ &= (x+3)(x^2 - 2x + 4)\\ &= x^3 - 2x^2 + 4x + 3x^2 - 6x + 12\\ &= x^3 + x^2 - 2x + 12\end{aligned}$$

61. $\sqrt{a^2 - 8a + 16} = \sqrt{(a-4)^2} = |a - 4|$

62.
$$\begin{aligned}\sqrt[4]{\frac{32x^5y^9}{2y^3z}} &= \sqrt[4]{\frac{16x^5y^6}{z}}\\ &= \frac{\sqrt[4]{16x^5y^6}}{\sqrt[4]{z}}\\ &= \frac{\sqrt[4]{16x^4y^4 \cdot xy^2}}{\sqrt[4]{z}}\\ &= \frac{2xy\sqrt[4]{xy^2}}{\sqrt[4]{z}} \cdot \frac{\sqrt[4]{z^3}}{\sqrt[4]{z^3}}\\ &= \frac{2xy\sqrt[4]{xy^2z^3}}{\sqrt[4]{z^4}}\\ &= \frac{2xy\sqrt[4]{xy^2z^3}}{z}\end{aligned}$$

Exercise Set 13.3

1. $y = \log_a x$

a. The base a must be positive and must not be equal to one.

b. The argument x represents a number that is greater than 0. Thus, the domain is $\{x|x \text{ is a real number and } x > 0\}$.

c. R

3. The functions $f(x) = a^x$ and $g(x) = \log_a x$ are inverse functions. Therefore, some of the points on the function are $g(x) = \log_a x$ are $\left(\frac{1}{27}, -3\right)\left(\frac{1}{9}, -2\right), \left(\frac{1}{3}, -1\right)$(1, 0), (3, 1), (9, 2), and (27, 3).

5. The functions $y = a^x$ and $y = \log_a x$ for $a \neq 1$ are inverses of each other, thus the graphs are symmetric with respect to the line $y = x$. For each ordered pair (x, y) on the graph of $y = a^x$, the ordered pair (y, x) is on the graph of $y = \log_a x$.

7. $y = \log_2 x$
Convert to exponential form.
$2^y = x$

x	$\frac{1}{4}$	$\frac{1}{2}$	1	2	4
y	-2	-1	0	1	2

Domain: $\{x|x > 0\}$
Range: R

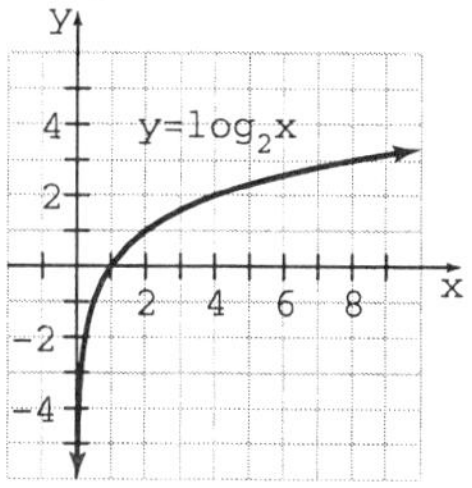

9. $y = \log_{1/2} x$
Convert to exponential form.
$x = \left(\frac{1}{2}\right)^y$

x	4	2	1	$\frac{1}{2}$	$\frac{1}{4}$
y	–2	–1	0	1	2

Domain: $\{x|x > 0\}$
Range: R

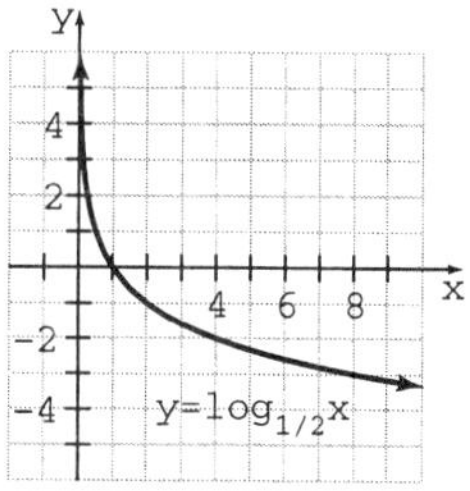

11. $y = \log_5 x$
Convert to the exponential form.
$x = 5^y$

x	$\frac{1}{25}$	$\frac{1}{5}$	1	5	25
y	–2	–1	0	1	2

Domain: $\{x|x > 0\}$
Range: R

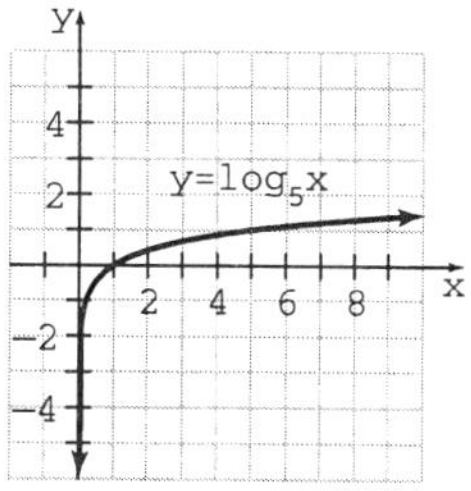

13. $y = \log_{1/5} x$ Convert to exponential form.
$x = \left(\frac{1}{5}\right)^y$

x	25	5	1	$\frac{1}{5}$	$\frac{1}{25}$
y	–2	–1	0	1	2

Domain: $\{x|x > 0\}$
Range: R

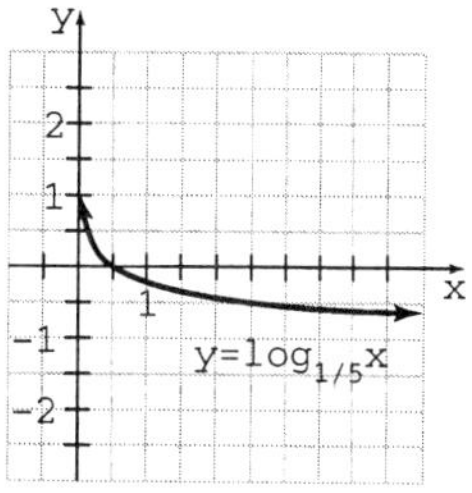

15. $y = 2^x$

x	–2	–1	0	1	2
y	$\frac{1}{4}$	$\frac{1}{2}$	1	2	4

$y = \log_{1/2} x$
Convert to exponential form.
$x = \left(\frac{1}{2}\right)^y$

x	4	2	1	$\frac{1}{2}$	$\frac{1}{4}$
y	–2	–1	0	1	2

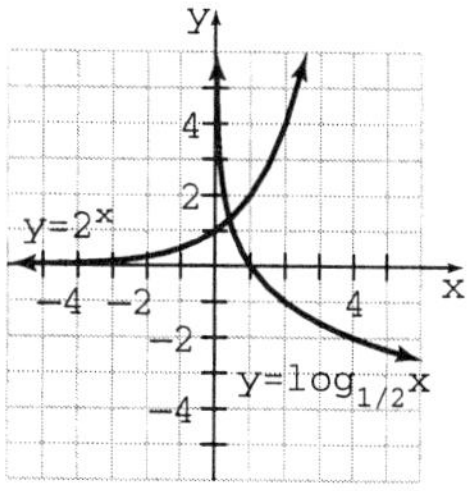

17. $y = 2^x$

x	-2	-1	0	1	2
y	$\frac{1}{4}$	$\frac{1}{2}$	1	2	4

$y = \log_2 x$
Convert to exponential form.
$x = 2^y$

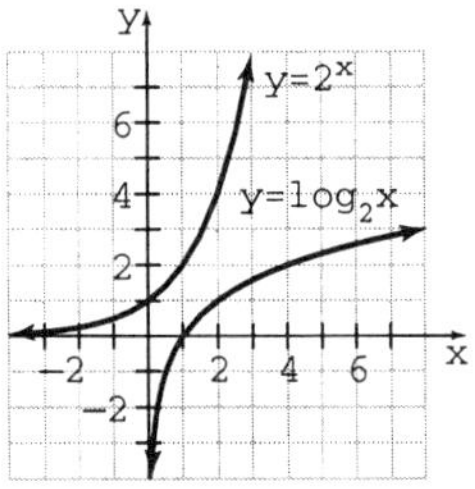

19. $2^3 = 8$
$\log_2 8 = 3$

21. $3^2 = 9$
$\log_3 9 = 2$

23. $16^{1/2} = 4$
$\log_{16} 4 = \frac{1}{2}$

25. $8^{1/3} = 2$
$\log_8 2 = \frac{1}{3}$

27. $\left(\frac{1}{2}\right)^5 = \frac{1}{32}$
$\log_{1/2}\left(\frac{1}{32}\right) = 5$

29. $2^{-3} = \frac{1}{8}$
$\log_2 \frac{1}{8} = -3$

31. $4^{-3} = \frac{1}{64}$
$\log_4 \frac{1}{64} = -3$

33. $64^{1/3} = 4$
$\log_{64} 4 = \frac{1}{3}$

35. $16^{-1/2} = \frac{1}{4}$
$\log_{16} \frac{1}{4} = -\frac{1}{2}$

37. $32^{-1/5} = \frac{1}{2}$
$\log_{32} \frac{1}{2} = -\frac{1}{5}$

39. $10^{0.6990} = 5$
$\log_{10} 5 = 0.6990$

41. $e^2 = 7.3891$
$\log_e 7.3891 = 2$

43. $c^b = w$
$\log_c w = b$

45. $\log_2 8 = 3$
$2^3 = 8$

47. $\log_{1/3} \frac{1}{27} = 3$
$\left(\frac{1}{3}\right)^3 = \frac{1}{27}$

49. $\log_5 \frac{1}{625} = -4$
$5^{-4} = \frac{1}{625}$

51. $\log_{81} 9 = \frac{1}{2}$
$81^{1/2} = 9$

53. $\log_8 \frac{1}{64} = -2$
$8^{-2} = \frac{1}{64}$

55. $\log_{27} \frac{1}{3} = -\frac{1}{3}$
$27^{-1/3} = \frac{1}{3}$

57. $\log_6 216 = 3$
$6^3 = 216$

59. $\log_{10} 8 = 0.9031$
$10^{0.9031} = 8$

61. $\log_e 6.52 = 1.8749$
$e^{1.8749} = 6.52$

63. $\log_r c = -a$
$r^{-a} = c$

65. $\log_4 16 = y$
$4^y = 16$
$4^y = 4^2$
$y = 2$

67. $\log_a 81 = 4$
$a^4 = 81$
$a^4 = 3^4$
$a = 3$

69. $\log_2 x = 5$
$2^5 = x$
$32 = x$

71. $\log_2 \frac{1}{8} = y$
$2^y = \frac{1}{8}$
$2^y = 2^{-3}$
$y = -3$

73. $\log_{1/2} x = 2$
$\left(\frac{1}{2}\right)^2 = x$
$\frac{1}{4} = x$

75. $\log_a \frac{1}{27} = -3$
$a^{-3} = \frac{1}{27}$
$a^{-3} = 3^{-3}$
$a = 3$

77. $\log_{10} 100 = 2$ because $10^2 = 100$

79. $\log_{10} 10 = 1$ because $10^1 = 10$

81. $\log_{10} \frac{1}{10} = -1$ because $10^{-1} = \frac{1}{10^1} = \frac{1}{10}$

83. $\log_{10} 10{,}000 = 4$ because $10^4 = 10{,}000$

85. $\log_4 64 = 3$ because $4^3 = 64$

87. $\log_3 \frac{1}{81} = -4$ because $3^{-4} = \frac{1}{3^4} = \frac{1}{81}$

89. $\log_8 \frac{1}{64} = -2$ because $8^{-2} = \frac{1}{8^2} = \frac{1}{64}$

91. $\log_7 1 = 0$ because $7^0 = 1$

93. $\log_9 9 = 1$ because $9^1 = 9$

95. $\log_4 1024 = 5$ because $4^5 = 1024$

97. If $f(x) = 4^x$, then $f^{-1}(x) = \log_4 x$.

99. $\log_{10} 425$ lies between 2 and 3 since 425 lies between $10^2 = 100$ and $10^3 = 1000$.

101. $\log_3 62$ lies between 3 and 4 since 62 lies between $3^3 = 27$ and $3^4 = 81$

103. For $x > 1$, 2^x will grow faster than $\log_{10} x$. Note that when $x = 10$, $2^x = 1024$ while $\log_{10} x = 1$.

105. $x = \log_{10} 10^5$
$10^x = 10^5$
$x = 5$

107. $x = \log_b b^3$
$b^x = b^3$
$x = 3$

109. $x = 10^{\log_{10} 8}$
$\log_{10} x = \log_{10} 8$
$x = 8$

111. $x = b^{\log_b 9}$
$\log_b x = \log_b 9$
$x = 9$

113. $R = \log_{10} I$
$7 = \log_{10} I$
$10^7 = I$
$I = 10{,}000{,}000$
The earthquake is 10,000,000 times more intense than the smallest measurable activity.

115.

$R = \log_{10} I$	$R = \log_{10} I$
$6 = \log_{10} I$	$2 = \log_{10} I$
$10^6 = I$	$10^2 = I$
$1{,}000{,}000 = I$	$100 = I$

$\frac{1{,}000{,}000}{100} = 10{,}000$
An earthquake that measures 6 is 10,000

times more intense than one that measures 2.

117. $y = \log_2(x-1)$ or $2^y = x - 1$

x	$1\frac{1}{4}$	$1\frac{1}{2}$	2	3	5
y	−2	−1	0	1	2

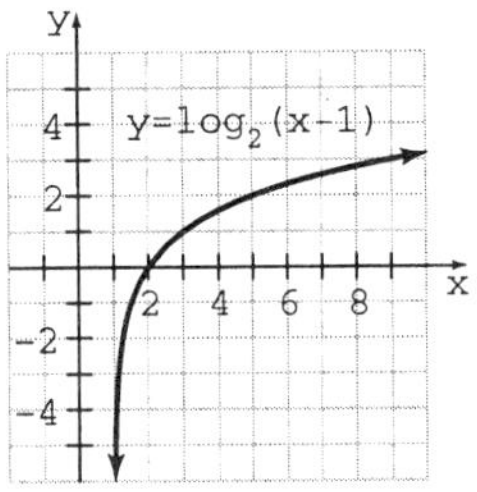

119. $3x^3 - 12x^2 - 36x = 3x(x^2 - 4x - 12)$
$= 3x(x-6)(x+2)$

120. $x^4 - 16 = (x^2+4)(x^2-4)$
$= (x^2+4)(x+2)(x-2)$

121. $40x^2 + 52x - 12 = 4(10x^2 + 13x - 3)$
$= 4(5x-1)(2x+3)$

122. $6r^2s^2 + rs - 1 = (3rs-1)(2rs+1)$

Exercise Set 13.4

1. Answers will vary.

3. Answers will vary.

5. Yes. This is true because of the product rule for logarithms.

7. $\log_4(3\cdot 10) = \log_4 3 + \log_4 10$

9. $\log_8 7(x+3) = \log_8 7 + \log_8(x+3)$

11. $\log_6 \frac{27}{5} = \log_6 27 - \log_6 5$

13.
$$\log_{10}\frac{\sqrt{x}}{x-9} = \log_{10}\frac{x^{1/2}}{x-9}$$
$$= \log_{10} x^{1/2} - \log_{10}(x-9)$$
$$= \frac{1}{2}\log_{10} x - \log_{10}(x-9)$$

15. $\log_6 x^7 = 7\log_6 x$

17.
$$\log_9 10(4^6) = \log_9 10 + \log_9 4^6$$
$$= \log_9 10 + 6\log_9 4$$

19.
$$\log_4\sqrt{\frac{a^3}{a+2}} = \log_4\left(\frac{a^3}{a+2}\right)^{1/2}$$
$$= \frac{1}{2}\log_4\frac{a^3}{a+2}$$
$$= \frac{1}{2}[\log_4 a^3 - \log_4(a+2)]$$
$$= \frac{1}{2}[3\log_4 a - \log_4(a+2)]$$
$$= \frac{3}{2}\log_4 a - \frac{1}{2}\log_4(a+2)$$

21.
$$\log_3\frac{d^6}{(a-5)^4} = \log_3 d^6 - \log_3(a-5)^4$$
$$= 6\log_3 d - 4\log_3(a-5)$$

23.
$$\log_8\frac{y(y+2)}{y^3} = \log_8 y + \log_8(y+2) - \log_8 y^3$$
$$= \log_8 y + \log_8(y+2) - 3\log_8 y$$
$$= -2\log_8 y + \log_8(y+2)$$

25.
$$\log_{10}\frac{2m}{3n} = \log_{10} 2m - \log_{10} 3n$$
$$= \log_{10} 2 + \log_{10} m - (\log_{10} 3 + \log_{10} n)$$
$$= \log_{10} 2 + \log_{10} m - \log_{10} 3 - \log_{10} n$$

27. $\log_5 2 + \log_5 3 = \log_5 6$

29. $\log_2 9 - \log_2 5 = \log_2\frac{9}{5}$

31. $5\log_4 2 = \log_4 2^5 = \log_4 32$

33. $\log_{10} x + \log_{10}(x+3) = \log_{10} x(x+3)$

35.
$$2\log_9 z - \log_9(z-2) = \log_9 z^2 - \log_9(z-2)$$
$$= \log_9\frac{z^2}{z-2}$$

37. $2(\log_5 p - \log_5 3) = 2\log_5 \frac{p}{3}$
$= \log_5\left(\frac{p}{3}\right)^2$

39. $\log_2 n + \log_2(n+4) - \log_2(n-3)$
$\log_2 n(n+4) - \log_2(n-3)$
$\log_2 \frac{n(n+4)}{n-3}$

41. $\frac{1}{2}[\log_5(x-4) - \log_5 x] = \frac{1}{2}\log_5 \frac{x-4}{x}$
$= \log_5\left[\frac{x-4}{x}\right]^{1/2}$
$= \log_5 \frac{\sqrt{x-4}}{x}$

43. $2\log_9 5 + \frac{1}{3}\log_9(r-6) - \frac{1}{2}\log_9 r$
$= \log_9 5^2 + \log_9(r-6)^{1/3} - \log_9 r^{1/2}$
$= \log_9 25 + \log_9 \sqrt[3]{r-6} - \log_9 \sqrt{r}$
$= \log_9 25\sqrt[3]{r-6} - \log_9 \sqrt{r}$
$= \log_9 \frac{25\sqrt[3]{r-6}}{\sqrt{r}}$

45. $4\log_6 3 - [2\log_6(x+3) + 4\log_6 x]$
$= \log_6 3^4 - [\log_6(x+3)^2 + \log_6 x^4]$
$= \log_6 3^4 - \log_6(x+3)^2 x^4$
$= \log_6 \frac{3^4}{(x+3)^2 x^4}$

47. $\log_a 10 = \log_a(2)(5)$
$= \log_a 2 + \log_a(5)$
$= 0.3010 + 0.6990$
$= 1$

49. $\log_a 2.5 = \log_a \frac{5}{2}$
$= \log_a 5 - \log_a 2$
$= 0.6990 - 0.3010$
$= 0.3980$

51. $\log_a 25 = \log_a 5^2$
$= 2(\log_a 5)$
$= 2(0.6990)$
$= 1.3980$

53. $5^{\log_5 10} = 10$

55. $(2^3)^{\log_8 5} = 8^{\log_8 5} = 5$

57. $\log_3 27 = \log_3 3^3 = 3$

59. $5(\sqrt[3]{27})^{\log_3 5} = 5(3)^{\log_3 5}$
$= 5(5)$
$= 25$

61. Yes

63. $\log_a \frac{x}{y} = \log_a xy^{-1}$
$= \log_a x + \log_a y^{-1}$
$= \log_a x + \log_a \frac{1}{y}$

65. $\log_a(x^2-4) - \log_a(x+2) = \log_a \frac{x^2-4}{x+2}$
$= \log_a \frac{(x+2)(x-2)}{x+2}$
$= \log_a(x-2)$

67. Yes;

$\log_a(x^2+8x+16) = \log_a(x+4)^2$
$= 2\log_a(x+4)$

69. $\log_{10} x^2 = 2\log_{10} x$
$= 2(0.4320)$
$= 0.8640$

71. $\log_{10}\sqrt[4]{x} = \log_{10} x^{1/4}$
$= \frac{1}{4}\log_{10} x$
$= \frac{1}{4}(0.4320) = 0.1080$

73. $\log_{10} xy = \log_{10} x + \log_{10} y$
$= 0.5000 + 0.2000$
$= 0.7000$

75. No; answers will vary.

77. $\log_2 \dfrac{\sqrt[4]{xy}\,\sqrt[3]{a}}{\sqrt[5]{a-b}} = \log_2 \sqrt[4]{xy}\,\sqrt[3]{a} - \log_2 \sqrt[5]{a-b}$

$= \log_2 (xy)^{1/4} + \log_2 a^{1/3} - \log_2 (a-b)^{1/5}$

$= \dfrac{1}{4}\log_2 xy + \dfrac{1}{3}\log_2 a - \dfrac{1}{5}\log_2 (a-b)$

$= \dfrac{1}{4}\log_2 x + \dfrac{1}{4}\log_2 y + \dfrac{1}{3}\log_2 a - \dfrac{1}{5}\log_2 (a-b)$

79. Let $\log_a x = m$ and $\log_a y = n$. Then

$a^m = x$ and $a^n = y$, so $\dfrac{x}{y} = \dfrac{a^m}{a^n} = a^{m-n}$.

Thus, $\log_a \dfrac{x}{y} = m - n = \log_a x - \log_a y$.

82. $\dfrac{2x+5}{x^2-7x+12} \div \dfrac{x-4}{2x^2-x-15}$

$= \dfrac{2x+5}{x^2-7x+12} \cdot \dfrac{2x^2-x-15}{x-4}$

$= \dfrac{2x+5}{(x-4)(x-3)} \cdot \dfrac{(2x+5)(x-3)}{x-4}$

$= \dfrac{(2x+5)^2}{(x-4)^2}$

83. $\dfrac{2x+5}{x^2-7x+12} - \dfrac{x-4}{2x^2-x-15}$

$= \dfrac{2x+5}{(x-4)(x-3)} - \dfrac{x-4}{(2x+5)(x-3)}$

$= \dfrac{(2x+5)(2x+5)}{(x-4)(x-3)(2x+5)} - \dfrac{(x-4)(x-4)}{(x-4)(x-3)(2x+5)}$

$= \dfrac{4x^2+20x+25}{(x-4)(x-3)(2x+5)} - \dfrac{x^2-8x+16}{(x-4)(x-3)(2x+5)}$

$= \dfrac{3x^2+28x+9}{(x-4)(x-3)(2x+5)}$

$= \dfrac{(3x+1)(x+9)}{(x-4)(x-3)(2x+5)}$

84. $2a - 7\sqrt{a} = 30$

$2a - 7\sqrt{a} - 30 = 0$

$2u^2 - 7u - 30 = 0$

$(2u+5)(u-6) = 0$

$2u + 5 = 0$ or $u - 6 = 0$

$u = -\dfrac{5}{2}$ $\quad u = 6$

$\sqrt{a} = -\dfrac{5}{2}$ $\quad \sqrt{a} = 6$

(no solution) $\quad a = 36$

The solution is $a = 36$.

85.

	Rate of Work	**Time Worked**	**Part of Task**
Mike	$\frac{1}{4}$	t	$\frac{t}{4}$
Jill	$\frac{1}{5}$	t	$\frac{t}{5}$

Let t be the time it would take for Mike and Jill to paint the house together.

$\dfrac{t}{4} + \dfrac{t}{5} = 1$

$20\left(\dfrac{t}{4} + \dfrac{t}{5}\right) = 20(1)$

$5t + 4t = 20$

$9t = 20$

$t = \dfrac{20}{9} = 2\dfrac{2}{9}$

It would take them $2\frac{2}{9}$ days to paint the house together.

86. $\sqrt[3]{4x^4y^7}\cdot\sqrt[3]{12x^7y^{10}}$
$=\sqrt[3]{48x^{11}y^{17}}$
$=\sqrt[3]{8\cdot 6x^9x^2y^{15}y^2}$
$=2x^3y^5\sqrt[3]{6x^2y^2}$

Exercise Set 13.5

1. Common logarithms are logarithms with base 10.

3. Antilogarithms are numbers obtained by taking 10 to the power of the logarithm.

5. $\log 45 = 1.6532$

7. $\log 19{,}200 = 4.2833$

9. $\log 0.0613 = -1.2125$

11. $\log 100 = 2.0000$

13. $\log 3.75 = 0.5740$

15. $\log 0.000472 = -3.3261$

17. antilog 0.4193 = 2.63

19. antilog 4.6283 = 42,500

21. antilog(−1.7086) = 0.0196

23. antilog 0.0000 = 1.00

25. antilog 2.7625= 579

27. antilog(−0.1543) = 0.701

29. $\log N = 2.0000$
$N =$ antilog 2.000
$N = 100$

31. $\log N = -2.103$
$N =$ antilog (−2.103)
$N = 0.00789$

33. $\log N = 4.1409$
$N =$ antilog 4.1409
$N = 13{,}800$

35. $\log N = -1.06$
$N =$ antilog (−1.06)
$N = 0.0871$

37. $\log N = -0.6218$
$N =$ antilog (−0.6218)
$N = 0.239$

39. $\log N = -0.3936$
$N =$ antilog (−0.3936)
$N = 0.404$

41. $\log 3560 = 3.5514$
Therefore, $10^{3.5514}\approx 3560$.

43. $\log 0.0727 = -1.1385$
Therefore, $10^{-1.1385}\approx 0.0727$

45. $\log 102 = 2.0086$
Therefore, $10^{2.0086}\approx 102$.

47. $\log 0.00128 = -2.8928$
Therefore, $10^{-2.8928}\approx 0.00128$.

49. $10^{2.9153} = 823$

51. $10^{-0.158} = 0.695$

53. $10^{-1.4802} = 0.0331$

55. $10^{1.3503} = 22.4$

57. $\log 1 = x$
$10^x = 1$
$10^x = 10^0$
$x = 0$
Therefore, $\log 1 = 0$.

59. $\log 0.1 = x$
$10^x = 0.1$
$10^x = \frac{1}{10}$
$10^x = 10^{-1}$
$x = -1$
Therefore, $\log 0.1 = -1$.

61. $\log 0.01 = x$
$10^x = 0.01$
$10^x = \frac{1}{100}$
$10^x = 10^{-2}$
$x = -2$
Therefore, $\log 0.01 = -2$.

63. $\log 0.001 = x$

$10^x = 0.001$

$10^x = \frac{1}{1000}$

$10^x = 10^{-3}$

$x = -3$

Therefore, $\log 0.001 = -3$.

65. $\log 10^7 = 7$

67. $10^{\log 7} = 7$

69. $6 \log 10^{5.2} = 6(5.2) = 31.2$

71. $5(10^{\log 9.4}) = 5(9.4) = 47$

73. No; $10^2 = 100$ and since $462 > 100$, log 462 must be greater than 2.

75. No; $10^0 = 1$ and $10^{-1} = 0.1$ and, since log 0.163 must be between 0 and −1.

77. No;

$$\log \frac{y}{3x} = \log y - \log 3x$$
$$= \log y - (\log 3 + \log x)$$
$$= \log y - \log 3 - \log x$$

79.
$$\log 125 = \log 5^3$$
$$= 3 \log 5$$
$$= 3(0.6990)$$
$$= 2.0970$$

81. log 30 is not possible given this information.

83.
$$\log \frac{1}{25} = \log 25^{-1}$$
$$= -\log 25$$
$$= -1(1.3979)$$
$$= -1.3979$$

85. $R = \log I$, $R = 3.41$

$3.41 = \log I$

$I = \text{anti}\log(3.41)$

$I \approx 2570$

This earthquake is about 2,570 times more intense than the smallest measurable activity.

87. $R = \log I$, $R = 6.37$

$6.37 = \log I$

$I = \text{anti}\log(6.37)$

$I \approx 2{,}340{,}000$

This earthquake is about 2,340,000 times more intense than the smallest measurable activity.

89. $\log d = 3.7 - 0.2g$

a. $g = 11$

$$\log d = 3.7 - 0.2(11)$$
$$= 3.7 - 2.2$$
$$= 1.5$$

$d = \text{antilog } 1.5 = 31.62$

A planet with absolute magnitude of 11 has a diameter of 31.62 kilometers.

b. $g = 20$

$$\log d = 3.7 - 0.2(20)$$
$$= 3.7 - 4$$
$$= -0.3$$

$d = \text{antilog } (-0.3) = 0.50$

A planet with absolute magnitude of 20 has a diameter of 0.50 kilometers.

c. $d = 5.8$

$$\log 5.8 = 3.7 - 0.2g$$
$$\log 5.8 - 3.7 = -0.2g$$
$$0.76343 - 3.7 = -0.2g$$
$$-2.93657 = -0.2g$$
$$\frac{-2.93657}{-0.2} = g$$
$$14.68 = g$$

A planet with diameter 5.8 kilometers has an absolute magnitude of 14.68.

91. $R(t) = 94 - 46.8 \log(t+1)$

a.
$$R(2) = 94 - 46.8 \log(2+1)$$
$$= 94 - 46.8 \log(3)$$
$$\approx 72$$

After two months, Sammy will remember about 72% of the course material.

b.
$$R(48) = 94 - 46.8 \log(2+48)$$
$$= 94 - 46.8 \log(50)$$
$$\approx 15$$

After forty-eight months, Sammy will remember about 15% of the course material.

93. $R = \log I$, $R = 4.6$

$4.6 = \log I$

$I = \text{anti}\log(4.6)$

$I \approx 39{,}800$

This earthquake is about 39,800 times more intense than the smallest measurable activity.

95. $\log E = 11.8 + 1.5m_s$

a. $\log E = 11.8 + 1.5(6)$
$\log E = 20.8$
$10^{20.8} = E$
$E = 6.31 \times 10^{20}$
The energy released is 6.31×10^{20}.

b. $\log(1.2 \times 10^{15}) = 11.8 + 1.5m_s$
$15.07918125 = 11.8 + 1.5m_s$
$3.27918125 = 1.5m_s$
$m_s \approx 2.19$
The surface wave has magnitude 2.19.

97. $M = \dfrac{\log E - 11.8}{1.5}$

$$M = \frac{\log(1.259 \times 10^{21}) - 11.8}{1.5}$$
$$= \frac{\log 1.259 + \log 10^{21} - 11.8}{1.5}$$
$$= \frac{\log 1.259 + 21 - 11.8}{1.5}$$
$$= \frac{\log 1.259 + 9.2}{1.5}$$
$$\approx \frac{0.1000 + 9.2}{1.5}$$
$$\approx \frac{9.3}{1.5}$$
$$\approx 6.2$$
The magnitude is about 6.2.

99. $R = \log I$

$\text{antilog}(R) = \text{antilog}(\log I)$

$\text{antilog}(R) = I$

101. $R = 85 - 41.9\log(t+1)$

$R - 85 = -41.9\log(t+1)$

$$\frac{R-85}{-41.9} = \frac{-41.9\log(t+1)}{-41.9}$$

$$\frac{85-R}{41.9} = \log(t+1)$$

$$\text{antilog}\left(\frac{85-R}{41.9}\right) = \text{antilog}(\log(t+1))$$

$$\text{antilog}\left(\frac{85-R}{41.9}\right) = t+1$$

$$\text{antilog}\left(\frac{85-R}{41.9}\right) - 1 = t$$

104. $3r = -4s - 6 \Rightarrow 3r + 4s = -6$ (1)
$3s = -5r + 1 \Rightarrow 5r + 3s = 1$ (2)

To eliminate the variable r, multiply equation (1) by –5 and equation (2) by 3 then add.

$-5(3r + 4s = -6) \Rightarrow -15r - 20s = 30$
$3(5r + 3s = 1) \Rightarrow \underline{15r + 9s = 3}$
$-11s = 33$
$s = -3$

Substitute –3 for s in equation (1) and solve for r.

$3r + 4(-3) = -6$

$3r - 12 = -6$

$3r = 6$

$r = 2$

The solution is (2, –3).

105. Let r equal the rate of car 2.

	d	r	t
Car 1	$4(r+5)$	$r+5$	4
Car 2	$4r$	r	4

The total distance was 420 miles.

$$4(r+5)+4r=420$$
$$4r+20+4r=420$$
$$8r+20=420$$
$$8r=400$$
$$r=50$$

The rate of car 2 is 50 mph and the rate of car 1 is 50 + 5 = 55 mph.

106. $-3x^2-4x-8=0$

$a=-3\ \ b=-4\ \ c=-8$

$$x=\frac{-b\pm\sqrt{b^2-4ac}}{2a}$$
$$=\frac{4\pm\sqrt{16-4(-3)(-8)}}{2(-3)}$$
$$=\frac{4\pm\sqrt{16-96}}{-6}$$
$$=\frac{4\pm\sqrt{-80}}{-6}$$
$$=\frac{4\pm\sqrt{-16}\sqrt{5}}{-6}$$
$$=\frac{4\pm 4i\sqrt{5}}{-6}$$
$$=\frac{-2(-2\pm 2i\sqrt{5})}{-2(3)}$$
$$=\frac{-2\pm 2i\sqrt{5}}{3}$$

107. Let r = the boat's speed in still water.

Direction	**d**	**r**	**t**
Downriver	15	$r+5$	$\frac{15}{r+5}$
Upriver	15	$r-5$	$\frac{15}{r-5}$

$$\frac{15}{r+5}+\frac{15}{r-5}=4$$
$$(r+5)(r-5)\left[\frac{15}{r+5}+\frac{15}{r-5}\right]=(r+5)(r-5)(4)$$
$$15(r-5)+15(r+5)=(r+5)(r-5)(4)$$
$$15r-75+15r+75=4r^2-100$$
$$30r=4r^2-100$$
$$0=4r^2+30r-100$$
$$0=2(2r^2+15r-50)$$
$$0=2(2r+5)(r-10)$$
$$2r+5=0 \qquad r-10=0$$
$$2r=-5 \qquad r=10$$
$$r=-\frac{5}{2}$$

Since r cannot be negative, the boat travels at 10 miles per hour.

108. In the function $f(x)=(x-2)^2+1$, h has a value of 2 and k has a value of 1. The graph of $f(x)$ will be the graph of $g(x)=x^2$ shifted 2 units to the right and 1 units up.

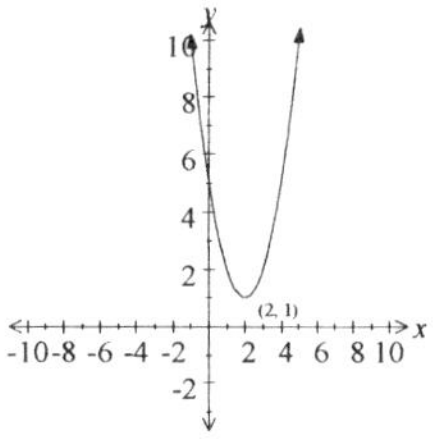

109. $\dfrac{2x-3}{5x+10} < 0$

$x \neq -2$

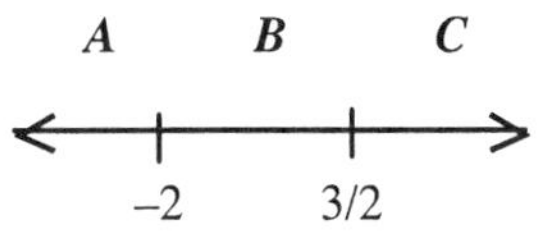

Interval A	Interval B	Interval C
Test Value, –4	Test Value, 0	Test Value, 6
$\dfrac{2x-3}{5x+10} < 0$	$\dfrac{2x-3}{5x+10} < 0$	$\dfrac{2x-3}{5x+10} < 0$
$\dfrac{-8-3}{-20+10} < 0$	$\dfrac{0-3}{0+10} < 0$	$\dfrac{12-3}{30+10} < 0$
$\dfrac{-11}{-10} < 0$	$\dfrac{-3}{10} < 0$ True	$\dfrac{9}{40} < 0$ True
$\dfrac{11}{10} < 0$ False		

The solution is $\left(-2, \dfrac{3}{2}\right)$.

Exercise Set 13.6

1. $c = d$

3. Check for extraneous roots.

5. log (–2) is not a real number

7.
$$5^x = 125$$
$$5^x = 5^3$$
$$x = 3$$

9.
$$3^x = 243$$
$$3^x = 3^5$$
$$x = 5$$

11.
$$49^x = 7$$
$$(7^2)^x = 7^1$$
$$7^{2x} = 7^1$$
$$2x = 1$$
$$x = \frac{1}{2}$$

13.
$$5^{-x} = \frac{1}{25}$$
$$5^{-x} = 5^{-2}$$
$$-x = -2$$
$$x = 2$$

15.
$$27^x = \frac{1}{3}$$
$$\left(3^3\right)^x = 3^{-1}$$
$$3^{3x} = 3^{-1}$$
$$3x = -1$$
$$x = -\frac{1}{3}$$

17.
$$2^{x+1} = 64$$
$$2^{x+1} = 2^6$$
$$x + 1 = 6$$
$$x = 5$$

19.
$$2^{3x-2} = 16$$
$$2^{3x-2} = 2^4$$
$$3x - 2 = 4$$
$$3x = 6$$
$$x = 2$$

21.
$$27^x = 3^{2x+3}$$
$$3^{3x} = 3^{2x+3}$$
$$3x = 2x + 3$$
$$x = 3$$

23.
$$7^x = 50$$
$$\log 7^x = \log 50$$
$$x \log 7 = \log 50$$
$$x = \frac{\log 50}{\log 7}$$
$$x \approx 2.01$$

25.
$$4^{x-1} = 20$$
$$\log 4^{x-1} = \log 20$$
$$(x-1)\log 4 = \log 20$$
$$x - 1 = \frac{\log 20}{\log 4}$$
$$x = \frac{\log 20}{\log 4} + 1$$
$$x \approx 3.16$$

27.
$$\begin{aligned} 1.63^{x+1} &= 25 \\ \log 1.63^{x+1} &= \log 25 \\ (x+1)\log 1.63 &= \log 25 \\ x+1 &= \frac{\log 25}{\log 1.63} \\ x+1 &\approx 6.59 \\ x &\approx 5.59 \end{aligned}$$

29.
$$\begin{aligned} 3^{x+4} &= 6^x \\ \log 3^{x+4} &= \log 6^x \\ (x+4)\log 3 &= x\log 6 \\ x\log 3 + 4\log 3 &= x\log 6 \\ 4\log 3 &= x\log 6 - x\log 3 \\ 4\log 3 &= x(\log 6 - \log 3) \\ \frac{4\log 3}{\log 6 - \log 3} &= x \\ 6.34 &\approx x \end{aligned}$$

31.
$$\begin{aligned} \log_{16} x &= \frac{1}{2} \\ 16^{1/2} &= x \\ \sqrt{16} &= x \\ 4 &= x \end{aligned}$$

33.
$$\begin{aligned} \log_{125} x &= \frac{1}{3} \\ 125^{1/3} &= x \\ \sqrt[3]{125} &= x \\ 5 &= x \end{aligned}$$

35.
$$\begin{aligned} \log_2 x &= -3 \\ 2^{-3} &= x \\ \frac{1}{2^3} &= x \\ \frac{1}{8} &= x \end{aligned}$$

37.
$$\begin{aligned} \log x &= 1 \\ \log_{10} x &= 1 \\ 10^1 &= x \\ 10 &= x \end{aligned}$$

39.
$$\begin{aligned} \log_2(5-3x) &= 3 \\ 2^3 &= 5-3x \\ 8 &= 5-3x \\ 3x &= -3 \\ x &= -1 \end{aligned}$$

41.
$$\begin{aligned} \log_5(x+2)^2 &= 2 \\ (x+2)^2 &= 5^2 \\ x+2 &= \pm\sqrt{25} \\ x+2 &= \pm 5 \end{aligned}$$
$$x+2=5 \quad \text{or} \quad x+2=-5$$
$$x=3 \qquad\qquad x=-7$$

Both values check. The solution is 3 and –7.

43.
$$\begin{aligned} \log_2(r+4)^2 &= 4 \\ (r+4)^2 &= 2^4 \\ r^2+8r+16 &= 16 \\ r^2+8r &= 0 \\ r(r+8) &= 0 \end{aligned}$$
$$r=0 \quad \text{or} \quad r+8=0$$
$$r=-8$$

45.
$$\begin{aligned} \log(x+3) &= 2 \\ \log_{10}(x+3) &= 2 \\ 10^2 &= x+3 \\ 100 &= x+3 \\ x &= 97 \end{aligned}$$

47.
$$\begin{aligned} \log_2 x + \log_2 5 &= 2 \\ \log_2 5x &= 2 \\ 5x &= 2^2 \\ x &= \frac{4}{5} \end{aligned}$$

49.
$$\begin{aligned} \log(r+2) &= \log(3r-1) \\ r+2 &= 3r-1 \\ 3 &= 2r \\ \frac{3}{2} &= r \end{aligned}$$

51.
$$\begin{aligned} \log(2x+1)+\log 4 &= \log(7x+8) \\ \log(8x+4) &= \log(7x+8) \\ 8x+4 &= 7x+8 \\ x &= 4 \end{aligned}$$

53. $\log n + \log(3n-5) = \log 2$
$\log(3n^2 - 5n) = \log 2$
$3n^2 - 5n = 2$
$3n^2 - 5n - 2 = 0$
$(3n+1)(n-2) = 0$

$3n + 1 = 0$ or $n - 2 = 0$
$3n = -1$ $\quad n = 2$
$n = -\frac{1}{3}$

Check: $n = -\frac{1}{3}$
$\log n + \log(3n-5) = \log 2$
$\log\left(-\frac{1}{3}\right) + \log\left[3\left(\frac{-1}{3}\right) - 5\right] = \log 2$
Logarithms of negative numbers are not real numbers.
Check: $n = 2$
$\log n + \log(3n-5) = \log 2$
$\log 2 + \log[3(2) - 5] = \log 2$
$\log 2 + \log 1 = \log 2$
$\log(2 \cdot 1) = \log 2$
$\log 2 = \log 2$
2 is the only solution.
$-\frac{1}{3}$ is an extraneous solution.

55. $\log 5 + \log y = 0.72$
$\log 5y = 0.72$
$5y \approx 5.2481$
$y \approx 1.05$

57. $2\log x - \log 4 = 2$
$\log x^2 - \log 4 = 2$
$\log \frac{x^2}{4} = 2$
$\frac{x^2}{4} = \text{antilog } 2$
$\frac{x^2}{4} = 100$
$x^2 = 400$
$x^2 - 400 = 0$
$(x+20)(x-20) = 0$

$x + 20 = 0$ or $x - 20 = 0$
$x = -20$ $\quad x = 20$
Check: $x = -20$
$2\log x - \log 4 = 2$
$2\log(-20) - \log 4 = 2$
Logarithms of negative numbers are not real numbers.
Check: $x = 20$
$2\log x - \log 4 = 2$
$2\log 20 - \log 4 = 2$
$\log \frac{400}{4} = 2$
$\log 100 = 2$
$100 = \text{antilog } 2$
$100 = 100$
Thus, 20 is the only solution.
–20 is an extraneous solution.

59. $\log x + \log(x-3) = 1$
$\log(x^2 - 3x) = 1$
$x^2 - 3x = \text{antilog } 1$
$x^2 - 3x = 10$
$x^2 - 3x - 10 = 0$
$(x-5)(x+2) = 0$
$x - 5 = 0$ or $x + 2 = 0$
$x = 5$ $\quad x = -2$
A check shows that 5 is the only solution.
–2 is an extraneous solution.

61. $\log x = \frac{1}{3}\log 27$
$\log x = \log 27^{1/3}$
$\log x = \log 3$
$x = 3$

63. $\log_8 x = 3\log_8 2 - \log_8 4$
$\log_8 x = \log_8 2^3 - \log_8 4$
$\log_8 x = \log_8 \frac{8}{4}$
$\log_8 x = \log_8 2$
$x = 2$

65. $\log_5(x+3) + \log_5(x-2) = \log_5 6$
$\log_5(x+3)(x-2) = \log_5 6$
$\log_5(x^2 + x - 6) = \log_5 6$
$x^2 + x - 6 = 6$
$x^2 + x - 12 = 0$
$(x+4)(x-3) = 0$
$x = -4$ or $x = 3$
Disregard $x = -4$ since
$\log(-4+3) = \log(-1)$.
Therefore, $x = 3$ is the only solution.

67. $\log_2(x+3) - \log_2(x-6) = \log_2 4$

$$\log_2 \frac{x+3}{x-6} = \log_2 4$$
$$\frac{x+3}{x-6} = 4$$
$$x+3 = 4x - 24$$
$$27 = 3x$$
$$9 = x$$

69.

$$50,000 = 4500(2^t)$$
$$\frac{50,000}{4500} = 2^t$$
$$\log \frac{50,000}{4500} = \log 2^t$$
$$\log \frac{50,000}{4500} = t \log 2$$
$$\log 50,000 - \log 4500 = t \log 2$$
$$\frac{\log 50,000 - \log 4500}{\log 2} = t$$
$$3.47 \approx t$$

There are 50,000 bacteria after about 3.47 hours.

71.

$$40 = 200(0.800)^t$$
$$0.2 = (0.800)^t$$
$$\log 0.2 = \log(0.800)^t$$
$$\log 0.2 = t \log 0.800$$
$$\frac{\log 0.2}{\log 0.800} = t$$
$$7.21 \approx t$$

40 grams remain after about 7.21 years.

73.

$$A = P(1+\frac{r}{n})^{nt}$$
$$4600 = 2000(1+\frac{0.05}{1})^{1 \cdot t}$$
$$4600 = 2000(1.05)^t$$
$$\frac{4600}{2000} = 1.05^t$$
$$2.3 = 1.05^t$$
$$\log 2.3 = \log 1.05^t$$
$$\log 2.3 = t \log 1.05$$
$$\frac{\log 2.3}{\log 1.05} = t \quad \Rightarrow \quad t \approx 17.07 \text{ years}$$

75. $f(t) = 26 - 12.1 \cdot \log(t+1)$

a. $x = 1990 - 1960 = 30$
$f(30) = 26 - 12.1 \log(30+1) = 7.95$
In 1990, the rate was 7.95 deaths per 1000 live births.

b. $x = 2005 - 1960 = 45$
$f(45) = 26 - 12.1 \log(45+1) \approx 5.88$
In 2005, the rate is expected to be 5.88 deaths per 1000 live births.

77. $c = 50,000$, $n = 12$, $r = 0.15$.

$$S = c(1-r)^n$$
$$S = 50,000(1-0.15)^{12}$$
$$S = 50,000(0.85)^{12}$$
$$S \approx 7112.09$$

The scrap value is about $7112.09.

79. $P_{out} = 12.6$ and $P_{in} = 0.146$

$$P = 10\log\left(\frac{12.6}{0.146}\right)$$
$$P \approx 10\log 86.30137$$
$$P \approx 10(1.936)$$
$$P \approx 19.36$$

The power gain is about 19.36.

81. a.

$$d = 120$$
$$d = 10 \log I$$
$$120 = 10 \log I$$
$$12 = \log I$$
$$I = \text{antilog } 12$$
$$I = 10^{12}$$
$$I = 1,000,000,000,000$$

The intensity is 1,000,000,000,000 times the minimum intensity of audible sound.

b.

$$d = 70$$
$$d = 10 \log I$$
$$70 = 10 \log I$$
$$7 = \log I$$
$$I = \text{antilog } 7$$
$$I = 10^7$$
$$I = 10,000,000$$
$$\frac{1,000,000,000,000}{10,000,000} = 100,000$$

The sound of an airplane engine is 100,000 times more intense than the noise in a busy city street.

83.

$$8^x = 16^{x-2}$$
$$2^{3x} = 2^{4(x-2)}$$
$$3x = 4(x-2)$$
$$3x = 4x - 8$$
$$8 = x$$

85. $2^{2x} - 6(2^x) + 8 = 0$

$(2^x)^2 - 6(2^x) + 8 = 0$

$y^2 - 6y + 8 = 0 \leftarrow$ Replace 2^x with y

$(y-4)(y-2) = 0$

$y - 4 = 0$ or $y - 2 = 0$

$y = 4 \qquad y = 2$

$2^x = 4 \qquad 2^x = 2 \leftarrow$ Replace y with 2^x

$2^x = 2^2 \qquad 2^x = 2^1$

$x = 2 \qquad x = 1$

The solutions are $x = 2$ and $x = 1$.

87. $2^x = 8^y$

$x + y = 4$

The first equation simplifies to

$2^x = (2^3)^y$

$2^x = 2^{3y}$

$x = 3y$

The system becomes

$x = 3y$

$x + y = 4$

Substitute $3y$ for x in the second equation.

$x + y = 4$

$3y + y = 4$

$4y = 4$

$y = 1$

Now, substitute 1 for y in the first equation.

$x = 3y$

$x = 3(1) = 3$

The solution is (3, 1).

89. $\log(x + y) = 2$

$x - y = 8$

The first equation can be written as

$x + y = 10^2$

$x + y = 100$

The system becomes

$$\begin{array}{rl} x + y = & 100 \\ x - y = & 8 \\ \hline \text{Add: } 2x \quad = & 108 \\ x \quad = & 54 \end{array}$$

Substitute 54 for x in the first equation.

$54 + y = 100$

$y = 46$

The solution is (54, 46).

91.

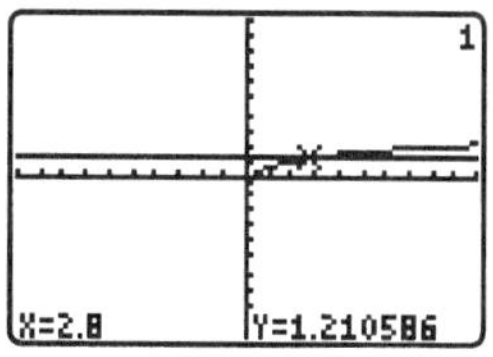

-10, 30, 5, -10, 10

The solution is $x \approx 2.8$.

93.

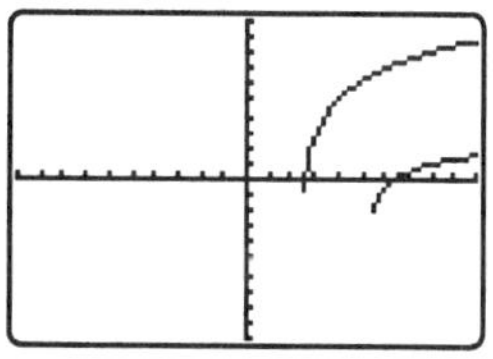

-10, 10, 1, -10, 10

There is no real-number solution.

95. Volume of cylinder:

$V_1 = \pi r^2 h = \pi\left(\frac{3}{2}\right)^2 \cdot 4 \approx 28.2743$ cubic feet

Volume of box:

$V_2 = l \cdot w \cdot h = (3)(3)(4) = 36$ cubic feet

The box has the greater volume.

Difference in volumes:

$V_2 - V_1 \approx 7.73$ cubic feet

96. $f(x) = x^2 - x,\ g(x) = x - 6$

$$\begin{aligned} (g-f)(x) &= (x-6) - (x^2 - x) \\ &= x - 6 - x^2 + x \\ &= -x^2 + 2x - 6 \end{aligned}$$

$$\begin{aligned} (g-f)(3) &= -(3)^2 + 2(3) - 6 \\ &= -9 + 6 - 6 \\ &= -9 \end{aligned}$$

97. Graph both inequalities on the same coordinate plane. The solution to the system is the double-shaded region.

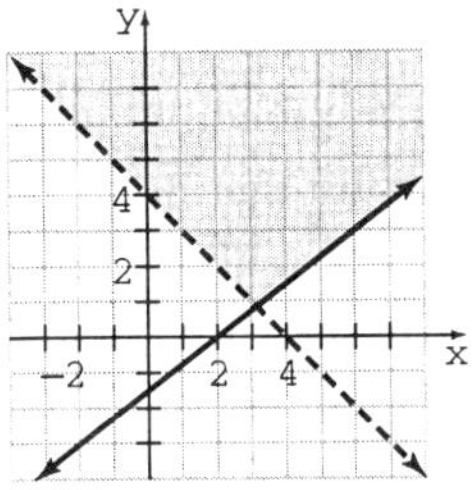

98.
$$\frac{2\sqrt{xy}-\sqrt{xy}}{\sqrt{x}+\sqrt{y}}=\frac{2\sqrt{xy}-\sqrt{xy}}{\sqrt{x}+\sqrt{y}}\cdot\frac{\sqrt{x}-\sqrt{y}}{\sqrt{x}-\sqrt{y}}$$
$$=\frac{\left(2\sqrt{xy}-\sqrt{xy}\right)\cdot\left(\sqrt{x}-\sqrt{y}\right)}{\left(\sqrt{x}+\sqrt{y}\right)\left(\sqrt{x}-\sqrt{y}\right)}$$
$$=\frac{2\sqrt{x^2y}-2\sqrt{xy^2}-\sqrt{x^2y}+\sqrt{xy^2}}{\sqrt{x^2}-\sqrt{xy}+\sqrt{xy}-\sqrt{y^2}}$$
$$=\frac{\sqrt{x^2y}-\sqrt{xy^2}}{\sqrt{x^2}-\sqrt{y^2}}$$
$$=\frac{x\sqrt{y}-y\sqrt{x}}{x-y}$$

99. $E=mc^2$, for c
$$\frac{E}{m}=\frac{mc^2}{m}$$
$$\frac{E}{m}=c^2$$
$$\sqrt{\frac{E}{m}}=c$$

100. Use $f(x)=a(x-h)^2+k$, where $a = 2$ and $(h, k) = (3, -2)$.
$$f(x)=2(x-3)^2-2$$

Exercise Set 13.7

1. **a.** The base in the natural exponential function is *e*.

   b. The approximate value of *e* is 2.7183.

3. The domain of $\ln x$ is $x > 0$.

5. $\log_a x=\dfrac{\log_b x}{\log_b a}$

7. $\ln e^x = x$

9. The inverse of $\ln x$ is e^x.

11. *P* decreases when *t* increases for $k < 0$.

13. $\ln 62 = 4.1271$

15. $\ln 0.813 = -0.2070$

17. $\ln N = 16$
$$e^{\ln N}=e^{1.6}$$
$$N=e^{1.6}\approx 4.95$$

19. $\ln N = -2.41$
$$e^{\ln N}=e^{-2.41}$$
$$N=e^{-2.41}\approx 0.0898$$

21. $\ln N = -0.0287$
$$e^{\ln N}=e^{-0.0287}$$
$$N=e^{-0.0287}\approx 0.972$$

23. $\log_3 56=\dfrac{\log 56}{\log 3}\approx 3.6640$

25. $\log_2 21=\dfrac{\log 21}{\log 2}\approx 4.3923$

27. $\log_4 11=\dfrac{\log 11}{\log 4}\approx 1.7297$

29. $\log_5 63=\dfrac{\log 63}{\log 5}\approx 2.5743$

31. $\log_6 123=\dfrac{\log 123}{\log 6}\approx 2.6857$

33. $\log_7 51=\dfrac{\log 51}{\log 7}\approx 2.0206$

35. $\log_5 0.463=\dfrac{\log 0.463}{\log 5}\approx -0.4784$

37. $\ln x + \ln(x-1) = \ln 12$

$\ln x(x-1) = \ln 12$

$e^{\ln[x(x-1)]} = e^{\ln 12}$

$x(x-1) = 12$

$x^2 - x - 12 = 0$

$(x-4)(x+3) = 0$

$x-4=0 \qquad x+3=0$

$x=4 \qquad x=-3$

Only $x=4$ checks. $x=-3$ is an extraneous solution since $\ln x$ becomes $\ln(-3)$ which is not a real number.

39. $\ln x + \ln(x+4) = \ln 5$

$\ln x(x+4) = \ln 5$

$e^{\ln(x^2+4x)} = e^{\ln 5}$

$x^2 + 4x = 5$

$x^2 + 4x - 5 = 0$

$(x+5)(x-1) = 0$

$x+5=0$ or $x-1=0$

$x=-5 \qquad x=1$

Only $x=1$ checks. $x=-5$ is an extraneous solution since $\ln x$ becomes $\ln(-5)$ which is not a real number.

41. $\ln x = 5\ln 2 - \ln 8$

$\ln x = \ln 2^5 - \ln 8$

$\ln x = \ln \frac{32}{8}$

$\ln x = \ln 4$

$e^{\ln x} = e^{\ln 4}$

$x = 4$

$x = 4$ checks.

43. $\ln(x^2-4) - \ln(x+2) = \ln 1$

$\ln(x^2-4) - \ln(x+2) = 0$

$\ln(x^2-4) = \ln(x+2)$

$e^{\ln(x^2-4)} = e^{\ln(x+2)}$

$x^2 - 4 = x + 2$

$x^2 - x - 6 = 0$

$(x-3)(x+2) = 0$

$x-3=0$ or $x+2=0$

$x=3 \qquad x=-2$

Only $x=3$ checks. $x=-2$ is an extraneous solution since $\ln(x+2)$ becomes $\ln(-2+2) = \ln(0)$ which is not a real number.

45. $P = 120e^{(2.3)(1.6)}$

$P = 120e^{3.68}$

$P \approx 4757.5673$

47. $50 = P_0 e^{-0.5(3)}$

$50 = P_0 e^{-1.5}$

$\frac{50}{e^{-1.5}} = P_0$

$P_0 \approx 224.0845$

49. $90 = 30e^{1.4t}$

$3 = e^{1.4t}$

$\ln 3 = \ln e^{1.4t}$

$\ln 3 = 1.4t$

$t = \frac{\ln 3}{1.4}$

$t \approx 0.7847$

51. $80 = 40e^{k(3)}$

$2 = e^{3k}$

$\ln 2 = \ln e^{3k}$

$\ln 2 = 3k$

$k = \frac{\ln 2}{3}$

$k \approx 0.2310$

53. $20 = 40e^{k(2.4)}$

$0.5 = e^{2.4k}$

$\ln 0.5 = \ln e^{2.4k}$

$\ln 0.5 = 2.4k$

$k = \frac{\ln 0.5}{2.4}$

$k \approx -0.2888$

55. $A = 6000e^{-0.08(3)}$

$A = 6000e^{-0.24}$

$A \approx 4719.77$

57. $V = V_0 e^{kt}$

$\frac{V}{e^{kt}} = V_0$ or $V_0 = \frac{V}{e^{kt}}$

59.
$$P = 150e^{4t}$$
$$\frac{P}{150} = e^{4t}$$
$$\ln\frac{P}{150} = \ln e^{4t}$$
$$\ln\frac{P}{150} = 4t$$
$$\frac{\ln P - \ln 150}{4} = t \text{ or } t = \frac{\ln P - \ln 150}{4}$$

61.
$$A = A_0 e^{kt}$$
$$\frac{A}{A_0} = e^{kt}$$
$$\ln\frac{A}{A_0} = \ln e^{kt}$$
$$\ln A - \ln A_0 = kt$$
$$\frac{\ln A - \ln A_0}{t} = k \text{ or } k = \frac{\ln A - \ln A_0}{t}$$

63.
$$\ln y - \ln x = 2.3$$
$$\ln\frac{y}{x} = 2.3$$
$$e^{\ln(y/x)} = e^{2.3}$$
$$\frac{y}{x} = e^{2.3}$$
$$y = xe^{2.3}$$

65. $\ln y - \ln(x+3) = 6$
$$\ln\frac{y}{x+3} = 6$$
$$e^{\ln\frac{y}{x+3}} = e^{6}$$
$$\frac{y}{x+3} = e^{6}$$
$$y = (x+3)e^{6}$$

67. $e^x = 12.183$

Take the natural logarithm of both sides of the equation.

$\ln e^x = \ln 12.183$

$x = \ln 12.183 \approx 2.5000$

69. $P = P_0 e^{kt}$

a. $P = 5000e^{0.08(2)}$

$= 5000e^{0.16}$

≈ 5867.55

The amount will be \$5867.55.

b. If the amount in the account is to double, then $P = 2(5000) = 10{,}000$.
$$10{,}000 = 5000e^{0.08t}$$
$$2 = e^{0.08t}$$
$$\ln 2 = \ln e^{0.08t}$$
$$\ln 2 = 0.08t$$
$$\frac{\ln 2}{0.08} = t$$
$$8.66 \approx t$$

It would take about 8.66 years for the value to double.

71. $P = P_0 e^{-0.028t}$

$P = 70e^{-0.028(20)}$

$P = 70e^{-0.56}$

$P \approx 39.98$

After 20 years, about 39.98 grams remain.

73. $f(t) = 1 - e^{-0.04t}$

a. $f(t) = 1 - e^{-0.04(50)} = 1 - e^{-2} \approx 0.8647$

About 86.47% of the target market buys the drink after 50 days of advertising.

b.
$$0.75 = 1 - e^{-0.04t}$$
$$-0.25 = -e^{-0.04t}$$
$$0.25 = e^{-0.04t}$$
$$\ln 0.25 = \ln e^{-0.04t}$$
$$\ln 0.25 = -0.04t$$

b.
$$t = \frac{\ln 0.25}{-0.04}$$
$$t \approx 34.66$$

About 34.66 days of advertising are needed if 75% of the target market is to buy the soft drink.

75. $f(P) = 0.37 \ln P + 0.05$

a.
$$f(972{,}000) = 0.37\ln(972{,}000) + 0.05$$
$$\approx 5.1012311 + 0.05$$
$$\approx 5.15$$

The average walking speed in Nashville, Tennessee is 5.15 feet per second.

b.
$$f(8{,}567{,}000) = 0.37\ln(8{,}567{,}000) + 0.05$$
$$\approx -5.906 + 0.05$$
$$\approx 5.96$$

The average walking speed in New York

City is 5.96 feet per second.

c.
$$5=0.37\ln P+0.05$$
$$4.95=0.37\ln P$$
$$13.378378=\ln P$$
$$e^{13.378378}=e^{\ln P}$$
$$P=e^{13378378}$$
$$P\approx 646{,}000$$
The population is about 646,000.

77. $V(t)=24e^{0.08t},\ t=2003-1626=377$
$$V(377)=24e^{0.08(377)}$$
$$\approx 300{,}977{,}000{,}000{,}000$$
The value of Manhattan in 2003 is about $300,977,000,000,000

79. $P(t)=6.30e^{0.013t}$

a. $t=2010-2003=7$
$$P(7)=6.30e^{0.013(7)}$$
$$=6.30e^{0.091}$$
$$\approx 6.9$$
The world's population in 2010 is expected to be about 6.9 billion.

b. 2(6.30 billion) $\Rightarrow$ 12.60 billion
$$12.60=6.30e^{0.013t}$$
$$\frac{12.60}{6.30}=\frac{6.30e^{0.013t}}{6.30}$$
$$2=e^{0.013t}$$
$$\ln 2=\ln e^{0.013t}$$
$$\ln 2=0.013t$$
$$t=\frac{\ln 2}{0.013}\approx 53$$
The world's population will double in about 53 years.

81. $y=15.29+5.93\ln x$

a. $y(18)=15.29+5.93\ln(18)\approx 32.43$ in.

b. $y(30)=15.29+5.93\ln(30)\approx 35.46$ in.

83. $f(t)=v_0e^{-0.0001205t}$

a. Use $f(t)=9$ and $v_0=20$.
$$9=20e^{-0.0001205t}$$
$$0.45=e^{-0.0001205t}$$
$$\ln 0.45=-0.0001205t$$
$$\frac{\ln 0.45}{-0.0001205}=t$$
$$t\approx 6626.62$$
The bone is about 6626.62 years old.

b. Let x equal the original amount of carbon 14 then $0.5x$ equals the remaining amount.
$$0.5x=xe^{-0.0001205t}$$
$$\frac{0.5x}{x}=\frac{xe^{-0.0001205t}}{x}$$
$$0.5=e^{-0.0001205t}$$
$$\ln 0.5=\ln e^{-0.0001205t}$$
$$\ln 0.5=-0.0001205t$$
$$\frac{\ln 0.5}{-0.0001205}=t$$
$$t\approx 5752.26$$
If 50% of the carbon 14 remains, the item is about 5752.26 years old.

85. Let P_0 be the initial investment, then $P=20{,}000$, $k=0.06$, and $t=18$.
$$P=P_0e^{kt}$$
$$20{,}000=P_0e^{0.06(18)}$$
$$20{,}000=P_0e^{1.08}$$
$$\frac{20{,}000}{e^{1.08}}=P_0$$
$$6791.91\approx P_0$$
The initial investment should be $6791.91.

87. a. Strontium 90 has a higher decay rate so it will decompose more quickly.

b. $P=P_0e^{-kt}$
$$P=P_0e^{-0.023(50)}$$
$$=P_0e^{-1.15}$$
$$\approx P_0(0.3166)$$
About 31.66% of the original amount will remain.

89. Answers will vary.

91. $e^{x-4} = 12\ln(x+2)$

$y_1 = e^{x-4}$

$y_2 = 12\ln(x+2)$

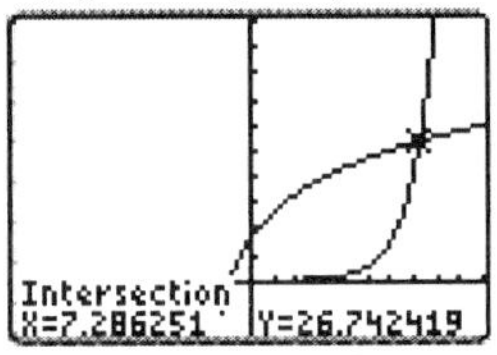

−10, 10, 1, −10, 50, 5

The intersection is approximately (7.286, 26.742). Therefore, $x \approx 7.286$.

93. $3x-6 = 2e^{0.2x} - 12$

$y_1 = 3x-6$

$y_2 = 2e^{0.2x} - 12$

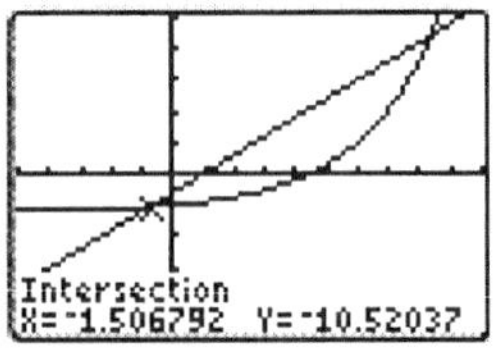

−10, 20, 2, −50, 50, 10

The intersections are approximately (–1.507, –10.520) and (16.659, 43.977). Therefore, $x \approx -1.507$ and 16.659.

95. $x = \frac{1}{k}\ln(kv_0t+1)$

$$xk = \ln(kv_0t+1)$$

$$e^{xk} = e^{\ln(kv_0t+1)}$$

$$e^{xk} = kv_0t+1$$

$$e^{xk} - 1 = kv_0t$$

$$\frac{e^{xk}-1}{kt} = v_0 \text{ or } v_0 = \frac{e^{xk}-1}{kt}$$

97. $\ln i - \ln I = \frac{-t}{RC}$

$$\ln\frac{i}{I} = \frac{-t}{RC}$$

$$e^{\ln(i/I)} = e^{-t/RC}$$

$$\frac{i}{I} = e^{-t/RC}$$

$$i = Ie^{-t/RC}$$

98. $(3xy^2+y)(4x-3xy)$

$$= 12x^2y^2 - 9x^2y^3 + 4xy - 3xy^2$$

$$= -9x^2y^3 + 12x^2y^2 - 3xy^2 + 4xy$$

99. $4x^2+bx+9 = (2x)^2+bx+(3)^2$ will be a perfect square trinomial if

$bx = \pm 2(2x)(3) \Rightarrow b = \pm 12$

100. $h(x) = \frac{x^2+4x}{x+6}$

a. $h(-3) = \frac{(-3)^2+4(-3)}{(-3)+6} = \frac{9-12}{3} = \frac{-3}{3} = -1$

b. $h\left(\frac{2}{5}\right) = \frac{\left(\frac{2}{5}\right)^2+4\left(\frac{2}{5}\right)}{\left(\frac{2}{5}\right)+6} = \frac{\frac{4}{25}+\frac{40}{25}}{\frac{2}{5}+\frac{30}{5}}$

$$= \frac{\frac{44}{25}}{\frac{32}{5}} = \frac{44}{25}\cdot\frac{5}{32} = \frac{11}{40} \text{ or } 0.275$$

101. Let x be the number of adult tickets sold and y be the number of children's tickets sold. The following system describes the situation.

$$x + y = 650$$

$$15x + 11y = 8790$$

Solve by elimination.

$-11(x + y = 650) \Rightarrow -11x - 11y = -7150$

$15x + 11y = 8790 \Rightarrow 15x + 11y = 8790$

$4x = 1640$

$x = 410$

Substitute $x = 410$ into the first equation to find y. $410 + y = 650 \Rightarrow y = 240$

410 adult tickets and 240 children's tickets must be sold.

102. $\sqrt[3]{x}\left(\sqrt[3]{x^2}+\sqrt[3]{x^5}\right) = \sqrt[3]{x}\cdot\sqrt[3]{x^2}+\sqrt[3]{x}\cdot\sqrt[3]{x^5}$

$$= \sqrt[3]{x^3}+\sqrt[3]{x^6}$$

$$= x + x^2$$

Review Exercises

1. $(f \circ g)(x) = (2x-5)^2 - 3(2x-5) + 4$

$$= 4x^2 - 20x + 25 - 6x + 15 + 4$$

$$= 4x^2 - 26x + 44$$

2. $(f \circ g)(x) = 4x^2 - 26x + 44$
$(f \circ g)(2) = 4(2)^2 - 26(2) + 44$
$= 16 - 52 + 44$
$= 8$

3. $(g \circ f)(x) = 2(x^2 - 3x + 4) - 5$
$= 2x^2 - 6x + 8 - 5$
$= 2x^2 - 6x + 3$

4. $(g \circ f)(x) = 2x^2 - 6x + 3$
$(g \circ f)(-3) = 2(-3)^2 - 6(-3) + 3$
$= 18 + 18 + 3$
$= 39$

5. $(f \circ g)(x) = 6\sqrt{x-3} + 1,\ x \geq 3$

6. $(g \circ f)(x) = \sqrt{(6x+1) - 3}$
$= \sqrt{6x - 2},\ x \geq \frac{1}{3}$

7. This function is one-to-one since it passes both the vertical line test and the horizontal line test.

8. The function is not one-to-one since the graph does not pass the horizontal line test.

9. Yes, the ordered pairs represent a one-to-one function. For each value of x, there is a unique value for y and each y-value has a unique x-value.

10. No, the ordered pairs do not represent a one-to-one function since the pairs (0, –2) and (3, –2) have different x-values but the same y-value.

11. Yes, $y = \sqrt{x+3},\ x \geq -3$, is a one-to-one function since it passes both the vertical line test and the horizontal line test.

12. No, $y = x^2 - 4$ is a parabola with vertex at (0, –4). It is not a one-to-one function since it does not pass the horizontal line test. Horizontal lines above $y = -4$ intersect the graph in two points.

13. $f(x)$: Domain: {–4, 0, 5, 6}
Range: {–3, 2, 3, 7}
$f^{-1}(x)$: Domain: {–3, 2, 3, 7}
Range: {–4, 0, 5, 6}

14. $f(x)$: Domain: $\{x \mid x \geq 0\}$
Range: $\{y \mid y \geq 2\}$
$f^{-1}(x)$: Domain: $\{x \mid x \geq 2\}$
Range: $\{y \mid y \geq 0\}$

15. $y = f(x) = 4x - 2$
$x = 4y - 2$
$x + 2 = 4y$
$\frac{x+2}{4} = y$
$f^{-1}(x) = \frac{x+2}{4}$

x	$f(x)$
0	2
$\frac{1}{2}$	0

x	$f^{-1}(x)$
0	$\frac{1}{2}$
–2	0

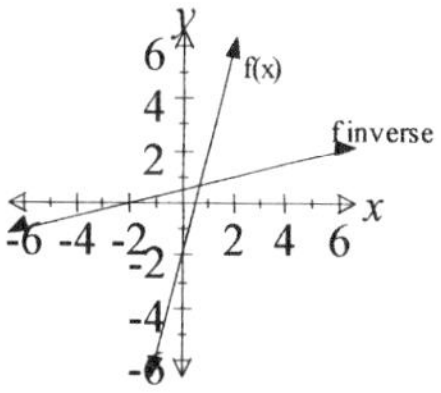

16. $y = f(x) = \sqrt[3]{x-1} = (x-1)^{1/3}$
$x = (y-1)^{1/3}$
$x^3 = [(y-1)^{1/3}]^3$
$x^3 = y - 1$
$x^3 + 1 = y$
$f^{-1}(x) = x^3 + 1$

x	$f(x)$
–7	–2
0	–1
1	0
2	1
9	2

x	$f^{-1}(x)$
–2	–7
–1	0
0	1
1	2
2	9

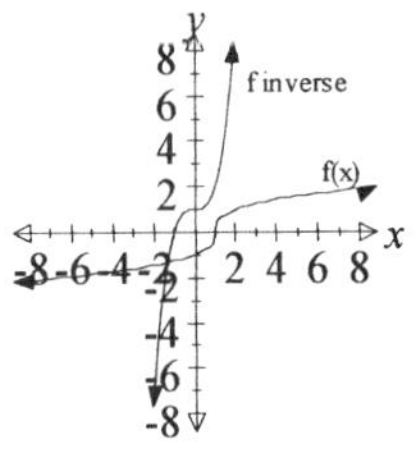

17. $f(x) = 36x \Rightarrow y = 36x \Rightarrow x = 36y \Rightarrow y = \frac{x}{36}$

$f^{-1}(x) = \frac{x}{36}$

$f^{-1}(x)$ represents yards and x represents inches.

18. $f(x) = 4x \Rightarrow y = 4x \Rightarrow x = 4y \Rightarrow y = \frac{x}{4}$

$f^{-1}(x) = \frac{x}{4}$

$f^{-1}(x)$ represents gallons and x represents quarts.

19. $y = 2^x$

x	−2	−1	0	1	2	3
y	$\frac{1}{4}$	$\frac{1}{2}$	1	2	4	8

Domain: R
Range: $\{y|y > 0\}$

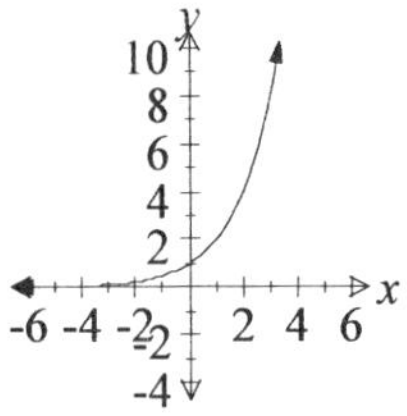

20. $y = \left(\frac{1}{2}\right)^x$

x	−2	−1	0	1	2
y	4	2	1	$\frac{1}{2}$	$\frac{1}{4}$

Domain: R
Range: $\{y|y > 0\}$

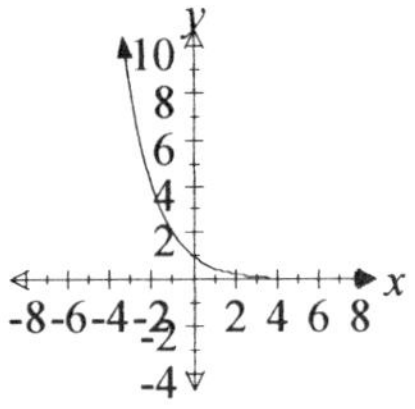

21. $f(t) = 7.02e^{0.365t}$

a. $t = 2003 - 1999 = 4$

$f(4) = 7.02e^{0.365(4)}$

≈ 30.23

The worldwide shipment in 2003 is about 30.23 million.

b. $t = 2005 - 1999 = 6$

$f(4) = 7.02e^{0.365(6)}$

≈ 62.73

The worldwide shipment in 2005 will be about 62.73 million.

c. $t = 2007 - 1999 = 8$

$f(8) = 7.02e^{0.365(8)}$

≈ 130.16

The worldwide shipment in 2007 will be about 130.16 million.

22. $7^2 = 49$

$\log_7 49 = 2$

23. $81^{1/4} = 3$

$\log_{81} 3 = \frac{1}{4}$

24. $5^{-2} = \frac{1}{25}$

$\log_5 \frac{1}{25} = -2$

25. $\log_2 16 = 4$

$2^4 = 16$

26. $\log_{1/3}\frac{1}{9}=2$

$\left(\frac{1}{3}\right)^2=\frac{1}{9}$

27. $\log_6\frac{1}{36}=-2$

$6^{-2}=\frac{1}{36}$

28. $3=\log_4 x$

$x=4^3$

$x=64$

29. $3=\log_a 8$

$a^3=8$

$a^3=2^3$

$a=2$

30. $-3=\log_{1/4} x$

$x=\left(\frac{1}{4}\right)^{-3}$

$x=\frac{1}{\left(\frac{1}{4}\right)^3}$

$x=\frac{1}{\frac{1}{64}}$

$x=64$

31. $y=\log_3 x$

$x=3^y$

x	$\frac{1}{9}$	$\frac{1}{3}$	1	3	9	27
y	−2	−1	0	1	2	3

Domain: $\{x|x>0\}$

Range: R

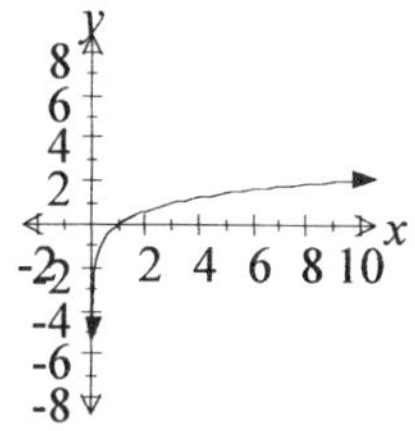

32. $y=\log_{1/2} x$

$x=\left(\frac{1}{2}\right)^y$

x	4	2	1	$\frac{1}{2}$	$\frac{1}{4}$
y	−2	−1	0	1	2

Domain: $\{x|x>0\}$

Range: R

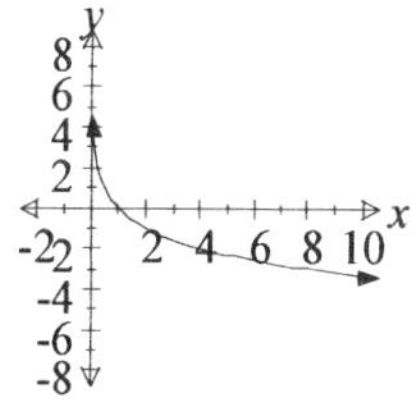

33. $\log_5 17^3=3\log_5 17$

34. $\log_3\sqrt{x-5}=\log_3(x-5)^{1/2}=\frac{1}{2}\log_3(x-5)$

35. $\log\frac{6(a+1)}{b}=\log 6+\log(a+1)-\log b$

36. $\log\frac{x^4}{9(2x+3)^5}=\log x^4-\log 9(2x+3)^5$

$=4\log x-[\log 9+\log(2x+3)^5]$

$=4\log x-[\log 9+5\log(2x+3)]$

$=4\log x-\log 9-5\log(2x+3)$

37. $2\log x-3\log(x+1)$

$=\log x^2-\log(x+1)^3$

$=\log\frac{x^2}{(x+1)^3}$

38. $3(\log 2+\log x)-\log y$

$=3(\log 2x)-\log y$

$=\log(2x)^3-\log y$

$=\log\frac{(2x)^3}{y}$

39. $\frac{1}{2}[\ln x-\ln(x+2)]-\ln 2$

$=\frac{1}{2}\left(\ln\frac{x}{x+2}\right)-\ln 2$

$=\ln\left(\frac{x}{x+2}\right)^{1/2}-\ln 2$

$=\ln\left(\frac{\sqrt{\frac{x}{x+2}}}{2}\right)$

40. $3\ln x + \frac{1}{2}\ln(x+1) - 3\ln(x+4)$
$= \ln x^3 + \ln(x+1)^{1/2} - \ln(x+4)^3$
$= \ln\frac{x^3\sqrt{x+1}}{(x+4)^3}$

41. $8^{\log_8 9} = 9$

42. $\log_4 4^5 = 5$

43. $7\log_9 81 = 7\log_9 9^2$
$= 7 \cdot 2$
$= 14$

44. $4^{\log_8 \sqrt{8}} = 4^{\log_8 8^{1/2}}$
$= 4^{1/2}$
$= \sqrt{4}$
$= 2$

45. log 763 = 2.8825

46. log 0.0281 = –3..5720

47. antilog 3.159 = 1440

48. antilog(–2.645) = 0.00226

49. $\log N = 2.3304$
$N = \text{antilog } 2.3304$
$N = 214$

50. $\log N = -1.2262$
$N = \text{antilog } (-1.2262)$
$N = 0.0594$

51. $\log 10^5 = 5$

52. $10^{\log 4} = 4$

53. $9\log 10^{3.2} = 9(3.2) = 28.8$

54. $2\left(10^{\log 4.7}\right) = 2(4.7) = 9.4$

55. $125 = 5^x$
$5^3 = 5^x$
$3 = x$

56. $81^x = \frac{1}{9}$
$(9^2)^x = 9^{-1}$
$9^{2x} = 9^{-1}$
$2x = -1$
$x = -\frac{1}{2}$

57. $2^{3x-1} = 32$
$2^{3x-1} = 2^5$
$3x - 1 = 5$
$3x = 6$
$x = 2$

58. $27^x = 3^{2x+5}$
$(3^3)^x = 3^{2x+5}$
$3^{3x} = 3^{2x+5}$
$3x = 2x + 5$
$x = 5$

59. $7^x = 89$
$\log 7^x = \log 89$
$x\log 7 = \log 89$
$x = \frac{\log 89}{\log 7}$
$x \approx 2.307$

60. $2.6^x = 714$
$\log 2.6^x = \log 714$
$x\log 2.6 = \log 714$
$x = \frac{\log 714}{\log 2.6}$
$x \approx 6.877$

61. $12.5^{x+1} = 381$
$\log 12.5^{x+1} = \log 381$
$(x+1)\log 12.5 = \log 381$
$x + 1 = \frac{\log 381}{\log 12.5}$
$x = \frac{\log 381}{\log 12.5} - 1$
$x \approx 1.353$

62.

$$3^{x+2} = 8^x$$
$$\log 3^{x+2} = \log 8^x$$
$$(x+2)\log 3 = x\log 8$$
$$x\log 3 + 2\log 3 = x\log 8$$
$$2\log 3 = x\log 8 - x\log 3$$
$$2\log 3 = x(\log 8 - \log 3)$$
$$\frac{2\log 3}{\log 8 - \log 3} = x$$
$$2.240 \approx x$$

63. $\log_3(5x+1) = 4$

$$3^4 = 5x+1$$
$$81 = 5x+1$$
$$80 = 5x$$
$$x = 16$$

64. $\log x + \log(4x-7) = \log 2$

$$\log(x(4x-7)) = \log 2$$
$$x(4x-7) = 2$$
$$4x^2 - 7x - 2 = 0$$
$$(4x+1)(x-2) = 0$$
$$4x+1 = 0 \quad \text{or} \quad x-2 = 0$$
$$x = -\frac{1}{4} \qquad x = 2$$

Only $x = 2$ checks. $x = -\frac{1}{4}$ is an extraneous solution .

65. $\log_3 x + \log_3(2x+1) = 1$

$$\log_3 x(2x+1) = 1$$
$$3^1 = x(2x+1)$$

is an e

$$0 = 2x^2 + x - 3$$
$$0 = (2x+3)(x-1)$$
$$2x+3 = 0 \quad \text{or} \quad x-1 = 0$$
$$x = -\frac{3}{2} \qquad x = 1$$

Only $x = 1$ checks. $x = -\frac{3}{2}$ is an extraneous solution since $\log_3 x$ becomes $\log_3\left(-\frac{3}{2}\right)$ which is not a real number.

66. $\ln(x+1) - \ln(x-2) = \ln 4$

$$\ln\frac{x+1}{x-2} = \ln 4$$
$$\frac{x+1}{x-2} = 4$$
$$x+1 = 4(x-2)$$
$$x+1 = 4x-8$$
$$1 = 3x-8$$
$$9 = 3x$$
$$x = 3$$

67.

$$40 = 20e^{0.6t}$$
$$2 = e^{0.6t}$$
$$\ln 2 = \ln e^{0.6t}$$
$$\ln 2 = 0.6t$$
$$\frac{\ln 2}{0.6} = t$$
$$1.155 \approx t$$

68.

$$100 = A_0 e^{-0.42(3)}$$
$$100 = A_0 e^{-1.26}$$
$$\frac{100}{e^{-1.26}} = A_0$$
$$352.542 \approx A_0$$

69.

$$A = A_0 e^{kt}$$
$$\frac{A}{A_0} = e^{kt}$$
$$\ln\frac{A}{A_0} = \ln e^{kt}$$
$$\ln\frac{A}{A_0} = kt$$
$$\frac{\ln\frac{A}{A_0}}{k} = t$$
$$\frac{\ln A - \ln A_0}{k} = t \text{ or } t = \frac{\ln A - \ln A_0}{k}$$

70.

$$150 = 600e^{kt}$$
$$\frac{150}{600} = e^{kt}$$
$$0.25 = e^{kt}$$
$$\ln 0.25 = \ln e^{kt}$$
$$\ln 0.25 = kt$$
$$\frac{\ln 0.25}{t} = k \text{ or } k = \frac{\ln 0.25}{t}$$

71. $\ln y - \ln x = 4$

$\ln \frac{y}{x} = 4$

$e^{\ln \frac{y}{x}} = e^4$

$\frac{y}{x} = e^4$

$y = xe^4$

72. $\ln(y+1) - \ln(x+5) = \ln 3$

$\ln \frac{y+1}{x+5} = \ln 3$

$\frac{y+1}{x+5} = 3$

$y+1 = 3(x+5)$

$y = 3(x+5) - 1$

$y = 3x + 15 - 1$

$y = 3x + 14$

73. $\log_2 196 = \frac{\log 196}{\log 2} \approx 7.6147$

74. $\log_3 74 = \frac{\log 74}{\log 3} \approx 3.9177$

75. $A = P(1+r)^n$

$= 12{,}000(1+0.1)^8$

$= 12{,}000(1.1)^8$

$\approx 25{,}723.07$

The amount is \$25,723.07.

76. $P = P_0 e^{kt}$

$P_0 = 10{,}000$, $k = 0.04$, and $P = 20{,}000$

$20{,}000 = 10{,}000 e^{(0.04)t}$

$2 = e^{0.04t}$

$\ln 2 = 0.04t$

$t = \frac{\ln 2}{0.04}$

$t \approx 17.3$

It will take about 17.3 years for the \$10,000 to double.

77. $N(t) = 2000(2)^{0.05t}$

a. Let $N(t) = 50{,}000$.

$50{,}000 = 2000(2)^{0.05t}$

$\frac{50{,}000}{2000} = 2^{0.05t}$

$25 = 2^{0.05t}$

$\log 25 = \log 2^{0.05t}$

$\log 25 = 0.05t \log 2$

$\frac{\log 25}{0.05 \log 2} = t$

$92.88 \approx t$

The time is 92.88 minutes.

b. Let $N(t) = 120{,}000$.

$120{,}000 = 2000(2)^{0.05t}$

$\frac{120{,}000}{2000} = 2^{0.05t}$

$60 = 2^{0.05t}$

$\log 60 = \log 2^{0.05t}$

$\log 60 = 0.05t \log 2$

$\frac{\log 60}{0.05 \log 2} = t$

118.14 Ut

The time is 118.14 minutes.

78. $P = 14.7e^{-0.00004x}$

$P = 14.7e^{-0.00004(8842)}$

$P = 14.7e^{-0.35368}$

$P \approx 14.7(0.7021)$

$P \approx 10.32$

The atmospheric pressure is 10.32 pounds per square inch at 8,842 feet above sea level.

79. $A(n) = 72 - 18\log(n+1)$

a. $A(0) = 72 - 18\log(0+1)$

$= 72 - 18\log(1)$

$= 72 - 18(0)$

$= 72$

The original class average was 72.

b. $A(2) = 72 - 18\log(2+1)$

$= 72 - 18\log(3)$

$\approx 72 - 8.6$

$= 63.4$

After 2 months, the class average was 63.4.

c. Let $A(n) = 59.4$.

$$59.4 = 72 - 18\log(n+1)$$
$$-12.6 = -18\log(n+1)$$
$$\frac{-12.6}{-18} = \log(n+1)$$
$$0.7 = \log(n+1)$$
$$10^{0.7} = 10^{\log(n+1)}$$
$$5.01 \approx n+1$$
$$4.01 \approx n$$

It takes about 4 months.

Practice Test

1. a. Yes, {(4, 2), (–3, 8), (–1, 3), (5, 7)} is one-to-one.

   b. {(2, 4), (8, –3), (3, –1), (7, 5)} is the inverse function.

2. a.
$$(f \circ g)(x) = f[g(x)] = f(x+2) = (x+2)^2 - 3 = x^2 + 4x + 4 - 3 = x^2 + 4x + 1$$

   b.
$$(f \circ g)(5) = 5^2 + 4(5) + 1 = 25 + 20 + 1 = 46$$

3. a.
$$(g \circ f)(x) = g[f(x)] = g(x^2+7) = \sqrt{x^2+7-5} = \sqrt{x^2+2}$$

   b.
$$(g \circ f)(4) = \sqrt{4^2+2} = \sqrt{16+2} = \sqrt{18} = 3\sqrt{2}$$

4. a.
$$y = f(x) = -3x - 5$$
$$x = -3y - 5$$
$$x + 5 = -3y$$
$$\frac{x+5}{-3} = y$$
$$-\frac{1}{3}(x+5) = y$$
$$f^{-1}(x) = -\frac{1}{3}(x+5)$$

   b.

x	$f(x)$
0	–5
$-\frac{5}{3}$	0

x	$f^{-1}(x)$
0	$-\frac{5}{3}$
–5	0

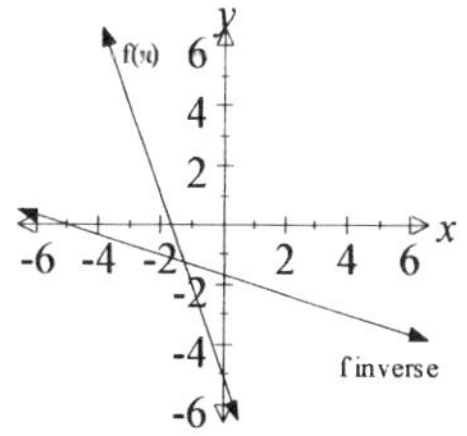

5. a.
$$y = f(x) = \sqrt{x-1},\ x \ge 1$$
$$x = (y-1)^{1/2}$$
$$x^2 = [(y-1)^{1/2}]^2$$
$$x^2 = y - 1$$
$$x^2 + 1 = y$$
$$f^{-1}(x) = x^2 + 1,\ x \ge 0$$

   b.

x	$f(x)$
1	0
2	1
5	2

x	$f^{-1}(x)$
0	1
1	2
2	5

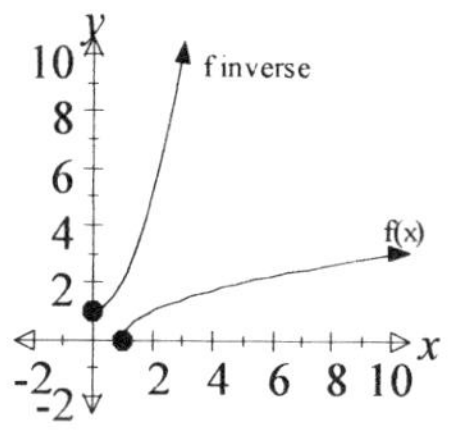

6. The domain of $y = \log_4(x)$ is $x > 0$.

7. $\log_3 \frac{1}{81} = \log_3 3^{-4} = -4$

8. $y = 3^x$

x	−2	−1	0	2	3
y	$\frac{1}{9}$	$\frac{1}{3}$	1	9	27

Domain: R

Range: $\{y | y > 0\}$

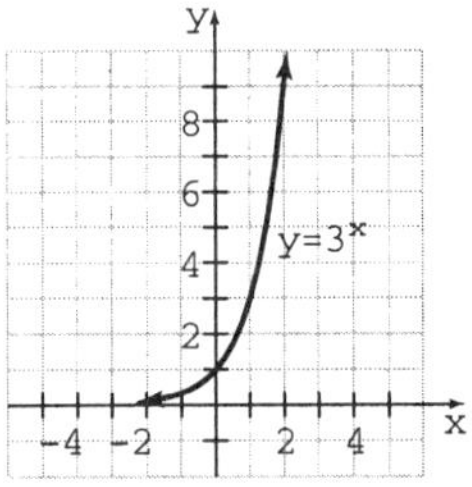

9. $y = \log_2 x$

$x = 2^y$

x	$\frac{1}{4}$	$\frac{1}{2}$	1	2	4
y	−2	−1	0	1	2

Domain: $\{x | x > 0\}$

Range: R

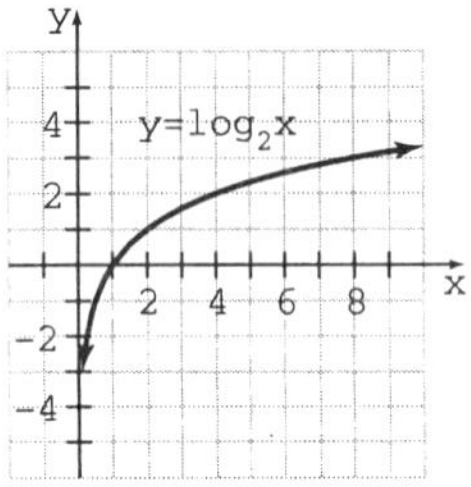

10. $2^{-5} = \frac{1}{32}$

$\log_2 \frac{1}{32} = -5$

11. $\log_5 625 = 4$

$5^4 = 625$

12. $4 = \log_2(x+1)$

$2^4 = x+1$

$16 = x+1$

$15 = x$

13. $y = \log_{64} 16$

$64^y = 16$

$4^{3y} = 4^2$

$3y = 2$

$y = \frac{2}{3}$

14. $\log_2 \frac{x^3(x-4)}{x+2}$

$= \log_2 x^3(x-4) - \log_2(x+2)$

$= \log_2 x^3 + \log_2(x-4) - \log_2(x+2)$

$= 3\log_2 x + \log_2(x-4) - \log_2(x+2)$

15. $5\log_6(x-4) + 2\log_6(x+3) - \frac{1}{2}\log_6 x$

$= \log_6(x-4)^5 + \log_6(x+3)^2 - \log_6 x^{1/2}$

$= \log_6 \frac{(x-4)^5(x+3)^2}{\sqrt{x}}$

16. $8\log_9\sqrt{9}$

$= 8\log_9 9^{1/2}$

$= 8 \cdot \frac{1}{2}$

$= 4$

17. a. $\log 4620 \approx 3.6646$

b. $\ln 0.0692 \approx -2.6708$

18. $3^x = 519$

$\log 3^x = \log 519$

$x\log 3 = \log 519$

$x = \frac{\log 519}{\log 3}$

$x \approx 5.69$

19. $\log 4x = \log(x+3) + \log 2$

$\log 4x = \log 2(x+3)$

$4x = 2(x+3)$

$4x = 2x+6$

$2x = 6$

$x = 3$

20. $\log(x+5)-\log(x-2)=\log 6$

$$\log\frac{x+5}{x-2}=\log 6$$
$$\frac{x+5}{x-2}=6$$
$$x+5=6x-12$$
$$17=5x$$
$$\frac{17}{5}=x$$

21.
$$\ln N=3.52$$
$$e^{3.52}=N$$
$$33.7844\approx N$$

22. $\log_6 40=\frac{\log 40}{\log 6}\approx 2.0588$

23.
$$200=500e^{-0.03t}$$
$$\frac{200}{500}=e^{-0.03t}$$
$$\ln\frac{200}{500}=-0.03t$$
$$\frac{\ln\frac{200}{500}}{-0.03}=t$$
$$t\approx 30.5430$$

24. $A=p\left(1+\frac{r}{n}\right)^{nt}$

Use $p=3500$, $r=0.06$ and $n=4$

$t=10$

$$A=3500\left(1+\frac{0.06}{4}\right)^{4\cdot 10}$$
$$=3500(1.015)^{40}$$
$$\approx 6349.06$$

The amount in the account is \$6349.06.

25. $v=v_0e^{-0.0001205t}$

Use $v=40$, and $v_0=60$.

$$40=60e^{-0.0001205t}$$
$$\frac{40}{60}=e^{-0.0001205t}$$
$$\frac{2}{3}=e^{-0.0001205t}$$
$$\ln\frac{2}{3}=\ln e^{-0.0001205t}$$
$$\ln\frac{2}{3}=-0.0001205t$$
$$\frac{\ln\frac{2}{3}}{-0.0001205}=t$$
$$3364.86\approx t$$

The fossil is approximately 3364.86 years old.

Cumulative Review Test

1.
$$\frac{6-|-18|\div 3^2-6}{4-|-8|\div 2^2}$$
$$=\frac{6-18\div 9-6}{4-8\div 4}$$
$$=\frac{6-2-6}{4-2}$$
$$=\frac{4-6}{2}$$
$$=\frac{-2}{2}$$
$$=-1$$

2.
$$4-(6x+6)=-(-2x+10)$$
$$4-6x-6=2x-10$$
$$-6x-2=2x-10$$
$$-6x+6x-2=2x+6x-10$$
$$-2=8x-10$$
$$-2+10=8x-10+10$$
$$8=8x$$
$$1=x$$

3.
$$2x-3y=5$$
$$-3y=5-2x$$
$$y=\frac{5-2x}{-3}\text{ or }y=\frac{2x-5}{3}$$

4. Let r be the tax rate. The tax rate is 7.5%.

$$22+22r=23.76$$
$$22r=1.76$$
$$r=\frac{1.76}{22}$$
$$\approx 0.08$$

The tax rate is 8%.

5. **a.** Let t = Kendra's jogging time.

Person	Rate	Time	Distance
Jason	4	$t+\frac{1}{2}$	$4(t+\frac{1}{2})$
Kendra	5	t	$5t$

$$4\left(t+\frac{1}{2}\right)=5t$$
$$4t+2=5t$$
$$2=t$$

They will meet 2 hours after Kendra starts jogging.

b. They will meet $5t = 5(2) = 10$ miles from the starting point.

6. Two points on the line are (0, 3) and (−2,−1).

$$m=\frac{y_2-y_1}{x_2-x_1}=\frac{-1-3}{-2-0}=\frac{-4}{-2}=2$$

Use point-slope form with $(0,3)$ and $m = 2$.

$$y-3=2(x-0)$$
$$y-3=2x$$
$$y=2x+3$$

7.
$$\left(\frac{3x^4y^{-3}}{6xy^4z^2}\right)^{-3}=\left(\frac{x^{4-1}y^{-3-4}}{2z^2}\right)^{-3}$$
$$=\left(\frac{x^3y^{-7}}{2z^2}\right)^{-3}$$
$$=\left(\frac{2z^2}{x^3y^{-7}}\right)^{3}$$
$$=\frac{2^3z^{2\cdot 3}}{x^{3\cdot 3}y^{-7\cdot 3}}$$
$$=\frac{8z^6}{x^9y^{-21}}$$
$$=\frac{8y^{21}z^6}{x^9}$$

8. $\dfrac{x^3+3x^2+5x+4}{x+1}$

Using synthetic division:

−1	1	3	5	4
		−1	−2	−3
	1	2	3	1

$$\frac{x^3+3x^2+5x+4}{x+1}=x^2+2x+3+\frac{1}{x+1}$$

9. $12x^2-5xy-3y^2=(4x-3y)(3x+y)$

10.
$$x^2-2xy+y^2-25$$
$$\left(x^2-2xy+y^2\right)-25$$
$$(x-y)^2-25$$
$$(x-y+5)(x-y-5)$$

11.
$$\frac{x+1}{x+2}+\frac{x-2}{x-3}=\frac{x^2-4}{x^2-x-6}$$
$$\frac{x+1}{x+2}+\frac{x-2}{x-3}=\frac{x^2-4}{(x+2)(x-3)}$$
$$(x+2)(x-3)\left(\frac{x+1}{x+2}+\frac{x-2}{x-3}\right)=(x+2)(x-3)\left(\frac{x^2-4}{(x+2)(x-3)}\right)$$
$$(x+1)(x-3)+(x-2)(x+2)=x^2-4$$
$$x^2-2x-3+x^2-4=x^2-4$$
$$2x^2-2x-7=x^2-4$$
$$x^2-2x-3=0$$
$$(x-3)(x+1)=0$$
$$x-3=0 \quad x+1=0$$
$$x=3 \quad x=-1$$

$x=-1$ because $x\neq 3$

12. The equation of variation is

$L=\dfrac{k}{P^2}$, with $P = 4$ and $k = 100$.

$$L=\frac{100}{4^2}=\frac{100}{16}=6.25$$

13. $h(x)=\dfrac{x^2+4x}{x+6}$

$$h(-3)=\frac{(-3)^2+4(-3)}{(-3)+6}$$
$$=\frac{9-12}{3}$$
$$=\frac{-3}{3} \text{ or } -1$$

14. $0.4x+0.6y=3.2 \quad (1)$

$\underline{1.4x-0.3y\ =1.6} \quad (2)$

Multiply (2) by 2, then add.

$0.4x+0.6y=3.2$

$\underline{2.8x \quad 0.6y=3.2}$

$3.2x \qquad =6.4$

$x=2$

Now substitute 2 for x in eq. (1) and solve for y.

$0.4(2)+0.6y=3.2$

$0.8+0.6y=3.2$

$0.6y=2.4$

$y=4$

The solution is (2, 4).

15. $\left[\begin{array}{cc|c} 1 & 1 & 6 \\ -2 & 1 & 3 \end{array}\right] \quad 2R_1+R_2=R_2$

$\left[\begin{array}{cc|c} 1 & 1 & 6 \\ 0 & 3 & 15 \end{array}\right] \quad \frac{1}{3}R_2$

$\left[\begin{array}{cc|c} 1 & 1 & 6 \\ 0 & 1 & 5 \end{array}\right] \quad -R_2+R_1=R_2$

$\left[\begin{array}{cc|c} 1 & 0 & 1 \\ 0 & 1 & 5 \end{array}\right]$

The solution is (1, 5).

16. $\begin{vmatrix} 3 & 0 & -1 \\ 2 & 5 & 3 \\ -1 & 4 & 6 \end{vmatrix} = 3\begin{vmatrix} 5 & 3 \\ 4 & 6 \end{vmatrix} - 2\begin{vmatrix} 0 & -1 \\ 4 & 6 \end{vmatrix} + (-1)\begin{vmatrix} 0 & -1 \\ 5 & 3 \end{vmatrix}$

$$=3(30-12)-2(0-(-4))-1(0-(-5))$$
$$=3(18)-2(4)-1(5)$$
$$=54-8-5$$
$$=46-5$$
$$=41$$

17. $y \le \frac{1}{3}x+6$

Plot a dashed line at $y \le \frac{1}{3}x+6$.

Use the point (0, 0) as the check point.

$y \le \frac{1}{3}x+6$

$0 \le \frac{1}{3}(0)+6$

$0 \le 6$ True

Therefore, shade the half-plane containing (0, 0)

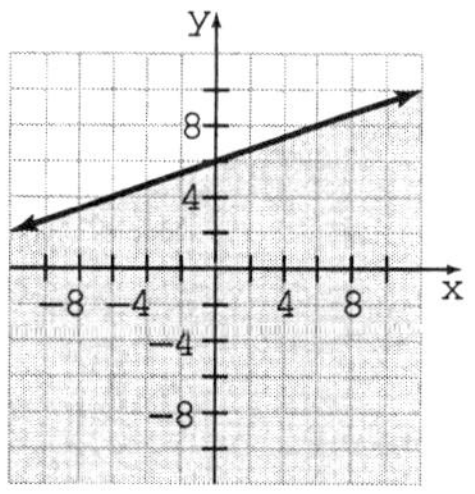

18. $g(x)=x^2-4x-5$

a. $g(x)=x^2-4x-5$

$g(x)=(x^2-4x+4)-5-4$

$g(x)=(x-2)^2-9$

b.

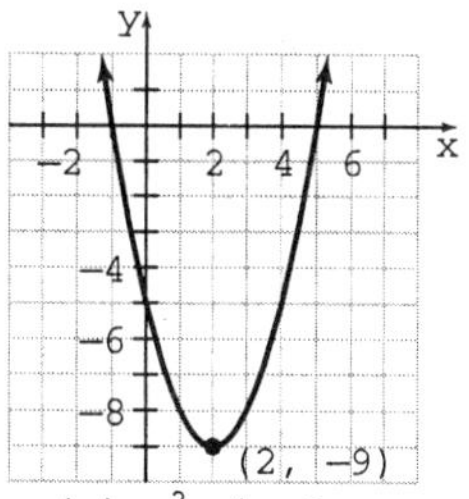

$g(x)=x^2-4x-5$

19. $f(x) = x^3 - 6x^2 + 5x$

$$x^3 - 6x^2 + 5x = 0$$
$$x(x^2 - 6x + 5) = 0$$
$$x(x-5)(x-1) = 0$$
$$x(x-5)(x-1) \geq 0$$

Interval *A*

Test Value, –2

$$x^3 - 6x^2 + 5x \geq 0$$
$$(-2)^3 - 6(-2)^2 + 5(-2) \geq 0$$
$$-8 - 6(4) - 10 \geq 0$$
$$-8 - 24 - 10 \geq 0$$
$$-42 \geq 0 \text{ False}$$

Interval *B*

Test Value, $\frac{1}{2}$

$$x^3 - 6x^2 + 5x \geq 0$$
$$\left(\frac{1}{2}\right)^3 - 6\left(\frac{1}{2}\right)^2 + 5\left(\frac{1}{2}\right) \geq 0$$
$$\frac{1}{8} - 6\left(\frac{1}{4}\right) + \frac{5}{2} \geq 0$$
$$\frac{1}{8} - \frac{6}{4} + \frac{5}{2} \geq 0$$
$$\frac{1}{8} - \frac{12}{8} + \frac{20}{8} \geq 0$$
$$\frac{9}{8} \geq 0 \text{ True}$$

Interval *C*

Test Value, 3

$$x^3 - 6x^2 + 5x \geq 0$$
$$(3)^3 - 6(3)^2 + 5(3) \geq 0$$
$$27 - 6(9) + 15 \geq 0$$
$$27 - 54 + 15 \geq 0$$
$$-12 \geq 0 \text{ False}$$

Interval *D*

Test Value, 6

$$x^3 - 6x^2 + 5x \geq 0$$
$$(6)^3 - 6(6)^2 + 5(6) \geq 0$$
$$216 - 6(36) + 30 \geq 0$$
$$216 - 216 + 30 \geq 0$$
$$30 \geq 0 \text{ True}$$

Thus, the solution is $[0, 1] \cup [5, \infty)$

A	*B*	*C*	*D*
False	True	False	True

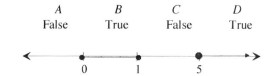

20. a. $P = P_0 e^{-0.028t}$

$= 600e^{-0.028(60)}$

$= 600e^{-1.68} = 600(0.1863739) = 111.82$

There will be 111.82 grams left after 60 years.

b. $P = P_0 e^{-0.028t}$

$300 = 600e^{-0.028t}$

$0.5 = e^{-0.028t}$

$\ln 0.5 = \ln e^{-0.028t}$

$-0.6931471 = -0.028t$

$24.755256 = t$

The half-life of strontium 90 is about 24.8 years.

Chapter 14

Exercise Set 14.1

1.

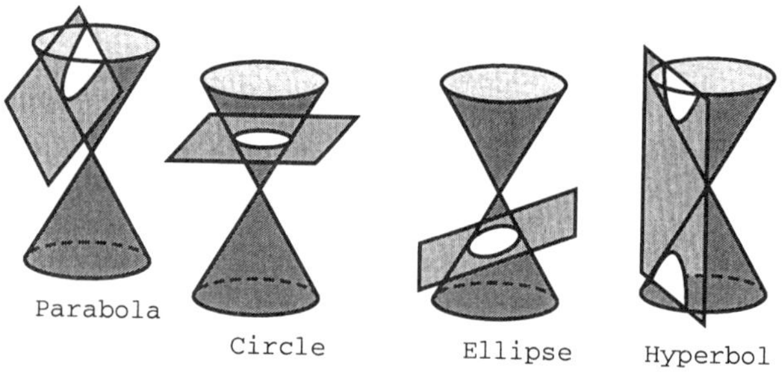

3. Yes, any parabola in the form $y = a(x - h)^2 + k$ is a function because each value of x corresponds to only one value of y. The domain is R, the set of all real numbers. Since the vertex is at (h, k) and $a > 0$, the range is $\{y | y \geq k\}$.

5. The graphs have the same vertex, (3, 4). The first graph opens upward, and the second one opens downward.

7. The distance is always a positive number because both distances are squared and we use the principal square root.

9. A circle is the set of all points in a plane that are the same distance from a fixed point.

11. No, the coefficients of the y^2- term and the x^2-term must both be the same.

13. No, the coefficients of the y^2- term and the x^2-term must both be the same.

15. No, equations of parabolas do not include both x^2 - and y^2- terms.

17. $y = (x-2)^2 + 3$

This is a parabola in the form $y = a(x - h)^2 + k$ with $a = 1$, $h = 2$ and $k = 3$. Since $a > 0$, the parabola opens upward. The vertex is (2, 3). The y-intercept is (0, 7). There are no x-intercepts.

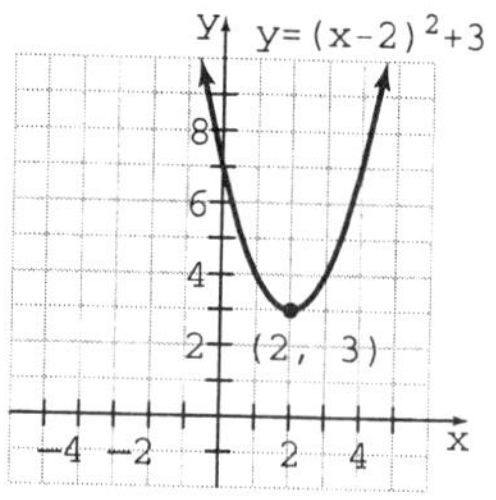

19. $y = (x+3)^2 + 2$

This is a parabola in the form $y = a(x - h)^2 + k$ with $a = 1$, $h = -3$ and $k = 2$. Since $a > 0$, the parabola opens upward. The vertex is (–3, 2). The y-intercept is (0, 11). There are no x-intercepts.

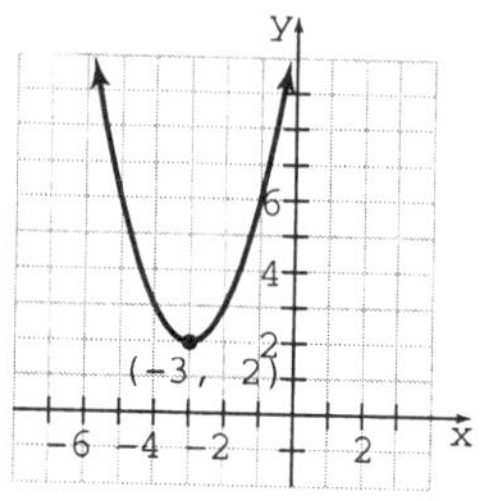

21. $y = (x-2)^2 - 1$

This is a parabola in the form $y = a(x - h)^2 + k$ with $a = 1$, $h = 2$ and $k = -1$. Since $a > 0$, the parabola opens upward. The vertex is (2, –1). The y-intercept is (0, 3). The x-intercepts are (1, 0) and (3, 0).

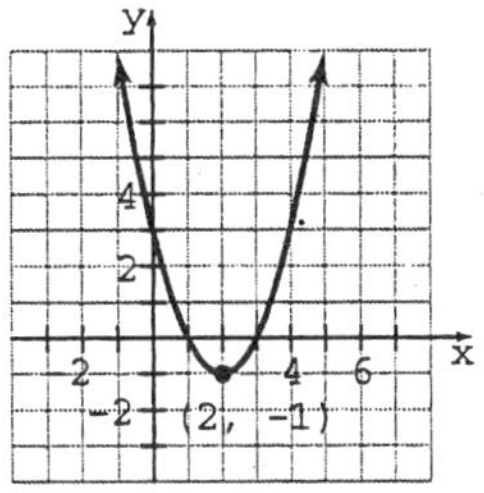

23. $y = -(x-1)^2 + 1$

This is a parabola in the form $y = a(x-h)^2 + k$ with $a = -1$, $h = 1$ and $k = 1$. Since $a < 0$, the parabola opens downward. The vertex is (1, 1). The y-intercept is (0, 0). The x-intercepts are (0, 0) and (2, 0).

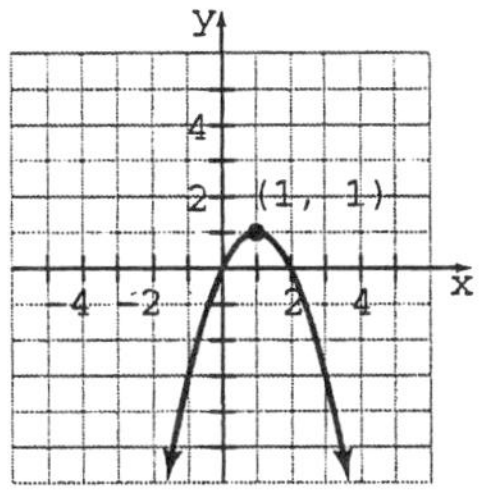

25. $y = -(x+3)^2 + 4$

This is a parabola in the form $y = a(x-h)^2 + k$ with $a = -1$, $h = -3$ and $k = 4$. Since $a < 0$, the parabola opens downward. The vertex is (−3, 4). The y-intercept is (0, −5). The x-intercepts are (−5, 0) and (−1, 0).

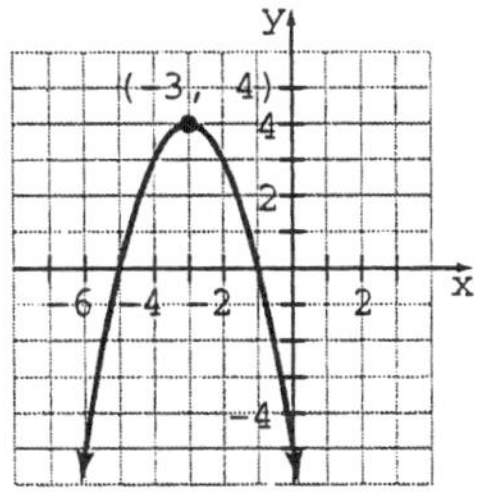

27. $y = -3(x-5)^2 + 3$

This is a parabola in the form $y = a(x-h)^2 + k$ with $a = -3$, $h = 5$ and $k = 3$. Since $a < 0$, the parabola opens downward. The vertex is (5, 3). The y-intercept is (0, −72). The x-intercepts are (4, 0) and (6, 0).

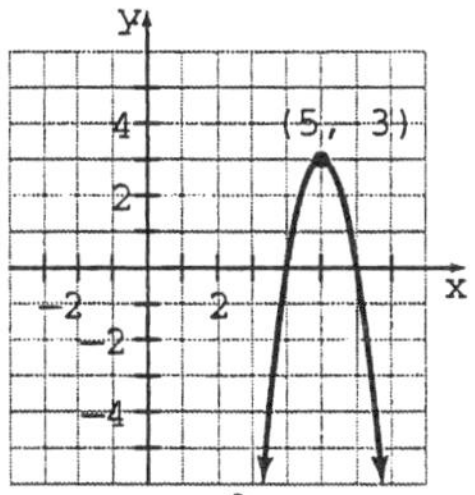

$y=-3(x-5)^2+3$

29. $x = (y-4)^2 - 3$

This is a parabola in the form $x = a(y-k)^2 + h$ with $a = 1$, $h = -3$ and $k = 4$. Since $a > 0$, the parabola opens to the right. The vertex is (−3, 4). The y-intercepts are about (0, 2.27) and (0, 5.73). The x-intercept is (13, 0).

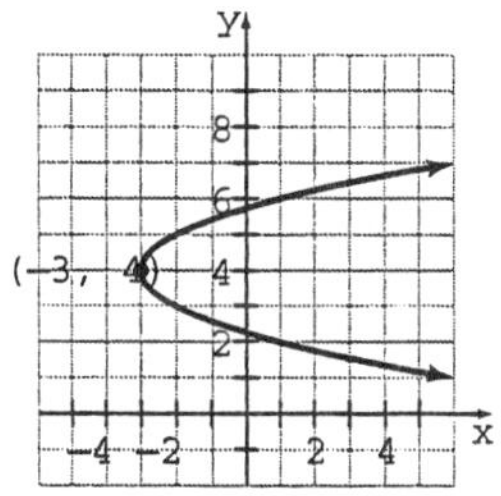

$x=(y-4)^2-3$

31. $x = -(y-5)^2 + 4$

This is a parabola in the form $x = a(y-k)^2 + h$ with $a = -1$, $h = 4$ and $k = 5$. Since $a < 0$, the parabola opens to the left. The vertex is (4, 5). The y-intercepts are (0, 3) and (0, 7). The x-intercept is (−21, 0).

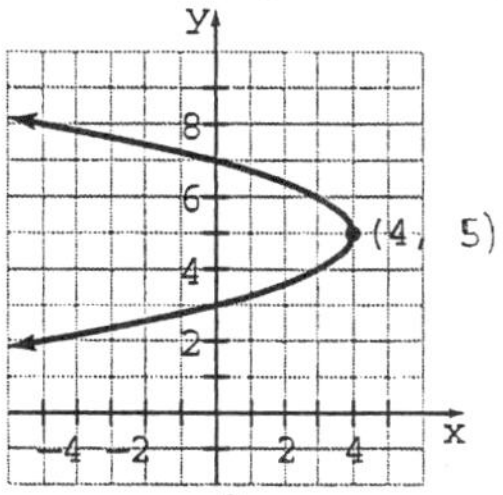

$x=-(y-5)^2+4$

33. $x = -5(y+3)^2 - 6$

This is a parabola in the form $x = a(y-k)^2 + h$ with $a = -5$, $h = -6$ and $k = -3$. Since $a < 0$, the parabola opens to the left. The vertex is (–6, –3). There are no y-intercepts. The x-intercept is (–51, 0)

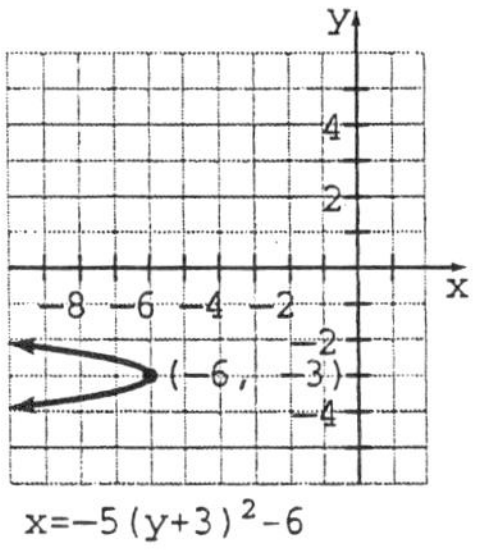

35.

This is a parabola in the form $y = a(x-h)^2 + k$ with $a = -2$, $h = -\frac{1}{2}$ and $k = 6$. Since $a < 0$, the parabola opens downward. The vertex is $\left(-\frac{1}{2}, 6\right)$. The y-intercept is $\left(0, \frac{11}{2}\right)$. The x-intercepts are about (–2.23, 0) and (1.23, 0).

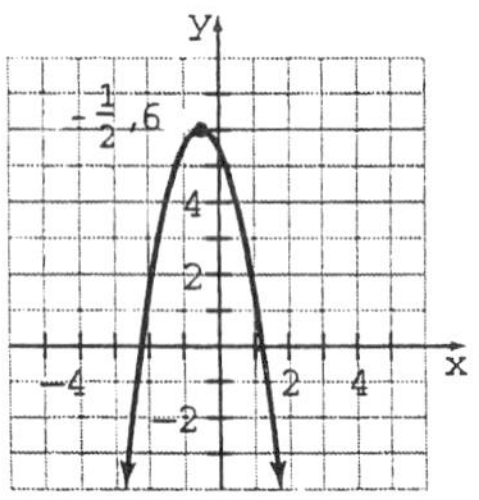

37. a. $y = x^2 + 2x$

$y = (x^2 + 2x + 1) - 1$

$y = (x+1)^2 - 1$

b. This is a parabola in the form $y = a(x-h)^2 + k$ with $a = 1$, $h = -1$ and $k = -1$. Since $a > 0$, the parabola opens upward. The vertex is (–1, –1). The y-intercept is (0, 0). The x-intercepts are (–2, 0) and (0, 0).

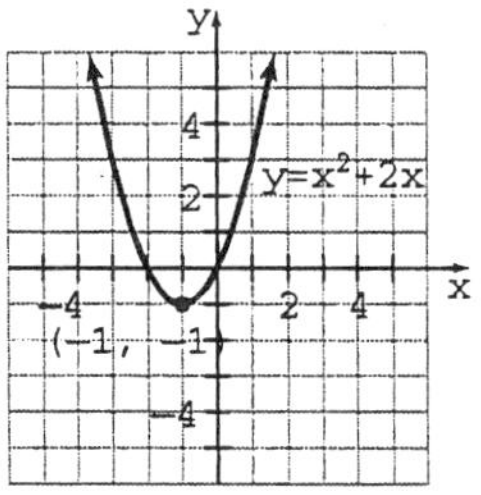

39. a. $y = x^2 + 6x$

$y = (x^2 + 6x + 9) - 9$

$y = (x+3)^2 - 9$

b. This is a parabola in the form $y = a(x-h)^2 + k$ with $a = 1$, $h = -3$ and $k = -9$. Since $a > 0$, the parabola opens upward. The vertex is (–3, –9). The y-intercept is (0, 0). The x-intercepts are (–6, 0) and (0, 0).

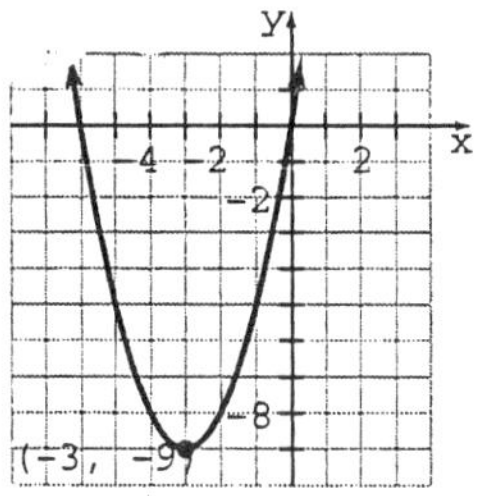

41. a. $x = y^2 + 4y$

$x = (y^2 + 4y + 4) - 4$

$x = (y+2)^2 - 4$

b. This is a parabola in the form $x = a(y-k)^2 + h$ with $a = 1$, $h = -4$ and $k = -2$. Since $a > 0$, the parabola opens to the right. The vertex is (–4, –2). The y-intercepts are (0, –4) and (0, 0). The x-intercept is (0, 0).

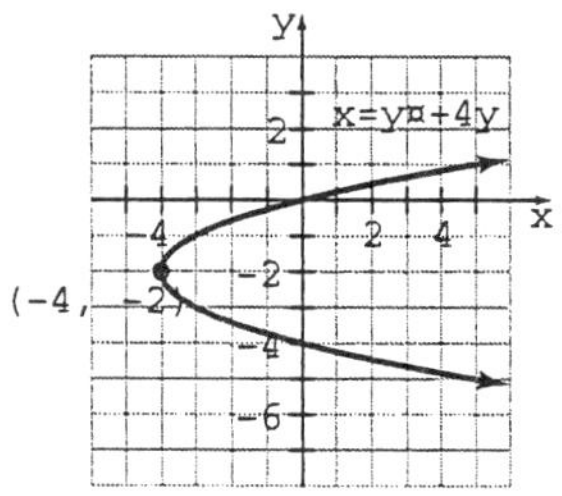

43. **a.** $y = x^2 + 7x + 10$

$y = \left(x^2 + 7x + \frac{49}{4}\right) - \frac{49}{4} + 10$

$y = \left(x + \frac{7}{2}\right)^2 - \frac{9}{4}$

b. This is a parabola in the form $y = a(x-h)^2 + k$ with $a = 1$, $h = -\frac{7}{2}$ and $k = -\frac{9}{4}$. Since $a > 0$, the parabola opens upward. The vertex is $\left(-\frac{7}{2}, -\frac{9}{4}\right)$. The y-intercept is (0, 10). The x-intercepts are (–5, 0) and (–2, 0).

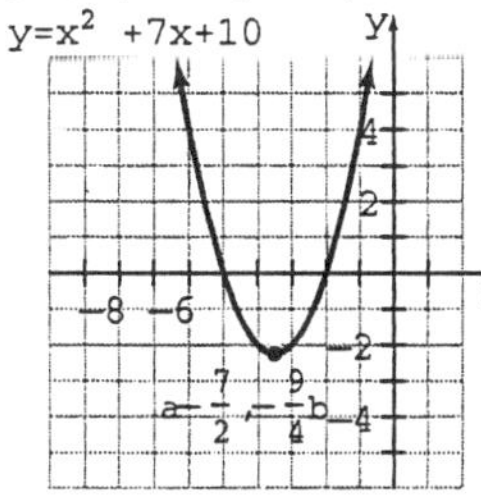

45. **a.** $x = -y^2 + 6y - 9$

$x = -(y^2 - 6y) - 9$

$x = -(y^2 - 6y + 9) + 9 - 9$

$x = -(y-3)^2$

b. This is a parabola in the form $x = a(y-k)^2 + h$ with $a = -1$, $h = 0$ and $k = 3$. Since $a < 0$, the parabola opens to the left. The vertex is (0, 3). The y-intercept is (0, 3). The x-intercept is (–9, 0).

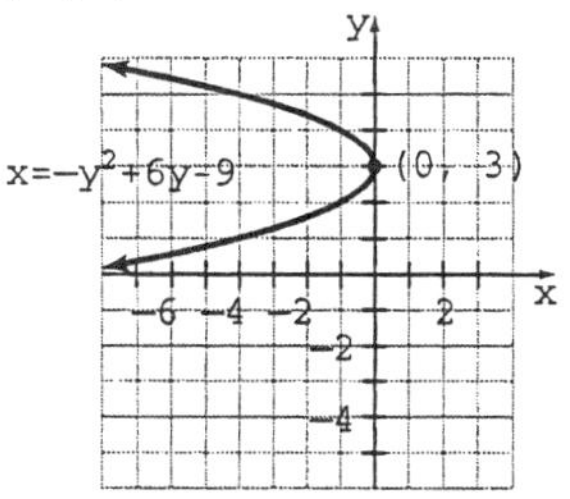

47. **a.** $y = -x^2 + 4x - 4$

$y = -(x^2 - 4x) - 4$

$y = -(x^2 - 4x + 4) + 4 - 4$

$y = -(x-2)^2$

b. This is a parabola in the form $y = a(x-h)^2 + k$ with $a = -1$, $h = 2$ and $k = 0$. Since $a < 0$, the parabola opens downward. The vertex is (2, 0). The y-intercept is (0, –4). The x-intercept is (2, 0).

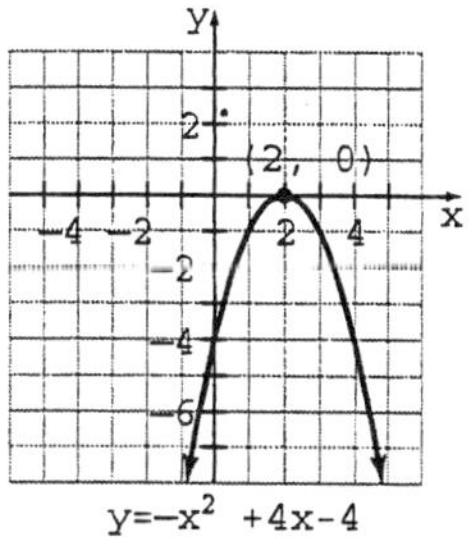

49. **a.** $x = -y^2 + 3y - 4$

$x = -(y^2 - 3y) - 4$

$x = -\left(y^2 - 3y + \frac{9}{4}\right) + \frac{9}{4} - 4$

$x = -\left(y - \frac{3}{2}\right)^2 - \frac{7}{4}$

b. This is a parabola in the form $x = a(y-k)^2 + h$ with $a = -1$, $h = -\frac{7}{4}$ and $k = \frac{3}{2}$. Since $a < 0$, the parabola opens to the left. The vertex is $\left(-\frac{7}{4}, \frac{3}{2}\right)$. There are no y-intercepts. The x-intercept is (–4, 0).

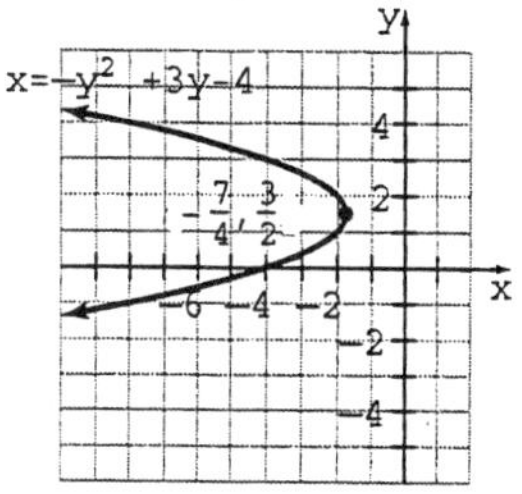

51. $$\begin{aligned} d &= \sqrt{(x_2-x_1)^2+(y_2-y_1)^2} \\ &= \sqrt{(3-3)^2+[-6-(-1)]^2} \\ &= \sqrt{0^2+(-5)^2} \\ &= \sqrt{0+25} \\ &= \sqrt{25} \\ &= 5 \end{aligned}$$

53. $$\begin{aligned} d &= \sqrt{(x_2-x_1)^2+(y_2-y_1)^2} \\ &= \sqrt{[5-(-4)]^2+(3-3)^2} \\ &= \sqrt{9^2+0^2} \\ &= \sqrt{81+0} \\ &= \sqrt{81} \\ &= 9 \end{aligned}$$

55. $$\begin{aligned} d &= \sqrt{(x_2-x_1)^2+(y_2-y_1)^2} \\ &= \sqrt{[4-(-1)]^2+[9-(-3)]^2} \\ &= \sqrt{5^2+12^2} \\ &= \sqrt{25+144} \\ &= \sqrt{169} \\ &= 13 \end{aligned}$$

57. $$\begin{aligned} d &= \sqrt{(x_2-x_1)^2+(y_2-y_1)^2} \\ &= \sqrt{[5-(-4)]^2+[-2-(-5)]^2} \\ &= \sqrt{9^2+3^2} \\ &= \sqrt{81+9} \\ &= \sqrt{90} \\ &\approx 9.49 \end{aligned}$$

59. $$\begin{aligned} d &= \sqrt{(x_2-x_1)^2+(y_2-y_1)^2} \\ &= \sqrt{\left(\frac{1}{2}-3\right)^2+[4-(-1)]^2} \\ &= \sqrt{\left(-\frac{5}{2}\right)^2+5^2} \\ &= \sqrt{\frac{25}{4}+25} \\ &= \sqrt{\frac{125}{4}} \\ &\approx 5.59 \end{aligned}$$

61. $$\begin{aligned} d &= \sqrt{(x_2-x_1)^2+(y_2-y_1)^2} \\ &= \sqrt{[-4.3-(-1.6)]^2+(-1.7-3.5)^2} \\ &= \sqrt{(-2.7)^2+(-5.2)^2} \\ &= \sqrt{7.29+27.04} \\ &= \sqrt{34.33} \\ &\approx 5.86 \end{aligned}$$

63. $$\begin{aligned} d &= \sqrt{(x_2-x_1)^2+(y_2-y_1)^2} \\ &= \sqrt{\left(0-\sqrt{7}\right)^2+\left[0-\sqrt{3}\right]^2} \\ &= \sqrt{\left(-\sqrt{7}\right)^2+\left(\sqrt{3}\right)^2} \\ &= \sqrt{7+3} \\ &= \sqrt{10} \\ &\approx 3.16 \end{aligned}$$

65. $$\begin{aligned} \text{Midpoint} &= \left(\frac{x_1+x_2}{2}, \frac{y_1+y_2}{2}\right) \\ &= \left(\frac{1+5}{2}, \frac{9+3}{2}\right) \\ &= (3, 6) \end{aligned}$$

67. $$\begin{aligned} \text{Midpoint} &= \left(\frac{x_1+x_2}{2}, \frac{y_1+y_2}{2}\right) \\ &= \left(\frac{-7+7}{2}, \frac{2+(-2)}{2}\right) \\ &= (0, 0) \end{aligned}$$

69. $\text{Midpoint} = \left(\frac{x_1+x_2}{2}, \frac{y_1+y_2}{2}\right) = \left(\frac{-1+4}{2}, \frac{4+6}{2}\right) = \left(\frac{3}{2}, 5\right)$

71. $\text{Midpoint} = \left(\frac{x_1+x_2}{2}, \frac{y_1+y_2}{2}\right) = \left(\frac{3+2}{2}, \frac{\frac{1}{2}+(-4)}{2}\right) = \left(\frac{5}{2}, -\frac{7}{4}\right)$

73. $\text{Midpoint} = \left(\frac{x_1+x_2}{2}, \frac{y_1+y_2}{2}\right) = \left(\frac{\sqrt{3}+\sqrt{2}}{2}, \frac{2+7}{2}\right) = \left(\frac{\sqrt{3}+\sqrt{2}}{2}, \frac{9}{2}\right)$

75.
$$(x-h)^2+(y-k)^2=r^2$$
$$(x-0)^2+(y-0)^2=6^2$$
$$x^2+y^2=36$$

77.
$$(x-h)^2+(y-k)^2=r^2$$
$$(x-2)^2+(y-0)^2=5^2$$
$$(x-2)^2+y^2=25$$

79.
$$(x-h)^2+(y-k)^2=r^2$$
$$(x-0)^2+[y-5]^2=2^2$$
$$x^2+(y-5)^2=4$$

81.
$$(x-h)^2+(y-k)^2=r^2$$
$$(x-3)^2+(y-4)^2=(9)^2$$
$$(x-3)^2+(y-4)^2=81$$

83.
$$(x-h)^2+(y-k)^2=r^2$$
$$[x-2]^2+[y-(-6)]^2=10^2$$
$$(x-2)^2+(y+6)^2=100$$

85.
$$(x-h)^2+(y-k)^2=r^2$$
$$(x-1)^2+(y-2)^2=\left(\sqrt{7}\right)^2$$
$$(x-1)^2+(y-2)^2=7$$

87. The center is (0, 0) and the radius is 4.
$$(x-h)^2+(y-k)^2=r^2$$
$$(x-0)^2+(y-0)^2=4^2$$
$$x^2+y^2=16$$

89. The center is (3, –2) and the radius is 3.
$$(x-h)^2+(y-k)^2=r^2$$
$$(x-3)^2+[y-(-2)]^2=3^2$$
$$(x-3)^2+(y+2)^2=9$$

91.
$$x^2+y^2=16$$
$$x^2+y^2=4^2$$
The graph is a circle with its center at the origin and radius 4.

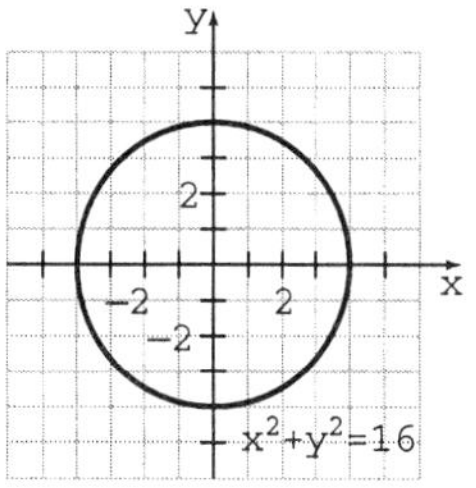

93.
$$x^2+y^2=10$$
$$x^2+y^2=\left(\sqrt{10}\right)^2$$
The graph is a circle with its center at the origin and radius $\sqrt{10}$.

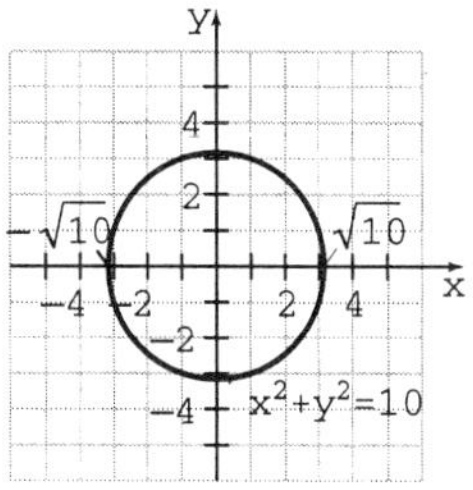

95. $(x+4)^2+y^2=25$

$(x+4)^2+(y-0)^2=5^2$

The graph is a circle with its center at (–4, 0) and radius 5.

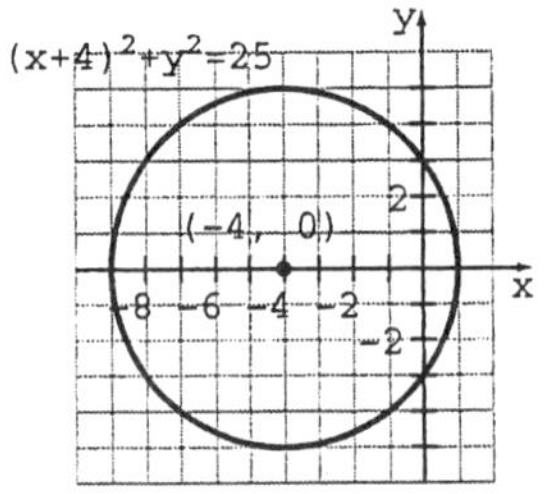

97. $x^2+(y+1)^2=9$

$(x-0)^2+(y+1)^2=(3)^2$

The graph is a circle with its center at (0, –1) and radius 3.

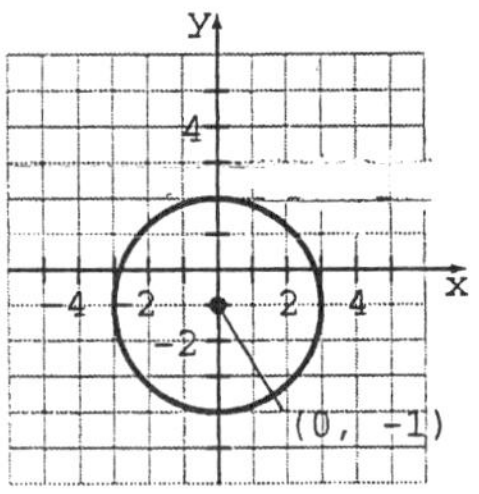

99. $(x+8)^2+(y+2)^2=9$

$(x+8)^2+(y+2)^2=3^2$

The graph is a circle with its center at (–8, –2) and radius 3.

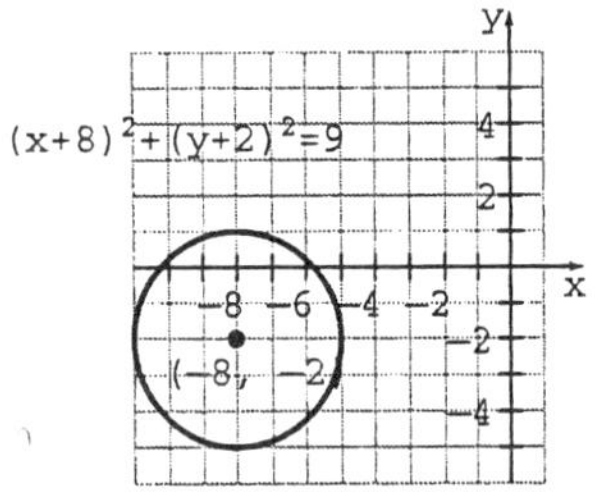

101. $y=\sqrt{25-x^2}$

If we solve $x^2+y^2=25$ for y, we obtain $y=\pm\sqrt{25-x^2}$. Therefore, the graph of $y=\sqrt{25-x^2}$ is the upper half ($y \geq 0$) of a circle with its center at the origin and radius 5.

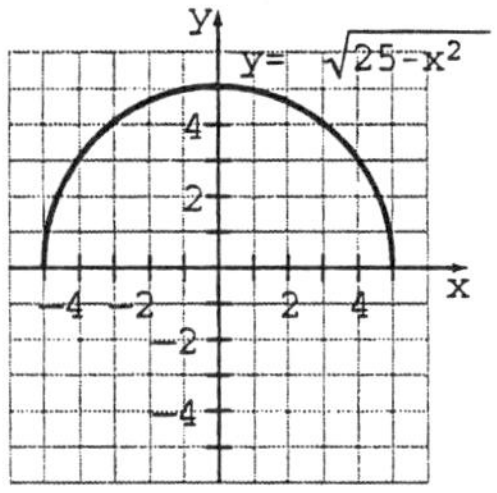

103. $y=-\sqrt{4-x^2}$

If we solve $x^2+y^2=4$ for y, we obtain $y=\pm\sqrt{4-x^2}$. Therefore, the graph of $y=-\sqrt{4-x^2}$ is the lower half ($y \leq 0$) of a circle with its center at the origin and radius 2.

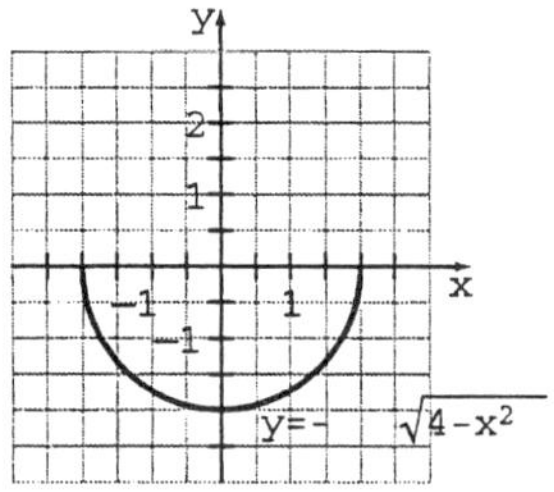

105. **a.**

$$x^2+y^2+8x+15=0$$
$$x^2+8x+y^2=-15$$
$$(x^2+8x+16)+y^2=-15+16$$
$$(x+4)^2+y^2=1$$
$$(x+4)^2+y^2=1^2$$

b. The graph is a circle with center (–4, 0) and radius 1.

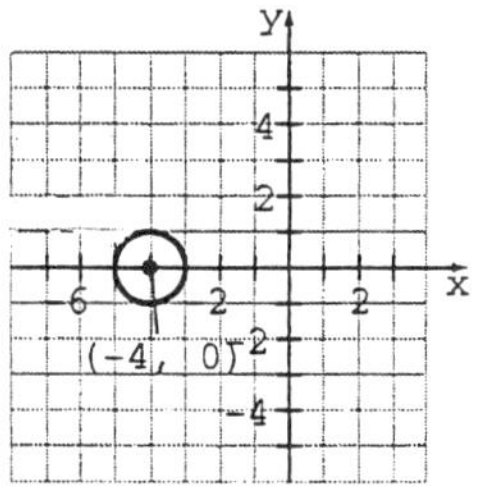

107. **a.**
$$x^2+y^2+6x-4y+9=0$$
$$x^2+6x+y^2-4y=-9$$
$$(x^2+6x+9)+(y^2-4y+4)=-9+9+4$$
$$(x+3)^2+(y-2)^2=4$$
$$(x+3)^2+(y-2)^2=2^2$$

b. The graph is a circle with center (–3, 2) and radius 2.

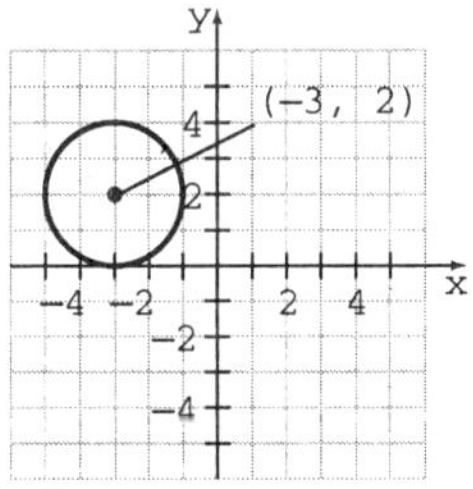

$x^2+y^2+6x-4y+9=0$

109. **a.**
$$x^2+y^2+6x-2y+6=0$$
$$x^2+6x+y^2-2y=-6$$
$$(x^2+6x+9)+(y^2-2y+1)=-6+9+1$$
$$(x+3)^2+(y-1)^2=4$$
$$(x+3)^2+(y-1)^2=2^2$$

b. The graph is a circle with center (–3, 1) and radius 2.

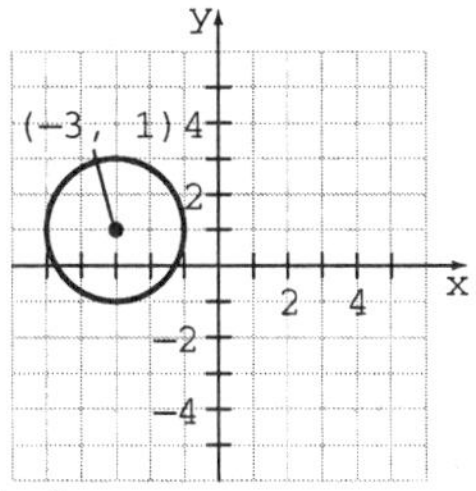

$x^2+y^2+6x-2y+6=0$

111. **a.**
$$x^2+y^2-8x+2y+13=0$$
$$x^2-8x+y^2+2y=-13$$
$$(x^2-8x+16)+(y^2+2y+1)=-13+16+1$$
$$(x-4)^2+(y+1)^2=4$$
$$(x-4)^2+(y+1)^2=2^2$$

b. The graph is a circle with center (4, –1) and radius 2.

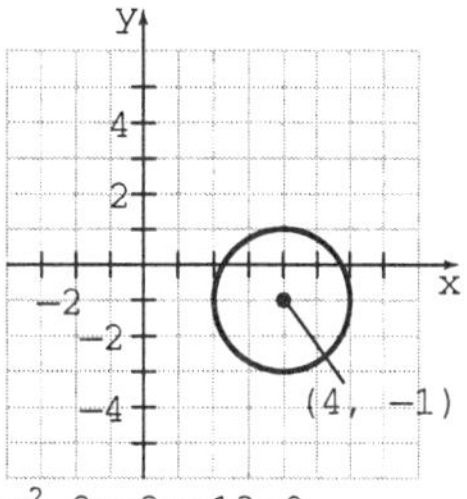

$x^2+y^2-8x+2y+13=0$

113. $x^2+y^2=16$
$x^2+y^2=4^2$
The graph is a circle with its center at the origin and radius 4.

Area $=\pi r^2=\pi(4)^2=16\pi\approx 50.3$ sq. units

115. x-intercept:
$x=0^2-6(0)-7$
$x=-7$
The x-intercept is (–7, 0)
y-intercepts:
$0=y^2-6y-7$
$0=(y+1)(y-7)$
$y=-1$ or $y=7$
The y-intercepts are (0, –1) and (0, 7).

117. x-intercept:
$x=2(0-5)^2+6$
$x=56$
The x-intercept is (56, 0).
Y-intercepts:
$0=2(y-5)^2+6$
Since $2(y-5)^2+6\geq 6$ for all real values of y, this equation has no real solutions.
There are no y-intercepts.

119. No. For example, the origin is the midpoint of both the segment from (1, 1) to (–1, –1) and the segment from (2, 2) to (–2, –2), but these segments have different lengths.

121. The distance from the midpoint (4, –6) to the endpoint (7, –2) is half the length of the line segment.

$$\frac{d}{2}=\sqrt{(7-4)^2+[-2-(-6)]^2}$$
$$=\sqrt{3^2+4^2}$$
$$=\sqrt{25}$$
$$=5$$

Since $\frac{d}{2}=5$, $d=10$. The length is 10 units.

123. Since (–5, 2) is 2 units above the x-axis, the radius is 2.

$$(x-h)^2+(y-k)^2=r^2$$
$$(x+5)^2+(y-2)^2=2^2$$
$$(x+5)^2+(y-2)^2=4$$

125. a. Diameter $=\sqrt{(x_2-x_1)^2+(y_2-y_1)^2}$

$$=\sqrt{(9-5)^2+(8-4)^2}$$
$$=\sqrt{4^2+4^2}$$
$$=\sqrt{16+16}$$
$$=\sqrt{32}$$
$$=4\sqrt{2}$$

Since the diameter is $4\sqrt{2}$ units, the radius is $2\sqrt{2}$ units.

b. Midpoint $=\left(\frac{x_1+x_2}{2},\ \frac{y_1+y_2}{2}\right)$

$$=\left(\frac{5+9}{2},\ \frac{4+8}{2}\right)$$
$$=(7,\ 6)$$

The center is (7, 6).

c.
$$(x-h)^2+(y-k)^2=r^2$$
$$(x-7)^2+(y-6)^2=\left(2\sqrt{2}\right)^2$$
$$(x-7)^2+(y-6)^2=8$$

127. The minimum number is 0 and the maximum number is 4 as shown in the diagrams.

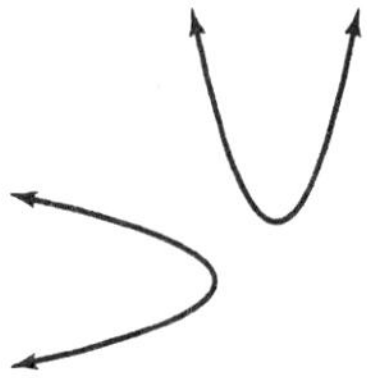

No points of intersection

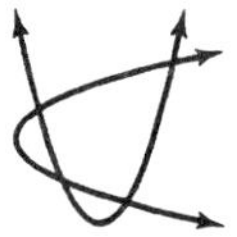

4 points of intersection

129. a. Since $150-2(68.2)=13.6$, the clearance is 13.6 feet.

b. Since $150-68.2=81.8$, the center of the wheel is 81.8 feet above the ground.

c.
$$(x-h)^2+(y-k)^2=r^2$$
$$(x-0)^2+(y-81.8)^2=68.2^2$$
$$x^2+(y-81.8)^2=68.2^2$$
$$x^2+(y-81.8)^2=4651.24$$

131. a. The center of the blue circle is the origin, and the radius is 4.

$$x^2+y^2=r^2$$
$$x^2+y^2=4^2$$
$$x^2+y^2=16$$

b. The center of the red circle is (2, 0), and the radius is 2.

$$(x-h)^2+(y-k)^2=r^2$$
$$(x-2)^2+(y-0)^2=2^2$$
$$(x-2)^2+y^2=4$$

c. The center of the green circle is (–2, 0), and the radius is 2.

$$(x-h)^2+(y-k)^2=r^2$$
$$[x-(-2)]^2+(y-0)^2=2^2$$
$$(x+2)^2+y^2=4$$

d. Shaded area = (blue circle area) – (red circle area) – (green circle area)

$$=\pi\left(4^2\right)-\pi\left(2^2\right)-\pi\left(2^2\right)$$
$$=16\pi-4\pi-4\pi$$
$$=8\pi$$

133. The radii are 4 and 8, respectively. So, the area between the circles is

$$\pi(8)^2-\pi(4)^2=64\pi-16\pi=48\pi.$$

136. $\dfrac{4x^{-3}y^4}{12x^{-2}y^3} = \dfrac{4}{12}x^{-3-(-2)}y^{4-3}$

$= \dfrac{1}{3}x^{-3+2}y^1$

$= \dfrac{1}{3}x^{-1}y$ or $\dfrac{y}{3x}$

137.

a. area 1: a^2 area 2: ab

area 3: ab area 4: b^2

b. $(a+b)^2$

138. $\begin{vmatrix} 4 & 0 & 3 \\ 5 & 2 & -1 \\ 3 & 6 & 4 \end{vmatrix} = 4\begin{vmatrix} 2 & -1 \\ 6 & 4 \end{vmatrix} - 5\begin{vmatrix} 0 & 3 \\ 6 & 4 \end{vmatrix} + 3\begin{vmatrix} 0 & 3 \\ 2 & -1 \end{vmatrix}$

$= 4(8+6) - 5(0-18) + 3(0-6)$

$= 4(14) - 5(-18) + 3(-6)$

$= 56 + 90 - 18$

$= 128$

139. $-4 < 3x - 4 < 8$

$-4 + 4 < 3x - 4 + 4 < 8 + 4$

$0 < 3x < 12$

$\dfrac{0}{3} < \dfrac{3x}{3} < \dfrac{12}{3}$

$0 < x < 4$

In interval notation: $(0, 4)$

140. $y = (x-4)^2 + 1$

This is a parabola in the form $y = a(x-h)^2 + k$ with $a = 1$, $h = 4$ and $k = 1$.

Since $a > 0$, the parabola opens upward. The vertex is (4, 1). The y-intercept is (0, 17). There are no x-intercepts.

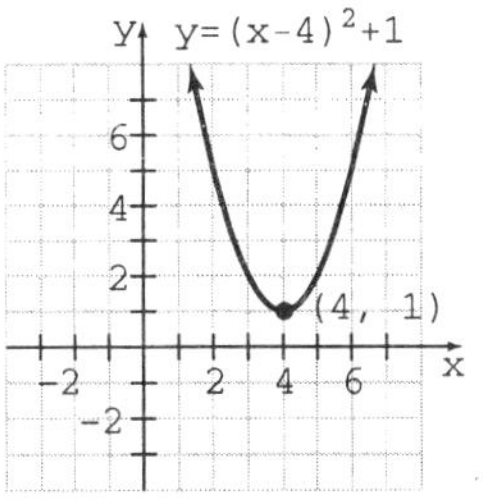

Exercise Set 14.2

1. An ellipse is a set of points in a plane, the sum of whose distances from two fixed points is constant.

3. $\dfrac{(x-h)^2}{a^2} + \dfrac{(y-k)^2}{b^2} = 1$

5. If $a = b$, the formula for a circle is obtained.

7. First divide both sides by 360.

9. No, the sign in front of the y^2 is negative.

11. $\dfrac{x^2}{4} + \dfrac{y^2}{1} = 1$

Since $a^2 = 4$, $a = 2$.

Since $b^2 = 1$, $b = 1$.

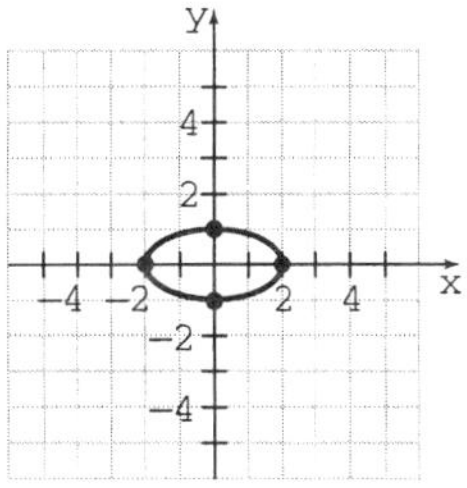

13. $\dfrac{x^2}{4} + \dfrac{y^2}{9} = 1$

Since $a^2 = 4$, $a = 2$.

Since $b^2 = 9$, $b = 3$.

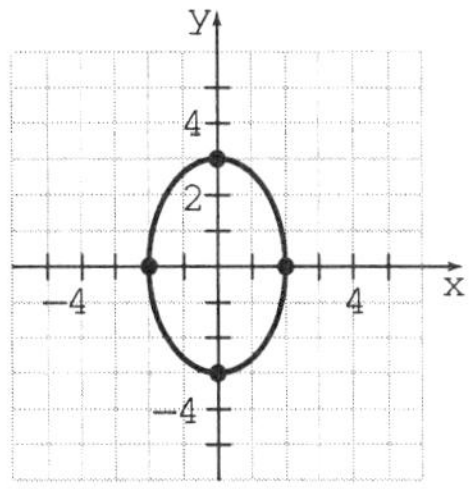

15. $\frac{x^2}{25}+\frac{y^2}{9}=1$

Since $a^2=25$, $a=5$.

Since $b^2=9$, $b=3$.

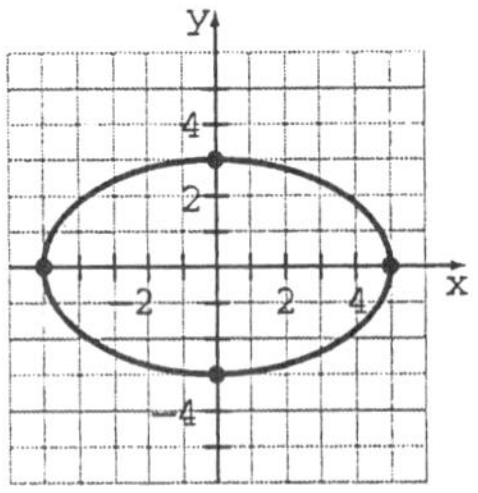

17. $\frac{x^2}{16}+\frac{y^2}{25}=1$

Since $a^2=16$, $a=4$.

Since $b^2=25$, $b=5$.

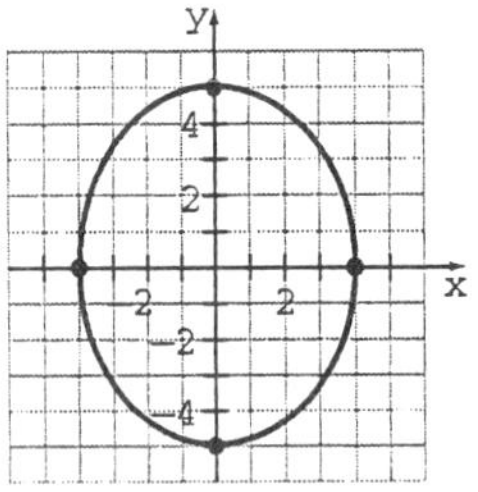

19. $x^2+36y^2=36$

$\frac{x^2}{36}+\frac{36y^2}{36}=1$

$\frac{x^2}{36}+\frac{y^2}{1}=1$

Since $a^2=36$, $a=6$.

Since $b^2=1$, $b=1$.

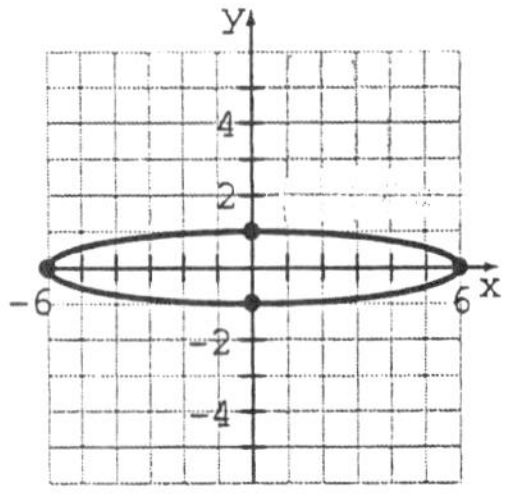

21. $9x^2+25y^2=225$

$\frac{9x^2}{225}+\frac{25y^2}{225}=1$

$\frac{x^2}{25}+\frac{y^2}{9}=1$

Since $a^2=25$, $a=5$.

Since $b^2=9$, $b=3$.

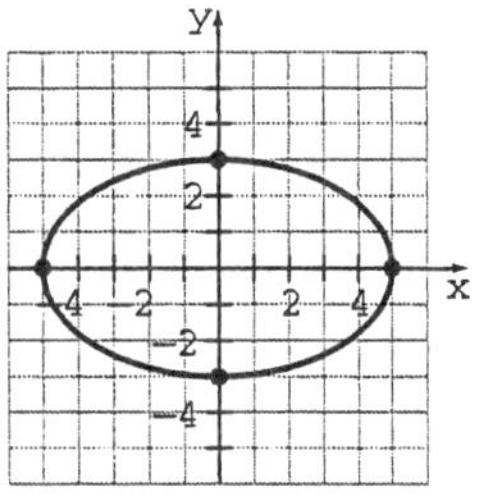

23. $9x^2+4y^2=36$

$\frac{9x^2}{36}+\frac{4y^2}{36}=1$

$\frac{x^2}{4}+\frac{y^2}{9}=1$

Since $a^2=4$, $a=2$.

Since $b^2=9$, $b=3$.

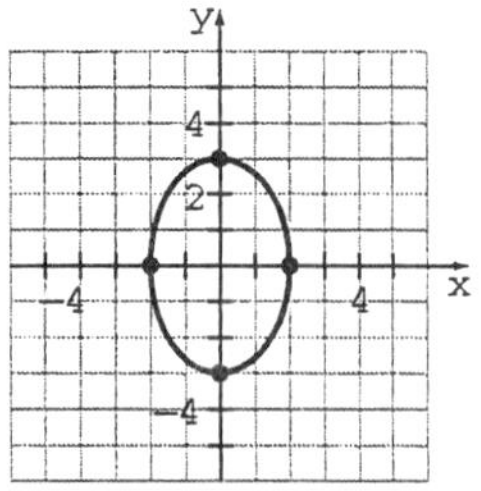

25. $100x^2+25y^2=400$

$4x^2+y^2=16$

$\frac{4x^2}{16}+\frac{y^2}{16}=1$

$\frac{x^2}{4}+\frac{y^2}{16}=1$

Since $a^2=4$, $a=2$.

Since $b^2=16$, $b=4$.

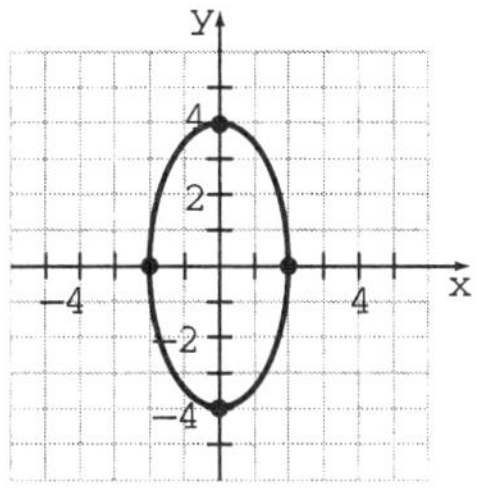

27. $x^2 + 2y^2 = 8$

$$\frac{x^2}{8} + \frac{2y^2}{8} = 1$$

$$\frac{x^2}{8} + \frac{y^2}{4} = 1$$

Since $a^2 = 8$, $a = \sqrt{8} = 2\sqrt{2} \approx 2.83$.

Since $b^2 = 4$, $b = 2$.

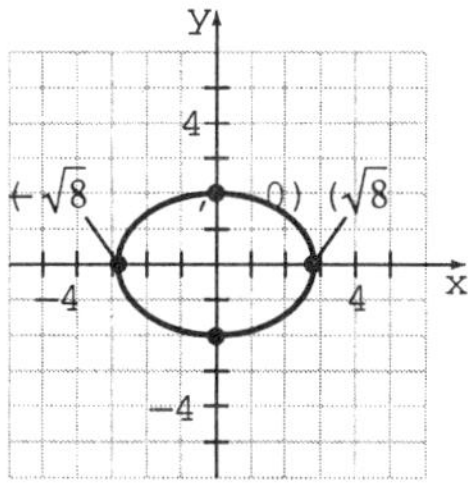

29. $\frac{x^2}{16} + \frac{(y-2)^2}{9} = 1$

The center is (0, 2).

Since $a^2 = 16$, $a = 4$.

Since $b^2 = 9$, $b = 3$.

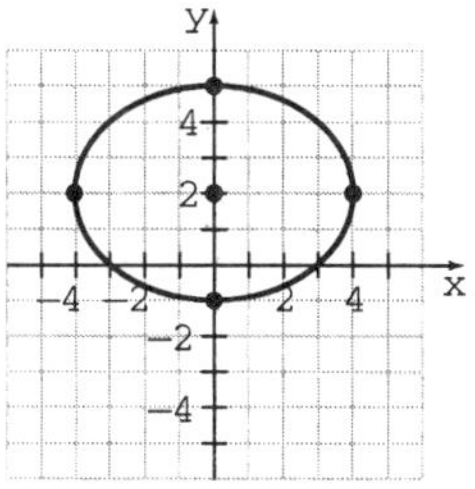

31. $\frac{(x-4)^2}{9} + \frac{(y+3)^2}{25} = 1$

The center is (4, −3).

Since $a^2 = 9$, $a = 3$.

Since $b^2 = 25$, $b = 5$.

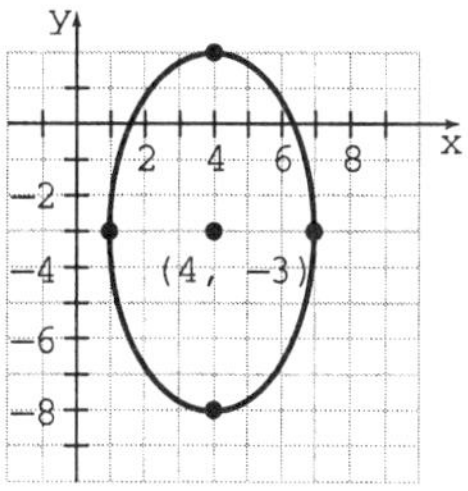

33. $\frac{(x+1)^2}{9} + \frac{(y-2)^2}{4} = 1$

The center is (−1, 2).

Since $a^2 = 9$, $a = 3$.

Since $b^2 = 4$, $b = 2$.

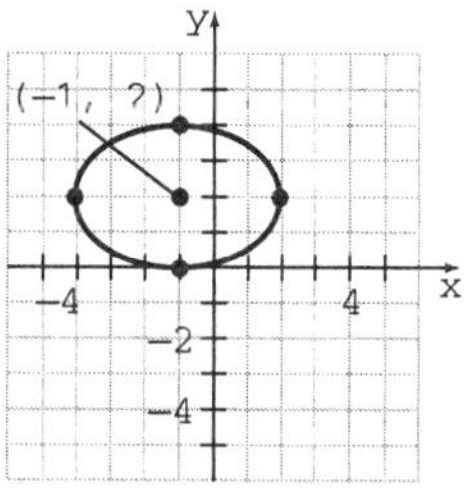

35. $(x+3)^2 + 9(y+1)^2 = 81$

$$\frac{(x+3)^2}{81} + \frac{9(y+1)^2}{81} = 1$$

$$\frac{(x+3)^2}{81} + \frac{(y+1)^2}{9} = 1$$

The center is (−3, −1).

Since $a^2 = 81$, $a = 9$.

Since $b^2 = 9$, $b = 3$.

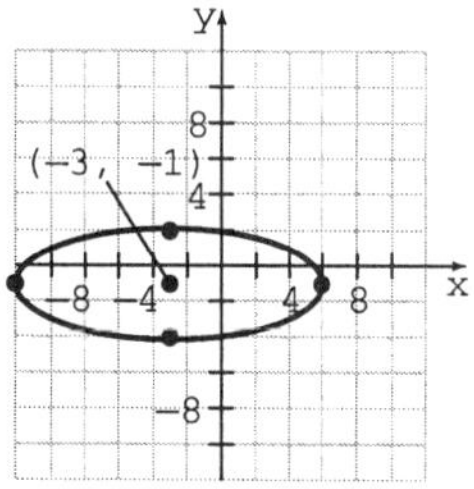

37. $4(x-2)^2 + 9(y+2)^2 = 36$

$$\frac{4(x-2)^2}{36} + \frac{9(y+2)^2}{36} = 1$$

$$\frac{(x-2)^2}{9} + \frac{(y+2)^2}{4} = 1$$

The center is (2, −2).

Since $a^2 = 9$, $a = 3$.

Since $b^2 = 4$, $b = 2$.

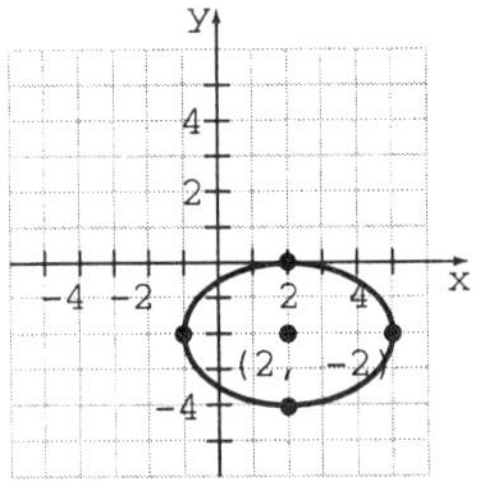

39. $$12(x+4)^2+3(y-1)^2=48$$
$$\frac{12(x+4)^2}{48}+\frac{3(y-1)^2}{48}=1$$
$$\frac{(x+4)^2}{4}+\frac{(y-1)^2}{16}=1$$
The center is (–4, 1).
Since $a^2=4$, $a=2$.
Since $b^2=16$, $b=4$.

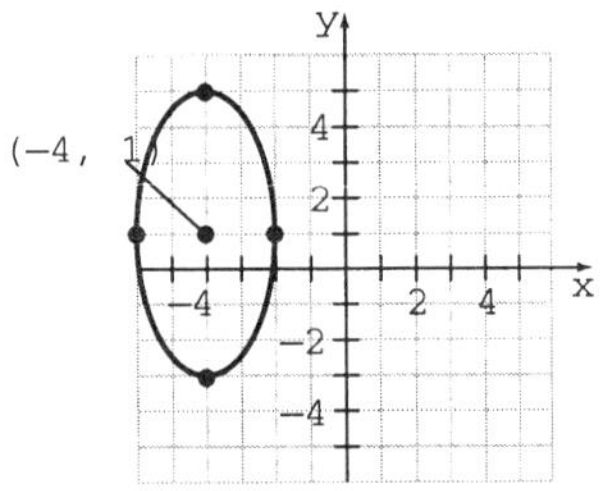

41. $$\frac{x^2}{4}+\frac{y^2}{1}=1$$
Since $a^2=4$, $a=2$.
Since $b^2=1$, $b=1$.

$$\begin{aligned}\text{Area} &= \pi ab\\ &= \pi(2)(1)\\ &= 2\pi\\ &\approx 6.3 \text{ square units}\end{aligned}$$

43. There is only one point, at (0, 0). The only way two non-negative numbers can sum to 0 is if they are both 0.

45. The center is the origin, $a = 3$, and $b = 5$.
$$\frac{x^2}{a^2}+\frac{y^2}{b^2}=1$$
$$\frac{x^2}{3^2}+\frac{y^2}{5^2}=1$$
$$\frac{x^2}{9}+\frac{y^2}{25}=1$$

47. The center is the origin, $a = 2$, and $b = 3$.
$$\frac{x^2}{a^2}+\frac{y^2}{b^2}=1$$
$$\frac{x^2}{2^2}+\frac{y^2}{3^2}=1$$
$$\frac{x^2}{4}+\frac{y^2}{9}=1$$

49. There are no points of intersection, because the ellipse with $a = 2$ and $b = 3$ is completely inside the circle of radius 4.

51. $$x^2+4y^2-4x-8y-92=0$$
$$x^2-4x+4y^2-8y=92$$
$$(x^2-4x+4)+4(y^2-2y+1)=92+4+4$$
$$(x-2)^2+4(y-1)^2=100$$
$$\frac{(x-2)^2}{100}+\frac{(y-1)^2}{25}=1$$
The center is (2, 1).

53. Since 90.2 – 20.7 = 69.5, the distance between the foci is 69.5 feet.

55. **a.** Consider the ellipse to be centered at the origin, (0, 0). Here, $a = 10$ and $b = 24$. The equation is
$$\frac{x^2}{10^2}+\frac{y^2}{24^2}=1 \Rightarrow \frac{x^2}{100}+\frac{y^2}{576}=1.$$

b. $$\begin{aligned}\text{Area} &= \pi ab\\ &= \pi(10)(24)\\ &= 240\pi \approx 753.98 \text{ square feet}\end{aligned}$$

c. Area of opening is half the area of ellipse
$$=\frac{\pi ab}{2}$$
$$=\frac{\pi(10)(24)}{2}$$
$$=120\pi \approx 376.99 \text{ square feet}$$

57. Using $a = 3$ and $b = 2$, we may assume that the ellipse has the equation $\frac{x^2}{9}+\frac{y^2}{4}=0$ and that the foci are located at $(\pm c,\ 0)$. Apply the definition of an ellipse using the points (3, 0) and (0, 2). That is, the distance from (3, 0) to (–c, 0) plus the distance from (3, 0) to (c, 0) is the same as the sum of the distance from (0, 2) to (–c, 0) and the distance from (0, 2) to (c, 0).

$$\sqrt{[3-(-c)]^2+(0-0)^2}+\sqrt{(3-c)^2+(0-0)^2}$$
$$=\sqrt{[0-(-c)]^2+(2-0)^2}+\sqrt{(0-c)^2+(2-0)^2}$$
$$|3+c|+|3-c|=\sqrt{c^2+4}+\sqrt{(-c)^2+4}$$

Note that the foci are inside the ellipse, so $3+c>0$ and $3-c>0$. So, $|3+c|=3+c$ and $|3-c|=3-c$.

$$(3+c)+(3-c)=2\sqrt{c^2+4}$$
$$6=2\sqrt{c^2+4}$$
$$3=\sqrt{c^2+4}$$
$$9=c^2+4$$
$$5=c^2$$
$$c=\pm\sqrt{5}$$

The foci are located at $\left(\pm\sqrt{5},\ 0\right)$.

That is, the foci are $\sqrt{5}\approx 2.24$ feet, in both directions, from the center of the ellipse, along the major axis.

59. Answers will vary.

61. Answers will vary.

63. The center is (4, 2).
$a=2$, $b=3$.

$$\frac{(x-h)^2}{a^2}+\frac{(y-k)^2}{b^2}=1$$
$$\frac{(x-4)^2}{2^2}+\frac{(y-2)^2}{3^2}=1$$
$$\frac{(x-4)^2}{4}+\frac{(y-2)^2}{9}=1$$

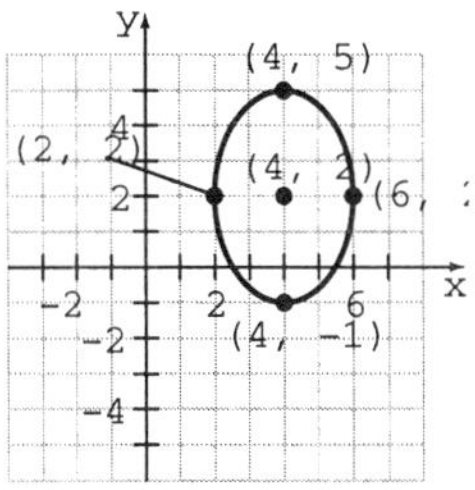

65.

$$S=\frac{n}{2}(f+l),\text{ for } l$$
$$S=\frac{nf}{2}+\frac{nl}{2}$$
$$S-\frac{nf}{2}=\frac{nl}{2}$$
$$2\left(S-\frac{nf}{2}\right)=2\left(\frac{nl}{2}\right)$$
$$2S-nf=nl$$
$$\frac{2S-nf}{n}=\frac{nl}{n}\Rightarrow l=\frac{2S-nf}{n}$$

66.

$$\begin{array}{r} x+\frac{5}{2} \\ 2x-3\overline{)\,2x^2+2x-2} \\ \underline{-2x^2+3x} \\ 5x-2 \\ \underline{-5x+\frac{15}{2}} \\ \frac{11}{2} \end{array}$$

$$\frac{2x^2+2x-2}{2x-3}=x+\frac{5}{2}+\frac{11}{2(2x-3)}$$

67.

$$\sqrt{8b-15}=10-b$$
$$8b-15=(10-b)^2$$
$$8b-15=100-20b+b^2$$
$$b^2-28b+115=0$$
$$(b-5)(b-23)=0$$
$$b-5=0 \quad\text{or}\quad b-23=0$$
$$b=5 \qquad\qquad b=23$$

Upon checking $b=23$ is extraneous. The solution is $b=5$.

68.

$$\frac{3x+5}{x-2}\le 0$$
$$3x+5=0\Rightarrow x=-\frac{5}{3}$$
$$x-2=0\Rightarrow x=2$$
$$\frac{3x+5}{x-2}\le 0\Rightarrow -\frac{5}{3}\le x<2\text{ or }\left[-\frac{5}{3},\ 2\right)$$

69.

$$\log_6 4000=\frac{\log 4000}{\log 6}\approx 4.6290$$

Exercise Set 14.3

1. A hyperbola is the set of points in a plane, the difference of whose distances from two fixed points is a constant.

3. The graph of $\frac{x^2}{a^2} - \frac{y^2}{b^2} = 1$ is a hyperbola with vertices at $(a, 0)$ and $(-a, 0)$. Its transverse axis lies along the x-axis. The asymptotes are $y = \pm\frac{b}{a}x$.

5. No, equations of hyperbolas have one positive square term and one negative square term. This equation has two positive square terms.

7. Yes, equations of hyperbolas have one positive square term and one negative square term. This equation satisfies that condition.

9. The first step is to divide both sides by 81 in order to make the right side equal to 1.

11. **a.** $\frac{x^2}{9} - \frac{y^2}{4} = 1$
Since $a^2 = 9$ and $b^2 = 4$, $a = 3$ and $b = 2$. The equations of the asymptotes are $y = \pm\frac{b}{a}x$, or $y = \pm\frac{2}{3}x$.

b. .To graph the asymptotes, plot the points $(3, 2)$, $(-3, 2)$, $(3, -2)$, and $(-3, -2)$. The graph intersects the x-axis at $(-3, 0)$ and $(3, 0)$.

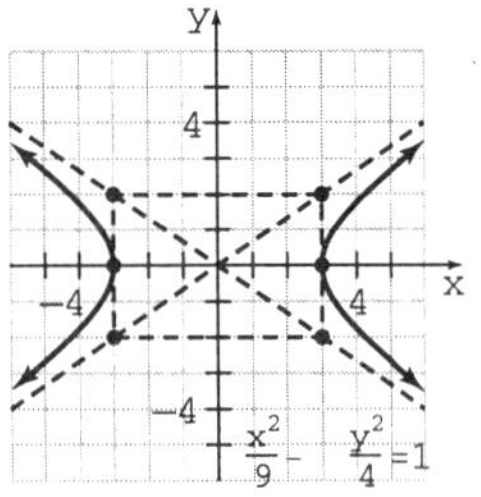

13. **a.** $\frac{x^2}{4} - \frac{y^2}{1} = 1$
Since $a^2 = 4$ and $b^2 = 1$, $a = 2$ and $b = 1$. The equations of the asymptotes are $y = \pm\frac{b}{a}x$, or $y = \pm\frac{1}{2}x$.

b. To graph the asymptotes, plot the points $(2, 1)$, $(-2, 1)$, $(2, -1)$, and $(-2, -1)$. The graph intersects the x-axis at $(-2, 0)$ and $(2, 0)$.

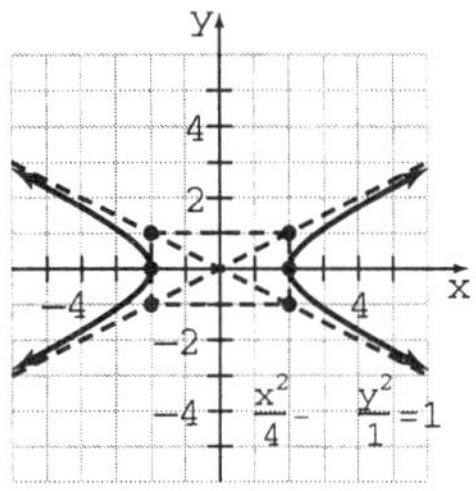

15. **a.** $\frac{x^2}{9} - \frac{y^2}{25} = 1$
Since $a^2 = 9$ and $b^2 = 25$, $a = 3$ and $b = 5$. The equations of the asymptotes are $y = \pm\frac{b}{a}x$, or $y = \pm\frac{5}{3}x$.

b. To graph the asymptotes, plot the points $(3, 5)$, $(-3, 5)$, $(3, -5)$, and $(-3, -5)$. The graph intersects the x-axis at $(-3, 0)$ and $(3, 0)$.

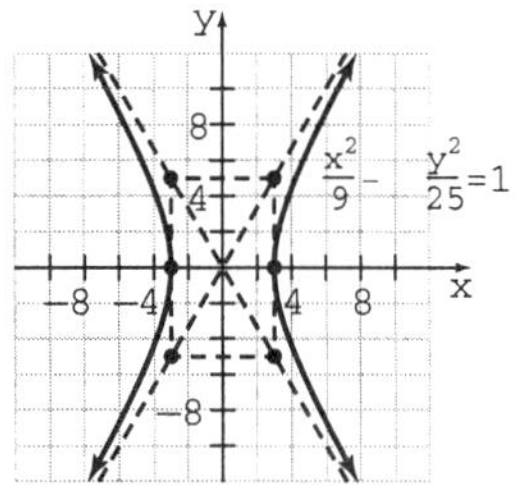

17. **a.** $\frac{x^2}{25} - \frac{y^2}{16} = 1$
Since $a^2 = 25$ and $b^2 = 16$, $a = 5$ and $b = 4$. The equations of the asymptotes are $y = \pm\frac{b}{a}x$, or $y = \pm\frac{4}{5}x$.

b. To graph the asymptotes, plot the points $(5, 4)$, $(-5, 4)$, $(5, -4)$, and $(-5, -4)$. The graph intersects the x-axis at $(-5, 0)$ and $(5, 0)$.

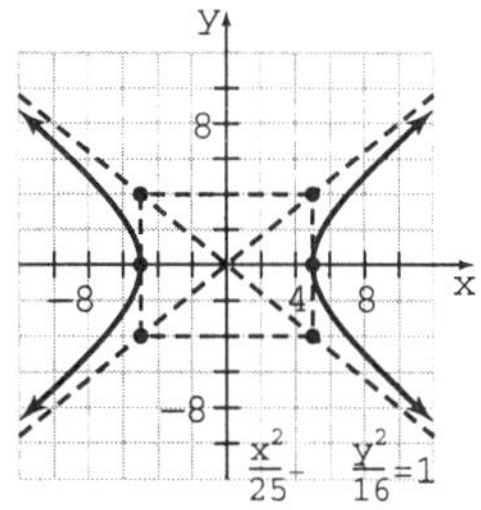

19. **a.** $\frac{y^2}{9} - \frac{x^2}{16} = 1$

Since $a^2 = 16$ and $b^2 = 9$, $a = 4$ and $b = 3$. The equations of the asymptotes are $y = \pm\frac{b}{a}x$, or $= y \pm \frac{3}{4}x$.

b. To graph the asymptotes, plot the points (4, 3), (–4, 3), (4, –3) and (–4, –3). The graph intersects the *y*-axis at (0, –3) and (0, 3).

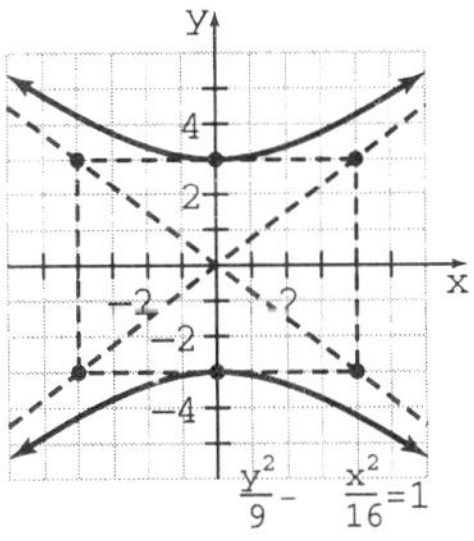

21. **a.** $\frac{y^2}{25} - \frac{x^2}{36} = 1$

Since $a^2 = 36$ and $b^2 = 25$, $a = 6$ and $b = 5$. The equations of the asymptotes are $y = \pm\frac{b}{a}x$, or $y = \pm\frac{5}{6}x$.

b. To graph the asymptotes, plot the points (6, 5), (–6, 5), (6, –5), and (–6, –5). The graph intersects the *y*-axis at (0, –5) and (0, 5).

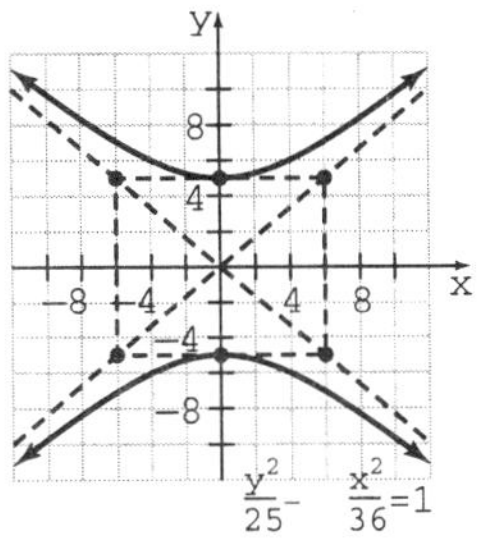

23. **a.** $\frac{y^2}{25} - \frac{x^2}{4} = 1$

Since $a^2 = 4$ and $b^2 = 25$, $a = 2$ and $b = 5$. The equations of the asymptotes are $y = \pm\frac{b}{a}x$ or $y = \pm\frac{5}{2}x$.

b. To graph the asymptotes, plot the points (2, 5), (–2, 5), (2, –5) and (–2, –5). The graph intersects the *y*-axis at (0, –5) and (0, 5).

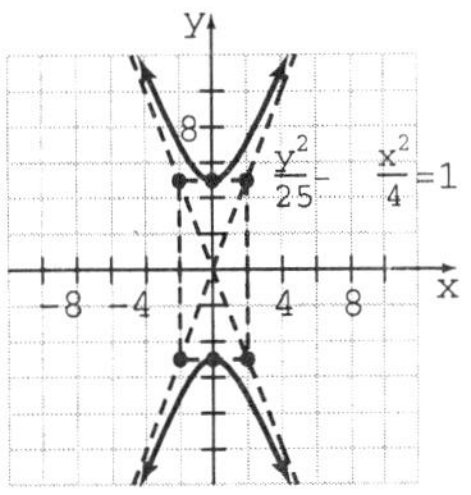

25. **a.** $\frac{y^2}{16} - \frac{x^2}{81} = 1$

Since $a^2 = 81$ and $b^2 = 16$, $a = 9$ and $b = 4$. The equations of the asymptotes are $y = \pm\frac{b}{a}x$, or $y = \pm\frac{4}{9}x$.

b. To graph the asymptotes, plot the points (9, 4), (–9, 4), (9, –4), and (–9, –4). The graph intersects the *y*-axis at (0, –4) and (0, 4).

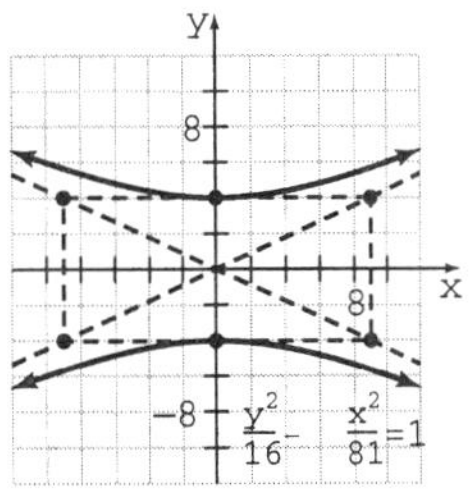

27. **a.** $x^2 - 25y^2 = 25$

$\frac{x^2}{25} - \frac{25y^2}{25} = 1$

$\frac{x^2}{25} - \frac{y^2}{1} = 1$

Since $a^2 = 25$ and $b^2 = 1$, $a = 5$ and $b = 1$. The equations of the asymptotes are $y = \pm\frac{b}{a}x$, or $y = \pm\frac{1}{5}x$.

b. To graph the asymptotes, plot the points (5, 1), (–5, 1), (5, –1), and (–5, –1). The graph intersects the *x*-axis at (–5, 0), and (5, 0).

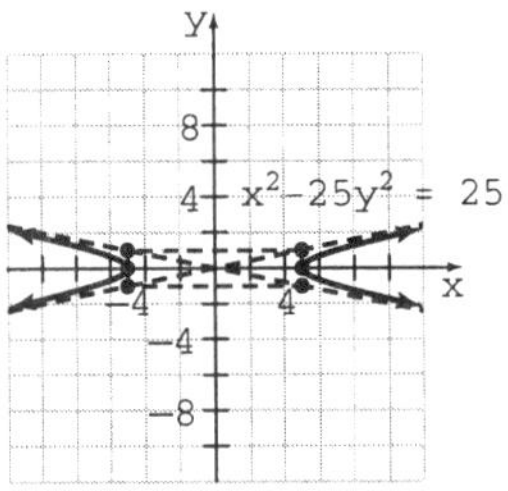

29. **a.** $16x^2 - 4y^2 = 64$

$$\frac{16x^2}{64} - \frac{4y^2}{64} = 1$$

$$\frac{x^2}{4} - \frac{y^2}{16} = 1$$

Since $a^2 = 4$ and $b^2 = 16$, $a = 2$ and $b =$ 4. The equations of the asymptotes are $y = \pm\frac{b}{a}x$, or $y = \pm 2x$.

b. To graph the asymptotes, plot the points (2, 4), (–2, 4), (2 –4), and (–2, –4). The graph intersects the x-axis at (–2, 0) and (2, 0).

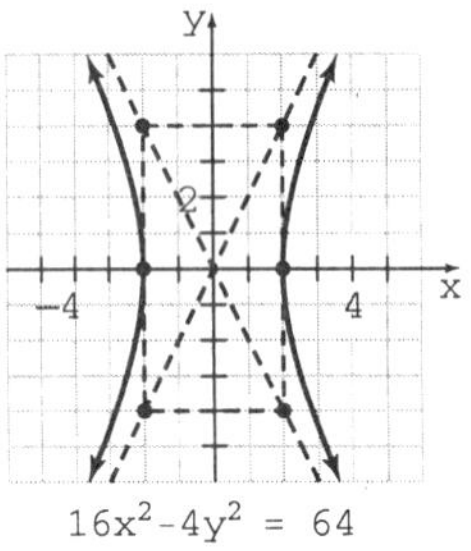

31. **a.** $9y^2 - x^2 = 9$

$$\frac{9y^2}{9} - \frac{x^2}{9} = 1$$

$$\frac{y^2}{1} - \frac{x^2}{9} = 1$$

Since $a^2 = 9$ and $b^2 = 1$, $a = 3$ and $b =$ 1. The equations of the asymptotes are $y = \pm\frac{b}{a}x$, or $\pm\frac{1}{3}x$.

b. To graph the asymptotes, plot the points (3, 1), (–3, 1), (3, –1), and (–3, –1). The graph intersects the y-axis at (0, –1) and (0, 1).

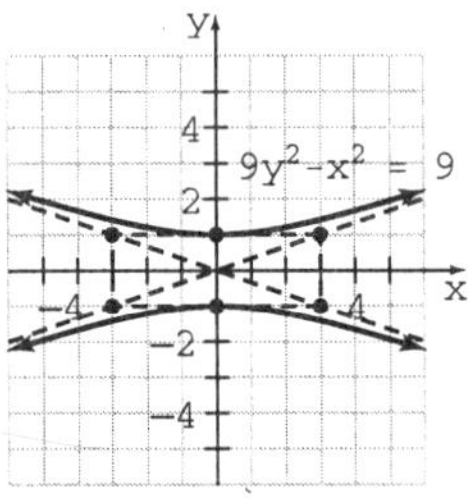

33. **a.** $25x^2 - 9y^2 = 225$

$$\frac{25x^2}{225} - \frac{9y^2}{225} = 1$$

$$\frac{x^2}{9} - \frac{y^2}{25} = 1$$

Since $a^2 = 9$ and $b^2 = 25$, $a = 3$ and $b =$ 5. The equations of the asymptotes are $y = \pm\frac{b}{a}x$, or $y = \pm\frac{5}{3}x$.

b. To graph the asymptotes, plot the points (3, 5), (–3, 5), (3, –5), and (–3, –5). The graph intersects the x-axis at (–3, 0) and (3, 0).

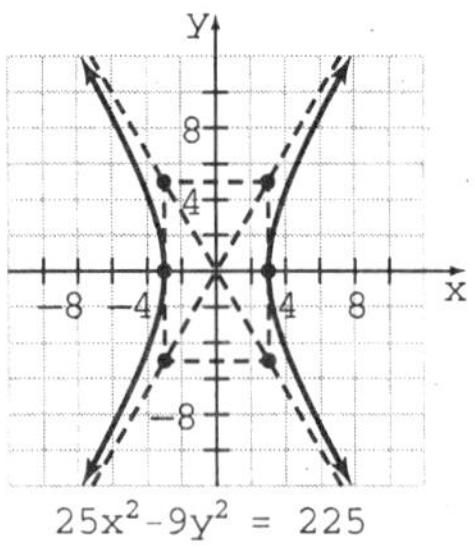

35. **a.** $4y^2 - 36x^2 = 144$

$$\frac{4y^2}{144} - \frac{36x^2}{144} = 1$$

$$\frac{y^2}{36} - \frac{x^2}{4} = 1$$

Since $a^2 = 4$ and $b^2 = 36$, $a = 2$ and $b = 6$. The equations of the asymptotes are $y = \pm\frac{b}{a}x$, or $y = \pm 3x$.

b. To graph the asymptotes, plot the points (2, 6), (–2, 6), (2, –6), and (–2, –6). The graph intersects the y-axis at (0, –6) and (0, 6).

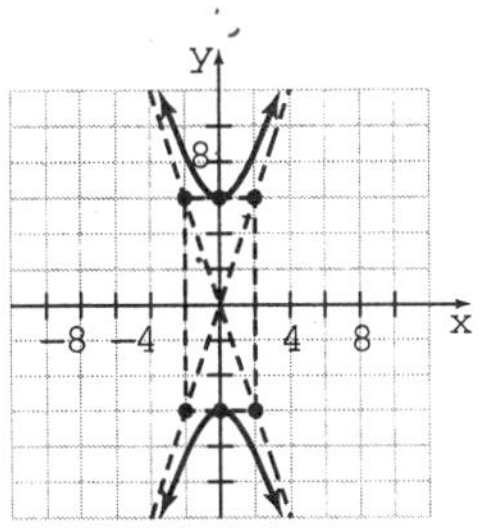

$4y^2-36x^2 = 144$

37. $5x^2+5y^2=125$

$\frac{5x^2}{5}+\frac{5y^2}{5}=\frac{125}{5}$

$x^2+y^2=25$

The graph is a circle.

39. $x^2+5y^2=125$

$\frac{x^2}{125}+\frac{5y^2}{125}=1$

$\frac{x^2}{125}+\frac{y^2}{25}=1$

The graph is an ellipse.

41. $4x^2-4y^2=29$

$\frac{4x^2}{29}-\frac{4y^2}{29}=1$

$\frac{x^2}{\frac{29}{4}}-\frac{y^2}{\frac{29}{4}}=1$

The graph is a hyperbola.

43. $2y=12x^2-8x+20$

$y=-6x^2-4x+10$

The graph is a parabola.

45. $6x^2+7y^2=42$

$\frac{6x^2}{42}+\frac{7y^2}{42}=1$

$\frac{x^2}{7}+\frac{y^2}{6}=1$

The graph is an ellipse.

47. $3x=-2y^2+9y-15$

$x=-\frac{2}{3}y^2+3y-5$

The graph is a parabola.

49. $6x^2+6y^2=36$

$\frac{6x^2}{6}+\frac{6y^2}{6}=36$

$x^2+y^2=6$

The graph is a circle.

51. $5y^2=7x^2+35$

$5y^2-7x^2=35$

$\frac{5y^2}{35}-\frac{7x^2}{35}=1$

$\frac{y^2}{7}-\frac{x^2}{5}=1$

The graph is a hyperbola.

53. $x+y=2y^2+$

$x=2y^2-y+5$

The graph is a parabola.

55. $3x^2=12y^2+48$

$3x^2-12y^2=48$

$\frac{3x^2}{48}-\frac{12y^2}{48}=1$

$\frac{x^2}{16}-\frac{y^2}{4}=1$

The graph is a hyperbola.

57. $y-x+2=x^2$

$y=x^2+x-2$

The graph is a parabola.

59. $-3x^2-3y^2=-243$

$\frac{-3x^2}{-3}+\frac{-3y^2}{-3}=\frac{-243}{-3}$

$x^2+y^2=81$

The graph is a circle.

61. Since the vertices are $(0, \pm 2)$, the hyperbola is of the form $\frac{y^2}{b^2}-\frac{x^2}{a^2}=1$ with $b = 2$. Since the asymptotes are $y=\pm\frac{1}{2}x$, we have $\frac{b}{a}=\frac{1}{2}$. Therefore, $\frac{2}{a}=\frac{1}{2}$, so $a = 4$. The equation of the hyperbola is $\frac{y^2}{2^2}-\frac{x^2}{4^2}=1$, or $\frac{y^2}{4}-\frac{x^2}{16}=1$.

63. Since the vertices are $(0, \pm 3)$, the hyperbola is of the form $\frac{x^2}{a^2} - \frac{y^2}{b^2} = 1$ with $a = 3$. Since the asymptotes are $y = \pm 2x$, we have $\frac{b}{a} = 2$. Therefore, $\frac{b}{3} = 2$, so $b = 6$. The equation of the hyperbola is $\frac{x^2}{3^2} - \frac{y^2}{6^2} = 1$, or $\frac{x^2}{9} - \frac{y^2}{36} = 1$.

65. Since the transverse axis is along the x-axis, the equation is of the form $\frac{x^2}{a^2} - \frac{y^2}{b^2} = 1$ Since the asymptotes are $y = \pm\frac{5}{3}x$, we require $\frac{b}{a} = \frac{5}{3}$.
Using $a = 3$ and $b = 5$, the equation of the hyperbola is $\frac{x^2}{3^2} - \frac{y^2}{5^2} = 1$, or $\frac{x^2}{9} - \frac{y^2}{25} = 1$.
No, this is not the only possible answer, because a and b are not uniquely determined.
$\frac{x^2}{18} - \frac{y^2}{50} = 1$ and others will also work.

67. No, for each value of x with $|x| > a$, there are 2 possible values of y.

69. $\frac{x^2}{9} - \frac{y^2}{4} = 1$. This hyperbola has its transverse axis along the x-axis with vertices at $(\pm 3, 0)$.
Domain: $(-\infty, -3] \cup [3, \infty)$
Range: R

71. The equation is changed from $\frac{x^2}{a^2} - \frac{y^2}{b^2} = 1$ to $\frac{x^2}{b^2} - \frac{y^2}{a^2} = 1$. Both graphs have a transverse axis along the x-axis. The vertices of the second graph will be closer to the origin, at $(\pm b, 0)$ instead of $(\pm a, 0)$. The second graph will open wider.

73. Answers will vary.

75. The points are $(-6, 4)$ and $(-2, 2)$.

$$m = \frac{y_2 - y_1}{x_2 - x_1} = \frac{2-4}{-2-(-6)} = \frac{-2}{4} = -\frac{1}{2}$$

Use $y - y_1 = m(x - x_1)$, with $m = -\frac{1}{2}, (-2, 2)$

$$y - 2 = -\frac{1}{2}(x-(-2))$$
$$y - 2 = -\frac{1}{2}x - 1$$
$$y = -\frac{1}{2}x + 1$$

76. $\frac{3x}{2x-3} + \frac{3x+6}{2x^2+x-6}$

$$= \frac{3x}{2x-3} + \frac{3x+6}{(2x-3)(x+2)}$$
$$= \frac{3x}{2x-3} \cdot \frac{x+2}{x+2} + \frac{3x+6}{(2x-3)(x+2)}$$
$$= \frac{3x^2+6x+3x+6}{(2x-3)(x+2)}$$
$$= \frac{3x^2+9x+6}{(2x-3)(x+2)}$$
$$= \frac{3(x+1)(x+2)}{(2x-3)(x+2)} \text{ or } \frac{3(x+1)}{2x-3}$$

77. $f(x) = 3x^2 - x + 4, \; g(x) = 6 - 4x^2$

$$(f+g)(x) = (3x^2 - x + 4) + (6 - 4x^2)$$
$$= -x^2 - x + 10$$

78.
$$5(-4x + 9y = 7) \Rightarrow -20x + 45y = 35$$
$$4(5x + 6y = -3) \Rightarrow 20x + 24y = -12$$
$$69y = 23$$
$$y = \frac{1}{3}$$

$$5x + 6\left(\frac{1}{3}\right) = -3 \Rightarrow 5x + 2 = -3 \Rightarrow x = -1$$

The solution is $\left(-1, \frac{1}{3}\right)$.

79. $E = \frac{1}{2}mv^2$, for v

$$2E = mv^2$$
$$\frac{2E}{m} = v^2$$
$$\sqrt{\frac{2E}{m}} = v \text{ or } v = \sqrt{\frac{2E}{m}}$$

80. $\log(x+3) = \log 4 - \log x$

$\log(x+3) = \log\frac{4}{x}$

$x+3 = \frac{4}{x}$

$x(x+3) = 4$

$x^2 + 3x - 4 = 0$

$(x+4)(x-1) = 0$

$x+4=0$ or $x-1=0$

$x=-4$ $\quad x=1$

Upon checking, $x=-4$ does not satisfy the equation. The solution is $x=1$.

Exercise Set 14.4

1. A nonlinear system of equations is a system in which at least one equation is nonlinear.

3. Yes

5. Yes

7. $x^2 + y^2 = 8$

$x + y = 0$

Solve $x+y=0$ for x: $x=-y$.

Substitute $x=-y$ for x in $x^2+y^2=8$.

$x^2 + y^2 = 8$

$(-y)^2 + y^2 = 8$

$y^2 + y^2 = 8$

$2y^2 = 8$

$y^2 = 4$

$y = \pm 2$

$y=2$ or $y=-2$

$x=-2$ $\quad x=2$

The solutions are (2, –2) and (–2, 2).

9. $x^2 + y^2 = 9$

$x + 2y = 3$

Solve $x+2y=3$ for x: $x=3-2y$.

Substitute $3-2y$ for x in $x^2+y^2=9$.

$x^2 + y^2 = 9$

$(3-2y)^2 + y^2 = 9$

$9 - 12y + 4y^2 + y^2 = 9$

$5y^2 - 12y = 0$

$y(5y-12) = 0$

$y=0$ or $y=\frac{12}{5}$

$x = 3-2y$
$x = 3-2(0)$
$x = 3$

$x = 3-2y$
$x = 3-2\left(\frac{12}{5}\right)$
$x = -\frac{9}{5}$

The solutions are (3, 0) and $\left(-\frac{9}{5}, \frac{12}{5}\right)$.

11. $y = x^2 - 5$

$3x + 2y = 10$

Substitute x^2-5 for y in $3x+2y=10$

$3x + 2y = 10$

$3x + 2(x^2-5) = 10$

$3x + 2x^2 - 10 = 10$

$2x^2 + 3x - 20 = 0$

$(x+4)(2x-5) = 0$

$x=-4$ or $x=\frac{5}{2}$

$y = x^2 - 5$
$y = (-4)^2 - 5$
$y = 11$

$y = x^2 - 5$
$y = \left(\frac{5}{2}\right)^2 - 5$
$y = \frac{5}{4}$

The solutions are $(-4, 11)$ and $\left(\frac{5}{2}, \frac{5}{4}\right)$.

13. $x + y = 4$

$x^2 - y^2 = 4$

Solve $x+y=4$ for y: $y=4-x$.

Substitute $4-x$ for y in $x^2-y^2=4$

$$x^2-y^2=4$$
$$x^2-(4-x)^2=4$$
$$x^2-(16-8x+x^2)=4$$
$$-16+8x=4$$
$$8x=20$$
$$x=\frac{5}{2}$$

$y=4-x$

$y=4-\frac{5}{2}$

$y=\frac{3}{2}$

The solution is $\left(\frac{5}{2}, \frac{3}{2}\right)$.

15. $x+y^2=4 \Rightarrow y^2=-x+4$

$x^2+y^2=6$

Substitute $-x+4$ for y^2 in $x^2+y^2=6$.

$$x^2+y^2=6$$
$$x^2+(-x+4)=6$$
$$x^2-x-2=0$$
$$(x-2)(x+1)=0$$

$x=2$ or $x=-1$

$y^2=-x+4$	$y^2=-x+4$
$y^2=-2+4$	$y^2=-(-1)+4$
$y^2=2$	$y^2=5$
$y=\pm\sqrt{2}$	$y=\pm\sqrt{5}$

The solutions are $(2, \sqrt{2})$, $(2, -\sqrt{2})$, $(-1, \sqrt{5})$, and $(-1, -\sqrt{5})$.

17. $x^2+y^2=4$

$y=x^2-12 \Rightarrow x^2=y+12$

Substitute $y+12$ for x^2 in $x^2+y^2=4$.

$$x^2+y^2=4$$
$$(y+12)+y^2=4$$
$$y^2+y+8=0$$
$$y=\frac{-1\pm\sqrt{1^2-4(1)(8)}}{2(1)}$$
$$=\frac{-1\pm\sqrt{-31}}{2}$$
$$=\frac{-1\pm i\sqrt{31}}{2}$$

There is no real solution.

19. $x^2+y^2=9$

$y=x^2-3$

Solve $y=x^2-3$ for x^2: $x^2=y+3$.

Substitute $y+3$ for x^2 in $x^2+y^2=9$.

$$x^2+y^2=9$$
$$(y+3)+y^2=9$$
$$y^2+y-6=0$$
$$(y-2)(y+3)=0$$

$y=2$ or $y=-3$

$x^2=y+3$	$x^2=y+3$
$x^2=2+3$	$x^2=-3+3$
$x^2=5$	$x^2=0$
$x=\pm\sqrt{5}$	$x=0$

The solutions are $(0, -3)$, $(\sqrt{5}, 2)$, and $(-\sqrt{5}, 2)$

21. $2x^2-y^2=-8$

$x-y=6$

Solve the second equation for y: $y=x-6$.

Substitute $x-6$ for y in $2x^2-y^2=-8$.

$$2x^2-y^2=-8$$
$$2x^2-(x-6)^2=-8$$
$$2x^2-(x^2-12x+36)=-8$$
$$2x^2-x^2+12x-36=-8$$
$$x^2+12x-28=0$$
$$(x-2)(x+14)=0$$

$x=2$ or $x=-14$

$y=x-6$	$y=x-6$
$y=2-6$	$y=-14-6$
$y=-4$	$y=-20$

The solutions are (2, –4) and (–14, –20)

23. $x^2 - y^2 = 4$

$\underline{2x^2 + y^2 = 8}$

$3x^2 \quad = 12$

$x^2 = 4$

$x = 2$ or $x = -2$

$x^2 - y^2 = 4$ $\quad$ $x^2 - y^2 = 4$

$2^2 - y^2 = 4$ $\quad$ $(-2)^2 - y^2 = 4$

$y^2 = 0$ $\quad$ $y^2 = 0$

$y = 0$ $\quad$ $y = 0$

The solutions are (2, 0) and (–2, 0).

25. $x^2 + y^2 = 13$ (1)

$2x^2 + 3y^2 = 30$ (2)

$-2x^2 - 2y^2 = -26$ (1) multiplied by – 2

$\underline{2x^2 + 3y^2 = 30}$ (2)

$y^2 = 4$

$y = 2$ or $y = -2$

$x^2 + y^2 = 13$ $\quad$ $x^2 + y^2 = 13$

$x^2 + 2^2 = 13$ $\quad$ $x^2 + (-2)^2 = 13$

$x^2 = 9$ $\quad$ $x^2 = 9$

$x = \pm 3$ $\quad$ $x = \pm 3$

The solutions are (3, 2), (3, –2), (–3, 2), and (–3, –2).

27. $x^2 + y^2 = 25$ (1)

$x^2 - 2y^2 = 7$ (2)

$2x^2 + 2y^2 = 50$ (1) Multiplied by 2

$\underline{x^2 - 2y^2 = 7}$ (2)

$3x^2 = 57$

$x^2 = 19$

$x = \sqrt{19}$ or $x = -\sqrt{19}$

$x^2 + y^2 = 25$ $\quad$ $x^2 + y^2 = 25$

$\left(\sqrt{19}\right)^2 + y^2 = 25$ $\quad$ $\left(-\sqrt{19}\right)^2 + y^2 = 25$

$y^2 = 6$ $\quad$ $y^2 = 6$

$y = \pm\sqrt{6}$ $\quad$ $y = \pm\sqrt{6}$

The solutions are $\left(\sqrt{19}, \sqrt{6}\right)$, $\left(\sqrt{19}, -\sqrt{6}\right)$, $\left(-\sqrt{19}, \sqrt{6}\right)$, and $\left(-\sqrt{19}, -\sqrt{6}\right)$.

29. $4x^2 + 9y^2 = 36$

$\underline{2x^2 - 9y^2 = 18}$

$6x^2 = 54$

$x^2 = 9$

$x = 3$ or $x = -3$

$4x^2 + 9y^2 = 36$ $\quad$ $4x^2 + 9y^2 = 36$

$4(3)^2 + 9y^2 = 36$ $\quad$ $4(-3)^2 + 9y^2 = 36$

$9y^2 = 0$ $\quad$ $9y^2 = 0$

$y^2 = 0$ $\quad$ $y^2 = 0$

$y = 0$ $\quad$ $y = 0$

The solutions are (3, 0) and (–3, 0).

31. $5x^2 - 2y^2 = -13$ (1)

$3x^2 + 4y^2 = 39$ (2)

$10x^2 - 4y^2 = -26$ (1) multiplied by 2

$\underline{3x^2 + 4y^2 = 39}$ (2)

$13x^2 = 13$

$x^2 = 1$

$x = 1$ or $x = -1$

$3x^2 + 4y^2 = 39$ $\quad$ $3x^2 + 4y^2 = 39$

$3(1)^2 + 4y^2 = 39$ $\quad$ $3(-1)^2 + 4y^2 = 39$

$4y^2 = 36$ $\quad$ $4y^2 = 36$

$y^2 = 9$ $\quad$ $y^2 = 9$

$y = \pm 3$ $\quad$ $y = \pm 3$

The solutions are (1, 3), (1, –3), (–1, 3), and (–1, –3).

33. $x^2 - 2y^2 = 7$ (1)

$x^2 + y^2 = 34$ (2)

$-x^2 + 2y^2 = -7$ (1) multiplied by – 1

$\underline{x^2 + y^2 = 34}$ (2)

$3y^2 = 27$

$y^2 = 9$

$y = 3$ or $y = -3$

$x^2 + y^2 = 34$ $\quad x^2 + y^2 = 34$
$x^2 + 3^2 = 34$ $\quad x^2 + (-3)^2 = 34$
$x^2 = 25$ $\quad x^2 = 25$
$x = \pm 5$ $\quad x = \pm 5$

The solutions are (5, 3), (5, –3), (–5, 3), and (–5, –3).

35. $x^2 + y^2 = 9$ (1)
$16x^2 - 4y^2 = 64$ (2)
$4x^2 + 4y^2 = 36$ (1) multiplied by 4
$16x^2 - 4y^2 = 64$ (2)
$20x^2 = 100$
$x^2 = 5$
$x = \sqrt{5}$ or $x = -\sqrt{5}$

$x^2 + y^2 = 9$ $\quad x^2 + y^2 = 9$
$\left(\sqrt{5}\right)^2 + y^2 = 9$ $\quad \left(-\sqrt{5}\right)^2 + y^2 = 9$
$y^2 = 4$ $\quad y^2 = 4$
$y = \pm 2$ $\quad y = \pm 2$

The solutions are $\left(\sqrt{5},\ 2\right)$, $\left(\sqrt{5},\ -2\right)$, $\left(-\sqrt{5},\ 2\right)$, and $\left(-\sqrt{5},\ -2\right)$.

37. Answers will vary.

39. Let x = length
y = width
$xy = 500$
$2x + 2y = 90$
Solve $2x + 2y = 90$ for y: $y = 45 - x$.
Substitute $45 - x$ for y in $xy = 500$.
$xy = 500$
$x(45 - x) = 500$
$45x - x^2 = 500$
$x^2 - 45x + 500 = 0$
$(x - 20)(x - 25) = 0$
$x - 20 = 0$ $\quad x - 25 = 0$
$x = 20$ or $x = 25$

$y = 45 - x$ $\quad y = 45 - 25$
$y = 45 - 20$ $\quad y = 45 - 25$
$y = 25$ $\quad y = 20$

The solutions are (20, 25) and (25, 20).
The dimensions of the dance floor are 20 m by 25 m.

41. $xy = 48$
$2x + y = 20$
Solve $2x + y = 20$ for y: $y = 20 - 2x$.
Substitute $20 - 2x$ for y in $yx = 48$.
$xy = 48$
$x(20 - 2x) = 48$
$20x - 2x^2 = 48$
$2x^2 - 20x + 48 = 0$
$x^2 - 10x + 24 = 0$
$(x - 4)(x - 6) = 0$
$x - 4 = 0$ or
$x = 4$

$y = 20 - 2x$
$y = 20 - 2(4)$
$y = 12$
$x - 6 = 0$
$x = 6$

$y = 20 - 2x$
$y = 20 - 2(6)$
$y = 8$

The solutions are (4, 12) and (6, 8).
The dimensions are 6 ft by 8 ft or 4 ft by 12 ft.

43. Let x = length
y = width
$xy = 112$
$x^2 + y^2 = \left(\sqrt{260}\right)^2$
Solve $xy = 112$ for y: $y = \frac{112}{x}$.
$y^2 + x^2 = 260$
$y^2 + \left(\frac{112}{y}\right)^2 = 260$
$y^2 + \frac{12{,}544}{y^2} = 260$
$y^4 + 12{,}544 = 260y^2$
$y^4 - 260y^2 + 12{,}544 = 0$
$(y^2 - 64)(y^2 - 196) = 0$
$x^2 - 64 = 0$ or
$x^2 = 64$
$x = \pm 8$

$x^2 - 196 = 0$
$x^2 = 196$
$x = \pm 14$
Since x must be positive, $x = 8$ or $x = 14$.
If $x = 8$, then $y = \dfrac{112}{8} = 14$.
If $x = 14$, then $y = \dfrac{112}{14} = 8$.
The dimensions of the new bill are 8 cm by 14 cm.

45. Let x = length
y = width
$x^2 + y^2 = 17^2$
$x + y + 17 = 40$
Solve $x + y + 17 = 40$ for y: $y = 23 - x$.
Substitute $23 - x$ for y in $x^2 + y^2 = 17^2$.

$$x^2 + y^2 = 17^2$$
$$x^2 + (23 - x)^2 = 17^2$$
$$x^2 + (529 - 46x + x^2) = 289$$
$$2x^2 - 46 + 529 = 289$$
$$2x^2 - 46x + 240 = 0$$
$$x^2 - 23x + 120 = 0$$
$$(x - 8)(x - 15) = 0$$

$x - 8 = 0$ or $x - 15 = 0$
$x = 8$ $\quad x = 15$

$y = 23 - x$ $\quad y = 23 - x$
$y = 23 - 8$ $\quad y = 23 - 15$
$y = 15$ $\quad y = 8$
The solutions are (8, 15) and (15, 8).
The dimensions of the piece of wood are 8 in. by 15 in.

47. $d = -16t^2 + 64t$
$d = -16t^2 + 16t + 80$
Substitute $-16t^2 + 64t$ for d in $d = -16t^2 + 16t + 80$.

$$d = -16t^2 + 16t + 80$$
$$-16t^2 + 64t = -16t^2 + 16t + 80$$
$$64t = 16t + 80$$
$$48t = 80$$
$$t = \frac{80}{48} = \frac{5}{3} \approx 1.67$$

The balls are the same height above the ground at $t \approx 1.67$ sec.

49. Since $t = 1$ year, we may write the formula as $i = pr$.
$7.50 = pr$
$7.50 = (p + 25)(r - 0.01)$
Rewrite the second equation by multiplying the binomials. Then substitute 7.50 for pr and solve for r.

$$7.50 = (p + 25)(r - 0.01)$$
$$7.50 = pr - 0.01p + 25r - 0.25$$
$$7.50 = 7.50 - 0.01p + 25r - 0.25$$
$$0 = -0.01p + 25r - 0.25$$
$$0.01p + 0.25 = 25r$$
$$\frac{0.01p}{25} + \frac{0.25}{25} = \frac{25r}{25}$$
$$r = 0.0004p + 0.01$$

Substitute $0.0004p + 0.01$ for r in $7.50 = pr$.
$7.50 = pr$
$7.50 = p(0.0004p + 0.01)$
$7.50 = 0.0004p^2 + 0.01p$
$0 = 0.0004p^2 + 0.01p - 7.50$
$0 = p^2 + 25p - 18,750$
$0 = (p - 125)(p + 150)$

$p - 125 = 0$ or
$p = 125$
$p + 150 = 0$
$p = -150$
Since the principal must be positive, use $p = 125$.
$r = 0.0004p + 0.01$
$r = 0.0004(125) + 0.01$
$r = 0.06$
The principal is \$125 and the interest rate is 6%.

51. $C = 10x + 300$
$R = 30x - 0.1x^2$

$$C = R$$
$$10x + 300 = 30x - 0.1x^2$$
$$0.1x^2 - 20x + 300 = 0$$
$$x = \frac{-b \pm \sqrt{b^2 - 4ac}}{2a}$$
$$= \frac{-(-20) \pm \sqrt{(-20)^2 - 4(0.1)(300)}}{2(0.1)}$$
$$= \frac{20 \pm \sqrt{280}}{0.2}$$
$$x = \frac{20 + \sqrt{280}}{0.2} \approx 183.7 \text{ or}$$

$x = \frac{20-\sqrt{280}}{0.2} \approx 16.3$

The break-even points are ≈ 16 and ≈ 184.

53. $C = 80x + 900$
$R = 120x - 0.2x^2$

$$C = R$$
$$80x + 900 = 120x - 0.2x^2$$
$$0.2x^2 - 40x + 900 = 0$$
$$x = \frac{-b \pm \sqrt{b^2 - 4ac}}{2a}$$
$$= \frac{-(-40) \pm \sqrt{(-40)^2 - 4(0.2)(900)}}{2(0.2)}$$
$$= \frac{40 \pm \sqrt{880}}{0.4}$$
$$x = \frac{40+\sqrt{880}}{0.4} \approx 174.2 \text{ or}$$
$$x = \frac{40-\sqrt{880}}{0.4} \approx 25.8$$

The break-even points are ≈ 26 and ≈ 174.

55. Solve each equation for y.

$$3x - 5y = 12$$
$$-5y = -3x + 12$$
$$y = \frac{3}{5}x - \frac{12}{5}$$
$$x^2 + y^2 = 10$$
$$y^2 = 10 - x^2$$
$$y = \pm\sqrt{10 - x^2}$$

Use $y_1 = \frac{3}{5}x - \frac{12}{5}$, $y_2 = \sqrt{10 - x^2}$, and $y_2 = -\sqrt{10 - x^2}$.

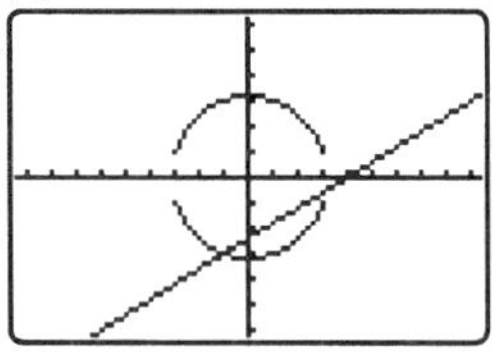

-9.4, 9.4, 1, -6.2,

Approximate solutions: (–1, –3), (3.12, –0.53)

57. Let x = length of one leg
y = length of other leg

$$x^2 + y^2 = 26^2$$
$$\frac{1}{2}xy = 120$$

Solve $\frac{1}{2}xy = 120$ for y: $y = \frac{240}{x}$.

Substitute $\frac{240}{x}$ for y in $x^2 + y^2 = 26^2$

$$x^2 + y^2 = 26^2$$
$$x^2 + \left(\frac{240}{x}\right)^2 = 676$$
$$x^2 + \frac{57,600}{x^2} = 676$$
$$x^4 + 57,600 = 676x^2$$
$$x^4 - 676x^2 + 57,600 = 0$$
$$(x^2 - 100)(x^2 - 576) = 0$$

$x^2 - 100 = 0$ or
$x^2 = 100$
$x = \pm 10$

$x^2 - 576 = 0$
$x^2 = 576$
$x = \pm 24$

Since x is a length, x must be positive.

If $x = 10$, then $y = \frac{240}{10} = 24$.

If $x = 24$, then $y = \frac{240}{24} = 10$.

The legs have lengths 10 yards and 24 yards.

59. The operations are evaluated in the following order: parentheses, exponents, multiplication or division, addition or subtraction.

60.
$$(x+1)^3 + 1$$
$$= (x+1)^3 + (1)^3$$
$$= (x+1+1)\left((x+1)^2 - (x+1)(1) + (1)^2\right)$$
$$= (x+2)\left(x^2 + 2x + 1 - x - 1 + 1\right)$$
$$= (x+2)\left(x^2 + x + 1\right)$$

61. $x = \frac{k}{P^2}$

$10 = \frac{k}{6^2} \Rightarrow k = 360 \Rightarrow x = \frac{360}{P^2}$

$x = \frac{360}{20^2} = \frac{360}{400} = \frac{9}{10}$ or 0.9

62.
$$\frac{2}{\sqrt{x+2}-3} = \frac{2}{\sqrt{x+2}-3} \cdot \frac{\sqrt{x+2}+3}{\sqrt{x+2}+3}$$
$$= \frac{2\sqrt{x+2}+6}{x+2+3\sqrt{x+2}-3\sqrt{x+2}-9}$$
$$= \frac{2\sqrt{x+2}+6}{x-7}$$

63. $A = A_0 e^{kt}$, for k

$\frac{A}{A_0} = e^{kt}$

$\ln\frac{A}{A_0} = \ln e^{kt}$

$\ln A - \ln A_0 = kt$

$\frac{\ln A - \ln A_0}{t} = k$

Review Exercises

1. $d = \sqrt{(x_2 - x_1)^2 + (y_2 - y_1)^2}$
$= \sqrt{(3 - 0)^2 + (-4 - 0)^2}$
$= \sqrt{3^2 + (-4)^2}$
$= \sqrt{9 + 16}$
$= \sqrt{25}$
$= 5$

Midpoint $= \left(\frac{x_1 + x_2}{2}, \frac{y_1 + y_2}{2}\right)$
$= \left(\frac{0 + 3}{2}, \frac{0 + (-4)}{2}\right)$
$= \left(\frac{3}{2}, -2\right)$

2. $d = \sqrt{(x_2 - x_1)^2 + (y_2 - y_1)^2}$
$= \sqrt{(-1-(-4))^2 + (5-1)^2}$
$= \sqrt{3^2 + 4^2}$
$= \sqrt{9+16}$
$= \sqrt{25}$
$= 5$

Midpoint $= \left(\frac{x_1 + x_2}{2}, \frac{y_1 + y_2}{2}\right)$
$= \left(\frac{-4+(-1)}{2}, \frac{1+5}{2}\right)$
$= \left(-\frac{5}{2}, 3\right)$

3. $d = \sqrt{(x_2 - x_1)^2 + (y_2 - y_1)^2}$
$= \sqrt{[-6-(-1)]^2 + [10-(-2)]^2}$
$= \sqrt{(-5)^2 + 12^2}$
$= \sqrt{25+144}$
$= \sqrt{169}$
$= 13$

Midpoint $= \left(\frac{x_1 + x_2}{2}, \frac{y_1 + y_2}{2}\right)$
$= \left(\frac{-6+(-1)}{2}, \frac{-2+10}{2}\right)$
$= \left(-\frac{7}{2}, 4\right)$

4. $d = \sqrt{(x_2 - x_1)^2 + (y_2 - y_1)^2}$
$= \sqrt{[-2-(-4)]^2 + (5-3)^2}$
$= \sqrt{2^2 + 2^2}$
$= \sqrt{4+4}$
$= \sqrt{8}$
≈ 2.83

Midpoint $= \left(\frac{x_1 + x_2}{2}, \frac{y_1 + y_2}{2}\right)$
$= \left(\frac{-4+(-2)}{2}, \frac{3+5}{2}\right)$
$= (-3, 4)$

5. $y = (x-2)^2 + 1$
This is a parabola in the form $y = a(x-h)^2 + k$ with $a = 1$, $h = 2$, and $k = 1$. Since $a > 0$, the parabola opens upward. The vertex is (2, 1). The y-intercept is (0, 5). There are no x-intercepts.

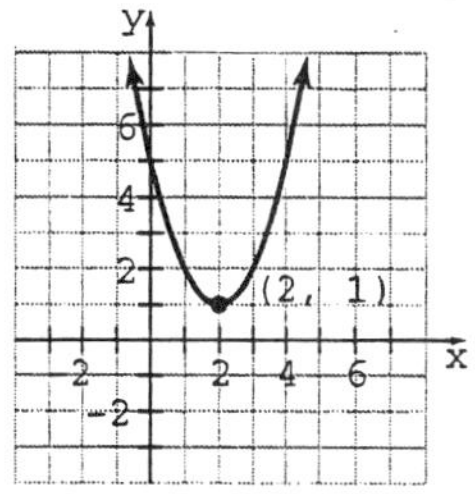

6. $y = (x+3)^2 - 4$

This is a parabola in the form $y = a(x-h)^2 + k$ with $a = 1$, $h = -3$, and $k = -4$. Since $a > 0$, the parabola opens upward. The vertex is (–3, –4). The y-intercept is (0, 5).

The x-intercepts are about (–5, 0) and (–1, 0).

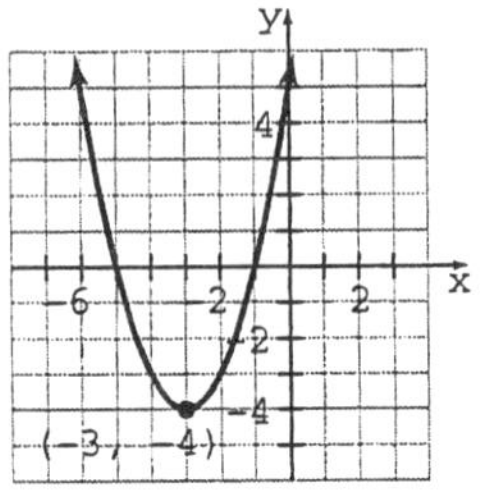

7. $x = (y-1)^2 + 4$

This is a parabola in the form $x = a(y-k)^2 + h$ with $a = 1$, $h = 4$, and $k = 1$. Since $a > 0$, the parabola opens to the right. The vertex is (4, 1). There are no y-intercepts.

The x-intercept is (5, 0).

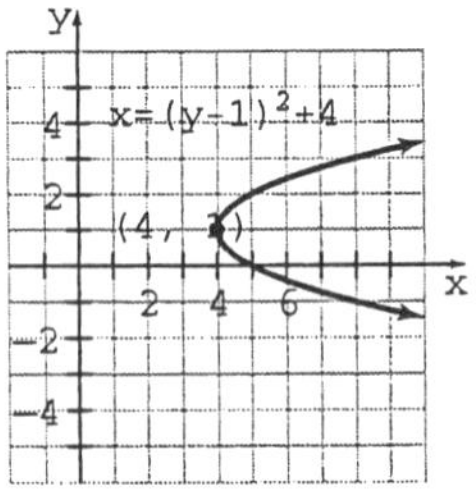

8. $x = -2(y+4)^2 - 3$

This is a parabola in the form $x = a(y-k)^2 + h$ with $a = -2$, $h = -3$, and $k = -4$. Since $a < 0$, the parabola opens to the left. The vertex is (–3, –4). There are no y-intercepts.

The x-intercept is (–35, 0).

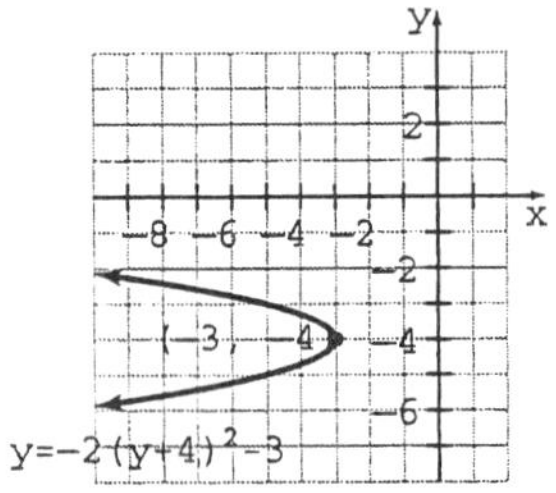

9. **a.** $y = x^2 - 8x + 22$

$y = \left(x^2 - 8x + 16\right) + 22 - 16$

$y = (x-4)^2 + 6$

b. This is a parabola in the form $y = a(x-h)^2 + k$ with $a = 1$, $h = 4$, and $k = 6$. Since $a > 0$, the parabola opens upward. The vertex is (4, 6). The y-intercept is (0, 22). There are no y-intercepts.

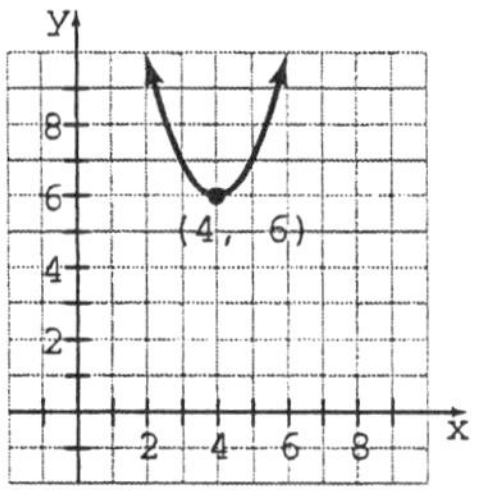

10. **a.** $x = -y^2 - 2y + 5$

$x = -(y^2 + 2y) + 5$

$x = -(y^2 + 2y + 1) + 1 + 5$

$x = -(y+1)^2 + 6$

b. This is a parabola in the form $x = a(y-k)^2 + h$ with $a = -1$, $h = 6$, and $k = -1$. Since $a < 0$, the parabola opens to the left. The vertex is (6, –1). The y-intercepts are about (0, –3.45) and (0, 1.45). The x-intercept is (5, 0).

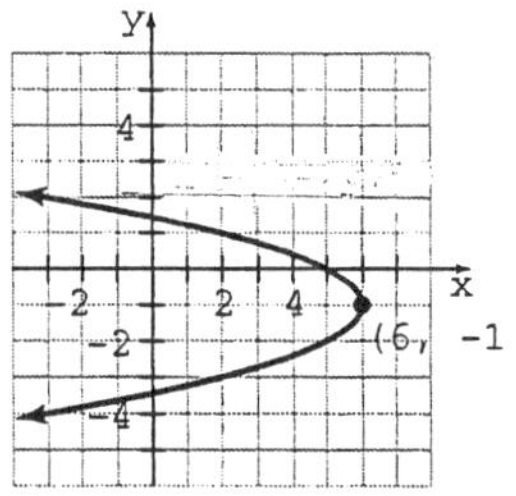

11. **a.** $x = y^2 + 5y + 4$

$x = \left(y^2 + 5y + \frac{25}{4}\right) - \frac{25}{4} + 4$

$x = \left(y + \frac{5}{2}\right)^2 - \frac{9}{4}$

b. This is a parabola in the form $x = a(y-k)^2 + h$ with $a = 1$, $h = -\frac{9}{4}$,

and $k = -\frac{5}{2}$. Since $a > 0$, the parabola opens to the right. The vertex is $\left(-\frac{9}{4}, -\frac{5}{2}\right)$.
The y-intercepts are (0, –4) and (0, –1). The x-intercept is (4, 0).

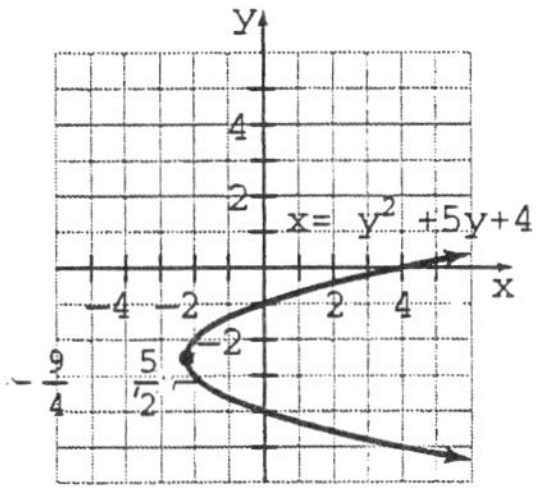

12. **a.**
$$y = 2x^2 - 8x - 24$$
$$y = 2(x^2 - 4x) - 24$$
$$y = 2(x^2 - 4x + 4) - 8 - 24$$
$$y = 2(x - 2)^2 - 32$$

b. This is a parabola in the form $y = a(x-h)^2 + k$ with $a = 2$, $h = 2$, and $k = -32$. Since $a > 0$, the parabola opens upward. The vertex is (2, –32). The
y-intercept is (0, –24). The x-intercepts are (–2, 0) and (6, 0).

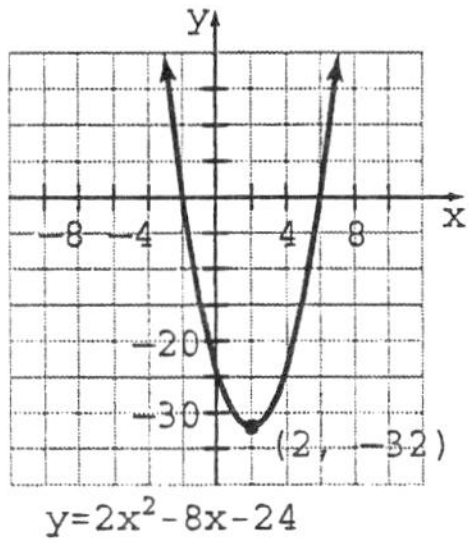

13. **a.**
$$(x-h)^2 + (y-k)^2 = r^2$$
$$(x-0)^2 + (y-0)^2 = 4^2$$
$$x^2 + y^2 = 4^2$$

b.

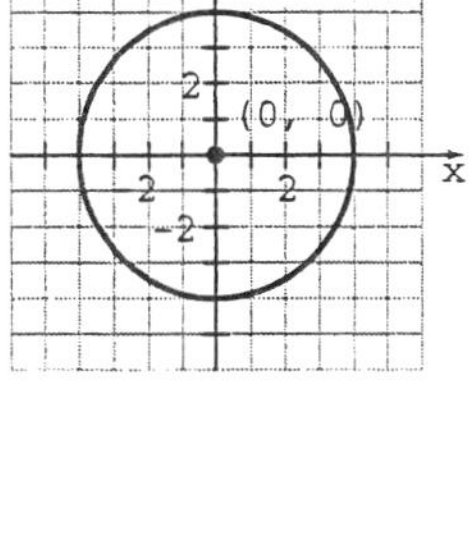

14. **a.**
$$(x-h)^2 + (y-k)^2 = r^2$$
$$[x-(-3)]^2 + (y-4)^2 = 1^2$$
$$(x+3)^2 + (y-4)^2 = 1^2$$

b.

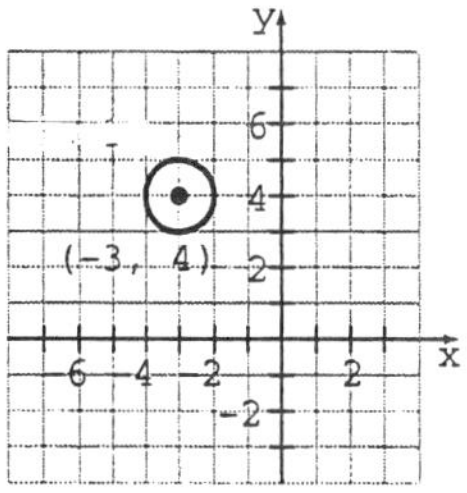

15. **a.**
$$x^2 + y^2 - 4y = 0$$
$$x^2 + (y^2 - 4y + 4) = 4$$
$$x^2 + (y-2)^2 = 2^2$$

b. The graph is a circle with center (0, 2) and radius 2.

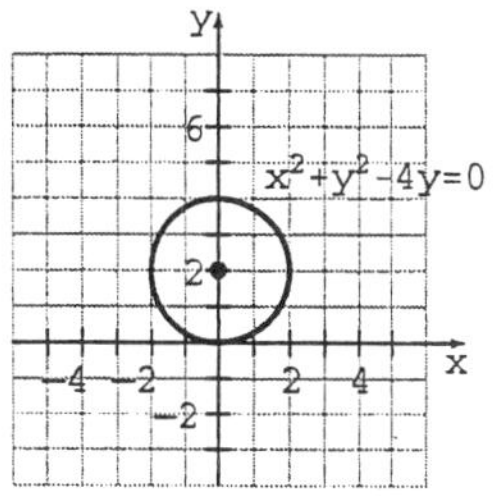

16. **a.**
$$x^2 + y^2 - 2x + 6y + 1 = 0$$
$$x^2 - 2x + y^2 + 6y = -1$$
$$(x^2 - 2x + 1) + (y^2 + 6y + 9) = -1 + 1 + 9$$
$$(x-1)^2 + (y+3)^2 = 9$$
$$(x-1)^2 + (y+3)^2 = 3^2$$

b. The graph is a circle with center (1, –3) and radius 3.

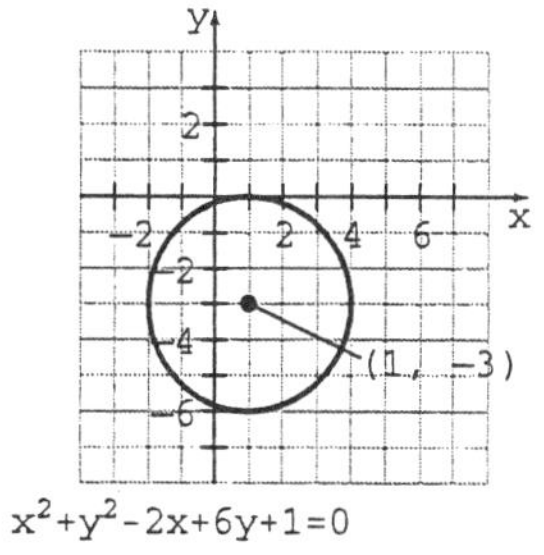

17. **a.**
$$x^2-8x+y^2-10y+40=0$$
$$(x^2-8x+16)+(y^2-10y+25)=-40+16+25$$
$$(x-4)^2+(y-5)^2=1$$
$$(x-4)^2+(y-5)^2=1^2$$

b. The graph is a circle with center (4, 5) and radius 1.

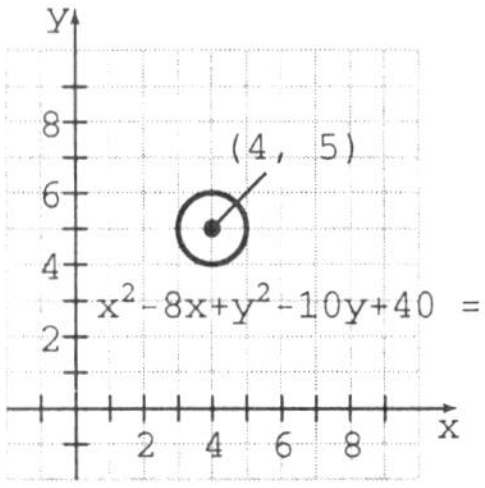

18. **a.**
$$x^2+y^2-4x+10y+17=0$$
$$x^2-4x+y^2+10y=-17$$
$$(x^2-4x+4)+(y^2+10y+25)=-17+4+25$$
$$(x-2)^2+(y+5)^2=12$$
$$(x-2)^2+(y+5)^2=\left(\sqrt{12}\right)^2$$

b. The graph is a circle with center (2, –5) and radius $\sqrt{12} \approx 3.46$.

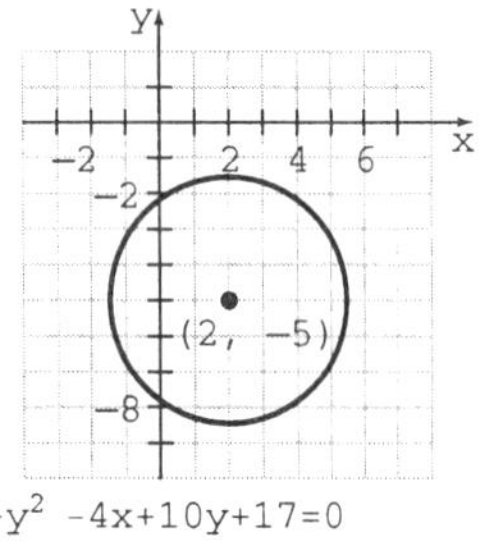

19. $y=\sqrt{9-x^2}$

If we solve $x^2+y^2=9$ for *y*, we obtain $y=\pm\sqrt{9-x^2}$. Therefore, the graph of $y=\sqrt{9-x^2}$ is the upper half $(y \geq 0)$ of a circle with its center at the origin and radius 4.

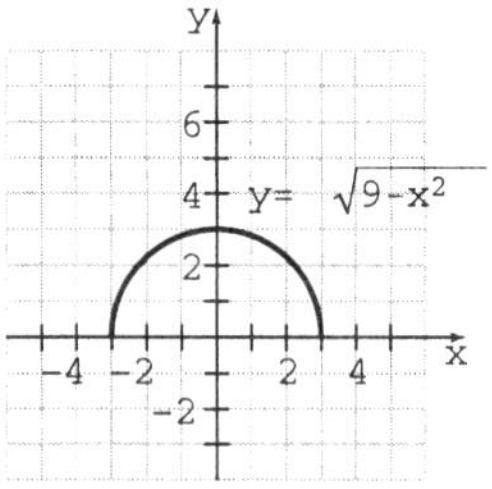

20. $y=-\sqrt{25-x^2}$

If we solve $x^2+y^2=25$ for *y*, we obtain $y=\pm\sqrt{25-x^2}$. Therefore, the graph of $y=-\sqrt{25-x^2}$ is the lower half $(y \leq 0)$ of a circle with its center at the origin and radius 5.

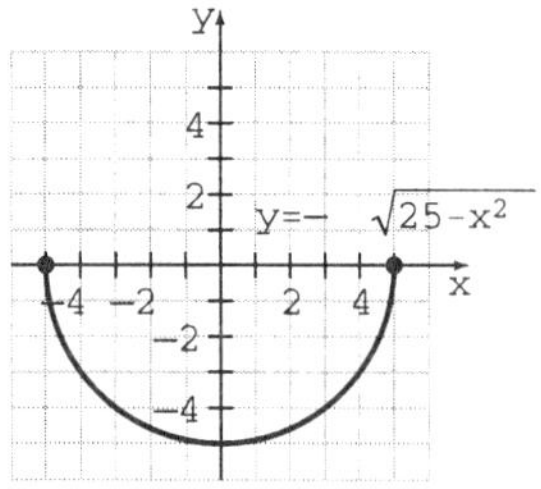

21. The center is (–1, 1) and the radius is 2.
$$(x-h)^2+(y-k)^2=r^2$$
$$[x-(-1)]^2+(y-1)^2=2^2$$
$$(x+1)^2+(y-1)^2=4$$

22. The center is (5, –3) and the radius is 3.
$$(x-h)^2+(y-k)^2=r^2$$
$$(x-5)^2+[y-(-3)]^2=3^2$$
$$(x-5)^2+(y+3)^2=9$$

23. $\frac{x^2}{4}+\frac{y^2}{9}=1$

Since $a^2=4$, $a=2$.

Since $b^2=9$, $b=3$.

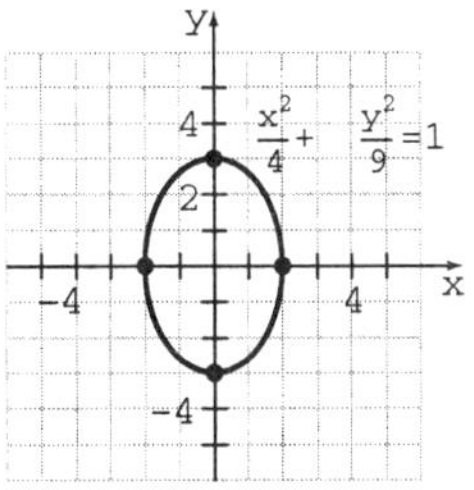

24. $\frac{x^2}{36}+\frac{y^2}{64}=1$

Since $a^2=36$, $a=6$.

Since $b^2=64$, $b=8$.

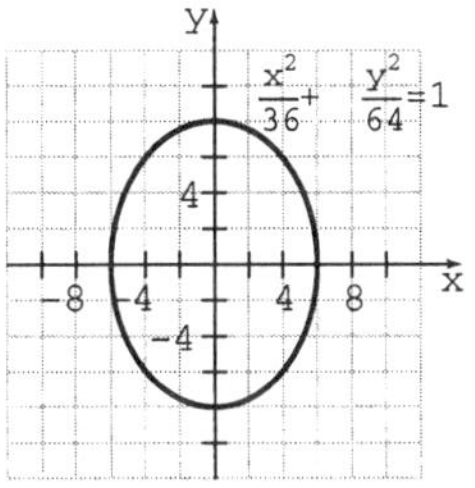

25. $4x^2+9y^2=36$

$\frac{4x^2}{36}+\frac{9y^2}{36}=1$

$\frac{x^2}{9}+\frac{y^2}{4}=1$

Since $a^2=9$, $a=3$.

Since $b^2=4$, $b=2$.

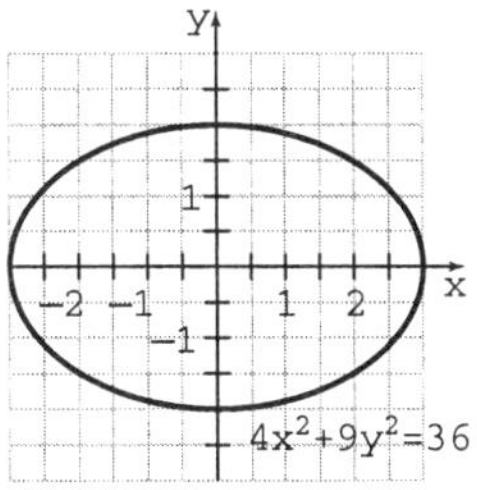

26. $9x^2+16y^2=144$

$\frac{9x^2}{144}+\frac{16y^2}{144}=1$

$\frac{x^2}{16}+\frac{y^2}{9}=1$

Since $a^2=16$, $a=4$.

Since $b^2=9$, $b=3$.

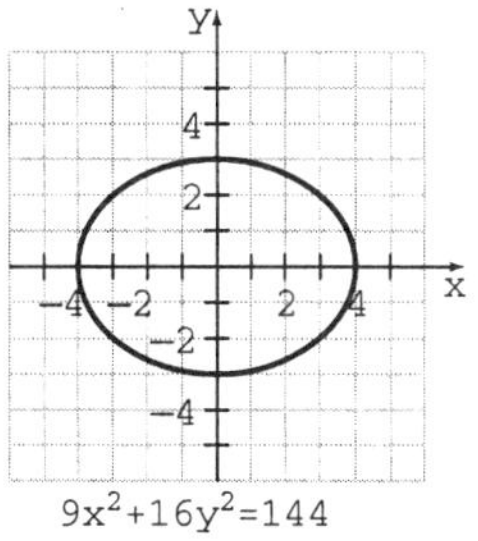

27. $\frac{(x-3)^2}{16}+\frac{(y+2)^2}{4}=1$

The center is (3, –2).

Since $a^2=16$, $a=4$.

Since $b^2=4$, $b=2$.

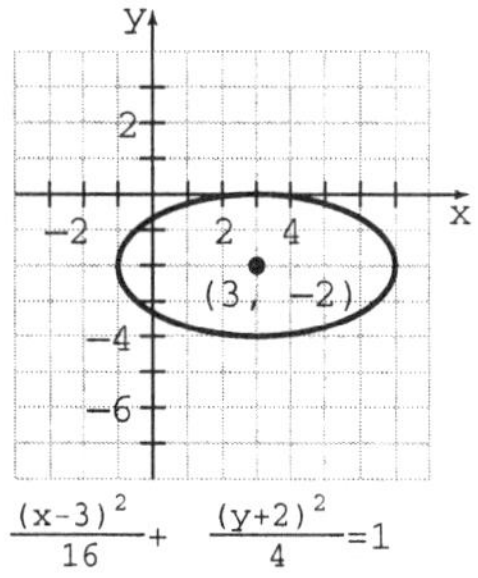

28. $\frac{(x+3)^2}{9}+\frac{y^2}{25}=1$

The center is (–3, 0).

Since $a^2=9$, $a=3$.

Since $b^2=25$, $b=5$.

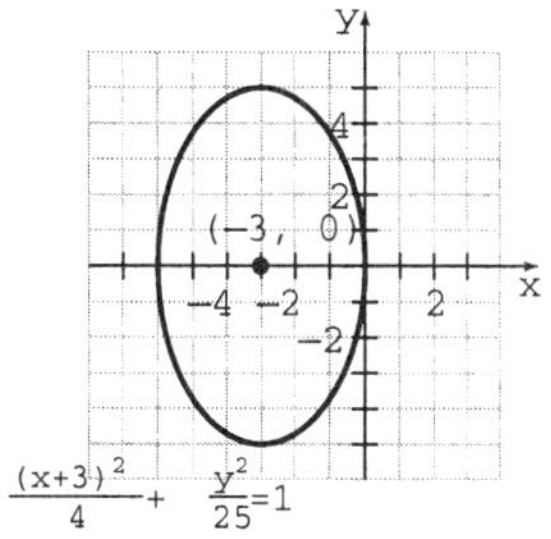

29. $25(x-2)^2 + 9(y-1)^2 = 225$

$\frac{25(x-2)^2}{225} + \frac{9(y-1)^2}{225} = 1$

$\frac{(x-2)^2}{9} + \frac{(y-1)^2}{25} = 1$

The center is (2, 1).
Since $a^2 = 9$, $a = 3$.
Since $b^2 = 25$, $b = 5$.

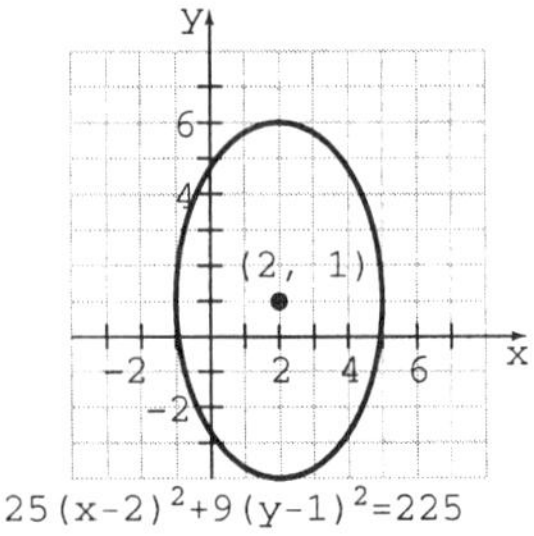

30. $\frac{x^2}{4} + \frac{y^2}{9} = 1$

Since $a^2 = 4$, $a = 2$.
Since $b^2 = 9$, $b = 3$.

Area $= \pi ab = \pi(2)(3) = 6\pi \approx 18.85$ sq. units

31. **a.** $\frac{x^2}{4} - \frac{y^2}{16} = 1$

Since $a^2 = 4$ and $b^2 = 16$, $a = 2$ and $b = 4$. The equations of the asymptotes are $y = \pm\frac{b}{a}x$, or $y = \pm 2x$.

b. To graph the asymptotes, plot the points (2, 4), (–2, 4), (2, –4), and (–2, –4). The graph intersects the *x*-axis at (–2, 0) and (2, 0).

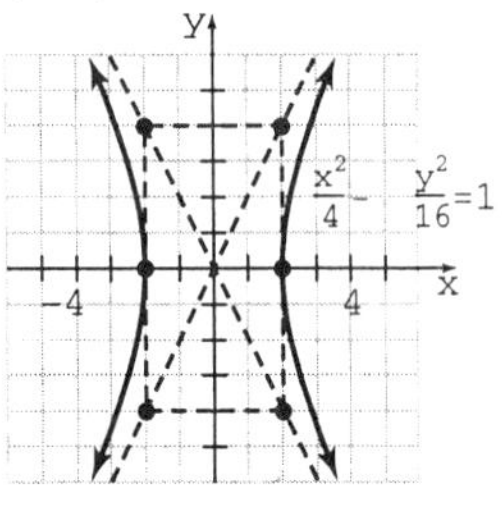

32. **a.** $\frac{x^2}{4} - \frac{y^2}{4} = 1$

Since $a^2 = 4$ and $b^2 = 4$, $a = 2$ and $b = 2$. The equations of the asymptotes are $y = \pm\frac{b}{a}x$, or $y = \pm\frac{2}{2}x = \pm x$.

b. To graph the asymptotes, plot the points (2, 2), (–2, 2), (2, –2), and (–2, –2). The graph intersects the *x*-axis at (–2, 0) and (2, 0).

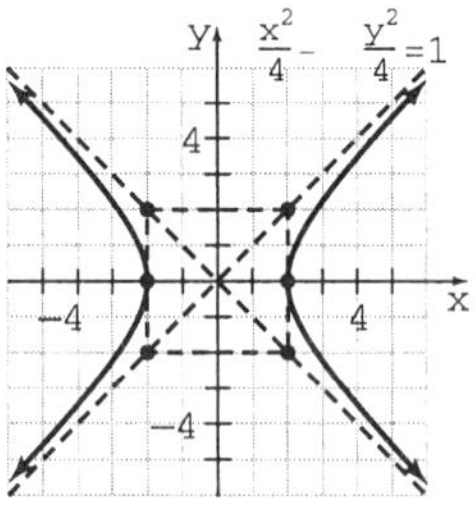

33. **a.** $\frac{y^2}{4} - \frac{x^2}{36} = 1$

Since $a^2 = 36$ and $b^2 = 4$, $a = 6$ and $b = 2$. The equations of the asymptotes are $y = \pm\frac{b}{a}x$, or $y = \pm\frac{1}{3}$.

b. To graph the asymptotes, plot the points (6, 2), (6, –2), (–6, 2), and (–6, –2). The graph intersects the *y*-axis at (0,–2) and (0, 2).

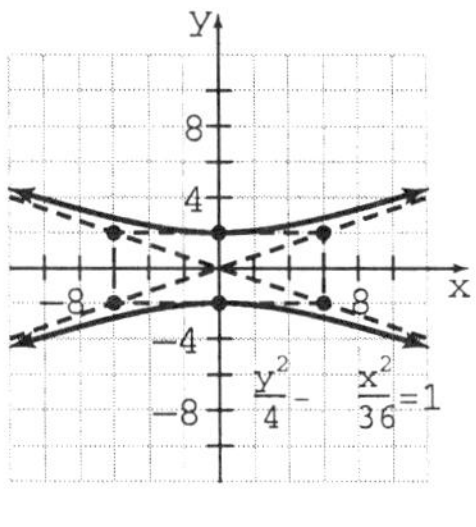

34. **a.** $\frac{y^2}{25} - \frac{x^2}{16} = 1$

Since $a^2 = 16$ and $b^2 = 25$, $a = 4$ and $b = 5$. The equations of the asymptotes are $y = \pm\frac{b}{a}x$, or $y = \pm\frac{5}{4}x$.

b. To graph the asymptotes, plot the points (4, 5), (4, –5), (–4, 5), and (–4, –5). The graph intersects the *y*-axis at (0, –5) and (0, 5).

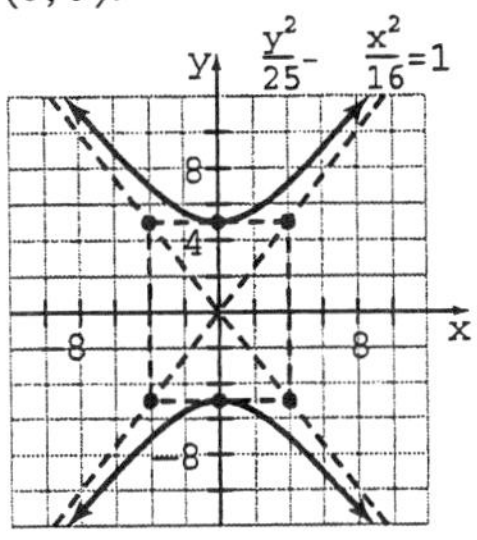

35. **a.** $x^2-9y^2=9$

$$\frac{x^2}{9}-\frac{9y^2}{9}=1$$

$$\frac{x^2}{9}-\frac{y^2}{1}=1$$

b. Since $a^2=9$ and $b^2=1$, $a=3$ and $b=1$.The equations of the asymptotes are $y=\pm\frac{b}{a}x$, or $y=\pm\frac{1}{3}x$.

c. To graph the asymptotes, plot the points (3, 1), (–3, 1), (3, –1), and (–3, –1). The graph intersects the *x*-axis at (–3, 0) and (3, 0).

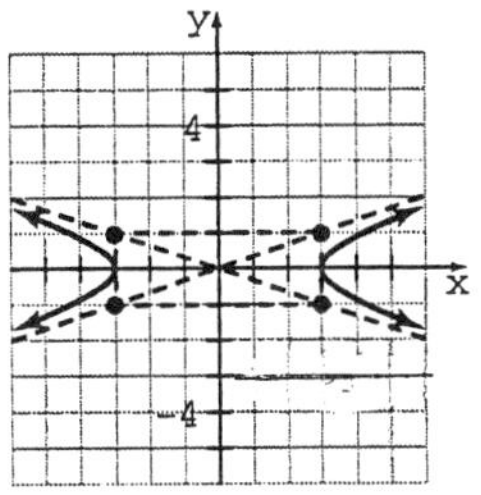

36. **a.** $25x^2-16y^2=400$

$$\frac{25x^2}{400}-\frac{16y^2}{400}=1$$

$$\frac{x^2}{16}-\frac{y^2}{25}=1$$

b. Since $a^2=16$ and $b^2=25$, $a=4$ and $b=5$. The equations of the asymptotes are $y=\pm\frac{b}{a}x$, or $y=\pm\frac{5}{4}x$.

c. To graph the asymptotes, plot the points (4, 5), (–4, 5), (4, –5), and (–4, –5). The graph intersects the *x*-axis at (–4, 0) and (4, 0).

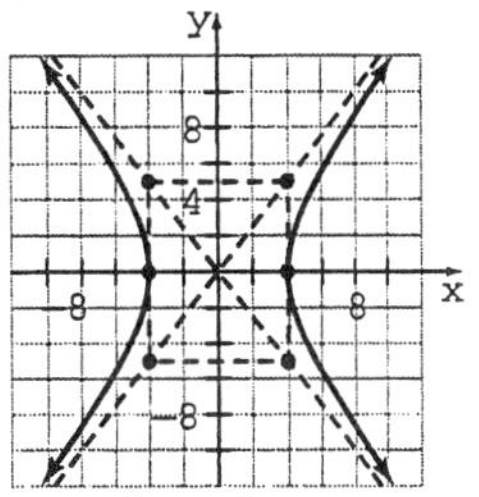

$25x^2-16y^2=400$

37. **a.** $4y^2-25x^2=100$

$$\frac{4y^2}{100}-\frac{25x^2}{100}=1$$

$$\frac{y^2}{25}-\frac{x^2}{4}=1$$

b. Since $a^2=4$ and $b^2=25$, $a=2$ and $b=5$. The equations of the asymptotes are $y=\pm\frac{b}{a}x$, or $y=\pm\frac{5}{2}x$.

c. To graph the asymptotes, plot the points (2, 5), (2, –5), (–2, 5), and (–2, –5). The graph intersects the *y*-axis at (0, –5) and (0, 5).

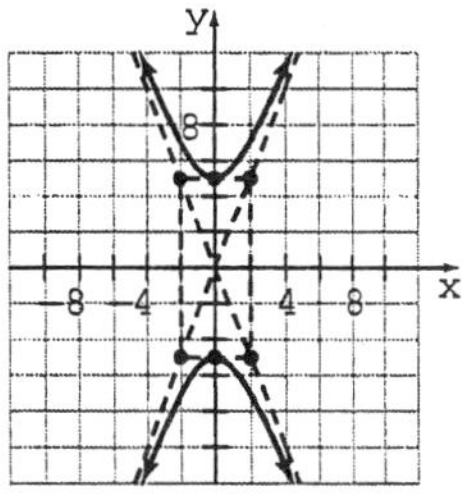

$4y^2-25x^2=100$

38. **a.** $49y^2-9x^2=441$

$$\frac{49y^2}{441}-\frac{9x^2}{441}=1$$

$$\frac{y^2}{9}-\frac{x^2}{49}=1$$

b. Since $a^2=49$ and $b^2=9$, $a=7$ and $b=3$. The equations of the asymptotes are $y=\pm\frac{b}{a}x$, or $y=\pm\frac{3}{7}x$.

c. To graph the asymptotes, plot the points (7, 3), (–7, 3), (7, –3), and (–7, –3). The graph intersects the *y*-axis at (0, –3) and (0, 3).

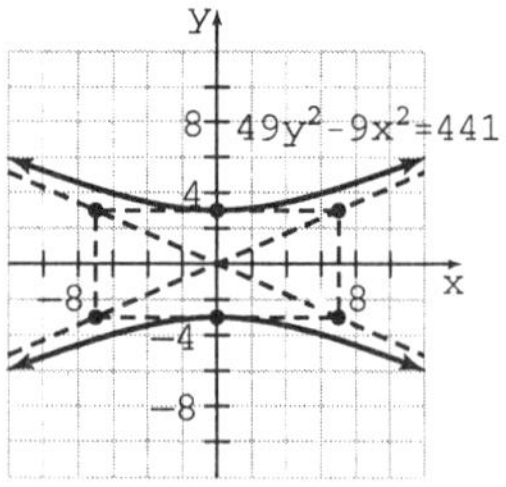

39. $\frac{x^2}{64}-\frac{y^2}{25}=1$
The graph is a hyperbola.

40. $3x^2+7y^2=21$

$\frac{3x^2}{21}+\frac{7y^2}{21}=1$

$\frac{x^2}{7}+\frac{y^2}{3}=1$

The graph is an ellipse.

41. $4x^2+4y^2=100$

$\frac{4x^2}{4}+\frac{4y^2}{4}=\frac{100}{4}$

$x^2+y^2=25$

The graph is a circle.

42. $4x^2-25y^2=25$

$\frac{4x^2}{25}-\frac{25y^2}{25}=1$

$\frac{x^2}{6.25}-\frac{y^2}{1}=1$

The graph is a hyperbola.

43. $\frac{x^2}{7}+\frac{y^2}{9}=1$
The graph is an ellipse.

44. $y=(x-3)^2+1$
The graph is a parabola.

45. $4x^2+9y^2=36$

$\frac{4x^2}{36}+\frac{9y^2}{36}=1$

$\frac{x^2}{9}+\frac{y^2}{4}=1$

The graph is an ellipse.

46. $x=-y^2+6y-7$
The graph is a parabola.

47. $x^2+y^2=16$

$x^2-y^2=16 \Rightarrow x^2=y^2+16$

Substitute y^2+16 for x^2 in $x^2+y^2=16$.

$x^2+y^2=16$

$y^2+16+y^2=16$

$2y^2=0$

$y^2=0$

Substitute 0 for y^2 in $x^2=y^2+16$

$x^2+0=16 \Rightarrow x=\pm 4$

The solution is (4, 0), and (–4, 0).

48. $x^2=y^2+4$

$x+y=4$

Solve $x+y=4$ for *y*: $y=4-x$.

Substitute $4-x$ for *y* in $x^2=y^2+4$.

$x^2=y^2+4$

$x^2=(4-x)^2+4$

$x^2=(16-8x+x^2)+4$

$8x-16=4$

$8x=20$

$x=\frac{5}{2}$

$y=4-x$

$y=4-\frac{5}{2}$

$y=\frac{3}{2}$

The solution is $\left(\frac{5}{2}, \frac{3}{2}\right)$.

49. $x^2 + y^2 = 9$

$y = 3x + 9$

Substitute $3x + 9$ for y in $x^2 + y^2 = 9$.

$$x^2 + y^2 = 9$$
$$x^2 + (3x+9)^2 = 9$$
$$x^2 + 9x^2 + 54x + 81 = 9$$
$$10x^2 + 54x + 72 = 0$$
$$5x^2 + 27x + 36 = 0$$
$$(x+3)(5x+12) = 0$$

$x + 3 = 0$ or $5x + 12 = 0$

$x = -3$ $\quad$ $x = -\frac{12}{5}$

$y = 3x + 9$

$y = 3(-3) + 9$

$y - 0$

$y = 3x + 9$

$y = 3\left(-\frac{12}{5}\right) + 9$

$y = \frac{9}{5}$

The solutions are (–3, 0) and $\left(-\frac{12}{5}, \frac{9}{5}\right)$.

50. $x^2 + 2y^2 = 7$

$x^2 - 6y^2 = 29$

Solve $x^2 + 2y^2 = 7$ for x^2: $x^2 = 7 - 2y^2$.

Substitute $7 - 2y^2$ for x^2 in $x^2 - 6y^2 = 29$.

$$x^2 - 6y^2 = 29$$
$$(7 - 2y^2) - 6y^2 = 29$$
$$7 - 8y^2 = 29$$
$$-8y^2 = 22$$
$$y^2 = -\frac{22}{8} \text{ or } -\frac{11}{4}$$
$$y = \pm i\sqrt{\frac{11}{4}}$$

There is no real solution.

51. $x^2 + y^2 = 49$

$x^2 - y^2 = 49$

$2x^2 = 98$

$x^2 = 49$

$x = 7$ $\quad$ $x = -7$

or

$x^2 + y^2 = 49$ $\quad$ $x^2 + y^2 = 49$

$7^2 + y^2 = 49$ $\quad$ $(-7)^2 + y^2 = 49$

$y^2 = 0$ $\quad$ $y^2 = 0$

$y = 0$ $\quad$ $y = 0$

The solutions are (7, 0) and (–7, 0).

52. $x^2 + y^2 = 25$ (1)

$x^2 - 2y^2 = -2$ (2)

$2x^2 + 2y^2 = 50$ (1) multiplied by 2

$x^2 - 2y^2 = -2$ (2)

$3x^2 = 48$

$x^2 = 16$

$x = 4$ or

$x^2 + y^2 = 25$

$4^2 + y^2 = 25$

$y^2 = 9$

$y = \pm 3$

$x = -4$

$x^2 + y^2 = 25$

$(-4)^2 + y^2 = 25$

$y^2 = 9$

$y = \pm 3$

The solutions are (4, 3), (4, –3), (–4, 3) and (–4, –3).

53. $-4x^2 + y^2 = -12$ (1)

$8x^2 + 2y^2 = -8$ (2)

$-4x^2 + y^2 = -12$ (1)

$4x^2 + y^2 = -4$ (2) divided by 2

$2y^2 = -16$

$y^2 = -8$

$y = \pm i\sqrt{8}$

$= \pm 2i\sqrt{2}$

There is no real solution.

54.
$-2x^2 - 3y^2 = -6$ (1)
$5x^2 + 4y^2 = 15$ (2)
$-10x^2 - 15y^2 = -30$ (1) multiplied by 5
$10x^2 + 8y^2 = 30$ (2) multiplied by 2
$-7y^2 = 0$
$y^2 = 0$
$y = 0$
$5x^2 + 4y^2 = 15$
$5x^2 + 4(0)^2 = 15$
$5x^2 = 15$
$x^2 = 3$
$x = \pm\sqrt{3}$
The solutions are $(\sqrt{3}, 0)$ and $(-\sqrt{3}, 0)$.

55. Let x = length
y = width
$xy = 36$
$2x + 2y = 26$
Solve $2x + 2y = 26$ for y: $y = 13 - x$.
Substitute $13 - x$ for y in $xy = 36$.
$xy = 36$
$x(13 - x) = 36$
$13x - x^2 = 36$
$x^2 - 13x + 36 = 0$
$(x - 4)(x - 9) = 0$
$x - 4 = 0$ $x - 9 = 0$
$x = 4$ or $x = 9$

$y = 13 - x$ $y = 12 - x$
$y = 13 - 4$ $y = 12 - 9$
$y = 9$ $y = 3$
The solutions are (4, 9) and (9, 4).
The dimensions of the pool table are 4 feet by 9 feet.

56. $C = 20.3x + 120$
$R = 50.2x - 0.2x^2$
$C = R$
$20.3x + 120 = 50.2x - 0.2x^2$
$0.2x^2 - 29.9 + 120 = 0$
$$x = \frac{29.9 \pm \sqrt{(-29.9)^2 - 4(0.2)(120)}}{2(0.2)}$$
$$x = \frac{29.9 \pm \sqrt{798.01}}{0.4}$$
$x \approx 145$ or 4
The company must sell either 4 bottles or 145 bottles to break even.

57. Since $t = 1$ year, we may rewrite the formula $i = prt$ as $i = pr$.
$250 = pr$
$250 = (p + 1250)(r - 0.01)$
Rewrite the second equation by multiplying the binomials. Then substitute 250 for pr and solve for r.
$250 = pr - 0.01p + 1250r - 12.5$
$250 = 250 - 0.01p + 1250r - 12.5$
$0 = -0.01p + 1250r - 12.5$
$0.01p + 12.5 = 1250r$
$$\frac{0.01p}{1250} + \frac{12.5}{1250} = r$$
$r = 0.000008p + 0.01$
Substitute $0.000008p + 0.01$ for r in $250 = pr$.
$250 = pr$
$250 = p(0.000008p + 0.01)$
$250 = 0.000008p^2 + 0.01p$
$0 = 0.000008p^2 + 0.01p - 250$
$0 = p^2 + 1250p - 31,250,000$
$0 = (p - 5000)(p + 6250)$
$p - 5000 = 0$ or $p + 6250 = 0$
$p = 5000$ $p = -6250$
The principal must be positive, so use $p = 5000$.
$r = 0.000008p + 0.01$
$r = 0.000008(5000) + 0.01$
$r = 0.05$
The principal is \$5000 and the rate is 5%.

Practice Test

1. They are formed by cutting a cone or pair of cones.

2. $d = \sqrt{(x_2 - x_1)^2 + (y_2 - y_1)^2}$
$= \sqrt{[3-(-4)]^2 + (4-5)^2}$
$= \sqrt{7^2 + (-1)^2}$
$= \sqrt{49+1}$
$= \sqrt{50}$
The length is $\sqrt{50} \approx 7.07$ units.

3. $\text{Midpoint} = \left(\frac{x_1+x_2}{2}, \frac{y_1+y_2}{2}\right)$
$= \left(\frac{-3+7}{2}, \frac{4+(-1)}{2}\right)$
$= \left(2, \frac{3}{2}\right)$

4. $y = -2(x+3)^2 + 1$
This is a parabola in the form
$y = a(x-h)^2 + k$
with $a = -2$, $h = -3$, and $k = 1$. Since $a < 0$, the parabola opens downward. The vertex is $(-3, 1)$. The y-intercept is $(0, -17)$. The x-intercepts are about $(-3.71, 0)$ and $(-2.29, 0)$.

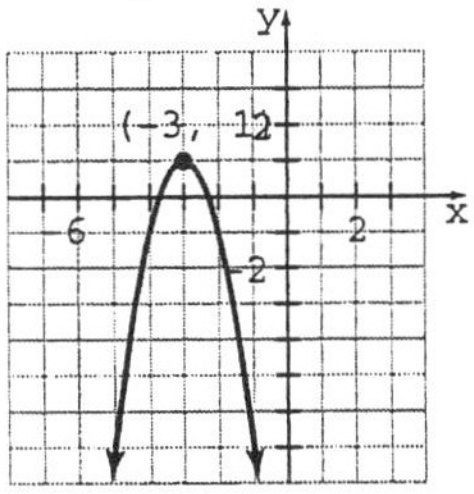

5. $x = y^2 - 2y + 4$
$x = (y^2 - 2y + 1) - 1 + 4$
$x = (y-1)^2 + 3$
This is a parabola in the form
$x = a(y-k)^2 + h$
with $a = 1$, $h = 3$ and $k = 1$. Since $a > 0$, the parabola opens to the right. The vertex is (3, 1).
There is no y-intercept. The x-intercept is (4, 0).

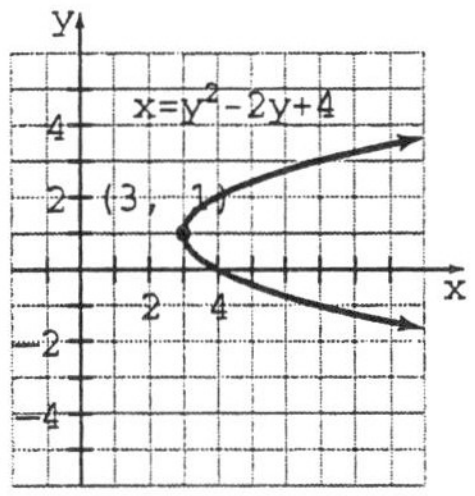

6. $x = -y^2 - 4y - 5$
$x = -(y^2 + 4y) - 5$
$x = -(y^2 + 4y + 4) + 4 - 5$
$x = -(y+2)^2 - 1$
This is a parabola in the form
$x = a(y-k)^2 + h$
with $a = -1$, $h = -1$, and $k = -2$. Since $a < 0$, the parabola opens to the left. The vertex is $(-1, -2)$. There are no y-intercepts. The x-intercept is $(-5,0)$.

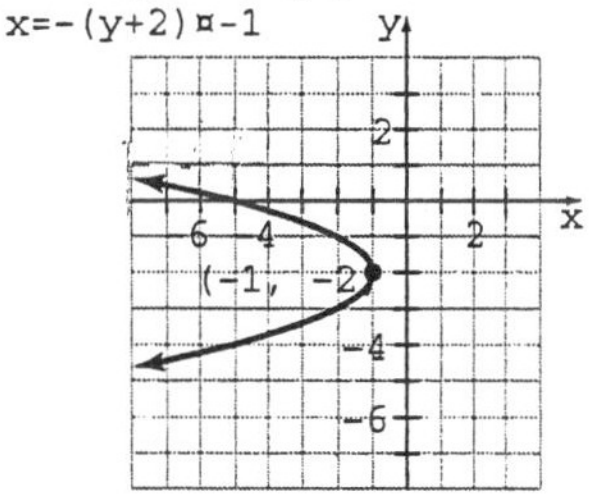

7. $(x-h)^2 + (y-k)^2 = r^2$
$[x-2]^2 + [y-4]^2 = 3^2$
$(x-2)^2 + (y-4)^2 = 9$

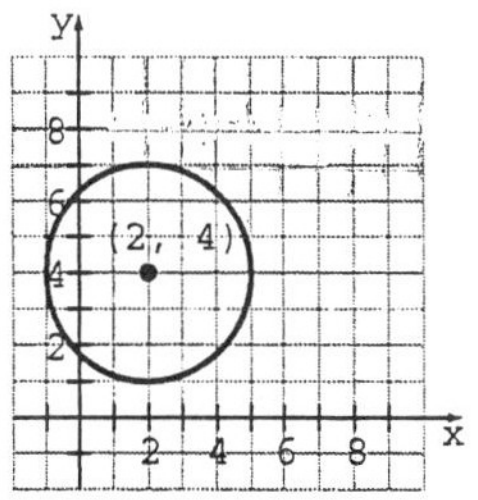

8. $(x+5)^2 + (y-5)^2 = 9$. The graph of this equation is a circle with center $(-5, 5)$ and radius 3.
$\text{Area} = \pi r^2$
$= \pi 3^2 = 9\pi \approx 28.27$ sq. units

9. The center is (3, –1) and the radius is 4.

$$(x-h)^2+(y-k)^2=r^2$$
$$(x-3)^2+[y-(-1)]^2=4^2$$
$$(x-3)^2+(y+1)^2=16$$

10. $y=-\sqrt{16-x^2}$

If we solve $x^2+y^2=16$ for *y*, we obtain $y=\pm\sqrt{16-x^2}$. Therefore, the graph of $y=-\sqrt{16-x^2}$ is the lower half ($y \le 0$) of a circle with its center at the origin and radius 4.

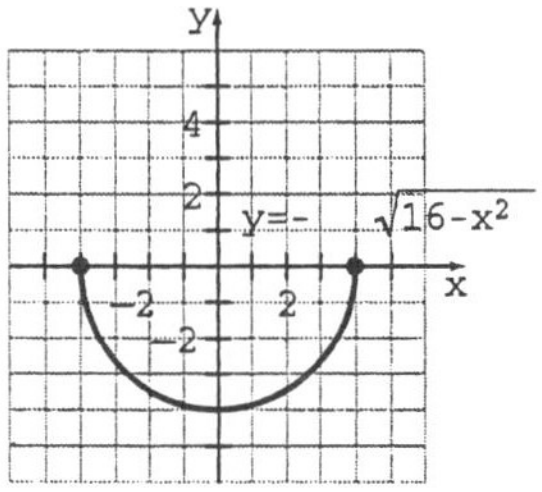

11.

$$x^2+y^2+2x-6y+1=0$$
$$x^2+2x+y^2-6y=-1$$
$$(x^2+2x+1)+(y^2-6y+9)=-1+1+9$$
$$(x+1)^2+(y-3)^2=9$$

The graph is a circle with center (1, 3) and radius 3.

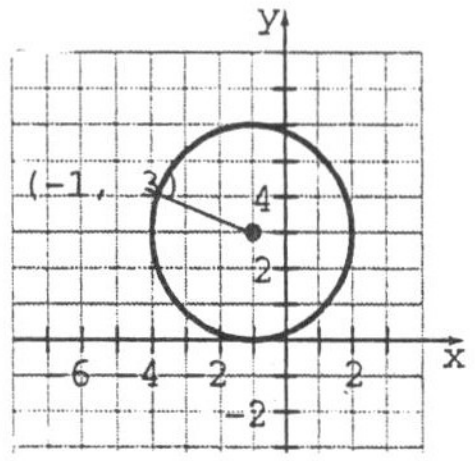

12. $4x^2+25y^2=100$

$$\frac{4x^2}{100}+\frac{25y^2}{100}=1$$
$$\frac{x^2}{25}+\frac{y^2}{4}=1$$

Since $a^2=25, a=5$

Since $b^2=4, b=2$.

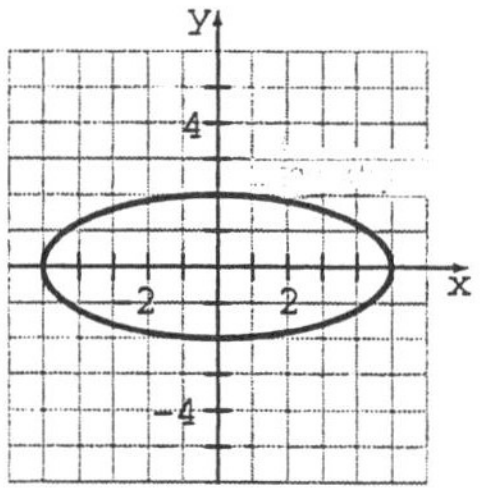

13. The center is (–2, –1), $a = 4$, and $b = 2$.

$$\frac{(x-h)^2}{a^2}+\frac{(y-k)^2}{b^2}=1$$
$$\frac{[x-(-2)]^2}{4^2}+\frac{[y-(-1)]^2}{2^2}=1$$
$$\frac{(x+2)^2}{16}+\frac{(y+1)^2}{4}=1$$

The values of a^2 and b^2 are switched, so this is not the graph of the given equation.

14. $4(x-4)^2+36(y+2)^2=36$

$$\frac{4(x-4)^2}{36}+\frac{36(y+2)^2}{36}=1$$
$$\frac{(x-4)^2}{9}+\frac{(y+2)^2}{1}=1$$

The center is (4, –2). Since $a^2=9, a=3$

Since $b^2=1$, $b=1$

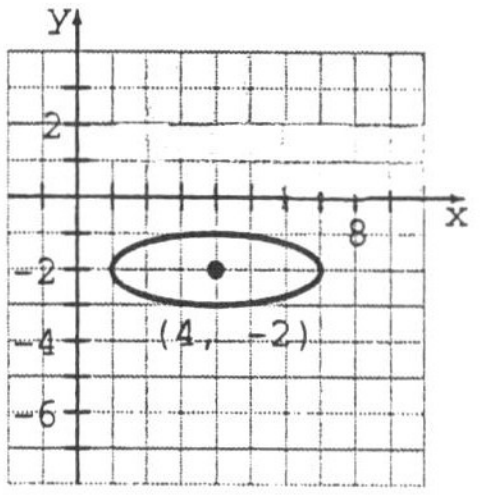

15. $3(x-5)^2+6(y+3)^2=18$

$$\frac{3(x-5)^2}{18}+\frac{6(y+3)^2}{18}=1$$
$$\frac{(x-5)^2}{6}+\frac{(y+3)^2}{3}=1$$

The center is $(5,-3.)$

16. The transverse axis lies along the axis corresponding to the positive term of the equation in standard form.